Optimising pig herd health and production

It is widely recognised that agriculture is a significant contributor to global warming and climate change. Agriculture needs to reduce its environmental impact and adapt to current climate change whilst still feeding a growing population, i.e. become more 'climate-smart'. Burleigh Dodds Science Publishing is playing its part in achieving this by bringing together key research on making the production of the world's most important crops and livestock products more sustainable.

Based on extensive research, our publications specifically target the challenge of climate-smart agriculture. In this way we are using 'smart publishing' to help achieve climate-smart agriculture.

Burleigh Dodds Science Publishing is an independent and innovative publisher delivering high quality customer-focused agricultural science content in both print and online formats for the academic and research communities. Our aim is to build a foundation of knowledge on which researchers can build to meet the challenge of climate-smart agriculture.

For more information about Burleigh Dodds Science Publishing simply call us on +44 (0) 1223 839365, email info@bdspublishing.com or alternatively please visit our website at www.bdspublishing.com.

Related titles:

Achieving sustainable production of pig meat Volume 2: Animal breeding and nutrition
Print (ISBN 978-1-78676-092-0); Online (ISBN 978-1-78676-094-4, 978-1-78676-095-1)

Achieving sustainable production of pig meat Volume 3: Animal health and welfare
Print (ISBN 978-1-78676-096-8); Online (ISBN 978-1-78676-099-9, 978-1-78676-098-2)

Understanding the behaviour and improving the welfare of pigs
Print (ISBN 978-1-78676-443-0); Online (ISBN 978-1-78676-444-7, 978-1-78676-445-4)

Chapters are available individually from our online bookshop:
https://shop.bdspublishing.com

climate•SMART•publishing
IN AGRICULTURAL SCIENCE

BURLEIGH DODDS SERIES IN AGRICULTURAL SCIENCE
NUMBER 118

Optimising pig herd health and production

Edited by Professor Dominiek Maes, Ghent University, Belgium and Professor Joaquim Segalés, Universitat Autònoma de Barcelona and IRTA-CReSA, Spain

burleigh dodds
SCIENCE PUBLISHING

Published by Burleigh Dodds Science Publishing Limited
82 High Street, Sawston, Cambridge CB22 3HJ, UK
www.bdspublishing.com

Burleigh Dodds Science Publishing, 1518 Walnut Street, Suite 900, Philadelphia, PA 19102-3406, USA

First published 2023 by Burleigh Dodds Science Publishing Limited

Library of Congress Control Number: 2022934576

British Library Cataloguing in Publication Data
A catalogue record for this book is available from the British Library

ISBN 978-1-78676-883-4 (Print)
ISBN 978-1-78676-886-5 (PDF)
ISBN 978-1-78676-885-8 (ePub)
ISSN 2059-6936 (print)
ISSN 2059-6944 (online)

DOI 10.19103/AS.2022.0103

Typeset by Deanta Global Publishing Services, Dublin, Ireland

Contents

Part 2 Understanding immunity and disease resistance

Series list

Title	Series number
Achieving sustainable cultivation of maize - Vol 1 From improved varieties to local applications Edited by: Dr Dave Watson, CGIAR Maize Research Program Manager, CIMMYT, Mexico	001
Achieving sustainable cultivation of maize - Vol 2 Cultivation techniques, pest and disease control Edited by: Dr Dave Watson, CGIAR Maize Research Program Manager, CIMMYT, Mexico	002
Achieving sustainable cultivation of rice - Vol 1 Breeding for higher yield and quality Edited by: Prof. Takuji Sasaki, Tokyo University of Agriculture, Japan	003
Achieving sustainable cultivation of rice - Vol 2 Cultivation, pest and disease management Edited by: Prof. Takuji Sasaki, Tokyo University of Agriculture, Japan	004
Achieving sustainable cultivation of wheat - Vol 1 Breeding, quality traits, pests and diseases Edited by: Prof. Peter Langridge, The University of Adelaide, Australia	005
Achieving sustainable cultivation of wheat - Vol 2 Cultivation techniques Edited by: Prof. Peter Langridge, The University of Adelaide, Australia	006
Achieving sustainable cultivation of tomatoes Edited by: Dr Autar Mattoo, USDA-ARS, USA and Prof. Avtar Handa, Purdue University, USA	007
Achieving sustainable production of milk - Vol 1 Milk composition, genetics and breeding Edited by: Dr Nico van Belzen, International Dairy Federation (IDF), Belgium	008
Achieving sustainable production of milk - Vol 2 Safety, quality and sustainability Edited by: Dr Nico van Belzen, International Dairy Federation (IDF), Belgium	009
Achieving sustainable production of milk - Vol 3 Dairy herd management and welfare Edited by: Prof. John Webster, University of Bristol, UK	010
Ensuring safety and quality in the production of beef - Vol 1 Safety Edited by: Prof. Gary Acuff, Texas A&M University, USA and Prof. James Dickson, Iowa State University, USA	011
Ensuring safety and quality in the production of beef - Vol 2 Quality Edited by: Prof. Michael Dikeman, Kansas State University, USA	012
Achieving sustainable production of poultry meat - Vol 1 Safety, quality and sustainability Edited by: Prof. Steven C. Ricke, University of Arkansas, USA	013
Achieving sustainable production of poultry meat - Vol 2 Breeding and nutrition Edited by: Prof. Todd Applegate, University of Georgia, USA	014

Introduction

This volume highlights the need to develop more preventive measures that can be implemented to tackle the increasing threat of disease in pig production. The book addresses recent developments in disease prevention, focusing on how veterinarians, farmers and producers can implement feed management and housing to optimise pig health and welfare, as well as the role of vaccine development in preventing the onset of endemic and emerging diseases in pigs and how to treat animals to comply with a judicious use of antimicrobials.

The chapters of this book are split into four parts: Part 1 chapters focus on understanding and identifying diseases in pigs, focusing specifically on viral and bacterial disease and transmission and epidemiology. Chapters on disease surveillance and monitoring are also included. Chapters in Part 2 discuss understanding immunity and disease resistance, drawing attention to the development of immune function in pigs, the effect of gut function and the nutritional strategies that can be used to boost immune function as well. Chapters also review mechanisms of resistance to respiratory disease. Part 3 chapters examine disease prevention, discussing on-farm strategies to prevent diseases as well as managing feed, housing and stocking density to help optimise pig health and prevent disease. Chapters on vaccine development to prevent both endemic and emerging diseases are also included. Part 4 concludes the book by providing chapters that focus on optimising health through the life cycle of pigs, specifically focusing on gilts, pregnant sows, weaned piglets and finisher pigs.

Part 1 Understanding and identifying disease

Chapter 1 opens the book with a discussion on the advances in understanding the mechanisms of porcine viral disease transmission and epidemiology. The chapter examines the basic features of swine viruses, including aspects of pathogenesis and transmission mechanisms. In certain instances, clinical signs caused by swine viruses are not specific enough to establish a sound clinical and/or pathological diagnosis, and laboratory investigations are required to ascertain which virus or viruses are causing disease. A summary of the most frequent clinical signs, lesions and diagnostic methods for relevant swine viral infections will be described. The chapter concludes with commonly used measures for the prevention and control of viral pig diseases, emphasizing the importance of vaccination as the cornerstone of swine health programmes.

Moving on from Chapter 1, the next chapter focuses on understanding and identifying bacterial disease in swine. Chapter 2 first discusses classical phenotypic characteristics as well as more advanced molecular techniques to identify and classify bacteria. It then reviews the pathogenicity of porcine bacteria, their transmission and the clinical signs and lesions induced by swine bacterial pathogens. The chapter also reviews how these infections can be diagnosed, before concluding with a discussion of the commonly used measures for prevention and control, including strengths and weaknesses of vaccination.

Expanding on topics previously discussed in Chapters 1 and 2, Chapter 3 examines the ways to improve disease surveillance and the use of monitoring systems in pig herds. The chapter begins by discussing identifying pathogens of concern, which is then followed by an overview of the importance of defining the goals and objectives of a monitoring or surveillance system. The chapter moves on to discuss the different factors that need to be evaluated in order to decide on the detection target (or analyte) and examines the available sample types. Sections on sample collection approaches, their collection frequency and the importance of proper sample size calculations are also included. The chapter also reviews the pooling of samples, production parameters, interpreting results and classifying herd health status. Two case studies on monitoring disease in sow farms are also included.

The final chapter of Part 1 provides an analysis of the surveillance on swine farms using antemortem specimens. Chapter 4 outlines the elements of designing and implementing infectious disease surveillance in commercial swine herds using diagnostic testing of antemortem samples. It first provides a general overview of infectious disease surveillance on swine farms, it then goes on to discuss collecting production and surveillance data, focusing specifically on testing technology and the various samples and specimens that can be taken. A section on test performance is also provided, which is then followed by an example of a surveillance planning checklist.

Part 2 Understanding immunity and disease resistance

The first chapter of Part 2 examines the advances in understanding the development of immune function in pigs. Chapter 5 begins by looking at colostrum and milk and how its consumption can influence piglet survival, then moves on to examine the transfer of humoral and cellular immunity to piglets. A section on the strategies to improve the transfer of passive immunity to piglets is also provided, focusing specifically on the vaccination of sows during gestation and ensuring a good colostrum ingestion in piglets. The chapter then reviews the development of active immunity in piglets during the neonatal period and a section on the strategies to overcome the interference with maternally derived acquired immunity is also provided.

Chapter 6 focuses on advances in understanding gut function and immunity in pigs. The gut tissues are important for feed intake and at the same time provide a niche for the microbiota. As a consequence, the gut is exposed to a vast variety of foreign antigens and has evolved a cellular network to ensure gut homeostasis. Maintaining this balance is important for optimal growth of pigs, especially during the critical periods of life. This chapter explores the latest findings in gut function and immunity in pigs as well as provides insights into how the microbiota can impact on these functions and shape intestinal immune responses. Finally, it discusses recent advances in the development of novel tools to investigate the functions of the gut tissues and provides an outlook to the future of gut health research in swine.

Expanding on topics previously touched upon in Chapters 5 and 6, Chapter 7 focuses on the nutritional strategies that can be implemented to boost immune function in pigs. The chapter begins by analysing the possibility that immune challenge can affect pig growth, then goes on to look at pregnancy, lactation and pre-weaned piglets. The importance of proteins and amino acids in the pig's diet is discussed, which is then followed by an analysis of various other nutrients such as fibre, glucans and carbohydrates. The importance of minerals such as copper, zinc, selenium and iron in immunity development is also addressed. The chapter moves on to review vitamin supplementation and their benefit, focusing on vitamins A, D, C and E specifically. A section on non-nutritive feed additives is also included, examining probiotics, prebiotics, synbiotics and proteobiotics.

The final chapter of Part 2 draws attention to host-pathogen interactions and genetic tools for resistance to porcine reproductive and respiratory syndrome virus (PRRSV). Chapter 8 first examines PRRS and the porcine respiratory disease complex. The chapter moves on to analyse the lung macrophage as the target for PRRSV infection, then examines two genetic approaches that can be used to lessen the impact of PRRSV. The PRRS Host Genetics Consortium and the search for host genes affecting the response to PRRSV infection is also touched upon. Sections dedicated to several markers are included, specifically focusing on the CD163 marker and the WUR marker. Sections on PRRS resistance through modification of CD163 and genetic association between the pig genome and PCV2 and swine influenza A viruses are also included.

Part 3 Disease prevention

Part 3 opens with a chapter that focuses on on-farm strategies for preventing pig diseases and improving biosecurity. Chapter 9 discusses the five principles of farm-level biosecurity and illustrates the application of these principles both for external biosecurity (purchasing and quarantine policy, the transport of animals and removal of manure and carcasses, the supply of feed, water, and equipment, the access of persons, vermin and bird control, environmental risks) and for internal biosecurity (compartmentalization, working lines, and equipment, disease management, farrowing and suckling period, weaning and fattening period, and cleaning and disinfection). Finally, ways to measure and quantify the biosecurity status of a pig farm are discussed.

The subject of Chapter 10 is managing feed to optimise pig health. The chapter first discusses the essential contribution of balanced energy and protein levels to achieving optimal health and performance. It then examines several mineral, vitamins and types of fibre and concludes with the essential subject of feed ingredients. A case study illustrating how characteristics of feed form and structure may have contradictory impacts on the gastric health and performance of pigs is also included. Proper nutrition depends not only on the feed, but also on the farm situation. Feeding for optimal performance can increase the risk of health problems, necessitating a delicate balancing act.

Chapter 11 draws attention to managing housing and stocking density to optimise health, welfare and production in pig herds. The chapter describes major husbandry- and management-related factors affecting animal health, welfare and production efficiency in different swine production systems. Given the current transformation in pig production, there are several conflicting goals which need to be addressed during design and adaptation of husbandry systems in the future. Chapter sections deal with the most important areas impacting pig health and welfare such as space, climate, flooring, water and feed supply. Since it is not possible to give a comprehensive overview of all systems the authors summarise fundamental housing requirements which must be achieved in all systems. The chapter also reviews basic principles in assessing the quality of housing systems. The very important human factor is addressed at the end of the chapter.

The next chapter focuses on advances in the development and use of vaccines for prevention of endemic diseases in pigs. Chapter 12 begins by highlighting how infectious diseases become endemic in pig farms, describing the main forces driving endemic diseases. The chapter then moves on to review ways to control infections, first emphasising how targeting key populations at the right time can help to implement infection control more efficiently. The chapter also examines the aims of vaccination in endemically infected farms and potential vaccine efficacy and the immunological basis of vaccination. The types of vaccines available are also described, as well as the use of adjuvants to enhance a vaccine's effectiveness. A section on the limitations of vaccines and ways to overcome these limitations is also provided.

Chapter 13 expands on the topics previously discussed in Chapter 12 by instead focusing on developing vaccines for emerging diseases in pigs. The chapter begins by first highlighting the main emerging and re-emerging infectious diseases affecting pigs, drawing specific attention to PRRS, porcine circovirus diseases, porcine epidemic diarrhea and African swine fever. Other agents that cause emerging infectious diseases are also touched upon. The chapter moves on to review vaccine platforms and new technologies to develop vaccines against emerging swine diseases, then goes on to discuss the development and use of nucleic acid vaccines. The chapter emphasises the importance of developing and using vaccines as a tool for controlling diseases and improving health.

Part 4 Optimising health through the life cycle

The first chapter of Part 4 focuses on optimising the health of gilts and sows during pregnancy and parturition. Chapter 14 begins by highlighting current biosecurity issues, the importance of routine vaccination and deworming programmes in sows and gilts. The chapter moves on to highlight production diseases that cause fertility problems and reduced reproductive performance, such as vulvar discharge syndrome, postpartum dysgalactia syndrome, mastitis and severe mammary oedema. The chapter also discusses infectious diseases causing fertility problems and reduced reproductive performance, focusing specifically on viral causes for abortions, bacterial infections as well as parasites. Non-infectious factors causing fertility problems and reduced reproductive performance are also mentioned, which are then followed by an examination of the stages leading up to parturition. Sections on the transfer of immunity and microbiota involvement during pregnancy, parturition and lactation are also included.

Chapter 15 analyses optimising the health of weaned piglets. It firstly discusses the transition from lactation to nursery, focusing on the factors that influence health and performance of weaned pig, as well as the intervention and prevention strategies that can be implemented during this transition. This is then followed by a section that addresses the infectious agents that can emerge during this stage. A review of the viruses that can emerge and act as immunosuppressive agents, such as PRRSV and porcine circovirus 2, are also discussed. The chapter then provides examples of diseases recognising risk factors and prevention strategies in the transition from weaning.

The final chapter of the book reviews optimising the health of finisher pigs. Chapter 16 first discusses how to measure pig health in the grower-finisher phase, then discusses the types of farms that should be considered when it comes to optimising finisher pig health. This is then followed by a review of the importance of the pig flow through the system to improve health as well. Stocking density (or space allowance) is also mentioned, as is the importance of transfer weight and the transition diet. The chapter also reviews how general control of infectious diseases can also help to optimise finisher pig health, as well as the importance of focusing on both infectious and non-infectious disease control, as non-infectious diseases are often forgotten about.

Part 1

Understanding and identifying disease

Chapter 1

Advances in understanding mechanisms of porcine viral disease transmission/epidemiology

Joaquim Segalés, Departament de Sanitat i Anatomia Animals, Universitat Autònoma de Barcelona, Centre de Recerca en Sanitat Animal (CReSA) - Institut de Recerca i Tecnologia Agroalimentàries (IRTA) and OIE Collaborating Centre for the Research and Control of Emerging and Re-Emerging Pig Diseases in Europe (CReSA-IRTA), Spain

1 Introduction

In 1898, Dr Martinus W. Beijerinck (1851–1931) was the first to discover and name *contagium vivum* fluidum, now known as Tobacco mosaic virus (van Kammen, 1999). More than 9000 viral species have subsequently been classified by the International Committee for the Taxonomy of Viruses (www.ictvonline.org, Virus Taxonomy: 2020 Release). Modelling studies, however, estimate that there are approximately 40 000 virus species in mammals alone (including around 10 000 viruses with zoonotic potential) (Carlson et al., 2019), suggesting that a significant number of viral species are yet to be discovered. Moreover, it has been calculated that the number of virus particles existing globally at any given time exceeds 10^{31}, meaning that viruses exceed the number of cells by 10–100-fold (revised in Simmonds et al., 2017).

http://dx.doi.org/10.19103/AS.2022.0103.01

Approximately 90 virus species are able to infect swine (Zhang et al., 2019), but evidence of overt disease causality is recognised for half of them. It is very likely that the number of viruses infecting swine will increase in the future, based on the above-mentioned modelling studies. The advent of very sensitive molecular biology techniques that do not require previous knowledge of the sequence of the pathogen (generally referred to as next-generation sequencing (NGS) methods) will ensure these new discoveries (Blomström, 2011). As an example, Fournié et al. (2015) investigated the spatio-temporal trend and diversity of infectious agents newly detected in pigs from 1985 to 2010. In this literature review study, 173 new variants (from 91 species) were found, of which 73 species had not been previously described in swine (Fournié et al., 2015). Forty-three percent of these new species were viruses, with 77% of these variants being RNA viruses. These novel viral agents were mainly identified in important pork-producing countries, with the rate of discovery of new virus variants doubling within the last ten years of the study period (Fournié et al., 2015). The concept of virosphere includes the wide and dynamic scope of virus diversity (Gorbalenya et al., 2020). Since 2015, several new pathogens have been described in the swine virosphere, further increasing the list of viruses causing disease and infection in pigs. In most cases, the newly discovered agents were not new or causative of a novel disease; good examples would be the atypical porcine pestivirus (APPV) as the cause of congenital tremors type AII (de Groof et al., 2016) and porcine circovirus 3 (PCV-3) as the cause of systemic periarteritis (Saporiti et al., 2021). In both cases, the disease or lesions they cause have been known for decades.

Although the trend of newly discovered viruses is influenced by factors facilitating the detection of infectious agents (such as diagnostic methods, surveillance efforts and research interests), it has been suggested that different scales and types of production systems may promote the emergence of certain types of infectious agents (Fournié et al., 2015). Considering the fast transformation of the swine industry and its globalised trade, increased effort is needed to improve our understanding of factors influencing the emergence and transmission of infectious agents, including viruses. However, this effort should not only be directed to novel agents but also to those important viral agents that have been causing significant problems, both currently and historically. Among them, endemic viruses such as porcine reproductive and respiratory syndrome virus (PRRSV), porcine circovirus 2 (PCV-2), swine influenza viruses (SIV) or porcine epidemic diarrhoea virus (PEDV) are still major sources of economic losses globally. In addition, one virus, that is, the African swine fever virus (ASFV), which was considered an almost exclusively African pathogen just 15 years ago, is currently one of the most threatening viruses worldwide (Blome et al., 2020).

This chapter provides an overview of important viral swine pathogens, their pathogenic mechanisms, transmission routes and clinical signs and lesions, as well as prevention and control measures.

2 Phenotypic characteristics of swine viruses

Viruses are considered to be obligate intracellular microorganisms as they cannot be cultivated in artificial media (Zhang et al., 2019). In fact, their replication capabilities rely mostly on the host cell machinery. Viruses also have unique characteristics as they cannot multiply via binary fission, they contain DNA or RNA (but not both at the same time) and they are sensitive to interferons but not antibiotics (Zhang et al., 2019).

The infective form of a virus is called the virion and represents a virus in its extracellular phase, as opposed to the different intracellular viral structures observed during replication (Burrell et al., 2017). The virion consists of genomic DNA or RNA surrounded by a protein coat (capsid) formed by viral protein subunits (capsomeres); the viral genome and capsid form the nucleocapsid. The nucleocapsid can, in turn, be covered by an outer layer (envelope) that is acquired and maintained when the nucleocapsid crosses (budding) a cell membrane (cytoplasmic, intracytoplasmic or nuclear); in contrast, some nucleocapsids emerge from the cells via lysis, becoming non-enveloped virions (Zhang et al., 2019). Importantly, the envelope is usually a lipid bilayer with some virus-encoded proteins on its surface; these proteins are attached to the matrix protein located in between the envelope (inner surface) and the nucleocapsid (Burrell et al., 2017). A list of the most significant viral swine pathogens, depending on their size, genome characteristics and the presence or absence of an envelope, is displayed in Table 1.

Due to the size of viruses in contrast to bacteria, they cannot be studied morphologically using optical microscopy (they are smaller than the limit of resolution of the light microscope, approximately 0.3 μm or 300 nm); therefore, electron microscopy (transmission or scanning) is required. Over the past few years, new microscopy technologies such as cryo-electron microscopy have been developed, which allow visualising viruses in a hydrated state – in contrast with the desiccated conditions of negative staining for electron microscopy (Burrell et al., 2017). In fact, negative staining for transmission electron microscopy (TEM) was a technique used in several diagnostic laboratories to detect faecal viral infections of pigs, mainly caused by coronaviruses and rotaviruses. However, this diagnostic technique has been completely displaced by the advent of molecular biology methods. Consequently, electron microscopy is now mainly used for research purposes and not for diagnosis. The most advanced method to study viruses structurally is X-ray crystallography, which allows visualising the virion's organisation with structural details near to

Table 1 Size, genome characteristics and presence/absence of envelope of the most significant swine viruses (including their family).

Virus name (acronym)	Family	Virion size (nm)	Genome type	Envelope
Adenoviruses	*Adenoviridae*	80–110	DNA	No
African swine fever virus (ASFV)	*Asfarviridae*	175–215	DNA	Yes
Caliciviruses • Porcine noroviruses • Porcine sapoviruses • Vesicular exanthema of swine virus (VESV)	*Caliciviridae*	35–39	RNA	No
Circoviruses • Porcine circovirus 1 (PCV-1) • Porcine circovirus 2 (PCV-2) • Porcine circovirus 3 (PCV-3) • Porcine circovirus 4 (PCV-4)	*Circoviridae*	17–24	DNA	No
Coronaviruses • Transmissible gastroenteritis virus (TGEV) • Porcine respiratory coronavirus (PRCV) • Porcine diarrhoea disease virus (PEDV) • Swine enteric coronavirus (SeCoV) • Severe acute diarrhoea syndrome virus (SADSV) • Haemagglutinating encephalomyelitis virus (PHEV) • Porcine deltacoronavirus (PDCoV)	*Coronaviridae*	80–160	RNA	Yes
Filoviruses • Reston ebolavirus (RESTV) • Zaire ebolavirus (EBOV)	*Filoviridae*	Pleomorphic	RNA	Yes
Flaviviruses • Japanese encephalitis virus (JEV) • West Nile virus (WNV)	*Flaviviridae*	40–60	RNA	Yes
Hepatitis E virus (HEV)	*Hepeviridae*	27–34	RNA	No

Herpesviruses • Pseudorabies (Aujeszky's disease) virus (PRV/ADV) • Porcine cytomegalovirus (PCMV) • Malignant catharral fever virus/Ovine herpesvirus 2 (OvHV-2)	*Herpesviridae*	150–200	DNA	Yes
Influenza viruses (SIV) • Subtypes H1N1, H1N2, H3N2 and pandemic H1N1	*Orthomyxoviridae*	80–120	RNA	Yes
Paramyxoviruses • Nipah virus • Porcine rubulavirus • Menangle virus	*Paramyxoviridae*	150–600	RNA	Yes
Parvoviruses (PPV)	*Parvoviridae*	18–26	DNA	No
Pestiviruses • Classical swine fever virus (CSFV) • Atypical porcine pestivirus (APPV) • Border disease virus (BDV) • Bovine viral diarrhoea virus (BVDV)	*Flaviviridae*	40–60	RNA	Yes
Picornaviruses • Foot-and-mouth disease virus (FMDV) • Swine vesicular disease virus (SVDV) • Encephalomyocarditis virus (EMCV) • Senecavirus A (SVA)	*Picornaviridae*	22–30	RNA	No
Porcine reproductive and respiratory syndrome virus (PRRSV)	*Arteriviridae*	50–70	RNA	Yes
Swinepox virus (SwPV)	*Poxviridae*	200–300	DNA	Yes
Reoviruses • Rotaviruses (RV) • Reoviruses	*Reoviridae*	60–80	RNA	No
Retroviruses	*Retroviridae*	80–100	RNA	Yes
Rhabdoviruses • Rabies virus • Vesicular stomatitis virus (VSV)	*Rhabdoviridae*	75x180 (bullet shaped)	RNA	Yes

Adapted from Zhang et al. (2019).

atomic resolution, the location of antigenic sites on the surface of virions as well as aspects of virion attachment and cell penetration (Burrell et al., 2017).

3 Methods to identify and classify swine viruses

To obtain an accurate diagnosis of a viral infection based exclusively on a clinical approach is not feasible in most cases. The epidemiological context and clinical-pathological findings are cornerstones for the diagnosis of any disease, but laboratory testing is as important as field work and compulsory in many disease scenarios.

Virus detection methods have been rapidly evolving from a technological point of view. The pioneering techniques were based on virus isolation in cell culture or propagation in live animals, as well as negative staining for TEM, whereas current methods are focused on detecting viral antigens by means of immunological reactions and viral nucleic acids (with or without previous knowledge of the nucleotide sequence of the infectious agent) (Burrell et al., 2017). In addition, the detection of virus antibodies has become one of the major targets in the surveillance of swine viruses.

3.1 Viral antigen detection techniques

An antigen is a molecule or molecular structure that can bind to a specific antibody or T-cell receptor; its name refers to a substance that is an 'antibody generator' (Male, 2006). In virology, antigens are proteins that can be derived from any part of the virion, but their detection is usually proof of a replicating agent – in other words, equivalent to an active viral infection.

Various methods can be used to detect viral antigens in different samples, including fluorescent antibody staining, immunohistochemistry (IHC), antigen-capture enzyme-linked immunoassays (in particular, antigen-capture enzyme-linked immunosorbent assay, ag-ELISA) or immunochromatographic assays (Zhang et al., 2019). The immunofluorescence test and IHC are particularly valuable as they confirm the presence of the viral antigen in combination with existing microscopic lesions in the corresponding organ or tissue. Therefore, the association of viral proteins together with histopathological findings gives very strong evidence of causality (Fig. 1); it is not only the detection of the virus or part of the virus in a tissue, but in relation to the associated lesions caused by the studied virus.

3.2 Viral nucleic acid detection methods

Molecular biology methods applied to virus genome detection have rapidly evolved over the last 20 years. Polymerase chain reaction (PCR) methods have

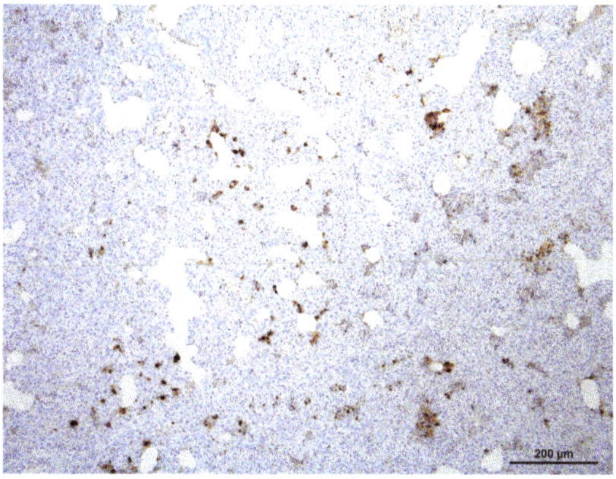

Figure 1 Example of immunohistochemistry. Detection of N protein of PRRSV in the lung of a pig displaying severe interstitial pneumonia; the viral antigen is shown as a granular brown staining in the cytoplasm of interstitial macrophages. Immunohistochemistry to detect PRRSV, haematoxylin counterstained.

been designed for virtually all viruses of interest in swine health (Zhang et al., 2019). Consequently, a significant number of diagnostic laboratories use PCR as the basis for quick, specific and sensitive detection of viral RNA or DNA. PCR (for DNA) or reverse transcriptase (RT)-PCR (for RNA) currently represent the most widely used technology to detect viral genomes. The high sensitivity of these methods, however, can be detrimental from a diagnostic point of view, as detection of the viral nucleic acid does not discriminate between a viable or non-viable microorganism (Belák, 2005). Therefore, it is important to couple PCR with other methods (histopathology, for example) that allow linking the agent with the disease condition. The advent of real-time quantitative PCR (qPCR for DNA and RT-qPCR for RNA) methods improved this latter point, as quantifying the viral amount enabled, at least for several diseases such as PCV-2 systemic disease (Olvera et al., 2004), establishing a relationship between the cause of disease and viral genome load.

More sophisticated methods such as microarrays have also been developed. This technique allows constructing microchips with dozens to thousands of oligonucleotide probes that can hybridise with viral genomes attached to a solid matrix. Therefore, microarrays can detect a huge number of viruses in one single sample, but they are not as sensitive as qPCR methods (Burrell et al., 2017). Despite promising capabilities, microarrays have not been as successful as PCR as the most convenient and practical technique to detect swine pathogens for diagnostic purposes.

In situ hybridisation (ISH) is an alternative method that allows detecting the viral genome within the microscopic lesions, like IHC. Consequently, it is an excellent technique to correlate the presence of the virus with the existing

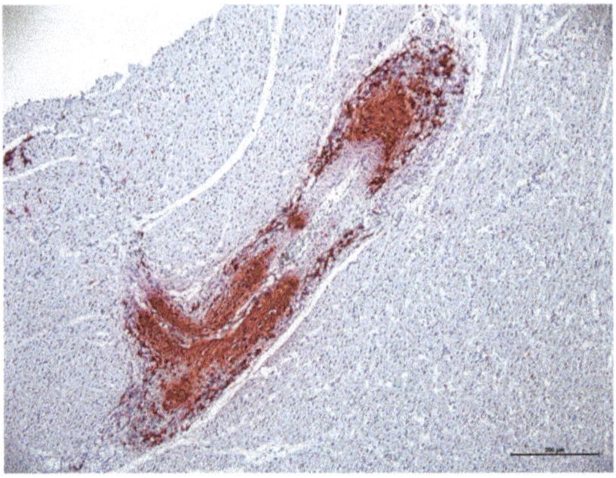

Figure 2 Example of *in situ* hybridisation. The detection of PCV-3 nucleic acid in the heart of a post-weaning pig suffering from wasting; labelling is mainly located in the arterial wall, which showed non-suppurative periarteritis and arteritis. *In situ* hybridisation to detect PCV-3, haematoxylin counterstained.

pathology (Fig. 2). ISH is particularly interesting for those viruses against which there is no validated antibody to perform IHC or the immunofluorescent test (Zhang et al., 2019).

All molecular biology methodologies described so far require prior genomic sequence information on the virus genera/families being investigated (Zhang et al., 2019). Metagenomics is a recently described platform that allows detection of all the genetic material present in each sample, independently of the previous existence of sequences for a given pathogen (Handelsman, 2004). Therefore, it represents an excellent tool for discovering new viruses as well as providing an unbiased approach to detect multiple pathogens in the sample (Belák et al., 2013). Virus sequencing is convenient as it allows differentiating variants within a virus species as well as discriminating sequences of wild viruses from vaccine-based attenuated viruses. The novel sequencing technologies based on metagenomics (NGS) allow the generation of millions of sequences in a single run, which implies a lower cost and faster speed. Although NGS is not yet a common diagnostic method for swine viruses, this technology shows good potential for future applications. So far, NGS has been successfully used to discover new porcine viruses, such as APPV (Hause et al., 2015) and PCV-3 (Phan et al., 2016).

4 Pathogenesis of swine viruses

Viral pathogenesis is the study of processes and mechanisms by which viruses cause diseases in their target hosts (Payne, 2017). However, the swine virosphere includes viruses able to cause disease via their sole infection of naïve pigs (such as ASFV, CSFV, ADV/PRV, PRRSV causing reproductive disease or FMDV), viruses able to cause disease as long as co-infecting with other agents (PCV-2 or PRRSV causing respiratory disease), viruses able to cause disease depending on the age or physiological stage of the animals (PPV, PEDV or TGEV), viruses that do not cause an apparent clinical outcome (HEV, Torque teno sus viruses or porcine bocaviruses) and viruses with unknown or poorly known clinical consequences (PCV-3, PCV-4, porcine kobuvirus or porcine astroviruses). Moreover, metagenomic studies have also proposed the existence of resident viruses, which would be defined as those viruses that are considered stably associated with healthy host tissues (Duerkop and Hooper, 2013). In relation to the latter point, a recent study performed in China with nearly 2000 healthy pigs found non-porcine parvovirus 1 parvoviruses, non-PCV circular DNA viruses, anelloviruses, sapeloviruses, toroviruses, mamastroviruses, picobirnaviruses, non-group A rotaviruses and enteroviruses in over 80% of farms, and were considered resident viruses (He et al., 2021). In this chapter, the so-called swine virobiota (viral microbiota of healthy subjects) and its virome (viral genes of the pig virobiota) will not be discussed; the focus will be the pathogenesis of viruses causing overt disease in pigs.

Globally, a viral disease consists of the viral replication effects on the host and the subsequent host immune response against the pathogen. There are several factors that modulate the pathogenesis of a given infectious agent, which include: (a) transmission, entry and spread within the host, (b) tissue or organ tropism, (c) virus intrinsic virulence and disease mechanisms, and (d) host factors and defences (Payne, 2017).

The first important step in pathogenesis is the introduction of the viral genome into the cytoplasm of the host cell. For this, the virus must attach to the cell surface and then penetrate the cellular membrane to deliver its nucleic acid along with any viral proteins needed for subsequent replication (Doms, 2016). Initial viral attachment often occurs via relatively non-specific, low-affinity interactions with cell surface molecules that support virus binding but are not required for virus infection (the so-called virus attachment factors). Although not strictly needed for infection, they can make virus attachment and infection more efficient. In contrast, virus receptors are host cell molecules (usually glycoproteins on the plasma membrane) that bind virions and are essential for subsequent infection. Viruses usually bind to their receptors with higher affinity than they do to attachment factors (Doms, 2016).

Virion attachment to its cellular receptor occurs through the viral attachment protein (VAP); in several cases, co-receptors are needed to increase infection efficiency. Importantly, the ability of a virus to enter one or various cell types is what determines the viral tropism. For example, the main cell target of PRRSV is the macrophage and can be internalised into the cells by means of sialoadhesin (CD169/Siglec-1) via interaction with the viral protein GP5/M heterodimer;

whereas, the CD163 receptor plays a role in viral internalisation and disassembly, interacting with GP2 and GP4 viral proteins (Van Gorp et al., 2008). As another example, PEDV and TGEV use their S protein to bind N aminopeptidase, an ectoenzyme abundantly expressed on the apical membrane of enterocytes, which is the functional receptor for both viruses (Delmas et al., 1992; Li et al., 2007). Consequently, the specific receptor/s for these viruses drive the systemic tropism of PRRSV and the enteric tropism of TGEV/PEDV.

After attachment, viruses enter the target cells following a multi-step process that results in the delivery of the genome to the site of replication, that is, the cytoplasm or nucleus (Doms, 2016). For this, the virus must cross the cell membrane, which is achieved by a membrane fusion reaction (typical of enveloped viruses) or the viral genome is delivered across the membrane via a pore formed by proteins of the viral capsid (typical of non-enveloped viruses). In general, RNA viruses replicate in the cytoplasm and DNA viruses in the nucleus, although there are some exceptions (i.e. swine poxvirus, one of the largest pig DNA viruses, replicates in the cytoplasm). Viruses use viral or host enzymes for genome replication and for the transcription and translation of viral proteins (Doms, 2016). The bigger the genome, the more likely that viral enzymes are involved in replication. As an example, PCV-2, one of the smallest DNA viruses of mammals, almost exclusively uses cell machinery for replication (Mankertz et al., 1998). Conversely, porcine coronavirus replication involves various viral and cellular proteins; the first open reading frame (ORF) of the coronavirus genomes encodes a large polyprotein, which is processed into several viral proteins that are required directly or indirectly for viral replication (Shi and Lai, 2005).

The last step in viral pathogenesis is the release of the virus from the infected cells. This can be achieved by lysis of the infected cell to release encapsidated virions (direct cytopathic effect) or budding from the plasma membrane as enveloped virions (Doms, 2016). Table 1 summarises which swine viruses are enveloped and non-enveloped. In addition to cellular budding to the extracellular media, some viruses can also bud at the site of cell-to-cell contact. While cell-free viral particles released into the extracellular media are compulsory for long-distance spread, this also implies that the presence of immune system components (antibodies and cells), the low success rate of cell infection by single particles and the relative fragility of viral particles in the extracellular environment may jeopardise viral dissemination into the host (Cifuentes-Muñoz et al., 2020). Such budding at the cell-to-cell contact site results in the formation of a structure called the 'virologic synapse,' making the transfer of the virus into the new host cell a very efficient process (Zhong et al., 2013). Some swine viruses such as ADV/PRV (Hanssens et al., 1995), PEDV (Yuan et al., 2021), porcine deltacoronavirus (Yang et al., 2020), SVDV (Vassalli et al., 1986), PRRSV (Wang et al., 2018) and others can also use the cell-to-cell transmission strategy to spread into an organ or tissue.

Viruses can disseminate to significant distances within the host using the lymphatic system, the blood or the peripheral nervous system (Doms, 2016). The most common mode of dissemination is via blood (viraemia), and a blood-borne virus can circulate either cell-free and/or as cell-associated virions. The ADV/PRV is a good example of a virus that can be found free in the blood or associated with cells, that is, monocytes (Nauwynck and Pensaert, 1995). Most swine viruses causing systemic disease have a viraemic period, although the duration can vary. For example, FMDV displays a relatively short viraemia of 1–3 days (Toka and Golde, 2013) but other viruses such as PCV-2 (Rodríguez-Arrioja et al., 2001) can show persistent, intermittent viraemia for several months. Some pig viruses can spread via the peripheral nervous system; classic examples include PRV/ADV (Taylor et al., 2012) and the rabies virus (Swenson et al., 2019). Viruses that spread via peripheral nerves use intracellular transport mechanisms within neurons.

The terms 'pathogenicity' and 'virulence' refer to the capacity of a virus to cause disease in its host and the degree of severity of the disease, respectively. A significant proportion of viral infections do not lead to clinical disease, being subclinical or inapparent; however, others result in diseases of variable severity that are accompanied by clinical signs in the pig. The outcome of the virus-host interaction is essentially the balance between the virulence of the infecting virus and the susceptibility of the host (Maclachlan and Dubovi, 2017).

Disease is caused by direct viral lysis of the infected cells or is immune-mediated by immune complexes and cytokines. Cytopathic viruses kill the cells in which they replicate, by preventing synthesis of host macromolecules, by producing degradative enzymes or toxic products or by inducing apoptosis. Several mechanisms of viral cell injury and death are described below (Maclachlan and Dubovi, 2017), including examples of the viruses responsible:

- inhibition of host-cell nucleic acid synthesis (porcine adenovirus);
- inhibition of host-cell RNA transcription (swine poxvirus, rabies virus);
- inhibition of processing of host-cell messenger RNAs (VSV, ADV/PRV, SIV);
- inhibition of host-cell protein synthesis (ADV/PRV, SIV, swine poxvirus, rabies virus);
- cytopathic effects of 'toxic' viral proteins (porcine adenovirus); and
- interference with cellular membrane function (porcine coronaviruses, rabies virus, porcine adenovirus).

Non-cytopathic viruses often cause persistent infection in which infected cells produce and release virions, but overall cellular metabolism is affected in a limited way (Maclachlan and Dubovi, 2017). A good example would be the various pestiviruses infecting swine.

As indicated previously, viruses typically cause direct damage to the host cells by subverting their metabolic machinery. In addition, inflammatory responses that occur in most viral infections contribute to disease pathogenesis. However, very often, the host immune response triggered by the viral infection mediates tissue injury and disease, particularly in those viruses that cause persistent, non-cytocidal infections (Maclachlan and Dubovi, 2017). The infiltration of virus-infected tissues by inflammatory cells, the release of cytokines and other mediators, and the resultant inflammation are characteristic of viral infections. These responses are fundamental for infection control as well as eventual virus clearance and induction of protective immunity. However, there is a delicate balance between the protective and destructive effects of host antiviral immune responses and between the ability of the virus to replicate and spread prior to the host protective response (Maclachlan and Dubovi, 2017). In fact, the onset and progression of the disease and its clinical signs can be mainly due to the host immune response.

5 Transmission of swine viruses

To establish infection, a virus must be exposed to host cells that are susceptible and permissive to infection. There are several different portals of entry that are used by different viruses for horizontal transmission. Most viruses interact with the cells of the host epithelium, which acts as the main barrier between the outer and inner environment of the body. The mucosal epithelium of the respiratory, gastrointestinal and genital tracts is the entry portal of many viruses (Louten, 2017). In addition, vertical transmission may occur, referring to the generational transmission of viruses from parents to their offspring, which can take several routes, including intrauterine (systemic viral infection of pregnant sows that cross the placental barrier), at farrowing during suckling, or through direct contact from the sow to the piglet during the lactating period.

As swine viral diseases can be spread through several routes, both external and internal biosecurity represent a high priority for pig farming. Transmission by direct contact takes place for almost all viral agents, but indirect transmission depends on the specific virus, its ability to persist in the environment and its capabilities of being carried by aerosols, wind or fomites. The different routes of transmission of a selected number of important swine viral pathogens are shown in Table 2.

6 Clinical signs and lesions of swine viral infections

An overview of the clinical signs, gross lesions and histopathological findings of the most significant swine viral agents are summarised in Table 3.

Table 2 Direct and indirect transmission routes of selected important viruses affecting swine (adapted from Diseases of Swine, 11th Edition). Described and contrasted routes of transmission are labelled with an X. The label (X) indicates a likely transmission route.

Viral agent[1]	Direct contact	Indirect contact										
		People	Manure/faeces	Semen	Intrauterine	Domestic/ feral animals	Rodents	Insects (vectors)	Aerosol	Animal feed	Water	Fomites
ASFV	X					X		X		X		
PCV-2	X			(X)*	X		(X)	(X)	X			X
PEDV	X	X	X						(X)	X		X
TGEV	X		X			X		(X)				X
JEV					X	X		X				
HEV	X		X							X	X	
ADV/PRV	X			X		X			X		X	X
PCMV	X				X							
SIV	X								X			
Nipah virus	X					(X)			(X)			
PPV	X			X	X	X	X					X
CSFV	X	X		X	X	X	(X)		X	X		
APPV	X			(X)	X							
FMDV	X	X							X	X		X
SVDV	X									X		X
SVA	X				(X)		(X)	(X)				
PRRSV	X			X	X			(X)	X	X	X	X
SwPV	X				X			X				
RV	X		X									X

[1]See abbreviations in Table 1.
*PCV-2 can be found in semen, but its amount is considered insufficient to cause infection in the recipient gilt/sow.

Table 3 Most frequent clinical signs and gross and microscopic lesions associated with diseases caused by selected important viruses affecting swine.

Viral agent[1]	Disease	Major clinical signs	Most significant gross lesions	Most significant microscopic lesions
ASFV	African swine fever	Hyperacute form: hyperthermia, depression and death Acute form: hyperthermia, cutaneous and systemic haemorrhages, dyspnoea, death Subacute form: as acute, but less severe Chronic form: growth retardation, dyspnoea	Splenomegaly, haemorrhagic diathesis, lymphadenomegaly (gastro-hepatic lymph node one of the major targets)	Haemorrhages, thrombosis, necrosis of lymphoid tissues, lymphocyte depletion
PCV-2	PCV-2-systemic disease	Wasting, respiratory distress, diarrhoea, jaundice, increased mortality	Generalised lymphadenomegaly, lack of pulmonary collapse with interstitial oedema, interstitial nephritis	Lymphocyte depletion with granulatomatous inflammation of lymphoid tissues, interstitial pneumonia, lymphohistiocytic infiltration can be present at any tissue
PEDV	Porcine epidemic diarrhoea	Vomiting, watery diarrhoea, anorexia, depression, high morbidity and lethality in young piglets	Catarrhal enteritis (apparent thinner small intestine wall)	Fusion and atrophy of intestinal villi
TGEV	Transmissible gastro-enteritis	Vomiting, profuse watery yellowish diarrhoea, high morbidity and lethality in young piglets	Catarrhal enteritis (apparent thinner small intestine wall)	Fusion and atrophy of intestinal villi
JEV	Japanese encephalitis	Reproductive failure, abortion, SMEDI-like signs; boar infertility; rare wasting in lactating/nursery pigs	Mummified foetuses, stillbirth, weak-born piglets often with malformations; vaginitis and epididymitis in boar	Stillborn and weak-born piglets with hydrocephalus, cerebellar hypoplasia, necrosis foci in liver and spleen; non-suppurative meningoencephalitis

HEV	Hepatitis E	No clinical signs	No gross lesions	Mild to moderate non-suppurative hepatitis
ADV/PRV	Aujeszky's disease/ Pseudorabies	Hyperthermia, anorexia, incoordination, vomiting, trembling,, dyspnoea, sudden death of young piglets, increased mortality; in pregnant animals, return-to-oestrus, abortion, SMEDI-like	Necrotic foci in liver, spleen and kidney in very young piglets; necrotising placentitis and endometritis in sows, stillborn and mummified piglets	Non-suppurative meningoencephalomyelitis and ganglioneuritis; systemic multifocal necrosis; necrotizing placentitis and endometritis
PCMV	Inclusion body rhinitis	Sudden death in neonates; SMEDI-like; sneezing and poor weight gain, nasal exudate	Stillborn and mummified piglets, interstitial pneumonia in neonates; catarrhal rhinitis	Lymphoplasmacytic inflammation in many organs in neonates with INIB in epithelial cells; lymphoplasmacytic rhinitis with INIB
SIV	Swine influenza	Hyperthermia, anorexia, tachypnoea and coughing	Craneo-ventral multifocal consolidation, sometimes extended to diaphragmatic lobes	Broncho-interstitial pneumonia, necrotising bronchitis and bronchiolitis
Nipah virus	Nipah virus infection	Hyperthermia, dyspnoea, paroxysmal non-productive ('barking') cough; muscle fasciculation, hind limb weakness, paresis	Lack of pulmonary collapse with interstitial oedema, generalised lymphadenomegaly; meningeal congestion	Interstitial pneumonia; vasculitis with fibrinoid necrosis, lymphocyte depletion; non-suppurative meningo-encephalitis
PPV	Porcine parvovirosis	SMEDI-like; mummified foetuses, weak-born piglets	Mummified foetuses with different crown-to-rump length; cutaneous vesicular lesions in piglets (very rare)	Non-suppurative endometritis and myometritis; foetuses: non-suppurative myocarditis and meningo-encephalitis, lymphocyte necrosis

(*Continued*)

Table 3 (Continued)

Viral agent[1]	Disease	Major clinical signs	Most significant gross lesions	Most significant microscopic lesions
CSFV	Classical swine fever	Peracute, acute, chronic and prenatal forms; hyperthermia, cutaneous and systemic haemorrhages, anorexia, lethargy, dyspnoea, diarrhoea; SMEDI-like with foetal malformations; congenital tremor (type AI)	Splenomegaly (with marginal infarcts), haemorrhagic diathesis, lymphadenomegaly; fibrino-necrotising typhlitis and colitis*	Haemorrhages, thrombosis, necrosis of lymphoid tissues, lymphocyte depletion, non-suppurative meningoencephalitis; foetal malformations (dysmyelinogenesis, cerebellar hypoplasia, microencephaly)
APPV	Atypical porcine pestivirus infection	Congenital tremor (type AII)	No gross lesions	Hypomyelinisation of the white matter of spinal cord and brain stem
FMDV	Foot and mouth disease	Hyperthermia, vesicle formation in mouth, snout and feet, lameness; high mortality (sudden mouth) in young piglets	Cutaneous vesicles, cutaneous erosions or ulcerations when rupturing; soft and flaccid heart in young piglets	Ballooning degeneration and necrosis of cutaneous stratum spinosum forming microvesicles, dermis oedema, cutaneous necrosis; non-suppurative myocarditis in young piglets
SVDV	Swine vesicular disease	Vesicle formation mainly around coronary bands, lameness (rare)	Vesicles mainly in the junction of the heel and the coronary band	Ballooning degeneration and necrosis of cutaneous stratum spinosum forming microvesicles, dermis oedema
SVA	Senecavirus A infection	Vesicle formation in mouth, snout and feet, lameness; neonatal pigs with diarrhoea	Cutaneous vesicles, cutaneous erosions or ulcerations when rupturing	Ballooning degeneration and necrosis of cutaneous stratum spinosum forming microvesicles, dermis oedema: villi atrophy in small intestine of piglets
PRRSV	Porcine reproductive and respiratory syndrome	Abortions, return-to-oestrus, sow mortality**, SMEDI-like, anorexia, lethargy; reduction in semen quality; high preweaning mortality, dyspnoea, diarrhoea, lethargy	Generalised lymphadenomegaly, lack of pulmonary collapse with interstitial oedema; mummified foetuses and stillborn piglets	Interstitial pneumonia, lymphocyte necrosis, lymphocyte depletion followed by hyperplasia; mild lymphohistiocytic endometritis and myometritis; atrophy of seminiferous tubules and depletion of germ cells

SwPV	Swinepox	Cutaneous crusty lesions mainly in flanks, ventral abdomen, legs, inguinal areas and ears; also, congenital form	Cutaneous crusty lesions	Hydropic degeneration of stratum spinosum of epidermis, intracytoplasmic inclusion bodies
RV	Rotavirosis	Vomiting, profuse watery yellowish diarrhoea, high morbidity and weight loss; lower mortality compared to coronavirus infections	Catarrhal enteritis (apparent thinner small intestine wall)	Fusion and atrophy of intestinal villi

[1]See abbreviations in Table 1.
SMEDI-like = presence of signs including stillborn and mummified piglets, embryonic death and/or infertility.
INIB = intranuclear inclusion bodies.
*These lesions in the intestine are caused by *Salmonella enterica* subsp. Typhimurium, not by the effect of CSFV replication. The immunosuppression caused by CSFV infection and viral replication in the intestine facilitates the subsequent effect of the bacteria.
**Especially high when infected by highly pathogenic PRRSV strains.

Recognition of a viral disease clinically or pathologically through necropsy can be challenging. As observed in Table 3, there are myriad different clinical signs caused by viral infections that are difficult to differentiate from diseases caused by other agents, toxins or nutritional deficiencies/toxicities. In general terms, a detailed clinical examination of each pig is not usually performed, although it is considered essential to establish the basis of the potential disease the animal is suffering from (Carr et al., 2018). Therefore, the clinical diagnostic approach must involve both group and individual examinations, including, eventually, the necropsy of representative animals showing the clinical signs of concern. At necropsy, there are some lesions or findings that can be potentially attributed to viral diseases and, consequently, help determine diagnosis of the condition, including:

- Interstitial pneumonia: tan-mottled lung with lack of collapse with eventual interstitial oedema (typical of PRRSV, PCV-2, PRCV); if a similar lesion is mainly located in cranio-ventral areas of the lung, it will probably be broncho-interstitial pneumonia, which would be compatible with SIV infection (also the basic lesion of a well-known bacterial agent of swine: *Mycoplasma hyopneumoniae*).
- Generalised lymphadenopathy: increase in size of most lymph nodes in a carcass is highly suggestive of a systemic viral infection, such as those produced by PRRSV, PCV-2, ASFV and CSFV. However, this can also be found in bacterial septicaemia.
- Catarrhal enteritis with very fluid content: atrophic enteritis due to shortening and fusion of villi (intestinal wall looks thinner than normal) is typical of enteric coronavirus and rotavirus infections. Very fluid content mainly in the small intestine.
- SMEDI (stillbirth, mummified, embryonic death and infertility): presence of stillborn piglets and mummified foetuses is usually found in viral reproductive infections such as those caused by PPV, swine enteroviruses and PCV-2. However, other causes of SMEDI findings exist that must be included in the differential diagnostic list.
- Generalised haemorrhages: haemorrhagic diathesis represents severe, acute damage of the cardiovascular system, and systemic viruses such as ASFV and CSFV must be suspected.

Of course, those lesions must be interpreted in the context of existing clinical signs and the epidemiological features of the condition being experienced in the herd, and further diagnostic efforts should confirm or rule out these potential viral infections.

7 Diagnosis of swine viral infections

A correct and precise disease diagnosis allows the establishment of adequate preventive and control measures at different levels (management, breeding, animal flow, therapy, vaccination schedule etc.), with high chances of success. Therefore, a global diagnostic approach including clinical history, epidemiological features, clinical assessment, gross pathological examination (necropsy) and, eventually, laboratory analyses, is paramount to deal with diseases.

This section offers a broad view of different types of laboratory analyses to help diagnosing swine viral diseases. It must be emphasised that the final diagnosis of a given condition is established by the farm-responsible veterinarian, since he/she is the person that compiles all the information that can be retrospectively and proactively gathered for a given disease case; laboratory analyses are support tools that allow establishing the participation/presence of a given pathogen in the pathological situation (Segalés, 2006).

Adequate sample selection is a crucial factor when testing to detect a virus. Although laboratory tests have particular sensitivity, specificity and predictive values, the submitting veterinarian has a major impact on the reliability of the result by selecting the right pig/s and samples/s to analyse (Segalés, 2006). Table 4 summarises the most frequent diagnostic laboratory tests used to detect swine viruses, including the type of samples to be used.

The type of sample used for diagnosing swine viral infections has expanded over time. Tissues fixed by immersion in 10% buffered formalin have been traditionally used for histopathological assessment. Microscopic lesions do not usually drive an aetiological diagnosis as most lesions only offer a differential diagnostic list of potential causal agents. However, histopathology can be coupled with alternative techniques such as IHC and ISH, which allow the detection of a specific antigen or nucleic acid of a particular virus within the microscopic lesions (section 3 of this chapter). Classic samples to detect a viral genome by different PCR (DNA) or RT-PCR (RNA) methods, quantitative or not, include the target tissues for the pathogen, such as serum in the case of agents causing viraemia. However, viral detection is not only used for specific disease diagnostic purposes but also for monitoring viral infections in farms. Over the last decade, sampling for herd monitoring has been simplified using oral and processing fluids. These sampling techniques are inexpensive, easy to acquire and can be applied to individual pigs (as a diagnostic) or pens/litters (for herd surveillance) (Prickett and Zimmerman, 2010; Turlewicz-Podbielska et al., 2020; Baliellas et al., 2021; Pleguezuelos et al., 2021):

- Oral fluids: a mixture of saliva and oral mucosal transudate that is collected when pigs chew on a cotton rope suspended at pig shoulder height in

Table 4 Most frequent diagnostic laboratory tests used to detect swine viruses, including the type of samples to be used for each of them.

Viral agent[1]	Diagnostic laboratory test	Sample type
ASFV	Viral detection: PCR, haemadsorption test Antibody detection: ELISA, indirect immunofluorescence (IFA)	PCR: tonsil, gastro-hepatic lymph node, kidney, spleen, lung, bone marrow, serum, blood Haemadsorption test: blood ELISA/IFA: serum
PCV-2	Lesion detection: histopathology Virus detection: IHC or ISH in lymphoid tissues, PCR, qPCR Antibody detection: ELISA	Histopathology: lymphoid tissues IHC/ISH: lymphoid tissues PCR, qPCR: serum, blood, tissues (lymph nodes, lung) ELISA: serum
PEDV	Lesion detection: histopathology Virus detection: RT-PCR, IHC, IFT Antibody detection: ELISA	Histopathology, IHC and IFT: small intestine RT-PCR: intestinal contents, faeces, small intestine scrapping ELISA: serum
TGEV	Lesion detection: histopathology Virus detection: RT-PCR, IHC, IFT Antibody detection: ELISA	Histopathology, IHC and IFT: small intestine RT-PCR: intestinal contents, faeces, small intestine scrapping ELISA: serum
JEV	Virus detection: virus isolation, RT-PCR, IHC Antibody detection: ELISA, VNT	Virus isolation, RT-PCR, IHC: Placenta from foetuses, stillbirth and neonates, brain, spleen, liver Virus isolation, RT-PCR: blood and serum ELISA, VNT: serum
HEV	Virus detection: RT-PCR Antibody detection: ELISA	RT-PCR: liver, bile, serum, mesenteric lymph node ELISA: serum
ADV/PRV	Lesion detection: histopathology Virus detection: PCR, IHC, virus isolation, IFT Antibody detection: ELISA*, VNT	Histopathology, PCR, IHC, IFT, virus isolation: brain, lung, lymph nodes, tonsil, liver, spleen, kidney ELISA, VNT: serum
PCMV	Lesion detection: histopathology Virus detection: PCR Antibody detection: ELISA	Histopathology: nasal turbinates, kidney, liver, bone marrow PCR: nasal swab, blood, lung, kidney; also brain, liver and bone marrow of foetuses ELISA: serum
SIV	Virus detection: RT-PCR, virus isolation Antibody detection: ELISA, HI	RT-PCR, virus isolation: nasal swab, lung, bronchoalveolar lavage ELISA, HI: serum
Nipah virus**	Virus detection: RT-PCR, RT-qPCR, IHC, virus isolation Antibody detection: ELISA, VNT	RT-PCR, RT-qPCR, virus isolation: nasal and oral swabs, brain, trigeminal ganglion, lung, spleen, lymph nodes, kidney IHC: brain, trigeminal ganglion, lung, spleen, lymph nodes, kidney ELISA, VNT: serum

(Continued)

Table 4 (Continued)

Viral agent[1]	Diagnostic laboratory test	Sample type
PPV	Virus detection: PCR, IFT Antibody detection: ELISA, HI	PCR: foetal tissues (brain and lung), amniotic liquid, semen IFT: foetal tissues (brain and lung) ELISA, HI: serum
CSFV	Virus detection: RT-PCR, virus isolation, antigen capture ELISA, IFT Antibody detection: ELISA, VNT	RT-PCR, virus isolation: blood, serum, plasma, nasal and rectal swabs, tonsil, spleen, ileum, lymph nodes Antigen capture ELISA: serum, blood, tissue homogenates IFT: Tonsil, spleen, ileum, lymph nodes ELISA, VNT: serum
APPV	Virus detection: RT-PCR Antibody detection: ELISA	RT-PCR: brain, tonsil, lymph nodes, serum ELISA: serum
FMDV	Virus detection: RT-PCR, antigen capture ELISA, virus isolation Antibody detection: ELISA, VNT	RT-PCR, antigen capture ELISA, virus isolation: epithelial tissue suspension, vesicular material ELISA, VNT: serum
SVDV	Virus detection: RT-PCR, antigen capture ELISA, virus isolation Antibody detection: ELISA, VNT	RT-PCR, antigen capture ELISA, virus isolation: epithelial tissue suspension, vesicular material ELISA, VNT: serum
SVA	Virus detection: RT-PCR, antigen capture ELISA, virus isolation Antibody detection: ELISA, VNT	RT-PCR, antigen capture ELISA, virus isolation: epithelial tissue suspension, vesicular material ELISA, VNT: serum
PRRSV	Virus detection: RT-PCR, RT-qPCR, IHC, ISH, virus isolation Antibody detection: ELISA, VNT	RT-PCR, RT-qPCR: blood, serum, tonsil, lymph nodes, lung, oral fluids, processing fluids Virus isolation: blood, serum, tonsil, lymph nodes, lung IHC, ISH: tonsil, lymph nodes, lung ELISA, VNT: serum
SwPV	Lesion detection: histopathology Virus detection: PCR, IHC Antibody detection: ELISA	Histopathology, IHC: skin PCR: epithelial tissue, crusty cutaneous material ELISA: serum
RV	Lesion detection: histopathology Virus detection: RT-PCR Antibody detection: ELISA	Histopathology: small intestine RT-PCR: intestinal contents, faeces, small intestine scrapping ELISA: serum

[1]See abbreviations in Table 1.
ELISA, enzyme linked immunosorbent assay; PCR, polymerase chain reaction; qPCR, quantitative PCR; RT-PCR, reverse transcription PCR; RT-qPCR, reverse transcription quantitative PCR; IHC, immunohistochemistry; ISH, *in situ* hybridisation; IFT, immunofluorescence test; VNT, virus neutralisation test; HI, haemagglutination inhibition test.
*ADV/PRV ELISA tests have been commercially developed to detect mainly gE and gB proteins; since most currently used vaccines have deletion of the gE gene, evidence of antibodies against gE (and gB) is indicative of natural infection, while detection of only antibodies against gB is indicative of vaccination.
**High-risk (biosafety level 4) pathogen; extreme care should be taken during sample collection, using the appropriate personal protective equipment.

the pen, typically for 20–30 minutes. Used to detect a variety of viruses, including PRRSV, PCV-2, SIV, CSFV, ASFV and SVDV.

- Processing fluids: tissues obtained during processing (castration and tail docking); the testicles and tails from littermates are placed in a clean bucket and the fluids are filtered through cheesecloth into a plastic bag. The contents are transferred into a sterile tube and shipped on ice to a veterinary diagnostic laboratory.
- Placental umbilical cords: a mixture of blood and transudate obtained from pooled umbilical cords. It has been mainly used to detect PRRSV and PCV-2.
- Tongue tissues from dead pigs: the collection of approximately 2-cm pieces of tongue, which are then stored in bags or containers in the freezer. The fluid generated after thawing is submitted to the diagnostic laboratory. Mostly used to monitor PRRSV infection.
- Nasal wipes and udder skin wipes: nasal wipes are collected using cotton gauze, which is wiped across the snout, collecting secretions from the snout and nares. Secretions from the oral and nasal cavities of lactating piglets are deposited on the skin of the sow udder, and the wipes enable the collection of these secretions. It has been used more sporadically than other collection methods to detect PRRSV, PCV-2 and SIV.

These novel sample types are non-aggressive for the pig (animal welfare) and user-friendly for the operator (easy to obtain). However, besides oral fluids, most of them are still poorly validated. Moreover, most of these samples are used to detect the viral genome through PCR platforms, and these methods must be validated for the accurate detection of DNA or RNA, as the efficiency of viral nucleic acid extraction depends upon the sample used (Prickett and Zimmerman, 2010; Turlewicz-Podbielska et al., 2020). It is also important to note that most of the novel sampling practices are population-based, so in general terms, are especially useful for monitoring a viral infection, and not for establishing a particular disease diagnosis.

In addition to pathogen detection, most monitoring programmes for viral agents include antibody testing. The general framework of a humoral response to a pathogen consists of an initial phase (usually a few days) in which the animal does not have antibodies. Antibodies of the IgM class then appear, and after 10–14 days (sometimes up to 3 weeks), IgG levels increase significantly. High IgM levels are usually indicative of a recent infection, while the predominance of IgG means that the onset of infection took place several weeks earlier (Segalés et al., 2013). So far, many techniques have been developed to detect antibodies against viral pathogens of swine, including ELISA, haemagglutination inhibition, immunofluorescence antibody test, complement fixation test, viral neutralisation and agglutination test (Segalés et al., 2013). However, only

ELISAs can be performed in a quick and high-throughput fashion, with many ELISA kits commercially available and fully validated (Christopher-Hennings et al., 2019). Consequently, ELISAs are currently the most popular assays for antibody detection.

Two major types of ELISAs exist, namely indirect and competitive/blocking (Segalés et al., 2013). Indirect ELISAs allow the semi-quantification of antibodies if a positive control sample includes a known concentration of antibodies, so a sample-to-positive (S/P) ratio can be offered in most of the cases. Competitive or blocking ELISAs have the advantage of not requiring species-specific conjugates, but readouts are in terms of positive/negative.

Commercial ELISAs have been mainly validated for the detection of antibodies in serum. However, only some of the tests have been validated to detect antibodies in oral fluids and meat juice, although the list of in-house or commercially modified ELISAs is much larger in research (Turlewicz-Podbielska et al., 2020). For example, oral fluids have been used to detect antibodies against PRRSV, PCV-2, SIV, ASFV, CSFV and SVDV, while meat juice has been used to analyse antibodies against PRRSV, PCV-2, ADV/PRV, CSFV and PEDV.

8 Prevention and control of swine viral infections

Several influencing factors determine whether a pig develops clinical illness after exposure to a virus. Factors affecting host susceptibility include age, immunocompetency, immunological status (due to previous natural infection, vaccination or the presence of maternally derived immunity), genetic predisposition, concurrent diseases/infections, stress, environment, management and nutrition. Importantly, there are also factors that can impact the risk of infection, which are the source of infection and the transmission route (Amass and Baysinger, 2006). Consequently, most of the prevention and control measures to be applied on a farm will be oriented to eliminate or minimise the source of infection and routes of transmission.

In the first instance, the most important task for producers and veterinarians is to avoid the entrance of the pathogen into the herd, which is mainly driven by external biosecurity measures (see chapter 10). This objective is fundamental in all cases, but especially crucial in the case of notifiable diseases from the World Organization of Animal Health (WOAH, 2021). Current viral infections included in the WOAH list are ASFV, CSFV, Nipah virus, PRRSV, TGEV, ADV/PRV, FMDV, Japanese encephalitis virus and rabies virus. Most of these viral infections are subjected to specific legislation worldwide, with the policy of stamping-out and limitations in national and international trading in force in many countries.

However, it is possible that the viral infection is already present in the herd, as happens with a significant number of endemic viral diseases. Although the world map of such endemic diseases is very diverse among countries and regions (i.e. PRRSV is endemic in many countries worldwide, but Brazil, Chile, Switzerland, Sweden, Norway and Finland are free from this viral infection), there are different generic principles of prevention and control that should be considered, including:

- Production system: closed, farrow-to-finish herds tend to maintain the infection endemically with a continuous exposure of the pathogen from one age group to others susceptible to the virus. Multi-site production systems have a higher risk of creating subpopulations of naïve animals and moving subclinically infected animals between sites.
- Replacement rates and acclimation: an important external biosecurity asset; it is crucial to know the sanitary status of the replacement stock and guarantee a sufficiently large quarantine or acclimation to ensure that an eventual active infection is not spread to the production site. It is particularly important to test that the acclimation period has been effective in terms of immunisation or deliberate infection of the gilts.
- Management and hygiene: general management practices that are of benefit for controlling viral diseases can be applied at all pig sites. It is imperative to ensure a good colostrum intake and limit cross-fostering in the farrowing crates within the first 24 hours after birth. Strict all in/all out procedures at all sites accompanied by cleaning, disinfection and downtime between batches are also measures that reduce infection transmission. Most of these measures were considered in the so-called 20-point plan of Madec (Madec et al., 2001), initially conceived to control PCV-2 systemic disease when no vaccines were available, and the MCREBEL (Management Changes to Reduce Exposure to Bacteria to Eliminate Losses) system (McCaw, 1995), which was developed to control PRRS in large herds. Another management measure that usually impacts positively on disease burden in the post-weaning area is increasing the weaning age (Faccin et al., 2020).
- Nutrition: feed composition and form mostly modulate the clinical expression of enteric bacterial diseases by significantly modifying the microbiota of the intestinal tract. However, the effect of the gut microbiota on viral disease outcome is a field of expanding research (Niederwerder, 2017). For example, in experimental models of PRRSV and PCV-2 infection, it was observed that pigs with lower microbial diversity of the microbiota were those exhibiting the worst clinical outcome (Niederwerder et al., 2016). In addition, the use of nutritional components as spray-dried plasma in nursery diets has been shown to be related with lower levels of mortality in the context of PCV-2 systemic disease (Dewey et al., 2006) or the use of isoflavones, like puerarin, showed improved intestinal function in piglets infected by PEDV (Wu et al., 2021). A final important point is to

highlight that nutritional components may serve as potential sources of viral contamination entry to a farm (Dee et al., 2021). In this latter study, samples of soya bean meal, lysine, choline and vitamin A were spiked with a mixture of PRRSV, PEDV and SVA and transported for 21 days in the trailer of a commercial transport vehicle for almost 10 000 km. After the transit period, infectious PRRSV, PEDV and SVA were detected by swine bioassay in soya products, PRRSV and SVA in vitamin A, and SVA in lysine and choline.

- Climate and housing conditions: different studies highlight that certain climate conditions (temperature, wind velocity and sunlight) may favour the transmission of several viral swine diseases such as PRRSV (Pitkin et al., 2009), FMDV (Hagerman et al., 2018) and Japanese encephalitis virus (Diallo et al., 2018) between farms. Environmental conditions within the barns (temperature, humidity and airspeed) can also affect virus transmission between animals, such as PRRSV (Fablet et al., 2016) or SIV (Ferreira et al., 2018).

In addition to the potential control of the risk factors within and between herds that facilitate viral transmission, the most effective way of preventing viral diseases in swine farms is by means of vaccination.

Vaccination can be administered to the piglet, but also to the sow to provide protection to the piglet and/or to protect the pregnancy. Boar vaccination is also recommended to avoid pathogen transmission. Several factors must be considered when designing a vaccination schedule for a herd (Chase and Lunney, 2019). First, the vaccination programme to be implemented in a farm should be based on the disease risks of the herd. Second, the effect of maternally derived immunity and age of the pig at vaccination must be balanced to avoid potential interference with vaccine efficacy. Third, vaccines are not intended to counteract deficiencies in management and biosecurity practices, that is, vaccination is an additional tool for the overall control of disease burden at the farm.

Most, if not all, vaccines used in swine should be considered as 'imperfect' vaccines, as they are only able to protect against moderate and severe clinical signs as well as mortality (Rose and Andraud, 2017). In other words, swine vaccines may show different degrees of efficiency in controlling a given viral disease, but none of them offer sterilising immunity, that is, they cannot prevent infection. The final objective of vaccinating every pig is to acquire herd immunity. This latter concept describes 'the threshold proportion of immune individuals that should lead to a decline in incidence of infection attributable to immunity acquired via infection and/or vaccination' (Chase and Lunney, 2019). Therefore, vaccination should ideally reach the critical threshold (in terms of population proportion) needed to mitigate the effects of viral infection in the herd. Furthermore, viral vaccines tend to be rather efficient in terms of reducing transmission, as demonstrated for PRRSV, PCV-2, SIV, CSFV and ADV/PRV (Rose and Andraud, 2017). Importantly, when multiple vaccine studies focusing on transmission reduction against some of these viruses are obtained, the results show good agreement among them, emphasising the efficacy of these vaccines.

For most of the viral pathogens infecting swine, it is impossible to distinguish the humoral immunity (antibodies) provided by vaccines from that originating from natural infection or colostrum. There are a couple of exceptions, which are configured by ADV/PRV vaccines containing a virus with deletion of the *gE* gene and CSFV subunit products based on the E2 recombinant protein. These vaccines are good examples of the so-called DIVA concept (Vannier et al., 2007), which allows differentiating infected from vaccinated animals. This strategy helps eliminate potential restrictions (national and international trading, quarantines) needed in the case of animals naturally infected by the wild-type virus.

There are several scenarios in which animals may develop disease despite vaccination (Roth, 1999):

- Vaccine administration in the face of maternal immunity: while this fact is evident for ADV/PRV, PPV and SIV vaccine intake, resulting in vaccine efficacy interference, others can overcome a certain level of maternally derived immunity, for example for PRRSV and PCV-2, and not lose efficacy.
- Vaccine administration after natural infection has occurred: the expected vaccine efficacy in this scenario is reduced, but for some viruses such as ADV/PRV, CSFV and FMDV, emergency vaccination in the case of a disease outbreak is recommended (Miller et al., 2015). As another example, PCV-2 vaccination also offers some beneficial effects when vaccinating already infected animals (Seo et al., 2014), but should never be the preferred option.
- Improper handling of vaccines (incorrect storage, sunlight exposure, expired, delivered in fat tissue) and/or administration equipment.
- Improper timing of vaccination and/or age when administered.
- Poor cross-protection against heterologous viral strains/variants.
- Immune dysfunction or compromise at the time of vaccination: a very good example would be PCV-2 vaccination at the time of PRRSV circulation, in which the vaccine is able to generate a normal humoral immune response, but the cellular immune response is impaired (Canelli et al., 2016).

Table 5 summarises the most frequent viral pathogens of swine for which vaccination with commercial products is practised all over the world, indicating the usual vaccination timing and schedule as well as vaccine type. Autogenous vaccines based on viruses are not currently allowed in Europe but are permitted in other parts of the world. This type of vaccine is a well-established tool of veterinary medicine to fill the gaps left by licensed vaccines and must strictly comply with basic quality and safety requirements (Hera and Bures, 2004).

Table 5 Most frequent viral pathogens of swine for which vaccination using commercial products is practised all over the world, indicating the usual vaccination timing and schedule as well as vaccine type.

Viral agent[1]	Animal group	Vaccination timing	Vaccine types
PCV-2	Sow	Any time before or during gestation	KV, subunit vaccine, KV chimeric
	Gilt	During acclimation	
	Piglet	Around weaning	
PEDV	Gilts/sows	Pre-farrowing	KV, MLV, RNA-based
ADV/PRV	Sow	Any time before or during gestation	KV, MLV (DIVA vaccines)
	Gilt	During acclimation	
	Pig	Late nursery and early fattening periods	
SIV	Sow	Any time before or during gestation	KV, MLV
	Gilt	During acclimation	
	Pig	Mid nursery and early fattening periods	
PPV	Gilts	During acclimation	KV
	Sows	Any time before or during gestation	
CSFV*	Sow	Any time before or during gestation	KV, MLV, subunit vaccine (DIVA vaccine)
	Gilt	During acclimation	
	Pig	Mid to late nursery and early fattening periods	
FMDV*	Gilts/sows	During gestation	KV
	Pigs	Mid to late nursery and early fattening periods	
PRRSV	Sow	Any time before or during gestation	KV, MLV
	Gilt	During acclimation	
	Piglet	Around weaning	
RV	Gilts/sows	Before farrowing	KV, MLV
	Pigs	Before weaning	

[1]See abbreviations in Table 1.
KV, killed, inactivated vaccine; MLV, modified live vaccine; DIVA, vaccines with genome deletion of the included virus that allow differentiating from wild-type viruses.
*In many parts of the world, stamping-out policies apply in front of these notifiable diseases; so, no vaccination is allowed.

9 Conclusion

Swine production contributes to one-third of meat consumed throughout the world, representing an important component of global food security, agricultural economies and trade (Van der Waal and Deen, 2018). In this context, infectious viral diseases of pigs may have dramatic consequences on the food chain supply, which is well illustrated with the outbreaks of ASF in Southeast

Asia in the last four years (Mason-D'Croz et al., 2020). Besides ASFV infection, the global swine industry is affected by several enzootic viral diseases, which exert detrimental effects on farm productivity with an important economic impact, for example, PRRSV, PCV-2, PEDV, ADV/PRV or CSFV. Moreover, some viral infections are considered to be a threat to public health due to their zoonotic nature, such as SIV and HEV (Andraud and Rose, 2020). In all cases, understanding viral infection dynamics, transmission modes, novel and faster diagnostic capabilities, as well as the development of more efficient vaccines are paramount to continue advancing towards sustainable, profitable and smart swine farming and production.

Importantly, understanding the factors influencing the emergence of novel infectious agents is key for the future of swine production, demanding anticipation and preparedness prior to such potential novel emerging infections. Fournié et al. (2015) found that from 1985 to 2010, an average annual rate of three novel swine pathogen species were identified. These numbers have further increased with the advent of more sophisticated research and diagnostic tools, so it is highly likely that the swine industry will have to deal with novel important swine viral pathogens in the future.

10 References

Amass, S. F. and Baysinger, A. (2006). Chapter 68: Swine disease transmission and prevention. In: Straw, B. E., Zimmerman, J. J., D'Allaire, S. and Taylor, D. J. (Eds). *Diseases of Swine* (9th edn.). Ames, IA: Blackwell Publishing, 1075–1098.

Andraud, M. and Rose, N. (2020). Modelling infectious viral diseases in swine populations: A state of the art. *Porcine Health Manag.* 6, 22.

Baliellas, J., Novell, E., Enric-Tarancon, V., Vilalta, C. and Fraile, L. (2021). Porcine reproductive and respiratory syndrome surveillance in breeding herds and nurseries using tongue tips from dead animals. *Vet. Sci.* 8(11), 259.

Belák, S. (2005). The molecular diagnosis of porcine viral diseases: A review. *Acta Vet. Hung.* 53(1), 113–124.

Belák, S., Karlsson, O. E., Blomström, A. L., Berg, M. and Granberg, F. (2013). New viruses in veterinary medicine, detected by metagenomic approaches. *Vet. Microbiol.* 165(1–2), 95–101.

Blome, S., Franzke, K. and Beer, M. (2020). African swine fever – A review of current knowledge. *Virus Res.* 287, 198099.

Blomström, A. L. (2011). Viral metagenomics as an emerging and powerful tool in veterinary medicine. *Vet. Q.* 31(3), 107–114.

Burrell, C. J., Howard, C. R. and Murphy, F. A. (2017). Chapter 3: Virion structure and composition. In: *Fenner and White's Medical Virology* (5th edn.). London: Academic Press, 27–37.

Canelli, E., Borghetti, P., Ferrari, L., De Angelis, E., Ferrarini, G., Catella, A., Ogno, G. and Martelli, P. (2016). Immune response to PCV2 vaccination in PRRSV viraemic piglets. *Vet. Rec.* 178(8), 193.

Carlson, C. J., Zipfel, C. M., Garnier, R. and Bansal, S. (2019). Global estimates of mammalian viral diversity accounting for host sharing. *Nat. Ecol. Evol.* 3(7), 1070–1075.

Carr, J., Chen, S. P., Connor, J. F., Kirkwood, R. and Segalés, J. (2018). Chapter 1: Clinical examination. In: Carr, J., Chen, S.-P. and Connor, J. F. (Eds). *Pig Health*. CRC Press, Taylor and Francis Group, 1-52.

Chase, C. and Lunney, J. K. (2019). Chapter 16: Immune system. In: Zimmerman, J. J., Karriker, L. A., Ramirez, A., Schwartz, K. J., Stevenson, G. W. and Zhang, J. (Eds). *Diseases of Swine* (11th edn.). Hoboken, NJ: John Wiley & Sons, Inc., 265–291.

Christopher-Hennings, J., Erickson, G. A., Hesse, R. A., Nelson, E. A., Rossow, S., Scaria, J. and Slavic, D. (2019). Chapter 6: Diagnostic tests, test performance, and considerations for interpretation. In: Zimmerman, J. J., Karriker, L. A., Ramirez, A., Schwartz, K. J., Stevenson, G. W. and Zhang, J. (Eds). *Diseases of Swine* (11th edn.). Hoboken, NJ: John Wiley & Sons, Inc., 75–97.

Cifuentes-Muñoz, N., El Najjar, F. and Dutch, R. E. (2020). Viral cell-to-cell spread: Conventional and non-conventional ways. *Adv. Virus Res.* 108, 85–125.

Dee, S., Shah, A., Cochrane, R., Clement, T., Singrey, A., Edler, R., Spronk, G., Niederwerder, M. and Nelson, E. (2021). Use of a demonstration project to evaluate viral survival in feed: Proof of concept. *Transbound. Emerg. Dis.* 68(2), 248–252.

de Groof, A., Deijs, M., Guelen, L., van Grinsven, L., van Os-Galdos, L., Vogels, W., Derks, C., Cruijsen, T., Geurts, V., Vrijenhoek, M., Suijskens, J., van Doorn, P., van Leengoed, L., Schrier, C. and van der Hoek, L. (2016). Atypical porcine pestivirus: A possible cause of congenital tremor Type A-II in newborn piglets. *Viruses* 8(10), 271.

Delmas, B., Gelfi, J., L'Haridon, R., Vogel, L. K., Sjöström, H., Norén, O. and Laude, H. (1992). Aminopeptidase N is a major receptor for the entero-pathogenic coronavirus TGEV. *Nature* 357(6377), 417–420.

Dewey, C. E., Johnston, W. T., Gould, L. and Whiting, T. L. (2006). Postweaning mortality in Manitoba swine. *Can. J. Vet. Res.* 70(3), 161–167.

Diallo, A. O. I., Chevalier, V., Cappelle, J., Duong, V., Fontenille, D. and Duboz, R. (2018). How much does direct transmission between pigs contribute to Japanese encephalitis virus circulation? A modelling approach in Cambodia. *PLoS ONE* 13(8), e0201209.

Doms, R. W. (2016). Chapter 3: Basic concepts: A step-by-step guide to viral infection. In: Katze, M. G., Korth, M. J. and Lynn, G. (Eds). *Viral Pathogenesis: From Basics to Systems Biology* (3rd edn.). Boston, MA: Academic Press, 29–40.

Duerkop, B. A. and Hooper, L. V. (2013). Resident viruses and their interactions with the immune system. *Nat. Immunol.* 14(7), 654–659.

Fablet, C., Marois-Créhan, C., Grasland, B., Simon, G. and Rose, N. (2016). Factors associated with herd-level PRRSV infection and age-time to seroconversion in farrow-to-finish herds. *Vet. Microbiol.* 192, 10–20.

Faccin, J. E. G., Laskoski, F., Hernig, L. F., Kummer, R., Lima, G. F. R., Orlando, U. A. D., Gonçalves, M. A. D., Mellagi, A. P. G., Ulguim, R. R. and Bortolozzo, F. P. (2020). Impact of increasing weaning age on pig performance and belly nosing prevalence in a commercial multisite production system. *J. Anim. Sci.* 98(4), skaa031.

Ferreira, J. B., Grgić, H., Friendship, R., Nagy, É. and Poljak, Z. (2018). Influence of microclimate conditions on the cumulative exposure of nursery pigs to swine influenza A viruses. *Transbound. Emerg. Dis.* 65(1), e145–e154.

Fournié, G., Kearsley-Fleet, L., Otte, J. and Pfeiffer, D. U. (2015). Spatiotemporal trends in the discovery of new swine infectious agents. *Vet. Res.* 46, 114.

Gorbalenya, A.E., Krupovic, M., Mushegian, A., Kropinski, A. M., Siddell, S. G., Varsani, A., Adams, M. J., Davison, A. J., Dutilh, B. E., Harrach, B., Harrison, R. L., Junglen, S., King, A. M. Q., Knowles, N. J., Lefkowitz, E. J., Nibert, M. L., Rubino, L., Sabanadzovic, S., Sanfaçon, H., Simmonds, P., Walker, P. J., Murilo Zerbini, F. and Kuhn, J. H. (2020). The new scope of virus taxonomy: Partitioning the virosphere into 15 hierarchical ranks. *Nat. Microbiol.* 5(5), 668–674.

Hagerman, A. D., South, D. D., Sondgerath, T. C., Patyk, K. A., Sanson, R. L., Schumacher, R. S., Delgado, A. H. and Magzamen, S. (2018). Temporal and geographic distribution of weather conditions favorable to airborne spread of foot-and-mouth disease in the coterminous United States. *Prev. Vet. Med.* 161, 41–49.

Handelsman, J. (2004). Metagenomics: application of genomics to uncultured microorganisms. *Microbiol. Mol. Biol. Rev.* 68(4), 669–685.

Hanssens, F. P., Nauwynck, H. J. and Mettenleiter, T. C. (1995). Role of glycoprotein gD in the adhesion of pseudorabies virus infected cells and subsequent cell-associated virus spread. *Arch. Virol.* 140(10), 1855–1862.

Hause, B. M., Collin, E. A., Peddireddi, L., Yuan, F., Chen, Z., Hesse, R. A., Gauger, P. C., Clement, T., Fang, Y. and Anderson, G. (2015). Discovery of a novel putative atypical porcine pestivirus in pigs in the USA. *J. Gen. Virol.* 96(10), 2994–2998.

He, B., Gong, W., Yan, X., Zhao, Z., Yang, L., Tan, Z., Xu, L., Zhu, A., Zhang, J., Rao, J., Yu, X., Jiang, J., Lu, Z., Zhang, Y., Wu, J., Li, Y., Shi, Y., Jiang, Q., Chen, X. and Tu, C. (2021). Viral metagenome-based precision surveillance of pig population at large scale reveals Viromic signatures of sample types and influence of farming management on pig Virome. *mSystems* 6(3), e0042021.

Hera, A. and Bures, J. (2004). Veterinary autogenous vaccines. *Dev. Biol. (Basel)* 117, 19–25.

Li, B. X., Ge, J. W. and Li, Y. J. (2007). Porcine aminopeptidase N is a functional receptor for the PEDV coronavirus. *Virology* 365(1), 166–172.

Louten, J. (2017). Chapter 5: Virus transmission and epidemiology. In: Louten, J. (Ed.). *Essential Human Virology.* Elsevier Inc., 71–92.

Maclachlan, N. J. and Dubovi, E. J. (2017). Chapter 3: Pathogenesis of viral infections and diseases. In: Maclachlan, N. J., Dubovi, E. J. (Eds). *Fenner Veterinary Virology* (5th edn.). London: Elsevier Inc., 47–78.

Madec, F., Rose, N., Eveno, E., Morvan, P., Larour, P., Jolly, J. P., Le Diguerher, G., Cariolet, R., Le Dimna, M., Blanchard, P. and Jestin, A. (2001). PMWS: on-farm observations and preliminary analytic epidemiology. In: *Proceedings of the ssDNA Viruses of Plants, Pigs and Primates*, Saint Maló, France, 86–88.

Male, D. (2006). Chapter 1: Introduction to the immune system. In: Male, D., Brostoff, J., Roth, D. B. and Roitt, I. (Eds). *Immunology* (Section 1) (7th edn.). Mosby: Elsevier, 3–18.

Mankertz, A., Mankertz, J., Wolf, K. and Buhk, H. J. (1998). Identification of a protein essential for replication of porcine circovirus. *J. Gen. Virol.* 79(2), 381–384.

Mason-D'Croz, D., Bogard, J. R., Herrero, M., Robinson, S., Sulser, T. B., Wiebe, K., Willenbockel, D. and Godfray, H. C. J. (2020). Modelling the global economic consequences of a major African swine fever outbreak in China. *Nat. Food* 1(4), 221–228.

McCaw, M. B. (1995). MCREBEL PRRS: Management procedures for PRRS control in large herd nurseries. In: *Proceedings of the Allen D. Leman Conference*, Vol. 22, 161–162.

Miller, G., Gale, S. B., Eshelman, C. E. and Wells, S. J. (2015). Emergency vaccination use in a modelled foot and mouth disease outbreak in Minnesota. *Rev. Sci. Tech.* 34(3), 729–740.

Nauwynck, H. J. and Pensaert, M. B. (1995). Cell-free and cell-associated viremia in pigs after oronasal infection with Aujeszky's disease virus. *Vet. Microbiol.* 43(4), 307–314.

Niederwerder, M. C. (2017). Role of the microbiome in swine respiratory disease. *Vet. Microbiol.* 209, 97–106.

Niederwerder, M. C., Jaing, C. J., Thissen, J. B., Cino-Ozuna, A. G., McLoughlin, K. S. and Rowland, R. R. (2016). Microbiome associations in pigs with the best and worst clinical outcomes following co-infection with porcine reproductive and respiratory syndrome virus (PRRSV) and porcine circovirus type 2 (PCV2). *Vet. Microbiol.* 188, 1–11.

WOAH, World Organization for Animal Health. (2021). Available at: https://www.woah.org/en/what-we-do/animal-health-and-welfare/animal-diseases/?tax%20animal=terrestrials%2Cswine (revised on 28 June 2022).

Olvera, A., Sibila, M., Calsamiglia, M., Segalés, J. and Domingo, M. (2004). Comparison of porcine circovirus type 2 load in serum quantified by a real time PCR in postweaning multisystemic wasting syndrome and porcine dermatitis and nephropathy syndrome naturally affected pigs. *J. Virol. Methods* 117(1), 75–80.

Payne, S. (2017). Chapter 9: Viral pathogenesis. In: *Viruses: From Understanding to Investigation*. Academic Press, 87–95.

Phan, T. G., Giannitti, F., Rossow, S., Marthaler, D., Knutson, T. P., Li, L., Deng, X., Resende, T., Vannucci, F. and Delwart, E. (2016). Detection of a novel circovirus PCV3 in pigs with cardiac and multi-systemic inflammation. *Virol. J.* 13(1), 184.

Pitkin, A., Deen, J. and Dee, S. (2009). Use of a production region model to assess the airborne spread of porcine reproductive and respiratory syndrome virus. *Vet. Microbiol.* 136(1–2), 1–7.

Pleguezuelos, P., Sibila, M., Cuadrado, R., López-Jiménez, R., Pérez, D., Huerta, E., Llorens, A. M., Núñez, J. I., Segalés, J. and López-Soria, S. (2021). Exploratory field study on the effects of porcine circovirus 2 (PCV-2) sow vaccination at different physiological stages mimicking blanket vaccination. *Porcine Health Manag.* 7(1), 35.

Prickett, J. R. and Zimmerman, J. J. (2010). The development of oral fluid-based diagnostics and applications in veterinary medicine. *Anim. Health Res. Rev.* 11(2), 207–216.

Rodríguez-Arrioja, G. M., Segalés, J., Balasch, M., Rosell, C., Quintana, J., Folch, J. M., Plana-Durán, J., Mankertz, A., Domingo, M. (2001). Serum antibodies to porcine circovirus type 1 (PCV1) and type 2 (PCV2) in pigs. *Vet. Rec.* 146, 762–764.

Rose, N. and Andraud, M. (2017). The use of vaccines to control pathogen spread in pig populations. *Porcine Health Manag.* 3, 8.

Roth, J. A. (1999). Mechanistic bases for adverse vaccine reactions and vaccine failures. *Adv. Vet. Med.* 41, 681–700.

Saporiti, V., Franzo, G., Sibila, M. and Segalés, J. (2021). Porcine circovirus 3 (PCV-3) as a causal agent of disease in swine and a proposal of PCV-3 associated disease case definition. *Transbound. Emerg. Dis.* 68(6), 2936–2948. doi: 10.1111/tbed.14204.

Segalés, J. (2006). Chapter 5: Selection of tests to be done. In: Virbac Animal Health (Ed.). *Pig Respiratory Disease Diagnosis: An Integrated Approach*. Virbac Animal Health, 68–85.

Segalés, J., Martínez, J., Castellà, J., Darwich, L., Domingo, M., Mateu, E., Martín, M. and Sibila, M. (2013). Chapter 4: Diagnostic methods. In: *Handbook of Laboratory Diagnosis in Swine*. Grupo Asis Biomedica S.L., 21-51.

Seo, H. W., Park, C., Han, K. and Chae, C. (2014). Effect of porcine circovirus type 2 (PCV2) vaccination on PCV2-viremic piglets after experimental PCV2 challenge. *Vet. Res.* 45(1), 13.

Shi, S. T. and Lai, M. M. (2005). Viral and cellular proteins involved in coronavirus replication. *Curr. Top. Microbiol. Immunol.* 287, 95-131.

Simmonds, P., Adams, M. J., Benkő, M., Breitbart, M., Brister, J. R., Carstens, E. B., Davison, A. J., Delwart, E., Gorbalenya, A. E., Harrach, B., Hull, R., King, A. M., Koonin, E. V., Krupovic, M., Kuhn, J. H., Lefkowitz, E. J., Nibert, M. L., Orton, R., Roossinck, M. J., Sabanadzovic, S., Sullivan, M. B., Suttle, C. A., Tesh, R. B., van der Vlugt, R. A., Varsani, A. and Zerbini, F. M. (2017). Consensus statement: Virus taxonomy in the age of metagenomics. *Nat. Rev. Microbiol.* 15(3), 161-168.

Swenson, S. L., Mead, D. G. and Kinker, D. R. (2019). Chapter 45: Rhabdoviruses. In: Zimmerman, J. J., Karriker, L. A., Ramirez, A., Schwartz, K. J., Stevenson, G. W. and Zhang, J. (Eds). *Diseases of Swine* (11th edn.). Hoboken, NJ: John Wiley & Sons, Inc., 733-742.

Taylor, M. P., Kobiler, O. and Enquist, L. W. (2012). Alphaherpesvirus axon-to-cell spread involves limited virion transmission. *Proc. Natl. Acad. Sci. U. S. A.* 109(42), 17046-17051.

Toka, F. N. and Golde, W. T. (2013). Cell-mediated innate responses of cattle and swine are diverse during foot-and-mouth disease virus (FMDV) infection: A unique landscape of innate immunity. *Immunol. Lett.* 152(2), 135-143.

Turlewicz-Podbielska, H., Włodarek, J. and Pomorska-Mól, M. (2020). Noninvasive strategies for surveillance of swine viral diseases: A review. *J. Vet. Diagn. Invest.* 32(4), 503-512.

Van der Waal, K. and Deen, J. (2018). Global trends in infectious diseases of swine. *Proc. Natl. Acad. Sci. U. S. A.* 115(45), 11495-11500.

Van Gorp, H., Van Breedam, W., Delputte, P. L. and Nauwynck, H. J. (2008). Sialoadhesin and CD163 join forces during entry of the porcine reproductive and respiratory syndrome virus. *J. Gen. Virol.* 89(12), 2943-2953.

van Kammen, A. (1999). Beijerinck's contribution to the virus concept - An introduction. *Arch. Virol. Suppl.* 15, 1-8.

Vannier, P., Capua, I., Le Potier, M. F., Mackay, D. K., Muylkens, B., Parida, S., Paton, D. J. and Thiry, E. (2007). Marker vaccines and the impact of their use on diagnosis and prophylactic measures. *Rev. Sci. Tech.* 26(2), 351-372.

Vassalli, J. D., Lombardi, T., Wohlwend, A., Montesano, R. and Orci, L. (1986). Direct cell-to-cell transmission of vesicular stomatitis virus. *J. Cell Sci.* 85, 125-131.

Wang, T., Fang, L., Zhao, F., Wang, D. and Xiao, S. (2018). Exosomes mediate intercellular transmission of porcine reproductive and respiratory syndrome virus. *J. Virol.* 92(4), e01734-17.

Wu, M., Yi, D., Zhang, Q., Wu, T., Yu, K., Peng, M., Wang, L., Zhao, D., Hou, Y. and Wu, G. (2021). Puerarin enhances intestinal function in piglets infected with porcine epidemic diarrhea virus. *Sci. Rep.* 11(1), 6552.

Yang, Y. L., Meng, F., Qin, P., Herrler, G., Huang, Y. W. and Tang, Y. D. (2020). Trypsin promotes porcine Deltacoronavirus mediating cell-to-cell fusion in a cell type-dependent manner. *Emerg. Microbes Infect.* 9(1), 457–468.

Yuan, C., Li, Y., Zhang, E., Jin, Y. and Yang, Q. (2021). The mechanism of PEDV-carrying CD3+ T cells migrate into the intestinal mucosa of neonatal piglets. *Viruses* 13(3), 469.

Zhang, J. Q., Yoon, K. J. and Zimmerman, J. J. (2019). Chapter 23: Overview of viruses. In: Zimmerman, J. J., Karriker, L. A., Ramirez, A., Schwartz, K. J., Stevenson, G. W. and Zhang, J. (Eds). *Diseases of Swine* (11th edn.). Hoboken, NJ: John Wiley & Sons, Inc., 427-437.

Zhong, P., Agosto, L. M., Munro, J. B. and Mothes, W. (2013). Cell-to-cell transmission of viruses. *Curr. Opin. Virol.* 3(1), 44–50.

Chapter 2

Understanding and identifying bacterial disease in swine

Dominiek Maes, Filip Boyen and Freddy Haesebrouck, Ghent University, Belgium

1 Introduction

Advances in the development of diagnostic tools during the last decades such as polymerase chain reaction (PCR) have allowed to diagnose different viral infections in pigs faster and with a higher sensitivity compared to previous cumbersome techniques such as virus isolation. This increased detection rate of viruses might have contributed to the perception that bacteria are considered as pathogens that are only or mainly important following previous viral infection. However, infections with bacterial pathogens have a significant impact on pork production worldwide and may lead to clinical disease even in the absence of prior viral infections (Ramirez, 2018). They continue to significantly affect the health, well-being, and performance of pigs in all types of production systems. Infections with bacterial respiratory and enteric pathogens are among the most common and economically significant diseases in swine. *Mycoplasma hyopneumoniae* is the primary pathogen that causes enzootic pneumonia in pigs and an important pathogen within the porcine respiratory disease complex (PRDC) (Rycroft, 2020). *Escherichia coli* (*E. coli*) is among the most important

http://dx.doi.org/10.19103/AS.2022.0103.02

causes of diarrhea in suckling and recently weaned pigs (Fairbrother and Nadeau, 2019). Some swine bacterial diseases are also on the Office International des Epizooties (OIE) list (2021) of notifiable animal diseases such as anthrax, infections with *Brucella suis*, and the *Mycobacterium tuberculosis* complex.

Infections with bacterial pathogens are also responsible for a major part of the antimicrobial use and, therefore, indirectly increase the risk for the development of antimicrobial resistance. Worldwide, pig farmers and veterinarians are urged to use less antimicrobials responsibly and judiciously. This implies that antimicrobials should not be used for prophylaxis and that antimicrobials that are critically important for human medicine should not be used anymore in animals or only under very strict conditions.

Finally, some bacterial pathogens in swine, such as *Salmonella enterica*, are important zoonotic agents. In the European Union (EU), salmonellosis is the second most reported zoonosis in humans (after campylobacteriosis), and *Salmonella* is the most frequently reported causative agent isolated in foodborne outbreaks (EFSA, 2018). Pigs and pork are considered as the source of 28% of the human infections caused by *Salmonella* Typhimurium and of 50% of the human infections caused by *Salmonella* Typhimurium monophasic variants (EFSA, 2018). Also in other parts of the world such as North America, *Campylobacter* and *Salmonella* cause the largest proportion of foodborne illnesses (Tack et al., 2020).

The present chapter provides an overview of important swine bacterial pathogens, the mechanisms of pathogenicity, transmission routes, clinical signs and lesions caused by the pathogens, and control measures.

2 Some phenotypic characteristics of swine bacterial pathogens

Bacteria can be classified according to some basic phenotypic characteristics such as (1) Gram stain reaction, (2) morphology, (3) oxygen requirements, and (4) ability to form endospores (Table 1) (Post, 2019).

Gram staining differentiates bacteria by the chemical and physical properties of their cell walls. Gram-positive cells have a thick layer of peptidoglycan in the cell wall that retains the primary stain, crystal violet. They appear as purple bacteria. Gram-negative cells have a thinner peptidoglycan layer that allows the crystal violet to wash out upon the addition of ethanol. They are stained pink or red by the counterstain, commonly safranin or fuchsine. Mycoplasmas are of the gram-positive lineage, but due to the lack of a cell wall, they do not retain the crystal violet stain. Treponemes and leptospires and also *Mycobacterium* species with their high lipid content in the cell walls cannot be properly stained by this method, but other methods such as Ziehl–Neelsen staining for *Mycobacterium* species are more appropriate.

Table 1 Classification of the principal bacterial pathogens of swine

Gram-staining	Oxygen requirements	Spore formation	Shape	Genus/genera
Gram-positive	Aerobic to microaerophilic	No	Bacilli	*Trueperella*, *Listeria*, *Mycobacterium*, *Rhodococcus*
	Aerobic or facultatively anaerobic	Yes	Bacilli	*Bacillus*
	Facultatively anaerobic	No	Cocci	*Enterococcus*, *Staphylococcus*, *Streptococcus*
		No	Bacilli	*Erysipelothrix*
	Anaerobic	Yes	Bacilli	*Clostridium*
		No	Bacilli	*Actinobaculum*
Gram-negative	Aerobic or facultatively anaerobic	No	Bacilli	*Actinobacillus*, *Bordetella*, *Brucella*, *Burkholderia*, *Escherichia*, *Glaesserella*, *Pasteurella*, *Salmonella*, *Yersinia*
	Microaerophilic or anaerobic	No	Curved to spiral-shaped bacilli	*Brachyspira*, *Leptospira*, *Treponema*, *Borrelia*, *Campylobacter*, *Helicobacter*, *Lawsonia*
Bacteria without cell wall and obligately intracellular bacteria				*Mycoplasma*, *Chlamydia*

Source: Adapted from Post (2019).

The cellular morphology of bacteria comprises cocci, bacilli, and spiral-shaped bacteria. Some cocci, such as *Staphylococcus* species, are perfectly rounded, while others are somewhat elongated such as many *Streptococcus* and *Enterococcus* species, or may even look like rods (coccobacilli). Staphylococci usually occur in irregular groups and less frequently in the classical 'grapes' and also in short chains and in pairs. Only some streptococci form clear chains. Many streptococci and all enterococci occur more often in groups, grapes, in duplo, as singles, or in short chains.

Bacilli may appear as regular rods (members of the genera *Escherichia*, *Salmonella*, and *Listeria*), others as coccobacillary (*Pasteurella*), pleomorphic (*Trueperella pyogenes* [*T. pyogenes*] and *Actinobacillus* species), curved (*Campylobacter* species and *Lawsonia intracellularis* [*L. intracellularis*]) or filamentous in shape (*Actinomyces* species and *Nocardia* species). Bacteria with a spiral shape include *Brachyspira*, *Treponema*, *Borrelia*, and *Leptospira* species, and several *Helicobacter* species. The size of the bacterium depends upon its growth phase and the type of medium used for cultivation. In general, most spirochetes, *Bacillus* species, and clostridia are regarded as large pseudomonads; Enterobacteriaceae (*Salmonella* species and *E. coli*) as medium-sized; *Brucella*, *Glaesserella*, and *Pasteurella* species as small; and mycoplasmas and chlamydias as very small.

Bacteria can also be classified by their ability to utilize or tolerate oxygen. Bacteria that require oxygen and can tolerate levels of oxygen that are present in the atmosphere (21%) are called 'obligate aerobes.' Facultatively anaerobic bacteria can multiply in both the presence and absence of oxygen. This group represents the majority of the swine bacterial pathogens. Microaerophilic bacteria such as the genera *Campylobacter* and *Helicobacter* require small amounts of oxygen but may be killed by normal atmospheric concentrations. Obligate anaerobes can only grow in the absence of oxygen and several of these bacteria are killed by even small amounts of oxygen. Many clostridia are obligate anaerobes. Some obligate anaerobic species, such as *Brachyspira hyodysenteriae*, can be more tolerant to oxygen and can therefore survive longer in the environment than other strictly anaerobic bacteria.

Bacteria may also be classified based on their ability to produce spores. Spores are formed when vegetative cells are deprived of necessary nutrition or environmental conditions. Spores are extremely resistant to harsh environmental conditions and disinfectants. Two important genera of spore-forming bacteria in swine are *Bacillus* and *Clostridium*.

The phenotypic characteristics of the bacteria may be important for the epidemiology, diagnostics, treatment, and control measures. The outcome of the Gram staining, for instance, may sometimes help to determine the choice of antimicrobial for treatment. Knowledge of a bacterium's oxygen requirements may be important for the storage and transport conditions of diagnostic samples to the laboratory. Organisms that produce spores, such as *Clostridia*, may be difficult to eliminate from the barns because the spores are resistant to most cleaning and disinfectant agents.

3 Advanced methods to identify and classify swine bacterial pathogens

Matrix-assisted laser desorption/ionization time-of-flight mass spectrometry (MALDI-TOF MS) is a technique which has become very popular in veterinary diagnostics in the last few years (Randall et al., 2015). An easy, cheap, and fast preparation protocol following a culture step allows identifying most bacterial species with high accuracy. The technique results in a peak spectrum representing the most abundant bacterial proteins, which is compared to a commercial and/or in-house library (Pavlovic et al., 2013).

The mentioned phenotypic characteristics and the traditional methods for identifying microbes are still used but have some drawbacks. First, they are applicable mainly to organisms that can be cultured *in vitro*. As the growth requirements of various species of bacteria are unknown, methods based on cultivation are not sufficient to explore specific microbial populations like the microbiome in the gastrointestinal tract (Lamendella et al., 2011). Second, some bacterial strains might exhibit unique biochemical characteristics that do not fit the pattern of any known genus and species. With the advent of molecular phylogeny and taxonomy, more advanced methods are used to identify and classify bacterial pathogens. They are not dependent on live cultures, and they can often reveal minute differences between organisms that escape detection by traditional means.

Several genes are used to identify species, the most important being the *16S* rRNA gene, followed by *23S*, internal transcribed spacer (ITS) region, gyrase B (*gyrB*), and other, sometimes species-specific, genes to attain a better resolution between species (Klindworth et al., 2013). In addition to MALDI-TOF MS, one of the quickest ways to match an isolated strain to a species or genus is done by amplifying its *16S* gene with universal primers, sequencing the 1.4 kb amplicon, and submitting it to a specialized web-based identification database. Several identification methods exist such as phenotypic analyses (fatty acid analyses, growth conditions), genetic analyses (DNA–DNA hybridization, DNA profiling, sequence, GC ratios), and phylogenetic analyses (*16S*-based phylogeny, phylogeny based on other genes, multi-gene sequence analysis, and whole-genome sequence-based analysis). Whole-genome sequencing is particularly useful for characterizing isolates, predicting phenotypes, tracking outbreaks, and identifying infection sources (Balloux et al., 2018).

4 Pathogenicity of swine bacteria

Most bacteria associated with pigs do not cause disease. These nonpathogenic, commensal, or mutualistic bacteria belong to the normal microbiota present on the skin and mucosae. Not only are they, generally speaking, harmless, but they can also be protective for the host. Most of the bacteria in the porcine alimentary tract, for instance, are commensals. Some bacteria belonging to the microbiota may cause disease under certain circumstances, for instance, when the host's immunity is compromised in some way or when the normal microbiota is disturbed. They are sometimes referred to as facultative pathogens. Actually, most bacteria that cause disease in pigs belong to this group of bacteria. *Trueperella pyogenes*, *Pasteurella multocida*, *Streptococcus suis*, *Escherichia coli*, *Fusobacterium necrophorum*, *Clostridium perfringens* type A (*C. perfringens*), and *Glaesserella parasuis* are examples of such bacteria. Obligate pathogens, on the other hand, do not belong to the normal microbiota. This terminology is sometimes also used to refer to bacteria that may cause disease in healthy, immunocompetent hosts, such as *Bacillus anthracis*. Opportunistic pathogenic bacteria generally exhibit low virulence, but if the host becomes injured or is seriously immunocompromised, they may cause disease.

Pathogenicity refers to the potential ability of an organism to cause disease and harm the host. This ability represents a genetic component of the pathogen, and the overt damage done to the host is a property of the host-pathogen interactions (Thomas and Elkinton, 2004). Mutualistic and most commensal bacteria belonging to the normal microbiota lack this inherent ability to cause disease. However, disease is not an inevitable outcome of the host-pathogen interaction and, furthermore, pathogens can express a wide range of virulence. Host immune factors, the number of organisms present in the initial exposure, and the virulence attributes of a bacterium all play a role in the development of disease. Virulence, a term often used interchangeably with pathogenicity, refers to the degree of disease caused by the organism. The extent of the virulence is usually correlated with the ability of the pathogen to multiply within the host and may be affected by other factors. Pathogenicity is a qualitative term, an 'all-or-none' concept, whereas virulence is a term that quantifies pathogenicity. An organism (species or strain) is defined as being pathogenic (or not) and, depending upon conditions, may exhibit different levels of virulence.

Bacteria may cause disease by different mechanisms; two important ones are tissue invasion and toxin production (Songer and Post, 2005). Tissue invasion can be accomplished by adhering to and/or penetrating cells, producing extracellular substances to facilitate invasion, and overcoming host defenses. Adhesins are surface proteins that facilitate adhesion to host cells. Many strains of pathogenic *E. coli* possess such surface adhesins. *Listeria monocytogenes*, *Mycobacterium* species, *Brucella* species, *Rhodococcus equi*, and *Salmonella enterica* are examples of facultatively intracellular bacteria that are able to invade and multiply in host cells (Niemann et al., 2004). Some bacteria produce extracellular enzymes such as coagulase produced by *Staphylococcus hyicus* and streptokinase produced by beta-hemolytic streptococci, enabling them to spread within the host tissue. Other bacteria mainly cause disease through toxin production. Exotoxins are metabolites that are produced in the bacterial cytoplasm and are mostly released into the extracellular environment by active secretion. Some exotoxins are only released in the environment after the lysis of the bacterial cell. Most exotoxins are proteins that are highly antigenic, and it may be useful to include inactivated exotoxin (toxoid) or a nontoxic but antigenic recombinant protein derived from the exotoxin in vaccines. The thermostable enterotoxins produced by some enterotoxigenic *E. coli* (ETEC) strains are polypeptides that are poorly immunogenic (Haesebrouck et al.,

2004). Exotoxins vary greatly in their potency, ranging from the highly toxic botulinum toxin, the cytolysins (Apx toxins) of *Actinobacillus pleuropneumoniae* to the weakly toxic product released by *T. pyogenes*. Other swine pathogens that may produce exotoxins are *C. perfringens*, enterotoxigenic strains of *E. coli*, *Pasteurella multocida*, and *Staphylococcus hyicus*. Endotoxins are the lipopolysaccharides found in the cell wall of gram-negative bacteria. These may be released from both actively growing bacterial cells and dead bacteria. The release of endotoxins is very important in the pathogenesis of gram-negative bacteria and is partially responsible for the clinical signs produced by these pathogens. Some cell wall components of Gram-positive bacteria, such as lipoteichoic acids and peptidoglycan, can cause similar host reactions as lipopolysaccharides from gram-negative bacteria (Middelveld and Alving, 2000).

Under field conditions, mixed infections with different pathogens (e.g. viruses, bacteria, and parasites) are common. A typical example is the PRDC. An overview of the interactions based on experimental infections with two respiratory pathogens (i.e. either two viruses, one virus and one bacterium, or two bacteria) has been published by Opriessnig et al. (2011). Although the different pathogens are capable of inducing disease independently, simultaneous infection with two or more pathogens often results in more severe disease. Some known mechanisms by which pig pathogens interact include damage to the muco-ciliated epithelium of the respiratory tract, inducing immune suppression, altering cytokine responses, or affecting macrophage function (Opriessnig et al., 2011). Also the sequence of infection with different pathogens likely plays a role in disease outcome.

5 Transmission of swine bacterial pathogens

The different routes of transmission of a selected number of important swine bacterial pathogens are shown in Table 2. The table is based on Amass (2005) and information from the literature published since then on the transmission of these pathogens has been added.

Transmission by direct contact occurs for all bacterial pathogens, whereas indirect transmission may vary depending on the pathogen. The indirect routes have been classified arbitrarily in the following categories: people, semen, manure, domestic/feral animals, rodents, insects (vectors), aerosol, animal feed, water, and fomites. Indirect transmission by close contact with infected droplets or feces and mechanical transfer by fomites or vectors are the ways in which bacteria are commonly spread.

Most bacterial pathogens can be transmitted from the sow to the offspring. For some pathogens like *Streptococcus suis*, piglets may be colonized already

Table 2 Direct and indirect transmission routes of swine pathogens and diseases

		Direct contact	People	Semen
Disease/pathogen	References	Purchasing policy, Transport/ removal of animals/manure/ cadavers, all subcategories of internal biosecurity[a]	Access check indirectly: transport/removal of animals/manure/ cadavers; supply of fodder, water, and equipment; cleaning and disinfection	Purchasing policy
Actinobacillus pleuropneumoniae	Assavacheep and Rycroft (2013); Loera-Muro et al. (2013); Tobias et al. (2014); Baroch et al. (2015); Marinou et al. (2015); Gottschalk and Broes (2019)	O		
Bordetella bronchiseptica	Nicholson et al. (2012); Brockmeier et al. (2019)	O		
Brachyspira hyodysenteriae	Desrosiers (2011); Hampson and Burrough (2019); Zeeh et al. (2020)	O	O	
Brucella suis	Wu et al. (2012); Lama and Bachoon (2018); Olsen et al. (2019)	O	X	X
Clostridium perfringens	Allaart et al. (2013); Uzal and Songer (2019)	O		
Erysipelothrix rhusiopathiae[c]	Bender et al. (2010); Boadella et al. (2012); Malmsten et al. (2018); Marinou et al. (2015); Opriessnig and Coutinho (2019)	O		
Escherichia coli	Cornick and VuKhac (2008); Callens et al. (2015); Pearson et al. (2016); Lama and Bachoon (2018); Fairbrother and Nadeau (2019)	O	X	
Glaesserella parasuis[1]	Brockmeier et al. (2013); Aragon et al. (2019)	O		
Lawsonia intracellularis[1]	Friedman et al. (2008); Baroch et al. (2015); Pearson et al. (2016); Vannucci et al. (2019)	O		
Leptospires	Maes et al. (2008); Althouse and Rossow (2011); Arent and Ellis (2019); Cilia et al. (2020)	O	X	O
Mycoplasma hyopneumoniae	Nathues et al. (2012); Baroch et al. (2015); Marinou et al. (2015); Pearson et al. (2016); Malmsten et al. (2018); Pieters and Maes (2019)	O	O	
Pasteurella multocida	Fablet et al. (2011); Desrosiers (2011); Register and Brockmeier (2019)	O	O	
Salmonella enterica.	Andres and Davies (2015); Martelli et al. (2018); Griffith et al. (2019)	O	X	
Streptococcus suis	Fablet et al. (2011); Gottschalk and Segura (2019); Giang et al. (2020).	O	X	

[a]As described in Biocheck.UgentTM.
[b]Transmitted to distances>2 km.
[c]Added to the original list of Amass (2005). 'X' represents possible transmission routes described by Amass (2005). 'O' describes new possible transmission routes based on studies published between 2005 and January 2021.
Source: Adapted from Amass (2005) and Filippitzi et al. (2018).

Indirect contact							
Manure	Domestic/ feral animals	Rodents	Insects (vectors)	Aerosol	Animal feed	Water	Fomites
Transport/ removal of animals/ manure/ cadavers	Vermin and bird control	Vermin and bird control	Location, environment	Purchasing policy, transport/ removal of animals/manure/ cadavers, location, environment, All	Supply of fodder, water and equipment	Supply of fodder, water and equipment	Cleaning and disinfection, indirectly: transport/removal of animals/manure/ cadavers; supply of fodder, water, and equipment; Access check; Vermin and bird control
	O			X		O	O
	X	X	O	X		X	O
X	X	X	O		O	X	O
O	X		O	O	O		
O			O	X		X	O
O	O	O			O	O	O
X	O	X	X	X	O	X	X
	O						
O	O	O	O				O
	X	X				X	
	O			X[b]		X	X
X	O			X		X	O
X	X	X	X	X	X	X	X
X	X		X	X		X	X

during or shortly after birth from vaginal fluids or while nursing. For other pathogens such as *Mycoplasma hyopneumoniae* and *A. pleuropneumoniae*, transmission from sow to offspring takes place later during lactation. For the latter pathogens, strategies such as (medicated) early weaning can be used to obtain piglets that are free from pathogens that are endemic in the sow population (Alexander et al., 1980).

6 Clinical signs and lesions induced by swine bacterial pathogens

An overview of the clinical signs and gross postmortem lesions of gram-positive and gram-negative swine bacterial pathogens are shown in Tables 3 and 4 (Muirhead and Alexander, 1997; Post, 2019).

A particular situation is the presence of bacteria in the ejaculates of breeding boars. Apart from the risk for the transmission of bacterial pathogens like *Brucella suis* to recipient sows inseminated with contaminated semen, bacterial contamination is also known to be detrimental to semen quality as it will cause sperm agglutination and reduce motility (Althouse and Lu, 2005). It may also decrease the longevity of the sperm during storage and the fertility potential (Sepúlveda et al., 2013, 2014).

Commonly isolated bacteria from extended semen and their effect on sperm quality are summarized in Table 5. They mostly belong to the Enterobacteriaceae family (Althouse and Lu, 2005; Úbeda et al., 2013). Within this family, *Serratia marcescens*, *Klebsiella* species, *Morganella* species, or *Proteus* species have been demonstrated to be present in a high percentage of samples and their presence is associated with reduced motility (Úbeda et al., 2013). *Enterobacter cloacae* at sperm:bacteria ratio of 1:5 and 1:10 reduced sperm motility and membrane integrity and resulted in sperm agglutination in semen doses stored at 15–17°C (Prieto-Martinez et al., 2014). *Clostridium perfringens* reduced sperm motility and viability after the inoculation of 10^8 cfu ml^{-1} in semen doses and 24 h incubation at 37°C or storage at 15°C (Sepúlveda et al., 2013). Similarly, experimental contamination of stored boar semen with *Pseudomonas aeruginosa* resulted in significant decreases in the percentages of sperm motility, viability, and acrosome integrity but did not affect pH (Sepúlveda et al., 2014). In a study by Althouse et al. (2000), extended semen samples were inoculated with colonies of the six most frequently isolated bacteria (*Enterobacter cloacae*, *E. coli*, *Serratia marcescens*, *Alcaligenes xylosoxidans*, *Burkholderia cepacia*, and *Stenotrophomonas maltophilia*). For all isolates, visual clumping, microscopic sperm-to-sperm agglutination, poor motility, and damaged acrosomes were observed after inoculation in a time-dependent manner. The exact interactions

Table 3 Gram-positive bacteria in swine and associated disease(s) and/or clinical signs

Bacterium	Disease(s)/clinical signs/lesions
Actinobaculum suis	Cystitis, pyelonephritis
Actinomyces hyovaginalis	Sporadic abortion, embolic pneumonia
Bacillus anthracis	Anthrax (pharyngeal, intestinal, septicemic form)
Clostridium botulinum	Botulism (paralysis, death) (pigs are not very sensitive)
Clostridium chauvoei	Blackleg (emphysematous necrotizing myositis) (pigs are not very sensitive)
Clostridioides (Clostridium) difficile	Neonatal colitis
Clostridium perfringens	Neonatal enteritis, pseudomembranous colitis
Clostridium novyi	Sudden death, hepatitis
Clostridium septicum	Malignant edema
Clostridium tetani	Tetanus (generalized skeletal muscle spasms)
Enterococcus sp.	Enteritis
Erysipelothrix rhusiopathiae	Erysipelas (acute, subacute, chronic form)
Listeria monocytogenes	Listeriosis, abortion, encephalitis, septicemia
Mycobacterium sp.	Tuberculosis (local, generalized form)
Mycoplasma hyopneumoniae	Enzootic pneumonia (broncho-interstitial pneumonia)
Mycoplasma hyorhinis	Arthritis, otitis, polyserositis
Mycoplasma hyosynoviae	Arthritis
Mycoplasma suis	Anemia, infertility, poor growth, pericarditis, unthriftiness
Rhodococcus equi	Granulomatous lymphadenitis
Staphylococcus aureus	Abscesses, arthritis, enteritis, mastitis, metritis, neonatal septicemia, vaginitis
Staphylococcus hyicus	Exudative epidermitis (greasy pig disease)
Streptococcus dysgalactiae spp. *equisimilis*	Arthritis, endocarditis, meningitis, septicemia
Streptococcus porcinus	Cervical lymphadenitis
Streptococcus suis	Septicemia, arthritis, endocarditis, meningitis, polyserositis
Trueperella abortisuis	Sporadic abortion
Trueperella pyogenes	Abscesses, arthritis, endocarditis, mastitis, osteomyelitis, pneumonia

Source: Adapted from Muirhead and Alexander (1997) and Post (2019).

of the different bacteria with sperm are not yet fully investigated. It has been suggested that *P. aeruginosa* decreases the ability of sperm to accomplish capacitation (Sepúlveda et al., 2016).

Table 4 Gram-negative bacteria in swine and associated disease(s) and/or clinical signs

Bacterium	Disease(s)/clinical signs/lesions
Actinobacillus pleuropneumoniae	Pleuropneumonia (peracute, acute, chronic form)
Actinobacillus suis	Pneumonia, septicemia
Bordetella bronchiseptica	Nonprogressive atrophic rhinitis, pneumonia
Burkholderia pseudomallei	Melioidosis, internal abscesses, lymph node abscesses
Brachyspira hyodysenteriae, Brachyspira hampsonii, or Brachyspira suanatina	Swine dysentery, mucohemorrhagic enteritis of the large intestine
Brachyspira pilosicoli	Porcine intestinal/colonic spirochetosis
Brachyspira murdochii or Brachyspira intermedia	Mild colitis, loose stools
Brucella suis	Brucellosis, abortion, orchitis, arthritis, infertility
Campylobacter coli/jejuni	Subclinical enterocolitis, mild diarrhea
Chlamydia suis	Conjunctivitis, neonatal diarrhea, pneumonia, possibly abortion
Chlamydia abortus	Early embryonic death, abortion
Escherichia coli	Colibacillosis, edema disease, cystitis, enteritis, mastitis, neonatal septicemia
Glaesserella parasuis	Glässer's disease, arthritis, polyserositis
Helicobacter suis	Gastritis, gastric ulcers
Klebsiella pneumoniae	Piglet septicemia
Lawsonia intracellularis	Porcine proliferative enteropathy
Leptospira sp.	Leptospirosis, infertility, stillbirths, weak piglets
Pasteurella multocida	Progressive atrophic rhinitis, pneumonia
Salmonella enterica	Enteric and septicemic salmonellosis
Treponema pedis	Cutaneous spirochetosis (granulomatous lymphadenitis)
Yersinia enterocolitica	Subclinical enterocolitis
Yersinia pseudotuberculosis	Diarrhea, enterocolitis

Source: Adapted from Muirhead and Alexander (1997) and Post (2019).

7 Diagnosis of swine bacterial pathogens

Developing an efficient diagnostic approach should start with gathering all relevant farm information and a detailed description of the problem. Further diagnostic workup is mostly necessary to establish a conclusive diagnosis, to elucidate the pathogens involved, and to assess their relative impact on the problem.

Table 5 Percentage of contaminated extended semen samples[a] in which different bacteria were isolated and their effect on sperm quality

Bacteria	Althouse and Lu (2005)	Schulze et al. (2015)	Úbeda et al. (2013)	Effect on sperm
Total contaminated samples	31.2% (78/250)	25.6% (88/344)	14.7% (263/1785)	Reduced sperm quality
Achromobacter xylosoxidans	10.3%	3.4%	ND	Agglutination, poor motility, damaged acrosomes, acidic pH (Althouse et al., 2000)
Burkholderia cepacia	2.6%	ND	ND	Agglutination, poor motility, damaged acrosomes, acidic pH (Althouse et al., 2000)
Clostridium perfringens	ND	ND	ND	Poor sperm viability and motility (Sepúlveda et al., 2013)
Enterobacter cloacae	2.6%	13.6%	ND	Agglutination, poor motility, damaged acrosomes, acidic pH, decreased the osmotic resistance (Althouse et al., 2000; Prieto-Martinez et al., 2014)
Enterococcus spp.	20.5%	8%	ND	ND
Escherichia coli	6.4%	ND	1.5%	Agglutination, poor motility, damaged acrosomes (Althouse et al., 2000)
Klebsiella spp.	3.8%	8%	11.8%	Poor motility (Úbeda et al., 2013)
Leifsonia aquatica	ND	20.5%	ND	ND
Morganella morganii	ND	ND	3.8%	Poor motility, damage acrosome, poorer response to the hypoosmoic swelling test (Úbeda et al., 2013)
Proteus mirabilis	1.3%	5.7%	1.9%	Poor motility, abnormal forms (Úbeda et al., 2013)
Providencia spp.	3.8%	ND	9.1 %	ND
*Pseudomonas*spp.	6.4%	5.7%	ND	*P. aeruginosa* reduced total and progressive sperm motility, sperm viability, and acrosome integrity
Ralstonia pickettii	ND	11.4%	ND	ND
Stenotrophomonas maltophilia	15.4%	ND	ND	Agglutination, poor motility, damaged acrosomes
Serratia marcescens	10.3%	2.3%	12.5%	Agglutination, poor motility, damaged acrosomes, acidic pH (Althouse et al., 2000)

[a]The percentage refers to the total contaminated samples. Bacteria present in a percentage of samples lower than 5% and identified only in one of the studies and for which no effect on sperm quality has been described are not included in the table. ND = not described. *Source*: Adapted from Lopéz et al. (2017).

Necropsies of a representative number of typically affected animals can be very helpful. The basic lesion caused by *Mycoplasma hyopneumoniae* and swine influenza viruses is a broncho-interstitial pneumonia; subsequent infection by other respiratory bacteria ends up with suppurative bronchopneumonia if the damage does not affect pleura and pleuropneumonia if it affects the pleura. It must be reminded that infections by *Glaesserella parasuis* and *Streptococcus suis* are systemic, and pleuritis usually occurs in the absence of pneumonia (Maes et al., 2001). However, mixed infections are generally the rule, and lesions may not appear very typical for one specific pathogen. Performing necropsies is mostly followed by taking samples for further diagnostic examinations such as histopathology, immunohistochemistry, pathogen isolation, or PCR testing.

In addition to investigating dead or euthanized animals, living animals from the affected age group can be sampled to detect antibodies and/or possible pathogens involved. Different sampling strategies can be used, depending on the age group, the type of sample, the disease, and the diagnostic test used. In the case of serology, paired serum samples will often be needed to establish the diagnosis. To save analysis costs, individual samples like fecal samples for *Brachyspira hyodysenteriae* may be pooled (Neirynck et al., 2020). For detection of *Salmonella* in subclinically infected pigs, overshoe samples may be used. They are collected at pen or compartment level and are more likely to be *Salmonella* positive than when samples from a limited number of individual animals are collected as the shedding of the bacteria in the feces by subclinically infected pigs is intermittent (Peeters et al., 2019). A detailed overview of sample selection, collection, and submission for different swine pathogens has been published by Arruda and Gauger (2019).

Examination of pigs in the slaughterhouse is an additional and valuable tool to obtain information on the health of the animals in an easy and cheap way (Alban et al., 2015). Different lesions can be monitored like lesions in the respiratory, gastrointestinal, or urogenital tract; tail biting lesions; joint lesions; and abscesses (Maes et al., 2001; de Jong et al., 2014). If the lesions are monitored from different successive batches of a herd, time evolution can be assessed. Slaughter checks are also useful to detect subclinical infections that produce gross lesions. Most scoring methods are based on a visual subjective estimation of the proportion of affected tissue. Therefore, it can involve possible errors. In addition, most lesions are not pathognomonic for a specific pathogen.

8 Control of swine bacterial pathogens

8.1 General control measures

Comprehensive reviews of the different control measures for particular swine bacterial pathogens have been published elsewhere (Luppi, 2017; Marco et al., 2020). They will not be repeated in detail here, even though they are of critical importance. Factors that should be considered to improve the overall biosecurity in pig farms are presented in a separate chapter in this book. A short overview of the most important control measures for swine bacterial disease will be discussed in the following sections. The measures are classified into the following categories: (1) type of production system, (2) replacement rates and gilt acclimation, (3) management, (4) hygiene, (5) nutrition, and (6) climate and housing conditions.

8.1.1 Type of production system

Control factors related to the type of production system include piglet source (one source lower risk than multiple sources), pig flow (not mixing of age groups) and batch system (more than 1-week systems lower risk than 1-week system), and parity segregation (separation of primiparous and multiparous sows). Closed herds or closed production systems that are geographically isolated and do not purchase pigs have the advantage that there is no risk for the introduction of bacterial pathogens via the introduction of new animals. However, in case closed farms are endemically infected, the infection is maintained in the farm, and breeding animals will continue to transmit the pathogen to the offspring. In case of multisite production systems, there is a risk to contract the disease by moving (subclinically) infected animals between the sites.

8.1.2 Replacement rates and gilt acclimation

Low replacement rates, purchasing breeding gilts from herds with similar or better health status, quarantine and implementing proper gilt acclimation practices can lower the risk of introducing a pathogenic bacterial species into the farm. As animals are often subclinically infected with bacterial pathogens (such as *Mycoplasma hyopneumoniae*, *A. pleuropneumoniae*, *Salmonella* Typhimurium, *Brachyspira hyodysenteriae*, and *Streptococcus suis*), the introduction of clinically healthy but infected pigs represents a great risk. Therefore, information from the origin farm, quarantine, and proper testing are needed.

Replacement gilts negative for specific bacterial pathogens have become available from breeding stock companies. However, also the introduction of naïve gilts into endemically infected farms may be a challenge. Naïve gilts will be exposed to positive sows and may eventually get infected and shed large numbers of bacteria. These gilts also become an infectious reservoir for newborn piglets. In addition, females of the recipient herd may be exposed to recirculation of the bacterium and become reinfected, generating an unstable disease status in the herd.

Proper gilt acclimation procedures, such as vaccination during the quarantine period, are therefore needed. Gilt acclimation will enhance immunity and alleviate or prevent introduction problems when gilts join an endemically infected breeding population. In the case of *Mycoplasma hyopneumoniae*, strategies have been used to purposefully expose incoming gilts to *Mycoplasma hyopneumoniae* at an early age, with the aim to immunize gilts and prevent shedding prior to entering the sow farm (Pieters and Fano, 2016).

8.1.3 Management

Management practices that are beneficial against bacterial disease include measures prior to weaning (sufficient colostrum intake, limiting cross-fostering, increased weaning age), strict all-in/all-out production, reducing stress rates (by maintaining optimal stocking densities and group size), prevention of other pathogens including viruses and parasites, and control of mycotoxins (Rhouma et al., 2017; Marco et al., 2020). Management strategies like weaning at a later age (28–33 instead of 21 days) ameliorate the stress of the piglets and reduce the risk for disease problems post-weaning (Faccin et al., 2020). However, more research is needed in this area including effects of changes in facilities like comingling systems, group lactations, or supplementary milk systems. All-in/all-out management with cleaning and disinfection between batches reduces the transmission of infection. It is of special importance to avoid any pig being delayed in the flow. All those pigs suffering the disease should be euthanized or moved forward with their group to avoid affecting the next batch, especially during an active outbreak. Stressful conditions such as pig handling, crowding, transportation, or severe weather conditions should be minimized to prevent clinical outbreaks in herds at risk.

8.1.4 Hygiene

A thorough cleaning to remove organic matter and disinfection of the pens should be practiced to limit or prevent carryover of pathogens from a previous batch of pigs to the next batch in the same barn (Dutkiewicz et al., 2017). In case of problems with a particular disease, a disinfectant that is active against the pathogen should be selected. Keeping the barns as dry as possible after disinfection may further lower the infection levels and minimize the effects of the disease. Some bacterial pathogens such as *Brachyspira hyodysenteriae* and *Salmonella* Typhimurium can be transmitted via rodents, domestic animals (e.g. cats and dogs), birds, and insects. Mice and rats can serve as potential reservoirs of these pathogens in pig herds, so implementation of effective rodent control is essential. Dogs and cats should not be allowed to have access to the pig stables. Also, stables should be preferably bird proof, so that birds do not have access to the area where the pigs are housed. Contact with wild boar or foxes should be avoided (Zeeh et al., 2020). Proper insect control is also needed to control the transmission of many bacterial pathogens such as *Brachyspira hyodysenteriae*, *Mycoplasma suis*, and *Streptococcus suis*. Unfortunately, in the case of outdoor pig production, it is almost impossible to prevent mechanical transmission of infectious material via birds, waterfowl, and other potential wildlife vectors. Peat used as bedding material for piglets might be contaminated by *Mycobacterium avium* and lead to infections in pigs (Johansen et al., 2014). Therefore, potentially contaminated bedding material should not be used or should be properly decontaminated.

8.1.5 Nutrition

Attention should also be paid to the composition and form of the diet as a clinical expression of enteric infections like *E. coli* infections post-weaning, *Brachyspira hyodysenteriae* and *Salmonella* infections is often influenced by dietary factors. Diet can influence the pH and the composition of the microbiota in the intestinal tract. Some diet components such as antimicrobial peptides, chitosan, lysozyme, short- and medium-chain fatty acids/triglycerides and plant extracts can have direct antibacterial effects as well (Vanrolleghem et al., 2019). Good quality ingredients, milk products, and digestible diets with a low protein content reduce the risk for post-weaning diarrhea (Luppi, 2017). Zinc oxide at high concentrations has been shown to be effective in controlling *E. coli* infections post-weaning. Its use will be banned in the EU from 2022 onward (EMA, 2017) because of the environmental concerns and because it has been associated with selection for zinc resistance, and other heavy metals like copper, and methicillin resistance in *Staphylococcus aureus* (Luppi, 2017). Zinc oxide is still commonly used in other parts of the world. Diets that reduce the amount of fermentable substrate entering the large intestine are beneficial against swine dysentery (Leser et al., 2000). The use of a coarse non-pelleted feed is associated with reduced *Salmonella* prevalence (O'Connor et al., 2008). Further research is warranted to assess the efficacy of nutritional supplements such as phytogenic ingredients, essential oils, prebiotics, probiotics, gut acidifiers, and mannan-oligosaccharides. Finally, good water quality is also important for pig health.

8.1.6 Climate and housing conditions

Optimal housing and climatic conditions are important for all bacterial infections but, in particular, for respiratory infections (Marco et al., 2020). Ventilation and hygiene are needed to maintain appropriate environmental thermal conditions in the stables (e.g. temperature, humidity, and airspeed) with minimal fluctuations and minimize air contaminant levels (e.g. dust, ammonia, and endotoxins). Season has an effect on respiratory disease, and generally there are less problems in summer when temperatures are higher and relative humidity is lower.

8.2 Antimicrobial medication

Antimicrobials are necessary for the treatment of bacterial infections and by doing so, we can safeguard animals. They should, however, be used judiciously and restrictively to avoid or limit the risk for antimicrobial resistance in both pathogens and the normal microbiota. High antimicrobial use and the resulting risk for antimicrobial resistance is an important concern worldwide. Prophylactic use or using antimicrobials as a substitute for improper management should be avoided. In some parts of the world, such as the EU, regulations to control antimicrobial resistance do not allow prophylactic use of antimicrobials (EMA, 2020). Animals that are clinically diseased due to a bacterial infection requiring antimicrobial treatment should preferably be treated by injection since they usually do not eat and might drink less. The treatment should be done following the principles stated in the Guidelines for the Prudent Use of Antimicrobials in Veterinary Medicine (2015/C 299/04). In addition, the use of critically important antimicrobials for human medicine such as fluoroquinolones, third- and fourth-generation cephalosporins, and recently also colistin should be restricted or prohibited. This is particularly important not only for infections with zoonotic bacterial pathogens but also for the treatment of other bacterial infections as there is an increasing resistance of *E. coli* strains isolated from pigs with diarrhea to colistin (Burow et al., 2019).

Judicious use of antimicrobials in animals in order to limit antimicrobial resistance is also receiving increased attention outside Europe, although there is substantial variation in legislation when viewed at an international level (Bright-Ponte, 2020; Hoque et al., 2020; Léger et al., 2019; Ogwuche et al., 2021). The US Food and Drug Administration (FDA) Center for Veterinary Medicine (CVM) has taken important steps to address potential antimicrobial resistance concerns surrounding the use of antimicrobials in food-producing animals, including changes to the approved use conditions of medically important antimicrobials to support their judicious use in food-producing animals (Bright-Ponte, 2020).

8.3 Vaccination

Improving the immunity status of susceptible animals by vaccination is a helpful tool to control infectious diseases, including swine bacterial disease. Vaccination is generally able to reduce the (risk for) clinical signs, lesions and performance losses due to disease. However, most vaccines only provide partial protection, do not prevent infection, and are not able to eliminate the pathogens from the herd. The decision of whether to implement vaccination is not straightforward. It depends on the pig producer (degree of being 'risk averse'), the type and severity of disease, the available vaccines and expected effects, and other factors such as the type of herd, flow of the pigs, infection level, and management and housing conditions. Bacterial diseases for which commercial vaccines are available are shown in Table 6. Some vaccines contain the inactivated toxin or a nontoxic but antigenic recombinant protein derived from the exotoxin. Examples of this group include vaccines against *A. pleuropneumoniae*, clostridial infections, enterotoxigenic *E. coli*, shiga-toxin-producing *E. coli* and dermonecrotic toxin *Pasteurella multocida*, and *Bordetella bronchiseptica* (progressive atrophic rhinitis). Other vaccines target extracellular bacteria, and protection is generally mediated by antibodies against their surface antigens and certain secreted antigens, but cell-mediated immunity may also be important. Examples of this group are vaccines against *Erysipelothrix rhusiopathiae* (*E. rhusiopathiae*) and *Mycoplasma hyopneumoniae*. Finally, there are also vaccines against (facultative) intracellular bacteria such as *L. intracellularis* and *Salmonella* Typhimurium. For protection against these bacteria, cell-mediated immune responses are important (Haesebrouck et al., 2004).

Vaccination schemes are highly variable between herds and also between countries. Gilts in the quarantine are often vaccinated against multiple bacterial pathogens (*A. pleuropneumoniae*, dermonecrotic toxin *Pasteurella multocida* and *Bordetella bronchiseptica*, *Clostridium* spp., *E. coli*, *E. rhusiopathiae*, *G. parasuis*, *Leptospira* spp., *L. intracellularis*, and *Mycoplasma hyopneumoniae*) as part of the acclimation practices. Breeding sows are often vaccinated against neonatal *E. coli* and clostridium infections and *E. rhusiopathiae*, and in some herds also against other bacterial pathogens such as dermonecrotic toxin *Pasteurella multocida* and *Bordetella bronchiseptica*, *G. parasuis*, *A. pleuropneumoniae*, and *Mycoplasma hyopneumoniae* infections, leptospirosis, salmonellosis, and so on. Some vaccinations in sows aim to provide protection to the piglets via maternally derived antibodies. Typical examples are neonatal *E. coli* and *Clostridium* infections and infections with dermonecrotic toxin *Pasteurella multocida* and *Bordetella bronchiseptica*. For these vaccinations to be successful, piglets should have sufficient colostrum intake (Declerck et al., 2015). Vaccination against *Mycoplasma hyopneumoniae* in piglets is very frequently practiced worldwide. Other bacterial vaccines that are also available for piglets

Table 6 Common bacterial infections and diseases in the pig and availability of commercial vaccines

	Bacterium		Disease/clinical sign/lesion	Age group	Commercial vaccine[a]
Enteric	*Escherichia coli* (ETEC, EPEC, STEC)[b]		Neonatal diarrhea	1–3 days	Yes
			Pre-weaning diarrhea	7–21 days	No
			Post-weaning diarrhea	5–14 days post weaning	Yes
			Edema disease	5–14 days post weaning	Yes
	Clostridium perfringens	Type C	Necrotic enteritis	1–14 days	Yes
		Type A	Diarrhea	3–28 days, weaned pigs	Yes
	Clostridioides (Clostridium) difficile		Diarrhea, ill thrift	5–21 days, rarely older	No
	Helicobacter suis		Gastritis, gastric ulcers	Fattening pigs, adults	No
	Salmonella enterica	Typhimurium Derby Choleraesuis	Occasional diarrhea, septicemia, death Occasional diarrhea Septicemia, diarrhea, death	All ages, usually after weaning	Yes No Yes
	Lawsonia intracellularis[c]		Porcine proliferative enteropathy (ileitis) Regional/necrotic ileitis Porcine hemorrhagic enteropathy	Nursery-fattening pigs Nursery-fattening pigs Fattening pigs, young adults	Yes[c]
	Strongly hemolytic *Brachyspira* spp.		Swine dysentery	All ages, mainly grower-fattening pigs	No
	Brachyspira pilosicoli		Porcine intestinal spirochaetosis (PIS)	5 weeks to 4 months	No
Respiratory	Dermonecrotic toxin *Pasteurella multocida* (D) *Bordetella bronchiseptica*		Progressive atrophic rhinitis (PAR)	1–8 weeks Nasal distortion lasts for life	Yes
	Mycoplasma hyopneumoniae		Enzootic pneumonia	Grower-fattening pigs	Yes
	Pasteurella multocida		Mycoplasma-induced respiratory disease	Grower-fattening pigs – secondary invader	No

	Actinobacillus pleuropneumoniae	Pleuropneumonia	Grower-fattening pigs	Yes
Septicemic/ bacteremic/ other infections	*E. coli*	Bacteremia, arthritis, navel infections Cystitis, nephritis	Pigs post-weaning Sows	No
	Streptococcus suis	Meningitis, endocarditis, arthritis, and peritonitis	2–10 weeks	No
	Glaesserella parasuis	Glässer's disease	2–10 weeks	Yes
	Mycoplasma hyosynoviae	Mycoplasmal arthritis	>16 weeks	No
	Staphylococcus aureus	Bacteremia, arthritis, osteomyelitis, mastitis and metritis	All age groups	No
	Staphylococcus hyicus	Exudative epidermitis	Pre- and post-weaning piglets	No
	Erysipelothrix rhusiopathiae	Erysipelas (dermatitis, arthritis, endocarditis)	Grower-fattening pigs, breeding pigs	Yes
	Leptospira spp.	Leptospirosis	Breeding animals	Yes

[a]Commercial vaccines available in only one or a few countries are not included in the table.
[b]ETEC: enterotoxic; EPEC enteropathogenic; STEC Shiga-toxin-producing *E. coli*.

[c]Infections with *Lawsonia intracellularis* may lead to different pathological/clinical forms depending on the age of the animals and other factors. Commercial vaccines are available against *L. intracellularis*. *Source*: Adapted from Aarestrup et al. (2008) and Thomson and Friendship (2019).

include *A. pleuropneumoniae*, *G. parasuis*, post-weaning enterotoxigenic *E. coli* and shiga-toxin-producing *E. coli*, *L. intracellularis*, *Salmonella* spp., and so on.

In general, vaccination is economically justified in clinically affected herds, especially if no major risk factors can be identified or changed in short term. However, vaccination may also be cost-efficient in subclinically infected herds as subclinical infections with pathogens like *Mycoplasma hyopneumoniae* or *A. pleuropneumoniae* also reduce performance (Regula et al., 2000; Maes et al., 2003). For some important pig pathogens (e.g. *Brachyspira hyodysenteriae* and *Streptococcus suis*), however, no or few commercial vaccines are available. Autogenous vaccines may also be used. These are mostly inactivated vaccines, bacterins based on the pathogen strain(s) isolated from animals affected by the disease. Autogenous vaccines against different pathogens such as *Streptococcus suis*, *A. pleuropneumoniae*, and *E. coli* have been described (Baums et al., 2010). One of the disadvantages of autogenous vaccines is that often limited data are available on vaccine efficacy and safety (Haesebrouck et al., 2004).

9 Conclusion

Infections with swine bacterial pathogens significantly affect the health, well-being, and performance of pigs in all types of production systems worldwide, and this will likely remain in the future. The pathogens can be classified according to different characteristics; some of them may determine their interaction with the host and virulence and/or may be important for the epidemiology, diagnostics, treatment, and control measures. Control should focus primarily on the improvement of management, nutrition and housing conditions, and also on vaccination. Most of these measures provide only partial protection against clinical signs, lesions, and production losses. Advances in research on pathogen characteristics, epidemiology, and pathogenesis may improve diagnostics and contribute to the development of new vaccines that provide better protection.

10 Where to look for further information

- *Diseases of Swine*, Eleventh Edition. Editors: Jeffrey J. Zimmerman, Locke A. Karriker, Alejandro Ramirez, Kent J. Schwartz, Gregory W. Stevenson, and Jianqiang Zhang. Published 2019 by John Wiley & Sons, Inc.
- *Porcine Health Management: The Official Journal of the European College of Porcine Health Management and the European Association of Porcine Health Management*: https://porcinehealthmanagement.biomedcentral.com/.
- *Journal of Swine Health and Production: The Official Journal of the American Association of Swine Veterinarians*: https://www.aasv.org/shap.html.

- *Book Mycoplasmas in Swine*. Editors: Dominiek Maes, Marina Sibila and Maria Pieters. Acco, Leuven Belgium, pp. 344.
- *Advancements and Technologies in Pig and Poultry Bacterial Disease Control*. Editors: Neil Foster, Ilias Kyriazakis and Paul Burrow. Elsevier, ISBN 978-0-12-818030-3, pp 278.
- *Achieving Sustainable Production of Pig Meat. Volume 3: Animal Health and Welfare*. Editor: Julian Wiseman. Burleigh Dodds Science Publishing, 328 pp.
- *Biosecurity in Animal Production and Veterinary Medicine: From Principles to Practice*. Editors: Jeroen Dewulf and Filip Van Immerseel. Acco Leuven, 524 pp.

11 References

Aarestrup, F. M., Oliver Duran, C. and Burch, D. G. (2008). Antimicrobial resistance in swine production. *Anim. Health Res. Rev.* 9(2), 135–148.

Alban, L., Petersen, J. V. and Busch, M. E. (2015). A comparison between lesions found during meat inspection of finishing pigs raised under organic/free-range conditions and conventional, indoor conditions. *Porcine Health Manag.* 1, 4.

Alexander, T. J., Thornton, K., Boon, G., Lysons, R. J. and Gush, A. F. (1980). Medicated early weaning to obtain pigs free from pathogens endemic in the herd of origin. *Vet. Rec.* 106(6), 114–119.

Allaart, J. G., van Asten, A. J. A. M. and Grone, A. (2013). Predisposing factors and prevention of *Clostridium perfringens*-associated enteritis. *Comp. Immunol. Microbiol. Infect. Dis.* 36(5), 449–464.

Althouse, G. C., Kuster, C. E., Clark, S. G. and Weisiger, R. M. (2000). Field investigations of bacterial contaminants and their effects on extended porcine semen. *Theriogenology* 53(5), 1167–1176.

Althouse, G. C. and Lu, K. G. (2005). Bacteriospermia in extended porcine semen. *Theriogenology* 63(2), 573–584.

Althouse, G. C. and Rossow, K. (2011). The potential risk of infectious disease dissemination via artificial insemination in swine. *Reprod. Domest. Anim.* 46(Suppl. 2), 64–67.

Amass, S. F. (2005). Biosecurity: Stopping the bugs from getting in. *Pig J.* 55, 104–114.

Andres, V. M. and Davies, R. H. (2015). Biosecurity measures to control salmonella and other infectious agents in pig farms: A review. *Compr. Rev. Food Sci. Food Saf.* 14(4), 317–335.

Aragon, V., Segalés, J. and Tucker, D. (2019). Chapter 54: Glässer's disease. In: Zimmerman, J. J., Karriker, L. A., Ramirez, A., Schwartz, K. J., Stevenson, G. W. and Zhang, J. (Eds) *Diseases of Swine* (11th edn, vol. 2019, pp. 844–853). Hoboken, NJ: John Wiley & Sons, Inc.

Arent, Z. and Ellis, W. (2019). Chapter 55: Leptospirosis. In: Zimmerman, J. J., Karriker, L. A., Ramirez, A., Schwartz, K. J., Stevenson, G. W. and Zhang, J. (Eds) *Diseases of Swine* (11th edn, vol. 2019, pp. 854–862). Hoboken, NJ: John Wiley & Sons, Inc.

Arruda, P. H. E. and Gauger, P. (2019). Chapter 7: Optimizing sample selection, collection and submission to optimize diagnostic value. In: Zimmerman, J. J., Karriker, L. A.,

Ramirez, A., Schwartz, K. J., Stevenson, G. W. and Zhang, J. (Eds) *Diseases of Swine* (11th edn, vol. 2019, pp. 98–111). John Wiley & Sons, Inc.

Assavacheep, P. and Rycroft, A. N. (2013). Survival of *Actinobacillus pleuropneumoniae* outside the pig. *Res. Vet. Sci.* 94(1), 22–26.

Balloux, F., Brønstad Brynildsrud, O., van Dorp, L., Shaw, L. P., Chen, H., Harris, K. A., Wang, H. and Eldholm, V. (2018). From theory to practice: Translating whole-genome sequencing (WGS) into the clinic. *Trends Microbiol.* 26(12), 1035–1048.

Baroch, J. A., Gagnon, C. A., Lacouture, S. and Gottschalk, M. (2015). Exposure of feral swine (*Sus scrofa*) in the United States to selected pathogens. *Can. J. Vet. Res.* 79(1), 74–78.

Baums, C. G., Brüggemann, C., Kock, C., Beineke, A., Waldmann, K. H. and Valentin-Weigand, P. (2010). Immunogenicity of an autogenous *Streptococcussuis* bacterin in preparturient sows and their piglets in relation to protection after weaning. *Clin. Vaccine Immunol.* 17(10), 1589–1597.

Bender, J. S., Shen, H. G., Irwin, C. K., Schwartz, K. J. and Opriessnig, T. (2010). Characterization of *Erysipelothrix* species isolates from clinically affected pigs, environmental samples and vaccine strains from six recent swine erysipelas outbreaks in the United States. *Clin. Vaccine Immunol.* 17(10), 1605–1611.

Boadella, M., Ruiz-Fons, J. F., Vicente, J., Martín, M., Segalés, J. and Gortazar, C. (2012). Seroprevalence evolution of selected pathogens in Iberian wild boar. *Transbound. Emerg. Dis.* 59(5), 395–404.

Bright-Ponte, S. J. (2020). Antimicrobial use data collection in animal agriculture. *Zoonoses Public Health* 67(Suppl. 1), 1–5.

Brockmeier, S., Register, K., Nicholson, T. and Loving, C. (2019). Chapter 49: Bordetellosis. In: Zimmerman, J. J., Karriker, L. A., Ramirez, A., Schwartz, K. J., Stevenson, G. W. and Zhang, J. (Eds) *Diseases of Swine* (11th edn, vol. 2019, pp. 767–777). Hoboken, NJ: John Wiley & Sons, Inc.

Brockmeier, S. L., Loving, C. L., Mullins, M. A., Register, K. B., Nicholson, T. L., Wiseman, B. S., Baker, R. B. and Kehrli, M. E. (2013). Virulence, transmission and heterologous protection of four isolates of *Haemophilus parasuis*. *Clin. Vaccine Immunol.* 20(9), 1466–1472.

Burow, E., Rostalski, A., Harlizius, J., Gangl, A., Simoneit, C., Grobbel, M., Kollas, C., Tenhagen, B. A. and Käsbohrer, A. (2019). Antibiotic resistance in *Escherichia coli* from birth to slaughter and its association with antibiotic treatment. *Prev. Vet. Med.* 165, 52–62.

Callens, B., Faes, C., Maes, D., Catry, B., Boyen, F., Francoys, D., de Jong, E., Haesebrouck, F. and Dewulf, J. (2015). Presence of antimicrobial resistance and antimicrobial use in sows are risk factors for antimicrobial resistance in their offspring. *Microb. Drug Resist.* 21(1), 50–58.

Cilia, G., Bertelloni, F., Piredda, I., Ponti, M. N., Turchi, B., Cantile, C., Parisi, F., Pinzauti, P., Armani, A., Palmas, B., Noworol, M., Cerri, D. and Fratini, F. (2020). Presence of pathogenic Leptospira spp. In the reproductive system and fetuses of wild boars (Sus scrofa) in Italy. *PLoS Negl. Trop. Dis.* 14(12), e0008982.

Cornick, N. A. and VuKhac, H. (2008). Indirect transmission of *Escherichia coli* O157:H7 occurs readily among swine but not among sheep. *Appl. Environ. Microbiol.* 74(8), 2488–2491.

de Jong, E., Appeltant, R., Cools, A., Beek, J., Boyen, F., Chiers, K. and Maes, D. (2014). Slaughterhouse examination of culled sows in commercial pig herds. *Livest. Sci.* 167, 362–369.

Declerck, I., Dewulf, J., Piepers, S., Decaluwé, R. and Maes, D. (2015). Sow and litter factors influencing colostrum yield and nutritional composition. *J. Anim. Sci.* 93(3), 1309–1317.

Desrosiers, R. (2011). Transmission of swine pathogens: Different means, different needs. *Anim. Health Res. Rev.* 12(1), 1–13.

Dutkiewicz, J., Sroka, J., Zając, V., Wasiński, B., Cisak, E., Sawczyn, A., Kloc, A. and Wójcik-Fatla, A. (2017). *Streptococcussuis*: A re-emerging pathogen associated with occupational exposure to pigs or pork products. Part I – Epidemiology. *Ann. Agric. Environ. Med.* 24(4), 683–695.

EMA (European Medicines Agency). (2017). Zinc oxide. Available at: https://www.ema.europa.eu/en/medicines/veterinary/referrals/zinc-oxide. Accessed: 21 March 2022.

EMA (European Medicines Agency). (2020). Categorisation of antibiotics used in animals promotes responsible use to protect public and animal health. Available at: https://www.ema.europa.eu/en/news/categorisation-antibiotics-used-animals-promotes-responsible-use-protect-public-animal-health. Accessed: 21 March 2022.

European Food Safety Authority and European Centre for Disease Prevention and Control (EFSA and ECDC). (2018). The European Union summary report on trends and sources of zoonoses, zoonotic agents and food-borne outbreaks in 2017. *EFSA J.* 16(12), e05500.

Fablet, C., Marois, C., Kuntz-Simon, G., Rose, N., Dorenlor, V., Eono, F., Eveno, E., Jolly, J. P., Le Devendec, L., Tocqueville, V., Queguiner, S., Gorin, S., Kobisch, M. and Madec, F. (2011). Longitudinal study of respiratory infection patterns of breeding sows in five farrow-to-finish herds. *Vet. Microbiol.* 147(3–4), 329–339.

Faccin, J. E. G., Laskoski, F., Hernig, L. F., Kummer, R., Lima, G. F. R., Orlando, U. A. D., Gonçalves, M. A. D., Mellagi, A. P. G., Ulguim, R. R. and Bortolozzo, F. P. (2020). Impact of increasing weaning age on pig performance and belly nosing prevalence in a commercial multisite production system. *J. Anim. Sci.* 98(4), skaa031.

Fairbrother, J. and Nadeau, E. (2019). Chapter 52: Colibacillosis. In: Zimmerman, J. J., Karriker, L. A., Ramirez, A., Schwartz, K. J., Stevenson, G. W. and Zhang, J. (Eds) *Diseases of Swine* (11th edn, vol. 2019, pp. 807–835). Hoboken, NJ: John Wiley & Sons, Inc.

Filippitzi, M. E., Brinch-Kruse, A., Postma, M., Sarrazin, S., Maes, D., Alban, L., Nielsen, L. R. and Dewulf, J. (2018). Review of transmission routes of 24 infectious diseases preventable by biosecurity measures and comparison of the implementation of these measures in pig herds in six European countries. *Transbound. Emerg. Dis.* 65(2), 381–398.

Friedman, M., Bednár, V., Klimeš, J., Smola, J., Mrlík, V. and Literák, I. (2008). *Lawsoniaintracellularis* in rodents from pig farms with the occurrence of porcine proliferative enteropathy. *Lett. Appl. Microbiol.* 47(2), 117–121.

Giang, E., Hetman, B. M., Sargeant, J. M., Poljak, Z. and Greer, A. L. (2020). Examining the effect of host recruitment rates on the transmission of *Streptococcussuis* in nursery swine populations. *Pathogens* 9(3), 174.

Gottschalk, M. and Broes, A. (2019). Chapter 48: Actinobacillosis. In: Zimmerman, J. J., Karriker, L. A., Ramirez, A., Schwartz, K. J., Stevenson, G. W. and Zhang, J. (Eds) *Diseases of Swine* (11th edn, vol. 2019, pp. 749–766). Hoboken, NJ: John Wiley & Sons, Inc.

Gottschalk, M. and Segura, M. (2019). Chapter 61: Streptococcosis. In: Zimmerman, J. J., Karriker, L. A., Ramirez, A., Schwartz, K. J., Stevenson, G. W. and Zhang, J. (Eds) *Diseases of Swine* (11th edn, vol. 2019, pp. 934–950). Hoboken, NJ: John Wiley & Sons, Inc.

Griffith, R., Carlson, S. and Krull, A. (2019). Chapter 59: Salmonellosis. In: Zimmerman, J. J., Karriker, L. A., Ramirez, A., Schwartz, K. J., Stevenson, G. W. and Zhang, J. (Eds) *Diseases of Swine* (11th edn, vol. 2019, pp. 912–925). Hoboken, NJ: John Wiley & Sons, Inc.

Haesebrouck, F., Pasmans, F., Chiers, K., Maes, D., Ducatelle, R. and Decostere, A. (2004). Efficacy of vaccines against bacterial diseases in swine: What can we expect? *Vet. Microbiol.* 100(3–4), 255–268.

Hampson, D. and Burrough, E. (2019). Chapter 62: Swine dysentery and Brachyspiral colitis. In: Zimmerman, J. J., Karriker, L. A., Ramirez, A., Schwartz, K. J., Stevenson, G. W. and Zhang, J. (Eds) *Diseases of Swine* (11th edn, vol. 2019, p. 951). Hoboken, NJ: John Wiley & Sons, Inc.

Hoque, R., Ahmed, S. M., Naher, N., Islam, M. A., Rousham, E. K., Islam, B. Z. and Hassan, S. (2020). Tackling antimicrobial resistance in Bangladesh: A scoping review of policy and practice in human, animal and environment sectors. *PLoS ONE* 15(1), e0227947.

Johansen, T., Agdestein, A., Lium, B., Jørgensen, J. A. and Djønne, B. (2014). *Mycobacterium aviumsubsp. hominissuis* infection in swine associated with peat used for bedding. *J. Biomed. Biotechnol.* 1, 189649.

Klindworth, A., Pruesse, E., Schweer, T., Peplies, J., Quast, C., Horn, M. and Glöckner, F. O. (2013). Evaluation of general 16S ribosomal RNA gene PCR primers for classical and next-generation sequencing-based diversity studies. *Nucleic Acids Res.* 41(1), e1.

Lama, J. K. and Bachoon, D. S. (2018). Detection of *Brucella suis*, *Campylobacter jejuni*, and *Escherichia coli* strains in feral pig (Sus scrofa) communities of Georgia. *Vector Borne Zoonotic Dis.* 18(7), 350–355.

Lamendella, R., Domingo, J. W., Ghosh, S., Martinson, J. and Oerther, D. B. (2011). Comparative fecal metagenomics unveils unique functional capacity of the swine gut. *BMC Microbiol.* 11, 103.

Léger, A., Alban, L., Veldhuis, A., van Schaik, G. and Stärk, K. D. C. (2019). Comparison of international legislation and standards on veterinary drug residues in food of animal origin. *J. Public Health Policy.* 40(3), 308–341.

Leser, T. D., Lindecrona, R. H., Jensen, T. K., Jensen, B. B. and Moller, K. (2000). Changes in bacterial community structure in the colon of pigs fed different experimental diets and after infection with *Brachyspira hyodysenteriae*. *Appl. Environ. Microbiol.* 66(8), 3290–3296.

Loera-Muro, V. M., Jacques, M., Tremblay, Y. D. N., Avelar-Gonzalez, F. J., Loera-Muro, A., Ramirez-Lopez, E. M., Medina-Figueroa, A., Gonzalez-Reynaga, H. M. and Guerrero Barrera, A. L. (2013). Detection of *Actinobacillus pleuropneumoniae* in drinking water from pig farms. *Microbiology (Reading)* 159(3), 536–544.

López, A., Van Soom, A., Arsenakis, I. and Maes, D. (2017). Boar management and semen handling factors affect the quality of boar extended semen. *Porcine Health Manag.* 3, 15.

Luppi, A. (2017). Swine enteric colibacillosis: Diagnosis, therapy and antimicrobial resistance. *Porcine Health Manag.* 3, 16.

Maes, D. G., Nauwynck, H., Rijsselaere, T., Mateusen, B., Vyt, Ph., de Kruif, A. and Van Soom, A. (2008). AI transmitted diseases in swine: An overview. *Theriogenology* 70(8), 1337–1345.

Maes, D. G., Verbeke, W., Vicca, J., Verdonck, M. and de Kruif, A. (2003). Benefit to cost of vaccination against *Mycoplasma hyopneumoniae* in pig herds under Belgian market conditions from 1996 to 2000. *Livest. Sci.* 83(1), 85–93.

Maes, D. G., Deluyker, H., Verdonck, M., Castryck, F., Miry, C., Vrijens, B., Ducatelle, R. and de Kruif, A. (2001). Non-infectious herd factors associated with macroscopic and microscopic lung lesions in slaughter pigs from farrow-to-finish pig herds. *Vet. Rec.* 148(2), 41–46.

Malmsten, A., Magnusson, U., Ruiz-Fons, F., González-Barrio, D. and Dalin, A. M. (2018). A serologic survey of pathogens in wild boar (Sus scrofa) in Sweden. *J. Wildl. Dis.* 54(2), 229–237.

Marco, E., Yeske, P. and Pieters, M. (2020). Chapter 9: General control measures against Mycoplasma hyopneumoniae infections. In: Maes, D., Sibila, M. and Pieters, M. (Eds) *Book Mycoplasmas in Swine* (pp. , 163–180). Leuven, Belgium: Acco Publishers. ISBN 978-94-6379-796-2.

Marinou, K. A., Papatsiros, V. G., Gkotsopoulos, E. K., Odatzoglou, P. K. and Athanasiou, L. V. (2015). Exposure of extensively farmed wild boars (Sus scrofa scrofa) to selected pig pathogens in Greece. *Vet. Q.* 35(2), 97–101.

Martelli, F., Andres, V. M., Davies, R. and Smith, R. P. (2018). Observations on the introduction and dissemination of *Salmonella* in three previously low prevalence status pig farms in the United Kingdom. *Food Microbiol.* 71, 129–134.

Middelveld, R. J. and Alving, K. (2000). Synergistic septicemic action of the gram-positive bacterial cell wall components peptidoglycan and lipoteichoic acid in the pig *in vivo*. *Shock* 13(4), 297–306.

Muirhead, M. R. and Alexander, T. J. L. (1997). Understanding disease. In: *Managing Pig Health and the Treatment of Disease, a Reference for the Farm (Chapter 2)* (pp. 21–54). Sheffield, UK: 5M Enterprises Limited UK.

Nathues, H., Woeste, H., Doehring, S., Fahrion, A. S., Doherr, M. G. and Grosse Beilage, E. (2012). Detection of *Mycoplasma hyopneumoniae* in nasal swabs sampled from pig farmers. *Vet. Rec.* 170(24), 623.

Neirynck, W., Boyen, F., Chantziaras, I., Vandersmissen, T., Vyt, P., Haesebrouck, F., Dewulf, J. and Maes, D. (2020). Implementation and evaluation of different eradication strategies for *Brachyspira hyodysenteriae*. *Porcine Health Manag.* 6, 27.

Niemann, H. H., Schubert, W.D., and Heinz, D. W. (2004). Adhesins and invasins of pathogenic bacteria: a structural view. *Microbes Infect.* 6(1), 101–112. doi:10.1016/j.micinf.2003.11.001. PMID: 14738899.

Nicholson, T. L., Brockmeier, S. L., Loving, C. L., Register, K. B., Kehrli, M. E., Stibitz, S. E. and Shore, S. M. (2012). Phenotypic modulation of the virulent Bvg phase is not required for pathogenesis and transmission of *Bordetella bronchiseptica* in swine. *Infect. Immun.* 80(3), 1025–1036.

O'Connor, A. M., Denagamage, T., Sargeant, J. M., Rajic, A. and McKean, J. (2008). Feeding management practices and feed characteristics associated with *Salmonella* prevalence in live and slaughtered market-weight finisher swine: A systematic review and summation of evidence from 1950 to 2005. *Prev. Vet. Med.* 87(3-4), 213-228.

Office International des Epizooties (OIE). (2021). Notifiable animal diseases. Available at: https://www.oie.int/en/what-we-do/animal-health-and-welfare/animal-diseases/?_tax_animal=terrestrials%2Cmultiple-species. Accessed: 21 March 2022.

Ogwuche, A., Ekiri, A. B., Endacott, I., Maikai, B. V., Idoga, E. S., Alafiatayo, R. and Cook, A. J. (2021). Antibiotic use practices of veterinarians and para-veterinarians and the implications for antibiotic stewardship in Nigeria. *J. S. Afr. Vet. Assoc.* 92, e1-e14.

Olsen, S., Boggiatto, P. and Samartino, L. (2019). Chapter 50: Bordetellosis. In: Zimmerman, J. J., Karriker, L. A., Ramirez, A., Schwartz, K. J., Stevenson, G. W. and Zhang, J. (Eds) *Diseases of Swine* (11th edn, vol. 2019, pp. 778-791). John Wiley & Sons, Inc.

Opriessnig, T. and Coutinho, T. (2019). Chapter 53: Erysipelas. In: Zimmerman, J. J., Karriker, L. A., Ramirez, A., Schwartz, K. J., Stevenson, G. W. and Zhang, J. (Eds) *Diseases of Swine* (11th edn, vol. 2019, pp. 835-843). Hoboken, NJ: John Wiley & Sons, Inc.

Opriessnig, T., Giménez-Lirola, L. G. and Halbur, P. G. (2011). Polymicrobial respiratory disease in pigs. *Anim. Health Res. Rev.* 12(2), 133-148.

Pavlovic, M., Huber, I., Konrad, R. and Busch, U. (2013). Application of MALDI-TOF MS for the identification of food borne bacteria. *Open Microbiol. J.* 7, 135-141.

Pearson, H. E., Toribio, J. L. M. L., Lapidge, S. J. and Hernandez-Jover, M. (2016). Evaluating the risk of pathogen transmission from wild animals to domestic pigs in Australia. *Prev. Vet. Med.* 123, 39-51.

Peeters, L., Dewulf, J., Boyen, F., Brossé, C., Vandersmissen, T., Rasschaert, G., Heyndrickx, M., Cargnel, M., Pasmans, F. and Maes, D. (2019). Effects of attenuated vaccine protocols against *Salmonella Typhimurium* on *Salmonella* serology in subclinically infected pig herds. *Vet. J.* 249, 67-72.

Pieters, M. and Fano, E. (2016). *Mycoplasma hyopneumoniae* management in gilts. *Vet. Rec.* 178(5), 122-123.

Pieters, M. and Maes, D. (2019). Chapter 56: Mycoplasmosis. In: Zimmerman, J. J., Karriker, L. A., Ramirez, A., Schwartz, K. J., Stevenson, G. W. and Zhang, J. (Eds) *Diseases of Swine* (11th edn, vol. 2019, pp. 863-883). Hoboken, NJ: John Wiley & Sons, Inc.

Post, K. (2019). Chapter 47: Overview of bacteria. In: Zimmerman, J. J., Karriker, L. A., Ramirez, A., Schwartz, K. J., Stevenson, G. W. and Zhang, J. (Eds) *Diseases of Swine* (11th edn, vol. 2019, pp. 745-749). Hoboken, NJ: John Wiley & Sons, Inc.

Prieto-Martinez, N., Bussalleu, E., Garcia-Bonavila, E., Bonet, S. and Yeste, M. (2014). Effects of Enterobacter cloacae on boar sperm quality during liquid storage at 17 degrees C. *Anim. Reprod. Sci.* 148(1-2), 72-82.

Ramirez, A. (2018). Chapter 1: Diseases affecting pigs: An overview of common bacterial, viral and parasitic pathogens of pigs. In: Wiseman, J. (Ed.) *Achieving Sustainable Production of Pig Meat: Animal Health and Welfare* (vol. 3, p. 328). Burleigh Dodds Science Publishing Limited.

Randall, L. P., Lemma, F., Koylass, M., Rogers, J., Ayling, R. D., Worth, D., Klita, M., Steventon, A., Line, K., Wragg, P., Muchowski, J., Kostrzewa, M. and Whatmore, A. M. (2015). Evaluation of MALDI-ToF as a method for the identification of bacteria in the veterinary diagnostic laboratory. *Res. Vet. Sci.* 101, 42-49.

Register, K. and Brockmeier, S. (2019). Chapter 57: Pasteurellosis. In: Zimmerman, J. J., Karriker, L. A., Ramirez, A., Schwartz, K. J., Stevenson, G. W. and Zhang, J. (Eds)

Diseases of Swine (11th edn, vol. 2019, pp. 884–897). Hoboken, NJ: John Wiley & Sons, Inc.
Regula, G., Lichtensteiger, C. A., Mateus-Pinilla, N. E., Scherba, G., Miller, G. Y. and Weigel, R. M. (2000). Comparison of serologic testing and slaughter evaluation for assessing the effects of subclinical infection on growth in pigs. *J. Am. Vet. Med. Assoc.* 217(6), 888–895.
Rhouma, M., Fairbrother, J. M., Beaudry, F. and Letellier, A. (2017). Post weaning diarrhea in pigs: Risk factors and non-colistin-based control strategies. *Acta Vet. Scand.* 59(1), 31.
Rycroft, A. (2020). Chapter 1: Overview of the general characteristics and classification of porcine *Mycoplasma* species. In: Maes, D., Sibila, M. and Pieters, M. (Eds) *Book Mycoplasmas in Swine* (pp. 25–46). Leuven, Belgium: Acco Publishers, ISBN 978-94-6379-796-2.
Schulze, M., Ammon, C., Rudiger, K., Jung, M. and Grobbel, M. (2015). Analysis of hygienic critical control points in boar semen production. *Theriogenology* 83(3), 430–437.
Sepúlveda, L., Bussalleu, E., Yeste, M. and Bonet, S. (2014). Effects of different concentrations of *Pseudomonas aeruginosa* on boar sperm quality. *Anim. Reprod. Sci.* 150(3–4), 96–106.
Sepúlveda, L., Bussalleu, E., Yeste, M. and Bonet, S. (2016). Effect of *Pseudomonas aeruginosa* on sperm capacitation and protein phosphorylation of boar spermatozoa. *Theriogenology* 85(8), 1421–1431.
Sepúlveda, L., Bussalleu, E., Yeste, M., Torner, E. and Bonet, S. (2013). How do different concentrations of *Clostridium perfringens* affect the quality of extended boar spermatozoa? *Anim. Reprod. Sci.* 140(1–2), 83–91.
Songer, J. G. and Post, K. W. (2005). Origin and evolution of virulence. In: *Veterinary Microbiology: Bacterial and Fungal Agents of Animal Disease* (pp. 3–9). St. Louis, MO: Elsevier Saunders.
Tack, D., Ray, L., Griffin, P., Cieslak, P., Dunn, J., Rissman, T., Jervis, R., Lathrop, S., Muse, A., Duwell, M., Smith, K., Tobin-D'Angelo, M., Vugia, D., Zablotsky-Kufel, J., Wolpert, B., Tauxe, R. and Payne, D. (2020). Preliminary incidence and trends of infections with pathogens transmitted commonly through food – Foodborne diseases active surveillance network. *MMWR Morb. Mortal. Wkly Rep. (MMWR)* 1069(17), 509–514.
Thomas, S. R. and Elkinton, J. S. (2004). Pathogenicity and virulence. *J. Invertebr. Pathol.* 85(3), 146–151.
Thomson, J. and Friendship, R. (2019). Chapter 15: Digestive system. In: Zimmerman, J. J., Karriker, L. A., Ramirez, A., Schwartz, K. J., Stevenson, G. W. and Zhang, J. (Eds) *Diseases of Swine* (11th edn, vol. 2019, pp. 234–263). Hoboken, NJ: John Wiley & Sons, Inc.
Tobias, T. J., Bouma, A., van den Broek, J., van Nes, A., Daemen, A. J. J. M., Wagenaar, J. A., Stegeman, J. A. and Klinkenberg, D. (2014). Transmission of *Actinobacillus pleuropneumoniae* among weaned piglets on endemically infected farms. *Prev. Vet. Med.* 117(1), 207–214.
Úbeda, J. L., Ausejo, R., Dahmani, Y., Falceto, M. V., Usan, A., Malo, C. and Perez-Martinez, F. C. (2013). Adverse effects of members of the Enterobacteriaceae family on boar sperm quality. *Theriogenology* 80(6), 565–570.
Uzal, F. and Songer, G. (2019). Chapter 51: Clostridial diseases. In: Zimmerman, J. J., Karriker, L. A., Ramirez, A., Schwartz, K. J., Stevenson, G. W. and Zhang, J. (Eds) *Diseases of Swine* (11th edn, vol. 2019, pp. 792–806). John Wiley & Sons, Inc.

Vannucci, F., Gebhart, C. and McOrist, S. (2019). Chapter 58: Proliferative enteropathy. In: Zimmerman, J. J., Karriker, L. A., Ramirez, A., Schwartz, K. J., Stevenson, G. W. and Zhang, J. (Eds) *Diseases of Swine* (11th edn, vol. 2019, pp. 898–911). John Wiley & Sons, Inc.

Vanrolleghem, W., Tanghe, S., Verstringe, S., Bruggeman, G., Papadopoulos, D., Trevisi, P., Zentek, J., Sarrazin, S. and Dewulf, J. (2019). Potential dietary feed additives with antibacterial effects and their impact on performance of weaned piglets: A meta-analysis. *Vet. J.* 249, 24–32.

Wu, N., Abril, C., Thomann, A., Grosclaude, E., Doherr, M. G., Boujon, P. and Ryser-Degiorgis, M. P. (2012). Risk factors for contact between wild boar and outdoor pigs in Switzerland and investigations into potential *Brucella suis* spill-over. *BMC Vet. Res.* 8, 116.

Zeeh, F., Vidondo, B. and Nathues, H. (2020). Risk factors for the infection with *Brachyspira hyodysenteriae* in pig herds. *Prev. Vet. Med.* 174, 104819.

Chapter 3

Improving disease surveillance and monitoring systems in pig herds

Alejandro Ramirez, University of Arizona, USA

1 Introduction

Although sometimes the terms 'disease monitoring' and 'surveillance' are used interchangeably, their epidemiological meanings are quite different. As Christensen (2001) states, disease monitoring and surveillance both involve the collection of data to evaluate the health and disease status of a given population over time. In the case of disease surveillance, it is an active and targeted approach that contains at least three components: (1) defined disease monitoring system, (2) a predefined disease intervention strategy, and (3) a defined threshold of disease frequency (Christensen, 2001). The *Manual on Livestock Disease Surveillance and Information Systems* (1999) from the Food and Agriculture Organization of

http://dx.doi.org/10.19103/AS.2022.0103.03

the United Nations has some slightly different definitions for these two terms. The manual defines monitoring as 'all activities aimed at detecting changes in the epidemiological parameters of a specified disease'. Disease surveillance is defined as 'all regular activities aimed at ascertaining the health status of a given population with the aim of early detection and control of animal diseases of importance to national economies, food security and trade'. A key component of surveillance is early detection. Early detection allows for early interventions which is a key component of disease control. Often, in pig production, disease monitoring is done in relation to known endemic diseases, while disease surveillance is done for diseases not known to be present (Burrough et al., 2019).

Surveillance systems can be passive or active. In passive surveillance, diseases are reported by pig farmers or veterinarians as they occur. In active surveillance, monitoring is targeted and specific, at a specific time, in order to actively find and identify more new cases. This distinction is important because in an active system, information collected is more complete and comprehensive.

For the purpose of this chapter, the term 'monitoring' will be used to imply both traditional monitoring and the monitoring phase of surveillance since the same concepts apply to both systems.

2 Identifying pathogens of concern

One of the first steps in the development of disease monitoring programs is the identification of the diseases of concern. Although there are many pig diseases of concern, the limitation on resources (time and money) along with logistics will dictate a focus on those diseases of greatest concern.

Some of the most common reasons for selecting diseases for monitoring/surveilling include (1) diseases with regulatory implications (especially foreign animal disease), (2) diseases with a high economic impact on production, or (3) diseases with high zoonotic potential. There are times - especially when implementing, or deciding to implement, a disease control or elimination program - that other diseases may be selected for monitoring even though they may not be of primary concern. As will be discussed later, incorporating multiple diseases into the same monitoring program can significantly impact the effectiveness of the program implemented. It is not the scope of this chapter to provide guidance on which diseases should be monitored or surveilled but to provide the necessary information to guide the pig farmers and veterinarians on improving their disease monitoring/surveillance systems in their herds.

3 Defining clear goals and objectives

The overall goal of a monitoring or surveillance system is to collect data for decision-making. Before any disease monitoring or surveillance system is established, the goals and objectives must be clearly stated and agreed upon by all those who are impacted by such decision-making. Although this seems obvious, often it is forgotten that many goals can be interpreted very differently by individuals.

We first start by clarifying the difference between a goal and an objective. Goals are the big picture expectations or vision, while objectives are the smaller, well-defined strategies that will help to achieve the overall goal. Let us look at an example to better understand the importance of differentiating goals and objectives. A pig farmer had a disease outbreak of porcine reproductive and respiratory syndrome (PRRS) 8 months ago in their breeding herd. Their goal is to test negative for PRRS again. This is a great goal, but unfortunately there are many different ways to interpret 'negative for PRRS' making it challenging to figure out how to monitor for such goal. In this case, it would be more beneficial to better clarify what is meant by 'PRRS negative' as well as create a series of objectives which can be monitored so that in time, the goal of testing PRRS negative again can be achieved. Some individual objectives may include the following:

- Stop abortions.
- Stop PRRS virus (PRRSV) transmission between sows.
- Return to normal sow productivity (pigs weaned/mated female/year).
- Eliminate PRRSV viremic piglets at birth.
- Wean pigs negative for PRRSV (no viremia).
- Re-establish breeding herd seronegative to PRRS.

Each of the above objectives will require completely different PRRSV monitoring programs. Being clearer about the specific objectives of the monitoring program will help to evaluate the effectiveness of different interventions by interpreting results under the same context. Later in this chapter, we will discuss the limitation in interpreting results, especially in the inability to prove a negative.

As monitoring programs are developed, it is critical to not forget the most important rule in diagnostics: 'Do not run a test unless you plan on how you will utilize the result'. This rule is very important in two contexts. First, data collected must serve to inform decision-making; otherwise it is a waste of time and resources. It is recognized that sometimes that decision-making may not be immediate, especially with passive monitoring systems, but ultimately the data collected should have a purpose and value for future decision-making. Second, the plan must include how to respond to either a positive or negative assay result. This can create a conundrum in deciding how to interpret an unanticipated result, especially in regards to a disease considered to have national and international phytosanitary implications. One such example is the desire to test imported feed ingredients or equipment for the presence of a foreign animal disease such as the African swine fever virus (ASFV). Intellectual curiosity and concern would suggest testing such imported products, but the challenge becomes what do you do with the result information? If living in a country where ASFV is not present, are able to test for ASFV, and you detect the virus on an imported metal flooring, what will that mean? Is the virus now confirmed to be present in that country? Is this reportable to the World Organization for Animal Health (OIE)? Even if it is not reportable, how will local media report this? What will be the impact on the local or national pig markets? Will the farm where that flooring was located need to be quarantined or even depopulated? If the solution to a positive result is the cleaning and disinfection of the flooring, why not just do this step regardless of testing? How will testing change what would have been done on the farm? There are many questions here, but the point is that one must be prepared to deal with any and all results that any monitoring program establishes.

As each objective is evaluated for possible monitoring, it is critical to make sure that the testing scheme established can provide data to answer the question on hand. In an academic setting, research questions are hypothesis driven and are best answered one at a time. That is, the best way to answer a question is to target the sampling specifically to answer that question. A challenge in the field is that often there are too many different questions that need to be answered. This creates a conundrum requiring compromise. Academic researchers have the advantage of having time on their side so detail and complexity of their projects can be focused on answering question one at a time. Pig farmers and veterinarians do not have that luxury and often want/need answers now. Again, compromise is needed as trying to move too fast and including multiple different objectives at one time are likely going to significantly compromise all results. Although these multiple results will seem to be answering multiple questions more quickly, if incorrect conclusions are made, the long-term damage might be worse than just having to wait for more accurate answers from a more focused monitoring program.

4 Detection target (analyte): agent or antibodies

Once clear objectives have been established, it will be necessary to evaluate several different factors that must be considered for the disease monitoring program. These factors will include recognizing the targeted analyte, disease ecology, age of pigs, management practices including vaccination programs, and antimicrobial usage, as well as assays available for that specific disease.

4.1 Analyte

Often, monitoring and surveillance systems incorporate some type of diagnostic testing. The very first consideration in assay selection involves selecting the desired target for the assay (analyte). This diagnostic testing can be broken down into just two main objectives (targets): (1) looking to detect the presence of an agent (e.g. bacterial, viral, fungal, and parasite) or parts of that agent or (2) looking for an immune response. For the purpose of this chapter, immune response will be focused on antibody production since that is the most common immune response measured by assays used for field diagnostics including disease monitoring programs. The progression of the disease usually starts with infection (presence of the agent), followed by tissue damage and clinical signs (depending on the severity of damage), then followed by an immune response. As a general rule, detection of antibodies usually takes at least 7–14 days after infection. Knowing which analyte to target requires extensive training and a full understanding of diagnostics, disease epidemiology, disease pathogenesis (how the disease develops within a pig), and disease ecology for each pathogen of concern. Veterinarians have extensive training in these areas and are key in helping develop the proper monitoring and surveillance programs.

4.2 Disease ecology

Disease ecology is defined as the ecological study of host-pathogen interactions within the context of their environment and evolution and at its foundation there are efforts to understand pathogen transmission and spread over space and time (Kilpatrick and Altizer, 2010). Knowledge of disease ecology continues to evolve over time as pig farms continue to change in size and structure. Early weaning, multi-site production, and all-in-all-out technology have all contributed to changes in disease transmission (Harris, 2000).

Veterinarians must work closely with pig farmers to understand their particular farm and pig flow (how groups of pigs move through the different buildings and phases of production). Pig flow has a significant impact on disease transmission and thus is a major influencer of disease ecology. Pig flow along with other management practices, vaccination, and antibiotic usage are all tools used to manage diseases and therefore will also impact monitoring planning. For most diseases, not all pigs get exposed to the pathogen at the same time (Rotolo et al., 2017). The spread of disease within a herd is impacted by the contagiousness (how many other pigs one pig can infect) as well as the infectiousness (how many bacteria or viruses are needed to cause infection) of the particular pathogen of concern. This spread can also be impacted by the overall immunity of the herd to that pathogen. The higher the herd immunity, the more limited and slower the spread. Within a pathogen, differences between different isolates (sometimes referred to as strains) can also dramatically change transmissibility within a herd. In larger herds, groups of pigs tend not to be homogenous and there often have a different subpopulation of pigs with different immunity or disease exposure levels that must be taken into consideration when planning monitoring programs (Dee et al., 1996; Rotolo et al., 2017).

Disease ecology and management factors not only impact where (what groups) and when (what age) disease may be found on a farm but will also impact the prevalence of the disease. Often farmers implement practices intended to either stop or slow down the spread of a particular pathogen. This is when the implementation of surveillance systems can be useful. In this case, active and routine surveillance will be done so that detection of a particular pathogen (at a preestablished level) can occur early enough to allow for an intervention to occur with hopes to mitigate the negative effects of the pathogen.

4.3 Pig age

The age of the animals to be monitored impacts monitoring plans in several ways. Disease occurrence and the level of prevalence in any age group of animals are highly dependent on the disease of concern along with the epidemiology of that particular disease. Often, the longer a pig is on the farm, the greater the probability that they encounter a pathogen. For some pathogens, animals develop resistance over time although the exact reasons for this resistance can be difficult to establish (Gelberg, 1992; Klinge et al., 2009).

The existence of maternally derived antibodies is also impacted by the pig's age. Pigs are not able to obtain antibodies from their mothers while in utero due to placental characteristics. This necessitates piglets to obtain all their passive antibodies through colostrum preferably within the first 6 h of birth (Miller et al., 1962). On the first day of life, piglets will consume colostrum that is rich in antibodies, especially IgG. These IgG antibodies are absorbed, intact, through the piglets' gut and enter their circulatory system, contributing to systemic immunity. This immunity is passive, meaning these antibodies will not increase but rather will decrease over time, lowering the piglet's protection and increasing their susceptibility to several diseases. This means that often, young piglets (say <10 or 12 weeks of age) can test positive for antibodies to a particular pathogen even though these piglets themselves have never been exposed to such a pathogen; these are maternally derived antibodies. Age is also important when considering the impact of vaccination on test results.

4.4 Vaccination

The use of vaccines is routine in most of today's pig farms worldwide. Vaccines are key components in mitigating, and sometimes even eliminating, diseases. As such they not only can change the prevalence of disease but can also impact test results and their interpretation.

Vaccines stimulate an immune response that includes antibody production. Regardless of the type of vaccine used (killed or modified live), with a few exceptions, these antibodies are usually indistinguishable from those produced through natural exposure to wild-type pathogens. This means that most diagnostic assays cannot differentiate a vaccinated animal from that of one that has been infected. The clinical implication of such differentiation is obvious.

There are also vaccines, such as those for *Mycoplasma hyopneumoniae*, which provide some level of protection to pigs but yet do not consistently produce antibodies measurable by the different commercially available *M. hyopneumoniae* antibody assays (Hodgins et al., 2004; Thacker et al., 1998).

The use of live or modified live vaccines will impact not only the interpretation of assays detecting antibodies (just like killed vaccines) but also assays detecting the disease agent (such as polymerase chain reaction [PCR] or culture of the organism). In these cases, a secondary diagnostic assay, such as genetic sequencing, may be necessary to differentiate vaccine isolates from wild-type isolates.

4.5 Antimicrobials

When looking to evaluate exposure to bacterial pathogens, the use of antimicrobials can change the dynamics of pathogen detection and pig immunity. Understanding the what, where, and when to use antimicrobials on the pig farm will be critical in designing the proper monitoring program to prevent the incorrect interpretation of results, especially negative results.

Antimicrobial use also relates to common pathogens encountered at the different stages of growth. Early on in life, enteric diseases are the primary cause of disease. Antimicrobials are then often used to minimize pathogen exposure with hopes to mitigate clinical disease. Often, antimicrobial use can change the dose and/or timing of pathogen exposure, resulting in a delay or prevention of the pig's immune response to the pathogen. As pigs continue to grow, they seem to develop resistance to many enteric pathogens, and respiratory pathogens become the primary diseases of concern.

4.6 Assay selection

There are a large number of assays available for pig diagnostics. It is not the scope of this chapter to cover all these. The goal is to highlight the big picture and the importance of selecting the right assay incorporating all the factors already discussed.

Assay availability and cost are quite variable locally, nationally, and worldwide. Assay availability is also quite variable depending on the disease of concern. Some diseases have a great diversity in assay availability, while other diseases may not have any options available. Understanding the pig's age, disease epidemiology, disease ecology, vaccination programs, and antimicrobial use on farm will decide the proper analyte (agent detection versus antibody response) for testing. It is important to recognize that some management practices can significantly impact assay selection. This is especially true when modified live vaccines are used compounding the interpretation of a positive assay result for antibodies or detection of the agent. In such cases, secondary testing may be available to differentiate infected animals from vaccinated ones. Unfortunately, such secondary testing capabilities are limited to just a handful of diseases and not for many diseases with major production impact for pig farmers.

The diagnostic assay sensitivity and specificity, as well as analytical sensitivity, must be considered. Diagnostic sensitivity is defined as the ability of an assay to correctly identify all positive samples in a group of known positives. Diagnostic specificity is defined as the ability of an assay to correctly identify all negative samples from a group of known negatives. Analytical sensitivity relates to the mechanistic detection of the assay, in other words, how many bacteria, virus particles, or antibodies must be present in the sample being tested in order for the assay to detect them.

Assays for antibody detection from high analytical sensitivity (detect low quantities) to low analytical sensitivity (require higher concentrations of antibodies) include the following (Stevenson, 1999): serum neutralization (SN) > enzyme-linked immunosorbent assay (ELISA) > hemagglutination inhibition > complement fixation > gel precipitation.

Assays for agent detection from high analytical sensitivity (detect low quantities) to low analytical sensitivity (require higher concentrations of agent) include the following (Stevenson, 1999): bioassay > polymerase chain reaction (PCR) > virus isolation > antigen ELISA (agELISA).

The differences in analytical sensitivity mentioned can result in differences in assay results (based on the stage of the disease). In addition, within a particular type of assay there can be great variation in the particular target of the assay. A good example of this is seen with the different ELISA assays available for *M. hyopneumoniae*. Some ELISAs are manufactured using whole cells (target many different proteins), while others may only target a specific internal protein. Even though both of these assays are based on ELISA technology, the results obtained by them can be very different (Erlandson et al., 2005; Thacker, 2004)

5 Sample types

Traditionally when discussing monitoring programs, the first sample type to think about collecting has been blood (serum). Serum samples have been very useful because many of the older diagnostic assays were for antibody detection. Tissue samples have also been another common sample collected over the years, especially for diagnostic purposes. With the development of PCR technology, there has been the opportunity to expand sample types for pathogen detection. Then in the late 2000s, swine veterinarians were introduced to the concept of oral fluids (Prickett et al., 2008). The advent and readily availability of PCR technology along with the acceptance of sampling using oral fluids have provided a path of creating an innovative variety of sampling times.

One of the initial challenges in accepting oral fluid samples was the change in paradigm in moving away from individual animal sampling to population-based sampling. This population-based sampling incorporated a new concept of an aggregate sample rather than the traditional 'pooling' associated with individual sampling. An aggregate sample is a sample that is composed of specimens from several animals, but these animals are not represented equally in the final sample. This is in comparison to the pooling of samples where the final 'mixture' contains equal amounts of all samples to be pooled.

Often the question becomes, what is the best sample for this monitoring program? The answer is complex and depends on a number of factors. It is important to understand the ecology and epidemiology of the disease of concern. For example, influenza A virus (IAV) in swine is a virus that stays in the lungs of pigs and does not go systemic (does not enter blood). Collecting a blood sample to test (PCR) or try and culture (virus isolation) for the presence of the virus is inappropriate since the virus is not found in that sample type. Blood can be used for testing antibodies for IAV, but nasal swabs, oral fluids, or lung tissue is needed for finding the actual virus.

The following is a list of different sample types that can be used in disease monitoring programs along with some brief, general comments provided to initiate discussion with the veterinary advisor (Almeida et al., 2020; Arruda and Gauger, 2019; Jones et al., 2020; Perkins et al., 2019; Schwartz, 1982):

- Serum - most common sample type. Can be used for testing antibodies for many diseases or for actual disease agents for pathogens that go systemic. Require animal restraining and technical training.
- Whole blood - requires the use of special blood tubes with an anticoagulant to ensure blood does not coagulate. Has a similar advantage as the serum but can also be helpful for diseases that infect red blood cells. Requires animal restraining and technical training.
- Nasal swabs - usually used for IAV or other diseases commonly present in the upper respiratory tract, especially the nose. It is usually important that a swab be taken from the middle one-third of the nose. Require animal restraining and minimal technical training.

- Nasal wipes - similar to nasal swabs but easier to do. Uses a gauze pad to swab over the nose of pigs. Can be used for IAV with good success. Greater chance for environmental contamination. Often can just confine an animal without needing to restrain. Minimal technical training is needed.
- Feces - commonly used for enteric pathogens. The best samples are those freshly obtained from a pig. No to minimal restraint and no technical training needed.
- Fecal swabs - commonly used for monitoring diarrhea, especially in young pigs. Requires restraint and no technical training.
- Tracheal swabs - commonly used for upper respiratory pathogens, especially *M. hyopneumoniae*. Requires restraint, special equipment (speculum and long swab) and high technical training.
- Laryngeal swabs - same as tracheal swabs but slightly lower diagnostic sensitivity (lower rate of detection). Requires restraint, special equipment (speculum and long swab), and minimal technical training.
- Fluid - swabs or syringes can be used to collect any fluid for testing. This includes fluid found in the chest, around the heart, abdomen, joints, urine, milk, and so on. Care should be taken to minimize or eliminate contamination of the sample. Usually collected on a dead animal and minimal technical training needed.
- Processing fluids - collection of testicles (from castrated males) and tails (from males and females) when piglets are processed after birth. This newer type of sample is used to monitor the status of newborn piglets. In countries where castration or tail docking is not done, sampling of tongues from dead piglets has been used instead. This is used as an aggregate sample of all tissues collected. No additional restraint and no technical training are needed.
- Oral fluids - usually an aggregate sample of a group of pigs (usually in a pen) that are allowed to chew on a cotton rope. No animal restraint and minimal technical training are needed.
- Family oral fluids - a variation of oral fluids done in the farrowing house, where a rope is placed in the location where the sow and her piglets are allowed to chew. Having the sow access the rope significantly increases the rate of successful oral fluid collection from the piglets. No animal restraint and minimal technical training are needed.
- Environmental - environmental sampling is gaining traction in monitoring programs. There are multiple ways that the environment can be sampled depending on the pathogen of interest and objectives. The biggest issue with environmental sampling is interpreting results and differentiating disease from just environmental contamination. No animal restraint and no training are needed.

- Placental umbilical cords (PUCs) - collecting blood from umbilical cords on placentas. Sampling was done at birth to establish the pre-colostrum status of piglets. No animal restraint and moderate technical training are needed.
- Tissues - tissue collection is usually limited to postmortem (after death) sampling. The quality of the sample is highly dependent on whether the animal was found dead or was collected immediately after euthanasia. No additional restraint and moderate technical training are needed.
- Feed/water - sampling of nutrition inputs can occur at different levels including sampling outside the farm, on arrival to the farm, in the bulk storage area, and at pig level. No animal restraint and minimal technical training are needed.
- Air - air sampling can be very complex and time consuming and is usually restricted to research only. No animal restraint and special equipment (expensive) and high technical training are needed.
- Others - As creativity in sample type continue to expand, there is likely to be new sample types added to this list.

Sometimes the proper monitoring program may require more than one sample type to be collected. Resources (time and money) are always limited on farms and sometimes the tendency is to focus only on those sample types which are 'easier' to collect. Each of the abovementioned sample types has its own advantages and disadvantages depending on the disease of interest. It is critical to ensure that the proper sample type is collected to properly answer the desired objective. Ease of collection is important, but the accuracy of the results as a whole is even more important.

6 Collection approach

It is much easier to sample individuals and be able to interpret individual results as compared to sampling a population and trying to infer the status of a group of pigs or entire herd from such sampling. On the other hand, from a logistic and resource (time and money) standpoint it is much more beneficial to sample part of a population rather than having to sample all individuals. Population medicine is focused on interpreting results from a herd or population perspective and not on an individual basis. What is important to recognize is that in the case of a population, we will interpret results from the perspective of the population and not individuals. The rare exception to this is when establishing monitoring or surveillance programs involving high-value animals such as replacement breeding animals (especially boars) and animals for international export.

6.1 Entire population or sample

From a group or herd perspective, sampling a subset of animals will always provide some uncertainty in the result. Sample size will be discussed in the next section of this chapter.

There are usually two main sampling objectives. Most often the time, the objective of the sampling is to detect diseases (especially in the case of surveillance), which means finding a single positive will classify the entire group of pigs (and possibly the entire farm) as positive. Again, the objective is to establish the status of the group or herd and not each individual. The other sampling objective relates to establishing prevalence or changes in prevalence.

It is then recognized that most monitoring programs will rely on sampling a subset of the population. There are many complexities in trying to decide which subgroups should be considered for testing. Selecting the correct subgroup will maximize the validity of the results. Often the subgroup selected is determined by identifying the groups of animals with the highest risk for disease exposure or those which the target analyte selected (agent or antibodies) is most likely to be present. This decision-making will also incorporate all the concepts discussed so far with special attention to the disease epidemiology for the pathogen of concern and the desired sample type selected.

6.2 Random or targeted sampling

When selecting the appropriate subgroup for monitoring, there are two general approaches to refine the sampling. Animals or groups of animals can be randomly selected or purposely targeted for sampling. Statistics and the scientific process emphasize the importance of randomization to minimize bias. In the case of most disease monitoring programs, we recognize that disease exposure on a farm is not homogenous; diarrhea is more common in younger animals and abortions only occur in the breeding herd. With such differences in the epidemiology of the disease, it is important to focus efforts through a targeted sampling of these higher-risk populations to maximize the opportunity for disease detection as well as minimize its impact on resources.

Work done by Rotolo et al. (2017) has also demonstrated the value of not randomizing the pen selection for oral fluid collection. Her team proposed the concept of fixed spatial sampling in which sampling was based on selecting pens equidistant to each other to ensure coverage of the entire barn. These same pens are then used consistently for the entire monitoring program. This approach is used as a disease within a barn spread from pen to pen over time in a spatial manner and is not impacting pigs or pens randomly. This approach allows for a greater probability of disease detection.

A random sampling of populations does have value. Random sampling is often used when doing cross-sectional monitoring or when doing monitoring from a larger system perspective (Wright et al., 2014). In these cases, although the particular group (e.g. pigs 4–6 months of age) or building may already be preestablished for targeted sampling, the animals or pens to be tested may be randomly selected. This is often the case when there are no clinical signs present in the herd or when no particular group within the herd is known to be at greater risk for disease exposure. In these cases, random sampling is used to increase the likelihood that the animals sampled are good representatives of the status of the population of interest.

6.3 Slaughter checks

Slaughter checks have been used as part of disease monitoring programs for years (Morrison et al., 1985; Scollo et al., 2017; Straw et al., 1989). These slaughter checks involved a systematic evaluation of animals going to slaughter allowing for the evaluation/quantification of lesions associated with ascarid infestation, progressive atrophic rhinitis, pneumonia, and pleural adhesions. This type of evaluation was specific for ascarids and atrophic rhinitis but was nonspecific in regard to identifying the cause for the lung lesions or pleurisy. Slaughter checks also focus on disease occurring in the late stage of growth and often do not represent disease seen throughout the entire production cycle. Unfortunately, the group of pigs selected for sampling has a great impact on the amount and severity of diseases detected in the herd. Selecting the fastest growing pigs that go to the market in late summer or early fall will show less disease present on the farm than the last group of tail-ends that go to slaughter in early spring after being housed indoors over a long winter with more challenged ventilation (Sanchez-Vazquez et al., 2012; Willeberg et al., 1984).

7 Frequency of sampling

There are multiple factors that should be considered when deciding the proper sample size and frequency of a monitoring program. As a general rule, the larger the sample size, the higher the confidence in the results of the sampling (see next section). If monitoring is used as the primary means for detecting a change (change in frequency or change in status), then the frequency of sampling dictates the frequency of detection. Monitoring once a year only allows disease detection once a year. If weekly detection is desired, monitoring must then occur at least on a weekly basis.

7.1 Laboratory access and reporting

Access to a diagnostic laboratory as well as the timeliness of reporting, also significantly impact the decision-making regarding sampling. The key for a successful monitoring program (especially surveillance) is having a trusted lab that is reputable and reports results on a timely basis. Focusing on a single lab or at least a single lab for each specific assay is critical as it can be challenging to compare results between different labs since often, the specific assays used can vary significantly between laboratories.

7.2 Internal versus external transmission

Monitor for endemic diseases is often easy. For these diseases, it is easy to know how the disease is spreading within the farm as well as identify which group of animals provides the most valuable information to achieve the desired objective. For diseases transmitted vertically (from the dam to the offspring) we target the offspring. Historical observations can also be used to target specific age groups on the farm.

When doing surveillance, the focus is usually an external introduction of a new disease. A risk assessment will need to be done to identify which pig populations are most at risk for a new disease introduction. Purchase replacement breeding stock is likely the highest group at risk for introducing a new disease into the farm. Active surveillance programs should involve every group of replacement breeding animals. If purchasing semen, this semen should be sourced only from boar studs that routinely do disease monitoring.

7.3 Target detection level

All monitoring programs require the establishment of a detection level that informs the sample size calculations. This detection level is usually expressed as a percentage of prevalence as it is based not only on a numerator (how many) but also on a denominator (population size). It is recognized that producers would like to detect diseases at the lowest prevalence possible. Unfortunately, it is important to remember that no amount of testing can prove a negative (0%) prevalence. This prevalence is the lowest prevalence of the target (agent or antibody) for testing and not necessarily the disease itself. This is important to remember as in the case of an acute outbreak of IAV in a group of finishing pigs, the number of infected animals (disease prevalence) will increase over time, but the number of animals shedding virus (prevalence of agent) at any one time will dramatically decrease after the first 2–3 weeks of the outbreak (Canini et al., 2020; Detmer et al., 2013; Simon-Grifé et al., 2012).

The epidemiology of each disease has a direct influence on informing the necessary frequency of monitoring needed to achieve the desired objectives. Herds that are at constant risk for disease spread or introduction will require a high frequency of testing. This high frequency of testing along with a targeted desire for detecting a low prevalence will require an extremely large number of samples (see next section) to be tested every time. As a compromise, more frequent testing can be done using a higher target for detection level, resulting in less monitoring samples needed at each sampling period understanding that a new disease introduction in a naïve population will result in quick changes in disease prevalence over time.

8 Sample size calculations

Sample size is defined as the number of animals to sample at each time period. As a general rule, as the sample size increases, the ability to detect disease also increases. Also, as the sample size increases the confidence of the results as a true representation of the status of the population increases.

The focus of this section is to provide the big picture overview when considering sample size calculations. When calculating samples sizes, it is important to consider the following factors: (1) objective of testing (detect the pathogen or estimate its prevalence); (2) collection approach (individual or group sampling); (3) herd size; (4) detection limit (prevalence); and (5) desired probability of detection (confidence).

8.1 Objectives of testing

Most of the monitoring done in pigs (especially surveillance) is targeted at pathogen detection. Pathogen detection is simply whether the pathogen is present or not. Such sample size calculations have the goal of finding just one positive sample in order to call the entire group of pigs positive. This sample size is not designed to allow for the comparison of the number of positive samples between groups of animals or between different time periods. This will be discussed in more detail under the section on result interpretation.

Sampling for prevalence estimation requires a completely different calculation and usually requires the use of a much large sample size. Prevalence estimation is simply wanting to know how many pigs in a population may be infected. In a population of 1200 pigs, the sample size for pathogen detection up to a 10% prevalence will require testing 30 individual animals, whereas the sample size for confirming a prevalence of 10% (within the range of 5–15%) will require sampling 164 individual animals.

8.2 Collection approach

The traditional books, papers, computer programs, and websites that can assist in calculating the proper sample size for any monitoring program focus on individual animal sampling. If group sampling is utilized (such as oral fluids), then completely different calculations need to be used. That is because a single sample represents an aggregate of multiple animals, but that aggregate is spatially correlated to those pigs in that group and not really representative of the entire population. A good way to think about this is sampling water in a large lake. Although a large volume of water can be collected from one location, this water sample likely does not fully represent the water in the entire lake. Guidelines for oral fluid sample size calculations can be found in publications by Henao-Diaz et al. (2020) and Rotolo et al. (2017).

8.3 Herd size

As herds become larger, sample size tends to increase up to a certain point. This is one of the big advantages of larger operations. This may not be intuitive, but it must be remembered that prevalences are a ratio so as herds get larger the ratio stays the same but the number of positive animals also gets larger. A 10% prevalence in a 100 head herd indicates 10 positive animals. In a herd of 10 000 animals a 10% prevalence indicates 1000 positive animals present. So, although there are more total animals in the herd, there are also more positive animals that can be found. The sample size for detecting a 10% prevalence level would require sampling 25 animals in a herd of 100 animals and just increasing the sample size to 30 animals in a herd of 1200.

8.4 Detection limit

The minimum detection limit is a key driver of sample size calculations. This is logical as when something is rare, it will require a larger number of animals to sample to find it. There are many books, papers, computer programs, and websites that can help to calculate the proper sample size for any monitoring program. For most herd sizes, when calculating sample size for disease detection some easy numbers to remember for different prevalence numbers are as follows: ≥50% = 5 animals; ≥33% = 8 animals; ≥25% = 11 animals; ≥20% = 14 animals; ≥10% = 30 animals; ≥5% = 60 animals; and ≥2% = 138 animals.

It is important to note in the above sample size calculations that as the prevalence drops below 10%, the sample size needed to find a positive sample dramatically increase. In those cases, where the disease is rare, there really is not much difference between sampling 30 animals compared to sampling

40 or 45. To go from a 10% detection limit to a 5% detection, the sample size needs to double from 30 to 60.

8.5 Confidence

The last factor to impact sample size calculations is the desired confidence for the sampling. This number is traditionally set at a 95% confidence in detecting a positive, which suggests that if sampling is done in a population with the desired targeted limit of detection, 95 out of 100 times at least one positive sample will be detected. It is important to recognize that in particular situations, even an 80% or 85% confidence may be sufficient depending on the balance between cost (time/money) in relation to missing detection. From a regulatory or gold-standard perspective, 95% confidence will continue to be the likely standard. A full understanding of the benefits and risks must be considered when making the final decision that deviates from the gold standard.

8.6 Finalizing sample size

In the end, sample size calculations guide the decision-making, but ultimately cost can impact the final plan. The final decision needs to be made by the pig farmer balancing cost with benefit as well as incorporating how much risk is willing to be taken.

One assumption that is often made is that when a new disease enters a population, especially one that is naïve to that disease, the disease prevalence will increase over time. As disease prevalence increases, the likelihood that a routine monitoring program will detect it increases as well. Here is where the frequency of monitoring can offset a higher detection limit. This assumption only holds true if in fact disease prevalence is expected to increase with time which is probably correct for a new disease introduction in a herd with no immunity, but incorrect for endemic diseases where the overall herd immunity may suppress the disease prevalence for a longer period of time.

9 Pooling of samples

Disease monitoring comes with a cost. Not only are time and other resources spent in collecting and processing samples, but the cost of running different assays can add to a financial burden, especially when the sample size is large. There is a cost associated with each pig disease. Some diseases have a greater financial impact on the herd than others. Some like ASF can even be disastrous to a pig farmer not only due to production losses but also the regulatory implication due to the presence of the disease both in a herd as well as in a country's entire pig industry.

One of the easiest ways to decrease the cost of disease monitoring is to pool samples for testing. Pooling is defined as combining samples from different individuals into a new single sample which contains equal amounts of all samples to be pooled. This new sample is what is then tested via the desired assay as a way to decrease the number of assays done. Dorfman (1943) was one of the first individuals to describe the economic benefits of pooling samples when testing for diseases.

Before pooling samples, it is important to consider the following four main points:

1. Assay used: Pooling should not be done for assays for which the concentration of analyte is important (e.g. antibody concentration such as in an ELISA). Pooling is most often done when selecting a PCR that involves active amplification of the genetic material found in the sample (Cortey et al., 2011; Laurin et al., 2019).
2. Expected concentration of analyte on sample: When evaluating the possibility of diluting a sample below the analytical detection level of an assay, it is helpful to assume a single positive sample will be pooled with negative samples. In a pool of five, assume one positive sample combined with four negative samples. If the concentration of the target analyte is already expected to be low on a positive sample, pooling should be avoided (Zimmerman et al., 2019).
    a. *Lawsonia intracellularis* - animals with no clinical signs of diarrhea or under treatment of antimicrobials are expected to have low concentrations of bacteria in their feces, so pulling too many samples together are likely to result in negative results (Burrough et al., 2015; Nathues and Grosse Beilage, 2008).
    b. PRRSV - aborted fetuses are known to contain a large amount of PRRSV and therefore even pooling 10 aborted fetuses in a single PCR assay will still produce a strong positive result (Althouse et al., 2019; Benson et al., 2002; Zimmerman et al., 2019).
3. Sample type: It is easier to justify pooling individual samples compared to pooling of aggregate samples that already contain multiple pigs in a single sample. Serum samples (individual) are easy to pool compared to the pooling of pen oral fluids (aggregate) (Zimmerman et al., 2019).
4. Result interpretation: Easier to justify pooling samples when results are interpreted as a group. If positive pool samples require retesting to identify individual animal results, pooling should only be done when prevalence is expected to be low otherwise too many pools will need to be retested resulting in higher assay cost than if samples were just tested individual. A formula for the proper pooling size when individual results are necessary is described by Dorfman (1943).

The biggest advantage of pooling comes with allowing for the increasing of sample size and/or frequency of monitoring and thus allows for greater confidence in detecting disease early while staying within the same budget (Kittawornrat et al., 2014; López et al., 2018).

10 Production parameters

Diagnostic monitoring programs can be strengthened when combined with the monitoring of production parameters (Silva et al., 2017). Often, these production parameters are readily available and easy to compile. They can serve as part of the routine, passive monitoring for detecting possible health changes that may warrant further investigation.

Some of the common production parameters monitored include number of abortions, pigs weaned, weaning and marketing weights, days on feed, average daily gain, and feed efficiency to just mention a few. Some other parameters that have gained interest in more recent years involve monitoring of clinical signs, morbidity, mortality, and daily water usage. These lists are mentioned not to be all inclusive but to highlight the variety in production parameters that can be monitored. For example, monitoring clinical signs can involve the daily temperature taking of boars, using some type of objective scales to quantify and score lameness or diarrhea, or even the use of technology to quantify and qualify the extent of coughing in a building.

It is important to recognize that all of these production parameters are not specific to any disease. Often they are used to identify areas of concern so that further, more specific monitoring or diagnostics can be implemented. Along with the lack of specificity of these parameters, it is good to differentiate whether the parameter is an indicator of throughput or efficiency (Silva et al., 2017).

Throughput measures are usually a number that represents a measurement and is not calculated. It does not involve a denominator and thus is an absolute and not relative. Most of the pig farmers are often driven by targeted throughputs. For example, pig farmer #1 targets to wean 500 pigs each week whereas pig farmer #2 is driven to produce 600 weaned pigs each week. Because we are not aware of the denominators (i.e. herd size) for these targeted throughputs, it is impossible to compare these two farms. Also, within a farm, there can be changes in group sizes from week to week, creating great variability each time these parameters are reported.

On the other hand, efficiency measures are numbers that are calculated rather than measured. They usually involve ratios, rates, or some other type of inclusion of a denominator. The advantage of efficiency measures is, they can be compared across groups and farms of different sizes since they are already standardized to a common denominator. It will be important to clarify and be consistent on which denominator is to be used (e.g. pigs weaned per inventory sow or pigs weaned per mated female). Even with a parameter such as feed efficiency, which seems to be simple and 'standard', is really depending on how it is calculated, there can be great variability in the reported numbers which can result in different farm implications that would need further clarification on weight range of pigs (start and end weights), energy concentration of diet residual feed, and even differences in pathogen exposure (Patience et al., 2015).

For production parameters to be useful, they must be standardized and summarized in a way that 'real' changes can be detected while accounting for the normal variation expected. As previously mentioned, the use of efficiency measures is usually preferred over throughput measures unless the parameters of the throughput measure (populations size, time period evaluated, etc.) remain constant. It is very helpful if trends are presented in a visual manner (line or bar graphs) and the incorporation of some type of statistical processes (from simple to complex depending) to better account for normal variation (De Vries and Reneau, 2010).

11 Interpreting results

Ultimately monitoring programs have been established and resources are invested in order to utilize the results in decision making. It is the results that dictate when and what type of action is needed to mitigate the impact of a disease. Regardless of the monitoring program established, success will be determined by the proper interpretation of the results.

To better understand the limitation of any monitoring program, it is critical to be reminded of some of the key factors that can impact the interpretation of the results. As previously discussed, all assays have limitations. A positive PCR only confirms the presence of genetic material and does not indicate whether the pathogen is alive or infectious. The presence of antibodies only confirms they are present and, when evaluated on their own, cannot indicate whether the animal is still infectious (i.e. shedding the pathogen) or even how long ago it was infected. Most of the time antibodies from wild-type infections cannot be differentiated from those derived from vaccination or even those acquired passively from colostrum. Environmental sampling is helpful in establishing the presence of a pathogen on the farm, but lacks the ability to clarify the impact of such a pathogen. That is, environmental sampling confirms the presence of the pathogen, but it does not indicate whether it is causing disease or not. As all these examples demonstrate, it is critical that all results be evaluated in the context of the clinical presentation including such factors as clinical signs, age, and vaccination history.

As mentioned previously, it is also important to remember that when sampling for detection, the focus is finding a single positive to classify the entire group/population as positive. This sampling size is not designed to compare differences in the rate of positivity. Let us say we randomly sample 30 heads in two different groups of pigs seeking a 95% confidence level that we can detect at least one positive when the prevalence ≥10%. In one group, we find 1 out of 30 samples positive and in the second group we detect 3 out of 30 samples as positive. Our only conclusion is that both groups are positive. It would be inappropriate to compare the number of positives and suggest that the second group has three times as many positives as the first group.

One of the challenges in interpreting results is that we actually do not know the true status of the population; often we are testing to try and figure this out. When unexpected results are obtained, some additional testing may be required. Two common approaches for such is incorporating the idea of serial and parallel testing.

Testing in serial is done in order to confirm a positive result (Burrough et al., 2019). This involves performing two separate assays (often at different times) on the same sample and requiring both assays to be positive in order to classify the sample as positive. One common example of this is the use of indirect immunofluorescent antibody assay (IFA) to confirm an unexpected ELISA positive sample (Holtkamp et al., 2011). The basis for this conclusion is that having two separate assays produce positive results decreases the likelihood that such results were actual false positives.

On the other hand, parallel testing is usually done to avoid false negatives (Burrough et al., 2019). In this case, two separate assays are usually done in order to increase the likelihood of detecting a disease. One such example is that commonly both ELISA and PCR testing are done on the same samples at the same time as each detects a different stage of the disease (early detection of agent versus later detection of antibody response). In this case, obtaining a single positive case will classify the sample as positive.

Veterinarians are highly trained in diagnostics and epidemiology. It is important to rely on their knowledge and expertise in conjunction with the understanding of the health history and pig flow of each individual farm, in arriving at the correct interpretations of the results. As mentioned at the beginning of this chapter, 'Do not run a test unless you plan on how you will utilize the result'. Under the proper scenario, the monitoring plan was designed to address specific objectives and thus by the time the monitoring program is implemented, everyone has anticipated and is ready to take action as soon as results are obtained.

12 Classifying herd health status

From a regulatory standpoint, there are many diseases that are continuously being monitored around the world in order to identify the disease status of herds and countries, especially in regards to international trade. Surprisingly, these monitoring programs are not really dictated from an international perspective but rather are designed to address specific concerns for specific countries. These regulatory sanitary programs often focus on obtaining results that impact the importation of pork and often involve not only disease monitoring but also antimicrobial residues.

As described multiple times in this chapter it is impossible to prove a herd is negative to a pathogen. Often, veterinarians and pig farmers talk about their herds as being negative for specific diseases. In reality, all that can be said about these herds is that they have been testing negative. In clarifying this, the obvious question becomes, based on what testing? Who was tested? What assays were used? How many animals? What was the frequency of such testing? and so on. In order to minimize misunderstandings, it is important for everyone to speak the same 'language'. As such, Holtkamp et al. (2011) developed terminology for classifying swine herds by PRRSV status and clarifying the necessary monitoring program needed to achieve each of the different categories described. Although such a program is not perfect, it created some common terminology to allow veterinarians and pig farmers to better communicate with each other. This current classification system is currently being updated to better incorporate new knowledge and diagnostics, especially in regards to new sample types available for testing. Such classification systems are gaining momentum in the swine industry. In the case of *M. hyopneumoniae*, Garza-Moreno et al. (2018) proposed a new herd classification system, while Clavijo et al. (2021) established recommendations for sampling guidelines for growing herds. There is no doubt that in the future, there will be more and more monitoring programs that will be developed to clarify the classification of the health status of herds for each of the diseases of great economic impact in pig production.

13 Case studies

Disease monitoring programs are complex. They need to incorporate knowledge on diagnostics, disease epidemiology, and a full understanding of the history and pig flow of each pig farm. This means that it is unlikely there will be a single monitoring program that will meet the objectives of all pig farmers. The following two scenarios are used to aid in incorporating all the information discussed in this chapter into a specific applied application with the hope to broaden the clinical implementation and interpretation of disease monitoring.

13.1 Case study 1 - Monitoring replacement breeding stock for Aujeszky's disease

A 350 farrow-to-finish sow farm is looking to change its herd's genetics. Their plan is to purchase 45–50 replacement gilts every 2 months and roll over their current herd to the new genetics.

- Objective 1: Ensure all replacement breeding stock have not been exposed to wild-type pseudorabies virus.
  - Target sample type = serum.
  - Number of samples = 30 head (sample) or 45–50 heads (entire group) of replacement animals.
  - Target Analyte = antibody.
  - Assay = SN.
    - If animals have been vaccinated with a gene-deleted vaccine, use an ELISA that targets the protein from the gene deletion.
    - Can use both serum naturalization and gene-deleted ELISA to ensure that animals were not exposed to wild-type virus.
  - Frequency of collection = Every 2 months.
    - Test every group of replacement animals on arrival at the isolation facility.
  - Desired outcome = 100% SN negative results if not vaccinated.
    - If vaccinated expect 100% SN positive and 100% gene-deleted ELISA negative.

13.2 Case study 2 - PRRS monitoring after disease outbreak

A 1250 head sow farm just underwent a severe PRRS outbreak. Their objective is to establish a comprehensive monitoring program to achieve their goal of weaning piglets at 3-weeks of age that are negative for PRRSV.

- Objective 1: Since the diagnosis of PRRSV has already been confirmed, there is no advantage to doing any type of diagnostic monitoring for PRRSV until the sow abortions have returned to baseline.
  - Piglets born are already expected to be positive during this time period.
  - Production records will be utilized for monitoring breeding herd abortions.
- Objective 2: Once the breeding herd is 'stabilized', the first objective would be to monitor the PRRSV status of newborn piglets.
  - Target sample type = processing fluids (testicles and tails).
  - Number of samples = Aggregate sample of all processing fluids from each farrowing room combined as a single sample.

- Target Analyte = detect virus genetic material.
- Assay = PCR.
- Frequency of collection = Weekly.
- Desired outcome = 8 consecutive weeks of negative PCR samples.
  - Need to wait for 8 weeks because disease prevalence (presence of viremia) will fluctuate significantly each week as herd stability is approached (Trevisan et al., 2019; López et al., 2020).

- Objective 3: Once pigs born are negative to PRRSV, the next question will be whether these pigs are able to stay negative until weaning time.
  - Target sample type = Family oral fluids.
  - Number of samples = one sample per farrowing crate.
  - Target Analyte = detect virus genetic material.
  - Assay = PCR.
  - Frequency of collection = Weekly.
  - Desired outcome = 2 or 3 weeks of negative PCR samples.

14 Conclusion

Disease monitoring and surveillance are complex topics. Unfortunately, there are no perfect monitoring systems. All diagnostic assays have advantages and disadvantages and resources (especially time and money) are always limited. All monitoring programs require the assumption of some risk by the pig farmer in order to be feasible for implementation in the field. The goal of this chapter was to highlight the many factors that need to be taken into consideration in order to design an effective and efficient monitoring system. It takes team work and collaboration between veterinarians and pig producers to ensure that the correct objectives are being achieved through the current monitoring program.

It is important to remember the golden rule of diagnostic testing 'Do not run a test unless you plan on how you will utilize the result'. Setting clear objectives for monitoring programs is critical. In the early part of a disease outbreak the detection of the agent may be beneficial, whereas later in the disease the target for testing may change to detecting antibodies. The presence of disease may require confirmation through tissue lesions.

The disease ecology of each pathogen of interest must be understood. The disease ecology can be impacted by current management practices (including vaccine our antimicrobial use) and pig flow. It must also be remembered that in larger herds not all pigs are homogenous in exposure or immunity. Pig's age can also have a significant impact on their susceptibility to different pathogens. Age will also impact the amount (if any) of maternal antibodies present and therefore, for endemic diseases, the probability of pathogen exposure usually increases with age.

Monitoring programs are usually designed to establish the health status of the population and not individuals. Aggregate samples (e.g. oral fluids) can be utilized to sample more individuals within a population while conserving diagnostic costs. Monitoring programs often focus on sampling groups of animals with the highest risk for exposure or likely to have the highest presence for the target analyte. This targeted sampling often increases the likelihood of detecting positive animals.

Sample size calculations for aggregate samples are different than those from individual animal samples. The lower the targeted level of disease detection the larger the sample size needed for monitoring. Another toll to minimize the cost of diagnostics is to utilize the pooling of samples. Sample pooling is more commonly done in PCR testing and is usually not recommended for ELISA testing. Often aggregate samples are not pooled for testing since they are already considered diluted.

When interpreting diagnostic results, it is important to remember that PCR positive results only confirm the presence of genetic material and do not clarify whether the pathogen is alive or infectious. ELISA positive results confirm the presence of antibodies and usually cannot differentiate whether the animal is infectious, exposed to wild-type infection, exposed to the vaccine, or maternally derives passive antibodies.

Knowledge and technology do change over time. It is essential that all monitoring programs be reviewed periodically and at least once a year. It is easy to just continue doing the same program year after year. Requiring an annual review provides an opportunity to make sure overall farm goals and objectives have not changed.

One of the biggest challenges with monitoring programs is the tendency to want to answer too many objectives without significantly changing the monitoring program. The temptation is to just tag along additional testing to the samples already collected. Although results will be obtained, the true challenge becomes in figuring out the proper interpretation of all the additional testing. Too often the incorrect sample size, incorrect sample type, or even incorrect animal group is being tested. The value of any monitoring program relies on the ability to make the correct interpretation of the results so early action can be implemented to mitigate negative effects from the disease monitored.

15 Where to look for further information

- Swine disease monitoring systems within different countries are increasing in popularity. Unfortunately, many of these resources are not openly available. The following resources are some of the few resources that are available to the public providing further insight into goals, objectives, and data available in different countries: Canadian Animal Health Surveillance System: https://www.cahss.ca/cahss-networks/swine.

- Great Britain pig quarterly report: disease surveillance and emerging threats: https://www.gov.uk/government/publications/pig-gb-disease-surveillance-and-emerging-threats-reports.
- United States Animal Health Surveillance System: https://www.aphis.usda.gov/aphis/ourfocus/animalhealth/animal-disease-information/swine-disease-information.
- United States National Animal Health Monitoring System: https://www.aphis.usda.gov/aphis/ourfocus/animalhealth/monitoring-and-surveillance/nahms
- United States Morrison Swine Health Monitoring Program: https://vetmed.umn.edu/centers-programs/swine-program/outreach-leman-mshmp/mshmp
- United States Swine Health Information System – Domestic Disease Monitoring Reports: https://www.swinehealth.org/domestic-disease-surveillance-reports/.
- United States Swine Health Information System – Global Disease Monitoring Reports: https://www.swinehealth.org/global-disease-surveillance-reports/.
- World Organization for Animal Health – World Animal Health Information System: https://www.oie.int/en/what-we-do/animal-health-and-welfare/disease-data-collection/world-animal-health-information-system/.

16 References

Almeida, M. N., Rotto, H., Schneider, P., Robb, C., Zimmerman, J. J., Holtkamp, D. J., Rademacher, C. J. and Linhares, D. C. L. (2020). Collecting oral fluid samples from due-to-wean litters, *Prev. Vet. Med.* 174, 104810.

Althouse, G. C., Kauffold, J. and Rossow, S. (2019). Diseases of the reproductive system. In: Zimmerman, J., Karriker, L., Ramirez, A., Schwartz, K. and Stevenson, G. (Eds) *Disease of Swine* (11th edn), Hoboken, NJ: Blackwell Publishing, pp. 373–392.

Arruda, H. E. P. and Gauger, P. (2019). Optimizing sample selection, collection, and submission to optimize diagnostic value. In: Zimmerman, J., Karriker, L., Ramirez, A., Schwartz, K. and Stevenson, G. (Eds) *Disease of Swine* (11th edn), Hoboken, NJ: Blackwell Publishing, pp. 98–111.

Benson, J. E., Yaeger, M. J., Christopher-Hennings, J., Lager, K. and Yoon, K. J. (2002). A comparison of virus isolation, immunohistochemistry, fetal serology, and reverse-transcription polymerase chain reaction assay for the identification of porcine reproductive and respiratory syndrome virus transplacental infection in the fetus, *J. Vet. Diagn. Invest.* 14(1), 8–14.

Burrough, E. R., Baum, D. H. and Schwartz, K. J. (2019). Collecting evidence and establishing causality. In: Zimmerman, J., Karriker, L., Ramirez, A., Schwartz, K. and Stevenson, G. (Eds) *Disease of Swine* (11th edn), Hoboken, NJ: Blackwell Publishing, pp. 112–122.

Burrough, E. R., Rotolo, M. L., Gauger, P. C. and Madson, D. M. (2015). Correlation of Lawsonia intracellularis semi-quantitative fecal polymerase chain reaction assay results with the presence of histologic lesions of proliferative enteropathy and positive immunohistochemical staining, *J. Swine Health Prod.* 23(4), 204–207.

Canini, L., Holzer, B., Morgan, S., Dinie Hemmink, J., Clark, B., sLoLa Dynamics Consortium, Woolhouse, M. E. J., Tchilian, E. and Charleston, B. (2020). Timelines of infection and transmission dynamics of H1N1pdm09 in swine, *PLoS Pathog.* 16(7), e1008628.

Christensen, J. (2001). Epidemiological concepts regarding disease monitoring and surveillance, *Acta. Vet. Scand. Suppl.* 94, 11–16.

Clavijo, M. J., Hu, D., Krantz, S., Cano, J. P., Pereira Maróstica, T., Henao-Diaz, A., Poeta Silva, A. P. S., Hemker, D., Tapia, E., Zimmerman, S., Fano, E., Polson, D., Fitzgerald, R., Tucker, A., Main, R., Wang, C., Zimmerman, J. J. and Rotolo, M. L. (2021). *Mycoplasma hyopneumoniae* surveillance in pig populations: Establishing sampling guidelines for detection in growing pigs, *J. Clin. Microbiol.* 59(5), e0351-20.

Cortey, M., Napp, S., Alba, A., Pileri, E., Grau-Roma, L., Sibila, M. and Segalés, J. (2011). Theoretical and experimental approaches to estimate the usefulness of pooled serum samples for the diagnosis of postweaning multisystemic wasting syndrome, *J. Vet. Diagn. Invest.* 23(2), 233–240.

De Vries, A. and Reneau, J. K. (2010). Application of statistical process control charts to monitor changes in animal production systems, *J. Anim. Sci.* 88(13 Suppl.), E11–E24.

Dee, S. A., Joo, H. S., Henry, S., Tokach, L., Park, B. K., Molitor, T. and Pijoan, C. (1996). Detecting subpopulations after PRRS virus infection in large breeding herds using multiple serologic tests, *J. Swine Health Prod.* 4(4), 181–184.

Detmer, S., Gramer, M., Goyal, S., Torremorell, M. and Torrison, J. (2013). Diagnostics and surveillance for swine influenza, *Curr. Top. Microbiol. Immunol.* 370, 85–112.

Dorfman, R. (1943). The detection of defective members of large populations, *Ann. Math. Statist.* 14(4), 436–440.

Erlandson, K. R., Evans, R. B., Thacker, B. J., Wegner, M. W. and Thacker, E. L. (2005). Evaluation of three serum antibody enzyme-linked immunosorbent assays for *Mycoplasma hyopneumoniae*, *J. Swine Health Prod.* 13(4), 198–203.

Food and Agriculture Organization of the United Nations. (1999). *Manual on Livestock Disease Surveillance and Information Systems*, Rome, Italy: Food and Agriculture Organization of the United Nations.

Garza-Moreno, L., Segalés, J., Pieters, M., Romagosa, A. and Sibila, M. (2018). Acclimation strategies in gilts to control *Mycoplasma hyopneumoniae* infection, *Vet. Microbiol.* 219, 23–29.

Gelberg, H. B. (1992). Studies on the age resistance of swine to group A rotavirus infection, *Vet. Pathol.* 29(2), 161–168.

Harris, D. L. (2000). *Multi-Site Pig Production*, Ames, IA: Iowa State University Press.

Henao-Diaz, A., Giménez-Lirola, L., Baum, D. H. and Zimmerman, J. (2020). Guidelines for oral fluid-based surveillance of viral pathogens in swine, *Porcine Health Manag.* 6, 28.

Hodgins, D. C., Shewen, P. E. and Dewey, C. E. (2004). Influence of age and maternal antibodies on antibody responses of neonatal piglets vaccinated against *Mycoplasma hyopneumoniae*, *J. Swine Health Prod.* 12(1), 10–16.

Holtkamp, D. J., Polson, D. D., Torremorell, M., Classen, D. M., Becton, L., Henry, S., Rodibaugh, M. T., Rowland, R. R., Snelson, H., Straw, B. and Yeske, P. (2011).

Terminology for classifying swine herds by porcine reproductive and respiratory syndrome virus status, *J. Swine Health Prod.* 19(1), 44–56.

Jones, C., Stewart, S., Woodworth, J., Dritz, S. and Paulk, C. (2020). Validation of sampling methods in bulk feed ingredients for detection of swine viruses, *Transbound. Emerg. Dis.* 67(1), 1–5.

Kilpatrick, A. M. and Altizer, S. (2010). Disease ecology, *Nat. Ed. Knowl.* 3(10), 55.

Kittawornrat, A., Panyasing, Y., Goodell, C., Wang, C., Gauger, P., Harmon, K., Rauh, R., Desfresne, L., Levis, I. and Zimmerman, J. (2014). Porcine reproductive and respiratory syndrome virus (PRRSV) surveillance using pre-weaning oral fluid samples detects circulation of wild-type PRRSV, *Vet. Microbiol.* 168(2–4), 331–339.

Klinge, K. L., Vaughn, E. M., Roof, M. B., Bautista, E. M. and Murtaugh, M. P. (2009). Age-dependent resistance to porcine reproductive and respiratory syndrome virus replication in swine, *Virol. J.* 6, 177.

Laurin, E., Thakur, K., Mohr, P. G., Hick, P., Crane, M. S. J., Gardner, I. A., Moody, N. J. G., Colling, A. and Ernst, I. (2019). To pool or not to pool? Guidelines for pooling samples for use in surveillance testing of infectious diseases in aquatic animals, *J. Fish Dis.* 42(11), 1471–1491.

López, W. A., Angulo, J., Zimmerman, J. J., and Linhares, D. C. L. (2018). Porcine reproductive and respiratory syndrome monitoring in breeding herds using processing fluids, *J. Swine Health Prod.* 26(3), 146–150.

López, W. A., Zimmerman, J. J., Gauger, P. C., Harmon, K. M., Bradner, L., Zhang, M., Giménez-Lirola, L., Ramirez, A., Cano, J. P. and Linhares, D. C. L. (2020). Practical aspects of PRRSV RNA detection in processing fluids collected in commercial swine farms, *Prev. Vet. Med.* 180, 105021.

Miller, E. R., Harmon, B. G., Ullrey, D. E., Schmidt, D. A., Luecke, R. W. and Hoefer, J. A. (1962). Antibody absorption, retention and production by the baby pig, *J. Anim. Sci.* 21(2), 309–314.

Morrison, R. B., Pijoan, C., Hilley, H. D. and Rapp, V. (1985). Microorganisms associated with pneumonia in slaughter weight swine, *Can. J. Comp. Med.* 49(2), 129–137.

Nathues, H. and Grosse Beilage, E. (2008). Diagnosis of *Lawsoniaintracellularis* infection in pigs after vaccination or antimicrobial treatment, *Dtsch Tierarztl. Wochenschr.* 115(11), 404–409.

Patience, J. F., Rossoni-Serão, M. C. and Gutiérrez, N. A. (2015). A review of feed efficiency in swine: Biology and application, *J. Anim. Sci. Biotechnol.* 6(1), 33.

Perkins, S., Davenport, J., Jablonski, E., Amondie, D. and Trevisan, J. (2019). Comparison of processing fluids and placental umbilical cord serum to detect porcine circovirus 2 in sow farms. *Proc. Amer. Assoc. Swine Vet.* 50, 263–264.

Prickett, K. W., Simer, R., Yoon, K. and Zimmerman, J. (2008). Oral-fluid samples for surveillance of commercial growing pigs for porcine reproductive and respiratory syndrome virus and porcine circovirus type 2 infections, *J. Swine Health Prod.* 16, 86–91.

Rotolo, M. L., Sun, Y., Wang, C., Giménez-Lirola, L., Baum, D. H., Gauger, P. C., Harmon, K. M., Hoogland, M., Main, R. and Zimmerman, J. J. (2017). Sampling guidelines for oral fluid-based surveys of group-housed animals, *Vet. Microbiol.* 209, 20–29.

Sanchez-Vazquez, M. J., Nielen, M., Gunn, G. J. and Lewis, F. I. (2012). Using seasonal-trend decomposition based on loess (STL) to explore temporal patterns of pneumonic lesions in finishing pigs slaughtered in England 2005–2011, *Prev. Vet. Med.* 104(1–2), 65–73.

Schwartz, W. L. (1982). Laboratory diagnosis of swine diseases, *Vet. Clin. North Am. Large Anim. Pract.* 4(2), 201–223.

Scollo, A., Gottardo, F., Contiero, B., Mazzoni, C., Leneveu, P. and Edwards, S. A. (2017). Benchmarking of pluck lesions at slaughter as a health monitoring tool for pigs slaughtered at 170kg (heavy pigs), *Prev. Vet. Med.* 144, 20–28.

Silva, G. S., Schwartz, M., Morriso, R. B. and Linhares, D. C. L. (2017). Monitoring breeding herd production data to detect PRRSV outbreaks, *Prev. Vet. Med.* 148, 89–93.

Simon-Grifé, M., Martín-Valls, G. E., Vilar, M. J., Busquets, N., Mora-Salvatierra, M., Bestebroer, T. M., Fouchier, R. A., Martín, M., Mateu, E. and Casal, J. (2012). Swine influenza virus infection dynamics in two pig farms; results of a longitudinal assessment, *Vet. Res.* 43(1), 24.

Stevenson, G. W. (1999). Common mistakes in interpretation of population serology. *Proc. Am. Assoc. Swine Pract.* 30, 339–343.

Straw, B. E., Tuovinen, V. K. and Bigraspouin, M. (1989). Estimation of the cost of pneumonia in swine herds, *J. Am. Vet. Med. Assoc.* 195(12), 1702–1706.

Thacker, E. L. (2004). Diagnosis of *Mycoplasma hyopneumoniae*, *J. Swine Health Prod.* 12(5), 252–254.

Thacker, E. L., Thacker, B. J., Boettcher, T. B. and Jayappa, H. (1998). Comparison of antibody production, lymphocyte stimulation, and protection induced by four commercial *Mycoplasma hyopneumoniae* bacterins, *J. Swine Health Prod.* 6(3), 107–112.

Trevisan, G., Jablonski, E., Angulo, J., Lopez, W. A. and Linhares, D. C. L. (2019). Use of processing fluid samples for longitudinal monitoring of PRRS virus in herds undergoing virus elimination, *Porcine Health Manag.* 5, 18.

Willeberg, P., Gerbola, M.-A., Kirkegaard-Petersen, B. K. and Andersen, J. B. (1984). The Danish pig health scheme: Nation-wide computer-based abattoir surveillance and follow-up at the herd level, *Prev. Vet. Med.* 3(1), 79–91.

Wright, L. J., Newson, S. E. and Noble, D. G. (2014). The value of a random sampling design for annual monitoring of national populations of larger British terrestrial mammals, *Eur. J. Wildl. Res.* 60(2), 213–221.

Zimmerman, J. J., Dee, S. A., Holtkamp, D. J., Murtaugh, M. P., Stadejek, T., Stevenson, G. W., Torremorell, M., Yang, H. and Zhang, J. (2019). Porcine reproductive and respiratory syndrome viruses (Porcine arteriviruses). In: Zimmerman, J., Karriker, L., Ramirez, A., Schwartz, K. and Stevenson, G. (Eds) *Disease of Swine* (11th edn), Hoboken, NJ: Blackwell Publishing, pp. 685–708.

Chapter 4

Surveillance on swine farms using antemortem specimens

Berenice Munguía-Ramírez, Betsy Armenta-Leyva, Luis Giménez-Lirola, Chong Wang and Jeffrey Zimmerman, Iowa State University, USA

1 Introduction

Surveillance is the process of tracking health and productivity parameters over time to understand population health dynamics and to make better decisions on disease control. To be thorough, we note that monitoring and surveillance are not quite the same. Specifically, monitoring is the systematic collection and evaluation of population data over time, whereas surveillance is monitoring along with a plan prepared and ready for implementation if a specific threshold or disease condition is identified (Salman, 2003). Although they are not synonymous, monitoring and surveillance are generally used interchangeably in daily life and, for simplicity, in this review (Paskins, 1999).

The aim of this chapter is to provide a general, non-mathematical overview of infectious disease surveillance on swine farms based on testing. Because farms vary in size, structure, management, and surveillance goals, there is no 'one-size-fits-all' surveillance plan that can fit all circumstances and meet all objectives. Rather, the design and implementation of a surveillance program should be driven by the producer, ideally working in conjunction with an animal health specialist, and tailored to meet the specific objectives of the production system. Bedrock principles should guide the design process,

http://dx.doi.org/10.19103/AS.2022.0103.04

which should be periodically reviewed after the surveillance program is initiated:

- Surveillance objective(s) should be clear and shared by all involved.
- The process - from sample collection to data interpretation - should be simple, clearly understood, and easily performed.
- The process should produce timely, accurate, interpretable and actionable results.
- The process must provide a return on investment through the reduction or avoidance of disease losses and/or enhancement of the value of the product.
- The process should be adaptable and able to meet new objectives as they are identified.

2 Overview

Representative sampling, testing a subset of randomly selected individuals to establish the status of the entire population, was the first step toward efficient surveillance. First described in 1895 (Kruskal and Mosteller, 1980), statistical sampling was rarely used in livestock surveillance until a synopsis by Cannon and Roe (1982) made the concepts accessible and understandable to field veterinarians. Subsequently, surveillance sample sizes based on binomial sampling distributions were routinely designed into swine disease control programs, for example, the U.S. pseudorabies (Aujeszky's disease) eradication program (Anderson et al., 2008), and became an integral part of the thought processes of swine health specialists.

The two key assumptions underlying binomial sampling are (1) the population is homogeneous, that is, randomly selected pigs in the population have an equal chance of being positive, and (2) the pigs in the population are 'independent,' that is, the infectious disease status of one pig is not predictive of the status of another (Wroughton and Cole, 2013). These assumptions were sometimes true in the smaller herds of the past, but are rarely true today because pigs on commercial production sites are separated into buildings, rooms, and pens by age, production stage, and/or function, with little interaction between groups. The result is the heterogeneous distribution (clustering) of disease within a production site. That is, some groups may be positive and others negative for the pathogen of interest, on the same farm at the same point in time. In addition, because infectious agents are most commonly spread from pig to pig, pigs in the same pen or barn are likely to be of similar status (Rotolo et al., 2017). It follows that, because pigs in physical proximity are likely to share the same disease status, they are not independent.

To account for this population structure and the non-random disease distribution, at least to the degree possible, it is useful to design surveillance based on 'epidemiological units,' that is, groups of animals on the site that share a common environment and/or a comparable risk of exposure to the pathogen of interest (OIE, 2021). For example, in the U.S. pseudorabies (Aujeszky's disease) eradication program, '*each segregated group of swine on an individual premises* ... (was) *considered a separate herd.*' Thus, according to the program guidelines, a farm could consist of one or more 'herds' (epidemiological units). Regardless of the number, each 'herd' was sampled according to the official (binomial sampling) protocol, for example, 29 pigs were sampled in each barn holding ≥ 1000 animals to achieve a 95% probability of detection at 10% prevalence (USDA, 2003).

A further complication to surveillance is the continual turnover of animals on swine farms. The production cycle is short for both market pigs (six months from birth to market) and breeding stock; that is, the turnover in finishing barns may approach 250% per year and breeding herds replace 40–50% of females annually (Stalder et al., 2004). As a point of contrast, human population turnover in 28 European countries for 2016 ranged from a low of 2.4% in Italy to a high of 8.5% in Luxembourg (Eurostat News, 2017). Further, as animals complete the production cycle, replacement animals are introduced, either through birth or from other farms. If replacements are immunologically susceptible to an infectious agent on the farm, they will eventually become infected and perpetuate the pathogen on the farm. If new replacements are infected with a pathogen not present on the farm, the risk is that it will spread to the remainder of the herd. This is a common scenario, that is, moving animals between herds is the most frequent route of PRRSV spread (Pileri and Mateu, 2016). Thus, sampling and testing must be sufficiently frequent in order to accommodate the rate of population turnover and the continual introduction of replacements.

3 Collecting production data

Pig producers have long recognized the value of data. For example, British pig producers in the 1920s used records to identify prolific sows that produced fast-growing, early-maturing progeny with good carcass characteristics (Woods, 2012). With the appearance of specialized pig farms in the mid-twentieth century, the goals of surveillance were broadened to include other health and productivity parameters (Alexander, 1971; Muirhead, 1976). This was, in part, a response to new disease challenges. That is, indoor housing alleviated health and welfare issues associated with outdoor pig production by providing better parasite control, nutrition and protection against extreme weather. However, confinement also changed the ecological balance among pigs, pathogens, and their environment and, in some circumstances, led to the emergence of new multifactorial disease syndromes (Muirhead, 1976; Woods, 2012). This problem remains with us, a contemporary example being the porcine respiratory disease complex (PRDC), '*a combination of primary and opportunistic infectious agents, often facilitated by adverse environmental and management conditions*' (Yaeger and Van Alstine, 2019). Responding to the challenge, swine health innovators of the time recommended a new approach based on the systematic collection and analysis of farm data to establish herd baselines, unravel complex disease causalities, and evaluate the effects of management decisions on health and productivity (Muirhead, 1976; Schwabe, 1982; Stein et al., 1987). Working at the cusp of this transition, Stein et al. (1987) aptly described this process: '*... just as the stethoscope and thermometer are fundamental tools for individual medicine, production and health recording systems are fundamental tools for effective population medicine.*'

Production and health recording systems relied on hand-written records into the 1970s and later, but producers readily adopted electronic data management as computer technology became accessible. Thus, Pepper et al. (1977a) described the use of a Fortran computer program for the analysis of reproductive parameters. On the farm, the data were written on 'sow cards.' The cards were then taken to a computer center for data entry and analysis, for example, litters per sow per year, weaning-to-service intervals, etc. Pepper and Taylor (1977b) used this system, in combination with necropsies of all pigs that died between Sunday 9:00 am and Friday 9:00 am, to achieve major improvements in performance in a 260-sow breeding herd. Electronic production records became increasingly commonplace after affordable desktop computers became available in the mid-1980s and continuous improvements in information technology have brought us to the point where it is now possible to collect and analyze data across entire production systems with thousands of animals distributed among multiple production sites (Magalhães et al., 2022).

4 Collecting surveillance data

In contrast to systems for the collection of production data, the tools needed for practical infectious disease surveillance required breakthroughs in diagnostic medicine, which have only been realized in recent years. Among these developments, we will focus on tests and specimens compatible with live animal surveillance in commercial swine herds, that is, assays able to detect pathogen-specific antigen, antibody or nucleic acid in specimens collected from live pigs or their environment. Diagnostic technology continues in a state of ongoing development and a periodic review of new commercially available tests and their performance is advised for the sake of keeping up with innovations.

4.1 Testing technology

The first major breakthrough in testing came with the development of the enzyme-linked immunosorbent assay (ELISA) in 1971 (Engvall and Perlmann, 1971; Van Weemen and Schuurs, 1971). This was followed quickly by the development of a variety of pathogen-specific antibody ELISAs (Schuurs and Van Weemen, 1977). ELISAs can be designed to measure either antigen or antibody, but both use an enzyme to detect antigen–antibody binding. Essentially, in the presence of antigen–antibody binding, the enzyme converts a colorless enzyme substrate (chromogen) to a colored product. The strength of the color reaction is measured using a spectrophotometer, with the measurement (optical density) directly proportional (indirect ELISA) or inversely proportional (blocking and competitive ELISAs) to the concentration of the target in the sample. Currently, ELISA remains the most widely used antibody detection method because of its simplicity, low cost, consistent performance and wide commercial availability.

In addition to ELISAs, a variety of other immunoassay technologies have been reported, but most have not been commercialized. These assays take on a wide range of formats, but most consist of an antigen or antibody that is immobilized on a surface (planar or microbead-based) and binds virus-specific antigens or antibodies present in a sample. By adding a reporter protein, it is possible to detect a pathogen-specific immune signal to confirm the presence of pathogen-specific antibody. Many of these systems show promise, point-of-care tests being one example. However, they need to meet regulatory, test performance and commercial expectations (Hobbs et al., 2021).

In surveillance, the role of antibody testing is to detect infection with a pathogen (in the absence of vaccination), track an outbreak over time, assess the response to vaccination or controlled exposure and/or assess the level of protective immunity within a population (Arnold and Chung, 2018; Cutts and Hanson, 2016; MacNeil et al., 2014). Depending on the agent, pathogen-specific antibody is usually detectable one to three weeks post infection, and thereafter for months to years in various diagnostic specimens, for example, serum, oral fluids, feces, mammary secretions or tissue exudate ('meat juice,' processing fluid). Notably, maternal antibody or antibody induced by vaccines cannot usually be differentiated from wild-type infections, with the exception of antibody induced by DIVA (differentiating infected from vaccinated individuals) vaccines (van Oirschot, 1999). DIVA vaccines induce protective immunity. However, they are missing one or more epitopes present in the wild-type pathogen. Thus, DIVA-vaccinated animals are negative for specific antibodies that are present in wild-type-infected animals. Pseudorabies virus DIVA vaccines and the accompanying ELISA made pseudorabies virus control and elimination possible (Mettenleiter, 2020).

The second major diagnostic breakthrough came with the publication of the principles underlying the polymerase chain reaction (PCR) (Saiki et al., 1985). As with the ELISA, PCRs were quickly adapted to swine diagnostic medicine, with pseudorabies virus DNA detection being among the first examples of its use (Belak et al., 1989; Jestin et al., 1990; Maes et al., 1990). PCR is based on the amplification of nucleic acid into quantities that can be detected and analyzed. PCR begins with the addition of short DNA sequences (primers) to denatured (single-stranded) DNA. Except that, in the case of RNA viruses, PCR is preceded by a step to produce a complementary DNA template (cDNA) from viral RNA by the addition of a reverse transcriptase enzyme. Binding of primers to the single-stranded DNA forms double-stranded DNA and triggers a thermostable polymerase enzyme to extend the sequence and produce a full, complementary strand of DNA. End-point methods then measure the amount of amplified product ('amplicon') accumulated over the course of the reaction. The development and commercialization of fluorometry-based real-time PCR, that is, quantitative real-time PCR (qPCR or RT-qPCR for RNA viruses), improved and simplified the process of amplifying and detecting nucleic acid sequences while allowing for simultaneous and more precise quantification of their concentrations (Klein, 2002). Thus, samples with a higher concentration of target nucleic acid will require fewer PCR amplification cycles to reach the cycle threshold (Ct) established as the assay cutoff for detection (Schmittgen and Livak, 2008).

Depending on the pathogen, detectable levels of pathogen-specific nucleic acid may be present hours to days after infection in various diagnostic specimens, for example, serum, oral fluids, feces, mammary secretions, or tissue exudate ('meat juice,' processing fluid). The ability of PCR to detect acute infections makes it particularly useful in verifying freedom from infection in animals scheduled for upcoming movement. PCR is also compatible with testing environmental samples, for example, surfaces, water and air, but it should be borne in mind that current qPCRs cannot differentiate between viable and non-viable (inactivated) pathogens. That is, a positive result indicates the presence of pathogen-specific genetic material, but not its viability or infectivity.

Although not strictly a breakthrough in diagnostic technology, the third significant development in surveillance was the creation of systems for managing, storing and analyzing testing data. In top-tier veterinary diagnostic laboratories, this would include laboratory information management systems (LIMS) that are capable of tracking samples, retrieving/storing electronic output from test devices, and reporting results to clients in any of the various electronic formats. In smaller laboratories or in on-farm testing, this may only include receiving results as electronic data. Regardless, these technologies have played a major role in improved throughput and timely data analysis/response.

Published by Burleigh Dodds Science Publishing Limited, 2023.

4.2 Samples and specimens

The antemortem specimens commonly used in surveillance are listed in Tables 1 and 2 for specific bacterial and viral pathogens, respectively. These tables are intended to provide an overview and should not be considered comprehensive. To begin this section, a brief overview of sampling terminology may be useful.

'Discrete' samples are defined as samples collected at one point in time from a specific source and location (Cameron et al., 2003; Patil et al., 2010). Without time and place identifiers, it is not possible to interpret test results in the context of the population under surveillance. Discrete samples may be specimens from one animal, for example, serum, nasal swabs, etc., or simultaneously collected from more than one animal ('aggregate' samples), for example, pen-based oral fluids or environmental samples.

Two or more discrete samples may be combined (composite or 'pooled' samples) into one for testing. The purpose of pooling is to reduce the number of test samples and, thereby, reduce cost. Pooling is not a new idea. For example, Dorfman (1943) tested pooled serum samples when screening World War II draftees for syphilis. The concept of detection at a lower cost is highly alluring and there is an extensive body of literature on the subject of pooling (Daniel et al., 2021). However, pooling should be used cautiously because it increases the rate of false negative results. That is, combining negative and positive samples must necessarily dilute the concentration of the test target in the pool. The risk is diluting the sample to a concentration below the threshold of detection. For example, Rovira et al. (2007) compared the detection of PRRSV in blood swab and serum samples during acute infection (1 to 15 days post inoculation) and found that pooling in groups of five resulted in 6% fewer RT-qPCR-positive serum samples and 8% fewer positive blood swab samples compared to testing the samples individually. Ultimately, the decision to pool (or not) depends on the cost of false-negative results versus the savings afforded by performing fewer tests. If pooling is done, it should be done in such a way as to preserve sampling location and date integrity. That is, pooling samples from different locations or dates will produce uninterpretable results.

4.2.1 Blood-derived specimens

Commonly used blood-derived specimens include whole blood, serum and blood swabs. Methods for collecting blood samples from pigs are described elsewhere (Ramirez and Karriker, 2019). Pig blood coagulates quickly, and when collecting whole blood, the sample should be collected directly into a vacutainer blood-collection tube containing an anticoagulant. Djordjevic et al. (2006) reported that sodium heparin inhibited PCR amplification, but other anticoagulants did not, that is, sodium citrate, K3 EDTA and lithium heparin.

Table 1 An overview of antemortem specimens commonly used in surveillance for bacterial pathogens (numbers correspond to references listed in Section 9)

Bacterial pathogens	Assay Target[a]	Serum	Whole blood	Blood swab	Oral fluid	Oral swab	Nasal swab	Oropharyngeal swab	Tonsil scraping	Tracheal swab	Feces	Rectal swab	Processing fluids	Colostrum/ milk	Vaginal swab	Semen	Environmental
Actinobacillus pleuropneumoniae	AB	295	-	-	103	159	25	-	-	-	-	-	-	197	-	-	-
	NA	-	-	-	44	-	274	178	274	-	-	-	-	-	-	-	158
Actinobacillus suis	AB	145	-	-	-	-	-	-	-	-	-	-	-	-	-	-	-
	NA	-	-	-	-	-	-	-	-	-	-	-	-	-	-	-	-
Bordetella bronchiseptica	AB	131	-	-	-	-	141	-	-	-	-	-	-	215	-	-	-
	NA	-	-	-	177	-	177	-	-	-	-	-	-	-	-	-	-
Brachyspira spp.	AB	147	-	-	-	-	-	-	-	-	176	-	-	-	-	-	-
	NA	-	-	-	293	-	-	-	-	-	125	214	-	-	-	-	95
Brucella suis	AB	273	-	-	-	-	-	-	-	-	-	-	-	-	-	-	-
	NA	-	-	-	-	-	-	-	-	-	-	-	-	-	-	-	-
Escherichia coli	AB	228	-	-	-	24	-	-	-	-	280	-	-	227	-	-	-
	NA	-	-	-	-	-	-	-	-	-	264	119	-	-	-	-	139
Erysipelothrix rhusopathiae	AB	98	-	-	99	-	-	-	-	-	-	-	-	126	-	-	-
	NA	-	-	-	99	98	-	-	-	-	-	-	-	-	-	-	110
Glaesserella parasuis	AB	185	-	-		89	203	-	-	-	-	-	-	164	-	-	-
	NA	-	-	-	44	-	11	168	-	-	-	-	-	-	-	-	157
Mycoplasma hyopneumoniae	AB	79	-	-	-	151	80	-	-	-	151	-	23	235	-	-	-
	NA	-	-	-	114	-	31	78	78	78	279	-		-	-	-	266
Lawsonia intracellularis	AB	137	-	-	16	-	-	-	-	-	53	-	-	-	-	-	-
	NA	-	-	-	86	-	-	-	-	-	137	192	-	-	-	-	-
Mycoplasma hyorhinis	AB	174	-	-	-	-	-	-	-	-	-	-	-	-	-	-	-
	NA	96	-	-	44	-	51	243	-	-	-	-	-	-	-	-	-

Mycoplasma hyosynoviae	AB	109	-	-	-	-	-	-	-	-	-	-	-	146	-	-	-
	NA	96	-	-	96	-	-	243	102	-	-	-	-	-	-	-	-
Mycoplasma suis	AB	117	-	-	-	-	-	-	-	-	-	-	-	-	-	-	-
	NA	-	116	-	-	69	69	-	-	-	-	-	-	-	69	-	69
Pasteurella multocida	AB	196	-	-	-	-	-	-	-	-	-	-	-	196	-	-	-
	NA	-	-	-	44	-	129	129	-	275	-	-	-	-	-	-	157
Salmonella enterica	AB	194	-	-	14	63	-	-	-	-	-	-	23	124	-	-	-
	NA	-	-	-	86	-	-	-	-	-	223	-	-	-	-	-	205
Streptococcus suis	AB	29	-	-	-	-	-	-	-	-	-	-	-	18	-	-	-
	NA	-	-	-	44	12	66	179	-	-	12	-	-	-	12	-	179

[a] AB = antibody NA = nucleic acid

Table 2 An overview of antemortem specimens commonly used in surveillance for viral pathogens (numbers correspond to references listed in Section 9)

Viral pathogens[a]	Assay Target[b]	Serum	Whole blood	Blood swab	Oral fluid	Oral swab	Nasal swab	Oropharyngeal swab	Tonsil scraping	Tracheal swab	Feces	Rectal swab	Processing fluids	Colostrum/ milk	Vaginal swab	Semen	Environmental
APPV	AB	229	-	-	-	-	-	-	-	-	-	-	-	-	-	-	-
	NA	229	-	-	304	255	190	-	-	-	62	190	67	-	-	92	-
ASFV	AB	56	-	37	97	-	-	-	-	-	198	-	-	-	-	-	-
	NA	1	308	37	105	105	308	60	308	-	61	201	-	-	-	-	43
CSFV	AB	292	-	-	210	-	-	-	-	-	-	-	-	263	-	-	-
	NA	2	2	220	105	105	240	70	240		47	54				83	47
FMDV	AB	87	-	-	258	-	-	-	-	-	-	-	-	88	-	-	-
	NA	81	81	-	287	268	5	269	-	-	-	-	-	-	-	282	52
HEV	AB	288	-	-	-	-	-	-	-	-	191	-	68	73	-	-	-
	NA	132	-	-	225	-	181	-	-	-	82	40	-	-	-	149	15
IAV	AB	135	-	-	207	-	-	-	-	-	-	-	23	-	-	-	-
	NA	21	-	-	21	-	91	91	-	84	9	-	-	-	-	-	91
PCMV	AB	155	-	-	-	-	-	-	-	-	-	-	-	-	-	-	-
	NA	188	188	-	-	187	74	-	-	-	290	187	-	-	-	154	-
PCV2	AB	36	-	-	232	-	-	-	-	-	-	-	160	93	-	-	-
	NA	27	259	204	195	297	259	259	252	256	259	85	160	108	239	204	163
PDCoV	AB	272	-	-	-	-	-	-	-	-	-	-	276	166	-	-	-
	NA	286	-	-	118	-	286	-	-	-	305	305	-	305	-	-	262
PEDV	AB	94	-	-	22	-	-	-	-	-	94	-	-	226	48	-	-
	NA	-	150	-	22	-	193	-	-	-	170	22	-	-	-	90	165
PPV	AB	57	-	-	-	-	-	-	-	-	-	-	-	-	-	-	-
	NA	57	41	-	183	-	206	-	252	-	184	206	-	-	-	104	-

PRRSV	AB	3	-	217	231	217	-	-	-	-	113	-	160	242	-	217	-
	NA	50	123	217	231	217	46	65	153	45	294	-	160	-	-	217	250
PRV	AB	156	-	-	208	-	-	241	-	-	143	-	-	247	-	-	-
	NA	238	-	-	208	-	152	-	26	-	-	-	-	-	-	-	148
SVDV	AB	237	-	-	257	-	-	-	-	-	-	-	-	-	-	-	-
	NA	81	81	-	257	257	81	81	-	-	81	122	-	-	-	282	-
SVV	AB	28	-	-	-	-	-	-	-	-	-	-	-	301	-	-	-
	NA	28	-	-	42	128	28	306	-	-	34	28	-	-	-	271	299
TGEV	AB	272	-	-	175	55	-	-	-	-	-	-	-	144	-	-	-
	NA	-	-	-	175	-	-	-	-	-	133	211	-	-	-	-	-
VESV	AB	138	-	-	-	-	-	-	-	-	-	-	-	-	-	-	-
	NA		-	-	-	167	-	-	-	-	-	-	-	-	-	-	-
VSV	AB	186	-	-	-	-	-	-	-	-	-	-	-	130	-	-	-
	NA	186	284	-	-	-	186	186	-	-	81	-	-	-	-	-	-

[a] See abbreviations in Section 7.0.
[b] AB = antibody NA = nucleic acid

Serum, the liquid portion of clotted blood, is recovered by centrifugation of blood samples collected without an anticoagulant. Blood swabs, a major innovation for the detection of pathogens that produce a pronounced viremia, for example, PRRSV, are collected by puncturing a superficial vein with a needle, saturating the swab with the blood that pools at the site of the puncture, and then placing the swab into a tube containing normal saline (Reicks et al., 2006). Under some circumstances, blood swabs can be collected without animal restraint, and therefore, may be a practical alternative to standard blood collection methods in boars or adult pigs (Carlson et al., 2018).

4.2.2 Oral cavity and upper respiratory tract specimens

Oral fluid, originally reported for the detection of PRRSV-specific nucleic acid and antibody (Prickett et al., 2008a), is the liquid collected by placing an absorptive device in the buccal cavity. Oral fluids are collected from individual pigs or from a group of pigs by suspending a length of cotton rope in their pen and allowing them to chew it for 15–30 minutes. The sample is recovered by squeezing the rope after placing it in a plastic bag. The fluid that accumulates in the bottom of the bag is then poured into a tube. Oral fluids are easily collected, and as reviewed by Henao-Diaz et al. (2020a), nucleic acid and/or antibody have been reported in oral fluids for essentially all common pathogens of swine.

Oral swabs (aka buccal swabs) are collected from the rostral portion of the oral cavity, that is, inner cheek mucosa, tongue, teeth and gums. They have been used for the isolation of swine vesicular disease virus, *Campylobacter* spp., and foot-and-mouth disease virus (Callahan et al., 2002; Kodama et al., 1980; McOrist and Lawson, 1989). For sample collection, the pig's mouth is held open with an oral speculum and a sterile swab is inserted into the oral cavity. The swab is moved in circular motion in the mouth while trying to avoid feed or other contaminants and then placed into a tube containing the appropriate medium (Arai et al., 2018). Oral swabs are sometimes considered equivalent to oral fluid, but as reviewed by Henao-Diaz et al (2020a), the simple act of restraining the pig to collect an oral swab induces a stress response that results in vasoconstriction of vessels supplying the buccal tissues. This then reduces the flow of fluids to the mouth and alters the composition of buccal fluids. This explains why, in a direct comparison, Prickett et al. (2008a) reported a lower rate of detection in oral swabs vs oral fluid samples from the same animals.

Nasal swab samples were first reported in the late 1960s and early 1970s for the isolation of *Mycoplasma* spp., *Bordetella bronchiseptica*, parvovirus, cytomegalovirus and influenza A virus (Gois et al., 1969; Kemeny and Amtower, 1973; Pirtle, 1975; Watt, 1978). To collect a nasal swab sample, a sterile pre-moistened swab is inserted into one naris of the pig and rotated while avoiding contact with the outside of the nostril. This process is repeated in the other naris with the same swab or with a new one and then the swab(s) is (are) placed in a tube containing an appropriate medium.

The caudal portion of the oral cavity, including the soft palate, throat, tonsils and the back of the tongue are an excellent sampling site because the tonsils are the site of primary replication and/or chronic persistent infection for many of the most troublesome bacterial and viral pathogens, for example, *Salmonella* spp., *Actinobacillus pleuropneumoniae*, PRRSV, foot-and-mouth disease virus and classical swine fever virus (Horter et al., 2003). Samples collected from the oropharynx include swabs, tonsil biopsies, and tonsil scrapings. Oropharyngeal swabs are collected by holding the pig's mouth open with an oral speculum, swabbing the oropharynx with a pre-moistened sterile swab, particularly targeting the tonsils, and placing the swab into a tube containing an appropriate medium. Biopsies of the tonsil of the soft palate were done routinely in the US 'hog cholera' eradication program (USDA, 1981), but the procedure requires considerable technical skill. Tonsil scrapings are a highly effective and less invasive alternative to tonsil biopsies. The sample is collected by scraping the tonsils of the soft palate with a stainless steel spoon while holding the pig's mouth open with an oral speculum. The material that pools in the bowl of the spoon is then collected with a sterile swab and placed in a tube containing the appropriate medium (Wills et al., 1997). Tracheal swabs were originally reported for the isolation of influenza A virus and later for the isolation of *Glasserella parasuis* and *Mycoplasma hyopneumoniae* (Carrou et al., 2006; Kirkwood et al., 2001; Shortridge and Webster, 1979). Tracheal swabs provide improved detection for some respiratory pathogens but the procedure requires technical skill (https://vetmed.iastate.edu/story/vdl-tracheal-sampling).

4.2.3 Miscellaneous specimens

Fecal samples are easily collected, but may present practical challenges due to stool consistency and/or the potential risk of contamination during handling (Choudhury et al., 2019). For fecal collection, restrain the pig, insert two fingers of a gloved hand into the rectum, recover the sample, and place it into a clean tube or plastic bag for testing or storage. Alternatively, fecal samples may be collected by inserting a sterile swab or fecal loop into the rectum and rotating gently against the bowel wall. The swab (or fecal loop sample) is then placed into a tube containing the appropriate medium. The specimens may serve for pathogen isolation, but false negatives can occur because the analyte concentration is lower than in the actual fecal samples (Choudhury et al., 2019).

Processing fluids are serosanguineous fluids recovered from testicles and tails collected at the time of piglet processing. They were originally proposed as a sow surveillance tool for the detection of PRRSV, influenza A virus, porcine circovirus 2, porcine parvovirus, *Mycoplasma hyopneumoniae* and *Salmonella enterica* (Boettcher et al., 2010). This original observation has been supported by subsequent research (Campler et al., 2021; López et al., 2018; Vilalta et al., 2018).

Colostrum and milk were first used to evaluate antibody responses in dams immunized against influenza A virus (Young and Underdahl, 1950). For collection, udders are cleaned, the first ten spurts discarded, and the sample collected in clean tubes. Fat and debris may be removed by centrifugation (13 000 x *g* for 15 minutes), with the middle layer and pellet collected for testing (Ha et al., 2009; Poonsuk et al., 2016). Mammary secretions represent the infectious disease status of the dam and may be tested either for pathogens or for antibodies (Wagstrom et al., 2001).

The use of vaginal swabs includes reports of the isolation of African swine fever virus and *Brucella abortus* (Greig and Plowright, 1970; Stuart et al., 1987). Vaginal swabs are collected by introducing a sterile swab ~8 cm into the vaginal tract, rotating gently in a circular motion while holding against the vaginal wall, and then placing in a tube containing an appropriate medium (Pena Cortes et al., 2018). Care should be taken to avoid fecal contamination (Gresham, 2003).

Semen has been used for the detection of sexually transmitted pathogens such as *Brucella abortus* and PRRSV (Christopher-Hennings et al., 1995; Hutchings and Andrews, 1946), but it generally does not provide good test performance and other specimens should be given consideration (Christopher-Hennings et al., 1995; Maes et al., 2008). In addition, semen collection is a technical procedure that requires training both the personnel and the boar.

Environmental samples (air, water, surfaces and feedstuffs) are commonly reported in the literature and may be useful as surveillance samples. However, the limitations inherent in environmental samples are significant. That is, if present in the environment, (1) the concentration of target collected in the sample is often below the assay's limit of detection (Girones et al., 2010; Garrido-Mantilla et al., 2019) and (2) targets are often non-uniformly distributed (O'Connor et al., 2006). Thus, negative results may represent sampling/testing error, and positive results, if based on nucleic acid detection, may be difficult to interpret because the result may represent non-infectious material.

5 Test performance

Yerushalmy (1947) originally defined test performance in terms of sensitivity or specificity, but the introduction of PCR-based tests necessitated classifiers (Saah and Hoover, 1997):

- 'Diagnostic' sensitivity - the probability that a test on a sample from a positive source will produce a positive test result. Note that 'source' could be a specimen collected from an animal or from the environment (air, water, surfaces and feedstuffs).
- 'Analytical' sensitivity - an estimate of the lowest concentration of target that an assay will detect.
- 'Diagnostic' specificity - the probability that a test on a sample from a negative source will produce a negative test result.
- 'Analytical' specificity - a measure of a test's reactivity with non-target substances.

Among the four, diagnostic specificity is the most important parameter in surveillance because any uncertainty about a positive result will quickly undermine participants' confidence in the program. Laboratory approaches to increasing diagnostic specificity (i.e by reducing false positives) include raising assay cutoffs, confirmatory testing of samples that screen positive, or re-sampling and re-testing procedures. Processes for dealing with unexpected positives should be established and agreed upon before initiating the program. Counterintuitively, diagnostic sensitivity is less important. While 'adequate' diagnostic sensitivity is necessary, detection does not have to be absolute because ongoing surveillance will reveal the true status of the population and lead to the agreed-upon response.

Although diagnostic specificity tends to be stable over time and across specimens, diagnostic sensitivity varies widely as a function of specimen and time post infection. For example, Table 3 shows that the choice of specimen has a major effect on the detection of early PRRSV infection in boars using RT-qPCR, with testing of semen samples producing many false negatives. The 'sampling effect' is even more profound in the case of chronic persistent

Table 3 Detection of early PRRSV infection in individual boars using RT-PCR by specimen and day post inoculation (Pepin et al., 2015)[a]

	Percent (%) positive by day post inoculation						
Specimen	1	2	3	4	5	6	7
Serum	36.5	79.1	89.5	93.8	95.2	97.4	99.9
Blood swab	30.3	73.3	79.4	86.7	87.9	99.9	99.9
Oral fluid	3.6	59.0	89.4	97.6	99.9	99.9	99.9
Whole semen or supernatant	0	12.8	17.1	26.5	22.0	28.9	30.8
Cell fraction semen	0	0	14.7	21.4	18.8	47.1	43.5

[a] Reprinted with the permission of John Wiley and Sons. Probability calculated from data available in published reports on the detection of PRRSV in boars under controlled settings and analyzed using a binomial logistic regression model with estimates obtained using the least square methods.

infections, for example, classical swine fever virus (Panyasing et al., 2018b), PRRSV (Henao-Diaz et al., 2020b), African swine fever virus (Carrillo et al., 1994) and latent persistent infections such pseudorabies virus (Beran et al., 1980). These pathogens often produce inapparent carriers, that is, animals in which the pathogen is present for weeks or months after their initial infection.

Using PRRSV as an example, viremia is observed beginning early in the infection. Resolution of viremia is not an indication of sterilizing immunity; rather the infectious virus remains in lymphoid tissues for an extended period. Thus, transmission of PRRSV to susceptible penmates was reported 99 days post inoculation, and isolation of PRRSV from tonsil-scraping samples for up to 157 days post inoculation (Wills et al., 1997; Zimmerman et al., 1992). Henao-Diaz et al. (2020b) estimated that 2% of the animals were viremic at 98 days post infection, but the infectious virus was present in 30% (Table 4). Thus, for chronic infections, a combination of antibody and RNA assays is typically required to achieve the highest rate of detection.

In the case of aggregate specimens, diagnostic sensitivity varies both as a function of the proportion of positive members in the pool and their stage of infection (Olsen et al., 2013). Logically, increasing the proportion of positive individuals contributing to the sample will produce a corresponding increase in the probability of detection, as shown for pen-based oral fluids (Table 5). Likewise, in Table 5, a comparison of oral fluid samples versus serum samples demonstrates the utility of aggregate samples in surveillance, that is, many more individual pig samples would need to be collected to match aggregate samples' probability of detection under the same circumstances.

Table 4 Predicted rate of PRRSV positivity over time in PRRSV-infected animals: serum RT-PCR, bioassay and serum antibody (Henao-Diaz et al., 2020b)[a]

		Day post exposure to PRRSV or modified live vaccine							
Diagnostic approach[b]	n	3	7	14	28	42	77	98	175
Detection of PRRSV RNA in serum	1973	94%	92%	87%	72%	48%	7%	2%	0%
Detection of infectious PRRSV[c]	468	ND	77%	74%	68%	61%	41%	30%	7%
PRRSV serum antibody[d]	1866	ND	ND	92%	92%	92%	91%	91%	90%

[a] Reprinted with the permission of Elsevier.
[b] Rates estimated based on prediction equations for PRRSV detection derived from aggregated data ($n = 4307$) from 19 refereed publications (1995–2018).
[c] Bioassay performed by inoculation of naïve pigs with lymphoid tissue homogenate or virus isolation in cell culture.
[d] Antibody detection in individual serum samples collected at ≥ 14 days post exposure using on IDEXX ELISA HerdChek® or IDEXX PRRS ELISA X3 Ab Test (IDEXX Laboratories, Inc., Westbrook, ME, USA).

Table 5 Effect of within-pen PRRSV prevalence on the probability of detecting RNA or antibody using one pen-based oral fluid sample. A comparison of the number of individual pig serum samples to achieve the oral fluid detection rate at the same prevalence (right half of Table 5).

Within-pen prevalence (%)	Probability of detecting PRRSV in a pen of pigs using one oral fluid sample (95% confidence interval)[a]		Number of pig serum samples needed to match the oral fluid probability of detection (95% confidence interval)[a]	
	PRRSV RNA	PRRSV antibody	PRRSV RNA	PRRSV antibody
5	0.31 (0.09, 0.67)	0.17 (0.06, 0.38)	8 (3, 17)	5 (2, 10)
10	0.79 (0.48, 0.94)	0.59 (0.37, 0.77)	11 (5, 16)	7 (4, 10)
15	0.94 (0.76, 0.99)	0.85 (0.67, 0.94)	12 (8, 16)	9 (6, 12)
20	0.98 (0.88, 1.00)	0.94 (0.82, 0.98)	13 (8, 16)	10 (7, 13)
25	0.99 (0.93, 1.00)	0.97 (0.90, 0.99)	13 (9, 16)	11 (8, 13)

[a] Detection of PRRSV RNA or antibody in pen-based oral fluids were analyzed by logistic regression to model the probability of a positive outcome by within-pen prevalence. The number of serum samples needed to match the probability of detection for one oral fluid sample was calculated from the probability of ≥ 1 positive test by prevalence and sample size assuming hypergeometric distribution (J. Zimmerman, personal communication).

6 Surveillance planning checklist and summary

Step 1 - Define the surveillance objective. Surveillance objective(s) should reflect the producer's goals and the conditions found in the production system. The more precise the question, the more likely the success of the surveillance program. The objectives should be completely clear to all involved. This is also the time to plan a response to a positive result, if a response is part of the program. In particular, unexpected positives are often highly disruptive to day-to-day operations. What is the action plan if an unexpected positive result is received?

Step 2 - Determine the sampling plan. Issues on sampling and testing are reviewed in Sections 4 and 5. Simplicity and consistency are key; a complex sampling plan either will not be carried out or will not be done correctly. For interpretable surveillance data, samples should be collected in the context of a population that shares a common environment (epidemiological unit), that is, the animals in a pen, room, or barn. Results from animals that do not share a direct epidemiological connection are often uninterpretable.

Determination of sample size is complicated by the complexity of farm populations. As discussed in Section 2, sample size calculations based on binomial sampling distributions assume that (1) the population is homogeneous, that is, randomly selected pigs in the population have an equal chance of being positive, and (2) the pigs in the population are 'independent', that is, the infectious disease status of one pig is not predictive of the status of another. These circumstances are only occasionally encountered on swine farms. Nevertheless, if collecting individual discrete samples, for example, serum, the best option is to use binomial sample size calculations based on the population within each epidemiological unit.

Sample size calculators are not available for aggregate samples, for example, oral fluids. Rotolo et al. (2017) reported the probability of detecting PRRSV in wean-to-finish barns using oral fluids as a function of the number of positive pens. In the field, Ramirez et al. (2012) collected six pen-based oral fluid samples at two-week intervals in ten barns (most holding ~1 100 pigs) from placement to market. Testing by PCR detected porcine circovirus 2 and Torque teno sus viruses at one or more samplings in all ten barns, PRRSV in nine of ten barns, and influenza A virus in seven of ten. Thus, field results suggested that collecting ~6 oral fluids per 1000 growing animals provided a reasonable assurance of detection over time.

In the absence of broadly generalizable sample size calculations for pen-based samples, it may be useful to conduct a sample-size sensitivity analysis (Rotolo et al., 2017). Assuming independence among epidemiological units, that is, rooms or barns, the overall probability of detection from sampling ≥ 2 units is:

$$P=(1-(1-p_1)(1-p_2)(1-p_3)\ldots(1-p_k)),$$

where p_i is the probability of detection in the ith ($i = 1,2, \ldots, k$) unit. If the units are similar in design and are sampled using the same plan, then p_i can be assumed equal among units and the formula simplified to:

$$P=\left(1-(1-p)^k\right),$$

where

P = cumulative probability of detecting the target in ≥ 1 of the epidemiological units sampled,

$p = pr*pse$, where pr is the probability of selecting an infected pen within a unit for sampling and pse is the probability that the pen-based sample will test positive (see Table 5), and

k = number of units sampled.

The impact of the number of pen-level samples per unit on the probability of detection is embedded in '*pr*.' That is, if we suppose that one of five pens is positive in each unit and the decision is to sample one pen per unit, then $pr = 1/5 = 0.2$. If there are five units, and the pen-level probability of detection is assumed to be 0.79 (this assumes within-pen prevalence of 10% - from Table 5), then the cumulative probability of detection among the five units is calculated as,

$$P=\left(1-(1-0.2*0.79)^5\right)=40.3\%.$$

If two pens per unit are sampled then $pr = 0.4$, and

$$P = \left(1 - \left(1 - 0.4 * 0.79\right)^{5}\right) = 68.0\%.$$

Sample size for breeding herds is more problematic. Typically, it is not practical to sample a statistically sufficient number of sows on a routine basis. Alternatively, oral fluid samples may be collected from group-housed sows, but behavioral and detection information is sparse (Pierdon et al., 2016). More frequently, the status of recently farrow litters or piglets within litters is used as a proxy for the sow herd itself using serum samples from suckling piglets, oral fluids, or processing fluids (Almeida et al., 2021a,b,c; López et al., 2021).

Although the information discussed above is all relevant to the question of sample size, the ultimate decider of sample size is the budget allocated to sampling.

Step 3: Sample allocation and frequency.

Infectious diseases in swine farms tend to cluster, that is, pigs physically close to each other are more likely to be of similar infection status versus pigs further apart. Conventional sampling strategies based on 'random' selection are usually less efficient and more expensive than spatially based methods because randomization tends to result in over- or under-sampling disease clusters. That is, the cluster is either not sampled (missed entirely) or sampled multiple times when a single sampling would suffice. In contrast, spatially based methods take spatial autocorrelation and heterogeneity into account. Thus, within rooms or barns, Rotolo et al. (2017) showed that 'fixed spatial' sampling was as good or better than random sampling and that repeatedly sampling the same pen(s) over time provided logical and interpretable data. Here, 'fixed spatial' sampling is defined as uniform spatial allocation such that samples are spread equidistantly to each other over the room or barn being sampled.

There are no formulas to determine sampling frequency. Ultimately, the driver is the urgency of detection (the more urgent, the greater the frequency). Based on the literature, one or two samplings per month provide interpretable and actionable data (Ramirez et al., 2012; Rotolo et al., 2017).

Step 4: Plan for data analysis.

Every surveillance program needs a plan for collecting, managing, and analyzing surveillance data. The particular analyses will depend on the specific objective(s). Regardless, statistical process control (SPC), an approach that originated from quality control in industry, is highly useful in monitoring disease trends, detecting change points and understanding normal variation in the population. A description of SPC is beyond the scope of this review, but resources and examples of its use are widely available (Baum et al., 2005; de Vries and Reneau, 2010).

7 Conclusion

In swine production, herd immunity is manipulated to reduce clinical losses, biosecurity implemented to stop the entry of pathogens, and surveillance performed to verify that herd immunity and biosecurity are functioning effectively. Historically, surveillance was rarely feasible because of logistical and cost considerations, that is, sampling a meaningful number of individual animals was cumbersome, testing was slow and the entire process was expensive relative to the return. These impediments have largely been removed by developments in diagnostic medicine that provide for efficient sampling/testing and advancements in computer technology that facilitate the collection, manipulation and analysis of data. Importantly, these technical improvements are only tools whose use must be guided by a clear surveillance objective; a sampling/testing plan that is easy to implement, affordable and strategic; a system for the on-going analysis and interpretation of test data; and an action plan for responding to testing results.

8 Abbreviations

AB	Antibody
APPV	Atypical porcine pestivirus
ASFV	African swine fever virus
CSFV	Classical swine fever virus
DIVA	Differentiating infected from vaccinated
ELISA	Enzyme-linked immunosorbent assay
FMDV	Foot-and-mouth disease virus
HEV	Hepatitis E virus
IAV	Influenza A virus
LIMS	Laboratory information management system
NA	Nucleic acid, either DNA or RNA
OIE	World Organisation for Animal Health, formerly Office International des Epizooties (OIE)
PCMV	Porcine cytomegalovirus
PCR	Polymerase chain reaction
qPCR	Quantitative PCR (starting material DNA or cDNA)
RT-qPCR	Reverse-transcriptase qPCR (starting material RNA)
PRDC	Porcine respiratory disease complex
PCV2	Porcine circovirus type 2

Published by Burleigh Dodds Science Publishing Limited, 2023.

PDCoV	Porcine deltacoronavirus
PEDV	Porcine epidemic diarrhea virus
PPV	Porcine parvovirus
PRRSV	Porcine reproductive and respiratory syndrome virus
PRV	Pseudorabies (Aujeszky's disease) virus
SVDV	Swine vesicular disease virus
SVV	Seneca valley virus
TGEV	Transmissible gastroenteritis virus
USDA	United States Department of Agriculture
VESV	Vesicular exanthema of swine virus
VSV	Vesicular stomatitis virus

9 References

1. Agüero, M., Fernández, J., Romero, L., Sánchez Mascaraque, C., Arias, M. and Sánchez-Vizcaíno, J. M. (2003). Highly sensitive PCR assay for routine diagnosis of African swine fever virus in clinical samples. *J. Clin. Microbiol.* 41(9), 4431-4434.
2. Agüero, M., Fernández, J., Romero, L. J., Zamora, M. J., Sánchez, C., Belák, S., Arias, M. and Sánchez-Vizcaíno, J. M. (2004). A highly sensitive and specific gel-based multiplex RT-PCR assay for the simultaneous and differential diagnosis of African swine fever and classical swine fever in clinical samples. *Vet. Res.* 35(5), 551-563.
3. Albina, E. and Buffereau, J. P. (1993). Development and use of an immunoenzyme technique for the routine diagnosis of the porcine reproductive and respiratory syndrome. *Rev. Sci. Tech.* 12(2), 523-535.
4. Alexander, T. J. L. (1971). Changes in pig production in Britain and their effect on the veterinary profession. *Vet. Rec.* 88(6), 138-141.
5. Alexandersen, S., Quan, M., Murphy, C., Knight, J. and Zhang, Z. (2003). Studies of quantitative parameters of virus excretion and transmission in pigs and cattle experimentally infected with foot-and-mouth disease virus. *J. Comp. Pathol.* 129(4), 268-282.
6. Almeida, M. N., Corzo, C. A., Zimmerman, J. J. and Linhares, D. C. L. (2021a). Longitudinal piglet sampling in commercial sow farms highlights the challenge of PRRSV detection. *Porcine Health Manag.* 7(1), 31.
7. Almeida, M. N., Zhang, M., Lopez, W. A. L., Vilalta, C., Sanhueza, J., Corzo, C. A., Zimmerman, J. J. and Linhares, D. C. L. (2021b). A comparison of three sampling approaches for detecting PRRSV in suckling piglets. *Prev. Vet. Med.* 194, 105427.
8. Almeida, M. N., Zhang, M., Zimmerman, J. J., Holtkamp, D. J. and Linhares, D. C. L. (2021c). Finding PRRSV in sow herds: Family oral fluids vs. serum samples from due-to-wean pigs. *Prev. Vet. Med.* 193, 105397.
9. Anderson, B. D., Ma, M. J., Wang, G. L., Bi, Z. Q., Lu, B., Wang, X. J., Wang, C. X., Chen, S. H., Qian, Y. H., Song, S. X., Li, M., Zhao, T., Wu, M. N., Borkenhagen, L. K., Cao, W. C. and Gray, G. C. (2018). Prospective surveillance for influenza A virus in Chinese swine farms. *Emerg. Microbes Infect.* 7(1), 87.
10. Anderson, L. A., Black, N., Hagerty, T. J., Kluge, J. P. and Sundberg, P. L. (2008). *Pseudorabies (Aujeszky's disease) and its eradication, USDA: APHIS Technical Bulletin #1923.*

11. Angen, O., Oliveira, S., Ahrens, P., Svensmark, B. and Leser, T. D. (2007). Development of an improved species specific PCR test for detection of *Haemophilus parasuis*. *Vet. Microbiol.* 119(2–4), 266–276.

12. Arai, S., Kim, H., Watanabe, T., Tohya, M., Suzuki, E., Ishida-Kuroki, K., Maruyama, F., Murase, K., Nakagawa, I. and Sekizaki, T. (2018). Assessment of pig saliva as a *Streptococcus suis* reservoir and potential source of infection on farms by use of a novel quantitative polymerase chain reaction assay. *Am. J. Vet. Res.* 79(9), 941–948.

13. Arnold, K. B. and Chung, A. W. (2018). Prospects from systems serology research. *Immunology* 153, 279–289.

14. Atkinson, B. M., Bearson, B. L., Loving, C. L., Zimmerman, J. J., Kich, J. D. and Bearson, S. M. D. (2019). Detection of *Salmonella*-specific antibody in swine oral fluids. *Porcine Health Manag.* 5, 29.

15. Baez, P. A., Lopez, M. C., Duque-Jaramillo, A., Pelaez, D., Molina, F. and Navas, M. C. (2017). First evidence of the hepatitis E virus in environmental waters in Colombia. *PLoS ONE* 12(5), e0177525.

16. Barrera-Zarate, J. A., Andrade, M. R., Pereira, C. E. R., Vasconcellos, A., Wagatsuma, M. M., Sato, J. P. H., Daniel, A. G. S., Rezende, L. A., Otoni, L. A. V., Laub, R. P., Macedo, N. R., Costa, C. M., Haddad, J. P. A. and Guedes, R. M. C. (2019). Oral fluid for detection of exposure to *Lawsonia intracellularis* in naturally infected pigs. *Vet. J.* 244, 34–36.

17. Baum, D. H., Ward, S., Baum, C. L., Lee, N., Polson, D. D., Harris, D. L. and Nielsen, B. (2005). Statistical process control methods used to evaluate the serologic responses of pigs infected with three *Salmonella* serovars. *J. Swine Health Prod.* 13, 304–313.

18. Baums, C. G., Brüggemann, C., Kock, C., Beineke, A., Waldmann, K. H. and Valentin-Weigand, P. (2010). Immunogenicity of an autogenous *Streptococcus suis* bacterin in preparturient sows and their piglets in relation to protection after weaning. *Clin. Vaccine Immunol.* 17(10), 1589–1597.

19. Belak, S., Ballagi-Pordany, A., Flensburg, J. and Virtanen, A. (1989). Detection of pseudorabies virus DNA sequences by the polymerase chain reaction. *Arch. Virol.* 108(3–4), 279–286.

20. Beran, G. W., Davies, E. B., Arambulo, P. V., Will, L. A., Hill, H. T. and Rock, D. L. (1980). Persistence of pseudorabies virus in infected swine. *J. Am. Vet. Med. Assoc.* 176(10 Pt 1), 998–1000.

21. Biernacka, K., Karbowiak, P., Wróbel, P., Charęza, T., Czopowicz, M., Balka, G., Goodell, C., Rauh, R. and Stadejek, T. (2016). Detection of porcine reproductive and respiratory syndrome virus (PRRSV) and influenza A virus (IAV) in oral fluid of pigs. *Res. Vet. Sci.* 109, 74–80.

22. Bjustrom-Kraft, J., Woodard, K., Giménez-Lirola, L., Rotolo, M., Wang, C., Sun, Y., Lasley, P., Zhang, J., Baum, D., Gauger, P., Main, R. and Zimmerman, J. (2016). Porcine epidemic diarrhea virus (PEDV) detection and antibody response in commercial growing pigs. *BMC Vet. Res.* 12, 99.

23. Boettcher, J., Hensch, A., Gangl, A., Janowetz, B., Alex, M., Niemeyer, H. and Wittkowsi, G. (2010). Testicles of castrated piglets and anachronism – But a powerful sample for disease surveillance in sow herds. In: *Proceedings of the 21st International Pig Veterinary Society Congress*, Vancouver, Canada, p. 304.

24. Bosi, P., Casini, L., Finamore, A., Cremokolini, C., Merialdi, G., Trevisi, P., Nobili, F. and Mengheri, E. (2004). Spray-dried plasma improves growth performance and

reduces inflammatory status of weaned pigs challenged with enterotoxigenic *Escherichia coli* K88. *J. Anim. Sci.* 82(6), 1764–1772.
25. Bossé, J. T., Johnson, R. P., Nemec, M. and Rosendal, S. (1992). Protective local and systemic antibody responses of swine exposed to an aerosol of *Actinobacillus pleuropneumoniae* serotype 1. *Infect. Immun.* 60(2), 479–484.
26. Brown Jr., T. T., Shin, K. O. and Fuller, F. J. (1995). Detection of pseudorabies viral DNA in tonsillar epithelial cells of latently infected pigs. *Am. J. Vet. Res.* 56(5), 587–594.
27. Brunborg, I. M., Moldal, T. and Jonassen, C. M. (2004). Quantitation of porcine circovirus type 2 isolated from serum/plasma and tissue samples of healthy pigs and pigs with postweaning multisystemic wasting syndrome using a Taqman-based real-time PCR. *J. Virol. Methods* 122(2), 171–178.
28. Buckley, A., Montiel, N., Guo, B., Kulshreshtha, V., van Geelen, A., Hoang, H., Rademacher, C., Yoon, K. J. and Lager, K. (2018). Dexamethasone treatment did not exacerbate Seneca Valley virus infection in nursery-age pigs. *BMC Vet. Res.* 14, 1.
29. Busque, P., Higgins, R., Caya, F. and Quessy, S. (1997). Immunization of pigs against *Streptococcus suis* serotype 2 infection using a live avirulent strain. *Can. J. Vet. Res.* 61(4), 275–279.
30. Callahan, J. D., Brown, F., Osorio, F. A., Sur, J. H., Kramer, E., Long, G. W., Lubroth, J., Ellis, S. J., Shoulars, K. S., Gaffney, K. L., Rock, D. L. and Nelson, W. M. (2002). Use of a portable real-time reverse transcriptase-polymerase chain reaction assay for rapid detection of foot-and-mouth disease virus. *J. Am. Vet. Med. Assoc.* 220(11), 1636–1642.
31. Calsamiglia, M., Pijoan, C. and Trigo, A. (1999). Application of a nested polymerase chain reaction assay to detect *Mycoplasma hyopneumoniae* from nasal swabs. *J. Vet. Diagn. Invest.* 11(3), 246–251.
32. Cameron, A., Gardner, I., Doherr, M. G. and Wagner, B. (2003). Sampling considerations in surveys and monitoring and surveillance systems. In: Salman, M. D. (Ed.). *Animal Disease Surveillance and Survey Systems. Methods and Applications.* Ames, IA: Blackwell Publishing Professional, pp. 47–66.
33. Campler, M. R., Cheng, T. Y., Schroeder, D. C., Yang, M., Mor, S. K., Ferreira, J. B. and Arruda, A. G. (2021). A longitudinal study of PRRSV detection in swine herds with different demographics and PRRSV management strategies. *Transbound. Emerg. Dis.* (in press).
34. Canning, P., Canon, A., Bates, J. L., Gerardy, K., Linhares, D. C., Piñeyro, P. E., Schwartz, K. J., Yoon, K. J., Rademacher, C. J., Holtkamp, D. and Karriker, L. (2016). Neonatal mortality, vesicular lesions and lameness associated with Senecavirus A in a US sow farm. *Transbound. Emerg. Dis.* 63(4), 373–378.
35. Cannon, R. M. and Roe, R. T. (1982). *Livestock Disease Surveys. A Field Manual for Veterinarians.* Canberra, Australia: Australian Government Publishing Service.
36. Carasova, P., Celer, V., Takacova, K., Trundova, M., Molinkova, D., Lobova, D. and Smola, J. (2007). The levels of PCV2 specific antibodies and viremia in pigs. *Res. Vet. Sci.* 83(2), 274–278.
37. Carlson, J., Zani, L., Schwaiger, T., Nurmoja, I., Viltrop, A., Vilem, A., Beer, M. and Blome, S. (2018). Simplifying sampling for African swine fever surveillance: Assessment of antibody and pathogen detection from blood swabs. *Transbound. Emerg. Dis.* 65(1), e165–e172.

38. Carrillo, C., Borca, M. V., Afonso, C. L., Onisk, D. V. and Rock, D. L. (1994). Long-term persistent infection of swine monocytes/macrophages with African swine fever virus. *J. Virol.* 68(1), 580–583.
39. Carrou, J. L., Laurentie, M., Kobisch, M. and Gautier-Bouchardon, A. V. (2006). Persistence of *Mycoplasma hyopneumoniae* in experimentally infected pigs after marbofloxacin treatment and detection of mutations in the parC gene, antimicrob. *Agents Chemoter* 50, 1959–1966.
40. Casas, M., Cortés, R., Pina, S., Peralta, B., Allepuz, A., Cortey, M., Casal, J. and Martín, M. (2011). Longitudinal study of hepatitis E virus infection in Spanish farrow-to-finish swine herds. *Vet. Microbiol.* 148(1), 27–34.
41. Chen, H. T., Zhang, J., Yang, S. H., Ma, L. N., Ma, Y. P., Liu, X. T., Cai, X. P., Zhang, Y. G. and Liu, Y. S. (2009). Rapid detection of porcine parvovirus DNA by sensitive loop-mediated isothermal amplification. *J. Virol. Methods* 158(1–2), 100–103.
42. Chen, Z., Yuan, F., Li, Y., Shang, P., Schroeder, R., Lechtenberg, K., Henningson, J., Hause, B., Bai, J., Rowland, R. R. R., Clavijo, A. and Fang, Y. (2016). Construction and characterization of a full-length cDNA infectious clone of emerging porcine Senecavirus A. *Virology* 497, 111–124.
43. Chenais, E., Sternberg-Lewerin, S., Boqvist, S., Liu, L., LeBlanc, N., Aliro, T., Masembe, C. and Ståhl, K. (2017). African swine fever outbreak on a medium-sized farm in Uganda: Biosecurity breaches and within-farm virus contamination. *Trop. Anim. Health Prod.* 49(2), 337–346.
44. Cheong, Y., Oh, C., Lee, K. and Cho, K. H. (2017). Survey of porcine respiratory disease complex-associated pathogens among commercial pig farms in Korea via oral fluid method. *J. Vet. Sci.* 18(3), 283–289.
45. Cho, J., Dee, S., Deen, J. and Faaberg, K. (2005). The influence of animal age, bacterial co-infection and porcine reproductive and respiratory syndrome virus (PRRSV) isolate pathogenicity on virus load in individual pigs. In: *Proceedings of the Allen D. Leman Swine Conference*, Minneapolis, Minnesota, USA, pp. 46–48.
46. Cho, J. G., Dee, S. A., Deen, J., Guedes, A., Trincado, C., Fano, E., Jiang, Y., Faaberg, K., Collins, J. E., Murtaugh, M. P. and Joo, H. S. (2006). Evaluation of the effects of animal age, concurrent bacterial infection, and pathogenicity of porcine reproductive and respiratory syndrome virus on virus concentration in pigs. *Am. J. Vet. Res.* 67(3), 489–493.
47. Choe, S., Kim, J. H., Kim, K. S., Song, S., Kang, W. C., Kim, H. J., Park, G. N., Cha, R. M., Cho, I. S., Hyun, B. H., Park, B. K. and An, D. J. (2019). Impact of a live attenuated classical swine fever virus introduced to Jeju island, a CSF-free area. *Pathogens* 8(4), 251.
48. Choudhary, P., Fourie, K. R., Ng, S., Hamonic, G., Bérubé, N., Popowych, Y. and Wilson, H. L. (2021). Intrauterine immunizations trigger antigen-specific mucosal and systemic immunity in pigs and passive protection in suckling piglets. *Vaccine* 39(42), 6322–6332.
49. Choudhury, R., Middelkoop, A., Bolhuis, J. E. and Kleerebezem, M. (2019). Legitimate and reliable determination of the age-related intestinal microbiome in young piglets; rectal swabs and fecal samples provide comparable insights. *Front. Microbiol.* 10, 1886.
50. Christopher-Hennings, J., Nelson, E. A., Hines, R. J., Nelson, J. K., Swenson, S. L., Zimmerman, J. J., Chase, C. L., Yaeger, M. J. and Benfield, D. A. (1995). Persistence

of porcine reproductive and respiratory syndrome virus in serum and semen of adult boars. *J. Vet. Diagn. Invest.* 7(4), 456–464.

51. Clavijo, M. J., Murray, D., Oliveira, S. and Rovira, A. (2017). Infection dynamic of *Mycoplasma hyorhinis* in three commercial pig populations. *Vet. Rec.* 181(3), 68.
52. Colenutt, C., Brown, E., Nelson, N., Wadsworth, J., Maud, J., Adhikari, B., Chapagain Kafle, S., Upadhyaya, M., Kafle Pandey, S., Paton, D. J., Sumption, K. and Gubbins, S. (2018). Environmental sampling as a low-technology method for surveillance of foot-and-mouth disease virus in an area of endemicity. *Appl. Environ. Microbiol.* 84(16), e00686-18.
53. Cordes, H., Riber, U., Jensen, T. K. and Jungersen, G. (2012). Cell-mediated and humoral immune responses in pigs following primary and challenge-exposure to *Lawsonia intracellularis*. *Vet. Res.* 43, 9.
54. Coronado, L., Bohórquez, J. A., Muñoz-González, S., Perez, L. J., Rosell, R., Fonseca, O., Delgado, L., Perera, C. L., Frías, M. T. and Ganges, L. (2019). Investigation of chronic and persistent classical swine fever infections under field conditions and their impact on vaccine efficacy. *BMC Vet. Res.* 15(1), 247.
55. Cox, E., Pensaert, M. B. and Goddeeris, B. M. (1997). TGEV-specific IgA at different mucosae following infection of pigs with transmissible gastroenteritis virus or the antigenically related porcine respiratory coronavirus. *Immunol. Lett.* 56, 178.
56. Cubillos, C., Gómez-Sebastian, S., Moreno, N., Nuñez, M. C., Mulumba-Mfumu, L. K., Quembo, C. J., Heath, L., Etter, E. M., Jori, F., Escribano, J. M. and Blanco, E. (2013). African swine fever virus serodiagnosis: A general review with a focus on the analyses of African serum samples. *Virus Res.* 173(1), 159–167.
57. Cui, J., Biernacka, K., Fan, J., Gerber, P. F., Stadejek, T. and Opriessnig, T. (2017). Circulation of porcine parvovirus types 1 through 6 in serum samples obtained from six commercial polish pig farms. *Transbound. Emerg. Dis.* 64(6), 1945–1952.
58. Cutts, F. T. and Hanson, M. (2016). Seroepidemiology: An underused tool for designing and monitoring vaccination programmes in low- and middle-income countries. *Trop. Med. Int. Health* 21(9), 1086–1098.
59. Daniel, E. A., Esakialraj L, B. H., S, A., Muthuramalingam, K., Karunaianantham, R., Karunakaran, L. P., Nesakumar, M., Selvachithiram, M., Pattabiraman, S., Natarajan, S., Tripathy, S. P. and Hanna, L. E. (2021). Pooled testing strategies for SARS-CoV-2 diagnosis: A comprehensive review. *Diagn. Microbiol. Infect. Dis.* 101(2), 115432.
60. de Carvalho Ferreira, H. C., Weesendorp, E., Elbers, A. R. W., Bouma, A., Quak, S., Stegeman, J. A. and Loeffen, W. L. (2012). African swine fever virus excretion patterns in persistently infected animals: A quantitative approach. *Vet. Microbiol.* 160(3–4), 327–340.
61. de Carvalho Ferreira, H. C., Weesendorp, E., Quak, S., Stegeman, J. A. and Loeffen, W. L. (2014). Suitability of faeces and tissue samples as a basis for non-invasive sampling for African swine fever in wild boar. *Vet. Microbiol.* 172(3–4), 449–454.
62. de Groof, A., Deijs, M., Guelen, L., van Grinsven, L., van Os-Galdos, L., Vogels, W., Derks, C., Cruijsen, T., Geurts, V., Vrijenhoek, M., Suijskens, J., van Doorn, P., van Leengoed, L., Schrier, C. and van der Hoek, Lia (2016). Atypical porcine Pestivirus: A possible cause of congenital tremor type A-II in newborn piglets. *Viruses* 8(10), 271.
63. de Lucia, A., Cawthraw, S., Davies, R., Smith, R. P., Bianco, C., Ostanello, F. and Martelli, F. (2020). Correlation of anti-*Salmonella* antibodies between serum and saliva samples collected from finisher pigs. *Front. Vet. Sci.* 6, 489.

64. de Vries, A. and Reneau, J. K. (2010). Application of statistical process control charts to monitor changes in animal production systems. *J. Anim. Sci.* 88(13 Suppl.), E11–E24.
65. Dee, S. A., Deen, J., Otake, S. and Pijoan, C. (2004). An experimental model to evaluate the role of transport vehicles as a source of transmission of porcine reproductive and respiratory syndrome virus to susceptible pigs. *Can. J. Vet. Res.* 68(2), 128–133.
66. Dekker, N., Daemen, I., Verstappe, K., de Greeff, A., Smith, H. and Duim, B. (2016). Simultaneous quantification and differentiation of *Streptococcus suis* serotypes 2 and 9 by quantitative real-time PCR, evaluated in tonsillar and nasal samples of pigs. *Pathogens* 5(3), 46.
67. Dénes, L., Ruedas-Torres, I., Szilasi, A. and Balka, G. (2021). Detection and localization of atypical porcine pestivirus in the testicles of naturally infected, congenital tremor affected piglets. *Transbound. Emerg. Dis.* (in press).
68. di Bartolo, I., de Sabato, L., Chelli, E., Alborali, G. L., Tonni, M., Monini, M., de Lucia, A. and Ostanello, F. (2020). Pilot investigation on the presence of anti-hepatitis E virus (HEV) antibodies in piglet processing fluids. *Animals* 10(7), 1168.
69. Dietz, S., Mack, S. L., Hoelzle, K., Becker, K., Jannasch, C., Stadler, J., Ritzmann, M. and Hoelzle, L. E. (2014). Quantitative PCR analysis of Mycoplasma suis shedding patterns during experimental infection. *Vet. Microbiol.* 172(3–4), 581–585.
70. Dietze, K., Tucakov, A., Engel, T., Wirtz, S., Depner, K., Globig, A., Kammerer, R. and Mouchantat, S. (2016). Rope-based oral fluid sampling for early detection of classical swine fever in domestic pigs at group level. *BMC Vet. Res.* 13(1), 5.
71. Djordjevic, V., Stankovic, M., Nikolic, A., Antonijevic, N., Rakicevic, L. J., Divac, A. and Radojkovic, M. (2006). PCR amplification on whole blood samples treated with different commonly used anticoagulants. *Pediatr. Hematol. Oncol.* 23(6), 517–521.
72. Dorfman, R. (1943). The detection of defective members of large populations. *Ann. Math. Statist.* 14(4), 436–440.
73. dos Santos, D. R. L., Vitral, C. L., de Paula, V. S., Marchevsky, R. S., Lopes, J. F., Gaspar, A. M., Saddi, T. M., Júnior, N. C., Guimarães, Fde R., Júnior, J. G., Ximenes, L. L., Souto, F. J. and Pinto, M. A. (2009). Serological and molecular evidence of hepatitis E virus in swine in Brazil. *Vet. J.* 182(3), 474–480.
74. Egerer, S., Fiebig, U., Kessler, B., Zakhartchenko, V., Kurome, M., Reichart, B., Kupatt, C., Klymiuk, N., Wolf, E., Denner, J. and Bähr, A. (2018). Early weaning completely eliminates porcine cytomegalovirus from a newly established pig donor facility for xenotransplantation. *Xenotransplantation* 25(4), e12449.
75. Engvall, E. and Perlmann, P. (1971). Enzyme-linked immunosorbent assay (ELISA) quantitative assay of immunoglobulin G. *Immunochemistry* 8(9), 871–874.
76. Eurostat News. (2017). *Popul. Turnover EU Member S.* https://ec.europa.eu/ eurostat/web/products-eurostat-news/-/ddn-20170907-1 (accessed 28 December 2021).
77. Evans, C. M., Medley, G. F., Creasey, S. J. and Green, L. E. (2010). A stochastic mathematical model of the within-herd transmission dynamics of porcine reproductive and respiratory syndrome virus (PRRSV): Fade-out and persistence. *Prev. Vet. Med.* 93(4), 248–257.
78. Fablet, C., Marois, C., Kobisch, M., Madec, F. and Rose, N. (2010). Estimation of the sensitivity of four sampling methods for *Mycoplasma hyopneumoniae* detection in live pigs using a Bayesian approach. *Vet. Microbiol.* 143(2–4), 238–245.

79. Feld, N. C., Qvist, P., Ahrens, P., Friis, N. F. and Meyling, A. (1992). A monoclonal blocking ELISA detecting serum antibodies to *Mycoplasma hyopneumoniae*. *Vet. Microbiol.* 30(1), 35-46.
80. Feng, Z. X., Shao, G. Q., Liu, M. J., Wang, H. Y., Gan, Y. and Wu, X. S. (2010). Development and validation of a SIgA-ELISA for the detection of *Mycoplasma hyopneumoniae* infection. *Vet. Microbiol.* 143(2-4), 410-416.
81. Fernández, J., Agüero, M., Romero, L., Sánchez, C., Belák, S., Arias, M. and Sánchez-Vizcaíno, J. M. (2008). Rapid and differential diagnosis of foot-and-mouth disease, swine vesicular disease, and vesicular stomatitis by a new multiplex RT-PCR assay. *J. Virol. Methods* 147(2), 301-311.
82. Fernández-Barredo, S., Galiana, C., García, A., Vega, S., Gómez, M. T. and Pérez-Gracia, M. T. (2006). Detection of hepatitis E virus shedding in feces of pigs at different stages of production using reverse transcription-polymerase chain reaction. *J. Vet. Diagn. Invest.* 18(5), 462-465.
83. Floegel, G., Wehrend, A., Depner, K. R., Fritzemeier, J., Waberski, D. and Moennig, V. (2000). Detection of classical swine fever virus in semen of infected boars. *Vet. Microbiol.* 77(1-2), 109-116.
84. Foni, E., Chiapponi, C., Fratta, E., Garbarino, C., Barigazzi, G. and Merenda, M. (2003). Detection of swine influenza virus by RT-PCR and standard methods. In: *Proceedings of the 4th International Symposium on Emerging and Re-emerging Pig Diseases*, Rome, Italy, pp. 270-271.
85. Fraile, L., Grau-Roma, L., Sarasola, P., Sinovas, N., Nofrarías, M., López-Jimenez, R., López-Soria, S., Sibila, M. and Segalés, J. (2012). Inactivated PCV2 one shot vaccine applied in 3-week-old piglets: Improvement of production parameters and interaction with maternally derived immunity. *Vaccine* 30(11), 1986-1992.
86. Frana, T., Warneke, H., Stensland, W., Kinyon, J., Bower, L. and Burrough, E. (2014). Comparative detection of Lawsonia intracellularis, Salmonella and Brachyspira from oral fluids and feces. In: *Proceedings of the 45th Annual Meeting American Association of Swine Veterinarians*, Dallas, Texas, USA, pp. 67-69.
87. Francis, M. J. and Black, L. (1983). Antibody response in pig nasal fluid and serum following foot-and-mouth disease infection or vaccination. *J. Hyg. (Lond)* 91(2), 329-334.
88. Francis, M. J. and Black, L. (1984). The effect of vaccination regimen on the transfer of foot and mouth disease antibodies from the sow to her piglets. *J. Hyg. (Lond)* 93(1), 123-131.
89. Frandoloso, R., Chaudhuri, S., Paraboni Frandoloso, G. C., Yu, R-H. and Schryvers, A. B. (2020). Proof of concept for prevention of natural colonization by oral needle-free administration of a microparticle vaccine. *Front. Immunol.* 11, 2744.
90. Gallien, S., Moro, A., Lediguerher, G., Catinot, V., Paboeuf, F., Bigault, L., Berri, M., Gauger, P. C., Pozzi, N., Authié, E., Rose, N. and Grasland, B. (2018). Evidence of porcine epidemic diarrhea virus (PEDV) shedding in semen from infected specific pathogen-free boars. *Vet. Res.* 49(1), 7.
91. Garrido-Mantilla, J., Alvarez, J., Culhane, M., Nirmala, J., Cano, J. P. and Torremorell, M. (2019). Comparison of individual, group and environmental sampling strategies to conduct influenza surveillance in pigs. *BMC Vet. Res.* 15(1), 61.
92. Gatto, I. R. H., Arruda, P. H., Visek, C. A., Victoria, J. G., Patterson, A. R., Krull, A. C., Schwartz, K. J., de Oliveira, L. G. and Arruda, B. L. (2018). Detection of atypical

porcine Pestivirus in semen from commercial boar studs in the United States. *Transbound. Emerg. Dis.* 65(2), e339–e343.

93. Gerber, P. F., Garrocho, F. M., Lana, A. M. and Lobato, Z. I. (2012). Fetal infections and antibody profiles in pigs naturally infected with porcine circovirus type 2 (PCV2). *Can. J. Vet. Res.* 76(1), 38–44.
94. Gerber, P. F. and Opriessnig, T. (2015). Detection of immunoglobulin (Ig) A antibodies against porcine epidemic diarrhea virus (PEDV) in fecal and serum samples. *MethodsX* 2, 368–373.
95. Giacomini, E., Gasparrini, S., Lazzaro, M., Scali, F., Boniotti, M. B., Corradi, A., Pasquali, P. and Alborali, G. L. (2018). The role of transportation in the spread of *Brachyspira hyodysenteriae* in fattening farms. *BMC Vet. Res.* 14(1), 10.
96. Giménez-Lirola, L. G., Meiroz-De-Souza-Almeida, H., Magtoto, R. L., McDaniel, A. J., Merodio, M. M., Matias Ferreyra, F. S., Poonsuk, K., Gatto, I. R. H., Baum, D. H., Ross, R. F., Arruda, P. H. E., Schwartz, K. J., Zimmerman, J. J., Derscheid, R. J. and Arruda, B. L. (2019). Early detection and differential serodiagnosis of *Mycoplasma hyorhinis* and *Mycoplasma hyosynoviae* infections under experimental conditions. *PLoS ONE* 14(10), e0223459.
97. Giménez-Lirola, L. G., Mur, L., Rivera, B., Mogler, M., Sun, Y., Lizano, S., Goodell, C., Harris, D. L., Rowland, R. R., Gallardo, C., Sánchez-Vizcaíno, J. M. and Zimmerman, J. (2016). Detection of African swine fever virus antibodies in serum and oral fluid specimens using a recombinant protein 30 (p30) dual matrix indirect ELISA. *PLoS ONE* 11(9), e0161230.
98. Giménez-Lirola, L. G., Xiao, C. T., Halbur, P. G. and Opriessnig, T. (2012). Development and evaluation of an enzyme-linked immunosorbent assay based on a recombinant SpaA protein (rSpaA415) for detection of anti-*Erysipelothrix* spp. IgG antibodies in pigs. *J. Microbiol. Methods* 91(1), 191–197.
99. Giménez-Lirola, L. G., Xiao, C. T., Zavala, M., Halbur, P. G. and Opriessnig, T. (2013). Improving ante mortem diagnosis of *Erysipelothrix rhusiopathiae* infection by use of oral fluids for bacterial, nucleic acid and antibody detection. *J. Microbiol. Methods* 92(2), 113–121.
100. Girones, R., Ferrus, M. A., Alonso, J. L., Rodriguez-Manzano, J., Calgua, B., Correa, A. de A., Hundesa, A., Carratala, A. and Bofill-Mas, S. (2010). Molecular detection of pathogens in water – The pros and cons of molecular techniques. *Water Res.* 44, 4325–4339.
101. Gois, M., Cerny, M., Rozkosny, V. and Sovadina, M. (1969). Studies on the epizootiological significance of some species of *Mycoplasma* isolated from nasal swabs and lungs of pigs. *Zentralbl. Veterinarmed. B* 16(3), 253–265.
102. Gomes Neto, J. C., Raymond, M., Bower, L., Ramirez, A., Madson, D. M., Strait, E. L., Rosey, E. L. and Rapp-Gabrielson, V. J. (2016). Two clinical isolates of *Mycoplasma hyosynoviae* showed differing pattern of lameness and pathogen detection in experimentally challenged pigs. *J. Vet. Sci.* 17(4), 489–496.
103. González, W., Giménez-Lirola, L. G., Holmes, A., Lizano, S., Goodell, C., Poonsuk, K., Sitthicharoenchai, P., Sun, Y. and Zimmerman, J. (2017). Detection of *Actinobacillus pleuropneumoniae* ApxIV toxin antibody in serum and oral fluid specimens from pigs inoculated under experimental conditions. *J. Vet. Res.* 61(2), 163-171.
104. Gradil, C., Molitor, T., Harding, M. and Crabo, B. (1990). Excretion of porcine parvovirus through the genital tract of boars. *Am. J. Vet. Res.* 51(3), 359–362.

105. Grau, F. R., Schroeder, M. E., Mulhern, E. L., McIntosh, M. T. and Bounpheng, M. A. (2015). Detection of African swine fever, classical swine fever, and foot-and-mouth disease viruses in swine oral fluids by multiplex reverse transcription real-time polymerase chain reaction. *J. Vet. Diagn. Invest.* 27(2), 140–149.
106. Greig, A. and Plowright, W. (1970). The excretion of two virulent strains of African swine fever virus by domestic pigs. *J. Hygiene* 68, 673–683.
107. Gresham, A. (2003). Infectious reproductive disease in pigs. *In Pract.* 25, 466–473.
108. Ha, Y., Ahn, K. K., Kim, B., Cho, K. D., Lee, B. H., Oh, Y. S., Kim, S. H. and Chae, C. (2009). Evidence of shedding of porcine circovirus type 2 in milk from experimentally infected sows. *Res. Vet. Sci.* 86(1), 108–110.
109. Hagerdon-Olsen, T., Nielsen, N. C., Friis, N. F. and Nielsen, J. (1999). Progression of *Mycoplasma hyosynoviae* infection in three pig herds. Development of tonsillar carrier state, arthritis and antibodies in serum and synovial fluid in pigs from birth to slaughter. *Zentralbl. Veterinarmed. A* 46(9), 555–564.
110. Han, I., Congeevaram, S., Ki, D. W., Oh, B. T. and Park, J. (2011). Bacterial community analysis of swine manure treated with autothermal thermophilic aerobic digestion. *Appl. Microbiol. Biotechnol.* 89(3), 835–842.
111. Henao-Diaz, A., Giménez-Lirola, L., Baum, D. H. and Zimmerman, J. (2020a). Guidelines for oral fluid-based surveillance of viral pathogens in swine. *Porcine Health Manag.* 6, 28.
112. Henao-Diaz, A., Ji, J., Giménez-Lirola, L., Baum, D. H. and Zimmerman, J. (2020b). Understanding and interpreting PRRSV diagnostics in the context of "disease transition stages." *Res. Vet. Sci.* 131, 173–176.
113. Henao-Diaz, Y. A., Wang, G., Poonsuk, K., Ji, J., Giménez-Lirola, L., Zimmerman, J. J. (2019). PRRSV-specific antibody in fecal samples - A pilot study. In: *Proceedings of the 19th International Symposium of World Association of Veterinary Laboratory Diagnosticians and OIE Seminar*, Chiang Mai, Thailand, pp. 251-252.
114. Hernández-García, J., Robben, N., Magnée, D., Eley, T., Dennis, I., Kayes, S. M., Thomson, J. R. and Tucker, A. W. (2017). The use of oral fluids to monitor key pathogens in porcine respiratory disease complex. *Porcine Health Manag.* 3, 7.
115. Hobbs, E. C., Colling, A., Gurung, R. B. and Allen, J. (2021). The potential of diagnostic point-of-care tests (POCTs) for infectious and zoonotic animal diseases in developing countries: Technical, regulatory and sociocultural considerations. *Transbound. Emerg. Dis.* 68(4), 1835–1849.
116. Hoelzle, L. E., Adelt, D., Hoelzle, K., Heinritzi, K. and Wittenbrink, M. M. (2003). Development of a diagnostic PCR assay based on novel DNA sequences for the detection of *Mycoplasma suis* (*Eperythrozoon suis*) in porcine blood. *Vet. Microbiol.* 93(3), 185–196.
117. Hoelzle, L. E., Hoelzle, K., Ritzmann, M., Heinritzi, K. and Wittenbrink, M. M. (2006). *Mycoplasma suis* antigens recognized during humoral immune response in experimentally infected pigs. *Clin. Vaccine Immunol.* 13(1), 116–122.
118. Homwong, N., Jarvis, M. C., Lam, H. C., Diaz, A., Rovira, A., Nelson, M. and Marthaler, D. (2016). Characterization and evolution of porcine Deltacoronavirus in the United States. *Prev. Vet. Med.* 123, 168–174.
119. Hornes, E., Wasteson, Y. and Olsvik, O. (1991). Detection of *Escherichia coli* heat-stable enterotoxin genes in pig stool specimens by an immobilized, colorimetric, nested polymerase chain reaction. *J. Clin. Microbiol.* 29(11), 2375-2379.

120. Horter, D. C., Yoon, K. J. and Zimmerman, J. J. (2003). A review of porcine tonsils in immunity and disease. *Anim. Health Res. Rev.* 4(2), 143-155.
121. Hutchings, L. M. and Andrews, F. N. (1946). Studies on brucellosis in swine; Brucella infection in the boar. *Am. J. Vet. Res.* 7(25), 379-384.
122. Inoue, T., Alexandersen, S., Clark, A. T., Murphy, C., Quan, M., Reid, S. M., Sakoda, Y., Johns, H. L. and Belsham, G. J. (2005). Importance of arginine 20 of the swine vesicular disease virus 2A protease for activity and virulence. *J. Virol.* 79(1), 428-440.
123. Inoue, R., Tsukahara, T., Sunaba, C., Itoh, M. and Ushida, K. (2007). Simple and rapid detection of the porcine reproductive and respiratory syndrome virus from pig whole blood using filter paper. *J. Virol. Methods* 141(1), 102-106.
124. Inoue, T., Inoue, K. and Kitano, K. (1980). Antibacterial antibodies in swine colostrum. *Am. J. Vet. Res.* 41(2), 272-273.
125. Jensen, N. S., Casey, T. A. and Stanton, T. B. (1990). Detection of *Treponema hyodysenteriae* by using oligodeoxynucleotide probes complementary to 16S rRNA. *J. Clin. Microbiol.* 28(12), 2717-2721.
126. Jenvey, C. J., Reichel, M. P. and Cockcroft, P. D. (2015). *Erysipelothrix rhusiopathiae* and *Mycoplasma hyopneumoniae*: The sensitivities of enzyme-linked immunosorbent assays for detecting vaccinated sows of unknown disease status using serum and colostrum, and the correlation of the results for sow serum, colostrum and piglet serum. *J. Vet. Diagn. Invest.* 27(2), 211-216.
127. Jestin, A., Foulon, T., Pertuiset, B., Blanchard, P. and Labourdet, M. (1990). Rapid detection of pseudorabies virus genomic sequences in biological samples from infected pigs using polymerase chain reaction DNA amplification. *Vet. Microbiol.* 23(1-4), 317-328.
128. Joshi, L. R., Fernandes, M. H. V., Clement, T., Lawson, S., Pillatzki, A., Resende, T. P., Vannucci, F. A., Kutish, G. F., Nelson, E. A. and Diel, D. G. (2016). Pathogenesis of Senecavirus A infection in finishing pigs. *J. Gen. Virol.* 97(12), 3267-3279.
129. Kamp, E. M., Bokken, G. C. A. M., Vermeulen, T. M. M., de Jong, M. F., Buys, H. E., Reek, F. H. and Smits, M. A. (1996). A specific and sensitive PCR assay suitable for large-scale detection of toxigenic *Pasteurella multocida* in nasal and tonsillar swabs specimens of pigs. *J. Vet. Diagn. Invest.* 8(3), 304-309.
130. Ke, Y., Yu, D., Zhang, F., Gao, J., Wang, X., Fang, X., Wang, H. and Sun, T. (2019). Recombinant vesicular stomatitis virus expressing the spike protein of genotype 2b porcine epidemic diarrhea virus: A platform for vaccine development against emerging epidemic isolates. *Virology* 533, 77-85.
131. Kemeny, L. J. and Amtower, W. C. (1973). *Bordetella* agglutinating antibody in swine. A herd survey. *Can. J. Comp. Med.* 37(4), 409-412.
132. Key, K. F., DiCristina, J., Gillespie, J., Guenette, D. K. and Meng, X. J. (2007). Direct inoculation of RNA transcripts from an infectious cDNA clone of porcine reproductive and respiratory syndrome virus (PRRSV) into the lymph nodes and tonsils of pigs initiates PRRSV infection in vivo. *Arch. Virol.* 152(7), 1383-1387.
133. Kim, S. Y., Song, D. S. and Park, B. K. (2001). Differential detection of transmissible gastroenteritis virus and porcine epidemic diarrhea virus by duplex RT-PCR. *J. Vet. Diagn. Invest.* 13(6), 516-520.
134. Kirkwood, R. N., Rawluk, S. A., Cegielski, A. C. and Otto, A. J. (2001). Effect of pig age and autogenous sow vaccination on nasal mucosal colonization of pigs by *Haemophilus parasuis*. *J. Swine Health Prod.* 9, 77-79.

135. Kitikoon, P., Nilubol, D., Erickson, B. J., Janke, B. H., Hoover, T. C., Sornsen, S. A. and Thacker, E. L. (2006). The immune response and maternal antibody interference to a heterologous H1N1 swine influenza virus infection following vaccination. *Vet. Immunol. Immunopathol.* 112(3-4), 117-128.
136. Klein, D. (2002). Quantification using real-time PCR technology: Applications and limitations. *Trends Mol. Med.* 8(6), 257-260.
137. Knittel, J. P., Jordan, D. M., Schwartz, K. J., Janke, B. H., Roof, M. B., McOrist, S. and Harris, D. L. (1998). Evaluation of antemortem polymerase chain reaction and serologic methods for detection of *Lawsonia intracellularis*-exposed pigs. *Am. J. Vet. Res.* 59(6), 722-726.
138. Knowles, N. J. and Reuter, G. (2019). Caliciviruses. In: Zimmerman, J. J., Karriker, L. A., Ramirez, A., Schwartz, K. J., Stevenson, G. W. and Zhang, J. (Eds). *Diseases of Swine* (11th edn.). Hoboken, NJ: John Wiley and Sons, Inc., pp. 464-472.
139. Kobayashi, H., Khai, L. T. L., Phan, T. T., Yamasaki, S. and Taniguchi, T. (2003). Prevalence of pathogenic *Escherichia coli* in a swine breeding environment in Can Tho Province, Vietnam. *JARQ* 37(1), 59-63.
140. Kodama, M., Ogawa, T., Saito, T., Tokuda, G. and Sasahara, J. (1980). Swine vesicular disease viruses isolated from healthy pigs in non-epizootic period. I. Isolation and identification. *Natl. Inst. Anim. Health Q. (Tokyo)* 20(1), 1-10.
141. Kono, Y., Suzuki, S., Mukai, T., Okazaki, K., Honda, E. and Yamashiro, T. (1994). Detection of specific systemic and local IgG and IgA antibodies of pigs after infection with *Bordetella bronchiseptica* by ELISA. *J. Vet. Med. Sci.* 56(2), 249-253.
142. Kruskal, W. and Mosteller, F. (1980). Representative sampling, IV: The history of the concept in statistics, 1895-1939. *Int. Sat Rev.* 48, 169-195.
143. Lan, D., Ji, W., Yang, S., Cui, L., Yang, Z., Yuan, C. and Hua, X. (2011). Isolation and characterization of the first Chinese porcine Sapelovirus strain. *Arch. Virol.* 156(9), 1567-1574.
144. Lanza, I., Shoup, D. I. and Saif, L. J. (1995). Lactogenic immunity and milk antibody isotypes to transmissible gastroenteritis virus in sows exposed to porcine respiratory coronavirus during pregnancy. *Am. J. Vet. Res.* 56(6), 739-748.
145. Lapointe, L., D'Allaire, S., Lacouture, S. and Gottschalk, M. (2001). Serologic profile of a cohort of pigs and antibody response to an autogenous vaccine for *Actinobacillus suis*. *Vet. Res.* 32(2), 175-183.
146. Lauritsen, K. T., Hagedorn-Olsen, T., Jungersen, G., Riber, U., Stryhn, H., Friis, N. F., Lind, P. and Kristensen, B. (2017). Transfer of maternal immunity to piglets is involved in early protection against *Mycoplasma hyosynoviae* infection. *Vet. Immunol. Immunopathol.* 183, 22-30.
147. Lee, C. H. and Olson, L. D. (1976). Immunofluorescence of spirochetes with serum from swine recovered from swine dysentery using an indirect fluorescent antibody test. *Can. J. Comp. Med.* 40(4), 404-407.
148. Li, H., Liang, R., Pang, Y., Shi, L., Cui, S. and Lin, W. (2020). Evidence for interspecies transmission route of pseudorabies virus via virally contaminated fomites. *Vet. Microbiol.* 251, 108912.
149. Li, H., Wu, J., Sheng, Y., Lu, Q., Liu, B., Chen, Y., Sun, Y., Zhou, E. M. and Zhao, Q. (2019). Prevalence of hepatitis E virus (HEV) infection in various pig farms from Shaanxi Province, China: first detection of HEV RNA in pig semen. *Transbound. Emerg. Dis.* 66(1), 72-82.

150. Li, J., Yuan, C., Liu, P., Li, Y., Zhang, P. and Yang, Q. (2021). Red blood cells serve as a vehicle for PEDV transmission. *Vet. Microbiol.* 257, 109081.
151. Lin, J. H., Weng, C. N., Liao, C. W., Yeh, K. S. and Pan, M. J. (2003). Protective effects of oral microencapsulated *Mycoplasma hyopneumoniae* vaccine prepared by co-spray drying method. *J. Vet. Med. Sci.* 65(1), 69–74.
152. Lin, Y., Qigai, H., Xiaolan, Y., Weicheng, B. and Huanchun, C. (2005). The co-administrating of recombinant porcine IL-2 could enhance protective immune responses to PRV inactivated vaccine in pigs. *Vaccine* 23(35), 4436–4441.
153. Linhares, D. C., Cano, J. P., Wetzell, T., Nerem, J., Torremorell, M. and Dee, S. A. (2012). Effect of modified-live porcine reproductive and respiratory syndrome virus (PRRSv) vaccine on the shedding of wild-type virus from an infected population of growing pigs. *Vaccine* 30(2), 407–413.
154. Liu, X., Liao, S., Zhu, L., Xu, Z. and Zhou, Y. (2013). Molecular epidemiology of porcine cytomegalovirus (PCMV) in Sichuan Province, China: 2010–2012. *PLoS ONE* 8(6), e64648.
155. Liu, X., Zhu, L., Shi, X., Xu, Z., Mei, M., Xu, W., Zhou, Y., Guo, W. and Wang, X. (2012). Indirect-blocking ELISA for detecting antibodies against glycoprotein B (gB) of porcine cytomegalovirus (PCMV). *J. Virol. Methods* 186(1–2), 30–35.
156. Liu, Y., Zhang, S., Xu, Q., Wu, J., Zhai, X., Li, S., Wang, J., Ni, J., Yuan, L., Song, X., Zhao, B., Zhou, Z., Wang, C. and Yang, L. (2018). Investigation on pseudorabies prevalence in Chinese swine breeding farms in 2013–2016. *Trop. Anim. Health Prod.* 50(6), 1279–1285.
157. Loera-Muro, A., Ramírez-Castillo, F. Y., Moreno-Flores, A. C., Martin, E. M., Avelar-González, F. J. and Guerrero-Barrera, A. L. (2021). *Actinobacillus pleuropneumoniae* surviving on environmental multi-species biofilms in swine farms. *Front. Vet. Sci.* 8, 722683.
158. Loera-Muro, V. M., Jacques, M., Tremblay, Y. D. N., Avelar-González, F. J., Loera Muro, A., Ramírez-López, E. M., Medina-Figueroa, A., González-Reynaga, H. M. and Guerrero-Barrera, A. L. (2013). Detection of *Actinobacillus pleuropneumoniae* in drinking water from pig farms. *Microbiology* 159, 536–544.
159. Loftager, M. K., Eriksen, L. and Nielsen, R. (1993). Antibodies against *Actinobacillus pleuropneumoniae* serotype 2 in mucosal secretions and sera of infected pigs as demonstrated by an enzyme-linked immunosorbent assay. *Res. Vet. Sci.* 54(1), 57–62.
160. López, W. and Linhares, D. (2017). Processing fluids, blood serum, and tail blood swabs to detect PRRSV RNA and PCV2 DNA by PCR-based assays. In: Proceedings of the 2017 ISU James D. McKean Swine Disease Conference, Ames, IA, p. 69.
161. López, W. A., Angulo, J., Zimmerman, J. J. and Linhares, D. C. L. (2018). Porcine reproductive and respiratory syndrome monitoring in breeding herds using processing fluids. *J. Swine Health Prod.* 26, 146–150.
162. López, W. A., Gauger, P. C., Harmon, K. M., Holtkamp, D. J., Cano, J. P., Macedo, N., Zhang, M., Silva, G. S., Angulo, J., Zimmerman, J. J. and Linhares, D. C. L. (2021). Probability of PRRS virus detection in pooled processing fluid samples. *Vet. Microbiol.* 261, 109190.
163. López-Lorenzo, G., Díaz-Cao, J. M., Prieto, A., López-Novo, C., López, C. M., Díaz, P., Rodríguez-Vega, V., Díez-Baños, P. and Fernández, G. (2019). Environmental distribution of porcine circovirus type 2 (PCV2) in swine herds with natural infection. *Sci. Rep.* 9(1), 14816.

164. López-Serrano, S., Neila-Ibáñez, C., Costa-Hurtado, M., Mahmmod, Y., Martínez-Martínez, J., Galindo-Cardiel, I. J., Darji, A., Rodríguez, F., Sibila, M. and Aragon, V. (2021). Sow vaccination with a protein fragment against virulent *Glaesserella (Haemophilus) parasuis* modulates immunity traits in their offspring. *Vaccines* 9(5), 534.
165. Lowe, J., Gauger, P., Harmon, K., Zhang, J., Connor, J., Yeske, P., Loula, T., Levis, I., Dufresne, L. and Main, R. (2014). Role of transportation in spread of porcine epidemic diarrhea virus infection, United States. *Emerg. Infect. Dis.* 20(5), 872-874.
166. Lu, M., Liu, Q., Wang, X., Zhang, J., Zhang, X., Shi, D., Liu, J., Shi, H., Chen, J. and Feng, L. (2020). Development of an indirect ELISA for detecting porcine Deltacoronavirus IgA antibodies. *Arch. Virol.* 165(4), 845-851.
167. Lung, O., Fisher, M., Beeston, A., Hughes, K. B., Clavijo, A., Goolia, M., Pasick, J., Mauro, W. and Deregt, D. (2011). Multiplex RT-PCR detection and microarray typing of vesicular disease viruses. *J. Virol. Methods* 175(2), 236-245.
168. Macedo, N., Rovira, A., Oliveira, S., Holtcamp, A. and Torremorell, M. (2014). Effect of enrofloxacin in the carrier stage of *Haemophilus parasuis* in naturally colonized pigs. *Can. J. Vet. Res.* 78(1), 17-22.
169. MacNeil, A., Lee, C. W. and Dietz, V. (2014). Issues and considerations in the use of serologic biomarkers for classifying vaccination history in household surveys. *Vaccine* 32(39), 4893-4900.
170. Madson, D. M., Magstadt, D. R., Arruda, P. H. E., Hoang, H., Sun, D., Bower, L. P., Bhandari, M., Burrough, E. R., Gauger, P. C., Pillatzki, A. E., Stevenson, G. W., Wilberts, B. L., Brodie, J., Harmon, K. M., Wang, C., Main, R. G., Zhang, J. and Yoon, K. J. (2014). Pathogenesis of porcine epidemic diarrhea virus isolate (US/Iowa/18984/2013) in 3-week-old weaned pigs. *Vet. Microbiol.* 174(1-2), 60-68.
171. Maes, D., Nauwynck, H., Rijsselaere, T., Mateusen, B., Vyt, P., de Kruif, A. and Van Soom, A. (2008). Disease in swine transmitted by artificial insemination: An overview. *Theriogenology* 70(8), 1337-1345.
172. Maes, R. K., Beisel, C. E., Spatz, S. J. and Thacker, B. J. (1990). Polymerase chain reaction amplification of pseudorabies virus DNA from acutely and latently infected cells. *Vet. Microbiol.* 24(3-4), 281-295.
173. Magalhães, E. S., Zimmerman, J. J., Thomas, P., Moura, C. A. A., Trevisan, G., Holtkamp, D. J., Wang, C., Rademacher, C., Silva, G. S. and Linhares, D. C. L. (2022). Whole-herd risk factors associated with wean-to-finish mortality under the conditions of a Midwestern USA swine production system. *Prev. Vet. Med.* 198, 105545.
174. Magnusson, U., Wilkie, B., Mallard, B., Rosendal, S. and Kennedy, B. (1998). *Mycoplasma hyorhinis* infection of pigs selectively bred for high and low immune response. *Vet. Immunol. Immunopathol.* 61(1), 83-96.
175. Magtoto, R., Poonsuk, K., Baum, D., Zhang, J., Chen, Q., Ji, J., Piñeyro, P., Zimmerman, J. and Giménez-Lirola, L. G. (2019). Evaluation of the serologic cross-reactivity between transmissible gastroenteritis coronavirus and porcine respiratory coronavirus using commercial blocking enzyme-linked immunosorbent assay kits. *mSphere* 4(2), e00017-e00019.
176. Mahu, M., Boyen, F., Canessa, S., Zavala Marchan, J., Haesebrouck, F., Martel, A. and Pasmans, F. (2017). An avirulent *Brachyspira hyodysenteriae* strain elicits intestinal IgA and slows down spread of swine dysentery. *Vet. Res.* 48(1), 59.
177. Maldonado, J., Valls, L., Camprodon, A., Riquelme, R. and Riera, P. (2014). Improvement of surveillance of anthropic rhinitis in pigs by using qPCR on oral

fluid samples from individual and grouped pigs. In: *Proceedings of the 23rd International Pig Veterinary Society Congress*, Cancun, Mexico, p. 223.
178. Marois, C., Cariolet, R., Morvan, H. and Kobisch, M. (2008). Transmission of pathogenic respiratory bacteria to specific pathogen free pigs at slaughter. *Vet. Microbiol.* 129(3-4), 325-332.
179. Marois, C., Le Devendec, L., Gottschalk, M. and Kobisch, M. (2007). Detection and molecular typing of *Streptococcus suis* in tonsils from live pigs in France. *Can. J. Vet. Res.* 71(1), 14-22.
180. McOrist, S. and Lawson, G. H. (1989). Reproduction of proliferative enteritis in gnotobiotic pigs. *Res. Vet. Sci.* 46(1), 27-33.
181. Meng, X. J., Halbur, P. G., Haynes, J. S., Tsareva, T. S., Bruna, J. D., Royer, R. L., Purcell, R. H. and Emerson, S. U. (1998). Experimental infection of pigs with the newly identified swine hepatitis E virus (swine HEV), but not with human strains of HEV. *Arch. Virol.* 143(7), 1405-1415.
182. Mettenleiter, T. C. (2020). Aujeszky's disease and the development of the marker/DIVA vaccination concept. *Pathogens* 9, 563.
183. Miłek, D., Woźniak, A., Guzowska, M. and Stadejek, T. (2019). Detection patterns of porcine parvovirus (PPV) and novel porcine parvoviruses 2 through 6 (PPV2-PPV6) in polish swine farms. *Viruses* 11(5), 474.
184. Miłek, D., Woźniak, A. and Stadejek, T. (2018). The detection and genetic diversity of novel porcine parvovirus 7 (PPV7) on Polish pig farms. *Res. Vet. Sci.* 120, 28-32.
185. Miniats, O. P., Smart, N. L. and Ewert, E. (1991). Vaccination of gnotobiotic primary specific pathogen-free pigs against *Haemophilus parasuis*. *Can. J. Vet. Res.* 55(1), 33-36.
186. Morozov, I., Davis, A. S., Ellsworth, S., Trujillo, J. D., McDowell, C., Shivanna, V., Dasen, E. J., Nichols, R., Martin, B. K., Monath, T. P. and Richt, J. A. (2019). Comparative evaluation of pathogenicity of three isolates of vesicular stomatitis virus (Indiana serotype) in pigs. *J. Gen. Virol.* 100(11), 1478-1490.
187. Morozov, V. A., Heinrichs, G. and Denner, J. (2017). Effective detection of porcine cytomegalovirus using non-invasively taken samples from piglets. *Viruses* 9(1), 9.
188. Morozov, V. A., Morozov, A. V. and Denner, J. (2016). New PCR diagnostic systems for the detection and quantification of porcine cytomegalovirus (PCMV). *Arch. Virol.* 161(5), 1159-1168.
189. Muirhead, M. R. (1976). Veterinary problems of intensive pig husbandry. *Vet. Rec.* 99(15), 288-292.
190. Muñoz-González, S., Canturri, A., Pérez-Simó, M., Bohórquez, J. A., Rosell, R., Cabezón, O., Segalés, J., Domingo, M. and Ganges, L. (2017). First report of the novel atypical porcine Pestivirus in Spain and a retrospective study. *Transbound. Emerg. Dis.* 64(6), 1645-1649.
191. Nakai, I., Kato, K., Miyazaki, A., Yoshii, M., Li, T. C., Takeda, N., Tsunemitsu, H. and Ikeda, H. (2006). Different fecal shedding patterns of two common strains of hepatitis E virus at three Japanese swine farms. *Am. J. Trop. Med. Hyg.* 75(6), 1171-1177.
192. Nascimento Chiriboga, A. E. C. N., Guimarães, W. V., Vanetti, M. C. D. and Araújo, E. F. (1999). Detection of *Lawsonia intracellularis* in faeces of swine from the main producing regions in Brazil. *Can. J. Microbiol.* 45(3), 230-234.
193. Niederwerder, M. C., Nietfeld, J. C., Bai, J., Peddireddi, L., Breazeale, B., Anderson, J., Kerrigan, M. A., An, B., Oberst, R. D., Crawford, K., Lager, K. M., Madson, D.

M., Rowland, R. R., Anderson, G. A. and Hesse, R. A. (2016). Tissue localization, shedding, virus carriage, antibody response, and aerosol transmission of porcine epidemic diarrhea virus following inoculation of 4-week-old feeder pigs. *J. Vet. Diagn. Invest.* 28(6), 671–678.

194. Nielsen, B., Baggesen, D., Bager, F., Haugegaard, J. and Lind, P. (1995). The serological response to *Salmonella* serovars typhimurium and infantis in experimentally infected pigs. The time course followed with an indirect anti-LPS ELISA and bacteriological examinations. *Vet. Microbiol.* 47(3–4), 205–218.
195. Nielsen, G. B., Nielsen, J. P., Haugegaard, J., Leth, S. C., Larsen, L. E., Kristensen, C. S., Pedersen, K. S., Stege, H., Hjulsager, C. K. and Houe, H. (2018). Comparison of serum pools and oral fluid samples for detection of porcine circovirus type 2 by quantitative real-time PCR in finisher pigs. *Porcine Health Manag.* 4, 2.
196. Nielsen, J. P., Foged, N. T., Sørensen, V., Barfod, K., Bording, A. and Petersen, S. K. (1991). Vaccination against progressive atrophic rhinitis with a recombinant *Pasteurella multocida* toxin derivative. *Can. J. Vet. Res.* 55(2), 128–138.
197. Nielsen, R. (1995). Detection of antibodies against *Actinobacillus pleuropneumoniae*, serotype 2 in porcine colostrum using a blocking enzyme-linked immunosorbent assay specific for serotype 2. *Vet. Microbiol.* 43(4), 277–281.
198. Nieto-Pelegrín, E., Rivera-Arroyo, B. and Sánchez-Vizcaíno, J. M. (2015). First detection of antibodies against African swine fever virus in faeces samples. *Transbound. Emerg. Dis.* 62(6), 594–602.
199. O'Connor, A. M., Gailey, J., McKean, J. D. and Hurd, H. S. (2006). Quantity and distribution of Salmonella recovered from three swine lairage pens. *J. Food Prot.* 69(7), 1717–1719.
200. OIE. (2021). Chapter 1.4: Animal health surveillance. In: *Terrestrial Animal Health Code*. https://www.oie.int/en/what-we-do/standards/codes-and-manuals/terrestrial-code-online-access/ (accessed 29 December 2021).
201. Olesen, A. S., Lohse, L., Boklund, A., Halasa, T., Gallardo, C., Pejsak, Z., Belsham, G. J., Rasmussen, T. B. and Bøtner, A. (2017). Transmission of African swine fever virus from infected pigs by direct contact and aerosol routes. *Vet. Microbiol.* 211, 92–102.
202. Olsen, C., Wang, C., Christopher-Hennings, J., Doolittle, K., Harmon, K. M., Abate, S., Kittawornrat, A., Lizano, S., Main, R., Nelson, E. A., Otterson, T., Panyasing, Y., Rademacher, C., Rauh, R., Shah, R. and Zimmerman, J. (2013). Probability of detecting porcine reproductive and respiratory syndrome virus infection using pen-based swine oral fluid specimens as a function of within-pen prevalence. *J. Vet. Diagn. Invest.* 25(3), 328–335.
203. Olvera, A., Pina, S., Pérez-Simó, M., Aragón, V., Segalés, J. and Bensaid, A. (2011). Immunogenicity and protection against *Haemophilus parasuis* infection after vaccination with recombinant virulence associated trimeric autotransporters (VtaA). *Vaccine* 29(15), 2797–2802.
204. Pala, C., Tedde, T., Salza, S., Uda, M. T., Lollai, S., Carboni, V., Fadda, A., Marongiu, E. and Virgilio, S. (2019). Epidemiological survey on the prevalence of *Salmonella* spp. in the Sardinian pig production chain, using real-time PCR screening method. *Ital. J. Food Saf.* 8(2), 7843.
205. Palinski, R. M., Mitra, N. and Hause, B. M. (2016). Discovery of a novel *Parvovirinae* virus, porcine parvovirus 7, by metagenomic sequencing of porcine rectal swabs. *Virus Genes* 52(4), 564–567.

206. Panyasing, Y., Goodell, C. K., Giménez-Lirola, L., Kittawornrat, A., Wang, C., Schwartz, K. J. and Zimmerman, J. J. (2013). Kinetics of influenza A virus nucleoprotein antibody (IgM, IgA, and IgG) in serum and oral fluid specimens from pigs infected under experimental conditions. *Vaccine* 31(52), 6210–6215.
207. Panyasing, Y., Kedkovid, R., Kittawornrat, A., Ji, J., Zimmerman, J. and Thanawongnuwech, R. (2018a). Detection of Aujeszky's disease virus DNA and antibody in swine oral fluid specimens. *Transbound. Emerg. Dis.* 65(6), 1828–1835.
208. Panyasing, Y., Kedkovid, R., Thanawongnuwech, R., Kittawornrat, A., Ji, J., Giménez-Lirola, L. and Zimmerman, J. (2018b). Effective surveillance for early classical swine fever virus detection will utilize both virus and antibody detection capabilities. *Vet. Microbiol.* 216, 72–78.
209. Panyasing, Y., Thanawongnuwech, R., Ji, J., Giménez-Lirola, L. and Zimmerman, J. (2018c). Detection of classical swine fever virus (CSFV) E2 and Erns antibody (IgG, IgA) in oral fluid specimens from inoculated (ALD strain) or vaccinated (LOM strain) pigs. *Vet. Microbiol.* 224, 70–77.
210. Papatsiros, V. G., Stylianaki, I., Papakonstantinou, G., Papaioannou, N. and Christodoulopoulos, G. (2019). Case report of transmissible gastroenteritis coronavirus infection associated with small intestine and brain lesions in piglets. *Viral Immunol.* 32(1), 63–67.
211. Paskins, R. (1999). *Manual on Livestock Disease Surveillance and Information Systems*. Rome: Food and Agriculture Organization of the United Nations.
212. Patil, G. P., Gore, S. D. and Taillie, C. (2010). *Composite Sampling*. Boston, MA: Springer.
213. Patterson, A. H., Rubin, J. E., Fernando, C., Costa, M. O., Harding, J. C. and Hill, J. E. (2013). Fecal shedding of *Brachyspyra* spp. on a farrow-to-finish swine farm with a clinical history of "*Brachyspira hampsonii*"-associated colitis. *BMC Vet. Res.* 9, 137.
214. Pedersen, K. B. and Barfod, K. (1977). Effect of vaccination of sows with *Bordetella bronchiseptica* on the incidence of atrophic rhinitis in swine. *Nord. Vet. Med.* 29(9), 369–375.
215. Pena Cortes, L. C., LeVeque, R. M., Funk, J., Marsh, T. L. and Mulks, M. H. (2018). Development of the tonsillar microbiome in pigs from newborn to weaning. *BMC Microbiol.* 18, 35. doi: 10.1186/s12866-018-1176-x.
216. Pepin, B. J., Kittawornrat, A., Liu, F., Gauger, P. C., Harmon, K., Abate, S., Main, R., Garton, C., Hargrove, J., Rademacher, C., Ramirez, A. and Zimmerman, J. (2015). Comparison of specimens for detection of porcine reproductive and respiratory syndrome virus infection in boar studs. *Transbound. Emerg. Dis.* 62(3), 295–304.
217. Pepper, T. A., Boyd, H. W. and Rosenberg, P. (1977a). Breeding record analysis in pig herds and its veterinary applications. 1: Development of a program to monitor reproductive efficiency and weaner production. *Vet. Rec.* 101(10), 177–180.
218. Pepper, T. A. and Taylor, D. J. (1977b). Breeding record analysis in pig herds and its veterinary applications. 2. Experience with a large commercial unit. *Vet. Rec.* 101(11), 196–199.
219. Petrov, A., Schotte, U., Pietschmann, J., Dräger, C., Beer, M., Anheyer-Behmenburg, H., Goller, K. V. and Blome, S. (2014). Alternative sampling strategies for passive classical and African swine fever surveillance in wild boar. *Vet. Microbiol.* 173(3–4), 360–365.
220. Pierdon, M. K., Martell, A. L. and Parsons, T. D. (2016). Use of ropes to collect oral fluids from gestating sows housed in dynamic groups and fed via electronic sow feeder. *Swine Health Prod.* 24, 246–252.

221. Pileri, E. and Mateu, E. (2016). Review on the transmission of porcine reproductive and respiratory syndrome virus between pigs and farms and impact on vaccination. *Vet. Res.* 47(1), 108.
222. Pires, A. F. A., Funk, J. A., Lim, A. and Bolin, S. R. (2013). Enumeration of *Salmonella* in feces of naturally infected pigs. *Foodborne Pathog. Dis.* 10(11), 933-937.
223. Pirtle, E. C. (1975). Surveillance of Iowa swine herds for influenza-like illness: Combined serologic and virus isolation method. *Am. J. Vet. Res.* 36(1), 121-122.
224. Plut, J., Jamnikar-Ciglenecki, U. and Stukelj, M. (2020). Molecular detection of porcine reproductive and respiratory syndrome virus, porcine circovirus 2 and hepatitis E virus in oral fluid compared to their detection in faeces and serum. *BMC Vet. Res.* 16(1), 164.
225. Poonsuk, K., Giménez-Lirola, L. G., Zhang, J., Arruda, P., Chen, Q., Correa da Silva Carrion, L., Magtoto, R., Pineyro, P., Sarmento, L., Wang, C., Sun, Y., Madson, D., Johnson, J., Yoon, K. J., Zimmerman, J. and Main, R. (2016). Does circulating antibody play a role in the protection of piglets against porcine epidemic diarrhea virus? *PLoS ONE* 11(4), e0153041.
226. Porter, P. (1969). Porcine colostral IgA and IgM antibodies to *Escherichia coli* and their intestinal absorption by the neonatal piglet. *Immunology* 17, 617-629.
227. Porter, P. and Hill, I. R. (1970). Serological changes in immunoglobulins IgG, IgA and IgM and *Escherichia coli* antibodies in the young pig. *Immunology* 18, 565-573.
228. Postel, A., Meyer, D., Cagatay, G. N., Feliziani, F., De Mia, G. M., Fischer, N., Grundhoff, A., Milićević, V., Deng, M. C., Chang, C. Y., Qiu, H. J., Sun, Y., Wendt, M. and Becher, P. (2017). High abundance and genetic variability of atypical porcine Pestivirus in pigs from Europe and Asia. *Emerg. Infect. Dis.* 23(12), 2104-2107.
229. Prickett, J., Simer, R., Christopher-Hennings, J., Yoon, K. J., Evans, R. B. and Zimmerman, J. J. (2008a). Detection of porcine reproductive and respiratory syndrome virus infection in porcine oral fluid samples: A longitudinal study under experimental conditions. *J. Vet. Diagn. Invest.* 20(2), 156-163.
230. Prickett, J. R., Kim, W., Simer, R., Yoon, K. J. and Zimmerman, J. (2008b). Oral-fluid samples for surveillance of commercial growing pigs for porcine reproductive and respiratory syndrome virus and porcine circovirus type 2 infections. *J. Swine Health Prod.* 16, 86-91.
231. Prickett, J. R., Johnson, J., Murtaugh, M. P., Puvanendiran, S., Wang, C., Zimmerman, J. J. and Opriessnig, T. (2011). Prolonged detection of PCV2 and anti-PCV2 antibody in oral fluids following experimental inoculation. *Transbound. Emerg. Dis.* 58(2), 121-127.
232. Ramirez, A. and Karriker, L. A. (2019). Herd evaluation. In: Zimmerman, J. J., Karriker, L. A., Ramirez, A., Schwartz, K. J., Stevenson, G. W. and Zhang, J. (Eds). *Diseases of Swine* (11th edn.). Hoboken, NJ: John Wiley and Sons, Inc., pp. 3-16.
233. Ramirez, A., Wang, C., Prickett, J. R., Pogranichniy, R., Yoon, K. J., Main, R., Johnson, J. K., Rademacher, C., Hoogland, M., Hoffmann, P., Kurtz, A., Kurtz, E. and Zimmerman, J. (2012). Efficient surveillance of pig populations using oral fluids. *Prev. Vet. Med.* 104(3-4), 292-300.
234. Rautiainen, E. and Wallgren, P. (2001). Aspects of the transmission of protection against *Mycoplasma hyopneumoniae* from sow to offspring. *J. Vet. Med. B Infect. Dis. Vet. Public Health* 48(1), 55-65.
235. Reicks, D. L., Muñoz-Zanzi, C. and Rossow, K. (2006). Sampling of adult boars during early infection with porcine reproductive and respiratory syndrome virus

for testing by polymerase chain reaction using a new blood collection technique (blood-swab method). *J. Swine Health Prod.* 14, 258–264.

236. Reid, S. M., Paton, D. J., Wilsden, G., Hutchings, G. H., King, D. P., Ferris, N. P. and Alexandersen, S. (2004). Use of automated real-time reverse transcription-polymerase chain reaction (RT-PCR) to monitor experimental swine vesicular disease virus infection in pigs. *J. Comp. Pathol.* 131(4), 308–317.
237. Ren, M., Lin, H., Chen, S., Yang, M., An, W., Wang, Y., Xue, C., Sun, Y., Yan, Y. and Hu, J. (2018). Detection of pseudorabies virus by duplex droplet digital PCR assay. *J. Vet. Diagn. Invest.* 30(1), 105–112.
238. Rincón, M., Mogollón, J., Ramírez, G., Vera, V. and Correa, J. (2015). Dynamics of porcine circovirus type 2 infection and neutralizing antibodies in subclinically infected gilts, and the effect on their litters. *Rev. Colomb. Cienc. Pecuarias* 28(3), 218–228.
239. Risatti, G., Holinka, L., Lu, Z., Kutish, G., Callahan, J. D., Nelson, W. M., Brea Tió, E. and Borca, M. V. (2005). Diagnostic evaluation of a real-time reverse transcriptase PCR assay for detection of classical swine fever virus. *J. Clin. Microbiol.* 43(1), 468–471.
240. Rodak, L., Šmíd, B., Valíček, L. and Jurak, E. (1987). Four-layer enzyme immunoassay (EIA) detection of differences in IgG, IgM and IgA antibody response to Aujeszky's disease virus in infected and vaccinated pigs. *Vet. Microbiol.* 13(2), 121–133.
241. Rolinec, M., Bíro, D., Gálik, B., Šimko, M. and Juráček, M. (2012). Immunoglobulins in colostrum of sows with porcine reproductive and respiratory syndrome-PRRS. *J. Cent. Eur. Agric.* 13(2), 301–308.
242. Roos, L. R., Surendran Nair, M., Redahl, A. K. and Pieters, M. (2019). *Mycoplasma hyorhinis* and *Mycoplasma hyosynoviae* dual detection patterns in dams and piglets. *PLoS ONE* 14(1), e0209975.
243. Rotolo, M. L., Sun, Y., Wang, C., Giménez-Lirola, L., Baum, D. H., Gauger, P. C., Harmon, K. M., Hoogland, M., Main, R. and Zimmerman, J. J. (2017). Sampling guidelines for oral fluid-based surveys of group-housed animals. *Vet. Microbiol.* 209, 20–29.
244. Rovira, A., Clement, T., Christopher-Hennings, J., Thompson, B., Engle, M., Reicks, D. and Muñoz-Zanzi, C. (2007). Evaluation of the sensitivity of reverse-transcription polymerase chain reaction to detect porcine reproductive and respiratory syndrome virus on individual and pooled samples from boars. *J. Vet. Diagn. Invest.* 19(5), 502–509.
245. Saah, A. J. and Hoover, D. R. (1997). "Sensitivity" and "specificity" reconsidered: the meaning of these terms in analytical and diagnostic settings. *Ann. Intern. Med.* 126(1), 91–94.
246. Saif, L. F. and Bohl, E. H. (1977). Immunoglobulin classes of antibodies in milk of swine after intranasal exposure to pseudorabies virus or transmissible gastroenteritis virus. *Infect. Immun.* 16(3), 961–966.
247. Saiki, R. K., Scharf, S., Faloona, F., Mullis, K. B., Horn, G. T., Erlich, H. A. and Arnheim, N. (1985). Enzymatic amplification of β-globin genomic sequences and restriction site analysis for diagnosis of sickle cell anemia. *Science* 230(4732), 1350–1354.
248. Salman, M. D. (2003). Surveillance and monitoring systems for animal health programs and disease surveys. In: Salman, M. D. (Ed.). *Animal Disease Surveillance and Survey Systems: Methods and Applications*. Ames, IA: Blackwell Publishing Professional, pp. 3–14.

249. Schelkopf, A., Nerem, J., Cowles, B., Amodie, D., Swalla, R. and Dee, S. (2014). Reproductive, productivity, and mortality outcomes in late-gestation gilts and their litters following simulation of inadvertent exposure to a modified-live vaccine strain of porcine reproductive and respiratory syndrome (PRRS) virus. *Vaccine* 32(36), 4639–4643.
250. Schmittgen, T. D. and Livak, K. J. (2008). Analyzing real-time PCR data by the comparative CT method. *Nat. Protoc.* 3(6), 1101–1108.
251. Schmoll, F., Voglmayr, T., Exel, B., Dastig, B., Truschner, K., Schuh, M., Trcka, M. and Sipos, W. (2003). Detection of porcine circovirus type 2 and porcine parvovirus in boar semen and tonsillar scrapings. In: *Proceedings of the 4th International Symposium on Emerging and Re-Emerging Pig Diseases*, Rome, Italy, p. 235.
252. Schuurs, A. H. W. M. and Van Weemen, B. K. (1977). Enzyme immunoassay. *Clin. Chim. Acta* 81(1), 1-40.
253. Schwabe, C. (1982). The current epidemiological revolution in veterinary medicine. Part I. *Prev. Vet. Med.* 1(1), 5–15.
254. Schwarz, L., Riedel, C., Högler, S., Sinn, L. J., Voglmayr, T., Wöchtl, B., Dinhopl, N., Rebel-Bauder, B., Weissenböck, H., Ladinig, A., Rümenapf, T. and Lamp, B. (2017). Congenital infection with atypical porcine pestivirus (APPV) is associated with disease and viral persistence. *Vet. Res.* 48(1), 1.
255. Segalés, J., Calsamiglia, M., Olvera, A., Sibila, M., Badiella, L. and Domingo, M. (2005). Quantification of porcine circovirus type 2 (PCV2) DNA in serum and tonsillar, nasal, tracheo-bronchial, urinary and faecal swabs of pigs with and without postweaning multisystemic wasting syndrome (PMWS). *Vet. Microbiol.* 111(3–4), 223–229.
256. Senthilkumaran, C., Bittner, H., Ambagala, A., Lung, O., Babiuk, S., Yang, M., Zimmerman, J., Giménez-Lirola, L. G. and Nfon, C. (2017a). Use of oral fluids for detection of virus and antibodies in pigs infected with swine vesicular disease virus. *Transbound. Emerg. Dis.* 64(6), 1762–1770.
257. Senthilkumaran, C., Yang, M., Bittner, H., Ambagala, A., Lung, O., Zimmerman, J., Giménez-Lirola, L. G. and Nfon, C. (2017b). Detection of genome, antigen, and antibodies in oral fluids from pigs infected with foot-and-mouth disease virus. *Can. J. Vet. Res.* 81(2), 82-90.
258. Shibata, I., Okuda, Y., Yazawa, S., Ono, M., Sasaki, T., Itagaki, M., Nakajima, N., Okabe, Y. and Hidejima, I. (2003). PCR detection of porcine circovirus type 2 DNA in whole blood, serum, oropharyngeal swab, nasal swab, and feces from experimentally infected pigs and field cases. *J. Vet. Med. Sci.* 65(3), 405–408.
259. Shortridge, K. F. and Webster, R. G. (1979). Geographical distribution of swine (Hsw1N1) and Hong Kong (H3N2) influenza virus variants in pigs in Southeast Asia. *Intervirology* 11(1), 9–15.
260. Silva, J., Rocha, D., Cunha, I., Rui Sales, L., Neto, F., Fontes, M. C. and Simões, J. (2015). Serological profile of offspring on an intensive pig farm affected by porcine reproductive and respiratory syndrome. *Asian Pac. J. Reprod.* 4(4), 317–321.
261. Sinha, A., Gauger, P., Zhang, J., Yoon, K. J. and Harmon, K. (2015). PCR-based retrospective evaluation of diagnostic samples for emergence of porcine Deltacoronavirus in US swine. *Vet. Microbiol.* 179(3–4), 296–298.
262. Soos, P., Mojzis, M., Pollner, A. and Sümeghy, L. (2001). Evaluation of vaccine-induced maternal immunity against classical swine fever. *Acta Vet. Hung.* 49(1), 17–24.

263. Ståhl, M., Kokotovic, B., Hjulsager, C. K., Breum, S. Ø. and Angen, Ø. (2011). The use of quantitative PCR for identification and quantification of *Brachyspira pilosicoli*, *Lawsonia intracellularis* and *Escherichia coli* fimbrial types F4 and F18 in pig feces. *Vet. Microbiol.* 151(3-4), 307-314.
264. Stalder, K. J., Knauer, M., Baas, T. J., Rothschild, M. F., Mabry, J. W. (2004). Sow longevity. *Pig News and Information* 25, 53N.
265. Stärk, K. D. C., Nicolet, J. and Frey, J. (1998). Detection of *Mycoplasma hyopneumoniae* by air sampling with a nested PCR assay. *Appl. Environ. Microbiol.* 64(2), 543-548.
266. Stein, T., Martineau, G. P., Morris, R. D. and Charette, R. (1987). A new approach to managing health in swine operations. *Can. Vet. J.* 28(6), 355-362.
267. Stenfeldt, C., Lohse, L. and Belsham, G. J. (2013). The comparative utility of oral swabs and probang samples for detection of foot-and-mouth disease virus infection in cattle and pigs. *Vet. Microbiol.* 162(2-4), 330-337.
268. Stenfeldt, C., Pacheco, J. M., Rodriguez, L. L. and Arzt, J. (2014). Early events in the pathogenesis of foot-and-mouth disease in pigs; identification of oropharyngeal tonsils as sites of primary and sustained viral replication. *PLoS ONE* 9(9), e106859.
269. Stuart, F. A., Corbel, M. J. and Brewer, R. A. (1987). Experimental Brucella abortus infection in pigs. *Vet. Microbiol.* 14(4): 365-379.
270. Sturos, M., Murray, D., Reicks, D., Cano, J. P., Rossow, S. and Vannucci, F. (2017). Natural and experimentallyinduced Senecavirus A infections in boars. https://www.vetmed.umn.edu/sites/vetmed;umn.edu/files/shmp_2017l18.18_svv_in_boars_-_science_page.pdf (accessed 7 January 2022).
271. Thachil, A., Gerber, P. F., Xiao, C. T., Huang, Y. W. and Opriessnig, T. (2015). Development and application of an ELISA for the detection of porcine Deltacoronavirus IgG antibodies. *PLoS ONE* 10(4), e0124363.
272. Thoen, C. O., Hopkins, M. P., Armbrust, A. L., Angus, R. D. and Pietz, D. E. (1980). Development of an enzyme-linked immunosorbent assay for detecting antibodies in sera of *Brucella suis*-infected swine. *Can. J. Comp. Med.* 44(3), 294-298.
273. Tobias, T. J., Bouma, A., Klinkenberg, D., Daemen, A. J., Stegeman, J. A., Wagenaar, J. A. and Duim, B. (2012). Detection of *Actinobacillus pleuropneumoniae* in pigs by real-time quantitative PCR for the apxIVA gene. *Vet. J.* 193(2), 557-560.
274. Tocqueville, V., Kempf, I., Paboeuf, F. and Marois-Créhan, C. (2017). Quantificatioin of *Pasteurella multocida* in experimentally infected pigs using a real-time PCR assay. *Res. Vet. Sci.* 112, 177-184.
275. Trevisan, G., Angulo, J., Johnson, C., Jensen M., Utrera V., Marostica T., Gauger, P. and Linhares, D. C. L. (2020). Increasing the functionality of your processing fluid toolbox beyond PRRSV monitoring: PCV2, PEDV, and PDCoV. In: *Proceedings of the 51st Annual Meeting American Association of Swine Veterinarians*, Atlanta, Georgia, USA, pp. 27-28.
276. USDA. (1981). *Hog Cholera and Its Eradication: A Review of U.S. Experience*. United States Department of Agriculture, Animal and Plant Inspection Service, APHIS, pp. 91-55.
277. USDA. (2003). *Pseudorabies Eradication Program Standards.* United States Department of Agriculture, Animal and Plant Inspection Service, APHIS 91-55-071.
278. Valeris-Chacin, R., Sponheim, A., Fano, E., Isaacson, R., Singer, R. S., Nerem, J., Leite, F. L. and Pieters, M. (2021). Relationships among fecal, air, raol and tracheal microbial

communities in pigs in a respiratory infection disease model. *Microorganisms* 9(2), 252.
279. van den Broeck, W., Cox, E. and Goddeeris, B. M. (1999). Receptor-dependent immune responses in pigs after oral immunization with F4 fimbriae. *Infect. Immun.* 67(2), 520–526.
280. van Oirschot, J. T. (1999). Diva vaccines that reduce virus transmission. *J. Biotechnol.* 73(2–3), 195–205.
281. van Rijn, P. A., Wellenberg, G. J., Hakze-van der Honing, R., Jacobs, L., Moonen, P. L. and Feitsma, H. (2004). Detection of economically important viruses in boar semen by quantitative RealTime PCR™ technology. *J. Virol. Methods* 120(2), 151–160.
282. Van Weemen, B. K. and Schuurs, A. H. W. M. (1971). Immunoassay using antigen-enzyme conjugates. *FEBS Lett.* 15(3), 232–236.
283. Velazquez-Salinas, L., Pauszek, S. J., Stenfeldt, C., O'Hearn, E. S., Pacheco, J. M., Borca, M. V., Verdugo-Rodriguez, A., Arzt, J. and Rodriguez, L. L. (2018). Increased virulence of an epidemic strain of vesicular stomatitis virus is associated with interference of the innate response in pigs. *Front. Microbiol.* 9, 1891.
284. Vilalta, C., Sanhueza, J., Alvarez, J., Murray, D., Torremorell, M., Corzo, C. and Morrison, R. (2018). Use of processing fluids and serum samples to characterize porcine reproductive and respiratory syndrome virus dynamics in 3 day-old pigs. *Vet. Microbiol.* 225, 149–156.
285. Vitosh-Sillman, S., Loy, J. D., Brodersen, B., Kelling, C., Doster, A., Topliff, C., Nelson, E., Bai, J., Schirtzinger, E., Poulsen, E., Meadors, B., Anderson, J., Hause, B., Anderson, G. and Hesse, R. (2016). Experimental infection of conventional nursing pigs and their dams with porcine Deltacoronavirus. *J. Vet. Diagn. Invest.* 28(5), 486–497.
286. Vosloo, W., Morris, J., Davis, A., Giles, M., Wang, J., Nguyen, H. T., Kim, P. V., Quach, N. V., Le, P. T., Nguyen, P. H., Dang, H., Tran, H. X., Vu, P. P., Hung, V. V., Le, Q. T., Tran, T. M., Mai, T. M., Le, Q. T. and Singanallur, N. B. (2015). Collection of oral fluids using cotton ropes as a sampling method to detect foot-and-mouth disease virus infection in pigs. *Transbound. Emerg. Dis.* 62(5), e71–e75.
287. Wacheck, S., Werres, C., Mohn, U., Dorn, S., Soutschek, E., Fredriksson-Ahomaa, M. and Märtlbauer, E. (2012). Detection of IgM and IgG against hepatitis E virus in serum and meat juice samples from pigs at slaughter in Bavaria, Germany. *Foodborne Pathog. Dis.* 9(7), 655–660.
288. Wagstrom, E. A., Chang, C. C., Yoon, K. J. and Zimmerman, J. J. (2001). Shedding of porcine reproductive and respiratory syndrome virus in mammary gland secretions of sows. *Am. J. Vet. Res.* 62(12), 1876–1880.
289. Wang, H. Y., Song, J. K., Shin, S., Choi, K. M. and Kim, H. (2020). One-tube nested real-time PCR assay for rapid screening of porcine cytomegalovirus in clinical samples. *Front. Vet. Sci.* 7, 586045.
290. Watt, R. G. (1978). Virological study of two commercial pig herds with respiratory disease. *Res. Vet. Sci.* 24(2), 147–153.
291. Wensvoort, G., Bloemraad, M. and Terpstra, C. (1988). An enzyme immunoassay employing monoclonal antibodies and detecting specifically antibodies to classical swine fever virus. *Vet. Microbiol.* 17(2), 129–140.
292. Wilberts, B. L., Arruda, P. H., Kinyon, J. M., Frana, T. S., Wang, C., Magstadt, D. R., Madson, D. M., Patience, J. F. and Burrough, E. R. (2014). Investigation of the impact of increased dietary insoluble fiber through the feeding of distillers dried grains

with solubles (DDGS) on the incidence and severity of *Brachyspira*-associated colitis in pigs. *PLoS ONE* 9(12), e114741.

293. Wills, R. W., Zimmerman, J. J., Yoon, K. J., Swenson, S. L., McGinley, M. J., Hill, H. T., Platt, K. B., Christopher-Hennings, J. and Nelson, E. A. (1997). Porcine reproductive and respiratory syndrome virus: A persistent infection. *Vet. Microbiol.* 55(1–4), 231–240.
294. Willson, P. J., Falk, G. and Klashinsky, S. (1987). Detection of *Actinobacillus pleuropneumoniae* infection in pigs. *Can. Vet. J.* 28(3), 111–116.
295. Woods, A. (2012). Rethinking the history of modern agriculture: British pig production, c.1910-65, 20 Century Br. *Historian* 23, 165–191.
296. Woźniak, A., Miłek, D., Matyba, P. and Stadejek, T. (2019). Real-time PCR detection patterns of porcine circovirus type 2 (PCV2) in Polish farms with different statuses of vaccination against PCV2. *Viruses* 11(12), 1135.
297. Wroughton, J. and Cole, T. (2013). Distinguishing between binomial, hypergeometric and negative binomial distributions. *J. Stat. Educ.* 21(1), 1–16.
298. Xu, W., Hole, K., Goolia, M., Pickering, B., Salo, T., Lung, O. and Nfon, C. (2017). Genome wide analysis of the evolution of Senecavirus A from swine clinical material and assembly yard environmental samples. *PLoS ONE* 12(5), e0176964.
299. Yaeger, M. J. and Van Alstine, W. G. (2019). Respiratory system. In: Zimmerman, J. J., Karriker, L. A., Ramirez, A., Schwartz, K. J., Stevenson, G. W. and Zhang, J. (Eds). *Diseases of Swine* (11th edn.). Hoboken, NJ: John Wiley and Sons, Inc., pp. 393–407.
300. Yang, F., Zhu, Z., Liu, H., Cao, W., Zhang, W., Wei, T., Zheng, M., Zhang, K., Tian, H., Zeng, Q., Cai, X. and Zheng, H. (2021). Evaluation of antibody response in sows after vaccination with Senecavirus A vaccine and the effect of maternal antibody transfer on antibody dynamics in offspring. *Vaccines* 9(10), 1066.
301. Yerushalmy, J. (1947). Statistical problems in assessing methods of medical diagnosis, with special reference to x-ray techniques. *Public Health Rep.* 62(40), 1432–1449.
302. Young, G. A. and Underdahl, N. R. (1950). Neutralization and hemagglutination inhibition of swine influenza virus by serum from suckling swine and by milk from their dams. *J. Immunol.* 65(4), 369–373.
303. Yuan, F., Feng, Y., Bai, J., Liu, X., Arruda, B., Anbalagan, S. and Peddireddi, L. (2022). Genetic diversity and prevalence of atypical porcine pestivirus in the Midwest of US swine herds during 2016–2018. *Transbound. Emerg. Dis.* 69(2), 753–763.
304. Zhang, J. (2016). Porcine deltacoronavirus: Overview of infection dynamics, diagnostic methods, prevalence and genetic evolution. *Virus Res.* 226, 71–84.
305. Zhang, J., Nfon, C., Tsai, C. F., Lee, C. H., Fredericks, L., Chen, Q., Sinha, A., Bade, S., Harmon, K., Piñeyro, P., Gauger, P., Tsai, Y. L., Wang, H. T. and Lee, P. A. (2019). Development and evaluation of a real-time RT-PCR and a field-deployable RT-insulated isothermal PCR for the detection of Seneca Valley virus. *BMC Vet. Res.* 15(1), 168.
306. Zimmerman, J., Sanderson, T., Eernisse, A. and Hill, H. (1992). Transmission of SIRS virus in convalescent animals to commingled penmates under experimental conditions. *Am. Assoc. Swine Pract. Newslett.* 4(4), 25.
307. Zsak, L., Borca, M. V., Risatti, G. R., Zsak, A., French, R. A., Lu, Z., Kutish, G. F., Neilan, J. G., Callahan, J. D., Nelson, W. M. and Rock, D. L. (2005). Preclinical diagnosis of African swine fever in contact-exposed swine by a real-time PCR assay. *J. Clin. Microbiol.* 43(1), 112–119.

Part 2

Understanding immunity and disease resistance

Chapter 5

Advances in understanding the development of immune function in pigs

Lorenzo Fraile, University of Lleida, Spain

1 Introduction

Vertebrates resist infection by pathogenic microorganisms in various ways. First, they prevent them from entering their bodies with physical barriers (e.g. skin and gut flora), but if they are overcome, the innate immune system will control and remove the infection completely or control it until the adaptive immune system comes into play. The innate immune response is based on the chemical differences between pathogenic organisms and components of the host's own body. This response consists of mechanisms triggered by a foreign agent (the pathogen itself or a product of the pathogen) through a danger signal to stop pathogen colonisation by causing an inflammatory process. The process of inflammation involves heat, pain, redness and swelling in the affected area. Two components of the innate immune system are involved in bringing about the inflammatory process: cellular (mainly phagocytic cells) and humoral (acute phase proteins, cytokines and complement system) components (Fig. 1). The innate immune system's ability to respond is immediate and develops in the days immediately after infection. The final phase of the immune response is the adaptive phase, which requires several days before it starts to develop, and it

http://dx.doi.org/10.19103/AS.2022.0103.05

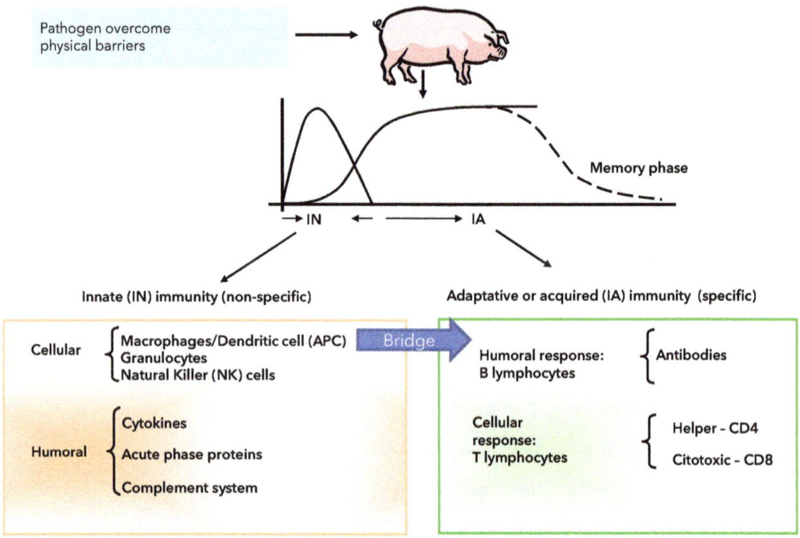

APC = Antigen presenting cells

Figure 1 General overview of the pig immune system.

begins when an antigen-presenting cell (APC) (often a dendritic cell) captures a pathogen in an infected tissue and migrates through the lymph system to the peripheral lymph nodes, where it interacts with lymphocytes. This APC forms a bridge between the innate and adaptive immune systems. The main cells of the adaptive immune system are B and T cells (Fig. 1). Both are derived from multipotent haematopoietic stem cells in the bone marrow. B cells are involved in the humoral immune response (immunoglobulin production), while T cells are mainly involved in the cell-mediated immune response (cellular component) (Tizard, 2013).

The porcine foetus begins to develop an immune response from 70 days of gestation onwards, but the newborn piglet's immune system is not completely developed until around 4 weeks of age (Butler et al., 2006, 2009; Salmon et al., 2009; Juul-Madsen et al., 2010). Neonates need to acquire passive immunity from their mothers before developing their own immunity against pathogens in order to decrease the probability to suffer from neonatal diseases (Holt and Jones, 2000; Declerck et al., 2016; Matias et al., 2017). Maternal immunity is transferred to the progeny through the placenta or by ingesting mammary secretions after birth in mammals (Salmon et al., 2009; Tizard, 2013). Serum antibodies are transferred through the placenta in some mammals such as humans, lagomorphs and rodents, but not in pigs, whose placenta is impermeable to immunoglobulins due to its anatomical structure. Therefore, piglets are hypo- or agammaglobulinaemic at birth. The mammary gland is a very specialised tissue that has evolved to feed the offspring, and milk production after parturition is a characteristic that differentiates mammals from other vertebrates (Poonsuk and Zimmerman, 2018).

2 Colostrum and milk: piglet survival

2.1 Colostrum and milk production

Mammary glands in sows are located in two parallel rows from the thoracic to the inguinal area on the ventral body wall. The total number of glands may vary between 12 and 18 depending on the genetic background of the individual (Labroue et al., 2001). Each mammary gland is composed of two complete gland systems that connect to one nipple (Martineau et al., 2012). Macroscopic and microscopic structures of the mammary gland differ between nulliparous and lactating sows. Thus, they are composed of cell buds distributed among fat and connective tissue in nulliparous sows versus glandular parenchyma in lactating sows. In these latter animals, the mammary gland is a highly organised system of ducts that terminate in the mammary alveoli surrounding the adipose tissue. The alveoli are lined with epithelial cells (lactocytes) whose function is milk synthesis, and they are surrounded by vascularised connective tissue and myoepithelial cells that function in milk ejection (Klopfenstein et al., 2006).

In the case of nulliparous sows, mammogenesis begins with the onset of puberty. Thus, their mammary gland parenchymal tissue increases by 51% and extraparenchymal tissue decreases by 17% (Farmer et al., 2004). After becoming pregnant, the rapid acceleration of mammary gland growth after day 75 of gestation coincides with an increase in the production of hormones that stimulate mammary development, including oestrogens, relaxin and prolactin (Poonsuk and Zimmerman, 2018). Thus, in the second half of gestation (days 75–90), the alveoli increase in number and size, and the amount of adipose tissue decreases, giving rise to the development of fully differentiated secretory lobules (day 105). Finally, lactogenesis is the process by which the components of colostrum and milk are created and secreted. This physiological process coincides with a reduction in oestradiol, cortisone and progesterone levels in the blood. Prolactin levels start to increase in the last quarter of gestation and continue after farrowing, inducing greater mammary gland development in response to suckling (Neville et al., 2002).

Mammary secretion in the sow gradually changes from colostrum, produced before farrowing, to transient milk and then to mature milk. Colostrum can be defined as the mammary secretion consumed by newborns in the first 24 h after the beginning of farrowing. By 34–36 h after parturition, the secretion produced until day 10 of lactation is termed transient milk, although the most significant changes are observed 72–96 h after farrowing. Finally, mature milk is defined as the milk secreted from day 10 of lactation onwards, and its composition remains virtually the same during this period (Bourne and Curtis, 1973; Theil et al., 2014a).

The composition of colostrum, transient milk and mature milk is detailed in Table 1. Colostrum is produced before the birth of the first piglet beginning around day 80 of gestation and finishing the last week of gestation. Colostrum is critical for the newborn piglet's survival due to its abundance of macronutrients and micronutrients, bioactive molecules like immunoglobulins, growth factors and enzymes (Hurley and Theil, 2011). Colostrum composition also changes significantly and rapidly, despite the extremely short period of colostrum production (Theil et al., 2014a). Hence, these levels change nearly every hour, but, for practical purposes, it is enough to differentiate between the composition of early (0 h), middle (12 h after farrowing) and late (24 h after farrowing) colostrum (Table 1). The protein fraction in colostrum not only provides energy but also serves other purposes, including the transference of immunity (immunoglobulins) and the stimulation of growth (growth factors). The most significant changes in the protein fraction occur gradually from early to late colostrum. In terms of overall nutritional value, including energy, late colostrum and transient milk are superior to early colostrum due to their higher levels of fat and lactose. Conversely, the immunological and bioactive properties of early colostrum are much greater than late colostrum. For this reason, colostrum should be ingested as soon as possible after birth to optimise the acquisition of passive immunity by the piglet (Le Dividich et al., 2005; Devillers et al., 2011; Ferrari et al., 2014).

2.2 Piglet survival

Newborn piglet requires large amounts of energy for physical activity (locomotion, suckling and competition for teats) and thermoregulation. Energy from liver glycogen, colostrum and transient milk (in the first few days) helps to maintain a constant body temperature and keeps the piglet alive. Prenatal glycogen stores provide enough energy for some 16 h postpartum, and the energy provided by colostrum should contribute significantly to piglet survival from 16 h to 34 h postpartum, as transient milk is not secreted until approximately 34 h after birth. Thus, newborn piglet survival depends on the ingestion of colostrum in the first hours of life as the piglet's energy reserves are low at birth (Decaluwé et al., 2014; Ferrari et al., 2014). Moreover, colostrum gives the newborn piglet passive immunity from the maternal origin that will be reviewed in later sections. In addition to the role of colostrum in neonatal nutrition and immunity, recent research suggests that it also plays a role in tissue development in the piglet, especially affecting the intestinal epithelium and reproductive tract. Thus, sow colostrum actually contains approximately 1500 ng/ml of epidermal

Table 1 Composition of colostrum, transient and mature milk in sows (Theil et al., 2014a)

		Colostrum			Transient milk		Mature milk
Postpartum time		Early (0 h)	Medium (12 h)	Late (24 h)	36 h	72 h	Day 10
Composition (g/100 g)	Fat	5.1[c]	5.3[c]	6.9[bc]	9.1[a]	9.8[a]	8.2[b]
	Protein	17.7[a]	12.2[b]	8.6[c]	7.3[cd]	6.1[d]	4.7[e]
	Lactose	3.5[d]	4.0[c]	4.4[bc]	4.6[b]	4.8[ab]	5.1[a]
	Dry matter	27.3[a]	22.4[b]	20.6[b]	21.4[b]	21.2[b]	18.9[c]
Energy (kJ/100 g)		260[d]	276[d]	346[c]	435[ab]	468[a]	409[b]
		Early colostrum			Mature milk		
IgG (total) (mg/ml)		61.8			1.6		
IgA (mg/ml)		11.3			4.1		
IgM (mg/ml)		3.8			1.5		

[a,b,c,d,e] Different letters in the same row means statistically significant differences ($P < 0.05$).

growth factor that accelerates intestinal growth. These 'lactocrine' effects are mediated, at least partially, by hormones in the colostrum. Curiously, there is a significant link between improved reproductive efficiency in sows and satisfactory colostrum consumption in newborn piglets (Vallet et al., 2015, 2016). These data show that colostrum consumption is critical to the animal's life, both in the short term (preweaning survival) and the long term (future reproductive performance) (Declerck et al., 2015, 2016).

Neonatal mortality is an important cause of preweaning production losses, and the first 24 h after birth is the most critical, with 28% of preweaning mortality occurring during this period (Devillers et al., 2011; Ferrari et al., 2014). Piglets without or with low colostrum consumption will be stunted, and they are more likely to be crushed or to suffer from deadly infectious diseases. Hence, consuming enough colostrum can reduce the preweaning mortality rate. Various studies have been carried out to evaluate the amount of colostrum that should be consumed. Mortality has been found to increase significantly in piglets that consumed <200 g of colostrum. In one study, the actual mortality rate was around 10% in piglets that consumed >200 g, 34% in those ingesting 100–200 g, and 63% in piglets consuming <100 g of colostrum (Ferrari et al., 2014). Thus, an intake of 200 g colostrum/piglet or 180 g/kg weight at birth seems to be the minimum required to minimise the risk of piglet mortality (Decaluwé et al., 2014; Hasan et al., 2019). These mortality results are also linked to the piglets' potential productivity, and an average consumption of 250 g colostrum/piglet is recommended for good body weight gain and to reduce the probability of poor productivity at 42 days of age (Ferrari et al., 2014; Vallet et al., 2015).

3 Transfer of humoral immunity to piglets

As previously detailed, colostrum is an extraordinarily complex mammary secretion that contains lactose, fat and proteins in proportions that change very quickly from birth (early colostrum) until 24 h postpartum (late colostrum) (Table 1). The proteins in colostrum are very important for the piglet, as they include immunoglobulins, which are essential elements of humoral acquired immunity. IgG is the most abundant immunoglobulin in colostrum (81%), and IgA is the most abundant in the sow's mature milk (57%) (Poonsuk and Zimmerman, 2018). The IgA to IgG ratio varies between 0.61 and 0.22 in colostrum and 2.1 and 6.96 in mature milk. Sow colostrum contains approximately 11 times more immunoglobulins (IgA, IgG and IgM) than mature milk because maternal immunoglobulins are concentrated from the sow blood to the colostrum in the mammary gland by a mechanism that depends on immunoglobulin receptors and these molecules are transcytosed into the gland alveoli via antibody isotype-dependent mechanisms, one for IgG and one for secretory IgA (S-IgA) and IgM (S-IgM) (Klobasa et al., 1987; Yoshida et al., 2006; Roopenian and Akilesh, 2007). Studies in several mammal species showed that the selective transfer of IgG from serum results in the high concentration of IgG in mammary secretions during late pregnancy that coincides with a period of high progesterone and low oestrogen levels (Schnulle and Hurley, 2003). On the day of parturition, the concentration of IgG in colostrum is three times the concentration present in sow serum. This fact may explain why sows that are seronegative for certain pathogens may have seropositive neonates for those pathogens after colostrum ingestion. The changes in immunoglobulin levels in mammary secretions reflect the ability of sows to transfer maternal immunity to their progeny. Scientific studies have demonstrated that almost 100% of the IgG, 24–54% of the IgA and 70–100% of the IgM in the colostrum comes from the sow's serum. In contrast, 20–38% of the IgG, 12–27% of the IgA and 4–32% of the IgM in the milk comes from the sow's serum. Obviously, the rest of the immunoglobulins are produced in the mammary gland. In particular, a very high proportion of IgA is synthesised in the mammary gland (Poonsuk and Zimmerman, 2018).

The immature structure of the neonatal intestine allows immunoglobulins to be absorbed during the first 24–36 h after birth. Thus, ingested maternal immunoglobulins remain intact within the neonatal digestive tract because the proteolytic activity is low due, among other factors, to the presence of sow colostrum trypsin inhibitor. Transcytosis of maternal immunoglobulins across neonatal intestinal epithelial cells is initiated by one of two mechanisms: non-selective endocytosis or antibody-specific receptor-mediated binding. These immunoglobulins transcytosed by the enterocytes are released in the lamina propia, diffuse to the intestinal lymphatic system and enter the circulation system (Moog, 1979). Moreover, levels of IgG and IgA in the piglet after colostrum ingestion suggest that absorption of colostrum immunoglobulins is non-selective, as serum levels of IgG and IgA in the piglet follow the ratios of IgG and IgA in the sow's colostrum (Bandrick et al., 2014a). After 24–36 h of birth, the mature intestine does not allow these macromolecules to be absorbed due to replacement of the intestinal cells for new ones during the first days after birth, and this phenomenon is called intestinal closure. Therefore, the IgGs ingested in colostrum are absorbed as intact molecules before intestinal closure (Lecce and Morgan, 1962; Klobasa et al., 1981). The administration of colostrum is clearly essential to give these animals humoral immunity to the pathogen. In addition, absorbed maternal IgM, IgA and IgG undergo transudation through the epithelial cells to the pig intestinal mucosal surfaces where it plays a major role in protecting neonates against enteric pathogens such as rotavirus and epidemic diarrhoea virus (Stirling et al., 2005; Poonsuk et al., 2016a,b). Colostrum consumption is also related to IgG levels 24 h after birth and remains linked to the piglet's IgG titre at weaning. In conclusion, sufficient colostrum must be consumed to guarantee the piglet derives a high IgG titre from passive immunity (Bandrick et al., 2014a).

After birth, IgG levels in the colostrum decrease rapidly, while IgA levels only decrease slightly (Theil et al., 2014a). IgAs are very important in establishing the intestinal microbiota and preventing food intolerances in the piglet. IgA ingested in both colostrum and mature milk is important to protect the gastrointestinal tract on a local level and IgA is essential to prevent digestive problems during the neonatal period. This lactogenic immunity is widely based, as there is 'communication' between the respiratory, digestive and genitourinary mucosae, so the immune response generated on a local level (basically of humoral and IgA type) in the respiratory, digestive and genitourinary mucosae appears in the mammary secretions due to migration of IgA-producing B cells through the regional lymphatic ganglia and blood to the mammary gland. This mechanism is mediated by a molecule (Mad-CAM-1) expressed in the venules of the Peyer's patches, the intestinal lamina propria and the mammary glands and by the ligand at the lymphocyte level (integrin $\alpha4/\beta7$) that binds to the MadCAM-1 and allows lymphocytes to pass from some compartments to others (Brandtzaeg, 2010). The presence of IgA in the milk could be due to selective transport of dimeric IgA from the induction site to the blood and from the blood by translocation through the mammary epithelium to the milk or by migration of IgA-producing B cells from the induction site (intestine and genitourinary and respiratory tracts) to the mammary gland, where it is transformed locally in plasma cells that produce IgA in the milk (Poonsuk and Zimmerman, 2018).

4 Transfer of cellular immunity to piglets

Cell-mediated immunity is very important in the control of many diseases, but the cellular contribution of colostrum has traditionally been overlooked in favour of non-cellular factors, such as immunoglobulins (Williams, 1993; Tuboly and Bernáth, 2002; Bandrick et al., 2008). Total cell yield was highest in colostrum than in milk because it decreased rapidly after farrowing. Mammary secretions of all species contain neutrophils, lymphocytes, epithelial cells and macrophages. The cell types and quantities depend on the stage of development of the mammary gland and other individual factors. Thus, the population of cells in colostrum consisted of 72% neutrophils, 26.4% lymphocytes, 1.3% macrophages, 0.4% epithelial cells and 0.2% eosinophils, whereas milk is enriched with epithelial cells (35–80%), and phagocytic cells (macrophages and neutrophils) only represent 32–55% of the cell population (Poonsuk and Zimmerman, 2018). In any case, these phagocytic cells from colostrum or milk origin are capable of phagocytosis but may lack some intracellular functions that could decrease their functionality versus these same blood cell types (Schollenberger et al., 1986). Finally, as previously commented, mammary alveolar epithelial cells play an important role in transferring antibody from the parenchyma into the alveoli, but their role for neonatal immunity is incomplete and merely descriptive, but several authors have described their role in cytokine production and as APCs. In any case, their number increases from day 1 to day 28 of lactation in the mammary secretions, and their phenotype, composition and function probably change over the course of lactation (Le Jan, 1996).

On average, 26% of the cells found in colostrum are lymphocytes, with B cells being less numerous (≈30%) than T cells (Nechvatalova et al., 2011). Research in other species suggests that B lymphocytes in mammary secretions could be plasma blasts or plasma cells, but their contribution to passive immunity in the pig remains unclear (Poonsuk and Zimmerman, 2018). On the other hand, the T cell subpopulations in colostrum are not representative of those found in the sow's blood, which suggests that the mononuclear cells in colostrum are selectively concentrated. The percentage of T cells (specifically CD8 and CD4) is significantly lower in the sow's colostrum than in her blood. In pigs, the αβ T cells that express CD4 and CD8 are effector memory cells, and stimulation by their specific antigen induces the proliferation of these double-positive (DP) lymphocytes. It is likely that many of the colostrum lymphocytes that are capable of generating an immune response in the piglet are DP effector memory cells. This DP CD4 and CD8 lymphocytes are rare during the first month of life in piglets (Stepanova et al., 2007). However, there are higher levels of γδ T cells in the sow's colostrum than in her blood; in fact, they make up nearly half of the mononuclear cells found in the colostrum (Bandrick et al., 2014b). γδ and αβT cells respond to antigens in different ways. γδ T cells are not major histocompatibility complex (MHC)-restricted, and they can recognise unprocessed nonprotein antigens and have natural killer (NK) activity and the ability to present antigens. This APC aspect is beneficial to the newborn, as neonatal APCs are immature and need greater stimulation to function in a similar way to mature animals. The ability of γδ T cells to evade destruction by the piglet's innate and adaptive immune responses allows it to complement the newborn's own immune system. Thus, NK cell activity in colostrum-fed piglets is higher than that in piglets before colostrum consumption (Bandrick et al., 2014b). Unfortunately, it is not known how maternal lymphocytes are transported across the neonatal intestinal epithelium, but there is evidence to suggest that there is selective lymphocyte transfer. If this is the case, there would be more selective absorption of γδ T cells. The transfer process is mediated by receptors that may be linked to phenotypic cellular changes produced within the colostrum (Tuboly et al., 1988; Tuboly and Bernáth, 2002; Reber et al., 2006).

It is worth noting that the process of cellular absorption is restricted to colostral cells from the piglet's own mother, and it does not occur with colostral cells that have been heat treated or that have come from another sow. In contrast, unlike these immune cells, colostral immunoglobulins cross the neonatal intestinal epithelium and enter the circulation independently of where they came from, so they can be immunoglobulins from another sow or even from another species. The mechanism behind the selective transfer of lymphoid cells in the neonatal intestinal epithelium is still unknown (Bandrick et al., 2011). After crossing the neonatal intestinal epithelium, lymphoid cells are transported to the mesenteric lymph nodes and other tissues (Williams, 1993). Classic immunological tests (e.g. immune response to a nonspecific mitogen) have demonstrated that these lymphocytes are functional. The high levels of IgG-secreting cells in the piglet's spleen, peripheral lymph nodes and bone marrow at 1 week of age could also be cells derived from colostral B cells that crossed the intestinal wall. There is limited published research on the role of colostral cells in protecting piglets against pathogens, but it could be very important because the development of the cellular part of the porcine immune system is not complete at birth (Stepanova et al., 2007). Thus, the transfer of peripheral blood mononuclear cells from adult pigs to neonates delayed the onset of the transmissible gastroenteritis virus (Cepica and Derbyshire, 1984). In conclusion, good piglet colostrum consumption was essential for optimum cellular and humoral passive immunity transfer (Blecha, 2001).

5 Strategies to improve the transfer of passive immunity to piglets

Piglet colostrum consumption depends on the sow's ability to produce and on the piglet's ability to ingest. It is worth noting that gilts usually produce colostrum of poorer quality than multiparous sows, and their piglets (usually smaller) are able to acquire less colostrum from their mothers (Quesnel, 2011). These two factors cause the level of passive immunity to be usually lower in piglets born to gilts. In any case, increasing the transfer of humoral and cellular passive immunity from the sow to the piglet is a very important objective that can be achieved by increasing the humoral and cellular immunity in sows through vaccination and/or assuring a good level of colostrum ingestion in piglets (Langel et al., 2016, 2020).

5.1 Vaccination of sows during gestation

Preventive medicine programmes usually involve the vaccination of gestating sows to induce a secondary or tertiary immune response and significantly increase the humoral and cellular immune response to the targeted pathogen(s) (Sibila et al., 2013; Hedegaard and Hedegaard, 2016; Matias et al., 2017). In this way, higher levels of antibodies and cellular immunity can be transferred to the piglet in the colostrum. Bandrick et al. (2014a,b) compared the humoral and cellular immunity acquired by piglets born to sows vaccinated twice against *M. hyo* during gestation and those born from non-vaccinated sows. Colostral antibody titres were clearly significantly higher in the vaccinated sows, and their piglets also acquired a better cellular immune response to the pathogen. The sows must be vaccinated correctly, and the vaccine must also be capable of generating the type of response required (helper T cells type Th1 or Th2) depending on the pathogen to fight against (Fraile et al., 2012a,b).

Maternal immunisation is commonly used in veterinary medicine for strengthening passive immunity by increasing the transfer of antibodies and cellular immunity from the mother to the piglets. In current production systems, sows are routinely vaccinated against a series of bacterial and viral pathogens (Poonsuk and Zimmerman, 2018). These vaccines induce production of high levels of IgG, IgM and IgA antibodies and cellular immunity that are transferred passively to the piglet, increasing the duration of both systemic and mucosal immune protection throughout the whole body against neonatal infections (Nechvatalova et al., 2011).

5.2 Assuring a good colostrum ingestion in piglets

It is essential to optimise the colostrum consumed by each piglet for its survival during the preweaning period. As previously described, piglets should ingest sufficient colostrum, estimated at approximately 250 g/piglet (Ferrari et al., 2014; Vallet et al., 2015). Moreover, this colostrum should be sourced from the piglet's own mother to guarantee that the animal acquires appropriate cellular immunity (Bandrick et al., 2011). Prolificacy has increased significantly on modern breeding farms due to intense genetic selection for this trait. Nowadays, it is now very common to find farms with an average prolificacy of over 15 piglets/sow. It means that the sow needs to produce at least 3750 g of colostrum to ensure that all her piglets can consume enough. The sow should be in good body condition to produce this amount of colostrum as colostrum production decreases when sows are excessively thin or fat. Finally, piglet management during the colostrum ingestion phase is extremely important for the acquisition of passive immunity (Vallet et al., 2013). The following conditions should be met for the piglet to be able to ingest sufficient colostrum:

1 Sows should enter their farrowing accommodation 1 week before the anticipated farrowing date, to get used to the new environment to prevent stress during farrowing (Papadopoulos et al., 2010).
2 It is advisable to feed a specific diet during the last week of gestation.
3 Sow accommodation should be clean, disinfected and fully suitable for farrowing.
4 Piglets should be kept warm and dried as soon as possible, so drying materials should be at hand in the farrowing accommodation.
5 Level/quality of stockmanship attending farrowings is very important: Staff should dry the piglets, guide them to the teats and assist during delivery if necessary.

Every farm should implement a strategy to ensure that all piglets consume sufficient colostrum. In particular, it is essential to allow piglets to remain with their mothers for at least 24 h after birth. These methods can differ from farm to farm, but monitoring the ingestion of enough colostrum is essential. In calves, a failure of passive transfer (FPT) at 24 h of age is determined by serum IgG concentrations <10 g/L (Weaver et al., 2000; Godden, 2008). Cabrera et al. (2012) showed that piglets seemed to have a 91% chance of survival until weaning (16–20 days), when their serum IgG concentration was between 23 and 25 g/L at 48–72 h of life. However, Devillers et al. (2011) showed that piglets that died within the first 3 days had an IgG concentration of 17 ± 2 g/L and piglets still alive at weaning had an average serum IgG level of 24 ± 1 g/L 24 h after birth. Surprisingly, in pigs so far, no such threshold value of FPT has been established probably due to the variability in the sampling time between studies among other factors. Anyway, it seems reasonable to achieve levels >20 g/L and preferably >30 g/L in piglet serum 24 h after birth (Schoos et al., 2021).

Several techniques can be applied in practice to estimate the piglet colostrum consumption. First, piglets can be weighed at birth and 24 h after birth. Any weight gain is obviously due to colostrum consumption in the first day of life. There is even a prediction equation that relates the colostrum consumed to the weight gain (Theil et al., 2014b). This technique is quite precise but is too laborious if there are many newborn piglets to be measured. Another way of evaluating colostrum ingestion is to measure total serum IgG levels in the piglets 24 h after farrowing. As discussed previously, piglets are born agammaglobulinaemic and the level of IgG in the serum is directly related to the ingestion of colostrum. There are several techniques that measure total IgG in pig serum, but it is not routinely performed by diagnostic laboratories and the different tests available can generate differences between laboratories (Schoos et al., 2021).

Other two techniques have been explored in the literature to estimate piglet colostrum consumption: immunocrit determination and Brix refractometry. Immunocrit determination enables estimating the amount of protein in the serum by chemically precipitating the protein and calculating the proportion of precipitate to non-precipitate serum. As previously discussed, most of the proteins in the piglet's blood in the first hours of life are immunoglobulins derived from the colostrum, especially IgG. An excellent correlation between immunocrit and plasma IgG levels has been described in newborn piglets (Vallet et al., 2013). Not surprisingly, low immunocrit values are linked to poor survival in the neonatal period, especially in piglets with low birth weights, and inmunocrit values lower than 10% has been associated with FTP to piglets (Vallet et al., 2013, 2015). This method has been used experimentally to measure the transfer of passive immunity in piglets, but very few studies have been carried out under field conditions, so more trials are needed under real production conditions to evaluate the usefulness of this technique and validate in a robust way the cut-off value proposed. Another technique is the Brix refractometry that measures in non-sucrose liquids the total solid percentage in a solution. This technique allows estimating the serum immunoglobulin concentration because immunoglobulins represent the major fraction of total protein in neonatal piglet serum after colostrum ingestion (Cabrera et al., 2012; Huang et al., 2012)., In a recent study, Schoos et al. (2021) used three different γ-globulin cut-off concentrations (10, 20 and 30 g/L γ-globulin) associated with piglet survival (Devillers et al., 2011; Cabrera et al., 2012) to set up a cut-off value for Brix refractometry. Thus, for a cut-off of 10 g/L γ-globulin, the best combination of sensitivity (100%) and specificity (98.5%) was at 5.4% Brix. The best combination of sensitivity (100%) and specificity (89.3%) for a cut-off of 20 g/L γ-globulin was at 7.0% Brix and for 30 g/L γ-globulin at 7.9% Brix with a sensitivity and specificity of 90.1% and 80.6%. Unfortunately, it is not possible to evaluate if those threshold values are accurate regarding FPT or piglet survival as there are no data about the clinical outcome of piglets and their association with Brix refractometry values. In any case, Brix refractometry level over 7% seems desirable in piglets 24 h after birth. Finally, monitoring of piglet colostrum consumption is a critical point to be developed at the farm level in parallel with the continuous increase in prolificacy due to genetic improvement. There is no doubt that pig production companies must integrate this monitoring system in sow farms to carry out the optimum differential diagnosis in outbreaks observed during the lactating period.

6 Development of active immunity in piglets during the neonatal period

Pigs are susceptible to a large number of bacterial, viral and parasitic diseases that can limit animal production efficiency. Because these diseases increase morbidity and/or mortality, production costs and reduce productivity, programmes of preventive medicine are required to minimise the negative effects of these diseases on swine production. It is very common to vaccinate piglets during the first 4 weeks of life as part of a preventive medicine programme and there are many vaccines that can be administered to piglets during this preweaning stage, following the recommendations issued by the pharmaceutical manufacturers (Segalés et al., 2009; Fraile et al., 2012a). Vaccination of newborn and suckling piglets is problematic for two reasons: the immune system is immature in the newborn piglet, and the animal has acquired humoral and cellular immunity through the colostrum (Parreño et al., 1999; Markowska-Daniel et al., 2011; Wood and Siegrist, 2011; Hedegaard and Hedegaard, 2016).

The neonatal immune system differs significantly from that of adult animals. Hence, newborns have fewer APCs, lower levels of costimulatory molecules and less surface expression of MHC II than adults. They also have a lower capacity to produce cytokines (particularly interleukin [IL]-12) in response to stimulation of the toll-like receptors (TLR) (Firth et al., 2005). CD4 T cells also show a natural bias towards a Th2 response rather than a Th1 response, so it is likely that the neonatal immune system will respond more effectively with a humoral adaptive immune response rather than a cellular one (Barrios et al., 1996; Chappuis, 1998; Butler et al., 2000, 2002, 2005, 2009). Newborns also possess a relative abundance of regulatory T cells (natural immunosuppressants). These cells secrete large amounts of IL-10 and transforming growth factor beta (TGF-β), which promote tolerance-inducing APCs and suppress the effector T cell response, the effects of which include defective plasma cell differentiation. In summary, newborns respond to the immune challenge with an adaptive humoral response which is both qualitatively and quantitatively weaker than the immunologically mature animal. Looking at these differences in the immune response between newborns and adult animals suggests that neonatal immunity is closely regulated and aims to encourage microbial colonisation and prevent adverse proinflammatory responses (Hodgins and Shewen, 2012).

Unfortunately, this also causes vulnerability to infection during the early stages of life. Finally, some other components of colostrum also have an important effect on the neonatal immune response. The corticosteroids in colostrum act as immunosuppressants, and the cytokines include proinflammatory factors (TNF-α, IL-1β and IL-6), anti-inflammatory factors (TGF-β and IL-10), growth promoters (erythropoietin, granulocyte colony-stimulating factor and macrophage colony-stimulating factor [IL-8/CXCL8 and CCL5]) and agents that promote a Th1-type response (interferon [IFN]-γ and IL-12) and a Th2 response (IL-6, IL-4 and IL-10). However, the predominant lactogenic cytokines (TGF-β1 and IL-4) are involved in strengthening the Th2-type response and reducing the immune response in newborns (Siegrist, 2001; Morein et al., 2007; Nguyen et al., 2007; Tizard, 2013).

Humoral and cellular immunity acquired from colostrum ingestion can modify the immune responses in newborns. However, there are significant differences in the action of passively transferred IgG and IgA antibodies. IgG antibodies are present in the newborn's bloodstream in small quantities that decline with time, and they could be responsible for suppressing the active immune responses that vaccines, mainly administered by the parenteral route, aim to generate in piglets (Chappuis, 1998; Endsley et al., 2003). Conversely, IgA antibodies are supplied continuously in the mother's milk and protect the gastrointestinal tract against pathogens without having any effect on the active immune response. In all mammalian species, maternal antibodies significantly reduce or eliminate antibody production following vaccination of the newborn progeny mainly for vaccines administered by the parenteral route. It is worth noting that this effect could be observed using common vaccines (live attenuated, inactivated and subunit vaccine) and it could reduce or eliminate protection against the disease (Hodgins and Shewen, 2012). The degree of the blocking effect of maternal antibodies is related to the titres of antibodies present in the newborn after colostrum consumption as the highest antibody titres persist for longer than the lowest antibody titres because their half-life is constant. Ultimately, maternal antibody titres decrease because the antibodies are metabolised and thus no longer offer protection after a period of time that is variable depending on the pathogen. However, in both humans and animals, even titres of maternal antibodies that are too low to be protective can still inhibit the response to vaccination against infectious diseases (Tizard, 2013). The antibody level at the moment of vaccination crucially affects the humoral immune response in young piglets. This effect could be observed when porcine circovirus type 2 (PCV2) vaccine was administered to piglets with variable titres of maternal antibodies to this virus. Animals with maternal antibody titres of 8 (using the immunoperoxidase monolayer assay technique) showed neither an increase nor a decrease in their antibody titres following vaccination against PCV2. However, PCV2 antibody titres of 4 showed consistently an increase, and levels in animals with PCV2 titres of over 10 showed a decrease, following PCV2 vaccination (Fraile et al., 2012a). In the case of Aujeszky's disease, various studies have demonstrated that vaccinating against this virus with live attenuated vaccines induced specific immune responses in newborn piglets, even in the presence of maternal antibodies. However, the size of the response depended on the quantity of antibodies present at the moment of vaccination, thus demonstrating the existence of interference from maternal antibodies to Aujeszky's disease virus. Clearly, these antibodies hinder optimum protection from vaccination (Pomorska-Mol et al., 2010a,b).

The interference with the development of humoral immunity when maternal antibodies are present involves a series of possible mechanisms, including some that neutralise live attenuated vaccines and others that involve feedback on antibody production (Kim et al., 2011). Maternal antibodies can neutralise an attenuated vaccine virus, which needs to reproduce itself to generate an immune response. This effect reduces the amount of viral antigen below a certain (undefined) threshold and interferes with the development of satisfactory immune recognition (Siegrist, 2001; Wood and Siegrist, 2011). Maternal antibodies suppress replication not only in live attenuated vaccines but also in subunit vaccines and vector vaccines (Klinkenberg et al., 2002). Non-neutralising antibodies also interfere with the development of immunity generated by live attenuated vaccines (Niewiesk, 2014; Tizard, 2013).

Regulation of antibody feedback is a phenomenon whereby injecting an antigen and a monoclonal antibody specific to that antigen at the same time reduces or inhibits the generation of antibodies specific to that antigen (Wiersma et al., 1989). This suppression increases with the quantity and affinity of the particular antibody and is independent of the immunoglobulin isotype. This phenomenon can be explained by two principal mechanisms: masking of the epitope and B cell regulation by a receptor called FcγRIIB. This masking of epitope theory states that antibodies can cover critical epitopes and hide them from recognition by B cells if there is enough antibody (Siegrist, 2003). Initially, it was thought that all the epitopes on an antigen were covered by antibodies to inhibit recognition by B cells (epitope-specific suppression). However, it has been demonstrated that a monoclonal antibody to an epitope can suppress B cell recognition of the entire antigen (non-epitope-specific suppression). Even low levels of antibodies are known to be able to inhibit antibody production, and the role of epitope masking in suppressing antibody production has always been the subject of debate. On the other hand, B cells express the FcγRIIB receptor on their surface, and this can bind to maternal antibodies at the constant domain of this shared region (Heyman, 2000; Karlsson et al., 2001; Getahun and Heyman, 2009). This receptor is in close proximity to the B-cell receptor (BCR) where the IgG specific to the BCR is just bound. This proximity inhibits activation of the B cells specific to the antigen because it is bound at the IgG-BCR complex and to the maternal IgG bounded to FcγRIIB receptor at the same time. This mechanism has been demonstrated in genetically modified mice that do not have FcγRI, FcγRIIB or FcγRIII. These animals have a stronger acquired humoral response than animals that are not genetically modified, which suggests the absence of a feedback mechanism to regulate B cell responses and antibody titres (Hjelm et al., 2006).

The immaturity of the neonatal immune system has already been described, and this is clearly apparent in the adaptive cellular immune response. The acquisition of a passive adaptive cellular immune response seems to be important for the newborn to face various challenges and may have a significant effect on the outcome of certain types of infection (Hedegaard and Hedegaard, 2016). However, maternal humoral and cellular immunity seems to interfere with the newborn's ability to develop this type of response. The neonatal cellular immune response does not seem to be affected by a high level of maternal humoral immunity, so maternal antibodies only have a minimal effect on T cell response, as seen with the Th1-type immune response generated in newborn babies after vaccination against the tuberculosis bacillus. This is believed to be because CD4 T cell responses are not inhibited, unlike the effect on B cells that has already been discussed (Jones et al., 2014). Finally, the cellular response in the newborn could in theory be inhibited by maternal cellular immunity, but in practice, it is quite rare (Endsley et al., 2003). In this respect, the cellular immune response to *M. hyo* in vaccinated piglets originating from sows vaccinated against this pathogen was very similar to those coming from non-vaccinated sows, when measured using two different methods (Bandrick et al., 2014a,b).

7 Strategies to overcome the interference with maternally derived acquired immunity

Despite the immaturity of the immune system in very young animals, an adequate immune response can be developed after vaccination because the immune system in these animals is sufficiently developed to generate, over a period of time, antibodies and memory T and B cells in response to appropriate combinations of vaccine antigens and adjuvants. In fact, many vaccines are designed to be administered to very young animals. A successful vaccine formulation should be able to evade or overcome the functional peculiarities of the neonatal period (Siegrist et al., 1998; Morris and Surendran, 2016). Research in this field is centred on the modulation of the neonatal innate immune responses to overcome the natural bias towards Th2 response adaptive immunity by incorporating safe but potent adjuvants into vaccines. Morris and Surendran (2016) demonstrated that a synergistic stimulation of multiple TLR by incorporating well-defined agonist-adjuvant combinations into vaccines is a promising strategy to ensure a protective vaccine response in neonates. In fact, adjuvants are a critical part of pharmaceutical companies' know-how and they are usually protected under patent (Leclerc, 2003; Morein et al., 2007; Polewicz et al., 2013). To test the efficacy of a vaccine in the presence of maternal antibodies, (a) maternal antibodies have to be high at the moment of vaccination, (b) the immunological response should be measured by the level of neutralising antibodies and (c) protection should be measured by clinical efficacy, as indicated by the absence of clinical signs, a significant reduction in viral/bacterial loads and/or reduction of severity of lesions associated to the pathogen. This is a requirement for pharmaceutical companies wishing to develop vaccines for very young animals. Thus, the summary of product characteristics (SPC) must be revised in detail because the potential interference of piglet vaccination with maternally derived immunity is described in the SPC because it is compulsory to be studied during the registration procedure of any vaccine and it is specific for each combination of pathogen and vaccine. All the requirements are clearly established in the European legislation in a concept paper on the provision of field efficacy studies in support of marketing authorisation applications for immunological veterinary medicinal products and on indications for veterinary vaccines (https://www.ema.europa.eu/en/documents/scientific-guideline/draft-concept-paper-provision-field-efficacy-studies-support-marketing-authorisation-applications_en.pdf 2021).

Various strategies to minimise the interference of maternal antibodies can be considered in the development of vaccines. Thus, it is known that inducing type I IFN can strengthen the B cell response in the presence of maternal antibodies via their bond with a surface receptor that stimulates antibody production, even when this pathway has previously been inhibited by maternal antibodies bound to the FcγRIIB receptor. Other approaches to developing an optimal vaccine for young piglets could be the continuous expression of an antigen even in the presence of maternal antibodies to obtain an antibody response through repeated exposure to the antigen. Thus, if a vector system can be designed that continuously expresses the antigen, the B cell responses should be stimulated when the maternal antibody titres have sufficiently reduced. Another strategy involves altering the antigen so that it continues to be immunogenic and does not inhibit antibody production by the B cells when binding to the FcγRIIB receptor (Lyubchenko et al., 2005).

Interference with maternal antibodies can be also minimised in piglets if the vaccine is administered in an anatomical location with a low IgG titre. For example, if a specific antibody isotype is found in respiratory, ocular fluids and in saliva at lower concentrations than in the blood, administering a vaccine via the mucosal route can minimise the interference (Griebel, 2009). Various authors have demonstrated that an intranasal Aujeszky's disease vaccine is more effective than a parenteral vaccine in the presence of maternal antibodies

(Salinas-Zacarias et al., 2020). Thus, early protection in susceptible animals could be induced using this immunization protocol, opening the possibility for its application against other viral pathogens of pigs and for translational studies in humans. In the case of the influenza virus, Pyo et al. (2015) also demonstrated that intranasal vaccination with attenuated live vaccines could be a good approach to overcome maternal immunity in pigs. Moreover, vector systems have also been used successfully in the presence of maternal antibodies in animal models by using several routes of administration for classical swine fever virus (Rangelova et al., 2012). Another strategy to overcome interference from maternal antibodies involves the development of DNA vaccines. Various authors have demonstrated that a single vaccination in 8-week-old piglets with a vaccine containing DNA fragments of the Aujeszky's disease virus, namely its glycoproteins gB, gC and gD, triggers a humoral response. It confirms that DNA vaccines can overcome interference from maternal antibodies (Fischer et al., 2003). A memory response has also been clearly demonstrated after secondary DNA vaccination at 11 weeks of age. These data suggest that DNA vaccination in animals with passive immunity successfully generates a strong response. This strategy has been also recently demonstrated for mRNA vaccines (Willis et al., 2020). Finally, another approach has been also proposed in the case of avian species. Thus, the idea is to vaccinate mothers and their progeny with different neutralising epitopes of the same pathogen. This will allow an effective immune response of the progeny towards neutralising epitopes while retaining maternally derived acquired immunity (MDA) protection until high levels of self-antibodies are produced. This concept is valid for various avian viruses that express two neutralizing proteins or have numerous neutralizing epitopes on the same protein, for example, the Newcastle disease virus. These may be used as protein subunit vaccines or live vaccines carried by a vector. This vaccination concept may overcome the gap in protection that occurs when MDAs decrease while self-immunity is still partial, to provide continuous protection at a young age. This approach could be also useful for swine diseases, but it is necessary to know different neutralising antibodies of the same pathogen (Pitcovski et al., 2017).

8 Conclusion and future trends

Preventive medicine programs in porcine health management are focused on minimising the negative effects of diseases on pig production. In the case of neonatal diseases, the acquisition of MDA is critical to providing enough humoral and cellular immunity to piglets before they can develop their own immune response against common pathogens. The acquisition of MDA is provided through the ingestion of enough colostrum during the first 24 h of life in piglets. To achieve this goal, good management procedures must be respected during the farrowing and postpartum periods. Moreover, the colostrum could be enriched with higher levels of immunoglobulins and immune cells if the sows are vaccinated against the main pathogens affecting piglets during the neonatal period. Finally, the development of a quick immune acquired response in piglets is a desirable goal in controlling many swine diseases during the rearing period. In this case, the immaturity of the neonatal immune system during the first 4 weeks of life and the presence of MDA could hinder the immune response generated in piglets with vaccines administered during the neonatal period. Thus, strategies to overcome MDA and the immaturity of the piglet are a necessary future trend in swine vaccinology that could be based on the use of different combinations of adjuvants, the use of different routes of vaccine administration and the development of new vaccine approaches for piglets.

9 Where to look for further information

9.1 Further reading

- Murphy, K., Travers, P. and Walport, M. (2008). *Janweys´s Immunobiology* (7th edition). Garland Science, New York.
- Tizard, I. (2016). Veterinary Immunology (10th edition). Elsevier.
- Articles in the journal Veterinary immunology and immunopathology: https://www.journals.elsevier.com/veterinary-immunology-and-immunopathology.

10 References

Bandrick, M., Theis, K. and Molitor, T. W. (2014a). Maternal immunity enhances Mycoplasma hyopneumoniae vaccination induced cell-mediated immune responses in piglets. *BMC Vet. Res.* 10:124.

Bandrick, M., Ariza-Nieto, C., Baidoo, S. K. and Molitor, T. W. (2014b). Colostral antibody-mediated and cell-mediated immunity contributes to innate and antigen-specific immunity in piglets. *Dev. Comp. Immunol.* 43(1):114–120.

Bandrick, M., Pieters, M., Pijoan, C., Baidoo, S. K. and Molitor, T. W. (2011). Effect of cross-fostering on transfer of maternal immunity to Mycoplasma hyopneumoniae to piglets. *Vet. Rec.* 168(4):100. doi: 10.1136/vr.c6163.

Bandrick, M., Pieters, M., Pijoan, C. and Molitor, T. W. (2008). Passive transfer of maternal Mycoplasma hyopneumoniae-specific cellular immunity to piglets. *Clin. Vaccine Immunol.* 15(3):540–543.

Barrios, C., Brawand, P., Berney, M., Brandt, C., Lambert, P. H. and Siegrist, C. A. (1996). Neonatal and early life immune responses to various forms of vaccine antigens qualitatively differ from adult responses: Predominance of a Th2-biased pattern which persists after adult boosting. *Eur. J. Immunol.* 26(7):1489–1496.

Blecha, F. (2001). Immunomodulators for prevention and treatment of infectious diseases in food-producing animals. *Vet. Clin. North Am. Food Anim. Pract.* 17(3):621-633.

Bourne, F. J. and Curtis, J. (1973). The transfer of immunoglobins IgG, IgA and IgM from serum to colostrum and milk in the sow. *Immunology* 24(1):157–162.

Brandtzaeg, P. (2010). The mucosal immune system and its integration with the mammary glands. *J. Pediatr.* 156(2) (Suppl.):S8–S15.

Butler, J. E., Francis, D. H., Freeling, J., Weber, P. and Krieg, A. M. (2005). Antibody repertoire development in fetal and newborn piglets. IX. Three pathogen-associated molecular patterns act synergistically to allow germfree piglets to respond to type 2 thymus-independent and thymus-dependent antigens. *J. Immunol.* 175(10):6772–6785.

Butler, J. E., Lager, K. M., Splichal, I., Francis, D., Kacskovics, I., Sinkora, M. (2009). The piglet as a model for B cell and immune system development. *Vet. Immunol. Immunopathol.* 128(1–3):147–170.

Butler, J. E., Sinkora, M., Wertz, N., Holtmeier, W. and Lemke, C. D. (2006). Development of the neonatal B and T cell repertoire in swine: Implications for comparative and veterinary immunology. *Vet. Res.* 37(3):417–441.

Butler, J. E., Sun, J., Weber, P., Navarro, P. and Francis, D. (2000). Antibody repertoire development in fetal and newborn piglets. III. Colonization of the gastrointestinal tract selectively diversifies the preimmune repertoire in mucosal lymphoid tissues. *Immunology* 100(1):119–130.

Butler, J. E., Weber, P., Sinkora, M., Baker, D., Schoenherr, A., Mayer, B. and Francis, D. (2002). Antibody repertoire development in fetal and newborn piglets. VIII. Colonization is required for newborn piglets to make serum antibodies to T-dependent and type 2 T-independent antigens. *J. Immunol.* 169(12):6822–6830.

Cabrera, R. A., Lin, X., Campbell, J. M., Moeser, A. J. and Odle, J. (2012). Influence of birth order, birth weight, colostrum and serum immunoglobulin G on neonatal piglet survival. *J. Anim. Sci. Biotechnol.* 3(1):42.

Cepica, A. and Derbyshire, J. B. (1984). Antibody-dependent and spontaneous cell-mediated cytoxicity against transmissible gastroenteritis virus infected cells by lymphocytes from sows fetuses and neonatal piglets. *Can. J. Comp. Med.* 48(3):258–261.

Chappuis, G. (1998). Neonatal immunity and immunisation in early age: Lessons from veterinary medicine. *Vaccine* 16(14–15):1468–1472.

Decaluwé, R., Maes, D., Wuyts, B., Cools, A., Piepers, S. and Janssens, G. P. J. (2014). Piglets' colostrum intake associates with daily weight gain and survival until weaning. *Livest. Sci.* 162:185–192.

Declerck, I., Dewulf, J., Piepers, S., Decaluwé, R. and Maes, D. (2015). Sow and litter factors influencing colostrum yield and nutritional composition. *J. Anim. Sci.* 93(3):1309–1317.

Declerck, I., Dewulf, J., Sarrazin, S. and Maes, D. (2016). Long-term effects of colostrum intake in piglet mortality and performance. *J. Anim. Sci.* 94(4):1633–1643.

Devillers, N., Le Dividich, J. and Prunier, A. (2011). Influence of colostrum intake on piglet survival and immunity. *Animal* 5(10):1605–1612.

Endsley, J. J., Roth, J. A., Ridpath, J. and Neill, J. (2003). Maternal antibody blocks humoral but not T cell responses to BVDV. *Biologicals* 31(2):123–151.

Farmer, C., Fisette, K., Robert, S., Quesnel, H. and Laforest, J. P. (2004). Use of recorded nursing grunts during lactation in two breeds of sows. II effects on sow performance and mammary development. *Can. J. Anim. Sci.* 84(4):581–587.

Ferrari, C. V., Sbardella, P. E., Bernardi, M. L., Coutinho, M. L., Vaz, I. S., Wentz, I. and Bortolozzo, F. P. (2014). Effect of birth weight and colostrum intake on mortality and performance of piglets after cross-fostering in sows of different parities. *Prev. Vet. Med.* 114(3-4):259-266.

Firth, M. A., Shewen, P. E. and Hodgins, D. C. (2005). Passive and active components of neonatal innate immune defenses. *Anim. Health Res. Rev.* 6(2):143-158.

Fischer, L., Barzu, S., Andreoni, C., Buisson, N., Brun, A. and Audonnet, J. C. (2003). DNA vaccination of neonate piglets in the face of maternal immunity induces humoral memory and protection against a virulent pseudorabies virus challenge. *Vaccine* 21(15):1732-1741.

Fraile, L., Sibila, M., Nofrarías, M., López-Jimenez, R., Huerta, E., Llorens, A., López-Soria, S., Pérez, D. and Segalés, J. (2012a). Effect of sow and piglet porcine circovirus type 2 (PCV2) vaccination on piglet mortality, viraemia, antibody titre and production parameters. *Vet. Microbiol.* 161(1-2):229-234.

Fraile, L., Grau-Roma, L., Sarasola, P., Sinovas, N., Nofrarías, M., López-Jimenez, R., López-Soria, S., Sibila, M. and Segalés, J. (2012b). Inactivated PCV2 one shot vaccine applied in 3-week-old piglets: Improvement of production parameters and interaction with maternally derived immunity. *Vaccine* 30(11):1986-1992.

Getahun, A. and Heyman, B. (2009). Studies on the mechanism by which antigen-specific IgG suppresses primary antibody responses: Evidence for epitope masking and decreased localization of antigen in the spleen. *Scand. J. Immunol.* 70(3):277-287.

Godden, S. (2008). Colostrum management for dairy calves. *Vet. Clin. North Am. Food Anim. Pract.* 24(1):19-39.

Griebel, P. J. (2009). Mucosal vaccination of the newborn: An unrealized opportunity. *Expert Rev. Vaccines* 8(1):1-3.

Hasan, S., Orro, T., Valros, A., Junnikkala, S., Peltoniemi, O. and Oliviero, C. (2019). Factors affecting sow colostrum yield and composition, and their impact on piglet growth and health. *Livest. Sci.* 227:60-67.

Hedegaard, C. J. and Hedegaard, P. M. (2016). Passive immunisation, an old idea revisited: Basic principles and application to modern animal production systems. *Vet. Immunol. Immunopathol.* 174:50-63.

Heyman, B. (2000). Regulation of antibody responses via antibodies, complement, and Fc receptors. *Annu. Rev. Immunol.* 18:709-737.

Hjelm, F., Carlsson, F., Getahun, A. and Herman, B. (2006). Antibody-mediated regulation of the immune response. *Scand. J. Immunol.* 64(3):177-184.

Hodgins, D. C. and Shewen, P. E. (2012). Vaccination of neonates: Problem and issues. *Vaccine* 30(9):1541-1559.

Holt, P. G. and Jones, C. A. (2000). The development of the immune system during pregnancy and early life. *Allergy* 55(8):688-697.

Huang, Y., Olson, D. J., Gordon, J. R., Middleton, D. M. and Simko, E. (2012). Plasma protein profiles of neonatal pigs before and after suckling. *Can. J. Vet. Res.* 76(1):1-7.

Hurley, W. L. and Theil, P. K. (2011). Perspectives on immunoglobulins in colostrum and milk. *Nutrients* 3(4):442-474.

Jones, C., Pollock, L., Barnett, S. M., Battersby, A. and Kampmann, B. (2014). The relationship between concentration of specific antibody at birth and subsequent response to primary immunization. *Vaccine* 32(8):996-1000.

Juul-Madsen, H. R., Jensen, K. H., Nielsen, J. and Damgaard, B. M. (2010). Ontogeny and characterization of blood leukocyte subsets and serum proteins in piglets before and after weaning. *Vet. Immunol. Immunopathol.* 133(2-4):95-108.

Karlsson, M. C., Getahun, A. and Heyman, B. (2001). FcgammaRIIB in IgG-mediated suppression of antibody responses: Different impact in vivo and in vitro. *J. Immunol.* 167(10):5558–5864.

Kim, D., Huey, D., Oglesbee, M. and Niewiesk, S. (2011). Insights into the regulatory mechanism controlling the inhibition of vaccine-induced seroconversion by maternal antibodies. *Blood* 117(23):6143–6151.

Klinkenberg, D., Moormann, R. J., De Smit, A. J., Bouma, A. and De Jong, M. C. (2002). Influence of maternal antibodies on efficacy of a subunit vaccine: Transmission of classical swine fever virus between pigs vaccinated at 2 weeks of age. *Vaccine* 20(23–24):3005–3013.

Klobasa, F., Werhahn, E. and Butler, J. E. (1981). Regulation of humoral immunity in the piglet by immunoglobulins of maternal origin. *Res. Vet. Sci.* 31(2):195–206.

Klobasa, F., Werhahn,E. and Butler, J. E. (1987). Composition of sow milk during lactation. *J. Anim. Sci.* 64(5):1458–1466.

Klopfenstein, C., Farmer, C. and Martineau, G. P. (2006). Diseases of the mammary glands and lactation problems. In: Straw, B. E., Zimmerman, J. J., D´Allaire, S. and Taylor, D. J. (Eds.), *Diseases of Swine* (9th edn.). Ames, IA: Iowa State University Press, pp. 833–860.

Labroue, F., Caugant, A., Ligonesche, B. and Gaudré, D. (2001). Etude de l´evolution des tétines d´apparence douteuse chez le cochette au cours de sa carrière. *J de La Recherche Porcine. En: France* (33). 143–150.

Langel, S. N., Paim, F. C., Lager, K. M., Vlasova, A. N. and Saif, L. J. (2016). Lactogenic immunity and vaccines for porcine epidemic diarrhea virus (PEDV): Historical and current concepts. *Virus Res.* 226:93–107.

Langel, S. N., Wang, Q., Vlasova, A. N. and Saif, L. J. (2020). Host factors affecting generation of immunity against porcine epidemic diarrhea virus in pregnant and lactating swine and passive protection of neonates. *Pathogens* 9(2):130.

Lecce, J. G. and Morgan, D. O. (1962). Effect of dietary regimen on cessation of intestinal absorption of large molecules (closure) in the neonatal pig and lamb. *J. Nutr.* 78:263–268.

Leclerc, C. (2003). New approaches in vaccine development. *Comp. Immunol. Microbiol. Infect. Dis.* 26(5–6):329–341.

Le Dividich, J., Rooke, J. A. and Herpin, P. (2005). Nutritional and immunological importance of colostrum for the newborn pig. *J. Agric. Sci.* 143(6):469–485.

Le Jan, C. (1996). Cellular components of mammary secretions and neonatal immunity: A review. *Vet. Res.* 27(4–5):403–417.

Lyubchenko, T., Dal Porto, J., Cambier, J. C. and Holers, V. M. (2005). Coligation of the B cell receptor with complement receptor type 2 (CR2/CD21) using its natural ligand C3dg: Activation without engagement of an inhibitory signaling pathway. *J. Immunol.* 174(6):3264–3272.

Markowska-Daniel, I., Pomorska-Mol, M. and Pejsak, Z. (2011). The influence of age and maternal antibodies on the postvaccinal response against swine influenza viruses in pigs. *Vet. Immunol. Immunopathol.* 142(1–2):81–86.

Martineau, G. P., Farmer, C. and Peltoniemi, O. (2012). Mammary system. In: Zimmerman, J. J., Karriker, L. A., Ramirez, A., Schwartz, K. J. and Stevenson, G. W. (Eds.), *Diseases of Swine* (10th edition). West Sussex, UK: Wiley-Blackwell, pp. 270–293.

Matías, J., Berzosa, M., Pastor, Y., Irache, J. M. and Gamazo, C. (2017). Maternal vaccination. Immunization of sows during pregnancy against ETEC infections. *Vaccines* 5(4):48.

Moog, F. (1979). Endocrine influences on the functional differentiation of the small intestine. *J. Anim. Sci.* 49(1):239–249.

Morein, B., Blomqvist, G. and Hu, K. (2007). Immune responsiveness in the neonatal period. *J. Comp. Pathol.* 137 (Suppl. 1):S27–S31.

Morris, M. C. and Surendran, N. (2016). Neonatal vaccination: Challenges and intervention strategies. *Neonatology* 109(3):161–169.

Nechvatalova, K., Kudlackova, H., Leva, L., Babickova, K. and Faldyna, M. (2011). Transfer of humoral and cell-mediated immunity via colostrum in pigs. *Vet. Immunol. Immunopathol.* 142(1-2):95–100.

Neville, M. C., McFadden, T. B. and Forsyth, I. (2002). Hormonal regulation of mammary differentiation and milk secretion. *J. Mammary Gland Biol. Neoplasia* 7(1):49–66.

Nguyen, T. V., Yuan, L., Azevedo, M. S. P., Jeong, K. I., Gonzalez, A. M. and Saif, L. J. (2007). Transfer of maternal cytokines to suckling piglets: In vivo and in vitro models with implications for immunomodulation of neonatal immunity. *Vet. Immunol. Immunopathol.* 117(3-4):236–248.

Niewiesk, S. (2014). Maternal antibodies: Clinical significance, mechanism of interference with immune responses, and possible vaccination strategies. *Front. Immunol.* 5:446.

Papadopoulos, G. A., Vanderhaeghe, C., Janssens, G. P. J., Dewulf, J. and Maes, D. G. (2010). Risk factors associated with postpartum dysgalactia syndrome in sows. *Vet. J.* 184(2):167–171.

Parreño, V., Hodgins, D. C., de Arriba, L., Kang, S. Y., Yuan, L., Ward, L. A., To, T. L. and Saif, L. J. (1999). Serum and intestinal isotype antibody responses to Wa human rotavirus in gnotobiotic pigs are modulated by maternal antibodies. *J. Gen. Virol.* 80(6):1417–1428.

Pitcovski, J., Pitcovski, E., Goldenberg, D. and Shahar, E. (2017). Pair-epitopes vaccination: Enabling offspring vaccination in the presence of maternal antibodies. *Avian Pathol.* 46(6):581–584.

Polewicz, M., Gracia, A., Garlapati, S., Van Kessel, J., Strom, S., Halperin, S. A., Hancock, R. E., Potter, A. A., Babiuk, L. A. and Gerdts, V. (2013). Novel vaccine formulations against pertussis offer earlier onset of immunity and provide protection in the presence of maternal antibodies. *Vaccine* 31(31):3148–3155.

Pomorska-Mol, M. and Markowska-Daniel, I. (2010a). Interferon-gamma secretion and proliferative responses of peripheral blood mononuclear cells after vaccination of pigs against Aujeszky's disease in the presence of maternal immunity. *FEMS Immunol. Med. Microbiol.* 58(3):405–411.

Pomorska-Mol, M., Markowska-Daniel, I. and Pejsak, Z. (2010b). Evaluation of humoral and antigen-specific T-cell responses after vaccination of pigs against pseudorabies in the presence of maternal antibodies. *Vet. Microbiol.* 144(3-4):450–454.

Poonsuk, K., Giménez-Lirola, L. G., Zhang, J., Arruda, P., Chen, Q., Correa da Silva Carrion, L., Magtoto, R., Pineyro, P., Sarmento, L., Wang, C., Sun, Y. (2016a). Does circulating antibody play a role in the protection of piglets against porcine epidemic diarrhea virus? *PLoS ONE* 11(4):e0153041.

Poonsuk, K., Zhang, J., Chen, Q., Gonzalez, W., da Silva Carrion, L. C., Sun, Y., Ji, J., Wang, C., Main, R., Zimmerman, J. and Giménez-Lirola, L. (2016b). Quantifying the effect of lactogenic antibody on porcine epidemic diarrhea virus infection in neonatal piglets. *Vet. Microbiol.* 197:83–92.

Poonsuk, K. and Zimmerman, J. (2018). Historical and contemporary aspects of maternal immunity in swine. *Anim. Health Res. Rev.* 19(1):31–45.

Pyo, H. M., Hlasny, M. and Zhou, Y. (2015). Influence of maternally-derived antibodies on live attenuated influenza vaccine efficacy in pigs.*Vaccine* 33(31):3667-3672.

Quesnel, H. (2011). Colostrum production by sows: Variability of colostrum yield and immunoglobulin G concentrations. *Animal* 5(10):1546-1553.

Rangelova, D., Nielsen, J., Strandbygaard, B., Koenen, F., Blome, S. and Uttenthal, A. (2012). Efficacy of marker vaccine candidate CP7_E2alf in piglets with maternally derived C-strain antibodies. *Vaccine* 30(45):6376-6381.

Reber, A. J., Lockwood, A., Hippen, A. R. and Hurley, D. J. (2006). Colostrum induced phenotypic and trafficking changes in maternal mononuclear cells in a peripheral blood leukocyte model for study of leukocyte transfer to the neonatal calf. *Vet. Immunol. Immunopathol.* 109(1-2):139-150.

Roopenian, D. C. and Akilesh, S. (2007). FcRn: The neonatal Fc receptor comes of age. *Nat. Rev. Immunol.* 7(9):715-725.

Salmon, H., Berri, M., Gerdts, V. and Meurens, F. (2009). Humoral and cellular factors of maternal immunity in swine. *Dev. Comp. Immunol.* 33(3):384-393.

Salinas-Zacarias, I., Guzman-Bautista , E. R., Ramírez-Estudillo, M. D. C., Chacón-Salinas, R. and Vega-López, M. A. (2020). Mucosal and systemic immune responses to Aujeszky's disease virus (ADV) in early vaccinated piglets. *Comp. Immunol. Microbiol. Infect. Dis.* 68:101400.

Schnulle, P. M. and Hurley, W. L. (2003). Sequence and expression of the FcRn in the porcine mammary gland. *Vet. Immunol. Immunopathol.* 91(3-4):227-231.

Schollenberger, A., Degorski, A., Frymus, T. and Schollenberger, A. (1986). Cells of sow mammary secretions. I. Morphology and differential counts during lactation. *Trans. Emerge. Dis.* 33(1):31-38.

Schoos, A., De Spiegelaere, W., Cools, A., Pardon, B., Van Audenhove, E., Bernaerdt, E., Janssens, G. P. J. and Maes, D. (2021). Evaluation of the agreement between Brix refractometry and serum immunoglobulin concentration in neonatal piglets. *Animal* 15(1):100041.

Segalés, J., Urniza, A., Alegre, A., Bru, T., Crisci, E., Nofrarías, M., López-Soria, S., Balasch, M., Sibila, M., Xu, Z., Chu, H. J., Fraile, L. and Plana-Duran, J. (2009). A genetically engineered chimeric vaccine against porcine circovirus type 2 (PCV2) improves clinical, pathological and virological outcomes in postweaning multisystemic wasting syndrome affected farms. *Vaccine* 27(52):7313-7321.

Sibila, M., Fraile, L., Ticó, G., López-Soria, S., Nofrarías, M., Huerta, E., Llorens, A., López-Jiménez, R., Pérez, D. and Segalés, J. (2013). Humoral response and colostral antibody transfer following 'one-dose' pre-mating vaccination of sows against porcine circovirus type-2. *Vet. J.* 197(3):881-883.

Siegrist, C. A. (2001). Neonatal and early life vaccinology. *Vaccine* 19(25-26):3331-3346.

Siegrist, C. A. (2003). Mechanisms by which maternal antibodies influence infant vaccine responses: Review of hypotheses and definition of main determinants. *Vaccine* 21(24):3406-3412.

Siegrist, C. A., Córdova, M., Brandt, C., Barrios, C., Berney, M., Tougne, C., Kovarik, J. and Lambert, P. H. (1998). Determinants of infant responses to vaccines in presence of maternal antibodies. *Vaccine* 16(14-15):1409-1414.

Stepanova, H., Samankova, P., Leva, L., Sinkora, J. and Faldyna, M. (2007). Early postnatal development of the immune system in piglets: The redistribution of T lymphocyte subsets. *Cell. Immunol.* 249(2):73-79.

Stirling, C. M., Charleston, B., Takamatsu, H., Claypool, S., Lencer, W., Blumberg, R. S. and Wileman, T. E. (2005). Characterization of the porcine neonatal Fc receptor - potential use for trans-epithelial protein delivery. *Immunology* 114(4):542-553.

Theil, P. K., Lauridsen, C. and Quesnel, H. (2014a). Neonatal piglet survival: Impact of sow nutrition around parturition on fetal glycogen deposition and production and composition of colostrum and transient milk. *Animal* 8(7):1021-1030.

Theil, P. K., Flummer, C., Hurley, W. L., Kristensen, N. B., Labouriau, R. L. and Sørensen, M. T. (2014b). Mechanistic model to predict colostrum intake based on deuterium oxide dilution technique data and impact of gestation and prefarrowing diets on piglet intake and sow yield of colostrum. *J. Anim. Sci.* 92(12):5507-5519.

Tizard, I. R. (2013). *Veterinary Immunology (9th edition)*. St Louis, MO: Saunders Elsevier.

Tuboly, S. and Bernáth, S. (2002). Intestinal absorption of colostral lymphoid cells in newborn animals. *Adv. Exp. Med. Biol.* 503:107-114.

Tuboly, S., Bernáth, S., Glávits, R. and Medveczky, I. (1988). Intestinal absorption of colostral lymphoid cells in newborn piglets. *Vet. Immunol. Immunopathol.* 20(1):75-85.

Vallet, J. L., Calderón-Díaz, J. A., Stalder, K. J., Phillips, C., Cushman, R. A., Miles, J. R., Rempel, L. A., Rohrer, G. A., Lents, C. A., Freking, B. A. and Nonneman, D. J. (2016). Litter-of-origin trait effects on gilt development. *J. Anim. Sci.* 94(1):96-105.

Vallet, J. L., Miles, J. R. and Rempel, L. A. (2013). A simple novel measure of passive transfer of maternal immunoglobulin is predictive of preweaning mortality in piglets. *Vet. J.* 195(1):91-97.

Vallet, J. L., Miles, J. R., Rempel, L. A., Nonneman, D. J. and Lents, C. A. (2015). Relationships between day one piglet serum immunoglobulin immunocrit and subsequent growth, puberty attainment, litter size, and lactation performance. *J. Anim. Sci.* 93(6):2722-2729.

Weaver, D. M., Tyler, J. W., VanMetre, D. C., Hostetler, D. E. and Barrington, G. M. (2000). Passive transfer of colostral immunoglobulins in calves. *J. Vet. Intern. Med.* 14(6):569-577.

Wiersma, E. J., Coulie, P. G. and Herman, B. (1989). Dual immunoregulatory effects of monoclonal IgG-antibodies: Suppression and enhancement of the antibody response. *Scand. J. Immunol.* 29(4):439-448.

Williams, P. P. (1993). Immunomodulating effects of intestinal absorbed maternal colostral leucocytes by neonatal pigs. *Can. J. Vet. Res.* 57(1):1-8.

Willis, E., Pardi, N., Parkhouse, K., Mui, B. L., Tam, Y. K., Weissman, D. and Hensley, S. E. (2020). Nucleoside-modified mRNA vaccination partially overcomes maternal antibody inhibition of de novo immune responses in mice. *Sci. Transl. Med.* 12(525):eaav5701.

Wood, N. and Siegrist, C. A. (2011). Neonatal immunization: Where do we stand? *Curr. Opin. Infect. Dis.* 24(3):190-195.

Yoshida, M., Masuda, A., Kuo, T. T., Kobayashi, K., Claypool, S. M., Takagawa, T., Kutsumi, H., Azuma, T., Lencer, W. I. and Blumberg, R. S. (2006). IgG transport across mucosal barriers by neonatal Fc receptor for IgG and mucosal immunity. *Springer Semin. Immunopathol.* 28(4):397-403.

Chapter 6

Advances in understanding gut function and immunity in pigs

Bert Devriendt, Ghent University, Belgium

1 Introduction

The gastrointestinal tract as an organ is responsible for feed intake, digestion of ingested feed, the absorption of feed-derived nutrients, minerals and water and the excretion of waste products. As such, it comes as no surprise that the gut constitutes the second largest surface area of the mammalian body. Accompanying its function of nutrient absorption, the gut is exposed daily to a plethora of foreign antigens derived from feed, the microbiota as well as enteric pathogens (Odenwald and Turner, 2017). To maintain gut homeostasis in the face of these foreign molecules, within the intestine, a complex cellular network has evolved to prevent unwanted immune responses to innocuous antigens, like feed or microbiota, while allowing swift protective responses against agents that cause infectious disease. In that context, the intestinal epithelium plays a crucial role and performs two seemingly opposing functions: absorption of nutrients and providing a barrier to protect against environmental insults (Allaire et al., 2018). In addition, the gut epithelium informs and educates intestinal immune cells to elicit responses tailored to the encountered microorganisms.

This network of intestinal epithelial cells and immune cells also responds and adapts to external cues originating from the microbiota and its metabolites. Especially during critical and stressful periods of life, like weaning, perturbations

http://dx.doi.org/10.19103/AS.2022.0103.06

of this complex network impair gut function and immunity, resulting in a growth lag. A significant factor in this post-weaning growth lag is the underdeveloped gut due to early weaning, resulting in a reduced ability of piglets to digest nutrients and raise immune responses.

As proper gut health is needed for optimal growth, insights into gut function and immunity will guide future research into dietary interventions aimed at improving gut health in pigs. In addition, these dietary interventions might also accelerate the development of the gut to ease the weaning transition as well as reduce the need for antibiotics during these critical periods of life.

This chapter will discuss the latest findings in gut function and immunity in pigs as well as provide insights into how the microbiota can impact these functions and shape the intestinal immune responses. Finally, it will discuss the recent advances in the development of novel tools to investigate the function of the gut and provide an outlook to the future of gut research in swine.

2 Architecture of the intestine

2.1 Macro-architecture of the intestine

The gut (intestine or bowel) extends from the pyloric sphincter of the stomach to the anus and is composed of different layers: the mucosa, submucosa, muscularis externa and serosa. The submucosa, muscularis externa and serosa mainly consist of muscular layers and connective tissue containing blood vessels, lymphatic vessels and nerves, supplying the mucosa with nutrients and immune cells. The mucosa is the innermost surface layer of the gastrointestinal tract and surrounds the lumen. It comprises three separate layers: the epithelium, lamina propria and muscularis mucosa, which is a thin layer of smooth muscle cells that guides the food through the gastrointestinal tract via peristalsis.

The gastrointestinal tract can be divided into the oral cavity, the esophagus, the stomach, small intestine, cecum, large intestine and rectum. The small intestine is further divided into different segments: duodenum, jejunum and ileum. These segments have different functions in nutrient digestion and absorption. In the duodenum, the first digestion of the gastric chyme takes place, using digestive enzymes delivered by the pancreas and bile acids from the liver for the emulsification of lipids. In the jejunum, a further digestion takes place and nutrients are absorbed, while in the ileum, the final digestion takes place together with the absorption of nutrients and vitamin B12 and reabsorption of bile acids (Kamaya et al., 2020). The large intestine mainly functions to reabsorb water and electrolytes, to provide a niche for the microbiota, and to absorb their metabolites resulting from the digestion of dietary fibers (Koh et al., 2016). To perform these functions, the different segments of the gut have a tailored micro-architecture and cellular composition of the intestinal epithelium.

2.2 Micro-architecture of the intestine

In both the small and large intestine, a single layer of columnar cells, the epithelium, separates the gut tissues from the external environment in the intestinal lumen. These intestinal epithelial cells are joined together by tight junctions, forming a continuous and relatively impermeable membrane. To perform its role in nutrient digestion and absorption, the intestinal epithelium in the small intestine is structured into crypts that invaginate the mesenchyme and villi that protrude into the intestinal lumen (Fig. 1). Together with the microvilli on the apical surface of the epithelium, this results in a several 100-fold increase in the absorptive surface as compared to a regular tube. In contrast to the small intestine, the large intestine lacks villi. At both sites, the intestinal epithelium is composed of many differentiated epithelial cells that are continuously renewed. For instance, in most adult mammals, the epithelial cells of the small intestinal villi have a lifespan of 3–5 days, which minimizes their exposure time to the high levels of mechanical stress and environmental hazards in the small intestinal lumen. To replenish these cells, the crypts contain crypt-based columnar stem cells, which give rise to the differentiated cells of the intestinal epithelium

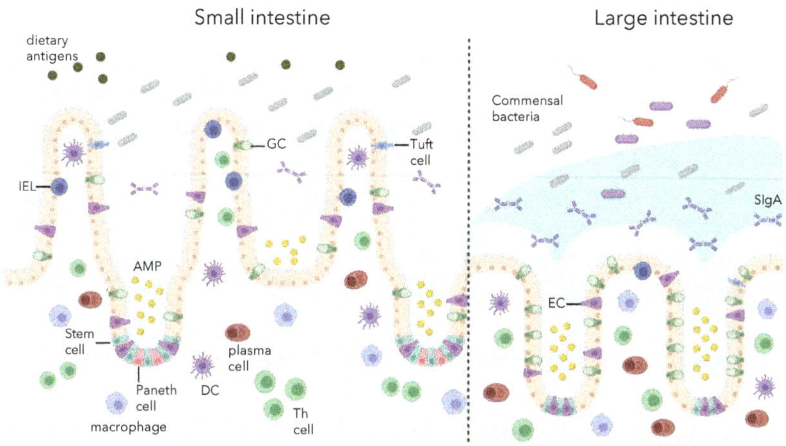

Figure 1 This graph illustrates the cellular complexity in the porcine small and large intestine. The intestine is continuously exposed to foreign antigens derived from feed and commensals. The density of dietary antigens decreases from proximal to the distal intestine, while the density of the microbiota follows the opposite direction. To cope with this, the intestine not only harbours a dense network of immune cells but also evolved a protective epithelial layer, composed of different cell types. The figure depicts the main characteristics of the small and large intestine together with an approximation of the region-specific abundance of the different epithelial cell types and immune cells. AMP, antimicrobial peptides; DC, dendritic cell; EC, enteroendocrine cell; GC, goblet cell; IEL, intraepithelial lymphocyte; SIgA, secretory IgA. Graphics were created with BioRender.com.

(Gehart and Clevers, 2019). The following sections provide an in-depth view of the function of the different epithelial cell types.

2.2.1 The intestinal epithelial stem cell

As mentioned in the previous section, the crypts in the small and large intestine contain epithelial stem cells, which express the adult stem cell marker, leucine-rich repeat-containing G-protein-coupled receptor 5 (Lgr5) (Barker et al., 2007; Wu et al., 2020). These crypt-based Lgr5+ epithelial stem cells depend on factors secreted by neighbouring stromal cells and contact-dependent cues to maintain their stemness, at least in mice (Sato et al., 2011; Gehart and Clevers, 2019). These stem cells continuously divide to give rise to daughter cells, which are pushed upwards along the crypt–villus axis. Upon leaving the crypt, these cells migrate towards the villus tip and differentiate into the two major epithelial cell lineages: the absorptive or secretory cell lineage. The former comprises enterocytes or colonocytes and microfold (M) cells, while the secretory lineage consists of mucus-producing goblet cells, a variety of hormone-secreting enteroendocrine cells, tuft cells and Paneth cells (Hooper, 2015). Recent efforts in murine models have unravelled the molecular networks controlling the maintenance and differentiation of the intestinal stem cells during steady state (Gehart and Clevers, 2019). In addition, these models have shown that nutritional interventions, inflammation and infection may rewire these signalling pathways to modulate the activity of the intestinal stem cells and hence the function of the gut (Yilmaz et al., 2012; Lindemans et al., 2015; Biton et al., 2018). If these signalling networks are conserved in swine and if similar perturbations impact intestinal stem cell activity in the pig gut remains to be investigated. Interestingly, a single-cell survey of the small intestinal epithelium in murine models revealed that each intestinal epithelial cell type comprises multiple subsets, further highlighting the previously unrecognized complexity of the gut epithelium (Haber et al., 2017). Currently, it is unknown if this complexity also exists in the porcine intestinal epithelium.

2.2.2 Absorptive intestinal epithelial cells

The absorptive intestinal epithelial cells are responsible for the final digestion of feed and absorption of nutrients, such as sugars, amino acids, lipids and vitamins via active and passive transport mechanisms. These tall, columnar-shaped cells are the most numerous cell type, constituting >90% of the intestinal epithelium. In the small intestine, these are called enterocytes, while in the large intestine, they are often referred to as colonocytes.

Enterocytes form a glycocalyx surface coat, which contains several digestive enzymes required for the degradation of polysaccharides, peptides and lipids.

These enzymes and receptors involved in nutrient absorption show a segment-specific expression. For instance, enterocytes within the distal ileum selectively express bile acid receptors, allowing reabsorption of bile acids that were not used to emulsify feed-derived lipids. Interestingly, glutamine, glutamate and aspartate are the main fuel sources for small intestinal epithelial cells, while colonocytes mainly use the short-chain fatty acid (SCFA) butyrate as an energy source (Duée et al., 1995; Roediger, 1982).

2.2.3 Goblet cells

Goblet cells are a secretory epithelial cell lineage in the small and large intestine, known for their production of mucus. Goblet cell numbers are higher in the large intestine as compared to the small intestine. The secreted mucus forms a gel-like layer over the apical surface of the epithelium, protecting it from mechanical damage and infection. Goblet cells have a typical morphology resembling a goblet and contain numerous vesicles that contain large quantities of stored mucinogen, a precursor of mucus (Hooper, 2015). In the colon, they also play an important role in maintaining immune homeostasis as sentinel cells by endocytosing and reacting to bacterial ligands in the crypts. This recognition of bacterial ligands by sentinel goblet cells promotes the mucus secretion of crypt-residing goblets cells via plumes and thereby aids in the protection of the crypt-based stem cells (Birchenough et al., 2016). Interestingly, in mice, intercrypt goblet cells in the colon produce mucus with a different glycosylation pattern than the crypt-residing goblet cells. This mucus produced by intercrypt goblet cells is important in the construction of the mucus layer in the large intestine and was more permeable for small molecules as compared to crypt mucus. This might be important for the absorption of ions and other small compounds by the epithelium, while the denser mucus in the crypts shields the crypt-based stem cell niche from potential perturbations (Nyström et al., 2021). In addition to mucus, goblet cells also secrete trefoil factors, which play a role in epithelial reconstitution upon injury, and resistin-like molecule β (RELMβ, mainly expressed in the colon), which affects T cell-mediated immunity (Huygelen et al., 2014; Bergstrom et al., 2015). RELMβ has been shown to recruit CD4+T cells to the intestine upon bacterial infection and instructs antigen-presenting cells to trigger IFNγ-producing T helper 1 (Th1) cells (Pine et al., 2018). In mice, goblet cells were also shown to acquire small molecules from the intestinal lumen and transfer these to subepithelial antigen-presenting cells, which then elicited tolerogenic responses (McDole et al., 2012).

2.2.4 Enteroendocrine cells

Enteroendocrine cells are scattered throughout the small and large intestine and during steady state constitute a minor epithelial cell population (<1%) (Sternini et al., 2008; Gonzalez et al., 2013). Nevertheless, together they form one of the largest endocrine systems of the body, producing over 30 different hormones. In general, one endocrine cell produces a single hormone, and depending on their hormone secretion profile, enteroendocrine cells are classified into different cell types: L cells (glucagon-like peptide 1), I cells (cholecystokinin), enterochromaffin (EC) cells (serotonin), X cells (ghrelin), S cells (secretin), K cells (gastric inhibitory peptide), delta cells (somatostatin) and N cells (neurotensin) (Gehart et al., 2019). However, a report on the existence of multihormonal cells in the murine gut makes this classification significantly more complicated with up to 20 different enteroendocrine cell types (Haber et al., 2017).

Enteroendocrine cells sense the contents of the intestinal lumen, mainly nutrients, via receptors on their apical membrane and feed that information via hormone secretion and neuropods to the nervous system to control gut physiology. Findings suggest that these receptors are directly related to those used in the taste buds (Sternini et al., 2008). Enteroendocrine cells through their secretion of hormones, like gastric inhibitory peptide, glucagon-like peptide and vasoactive intestinal peptide, play an important role in gut metabolism, intestinal motility, feed intake and appetite control (Worthington et al., 2018). As such, they are attractive targets for nutritional interventions to better control feed intake in young pigs.

2.2.5 Paneth cells

Paneth cells are a secretory epithelial cell lineage unique to the small intestine. Although their presence in pigs has long been debated due to the lack of tools and methods to detect lysozyme, advances in methodology have confirmed the presence of Paneth cells in the pig's small intestine (Van der Hee et al., 2018). Upon their differentiation, Paneth cells migrate back to the crypts and settle at the bottom of the crypts where these pyramidal-shaped cells can reside for up to 30 days (Clevers and Bevins, 2013). Within the crypts, Paneth cells nurture the epithelial stem cells by providing cell contact-dependent cues and secreting growth factors, which maintain stemness as well as sustain proliferation of the stem cells (Gehart and Clevers, 2019). As such, Paneth cells play an important role in the renewal of the intestinal epithelium.

On top of their nurturing role, Paneth cells also play an important role in protecting the crypt-based stem cells against microbial insults by secreting a variety of host defence proteins. Paneth cells contain abundant secretory granules, which contain host defence proteins, such as lysozyme, phospholipase A2 and α-defensins (Clevers and Bevins, 2013). In pigs, these α-defensins have not yet been identified (Sang and Blecha, 2009). Paneth cells continuously release these host defence proteins, forming a gradient along the crypt-villus axis. However, upon sensing microbial signals, Paneth cells further increase this output (Ayabe et al., 2000). Interestingly, transmissible gastroenteritis virus (TGEV) infects pig Paneth cells in an aminopeptidase N-dependent manner, resulting in their loss from the crypts. This in turn inhibited Lgr5+ intestinal stem cell renewal and affected their differentiation, as less absorptive enterocytes and more goblet cells were observed (Wu et al., 2020).

2.2.6 Tuft cells

Despite their discovery more than 60 years ago in rats, thanks to their unique tubulovesicular system and an apical bundle of microfilaments connected to a tuft of long and thick microvilli, the role of tuft cells within the intestinal epithelium has long been unclear, in particular, because this cell type constitutes a minor population (<1%) in the small and large intestine during steady-state conditions (Von Moltke et al., 2015; Gerbe and Jay, 2016). Tuft cells are chemosensory cells that dramatically increase in numbers upon helminth infection, indicating that these epithelial cells play an important role in the sensing and defence against parasitic helminths (Von Moltke et al., 2015; Gerbe et al., 2016). Interestingly, in murine models, deletion of the TRPM5 cation channel, which transduces signals from taste receptors that respond to bitter-, sweet- and umami-tasting substances, impaired intestinal tuft cell expansion and the clearance of helminths (Howitt et al., 2016). In pigs, infections with parasitic worms, like *Ascaris suum* and *Trichuris suis*, can result in performance losses due to a reduced feed conversion efficiency (Vlaminck et al., 2016). Similar to helminth infections in mice, *T. suis* infection in pigs resulted in increased Tuft cell numbers in the proximal colon (Myhill et al., 2018).

2.2.7 Microfold cells

Microfold (M) cells are epithelial cells specialised in the transport of macromolecules and particulate matter from the intestinal lumen to the subepithelial tissues. In pigs, M cells can be identified by cytokeratin-18 staining and are found in the follicle-associated epithelium covering the Peyer's patches (PPs) and isolated lymphoid follicles (Casteleyn et al., 2013). This follicle-associated epithelium harbours a filamentous brush border glycocalyx that lacks the membrane-associated hydrolytic enzymes characteristic of the absorptive epithelium (Neutra et al., 2001). M cells have some unique structural and functional features that facilitate the transepithelial transport of macromolecules and microorganisms. In contrast to enterocytes, M cells lack an extensive mucus layer, display a poorly organised brush border with irregular-shaped microvilli, have a thin glycocalyx and a reduced enzymatic activity. In addition, their basolateral membrane is invaginated to form intraepithelial pockets where lymphocytes and antigen-presenting cells are present

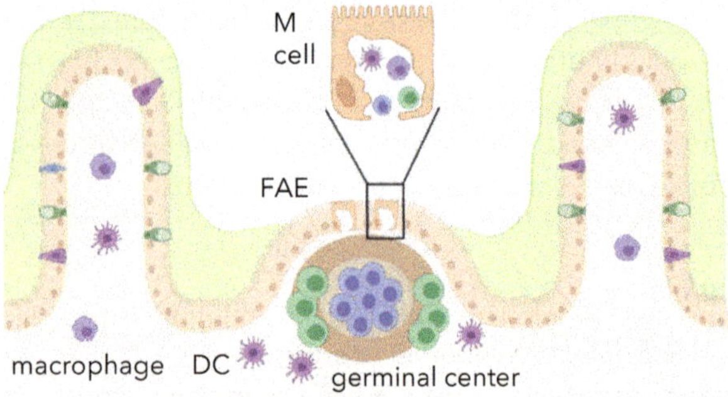

Figure 2 A schematic illustration of a Peyer's patch and microfold (M) cells. M cells are present within the follicle-associated epithelium of the porcine small intestine, which covers the follicles within the Peyer's patches. These follicles consist of a germinal centre comprising B cells (in blue) and interfollicular areas, which contain T cells (in green). M cells have an invagination of their basolateral membrane. Within this M cell pocket, antigen-presenting cells, like macrophages and dendritic cells, and lymphocytes reside. DC, dendritic cell; FAE, follicle-associated epithelium. Graphics were created with BioRender.com.

(Devriendt et al., 2012). Interestingly, M cells are also scattered throughout the absorptive epithelium, where they could transport antigens to immune cells residing in the lamina propria (Jang et al., 2004; Valpotić et al., 2010). Although M cells are important in generating gut immune responses, they also represent a weak point in the epithelial barrier and many pathogens exploit these cells as a portal of entry in the host (Fig. 2).

3 The intestinal epithelium control of gut immunity

This cellular complexity of the intestinal epithelium supports its two seemingly opposing functions. On the one hand, the epithelium needs to be permeable to allow efficient absorption of nutrients, while on the other hand, it provides a physical barrier to protect the underlying tissues against potential harmful molecules. This physical barrier is composed of closely connected neighbouring epithelial cells by tight junction complexes, the glycocalyx and the mucus layer. In addition to this physical barrier, the intestinal epithelium also takes an active role in defending the epithelial surfaces. It functions as a regulating nexus for intestinal immunity and crosstalk between bacteria and gut immune cells (Allaire et al., 2018; Nowarski et al., 2017). This occurs in several ways. First, intestinal epithelial cells were shown to sense microorganisms via a range of pattern recognition receptors (PRRs) and in response secrete a variety of host defence peptides to kill the bacteria. Second, upon recognition of microorganisms, intestinal epithelial cells can relay this information to the underlying immune cells so these can mount proper responses. Finally, gut epithelial cells transport dimeric IgA from their basolateral surface to the apical membrane, where it is released in the intestinal lumen as secretory IgA. This secretory IgA is the main antibody in the gut and plays an essential role to keep both microbiota and pathogens at bay (Devriendt et al., 2012). The following paragraphs will discuss the role of the intestinal epithelium in gut immunity in greater detail.

3.1 The gut epithelial barrier

The physical barrier formed by the gut epithelium consists of three layers: epithelial cells connected through tight junctions, the glycocalyx and the mucus layer. The tight junctions in gut epithelia are intercellular adhesion complexes, composed of occludin, zona occludens-1 and several members of the claudin family (claudin 2, 3, 4, 7, 8, 12 and 15) (Zihni et al., 2016). Tight junctions control the paracellular ion permeability of the gut epithelium, support the maintenance of cell polarity by restricting intermixing of apical and basolateral transmembrane proteins and control the transepithelial migration of immune cells. During infection, the composition of the tight junction complex can change to regulate the paracellular transport of molecules and immune cells (Dubreuil, 2017). The second layer (400–500 nm) constitutes the glycocalyx, a network of large, negatively charged, integral membrane mucin-like glycoproteins, which covers the tips of the microvilli. The glycocalyx not only contains digestive enzymes but also prevents direct contact of macromolecules and microorganisms with the membrane of the gut epithelial cells (Snoeck et al., 2005). On top of the glycocalyx lays another physical barrier, the mucus layer. The main component of the mucus layer in the small and large intestine is MUC2, secreted by goblet cells. In the small intestine, the mucus layer is a single, rather patchy layer, which varies in thickness depending on the segment, and has its smallest point in the jejunum. Changes in diet however can affect mucus secretion and the thickness of this layer (Swiech et al., 2019). The function of MUC2 within the mucus layer seems to stretch beyond providing a mere physical barrier. In the murine small intestine, MUC2 plays a role in maintaining immune homeostasis by delivering tolerogenic signals to gut dendritic cells (DCs) by activating the transcription factor β-catenin via a galectin-3-Dectin-1-FcγRIIB complex (Shan et al., 2013). Interestingly, some pathogens, such as *Escherichia coli*, can degrade MUC2 (Kumar et al., 2014). Whether this aids in tipping the balance of tolerance to immunity within the gut environment to protect against *E. coli* infections warrants further investigation.

In the large intestine, the mucus layer is thicker and in contrast to the mucus layer in the small intestine consists of an inner and an outer layer. While the inner layer is compact and essentially free of bacteria, the outer layer is more open and provides a niche for the microbiota. Interestingly, several gut bacteria, including butyrate-producing bacteria, can degrade mucins and use them as a carbon source when alternative sources in the environment are limited (Quintana-Hayashi et al., 2018). This results in extensive degradation of the colonic mucus layer, resulting in deeper penetration of this layer by the microbiota and ultimately might drive gut inflammation. Dietary changes and pathogens can alter the composition of the mucus layer in the colon (Quintana-Hayashi et al., 2018). For instance, *Brachyspira hyodysenteriae*, *Salmonella enterica* serovar Typhimurium and *Trichuris suis* trigger *de novo* synthesis of MUC5ac to increase their binding to the colonic mucus layer (Quintana-Hayashi et al., 2015; Kim et al., 2009; Kringel et al., 2006).

3.2 The interplay between intestinal epithelial cells and immune cells

The intestinal epithelium provides more than a physical barrier and is also involved in maintaining immune homeostasis by expressing various pattern recognition receptors (PRRs). Of these, Toll-like receptor (TLR) 2, TLR4 and TLR5 seem to be the most important PRRs to sense bacterial infections. In mice, the expression of these TLRs differs depending on age, gut location and cell type. For instance, TLR2 and TLR4 presence in epithelial cells was mainly restricted to the large intestine, while TLR5 was expressed by epithelial cells in the small intestine and proximal colon (Price et al., 2018). At present, it is unknown whether similar spatial patterns of epithelial TLR expression also occur in other species. The expression of PRRs allows intestinal epithelial cells to sense the presence of microorganisms and triggers a response to protect the epithelium from infection. In the case of bacterial infections, epithelial cells secrete an array of host defence proteins, such as β-defensins, cathelicidins and regenerating islet-derived III gamma (Reg3γ) (Sang and Blecha, 2009; Peng et al., 2021). Compared to other mammals, pigs have a very diverse array of host defence proteins, including 13 β-defensins and 11 cathelicidins, although not all are expressed in the gut (Sang and Blecha, 2009). This might compensate for the apparent lack of porcine α-defensins.

In addition to the secretion of host defence proteins, intestinal epithelial cells also relay information on the luminal content to the underlying immune cells through the secretion of inflammatory or tolerogenic mediators, allowing these cells to mount responses tailored to the encountered microorganism (Allaire et al., 2018) (Fig. 3). For instance, during the steady state, enterocytes produce thymic stromal lymphopoietin (TSLP) and transforming growth factor β (TGFβ), which imprint a tolerogenic phenotype on intestinal DCs (Iliev et al., 2009). During pathogen encounter, such as enterotoxigenic *E. coli,* intestinal epithelial cells secrete pro-inflammatory cytokines, such as interleukin (IL)6, and chemokines such as CCL20 and CXCL8 (or IL8) through recognition of flagella in a TLR5-dependent manner (Allaire et al., 2018; Devriendt et al.,

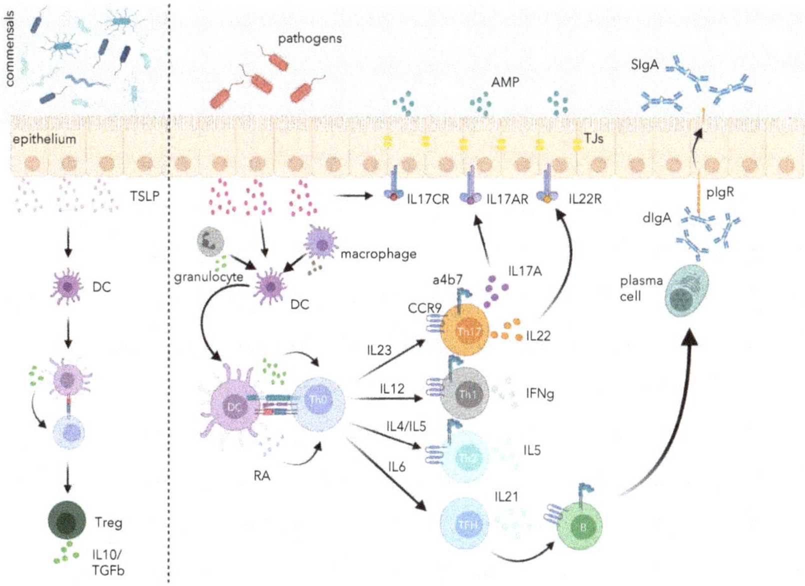

Figure 3 This graphic illustrates the crosstalk between gut epithelial cells and immune cells to maintain gut homeostasis. Under steady-state conditions, in the presence of commensals, intestinal epithelial cells secrete TSLP, which imprints a tolerogenic phenotype on gut dendritic cells, ultimately leading to the generation of commensal-specific regulatory T cells. In face of infection with pathogens, intestinal epithelial cells secrete proinflammatory cytokines, which activate innate immune cells, like dendritic cells, macrophages and granulocytes. The latter will in turn secrete pro-inflammatory cytokines, also leading to the activation and maturation of dendritic cells. These antigen-presenting cells have the unique ability to activate naïve T cells and drive their polarization to effector T cells. In addition, dendritic cells imprint in a retinoic acid-dependent manner a gut homing phenotype on these effector T cells. At the site of infection, effector T cells secrete many cytokines, like IL17A and IL22, which instruct intestinal epithelial cells to reinforce their barrier function by upregulating the expression of tight junction proteins and antimicrobial peptides. Besides these effector T cells, T_{FH} cells will activate B cells in an IL21-dependent manner. These B cells differentiate into IgA-producing plasma cells, which migrate back to the site of infection. Secreted IgA is then recognized by pIgR at the basolateral surface of the epithelial cells and transported to the apical side to release pathogen-specific secretory IgA in the intestinal lumen. AMP, antimicrobial peptide; DC, dendritic cell; dIgA, dimeric IgA; IL17AR, IL17A receptor; IL17CR, IL17C receptor; IL22R, IL22 receptor; pIgR, polymeric Ig receptor; RA, retinoic acid; T_{FH}, follicular helper T cells; Th, T helper cell; TJs, tight junctions; Treg, regulatory T cell; TSLP, thymic stromal lymphopoetin. Graphics were created with BioRender.com.

2010; Luo et al., 2019). CXCL8 is crucial for the recruitment of neutrophils, while CCL20 controls the migration of CCR6+ DCs. In addition, TLR5-mediated signalling triggered IL17C production by intestinal epithelial cells, which in turn promoted the expression of β-defensins and tight junction proteins by these cells (Luo et al., 2019). The TLR5–IL17C axis clearly plays an important role in the maintenance and fortification of the barrier function upon encountering flagellated pathogens. The secretion of these pro-inflammatory mediators by intestinal epithelial cells most likely facilitates a shift from a tolerogenic to an immune-inductive microenvironment, resulting in the proper activation of recruited immune cells at the site of infection. Upon their activation, these innate and adaptive immune cells in turn secrete IL17A and IL22, which triggers gut epithelial cells to fortify or restore their barrier function by upregulating their production of host defence proteins and tight junction proteins (Geremia and Aranciba-Carcamo, 2017; Brevi et al., 2020). During enterotoxigenic *E. coli* infection in pigs, IL17A, IL17F and IL22 transcript levels are increased in the small intestine, especially the ileal PPs. This transcriptional response is associated with an influx of IL17A+CD3+T cells in the small intestinal tissues of infected animals (Luo et al., 2015).

Another example of this feed-forward loop between epithelial cells and immune cells is the response of tuft cells to the luminal content. Tuft cells sense different luminal metabolites, such as succinate, derived from helminths as well as commensals and relay this information to immune cells through the secretion of IL25. This cytokine promotes the activation of innate immune cells and recruits T helper 2 cells, which in turn produce IL4, IL5 and IL13. These cytokines not only stimulate the expansion of eosinophils, but especially, IL13 will trigger further tuft cell expansion, induces hyperplasia of mucus-secreting goblet cells and smooth muscle hypercontractility, which are part of the so-called 'weep and sweep' response to clear parasitic helminths (Nadjsombati et al., 2018; Lei et al., 2018; Loke and Cadwell, 2018). It is tempting to speculate that nutritional interventions with molecules recognized by taste receptors on tuft cells could increase this cell population and aid in the clearance of parasitic infections. Interestingly, supplementing succinate to pigs increased IL18, IL25 and MUC2 production in the jejunum (Li et al., 2019).

Intestinal epithelial cells further support the maintenance of gut immune homeostasis by transporting dimeric IgA from their basolateral surface to the intestinal lumen (Fig. 3). Plasma cells at the basolateral surface of the epithelial cells secrete dimeric IgA, which then binds to the polymeric Ig receptor (pIgR) present at the basolateral membrane of the epithelial cells. This receptor then transports the dimeric IgA to the apical surface of the epithelium, where it is released as secretory IgA (SIgA) by proteolytic cleavage of pIgR. SIgA is composed of dimeric IgA and the secretory component, which is a part of the extracellular domain of pIgR (Corthésy, 2013). SIgA-mediated protection relies primarily on immune exclusion. SIgA has multiple complementary functions to create a barrier at the intestinal epithelium, to neutralize toxins or pathogens within epithelial cells, to keep microorganisms away from the gut epithelium and to shape the composition of the microbiota (Brandtzaeg et al., 2008).

These findings clearly indicate that the intestinal epithelium plays an important role in maintaining gut health in pigs and that intricate and complex signalling and cellular networks have evolved to balance and to fine-tune the response of the gut tissues to the microbiota and potential pathogens. The next section will focus on the role of immune cells within the gut-associated lymphoid tissues in maintaining intestinal health.

4 The gut-associated lymphoid tissues

4.1 Main lymphoid structures in the intestine

The gut constitutes the second largest surface area of the mammalian body and is exposed daily to a large variety of foreign substances. To protect that large surface area, the intestine harbours the largest immune system of the body. The mucosal immune system in the gut can in principle be divided into inductive sites, where sampled antigens stimulate antigen-presenting cells to activate naive T and B cells, and effector sites, where effector T and B cells perform their functions, such as assisting in the production of SIgA. In the small and large intestine, the inductive sites consist of the gut-associated lymphoid tissues (GALT), while the effector sites constitute the lamina propria and the surface epithelium. The GALT itself is composed of aggregated lymphoid follicles, called PPs, the lymphoglandular complex, isolated lymphoid follicles and the mesenteric lymph nodes. The presence of other GALT-like structures, such as lymphocyte-filled villi (rat, human) and cryptopatches (mouse), has not been observed in pigs (Brandtzaeg et al., 2008). Of note, the mesenteric lymph nodes have a specific, so-called inverted structure, which is also present in rhinoceros, dolphins and elephants. These inverted lymph nodes lack a larger medullary area and are preferentially composed of cortical areas and a paracortex. As a consequence, the majority of the lymphocytes exit the lymph node via the blood (Rothkötter, 2009).

In common with birds and most mammals, the pig has multiple, isolated PPs which are scattered throughout the jejunum. In adult animals, 20–30 PPs can be found in the jejunum. These jejunal PPs gradually increase in size from the proximal to the distal jejunum. Of note, based on the migration of lymphocytes within jejunal PPs, it has been suggested that more than one functional subset of PPs exists within the jejunum. Interestingly, pigs, like ruminants and dogs, have a single, large, continuous patch in the ileum. In contrast to ileal PPs in sheep, however, porcine ileal PPs are not primary lymphoid organs, as they are not a site of B cell lymphogenesis (Butler and Sinkora, 2013).

The structure of the ileal PPs differs from jejunal PPs and most likely also has a different function (Bailey, 2009; Liebler-Tenorio and Pabst, 2006). The latter is highlighted by a different immune cell composition, with a higher T/B cell ratio in the jejunal PPs as compared to ileal PPs and a predominant gene expression related to T cell differentiation in the jejunal PPs (Maroilley et al., 2018).

In the large intestine, the GALT is composed of isolated lymphoid follicles. In pigs, a patchy accumulation of lymphoid nodules is present at the ileocecal opening. Isolated lymphoid follicles on the other hand are scattered throughout the large intestine, although these are more abundant at the central colonic flexure (Liebler-Tenorio and Pabst, 2006). More than 1000 isolated lymphoid follicles are present in the adult large intestine. These isolated lymphoid follicles are sometimes organized in larger structures, consisting of several lymphoid follicles and interfollicular areas within the submucosa. These structures have been termed lymphoglandular complexes (Morfitt and Pohlenz, 1989).

4.2 Gut immunity

As mentioned in the previous section the intestinal epithelium forms the first barrier to protect the gut tissues from the external environment. This barrier is more than a physical barrier as the intestinal epithelial cells also inform gut-resident innate and adaptive immune cells on the luminal content. While innate immune cells form the second layer of defence to protect the gut tissues against pathogens, the final defence layer comprises adaptive immune cells, which are activated by intestinal antigen-presenting cells to mount pathogen-specific effector immune responses, like the production of SIgA. These antigen-presenting cells within the intestine have unique properties to enable them to educate pathogen-specific T and B cells to elicit gut immunity.

4.2.1 Innate immune cells

Just like other organs, the gut contains many tissue-resident innate immune cells, which are crucial to protect against infection with intestinal pathogens. An important immune cell population for maintaining immune homeostasis are T cells residing within the intestinal epithelium. These intraepithelial lymphocytes (IELs) are crucial for epithelial barrier integrity and provide the first line of defence against pathogens (Cheroutre et al., 2011). In humans and rodents, IELs can be further subdivided into natural and induced IELs. The latter are mainly CD8+TCRαβ+T cells that are recruited to the epithelium upon antigen exposure and reside within the epithelium as antigen-experienced effector or memory cells. This cell population increases with age together with increased antigen exposure. Natural IELs on the other hand are mainly TCRγδ+T cells which migrate to gut in an antigen-independent manner and are involved in tissue repair and lysis of malignant or infected epithelial cells (Luoma et al., 2014). Similar to their murine and human counterparts, porcine IEL numbers increase with age in the small and large intestines. Regional differences are however present as the large intestine harbours more IELs as compared to the small intestine, while small intestinal IELs are more activated than their counterparts in the colon, at least based on their expression of CD27 (Wiarda et al., 2020). Further studies should address if the abundance and activation status of these gut IELs are affected by dietary interventions, stress or disease. Another innate-like effector T cell population within the gut are mucosal-associated invariant T cells (MAITs), which are likely involved in the host defence against infectious diseases. MAITs have an invariant TCRα chain and biased use of TCRβ chains and are restricted by the highly conserved major histocompatibility complex class I (MHCI)-like molecule MR1 (Koay et al., 2019). MAITs were recently identified in pig and showed a higher relative abundance in the small and large intestine as compared to other tissues. Within the small intestine, MAITs were exclusively CD4−CD8+ (Xiao et al., 2019).

Other innate immune cells within the gut are mast cells, basophils and eosinophils. Mast cells are best known for their role in IgE-mediated immunity and their protection against helminth infections. At present, not much is known about the distribution of mast cells in the porcine gut. Some studies have reported their presence in the small intestine and cecum and showed that their numbers changed upon enrichment of feed with linseed oil or glycinin, the major storage protein of soybean (Sun et al., 2008; Boudry et al., 2009). The latter plays a role in the induction of soybean-related food allergies in pigs. In addition, increased intestinal mast cell activity is associated with stress induced by early weaning of piglets (Pohl et al., 2017). Basophils and eosinophils are also important to protect the host against helminth infections. While basophils have not been reported in the pig gut, eosinophil numbers were increased in the cecum of piglets infected with *A. suum* (Masure et al., 2013).

Upon infection of the intestine with bacteria or fungi, neutrophils are massively recruited from the blood to the site of infection. These neutrophils not only can phagocytose pathogens for intracellular killing but also are equipped to kill extracellular pathogens (Fournier and Parkos, 2012). Indeed, neutrophils are able to produce massive amounts of reactive oxygen species to destroy pathogens. In addition, they have several granules that contain a variety of host defence proteins (β-defensins, lactoferrin, cathelicidins) and enzymes (myeloperoxidase, hydrolytic enzymes, proteases and lipases) that are secreted upon contact with pathogens (Cassatella et al., 2019). On top of that, neutrophils form neutrophils extracellular traps, which are composed of nuclear contents and granule-derived anti-microbial proteins. The main function of these structures is to sequester and destroy pathogens, such as *Staphylococcus aureus* and *Pseudomonas aeruginosa* (Bréa et al., 2012; Liu et al., 2021). Emerging evidence in mice models has unravelled the role of neutrophils in tissue restoration during wound healing, even in the absence of infection (Phillipson and Kubes, 2019).

In addition to these innate immune cells, in mice and humans, another population of immune cells with important roles in maintaining immune homeostasis are the innate lymphoid cells (ILCs). Based on the cytokines they produce, these ILCs can be further subdivided into ILC1 (IFNγ), ILC2 (IL13) and ILC3 (IL22, IL17A) subsets. Surprisingly, these ILCs (except for natural killer cells, which are an ILC1 subset) still have to be identified in livestock species, including pigs.

4.2.2 Intestinal antigen-presenting cells

Another critical immune cell population within the intestine are the antigen-presenting cells residing in the gut tissues. Due to their antigen-presenting properties, these cells bridge innate and adaptive immunity. Abundant numbers of macrophages and DCs are present within the gut tissues. Gut macrophages derive from monocytes and are characterized in mice and humans by their poor migratory properties and their expression of the chemokine receptor CX3CR1. These cells play important roles not only in the induction of oral tolerance to harmless molecules and microorganisms but also to contain pathogens with their phagocytotic ability. During the steady state, CX3CR1+ macrophages can extend transepithelial dendrites through the epithelium to acquire luminal antigens and also transfer phagocytosed antigens via gap junctions to CD103+ DCs, which in turn elicit oral tolerance by triggering the polarization of naïve CD4+T cells into antigen-specific regulatory T cells (Tregs) (Mazzini et al., 2014).

DCs have the unique ability to activate naïve T cells and depending on the encountered molecules and external cues to drive the polarization of these cells into effector T cells. In most mammals, DCs can be further subdivided into conventional (cDCs) and plasmacytoid dendritic cells (pDCs). The latter are known for their production of IFNα and as such are important in mediating anti-viral responses (Lamote et al., 2017). In pigs, blood-derived pDCs can be phenotyped as CD14−CD172aloCD4+ and respond to poly I:C and CpG motifs via their expression of TLR3 and TLR9. cDCs are further divided into two subsets: cDC type 1 (cDC1) and type 2 (cDC2). Whereas the cDC1 subset is proficient at cross-presenting external antigens to CD4−CD8+ cytotoxic T cells via MHCI, cDC2 are specialized in the activation of CD4+CD8− Th cells and their differentiation into Th2 or Th17 effectors (Auray et al., 2016). In pigs, these DC subsets have been identified in blood, lung and tonsils (Auray et al., 2016, 2020; Maisonnasse et al., 2016). Although seminal work by Bimczok et al. (2006) provided valuable information on the presence of different mononuclear phagocytes within the porcine small intestine, at present, information is lacking on the abundance of the pDC, cDC1 and cDC2 subsets in the gut tissues and potential regional differences in their abundance and function. A major problem related to this issue is the lack of tools to unambiguously discriminate DCs from macrophages and other myeloid cell populations in the gut. Interestingly, dietary contaminants, such as mycotoxins, can abolish the antigen-presenting capacity of porcine small intestinal mononuclear phagocytes, leading to a reduced ability of the gut tissues to raise immune responses and to clear infections (Devriendt et al., 2009).

To activate and to drive the polarization of naïve CD4+T cells to the different effector phenotypes, DCs patrolling the tissues in an immature state will undergo a complex maturation process upon sensing danger signals (Rescigno and Di Sabatino, 2009) (Fig. 3). This maturation process entails the upregulation of surface MHCII and costimulatory molecules, such as CD40, CD80, CD86, to enable efficient antigen presentation to T cells, an enhanced antigen processing as well as a shift in their chemokine receptor profile. While immature DCs express CCR6 to migrate to the gut tissues, mature DCs downregulate CCR6 and upregulate CCR7, which is required for their migration to the mesenteric lymph nodes (Devriendt et al., 2011). Mature DCs also secrete cytokines to drive T cell differentiation, which can then control the ensuing immune responses to eradicate the encountered threat. IL12 secretion by DCs drives Th1 cell differentiation, IL23 secretion results in Th17 effector cells, while IL13 secretion aids Th2 differentiation (IL13, IL5). These T cell effector phenotypes are all present in pigs (Käser, 2021). Th1 cells are characterised by IFNγ production and play a crucial role in the protection against intracellular pathogens. Th2 cells are important for protection against extracellular pathogens, such as helminths and bacteria, by stimulating B cell activation and class switching to IgE and IgA, respectively. IL4, IL5 and IL13 are hallmark cytokines of Th2 cells. Finally, Th17 effector T cells produce IL17A, IL17F and IL22 and as such have a pivotal function in maintaining the epithelial barrier and in protecting the mucosal surfaces against extracellular bacteria, such enterotoxigenic *E. coli* and *Salmonella* typhimurium (Luo et al., 2015; Brevi et al., 2020; Schmidt et al., 2021). In the absence of danger signals, DCs will not complete their maturation process and secrete IL10 or TGFβ. This will trigger the differentiation of IL10- and TGFβ-producing FoxP3+Tregs. In the gut, DCs use retinoic acid (a vitamin A metabolite) and TGFβ to promote Treg differentiation from naïve CD4+T cells (Bakdash et al., 2015).

Besides the abovementioned CD4+CD8−T helper cells and CD4−CD8+cytotoxic T cells, also CD4+CD8+double-positive T-cells are present in pig. This CD4+CD8+T cell population is characterised by the expression of the CD8α chain and function as effector/memory T cells. These antigen-experienced T cells increase with age. Interestingly, regional differences are present in the abundance of this T cell subset within the small intestine. For instance, in the porcine jejunal PPs, CD4+CD8α+T cells comprise around 12% of the total T cell population, while in ileum, they only comprise 4% of the T cell pool (Zuckermann and Gaskins, 1996; Käser, 2021). This CD4+CD8α+T cell population increases upon infection of vaccination. For instance, oral vaccination of piglets with Salmoporc (a live-attenuated, histidine-adenine auxotrophic *Salmonella* vaccine strain; Ceva Santé Animale, Libourne, France) triggered the presence of vaccine-specific IL17A-producing CD4+CD8α+T cells in the lamina propria and gut-draining lymph nodes, which might contribute to protective immunity against *Salmonella* infections in pig (Schmidt et al., 2021).

On top of their role in T helper cell activation and differentiation, DCs also play an important role in the activation of B cells. They do so indirectly by driving the differentiation of naïve CD4+T cells within lymph nodes and PPs to the T follicular helper cell (Tfh) lineage in an IL6-dependent manner (Deenick et al., 2011). These activated Tfh cells are characterised by their expression of the chemokine receptor CXCR5, which allow them to migrate to the B cell zones within lymph nodes. Tfh play an essential role in the formation of germinal centres, structures within the B cell zone where B cells can differentiate into antibody-secreting plasma cells and memory B cells. Tfh drives this process by activation of B cells via TCR–MHCII, ICOS–ICOSL and CD40–CD40L interactions and induce B cell proliferation by the secretion of the Tfh hallmark cytokine IL21. The development of memory B cells by Tfh cells seems to be driven by their secretion of IL9 (Deenick et al., 2011; Wang et al., 2017). In pigs, a live-attenuated *Salmonella enterica* serovar typhimurium vaccine increased the frequency of IL21-producing Tfh cells and promoted germinal centre formation in the spleen (Ugolini et al., 2018).

4.2.3 Gut homing of effector T and B cells

In addition to driving T cell differentiation, intestinal DCs also have an important role in dictating the homing potential of recently activated lymphocytes. Small intestinal CD103+ DCs promoted the expression of the gut-homing receptors, the integrin α4β7 and the chemokine receptor CCR9 on T cells and IgA+ B cells (Johansson-Lindbom et al., 2005; Mora et al., 2006; Iwata et al., 2004). CCR9 binds to CCL25 produced by small intestinal epithelial cells, whereas α4β7 binds to MAdCAM-1 expressed by the high endothelial venules of the small intestine. Both MAdCAM-1 and CCL25 are constitutively expressed, implicating continuous migration of T and B cells to the small intestine. The ability to imprint the expression of these gut-homing receptors depends on the metabolism of vitamin A to retinoic acid by the DCs. In contrast to DCs at other tissues, intestinal DCs express retinal aldehyde dehydrogenases to convert vitamin A to its active metabolite retinoic acid (Habtezion et al., 2016; Agace and Persson, 2012). Thus, it appears that intestinal DCs utilize the vitamin A metabolic pathway to educate T and B cells for gut-homing migration. In pigs, tissue-specific homing of lymphocytes occurs and both the small intestine-specific expression of CCL25 and CCR9 as well as the strong expression of β7 on intestinal T and IgA+ B cells have been confirmed (Meurens et al., 2006; Chen et al., 2016). Porcine DCs use retinoic acid to imprint gut-homing receptors on T cells, and parenteral administration of vitamin A to piglets promoted migration of TGEV-specific CD8+ T cells to the lamina propria (Chen et al., 2016). Interestingly, vitamin A deficiency in gnotobiotic piglets impaired immune responses to a human rotavirus vaccine, while supplementing vitamin A to piglets enhanced IgA responses to porcine epidemic diarrhea virus (Kandasamy et al., 2014; Langel et al., 2019).

Apart from α4β7 and CCR9, gut homing of T cells to the large intestine is mainly driven by GPR15, while homing of IgA+ antibody-secreting cells (ASCs) to the gut also relies on CCR10. The latter recognizes the chemokine CCL28, which is not only expressed in the small intestinal epithelium but also in the trachea, the large intestine and the mammary gland (Habtezion et al., 2016). The expression of CCL28 in the mammary gland and its increased expression during lactation might play an important role in the recruitment of IgA+ ASCs to the mammary gland and the subsequent transfer of IgA to newborn piglets through colostrum (Bourges et al., 2008; Salmon et al., 2009).

Intestinal DCs clearly have some unique features through their expression of retinal aldehyde hydrogenases and the use of the vitamin A metabolite retinoic acid to induce Tregs and gut homing. The dietary component vitamin A thus plays an important role in shaping the gut immune system and indirectly in the transfer of passive immunity to neonates. In addition to dietary components, also the microbiota is known to control gut function and immunity. The following section delves deeper into our current understanding of the impact of the porcine microbiota on gut tissues.

5 The role of the microbiota in shaping gut function and immunity

The intestine is home to an abundant community of microorganisms, including bacteria, viruses and fungi, collectively called the microbiota. Some members of the microbiota form a symbiotic relationship with their host, while the latter provides a niche in which the microbiota can flourish. Intestinal bacteria confer essential benefits by fermenting dietary fibres to SCFAs, which are used as an energy source by colonocytes. In addition, the microbiota synthesizes essential vitamins, like vitamins B and K (Kamaya et al., 2020). In addition to these, other microbiota-produced metabolites can shape the function of the intestinal epithelium and gut immune system (Tremaroli and Bäckhed, 2012). Metabolites produced by the microbiota can reach the circulation and in mice and humans, and some of these metabolites are found at high concentrations in blood and might have an impact on tissues beyond the gut. Unsurprisingly, the microbiota and its metabolites have been associated with many diseases and disorders.

The intestinal microbiota are predominantly bacteria, and in post-weaning piglets, Bacteroidetes, Firmicutes and Proteobacteria are the most abundant phyla (Pajarillo et al., 2014). Similar to other mammals, the microbial richness and abundance differ between the intestinal segments. The microbial abundance is rather low in the small intestine but increases to tremendous densities in the large intestine. While the ileum is mainly colonized by firmicutes and Proteobacteria, the cecum and colon are dominated by Bacteroidetes. Interestingly, also in pigs, there is a difference between the composition of the planktonic microbiota in the intestinal lumen and the genera that dominate the mucus-adhering microbiota (Looft et al., 2014; Zhang et al., 2018).

Similar to other species, the microbiota is necessary for the development of the gut immune system in pigs (Stokes, 2017). In germ-free piglets, the ileum contains less IgA+ cells and the ratio of effector/memory T cells is lower as compared to conventional piglets (Potockova et al., 2015). This less-developed GALT in germ-free piglets is reflected by an IgM-dominated immune response upon oral immunisation as compared to the IgA responses in conventional piglets (Stuyven et al., 2010). In addition, rearing conditions during early life affect immune cell development at later ages. Under high-hygiene conditions, piglets around weaning had a different architecture of the gut antigen-presenting cells and less intestinal Tregs than outdoor reared piglets (Christoforidou et al., 2018). This study showed that exposure to environmental factors during early life impacts on the development of the gut immune system at later ages.

5.1 Microbiota metabolites and gut immunity

The composition of microbiota in pigs is affected by diet, antibiotics, physiological status, genetic background and the environment, although diet is the most important factor (Isaacson and Kim, 2012). Pig feed contains plant-derived fibres, such as lignin, hemicellulose, cellulose and other structural components, that are undigestable for pigs but are fermented by the colonic microbiota. This fermentation results in the production of SCFAs, such as acetate, propionate and butyrate. In the cecum of pigs, acetate is the most abundant SCFA followed by butyrate and propionate (Zaiss et al., 2015, Immunity). These SCFAs exert their beneficial effects through interactions with the free fatty acid receptors (FFAR) 2 (=GPR43), FFAR3 (=GPR41) and GPR109A (Venegas et al., 2019). In pigs, FFAR2 and FFAR3 mRNA expression has been reported in the ileum. In contrast to mice and humans, dietary fibre however does not affect transcript levels of these SCFA receptors in the pig intestine (Haenen et al., 2013). The best-studied SCFA is butyrate, which has many beneficial effects on the gut epithelium and immune cells in several species, including pigs (Bedford and Gong, 2018). For instance, butyrate promotes the epithelial barrier function by inducing the production of mucins, trefoil factors, tight junction proteins and host defence proteins by colonocytes. Butyrate also acts as an anti-inflammatory agent, inhibiting the expression of pro-inflammatory cytokines and chemokines (tumour necrosis factor (TNF)α, IL1β, IL6, IL8) (Liu et al., 2018). The mechanisms underlying the anti-inflammatory effect of butyrate rely in part on the inhibition of nuclear factor kappa B, a transcription factor that regulates the expression of a variety of genes involved in inflammation and immunity, and on the activation of peroxisome proliferator-activated receptor γ, a member of the nuclear hormone receptor family known to exert anti-inflammatory effects (Venkatraman et al., 2003; Marion-Letellier et al., 2009). Another mechanism by which butyrate might exert its anti-inflammatory activities could be attributed to its ability to inhibit histone deacetylases (HDACs) (Yang et al., 2020). These enzymes remove acetyl groups from lysines on the histones, resulting in a more condense and transcriptionally silent chromatin. Butyrate not only affects gut epithelial cells but also immune cells. In a GPR109a-dependent manner, butyrate imprints a tolerogenic function on intestinal DCs, resulting in the differentiation of regulatory T cells and increased production of the anti-inflammatory cytokines IL10 and TGFβ in the gut tissues (Atarashi et al., 2013). In addition, in mice, butyrate also promotes IL22 production by CD4+T cells through inhibiting HDACs in a GPR41-dependent manner (Yang et al., 2020). IL22 in turn acts upon epithelial cells to reinforce their barrier function. As mentioned, colonocytes use butyrate as their main energy source. However, in mice models, butyrate has opposing effects on the intestinal epithelial stem cells in the colon crypts, inhibiting their proliferation and delaying wound repair (Kaiko et al., 2016). Moreover, butyrate also inhibited proliferation of crypt-based epithelial stem cells in mouse and human small intestinal organoids (Pearce et al., 2020). This argues for a careful dose selection when supplementing feed with butyrate.

Other important metabolites produced by the microbiota are indolic compounds, which result from the catabolism of the aromatic amino acids phenylalanine, tyrosine and tryptophan, released by the digestion of dietary proteins (Roager and Licht, 2018). Several of these indolic compounds, such as indoleacetic acid, indolealdehyde (IAld) and indoleacrylic acid, serve as ligands for the aryl hydrocarbon receptor (AhR), a transcription factor commonly expressed by gut immune cells (Stockinger et al., 2014). However, the ligand specificity of AhR seems to depend on the species (Hubbard et al., 2015). AhR activation by indolic compounds not only triggers IL22 production but also favours the differentiation of FoxP3+Tregs, controls their gut homing by inducing the chemoattractant receptor GPR15 and inhibits Th17 differentiation from naïve CD4+T cells (Xiong et al., 2020; Swaminathan et al., 2021). Interestingly, AhR not only influences T cells differentiation but also plays a role in the differentiation of IL10-producing regulatory B cells (Bregs) (Piper et al., 2019). Moreover, AhR activation by indole derivatives produced by *Lactobacillus reuteri* reprograms intestinal epithelial lymphocytes (IELs) to suppressive effector cells (Cervantes-Barragan et al., 2017). In pigs, the transfer of fecal material from an adult pig to newborn piglets increased the concentration of indole derivatives in the gut. This was associated with a higher AhR activity and IL22 mRNA expression in the colon (Geng et al., 2018).

Finally, the microbiota also processes primary bile acids into secondary bile acids. Although most primary bile acids are reabsorbed in the ileum, a small fraction (1–5%) reaches the large intestine where they are transformed into secondary bile acids. The latter can exert immunomodulatory activities. For instance, in mice, the primary bile acid cholic acid is converted to deoxycholic acid, which increased the induction of peripheral FoxP3+ Tregs by diminishing the immunostimulatory ability of DCs (Campbell et al., 2020).

5.2 Dietary interventions to control the gut microbiota

The above examples clearly illustrate that intestinal health in swine is affected by metabolites produced by the microbiota and indicate that gut health might be modulated by dietary interventions aimed at changing the composition and the metabolic activity of the gut microbiota. Supplementation of feed with pre-, pro- and synbiotics might allow farmers to control the functional composition of the microbiota to improve gut health during the different stages of the production process. Especially, during weaning, dietary interventions to improve gut function might prove effective. Early weaning results in an underdeveloped gut, which is unable to properly digest and absorb nutrients. Many efforts are thus focused on feed additives to ease the transition from milk to solid feed in piglets. Prebiotics are defined as substances that the host is unable to digest and serve as nutrients for the microbiota. Prebiotics include functional oligo- and polysaccharides, polyols, protein hydrolysates and plant extracts (Markowiak and Slizewska, 2018). Supplementing feed with fructo-oligosaccharides (FOS), galacto-oligosaccharides and yeast cell wall (including mannan-oligosaccharides and β-glucans) promoted the intestinal barrier function and reduced gut inflammation in suckling and weaned piglets (Tian et al., 2019; Duan et al., 2019; Schokker et al., 2018). Also, more complex polysaccharides or fibres are intensely investigated for their impact on pig gut health (Van Hees et al., 2019). For instance, changing the fibre content in the feed to resistant starch increased the relative abundance of butyrate-producing microbiota in the pig colon and resulted in increased SCFA concentrations in the cecum and colon, which promoted intestinal immunity (Haenen et al., 2013; Trachsel et al., 2019). Interestingly, the choice of fibre source is important as different types of fibre have a different effect on the microbiota composition and gut physiology. The long-term intake of wheat bran fibre and pea fibre had beneficial effects on gut physiology in fattening pigs, while soybean fibre damaged the architecture of the pig intestine (Chen et al., 2014; Zhao et al., 2018).

Another method to change the composition of the gut microbiota to improve intestinal health is feeding probiotics. Several *Lactobacilli*, *Enterococci* and *Clostridia* species are under investigation for their beneficial effects on gut health in pigs (Markowiak and Slizewska, 2018). Although expectations are high, their implementation in pig production systems might not be straightforward due to difficulties with the production of probiotics at scale and the delivery of live bacteria to the large intestine.

Synbiotics are defined as the combination of pre- and probiotics, which have a synergistic effect on gut health when given combined. Some studies have investigated this synergism between pre- and probiotics. For example, a study showed a potential synergistic effect of FOS and *Lactobacillus plantarum* ZLP001 on the growth performance of weaned piglets and on systemic immune parameters (increased IFNγ and total IgG levels) (Wang et al., 2019). More studies are however warranted to determine the optimal combinations and their concentrations to improve gut health in swine.

In addition to pre- and probiotics, many studies have focused on the effect of postbiotics on gut health. Postbiotics are defined as the metabolites of beneficial microbiota with SCFAs as classical examples. Several studies assessed supplementation of feed with butyrate. In general, when fed to weaning piglets, butyrate supplementation did not result in increases in average daily weight gains (ADG), although it promoted gut health to some extent (Biagi et al., 2007; Mallo et al., 2012). The most pronounced effects of butyrate supplementation (free or encapsulated) were observed during artificial rearing of piglets (Kotunia et al., 2004) or intrauterine growth-restricted piglets (Dong et al., 2016). In these cases, butyrate supplementation accelerated the development of the small intestinal (increase in crypt depth, villus height and mucosa thickness) as compared to control littermates. In contrast, tributyrin supplementation of weaned piglets resulted in an increased ADG and gain-to-feed ratio, showing its potential as a feed additive (Sotira et al., 2020). These results on the beneficial effects of butyrate and pre-and probiotics on pig gut health provide a solid basis for future work to develop these compounds as valuable supplements in pig production and even as alternatives for antibiotics.

Another alternative to antibiotics is vaccination (Hoelzer et al., 2018). Given the crucial role of the microbiota in shaping gut immunity, it is not surprising that the composition of the gut microbiota can impact vaccine efficacy. For instance, in humans, antibiotic-driven bacterial imbalance can lead to impaired immune responses and can alter immune responses to rotavirus and seasonal influenza vaccines (Harris et al., 2018; Oh et al., 2014). Some studies have investigated the potential of probiotic supplementation to improve vaccine efficacy in humans and pigs (Zimmermann and Curtis, 2018; Vlasova et al., 2019). Despite these efforts, further research is needed to establish a causal relationship between microbiota composition and responses to vaccines as well as to identify microbiota-derived molecules or dietary components that can improve vaccine-induced responses.

6 Intestinal organoids as novel tools to explore the function of the gut epithelium

Mice models have been instrumental in advancing our understanding of the role the intestinal epithelium plays in regulating immune responses to commensals and invading pathogens. However, these models are largely lacking for livestock species, including a pig. Although much has been learned from porcine intestinal epithelial cell lines (e.g. IPEC-J2, IPEC-1, IPI-2i; Seeger, 2020), these lack the cellular heterogeneity of the gut epithelium, do not reproduce tissue functionality and lack the complex micro-architecture of the gut tissues (Beaumont et al., 2021). Gut-ligated loops, small intestinal segment perfusion models and *ex vivo*-cultured intestinal explants, such as precision-cut tissue slices, recapitulate these features but are only suitable for short-term experiments.

The development of intestinal organoids has solved many of these issues. Intestinal organoids derived from the small intestine are called enteroids, while those derived from the large intestine are often called colonoids. Since the initial development of murine enteroids, intestinal organoids have been established in many species, including pig (Powell and Behnke, 2017; Sato et al., 2009). Porcine enteroids were first described by Gonzalez et al. (2013), and since then, a handful of research groups have used these intestinal organoids to investigate the interaction of the small intestinal epithelium with dietary components, mycotoxins, commensals and gut pathogens (reviewed in Seeger, 2020; Beaumont et al., 2021). At present, large intestinal organoids have not been described for swine. Porcine small intestinal organoids are derived from Lgr5+ adult epithelial stem cells residing in the crypts. Upon their isolation from the intestine, these crypts are cultured in a hydrogel of extracellular matrix proteins (Matrigel™) in a complex intestinal epithelial stem cell culture medium containing several recombinant human or mouse growth factors (epidermal growth factor, R-spondin, Wnt3a and Noggin) and a variety of inhibitors (Y27632: inhibitor of ROCK; LY2157299, A8301: inhibitors of TGFβ receptor signalling; SB202190: a p38 MAPK inhibitor; CHIR99021: an inhibitor of glycogen synthase kinase 3) to promote cell proliferation and to inhibit cell death and differentiation (Fig. 4). An overview of the composition of

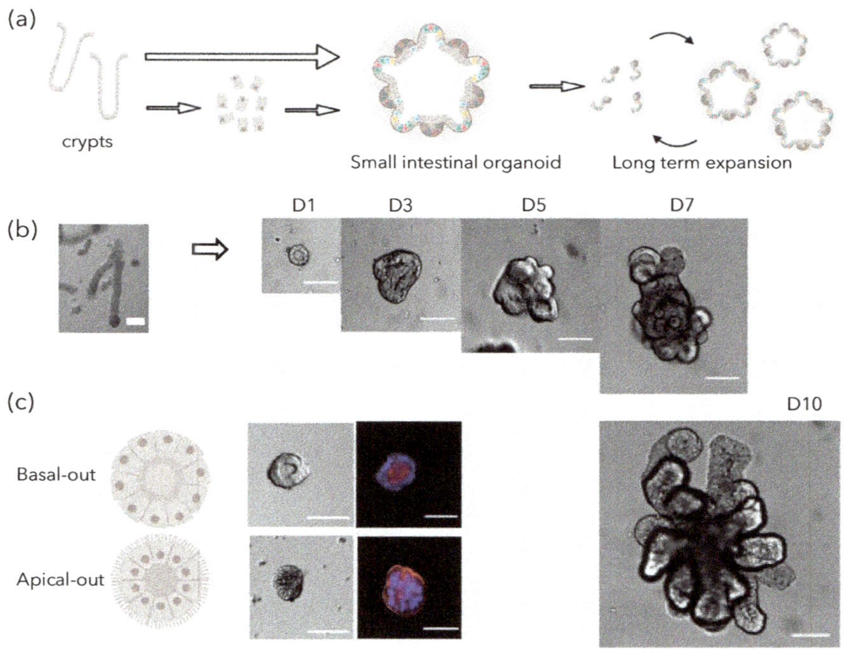

Figure 4 Growth of crypt-based stem cells into porcine small intestinal organoids. (a) Schematic presentation of small intestinal organoid culture and expansion. Crypts isolated from the intestine are cultured in intestinal epithelial stem cell (IESC) culture media embedded in Matrigel. The IESCs residing at the bottom of the crypts will grow into small intestinal organoids or enteroids. These enteroids can be fragmented (mechanical or enzymatic) and cultured again to expand the enteroid cultures. (b) An example of the growth and culture of porcine small intestinal crypts into enteroids over a period of 10 days. (c) Schematic presentation of basal-out and apical-out enteroids (left) and microscopic images of porcine basal-out and apical-out enteroids. Blue are nuclei, red shows a villin staining. Scale bars are 50 μm. Graphics were created with BioRender.com.

custom-made culture media used to grow porcine small intestinal organoids is given by Beaumont et al. (2021) and Seeger (2020). As an alternative to custom-made media, porcine intestinal organoids can be cultured in IntestiCult™, a media developed by StemCell Technologies (Vancouver, CA) to culture murine and human intestinal organoids. Its composition however needs further optimization, as the functionality of IntestiCult-grown porcine intestinal organoids does not fully resemble those grown in custom-made media (Vermeire et al., 2021).

When crypt-based intestinal stem cells are grown under these conditions for several days, they form self-organizing 3D structures with crypt-like domains, composed of a monolayer of polarized intestinal epithelial cells, which surrounds a central cavity or pseudolumen. Although physiologically relevant and useful to mimic diarrhoea in swelling assays, the apical surface of the epithelium in the intestinal organoids is not directly accessible (Vermeire et al., 2021). Although monolayers have been developed to circumvent this issue, these lack a complex 3D organization (Van der Hee et al., 2018). Alternatively, the epithelial polarity in human and porcine small intestinal organoids can be inverted using a straightforward method to generate so-called apical-out enteroids (Co et al., 2019; Li et al., 2020). In these apical-out enteroids, the apical surface of the epithelium is directly accessible, facilitating research investigating the interaction of the gut epithelium with dietary components and microorganisms.

Although intestinal organoids are a powerful tool to study the function of the gut epithelium in pigs, a further increase in their complexity and optimization of the composition of the culture media might be needed to unlock their full potential. For instance, the concentration of the different inhibitors and growth factors might be further fine-tuned for optimal growth and differentiation of porcine intestinal organoids. The use of porcine recombinant growth factors might be worth considering. Even though growth factors are usually highly conserved, small differences might influence receptor interactions and subsequent activation of signalling pathways for intestinal epithelial stem cell growth and differentiation. In addition, more defined, xeno-free hydrogels are much needed. Intestinal organoids are often cultured in Matrigel™ (derived from Engelbreth-Holm-Swarm sarcoma cells), which not only contains extracellular matrix components but also many other proteins, such as growth factors (Hughes et al., 2010), and is prone to batch-to-batch variations. Since the composition of the extracellular matrix and its mechanical properties (e.g. stiffness) define stem cell differentiation, the development of defined hydrogels or synthetic scaffolds is much needed (Kratochvil et al., 2019).

A further advancement should be an increased cellular complexity of the intestinal organoids, as immune, neuronal, mesenchymal and vascular cells are currently lacking. Moreover, mechanical forces and the microbiota are also not reproduced in current porcine intestinal organoid models, although these factors are key regulators of intestinal physiology. Further efforts should focus on the development of co-culture models to include other cell types, such as stromal cells and immune cells, like intraepithelial lymphocytes or macrophages, and microfluidic gut-on-a-chip models. Like intestinal organoids, the latter not only allow multi-lineage differentiation and the formation of an epithelial architecture but also more closely mimics cell–lumen interactions and the complex physical and biochemical microenvironment of the gut (Nikolaev et al., 2020).

7 Conclusion and future trends

A complex cellular network has evolved in the intestine to perform its function of feed digestion and nutrient absorption, while simultaneously preventing unwanted immune responses to innocuous antigens, like feed or microbiota, and allowing swift protective responses against pathogens. This cellular network within the intestine responds and adapts to these external cues, to which it is daily exposed. Perturbations of this network during critical and stressful periods, like weaning, result in a reduced ability to digest feed and impaired immune responses, which in turn lead to a growth lag. As a healthy gut is crucial for optimal growth and the microbiota play an important role in shaping gut function and immunity, many efforts are ongoing to design dietary interventions such as pre-, pro- and postbiotics to improve pig gut health. Future efforts should focus on answering one of the long-standing questions in gut research: what is a healthy microbiome? Most likely, the answer will be complex, as the microbiome composition and its function probably will shift depending on age and the different stages in pig production. Advances in sequencing, metabolomics and gut organoids might help to answer this question.

The use of intestinal organoids might be an alternative for animal models in gut research and could be used to further promote the 3Rs principle in research. To unlock its full potential, further advances in optimizing this model should focus on including intestinal stromal cells and immune cells as well as microbiota. Evidently, this will increase the complexity of the system; however, commonly used tools to explore cellular functions are inadequate to dissect the cellular networks that cooperate to maintain gut homeostasis during stress and infection in pigs. Translation of recent advances in single-cell RNA sequencing and other single-cell (multi)omics technology to livestock species will help to deepen our understanding of this interplay between stromal cells, immune cells and microbiota in gut health in pigs. The insights gained will inform on the design of dietary interventions aimed at improving gut health and accelerating its development in swine.

8 References

Agace, W. W. and Persson, E. K. 2012. How vitamin A metabolizing dendritic cells are generated in the gut mucosa. *Trends Immunol.* 33(1):42–48.

Allaire, J. M., Crowley, S. M., Law, H. T., Chang, S. Y., Ko, H. J. and Vallance, B. A. 2018. The intestinal epithelium: Central coordinator of mucosal immunity. *Trends Immunol.* 39(9):677–696.

Atarashi, K., Tanoue, T., Oshima, K., Suda, W., Nagano, Y., Nishikawa, H., Fukuda, S., Saito, T., Narushima, S., Hase, K., Kim, S., Fritz, J. V., Wilmes, P., Ueha, S., Matsushima, K., Ohno, H., Olle, B., Sakaguchi, S., Taniguchi, T., Morita, H., Hattori, M. and Honda, K. 2013. Treg induction by a rationally selected mixture of Clostridia strains from the human microbiota. *Nature* 500(7461):232–236.

Auray, G., Keller, I., Python, S., Gerber, M., Bruggmann, R., Ruggli, N. and Summerfield, A. 2016. Characterization and transcriptomic analysis of porcine blood conventional and plasmacytoid dendritic cells reveals striking species-specific differences. *J. Immunol.* 197(12):4791–4806.

Auray, G., Talker, S. C., Keller, I., Python, S., Gerber, M., Liniger, M., Ganges, L., Bruggmann, R., Ruggli, N. and Summerfield, A. 2020. High-resolution profiling of innate immune responses by porcine dendritic cell subsets *in vitro* and *in vivo*. *Front. Immunol.* 11:1429.

Ayabe, T., Satchell, D. P., Wilson, C. L., Parks, W. C., Selsted, M. E. and Ouellette, A. J. 2000. Secretion of microbicidal α-defensins by intestinal Paneth cells in response to bacteria. *Nat. Immunol.* 1(2):113–118.

Bailey, M. 2009. The mucosal immune system: Recent developments and future directions in the pig. *Dev. Comp. Immunol.* 33(3):375–383.

Bakdash, G., Vogelpoel, L. T. C., van Capel, T. M. M., Kapsenberg, M. L. and de Jong, E. C. 2015. Retinoic acid primes human dendritic cells to induce gut-homing, IL-10-producing regulatory T cells. *Mucosal Immunol.* 8(2):265–278.

Barker, N., Van Es, J. H., Kuipers, J., Kujala, P., Van Den Born, M., Cozijnsen, M., Haegebarth, A., Korving, J., Begthel, H., Peters, P. J. and Clevers, H. 2007. Identification of stem cells in small intestine and colon by marker gene Lgr5. *Nature* 449(7165):1003–1007.

Beaumont, M., Blanc, F., Cherbuy, C., Egidy, G., Giuffra, E., Lacroix-Lamandé, S. and Wiedemann, A. 2021. Intestinal organoids in Farm Animals. *Vet. Res.* 52(1):33.

Bedford, A. and Gong, J. 2018. Implications of butyrate and its derivatives for gut health and animal production. *Anim. Nutr.* 4(2):151–159.

Bergstrom, K. S. B., Morampudi, V., Chan, J. M., Bhinder, G., Lau, J., Yang, H., Ma, C., Huang, T., Ryz, N., Sham, H. P., Zarepour, M., Zaph, C., Artis, D., Nair, M. and Vallance, B. A. 2015. Goblet cell derived RELM-b recruits CD4+ T cells during infectious colitis to promote protective intestinal epithelial cell proliferation. *PLoS Pathog.* 11(8):e1005108.

Biagi, G., Piva, A., Moschini, M., Vezzali, E. and Roth, F. X. 2007. Performance, intestinal microflora, and wall morphology of weanling pigs fed sodium butyrate. *J. Anim. Sci.* 85(5):1184–1191.

Bimczok, D., Post, A., Tschernig, T. and Rothkötter, H. J. 2006. Phenotype and distribution of dendritic cells in the porcine small intestinal and tracheal mucosa and their spatial relationship to epithelial cells. *Cell Tissue Res.* 325(3):461–468.

Birchenough, G. M. H., Nyström, E. E. L., Johansson, M. E. V. and Hansson, G. C. 2016. A sentinel goblet cell guards the colonic crypt by triggering Nlrp6-dependent MUC2 secretion. *Science* 352(6293):1535–1542.

Biton, M., Haber, A. L., Rogel, N., Burgin, G., Beyaz, S., Schnell, A., Ashenberg, O., Su, C. W., Smillie, C., Shekhar, K., Chen, Z., Wu, C., Ordovas-Montanes, J., Alvarez, D., Herbst, R. H., Zhang, M., Tirosh, I., Dionne, D., Nguyen, L. T., Xifaras, M. E., Shalek, A. K., von Andrian, U. H., Graham, D. B., Rozenblatt-Rosen, O., Shi, H. N., Kuchroo, V., Yilmaz, O. H., Regev, A. and Xavier, R. J. 2018. T helper cell cytokines modulate intestinal stem cell renewal and differentiation. *Cell* 175(5):1307–1320.

Boudry, G., Douard, V., Mourot, J., Lallès, J. P. and Le Huërou-Luron, I. 2009. Linseed oil in the maternal diet during gestation and lactation modifies fatty acid composition, mucosal architecture, and mast cell regulation of the ileal barrier in piglets. *J. Nutr.* 139(6):1110–1117.

Bourges, D., Meurens, F., Berri, M., Chevaleyre, C., Zanello, G., Levast, B., Melo, S., Gerdts, V. and Salmon, H. 2008. New insights into the dual recruitment of IgA+ B cells in the developing mammary gland. *Mol. Immunol.* 45(12):3354–3362.

Brandtzaeg, P., Kiyono, H., Pabst, R. and Russell, M. W. 2008. Terminology: Nomenclature of mucosa-associated lymphoid tissue. *Mucosal Immunol.* 1(1):31–37.

Bréa, D., Meurens, F., Dubois, A. V., Gaillard, J., Chevaleyre, C., Jourdan, M. L., Winter, N., Arbeille, B., Si-Tahar, M., Gauthier, F. and Attucci, S. 2012. The pig as a model for investigating the role of neutrophil serine proteases in human inflammatory lung diseases. *Biochem. J.* 447(3):363–370.

Brevi, A., Cogrossi, L. L., Grazia, G., Masciovecchio, D., Impellizzieri, D., Lacanfora, L., Grioni, M. and Bellone, M. 2020. Much more than IL-17A: Cytokines of the IL-17 family between microbiota and cancer. *Front. Immunol.* 11:565470.

Butler, J. E. and Sinkora, M. 2013. The enigma of the lower gut-associated lymphoid tissue (GALT). *J. Leukoc. Biol.* 94(2):259–270.

Campbell, C., McKenney, P. T., Konstantinovsky, D., Isaeva, O. I., Schizas, M., Verter, J., Mai, C., Jin, W. B., Gua, C. J., Violante, S., Ramos, R. J., Cross, J. R., Kadaveru, K., Hambor, J. and Rudensky, A. Y. 2020. Bacterial metabolism of bile acids promotes generation of peripheral regulatory T cells. *Nature* 581(7809):475–479.

Cassatella, M. A., Östberg, N. K., Tamassia, N. and Soehnlein, O. 2019. Biological roles of neutrophil-derived granule proteinS and cytokines. *Trends Immunol.* 40(7):648–664.

Casteleyn, C., Van den Broeck, W., Gebert, A., Tambuyzer, B. R., Van Cruchten, S. and Van Ginneken, C. 2013. M cell specific markers in man and domestic animals: Valuable tools in vaccine development. *Comp. Immunol. Microbiol. Infect. Dis.* 36(4):353–364.

Cervantes-Barragan, L., Chai, J. N., Tianero, M. D., Di Luccia, B., Ahern, P. P., Merriman, J., Cortez, V. S., Caparon, M. G., Donia, M. S., Gilfillan, S., Cella, M., Gordon, J. I., Hsieh, C. S. and Colonna, M. 2017. *Lactobacillus reuteri* induces gut intraepithelial $CD4^{+}CD8\alpha\alpha^{+}$ T cells. *Science* 357(6353):806–810.

Chen, H., Mao, X. B., Che, L. Q., Yu, B., He, J., Yu, J., Han, G. Q., Huang, Z. Q., Zheng, P. and Chen, D. W. 2014. Impact of fiber types on gut microbiota, gut environment and gut function in fattening pigs. *Anim. Feed Sci. Technol.* 195:101–111.

Chen, X., Tu, C., Qin, T., Zhu, L., Yin, Y. and Yang, Q. 2016. Retinoic acid facilitates inactivated transmissible gastroenteritis virus induction of CD8(+) T-cell migration to the porcine gut. *Sci. Rep.* 6:24152.

Cheroutre, H., Lambolez, F. and Mucida, D. 2011. The light and dark sides of intestinal intraepithelial lymphocytes. *Nat. Rev. Immunol.* 11(7):445–456.

Christoforidou, Z., Burt, R., Mulder, I., Gil, B. P., Pluske, J., Kelly, D., Stokes, C. R., Bailey, M. and Lewis, M. C. 2018. Development of immune cells in the intestinal mucosa can be affected by intensive and extensive farm environments, and antibiotic use. *Front. Immunol.* 9:1061.

Clevers, H. C. and Bevins, C. L. 2013. Paneth cells: Maestros of the small intestinal crypts. *Annu. Rev. Physiol.* 75:289–311.

Co, J. Y., Margalef-Català, M., Li, X., Mah, A. T., Kuo, C. J., Monack, D. M. and Amieva, M. R. 2019 Controlling epithelial polarity: A human enteroid model for host-pathogen interactions. *Cell Rep.* 26(9):2509–2520.e4. doi: 10.1016/j.celrep.2019.01.108.

Corthésy, B. 2013. Multi-faceted functions of secretory IgA at mucosal surfaces. *Front. Immunol.* 4:185.

Deenick, E. K., Ma, C. S., Brink, R. and Tangye, S. G. 2011. Regulation of T follicular helper cell formation and function by antigen presenting cells. *Curr. Opin. Immunol.* 23(1):111–118.

Devriendt, B., De Geest, B. G. and Cox, E. 2011. Designing oral vaccines targeting intestinal dendritic cells. *Expert Opin. Drug Deliv.* 8(4):467–483.

Devriendt, B., De Geest, B. G., Goddeeris, B. M. and Cox, E. 2012. Crossing the barrier: Targeting epithelial receptors for enhanced oral vaccine delivery. *J. Control. Release* 160(3):431–439.

Devriendt, B., Gallois, M., Verdonck, F., Wache, Y., Bimczok, D., Oswald, I. P., Goddeeris, B. M. and Cox, E. 2009. The food contaminant fumonisin B_1 reduces the maturation of porcine CD11R1$^+$ antigen presenting cells and antigen-specific immune responses, leading to a prolonged intestinal ETEC infection. *Vet. Res.* 40(4):40.

Devriendt, B., Stuyven, E., Verdonck, F., Goddeeris, B. M. and Cox, E. 2010. Enterotoxigenic Escherichia coli (K88) induce proinflammatory responses in porcine intestinal epithelial cells. *Dev. Comp. Immunol.* 34(11):1175–1182.

Dong, L., Zhong, X., He, J., Zhang, L., Bai, K., Xu, W., Wang, T. and Huang, X. 2016. Supplementation of tributyrin improves the growth and intestinal digestive and barrier functions in intrauterine growth-restricted piglets. *Clin. Nutr.* 35(2):399–407.

Duan, X., Tian, G., Chen, D., Huang, L., Zhang, D., Zheng, P., Mao, X., Yu, J., He, J., Huang, Z. and Yu, B. 2019. Mannan oligosaccharide supplementation in diets of sow and (or) their offspring improved immunity and regulated intestinal bacteria in piglet1. *J. Anim. Sci.* 97(11):4548–4556.

Dubreuil, J. D. 2017. Enterotoxigenic Escherichia coli targeting intestinal epithelial tight junctions: An effective way to alter the barrier integrity. *Microb. Pathog.* 113:129–134.

Duée, P. H., Darcy-Vrillon, B., Blachier, F. and Morel, M. T. 1995. Fuel selection in intestinal cells. *Proc. Nutr. Soc.* 54(1):83–94.

Fournier, B. M. and Parkos, C. A. 2012. The role of neutrophils during intestinal inflammation. *Mucosal Immunol.* 5(4):354–366.

Gehart, H. and Clevers, H. 2019. Tales from the crypt: New insights into intestinal stem cells. *Nat. Rev. Gastroenterol. Hepatol.* 16(1):19–34.

Gehart, H., van Es, J. H., Hamer, K., Beumer, J., Kretzschmar, K., Dekkers, J. F., Rios, A. and Clevers, H. 2019. Identification of enteroendocrine regulators by real-time single cell differentiation mapping. *Cell* 176(5):1158–1173.e16.

Geng, S., Cheng, S., Li, Y., Wen, Z., Ma, X., Jiang, X., Wang, Y. and Han, X. 2018. Faecal microbiota transplantation reduces susceptibility to epithelial injury and modulates tryptophan metabolism of the microbial community in a piglet model. *J. Crohns. Colitis* 12(11):1359–1374.

Gerbe, F. and Jay, P. 2016. Intestinal tuft cells: Epithelial sentinels linking luminal cues to the immune system. *Mucosal Immunol.* 9(6):1353–1359.

Gerbe, F., Sidot, E., Smyth, D. J., Ohmoto, M., Matsumoto, I., Dardalhon, V., Cesses, P., Garnier, L., Pouzolles, M., Brulin, B., Bruschi, M., Harcus, Y., Zimmermann, V. S., Taylor, N., Maizels, R. M. and Philippe Jay, P. 2016. Intestinal epithelial tuft cells initiate type 2 mucosal immunity to helminth parasites. *Nature* 529(7585):226–230.

Geremia, A. and Aranciba-Carcamo, C. V. 2017. Innate lymphoid cells in intestinal inflammation. *Front. Immunol.* 8:1296.

Gonzalez, L. M., Williamson, I., Piedrahita, J. A., Blikslager, A. T. and Magness, S. T. 2013. Cell lineage identification and stem cell culture in a porcine model for the study of intestinal epithelial regeneration. *PLoS ONE* 8(6):e66465.

Haber, A. L., Biton, M., Rogel, N., Herbst, R. H., Shekhar, K., Smillie, C., Burgin, G., Delorey, T. M., Howitt, M. R., Katz, Y., Tirosh, I., Beyaz, S., Dionne, D., Zhang, M., Raychowdhury, R., Garrett, W. S., Rozenblatt-Rosen, O., Shi, H. N., Yilmaz, O., Xavier, R. J. and Regev, A. 2017. A single-cell survey of the small intestinal epithelium. *Nature* 551(7680):333–339.

Habtezion, A., Nguyen, L. P., Hadeiba, H. and Butcher, E. C. 2016. Leukocyte trafficking to the small intestine and colon. *Gastroenterology* 150(2):340–354.

Haenen, D., Zhang, J., Souza da Silva, C., Bosch, G., van der Meer, I. M., van Arkel, J., van den Borne, J. J. G. C., Pérez Gutiérrez, O., Smidt, H., Kemp, B., Müller, M. and Hooiveld, G. J. E. J. 2013. A diet high in resistant starch modulates microbiota composition, SCFA concentrations and gene expression in pig intestine. *J. Nutr.* 143(3):274–283.

Harris, V. C., Haak, B. W., Handley, S. A., Jiang, B., Velasquez, D. E., Hykes, B. L. Jr., Droit, L., Berbers, G. A. M., Kemper, E. M., van Leeuwen, E. M. M., Boele van Hensbroek, M. and Wiersinga, W. J. 2018. Effect of antibiotic-mediated microbiome modulation on rotavirus vaccine immunogenicity: A human, randomized-control proof-of-concept trial. *Cell Host Microbe* 24(2):197–207.e4.

Hoelzer, K., Bielke, L., Blake, D. P., Cox, E., Cutting, S. M., Devriendt, B., Erlacher-Vindel, E., Goossens, E., Karaca, K., Lemiere, S., Metzner, M., Raicek, M., Collell Suriñach, M., Wong, N. M., Gay, C. and Van Immerseel, F. 2018. Vaccines as alternatives to antibiotics for food producing animals. Part 1: Challenges and needs. *Vet. Res.* 49(1):64.

Hooper, L. V. 2015. Epithelial cell contributions to intestinal immunity. In: *Advances in Immunology*, Frederick W. Alt (Ed.), 126, Chapter 3, pp. 129–172, Elsevier.

Howitt, M. R., Lavoie, S., Michaud, M., Blum, A. M., Tran, S. V., Weinstock, J. V., Gallini, C. A., Redding, K., Margolskee, R. F., Osborne, L. C., Artis, D. and Garret, W. S. 2016. Tuft cells, taste-chemosensory cells, orchestrate parasite type 2 immunity in the gut. *Science* 351(6279):1329–1333.

Hubbard, T. D., Murray, I. A., Bisson, W. H., Lahoti, T. S., Gowda, K., Amin, S. G., Patterson, A. D. and Perdew, G. H. 2015. Adaptation of the human aryl hydrocarbon receptor to sense microbiota-derived indoles. *Sci. Rep.* 5:12689.

Hughes, C. S., Postovit, L. M. and Lajoie, G. A. 2010. Matrigel: A complex protein mixture required for optimal growth of cell culture. *Proteomics* 10(9):1886–1890.

Huygelen, V., De Vos, M., Willemen, S., Fransen, E., Casteleyn, C., Van Cruchten, S. and Van Ginneken, C. 2014. Age-related differences in mucosal barrier function and morphology of the small intestine in low and normal birth weight piglets. *J. Anim. Sci.* 92(8):3398–3406.

Iliev, I. D., Spadoni, I., Mileti, E., Matteoli, G., Sonzogni, A., Sampietro, G. M., Foschi, D., Caprioli, F., Viale, G. and Resigno, M. 2009. Human intestinal epithelial cells promote the differentiation of tolerogenic dendritic cells. *Gut* 58(11):1481–1489.

Isaacson, R. and Kim, H. B. 2012. The intestinal microbiome of the pig. *Anim. Health Res. Rev.* 13(1):100–109.

Iwata, M., Hirakiyama, A., Eshima, Y., Kagechika, H., Kato, C. and Song, S. Y. 2004. Retinoic acid imprints gut-homing specificity on T cells. *Immunity* 21(4):527–538.

Jang, M. H., Kweon, M. N., Iwatania, K., Yamamoto, M., Teraharad, K., Sasakawaf, C., Suzuki, T., Nochid, T., Yokotai, Y., Rennert, P. D., Hiroia, T., Tamagawaa, H., Iijimaa, H., Kunisawaa, J., Yukid, Y. and Kiyonoa, H. 2004. Intestinal villous M cells: An antigen entry site in the mucosal epithelium. *Proc. Natl. Acad. Sci. U. S. A.* 101(16):6110–6115.

Johansson-Lindbom, B., Svensson, M., Pabst, O., Palmqvist, C., Marquez, G., Forster, R. and Agace, W. W. 2005. Functional specialization of hut CD103+ dendritic cells in the regulation of tissue-selective T cell homing. *J. Exp. Med.* 202(8):1063–1073.

Kaiko, G. E., Ryu, S. H., Koues, O. I., Collins, P. L., Solnica-Krezel, L., Pearce, E. J., Pearce, E. L., Oltz, E. M. and Stappenbeck, T. S. 2016. The colonic crypt protects stem cells from microbiota-derived metabolites. *Cell* 165(7):1708–1720.

Kamaya, H., Okumura, R. and Takeda, K. 2020. Interaction between the microbiota, epithelia, and immune cells in the intestine. *Annu. Rev. Immunol.* 38:23–48.

Kandasamy, S., Chattha, K. S., Vlasova, A. N. and Saif, L. J. 2014. Prenatal vitamin A deficiency impairs adaptive immune responses to pentavalent rotavirus vaccine (RotaTeq®) in a neonatal gnotobiotic pig model. *Vaccine* 32(7):816–824.

Käser, T. 2021. Swine as biomedical animal model for T-cell research – success and potential for transmittable and non-transmittable human diseases. *Mol. Immunol.* 135:95–115.

Kim, C. H., Kim, D., Ha, Y., Cho, K. D., Lee, B. H., Seo, I. W., Kim, S. H. and Chae, C. 2009. Expression of mucins and trefoil factor family protein-1 in the colon of pigs naturally infected with *Salmonella typhimurium*. *J. Comp. Pathol.* 140(1):38–42.

Koay, H. F., Gherardin, N. A., Xu, C., Seneviratna, R., Zhao, Z., Chen, Z., Fairlie, D. P., McCluskey, J., Pellicci, D. G., Uldrich, A. P. and Godfrey, D. I. 2019. Diverse MR1-restricted T cells in mice and humans. *Nat. Commun.* 10(1):2243.

Koh, A., De Vadder, F., Kovatcheva-Datchary, P. and Bäckhed, F. 2016. From dietary fiber to host physiology: Short-chain fatty acids as key bacterial metabolites. *Cell* 165(6):1332–1345.

Kotunia, A., Woliński, J., Laubitz, D., Jurkowska, M., Romé, V., Guilloteau, P. and Zabielski, R. 2004. Effect of sodium butyrate on the small intestine development in neonatal piglets fed by artificial sow. *J. Physiol. Pharmacol.* 55 (Suppl. 2):59–68.

Kratochvil, M. J., Seymour, A. J., Li, T. L., Pasca, S. P., Kuo, C. J. and Heilshorn, S. C. 2019. Engineered materials for organoid systems. *Nat. Rev. Mater.* 4(9):606–622.

Kringel, H., Iburg, T., Dawson, H., Aasted, B. and Roepstorff, A. 2006. A time course study of immunological responses in *Trichuris suis* infected pigs demonstrates induction of a local type 2 response associated with worm burden. *Int. J. Parasitol.* 36(8):915–924.

Kumar, P., Luo, Q., Vickers, T. J., Seikh, A., Lewis, W. G. and Fleckenstein, J. M. 2014. EatA, an immunogenic protective antigen of enterotoxigenic E. coli, degrades intestinal mucin. *Infect. Immun.* 82(2):500–508.

Lamote, J. A. S., Kestens, M., Van Waesberghe, C., Delva, J., De Pelsmaeker, S., Devriendt, B. and Favoreel, H. W. 2017. The pseudorabies virus glycoprotein gE/gI complex suppresses type I interferon production by plasmacytoid dendritic cells. *J. Virol.* 91(7):e02276-16.

Langel, S. N., Paim, F. C., Alhamo, M. A., Lager, K. M., Vlasova, A. N. and Saif, L. J. 2019. Oral vitamin A supplementation of porcine epidemic diarrhea virus infected gilts enhances IgA and lactogenic immune protection of nursing piglets. *Vet. Res.* 50(1):101.

Lei, W., Ren, W., Ohmoto, M., Urban, J. F. Jr., Matsumoto, I., Margolskee, R. F. and Jiang, P. 2018. Activation of intestinal tuft cell-expressed Sucnr1 triggers type 2 immunity in the mouse small intestine. *Proc. Natl. Acad. Sci. U. S. A.* 115(21):5552–5557.

Li, X., Mao, M., Zhang, Y., Yu, K. and Zhu, W. 2019. Succinate modulates intestinal barrier function and inflammation response in pigs. *Biomolecules* 9(9):486.

Li, Y., Yang, N., Chen, J., Huang, X., Zhang, N., Yang, S., Liu, G. and Liu, G. 2020. Next-generation porcine intestinal organoids: an apical-out organoid model for swine enteric virus infection and immune response investigations. *J Virol.* 94(21):e01006-20. doi: 10.1128/JVI.01006-20.

Liebler-Tenorio, E. M. and Pabst, R. 2006. MALT structure and function in farm animals. *Vet. Res.* 37(3):257–280.

Lindemans, C. A., Calafiore, M. C., Mertelsmann, A. M., O'Connor, M. H., Dudakov, J. A., Jenq, R. R., Velardi, E., Young, L. F., Smith, O. M., Lawrence, G., Ivanov, J. A., Fu, Y. Y., Takashima, S., Hua, G., Martin, M. L., O'Rourke, K. P., Lo, Y. H., Mokry, M., Romera-Hernandez, M., Cupedo, T., Dow, L. E., Nieuwenhuis, E. E., Shroyer, N. F., Liu, C., Kolesnick, R., van der Brink, M. R. M. and Hanash, A. M. 2015. Interleukin-22 promotes intestinal-stem-cell-mediated epithelial regeneration. *Nature* 528(7583):560–564.

Liu, H., Wang, J., He, T., Becker, S., Zhang, G., Li, D. and Ma, X. 2018. Butyrate: A double-edged sword for health? *Adv. Nutr.* 9(1):21–29.

Liu, M. L., Lyu, X. and Werth, V. P. 2021. Recent progress in the mechanistic understanding of NET formation in neutrophils. *FEBS Journal*. doi: 10.1111/febs.16036.

Loke, P. and Cadwell, K. 2018. Getting a taste for parasites in the Gut. *Immunity* 49(1):16–18.

Looft, T., Allen, H. K., Cantarel, B. L., Levine, U. Y., Bayles, D. O., Alt, D. P., Henrissat, B. and Stanton, T. B. 2014. Bacteria, phages and pigs: The effects of in-feed antibiotics on the microbiome at different gut locations. *ISME J.* 8(8):1566–1576.

Luo, Y., Van Nguyen, U., de la Fe Rodriguez, P. Y., Devriendt, B. and Cox, E. 2015. F4^{+} ETEC infection and oral immunization with F4 fimbriae elicits an IL-17-dominated immune response. *Vet. Res.* 46:121.

Luo, Y., Xu, J., Zhang, C., Jiang, C., Ma, Y., He, H., Wu, Y., Devriendt, B., Cox, E. and Zhang, H. 2019. Toll-like receptor 5-mediated IL-17C expression in intestinal epithelial cells enhances epithelial host defense against F4^{+} ETEC infection. *Vet. Res.* 50(1):48.

Luoma, A. M., Castro, C. D. and Adams, E. J. 2014. Gammadelta T cell surveillance via CD1 molecules. *Trends Immunol.* 35(12):613–621.

Maisonnasse, P., Bouguyon, E., Piton, G., Ezquerra, A., Urien, C., Deloizy, C., Bourge, M., Leplat, J. J., Simon, G., Chevalier, C., Vincent-Naulleau, S., Crisci, E., Montoya, M., Schwartz-Cornil, I. and Bertho, N. 2016. The respiratory DC/macrophage network at steady-state and upon influenza infection in the swine biomedical model. *Mucosal Immunol.* 9(4):835–849.

Mallo, J. J., Balfagón, A., Gracia, M. I., Honrubia, P. and Puyalto, M. 2012. Evaluation of different protections of butyric acid aiming for release in the last part of the gastrointestinal tract of piglets. *J. Anim. Sci.* 90 (Suppl. 4):227–229.

Marion-Letellier, R., Déchelotte, P., Lacucci, M. and Ghosh, S. 2009. Dietary modulation of peroxisome proliferator-activated receptor gamma. *Gut* 58(4):586–593.

Markowiak, P. and Slizewska, K. 2018. The role of probiotics, prebiotics and synbiotics in animal nutrition. *Gut Pathog.* 10:21.

Maroilley, T., Berri, M., Lemonnier, G., Esquerré, D., Chevaleyre, C., Mélo, S., Meurens, F., Coville, J. L., Replat, J. J., Rau, A., Bed'hom, B., Vincent-Naulleau, S., Mercat, M. J., Billon, Y., Lepage, P., Rogel-Gaillard, C. and Estellé, J. 2018. Immunome differences between porcine ileal and jejunal Peyer's patches revealed by global transcriptome sequencing of gut- associated lymphoid tissues. *Sci. Rep.* 8(1):9077.

Masure, D., Vlaminck, J., Wang, T., Chiers, K., Van den Broeck, W., Vercruysse, J. and Geldhof, P. 2013. A role for eosinophils in the intestinal immunity against infective *Ascaris suum* larvae. *PLoS Negl. Trop. Dis.* 7(3):e2138.

Mazzini, E., Massimiliano, L., Penna, G. and Rescigno, M. 2014. Oral tolerance can be established via gap junction transfer of fed antigens from CX3CR1+ macrophages to CD103+ dendritic cells. *Immunity* 40(2):248–261.

McDole, J. R., Wheeler, L. W., McDonald, K. G., Wang, B., Konjufca, V., Knoop, K. A., Newberry, R. D. and Miller, M. J. 2012. Goblet cells deliver luminal antigen to CD103 + dendritic cells in the small intestine. *Nature* 483(7389):345–349.

Meurens, F., Berri, M., Whale, J., Dybvig, T., Strom, S., Thompson, D., Brownlie, R., Townsend, H. G., Salmon, H. and Gerdts, V. 2006. Expression of TECK/CCL25 and MEC/CCL28 chemokines and their respective receptors CCR9 and CCR10 in porcine mucosal tissues. *Vet. Immunol. Immunopathol.* 113(3–4):313–327.

Mora, J. R., Iwata, M., Eksteen, B., Song, S. Y., Junt, T., Senman, B., Otipoby, K. L., Yokota, A., Takeuchi, H., Ricciardi-Castagnoli, P., Rajewsky, K., Adams, D. H. and von Andrian, U. H. 2006. Generation of gut-homing IgA-secreting B cells by intestinal dendritic cells. *Science* 314(5802):1157–1160.

Morfitt, D. C. and Pohlenz, J. F. 1989. Porcine colonic lymphoglandular complex: Distribution, structure, and epithelium. *Am. J. Anat.* 184(1):41–51.

Myhill, L. J., Stolzenbach, S., Hansen, T. V. A., Skovgaard, K., Stensvold, C. R., Andersen, L. O., Nejsum, P., Mejer, H., Thamsborg, S. M. and Williams, A. R. 2018. Mucosal barrier and Th2 immune responses are enhanced by dietary inulin in pigs infected with *Trichuris suis*. *Front. Immunol.* 9:2557.

Nadjsombati, M. S., McGinty, J. W., Lyons-Cohen, M. R., Jaffe, J. B., DiPeso, L., Schneider, C., Miller, C. N., Pollack, J. L., Nagana Gowda, G. A., Fontana, M. F., Erle, D. J., Anderson, M. S., Locksley, R. M., Raftery, D. and Von Moltke, J. 2018. Detection of succinate by intestinal tuft cells triggers a type 2 innate immune circuit. *Immunity* 49(1):33–41.

Neutra, M. R., Mantis, N. J. and Kraehenbuhl, J. P. 2001. Collaboration of epithelial cells with organized mucosal lymphoid tissues. *Nat. Immunol.* 2(11):1004–1009.

Nikolaev, M., Mitrofanova, O., Broguiere, N., Geraldo, S., Dutta, D., Tabata, Y., Elci, B., Brandenberg, N., Kolotuev, I., Gjorevski, N., Clevers, H. and Lutolf, M. P. 2020. Homeostatic mini-intestines through scaffold-guided organoid morphogenesis. *Nature* 585(7826):574–578.

Nowarski, R., Jackson, R. and Flavell, R. A. 2017. The stromal intervention: Regulation of immunity and inflammation at the epithelial-mesenchymal barrier. *Cell* 168(3):362–375.

Nyström, E. E. L., Martinez-Abad, B., Arike, L., Birchenough, G. M. H., Nonnecke, E. B., Castillo, P. A., Svensson, F., Bevins, C. L., Hansson, G. C. and Johansson, M. E. V. 2021. An intercrypt subpopulation of goblet cells is essential for colonic mucus barrier function. *Science* 372(6539):eabb1590.

Odenwald, M. A. and Turner, J. R. 2017. The intestinal epithelial barrier: A therapeutic target? *Nat. Rev. Gastroenterol. Hepatol.* 14(1):9–21.

Oh, J. Z., Ravindran, R., Chassaing, B., Carvalho, F. A., Maddur, M. S., Bower, M., Hakimpour, P., Gill, K. P., Nakaya, H. I., Yarovinsky, F., Sartor, R. B., Gewirtz, A. T. and Pulendran, B. 2014. TLR5-mediated sensing of gut microbiota is necessary for antibody responses to seasonal influenza vaccination. *Immunity* 41(3):478–492.

Pajarillo, E. A. B., Chae, J. P., Balolong, M. P., Bum Kim, H. and Kang, D. K. 2014. Assessment of fecal bacterial diversity among healthy piglets during the weaning transition. *J Gen Appl Microbiol.* 60(4):140–146. doi: 10.2323/jgam.60.140.

Pearce, S. C., Weber, G. J., van Sambeek, D. M., Soares, J. W., Racicot, K. and Breault, D. T. 2020. Intestinal enteroids recapitulate the effects of short-chain fatty acids on the intestinal epithelium. *PLoS ONE* 15(4):e0230231.

Peng, J., Tang, Y. and Huang, Y. 2021. Gut health: The results of microbial and mucosal immune reactions in pigs. *Anim. Nutr.* 7(2):282–294, doi: 10.1016/j.aninu.2021.01.001.

Phillipson, M. and Kubes, P. 2019. The healing power of neutrophils. *Trends Immunol.* 40(7):635–647.

Pine, G. M., Batugedara, H. M. and Nair, M. G. 2018. Here, there and everywhere: Resistin-like molecules in infection, inflammation, and metabolic disorders. *Cytokine* 110:442–451.

Piper, C. J. M., Rosser, E. C., Oleinika, K., Nistala, K., Krausgruber, T., Rendeiro, A. F., Banos, A., Drozdov, I., Villa, M., Thomson, S., Xanthou, G., Bock, C., Stockinger, B. and Mauri, C. 2019. Aryl hydrocarbon receptor contributes to the transcriptional program of IL-10-producing regulatory B cells. *Cell Rep.* 29(7):1878–1892.

Pohl, C. S., Medland, J. E., Mackey, E., Edwards, L. L., Bagley, K. D., DeWilde, M. P., Williams, K. J. and Moeser, A. J. 2017. Early weaning stress induces chronic functional diarrhea, intestinal barrier defects, and increased mast cell activity in a porcine model of early life adversity. *Neurogastroenterol. Motil.* 29(11):e13118. doi: 10.1111/nmo.13118.

Potockova, H., Sinkorova, J., Karova, K. and Sinkora, M. 2015. The distribution of lymphoid cells in the small intestine of germ-free and conventional piglets. *Dev. Comp. Immunol.* 51(1):99–107.

Powell, R. H. and Behnke, M. S. 2017. WRN conditioned media is sufficient for in vitro propagation of intestinal organoids from large farm and small companion animals. *Biol. Open* 6(5):698–705.

Price, A. E., Shamardani, K., Lugo, K. A., Deguine, J., Roberts, A. W., Lee, B. L. and Barton, G. M. 2018. A map of toll-like receptor expression in the intestinal epithelium reveals distinct spatial, cell type-specific, and temporal patterns. *Immunity* 49(3):560–575.

Quintana-Hayashi, M. P., Mahu, M., De Pauw, N., Boyen, F., Pasmans, F., Martel, A., Premaratne, P., Fernandez, H. R., Teymournejad, O., Vande Maele, L., Haesebrouck, F. and Lindéna, S. K. 2015. The levels of *Brachyspira hyodysenteriae* binding to porcine colonic mucins differ between individuals, and binding is increased to mucins from infected pigs with de novo MUC5AC synthesis. *Infect. Immun.* 83(4):1610–1619.

Quintana-Hayashi, M. P., Padra, M., Padra, J. T., Benktander, J. and Linden, S. K. 2018. Mucus-pathogen interactions in the gastro-intestinal tract of farmed animals. *Microorganisms* 6(2):55.

Rescigno, M. and Di Sabatino, A. 2009. Dendritic cells in intestinal homeostasis and disease. *J. Clin. Invest.* 119(9):2441-2450.

Roager, H. M. and Licht, T. R. 2018. Microbial tryptophan catabolites in health and disease. *Nat. Commun.* 9(1):3294.

Roediger, W. E. W. 1982. Utilization of nutrients by isolated epithelial cells of the rat colon. *Gastroenterology* 83(2):424-429.

Rothkötter, H. J. 2009. Anatomical particularities of the porcine immune system - A physician's view. *Dev. Comp. Immunol.* 33(3):267-272.

Salmon, H., Berri, M., Gerdts, V. and Meurens, F. 2009. Humoral and cellular factors of maternal immunity in swine. *Dev. Comp. Immunol.* 33(3):384-393.

Sang, Y. and Blecha, F. 2009. Porcine host defense peptides: Expanding repertoire and functions. *Dev. Comp. Immunol.* 33(3):334-343.

Sato, T., van Es, J. H., Snippert, H. J., Stange, D. E., Vries, R. G., van de Born, M., Barker, N., Shroyer, N. F., van de Wetering, M. and Clevers, H. 2011. Paneth cells constitute the niche for Lgr5 stem cells in intestinal crypts. *Nature* 469(7330):415-418.

Sato, T., Vries, R. G., Snippert, H. J., van de Wetering, M., Barker, N., Stange, D. E., van Es, J. H., Abo, A., Kujala, P., Peters, P. J. and Clevers, H. 2009. Single Lgr5 stem cells build crypt-villus structures in vitro without a mesenchymal niche. *Nature* 459(7244):262-265.

Schmidt, S., Sassu, E. L., Vatzia, E., Pierron, A., Lagler, J., Mair, K. H., Stadler, M., Knecht, C., Spergser, J., Dolezal, M., Springer, S., Theuß, T., Fachinger, V., Ladinig, A., Saalmüller, A. and Gerner, W. 2021. Vaccination and infection of swine with *Salmonella* typhimurium induces a systemic and local multifunctional $CD4^+$ T-cell response. *Front. Immunol.* 11:603089.

Schokker, D., Fledderus, J., Jansen, R., Vastenhouw, S. A., de Bree, F. M., Smits, M. A. and Jansman, A. A. J. M. 2018. Supplementation of fructooligosaccharides to suckling piglets affects intestinal microbiota colonization and immune development. *J. Anim. Sci.* 96(6):2139-2153.

Seeger, B. 2020. Farm animal-derived models of the intestinal epithelium: Recent advances and future applications of intestinal organoids. *Altern Lab. Anim.* 48(5-6):215-233.

Shan, M., Gentile, M., Yeiser, J. R., Walland, A. C., Bornstein, V. U., Chen, K., He, B., Cassis, L., Bigas, A., Cols, M., Comerma, L., Huang, B., Blander, J. M., Xiong, H., Mayer, L., Berin, C., Augenlicht, L. H., Velcich, A. and Cerutti, A. 2013. Mucus enhances gut homeostasis and oral tolerance by delivering immunoregulatory signals. *Science* 342(6157):447-453.

Snoeck, V., Goddeeris, B. M. and Cox, E. 2005. The role of enterocytes in the intestinal barrier function and antigen uptake. *Microbes Infect.* 7(7-8):997-1004.

Sotira, S., Dell'Anno, M., Caprarulo, V., Hejna, M., Pirrone, F., Callegari, M. L., Tucci, T. V. and Rossi, L. 2020. Effects of tributyrin supplementation on growth performance, insulin, blood metabolites and gut microbiota in weaned piglets. *Animals (Basel)* 10(4):726.

Sternini, C., Anselmi, L. and Rozengurt, E. 2008. Enteroendocrine cells: A site of "taste" in gastrointestinal chemosensing. *Curr. Opin. Endocrinol. Diabetes Obes.* 15(1):73-78.

Stockinger, B., Di Meglio, P., Gialitakis, M. and Duarte, J. H. 2014. The aryl hydrocarbon receptor: Multitasking in the immune system. *Annu. Rev. Immunol.* 32:403-432.

Stokes, C. R. 2017. The development and role of microbial-host interactions in gut mucosal immune development. *J. Anim. Sci. Biotechnol.* 8:12.

Stuyven, E., Van de Broeck, W., Nauwynck, H., Goddeeris, B. M. and Cox, E. 2010. Oral administration of beta-1,3/1,6-glucan Macrogard fails to enhance the mucosal immune response following oral F4 fimbrial immunisation in gnotobiotic pigs. *Vet. Immunol. Immunopathol.* 137(3-4):291-297.

Sun, P., Li, D., Li, Z., Dong, B. and Wang, F. 2008. Effects of glycinin on IgE-mediated increase of mast cell numbers and histamine release in the small intestine. *J. Nutr. Biochem.* 19(9):627-633.

Swaminathan, G., Nguyen, L. P., Namkoong, H., Pan, J., Haileselassie, Y., Patel, A., Ji, A. R., Mikhail, D. M., Dinh, T. T., Singh, H., Liao, B., Vázquez-Montesino, L. M., Butcher, E. C. and Habtezion, A. 2021. The aryl hydrocarbon receptor regulates expression of mucosal trafficking receptor GPR15. *Mucosal Immunol.* 14(4):852-861, doi: 10.1038/s41385-021-00390-x.

Swiech, E., Tusnio, A., Barszcz, M., Taciak, M. and Siwiak, E. 2019. Goblet cells and mucus layer in the gut of young pigs: Response to dietary contents of threonine and non-essential amino acids. *J. Anim. Physiol. Anim. Nutr. (Berl)* 103(3):894-905.

Tian, S., Wang, J., Yu, H., Wang, J. and Zhu, W. 2019. Changes in ileal microbial composition and microbial metabolism by an early-life galacto-oligosaccharides intervention in a neonatal porcine model. *Nutrients* 11(8):1753.

Trachsel, J., Briggs, C., Gabler, N. K., Allen, H. K. and Loving, C. L. 2019. Dietary resistant starch alters intestinal microbial communisties and their metabolites, and marker of immune regulation and barrier function in swine. *Front. Immunol.* 10:1381.

Tremaroli, V. and Bäckhed, F. 2012. Functional interactions between the gut microbiota and host metabolism. *Nature* 489(7415):242-249.

Ugolini, M., Gerhard, J., Burkert, S., Jensen, K. J., Georg, P., Ebner, F., Volkers, S. M., Thada, S., Dietert, K., Bauer, L., Schäfer, A., Helbig, E. T., Opitz, B., Kurth, F., Sur, S., Dittrich, N., Gaddam, S., Conrad, M. L., Benn, C. S., Blohm, U., Gruber, A. D., Hutloff, A., Hartmann, S., Boekschoten, M. V., Müller, M., Jungersen, G., Schumann, R. R., Suttorp, N. and Sander, L. E. 2018. Recognition of microbial viability via TLR8 drives T_{FH} cell differentiation and vaccine responses. *Nat. Immunol.* 19(4):386-396.

Valpotić, H., Janjatović, A. K., Lacković, G., Božić, F., Dobranić, V., Svoboda, D., Valpotić, I. and Popović, M. 2010. Increased number of intestinal villous M cells in levamisole-pretreated weaned pigs experimentally infected with $F4ac^{+}$ enterotoxigenic Escherichia coli strain. *Eur. J. Histochem.* 54(2):e18.

Van der Hee, B., Loonen, L. M. P., Taverne, N., Taverne-Thiele, J. J., Smidt, H. and Wells, J. M. 2018. Optimized procedures for generating an enhanced, near physiological 2D culture system from porcine intestinal organoids. *Stem Cell Res.* 28:165-171.

Van Hees, H. M. J., Davids, M., Maes, D., Millet, S., Possemiers, S., den Hartog, L. A., van Kempen, T. A. T. G. and Janssens, G. P. J. 2019. Dietary fibre enrichment of supplemental feed modulates the development of the intestinal tract in suckling piglets. *J. Anim. Sci. Biotechnol.* 10:83.

Venegas, D. P., De la Fuente, M. K., Landskron, G., Gonzalez, M. J., Quera, R., Dijkstra, G., Harmsen, H. J. M., Faber, K. N. and Hermoso, M. A. 2019. Short chain fatty acids (SCFAs)-mediated gut epithelial and immune regulation and its relevance for inflammatory bowel diseases. *Front. Immunol.* 10:277.

Venkatraman, A., Ramakrishna, B. S., Shaji, R. V., Kumar, N. S. N., Pulimood, A. and Patra, S. 2003. Amerioration of dextran sulfate colitis by butyrate: Role of heat shock protein 70 and NF-kB. *Am. J. Physiol. Gastrointest. Liver Physiol.* 285(1):G177-G184.

Vermeire, B., Gonzalez, L. M., Jansens, R. J. J., Cox, E. and Devriendt, B. 2021. Porcine small intestinal enteroids as a model to explore ETEC-host interactions in the gut. *Vet. Res.* 52(1):94.

Vlaminck, J., Masure, D., Wang, T., Nejsum, P., Hokke, C. H. and Geldhof, P. 2016. A phosphorylcholine-containing glycolipid-like antigen present on the surface of infective stage larvae of *Ascaris* spp. Is a major antibody target in infected pigs and humans. *PLoS Negl. Trop. Dis.* 10(12):e0005166.

Vlasova, A. N., Takanashi, S., Miyazaki, A., Rajashekara, G. and Saif, L. J. 2019. How the gut microbiome regulates host immune responses to viral vaccines. *Curr. Opin. Virol.* 37:16–25.

Von Moltke, J., Ji, M., Liang, H. E. and Locksley, R. M. 2015. Tuft-cell-derived IL-25 regulates an intestinal ILC2-epithelial response circuit. *Nature* 529(7585):221–225.

Wang, J., Wang, S., Liu, H., Zhang, D., Wang, Y. and Ji, H. 2019. Effects of oligosaccharides on the growth and stress tolerance of *Lactobacillus plantarum* ZLP001 *in vitro*, and the potential symbiotic effects of *L. plantarum* ZLP001 and fructo-oligosaccharides in post-weaning piglets. *J. Anim. Sci.* 97(11):4588–4597.

Wang, Y., Shi, J., Yan, J., Xiao, Z., Hou, X., Lu, P., Hou, S., Mao, T., Liu, W., Ma, Y., Zhang, L., Yang, X. and Qi, H. 2017. Germinal center development of memory B cells driven by IL-9 from follicular helper T cells. *Nat. Immunol.* 18(8):921–930.

Wiarda, J. E., Trachsel, J. M., Bond, Z. F., Byrne, K. A., Gabler, N. K. and Loving, C. L. 2020. Intraepithelial T cells diverge by intestinal location as pigs age. *Front. Immunol.* 11:1139.

Worthington, J. J., Reimann, F. and Gribble, F. M. 2018. Enteroendocrine cells-sensory sentinels of the intestinal environment and orchestrators of mucosal immunity. *Mucosal Immunol.* 11(1):3–20. doi: 10.1038/mi.2017.73.

Wu, A., Yu, B., Zhang, K., Xu, Z., Wu, He, J., Luo, J., Luo, Y., Yu, J., Zheng, P., Che, L., Mao, X., Huang, Z., Wang, L., Zhao, J. and Chen, D. 2020. Transmissible gastroenteritis virus targets Paneth cells to inhibit the self-renewal and differentiation of Lgr5 intestinal stem cells via Notch signaling. *Cell Death Dis.* 11(1):40.

Xiao, X., Li, K., Ma, X., Liu, B., He, X., Yang, S., Wang, W., Jiang, B. and Cai, J. 2019. Mucosal-associated invariant T cells expressing the TRAV1-TRAJ33 chain are present in pigs. *Front. Immunol.* 10:2070.

Xiong, L., Dean, J. W., Fu, Z., Oliff, K. N., Bostick, J. W., Ye, J., Chen, Z. E., Mühlbauer, M. and Zhou, L. 2020. Ahr-Foxp3-RORγt axis controls gut homing of $CD4^+$ T cells by regulating GPR15. *Sci. Immunol.* 5(48):eaaz7277.

Yang, W., Yu, T., Huang, X., Bilotta, A. J., Xu, L., Lu, Y., Sun, J., Pan, F., Zhou, J., Zhang, W., Yao, S., Maynard, C. L., Singh, N., Dann, S. M., Liu, Z. and Cong, Y. 2020. Intestinal microbiota-derived short-chain fatty acids regulation of immune cell IL-22 production and gut immunity. *Nat. Commun.* 11(1):4457.

Yilmaz, Ö. H., Katajisto, P., Lamming, D. W., Gültekin, Y., Bauer-Rowe, K. E., Sengupta, S., Birsoy, K., Dursun, A., Yilmaz, V. O., Selig, M., Nielsen, G. P., Mino-Kenudson, M., Zukerberg, L. R., Bhan, A. K., Deshpande, V. and Sabatini, D. M. 2012. mTORC1 in the Paneth cell niche couples intestinal stem cell-function to calorie intake. *Nature* 486(7404):490–495.

Zaiss, M. M., Rapin, A., Lebon, L., Dubey, L. K., Mosconi, I., Sarter, K., Piersigilli, A., Menin, L., Walker, A. W., Rougemont, J., Parewyck, P., Geldhof, P., McCoy, K. D., Macpherson, A. J., Croese, J., Giacomin, P. R., Loukas, A., Junt, T., Marsland, B. J. and Harris, N. L.

2015. The intestinal microbiota contributes to the ability of helminths to modulate allergic inflammation. *Immunity* 43:1-2.

Zhang, L., Wu, W., Lee, Y. K., Wie, J. and Zhang, H. 2018. Spatial heterogeneity and co-occurrence of mucosal and luminal microbiome across swine intestinal tract. *Front. Microbiol.* 9:48.

Zhao, J., Liu, P., Wu, Y., Guo, P., Liu, L., Ma, N., Levesque, C., Chen, Y., Zhao, J., Zhang, J. and Ma, X. 2018. Dietary fiber increases butyrate-producing bacteria and improves the growth performance of weaned piglets. *J. Agric. Food Chem.* 66(30):7995-8004.

Zihni, C., Mills, C., Matter, K. and Balda, M. S. 2016. Tight junctions: From simple barriers to multifunctional molecular gates. *Nat. Rev. Mol. Cell Biol.* 17(9):564-580.

Zimmermann, P. and Curtis, N. 2018. The influence of probiotics on vaccine responses - A systematic review. *Vaccine* 36(2):207-213.

Zuckermann, F. A. and Gaskins, H. R. 1996. Distribution of porcine CD4/CD8 double-positive T lymphocytes in mucosa-associated lymphoid tissues. *Immunology* 87(3):493-499.

Chapter 7

Nutritional strategies to boost immune function in pigs

Glen W. Almond, North Carolina State University, USA; and Sara D. Hough, DSM Nutritional Products, USA

1 Introduction

Prior to discussing the influence of nutritional strategies on immune function, it is necessary to consider the changes in our knowledge of various nutritional constituents and the porcine immune system. A number of factors hindered the progress of our understanding of immunity in pigs. The lack of reagents and specific antibodies interfered with the study of T cell phenotypes and cytokines. Thus, some older studies were restricted to entire lymphocyte populations (Dalgaard et al., 2018). Rather than attempting to provide a review of porcine immunity, readers are encouraged to read excellent reviews (Saalmüller and Gerner, 2016; Chase and Lunney, 2019).

The number of papers examining the relationship between nutrition and porcine immunity has increased (Table 1), but the information often is confusing and confounded, particularly when assessing nutritional manipulation or immune system stimulation (ISS). Also, it is difficult to compare different animal

http://dx.doi.org/10.19103/AS.2022.0103.07

Table 1 Examples of how some nutrients may alter immunity in pigs. Please note that this is not an all-inclusive list

Nutrient	Application	Modification of immunity	Reference(s)
Astragalus polysaccharides	1 week prior to parturition	↑ IgG, IgM in colostrum	Tan et al. (2017)
Seaweed extract (10 g/day)	Day 109 of gestation until weaning	↑ IgG in colostrum and circulatory IgG in piglets	Leonard et al. (2010)
α-tocopherol acetate (500 mg/kg feed) and vitamin C (10 g/day)	Sow diets	↑ Total Ig and IgG in plasma of piglets	Pinelli-Saavedra et al. (2008)
Vitamin E (250 IU/kg)	Last week of gestation and lactation in sows	High levels of plasma IgG, IgA, total antioxidant capacity and catalase in piglets	Wang et al. (2017)
Dietary fully oxidized β-carotene supplementation	From gestation day 85 until day 21 of lactation	Enhanced colostrum immunoglobulin levels, and IgM and IgG content of 14-day milk. ↓ TNF-α and IL-8 levels in colostrum, as well as the TNF-α and IL-18 levels in 14-day milk	Chen et al. (2021)
Dietary conjugated linoleic acid	42 days prior to disease challenge	Early disappearance of antigen-stimulated proliferation and cytotoxicity, specific responses of CD8 αβ+ cells	Bassaganya-Riera et al. (2002)
Dietary ratios of n-6:n-3 PUFA of 1:1 and 5:1	During grow-finish stages of growth	Markedly suppressed inflammatory cytokines IL-1β, TNF-α and IL-6	Duan et al. (2014)
Supplementation with β(1-3)(1-6)-D-glucan (15 mg/kg/day)	Dietary supplement in piglets	Stimulated phagocytic activity, and IL-2 production, while lowering cortisol and TNF-α levels after LPS challenge	Vetvicka and Oliveira (2014)
β-glucans (*Saccharomyces cerevisiae* or *Sclerotium rolfsii*)	Fed for 2 weeks after weaning	Reduced susceptibility to an F4+ ETEC infection	Stuyven et al. (2009)
2500 ppm zinc oxide	Weaned pig diets	↑ numbers of CD3+ T cells, and CD3+CD4+Foxp3+ regulatory T cells, elevated IL-2 production and decreased IL-10 production	Kloubert et al. (2018)

Daily oral retinyl acetate (30 000 IU).	Starting at gestation day 76 and throughout lactation	↑ PEDV IgA antibody-secreting cells and PEDV IgA antibodies in serum pre-partum and IgA+ β7+ (gut homing) cells in milk	Langel et al. (2019)
Vitamin 25(OH)D3 supplementation (3500 IU/kg)	Weaned pig diets	↑ leukocyte cell numbers ↑ viability and phagocytic activity of leukocytes in serum and in bronchioalveolar lavage material	Konowalchuka et al. (2013)
Vitamin E (250 IU/kg)	Maternal diet	Enhanced humoral immune function and antioxidant activity in sows and piglets	Wang et al. (2017)
Wakame seaweed stalks as 1% seaweed powder	Diet supplementation for 20–30 kg and 70 kg BW pigs	↑ percentages of peripheral blood natural killer cells and CD3+CD4−CD8+ cells	Shimazu et al. (2019)

studies as confounding factors, such as diets, environmental conditions, feed types and ingredients, sex, age, pathogen strain, and dose influence the immune or disease parameters.

Interactions between the immune system and metabolism need to be considered since both systems have the potential to influence each other. Studies of immune system metabolism ('immunometabolism') consider the effects of immune cells or immune cell products on organs that regulate whole body metabolism. In addition, immunometabolism examines metabolic pathways *within* immune cells and how this may alter the immune response (Pearce and Pearce, 2013). Immunometabolism represents an entire field of study and is beyond the scope of this chapter. However, metabolism is influenced not only by substrate availability but also by signaling pathways elicited by metabolites (Pearce and Pearce, 2013; Igata et al., 2019; Davanzo et al., 2021). Immune cells reside in immune system organs and different tissues. These tissues have divergent metabolic rates (Caputa et al., 2019), and this raises the question of how nutrition interacts with immune cells. The obvious question is, 'What are the ideal methods to assess nutritional intake on immunity, particularly during pathogen challenge?'

2 How does immune challenge affect pig growth?

A recent study, using a Hierarchical Clustering on Principal Components analysis, evaluated different challenge models (CM) and their ability to compromise growth performance in pigs (Rodrigues et al., 2021). Lipopolysaccharide (LPS) challenge (*Escherichia coli* with doses ranging from 1 µg/kg BW to 200 µg/kg BW) was more detrimental to average daily gain (ADG) than environmental (e.g. elevated ambient temperature and crowding), respiratory challenge (*Actinobacillus pleuropneumoniae*, *Mycoplasma hyopneumoniae*, *Pasteurella multocida*, porcine circovirus type 2, porcine reproductive and respiratory syndrome virus [PRRSV], or others), and sanitary models (fecal accumulation without cleaning and disinfecting); LPS also lowered gain:feed ratios (GF) more than all the other CM. Injections with bacterial LPS were considered an appropriate model for investigating immune stimulation and growth impairment in pigs (Rodrigues et al., 2021). The aforementioned observations confirmed a previous study that demonstrated LPS treatment increased glucose consumption, presumably by immune cells, resulting in limited energy availability and thus diminished growth potential (Kvidera et al., 2017).

In general, *sub-clinical* diseases were shown to reduce lean tissue growth and feed efficiency in growing pigs (Williams et al., 1997a,b; Le Floc'h et al., 2009). Reduced growth in challenged pigs was attributed to changes in nutrient utilization efficiency (Rakhshandeh et al., 2012). A meta-analysis also showed effects on nutrient utilization and feed intake on the reduction in growth (Pastorelli et al., 2012). Perhaps some of the differing results were due to age or weight of the pigs at the time of challenge. Post-weaned piglets were more affected by a LPS challenge than heavier pigs in the grow-finish stages of production (Rodrigues et al., 2021). Younger pigs had sudden and dramatic decreases in ADG and average daily feed intake (ADFI). The older pigs had greater time for recovery (Rodrigues et al., 2021).

Gender also appears to influence the response of pigs to ISS. Williams et al. (2009) showed that the pro-inflammatory cytokine response was sex-dependent when barrows and gilts were injected with LPS. Female pigs had more robust local and systemic immune responses, which, in turn, could reflect altered nutrient requirements. Furthermore, growth and nutrient requirements differ among boars, barrows, and gilts and these differences often are more apparent as they reach heavier weights (Quiniou et al., 2010).

While several proteins may be synthesized in the body and contribute to inflammation, other proteins and amino acids (AAs) are secreted into the gut lumen and are not recovered by the body. When pigs experience reduced feed intake, energy supply may be more limiting for growth than a single nutrient. However, provision of specific AAs involved in the control of feed intake may improve the health status of animals (Trevisi et al., 2009). For example, dietary supplementation with 1 g L-tryptophan/kg increased feed intake of pigs challenged with enterotoxigenic *E. coli* K88 (Trevisi et al., 2009).

3 Pregnancy, lactation and pre-weaned piglets

The role of maternal nutrition cannot be overemphasized since it is critical to fetal development, growth of piglets, lactational performance, and production of immunoglobulins. After parturition, maternal nutrition continues to regulate the development of the immune system (Salmon et al., 2009). Specific essential fatty acids (FA) supplemented in gestation and lactation diets improved the quantity of sow colostrum immunoglobulins and weaning weight (Hasan et al., 2018; Yao et al., 2012).

Newborn pigs are immunologically competent, yet due to the epitheliochorial placentation, their adaptive immunity is naïve. Colostrum intake in the first 24 h is critical to early postnatal development of the mucosal and systemic immunity (Ogawa et al., 2016). An early study revealed that 70–90% of colostral lymphocytes were T lymphocytes, with CD8 lymphocytes predominating over CD4, and that the ratio CD4/CD8 was significantly lower in colostrum than in blood (Le Jan, 1994).

Diminished back fat thickness, low birth weights, or low colostrum intake increase the risk of piglet death (Hasan et al., 2019). It was previously shown that piglet serum IgG concentrations had a positive correlation with piglet weight, and piglets from larger litters had lower serum IgG (Nguyen et al., 2013). It is evident that sow body condition, colostrum quality and quantity, and piglet body condition contribute to piglet survivability and growth. Thus, it was not surprising that several studies were directed at improving colostrum quality and quantity.

Various sow or piglet dietary supplements were used to improve the quality of sow colostrum or the immunity of piglets. These supplements included *Astragalus* polysaccharides (APS) (Tan et al., 2017), seaweed extract or fish oil (Leonard et al., 2010), α-tocopherol acetate (ATA) and vitamin C (Pinelli-Saavedra et al., 2008), oxidized β-carotene (Chen et al., 2021), vitamin E (Wang et al., 2017), and vitamins A and D and copper (Verso et al., 2020).

Dietary supplementation with APS for 1 week prior to parturition increased IgG, IgM, and growth factors in colostrum collected at the time of parturition. It was concluded that the APS improved the immune components in sow colostrum and/or milk (Tan et al., 2017). Another study supplemented sows with seaweed extract (10 g/d) or fish oil (100 g/d) in late gestation (day > 109 d) onward until weaning. Seaweed extract increased colostral IgG concentrations and circulatory IgG concentrations in suckled piglets on days 5 and 12 of lactation, and fish oil supplement improved leukocyte and lymphocyte phagocytosis of *E. coli* at weaning (Leonard et al., 2010).

The dietary inclusion of ATA (500 mg/kg feed) and vitamin C (10 g/day) in sow diets increased the AT content of colostrum, milk, piglet plasma, liver, thymus, bone marrow, and spleen at weaning (Pinelli-Saavedra et al., 2008). The combined vitamin treatment increased the total Ig and IgG concentrations in piglet plasma. Vitamin C significantly increased lymphocyte response to ConA and PHA in sows, but not in piglets (Pinelli-Saavedra et al., 2008)

Additional investigations evaluated the impact of increased vitamin levels in sow diets. A high concentration (250 IU/kg) of vitamin E during the last week of gestation and lactation resulted in high levels of plasma IgG, IgA, total antioxidant capacity (T-AOC) and catalase (CAT) in piglets from supplemented sows (Wang et al., 2017). Dietary fully oxidised β-carotene (OxBC) supplementation from gestation day 85 until day 21 of lactation enhanced colostrum immunoglobin levels and IgM and IgG content of 14-d milk. The OxBC supplementation decreased TNF-α and IL-8 levels in colostrum, as well as the TNF-α and IL-18 levels in 14-d milk. Thus, dietary supplementation with OxBC improved the immune status of sows (Chen et al., 2021). Finally, it was shown that immune system development, including the percentages of lymphocytes, NK cells, and antigen-presenting cells, differed between low-birth-weight and high-birth-weight piglets (Verso et al., 2020). The supplementation of piglets with bovine colostrum and vitamins A (β-carotene; 30 000 IU) and D (25-hydroxyvitamin D; 4 000 IU) and copper sulfate (45 mg) could modify the development of immune functions (Verso et al., 2020).

4 Proteins and amino acids

The conventional classification of AAs includes the essential AA, nonessential AA, and conditionally essential AA. The essential AA cannot be synthesized in cells to meet the demands for productive functions; nonessential AA can be synthesized within the body or derived in the body from essential AA; and conditionally essential AA is usually not essential and depends on dietary and physiological conditions. This classification whether the AA is essential or nonessential for the appropriate response to immune challenges may be debatable.

Many contributions have been made to the study of nutrient supplementation on immunity-involved AA and the balance between calories, protein, and AA. A comprehensive review of the influence of AA and protein on immunity was previously published and it clearly illustrated the complexity of AA ratios and calorie:lysine ratios (Goodband et al., 2014). Previous studies documented that inflammation reduced N retention in the body and increased urinary N loss in growing pigs (van Heugten et al., 1994; Rakhshandeh et al., 2010). The altered N retention or loss potentially resulted from (a) decreased feed intake, (b) impaired digestion and absorption of nutrients, (c) enhanced catabolism of body proteins, and (d) increased oxidation of AAs and are orchestrated by inflammatory cytokines and hormones. In contrast to the initial studies on N metabolism, Goodband and coworkers (2014) stated, 'The evidence is sparse or equivocal for increasing nutrient requirements during an immune challenge. Nutritionists and swine producers should resist the pressure to alter the diet, limit feed, or add expensive feed additives during an immune challenge.' Despite this statement and comment, a vast number of research studies attempted to demonstrate the benefit of altering diets of immune-challenged pigs. Some of the recommendations have validity, others do not.

The influence of ISS on protein synthesis, protein degradation, and protein deposition was evaluated recently. Evidently, ISS caused a decrease in protein synthesis, degradation, and deposition, despite the failure to affect daily feed intake during the post-ISS period (McGilvray et al., 2019a). Other than clinical signs associated with ISS, a detailed description of the immunological response was lacking.

Rather than providing detailed descriptions and evaluations of all AA and their roles in immunity, which would be exhaustive and likely require a number of chapters, a few specific AA will be used as examples of previous research. Specific AA or combinations of AA were assessed for their role, and additional supplementation, in the response to ISS or specific pathogen challenge. The metabolism of threonine (Thr) during ISS provides an excellent example due to its role in the synthesis of Thr-rich immune system metabolites, such as immunoglobulins, acute phase proteins, and intestinal mucins (Faure et al., 2007; Rémond et al., 2009). The obvious question is whether Thr supplementation is required during and/or after ISS. Results of investigations were highly variable and contradictory at times. It is logical to assume that endogenous sources may meet the requirement for increased utilization of certain AA, and thus, dietary requirements may not be greater.

As described by Kampman-van de Hoek and coworkers (2016), 'the utilization of sulfur AAs, Thr and tryptophan (Trp) for growth was reduced when growing pigs were fed a diet with a low content of these AAs because of competition between skeletal muscle and the immune system.' Another study indicated that the physiological changes associated with ISS increased the dietary Thr requirements for protein deposition due to an increase in maintenance requirements (McGilvray et al., 2019b). Also, it was demonstrated that high dietary fiber (DF) and ISS independently increased the Thr requirement to maximize protein deposition (Wellington et al., 2018). Supplementation of Thr to meet requirements for high DF and ISS failed to sustain growth performance when pigs were fed high fiber (HF) diets and challenged with *Salmonella typhimurium* (Wellington et al., 2019).

In healthy pigs, injected with bovine serum albumin or classical swine fever-attenuated vaccine or ovalbumin (Wang et al., 2006), serum IgG and IgM concentrations increased with dietary Thr intake. In contrast, increased dietary Thr failed to improve IgG and IgM concentrations when piglets were challenged with pseudorabies live vaccine (Mao et al., 2014). Changing dietary Thr content from 8.5 g/kg to 9.0 g/kg resulted in a higher secretion of IgM in enterotoxigenic *Escherichia coli* (ETEC)-challenged pigs susceptible to ETEC. Evidently, health status influences the dietary effect of Thr supplementation on the humoral immune response.

Leucine (Leu) has a role in whole-body protein turnover and skeletal muscle, which may be affected by ISS. Supplemental Leu improved protein deposition (PD) but did not affect whole-body protein turnover during the ISS (Rudar et al., 2017). In addition, the efficiency of Trp utilization for PD was also reduced during ISS, thereby indicating that the Trp requirement for PD increased during an inflammatory state (de Ridder et al., 2012). A decrease in Lys, Phe, and Ile flux in pigs treated with LPS was attributed to a reduction in whole-body protein synthesis or decreased catabolism of these AA (Stuart et al., 2016). Additional studies involving PRRSV revealed that the viral infection increased plasma Met and Thr flux in growing pigs (McGilvray et al., 2019b). Collectively, these changes suggest modification(s) of AA utilization by LPS or PRRSV challenge in growing pigs.

An appropriate method to analyze plasma AA responses and changes in AA metabolism is to actually challenge pigs with pathogens or to compromise the health. For example, reduced plasma AA concentrations were reported in pigs with chronic lung inflammation (Melchior et al., 2004) or infectious peritonitis (Yoo et al., 1997). Further, co-infections with *Mycoplasma hyopneumoniae* and H1N1 influenza virus lowered postprandial concentrations of Thr and Arg in plasma (Le Floc'h et al., 2014). Plasma AA concentrations are influenced by nutritional status, dietary supply, and the metabolic fluxes of AAs entering or leaving the plasma pool. For example, Val and tyrosine (Tyr) fluxes were decreased by lung inflammation without concomitant changes in their pool size in plasma (Kampman-van de Hoek et al., 2015).

Modifications of the diet on intestinal mucosal function were the focus of numerous studies. Many proteinogenic AAs exert positive effects on gut villus morphology and improve intestinal mucosal barrier function when supplied above the estimated requirements (Jiao et al., 2015). The AAs Arg, Gln, Glu, Gly, sulfur AA, and Thr have the potential for managing the physical gut barrier, and as shown by Le Floc'h and colleagues, several AAs also are involved in defense proteins of the gut (Le Floc'h et al., 2018). Low-protein diets, currently used to reduce the production of nitrogenous wastes by swine farms, do not supply sufficient Arg or its AA precursors (Wu et al., 2014a).

How do producers prevent gut disorders and improve protein utilization in pigs? Gloaguen et al. (2014) confirmed the efficacy-reducing dietary protein while supplementing free AAs. The strategy was to formulate low-crude protein CP (13.5%) diets and include free AAs for the growth of 10–20 kg pigs. Other studies showed that 'moderate dietary protein restriction (CP 13–15.3%) was beneficial for a healthy balance in the gut microbiota and metabolic activity in the large intestine of pigs, and improved ileal mucosal barrier function' (Fan et al., 2017; Peng et al., 2017b). It was suggested that top dressing the commercial diet with functional AAs as powder might be beneficial during critical periods of production, such as weaning or pathogen exposure (Garcia-Launay et al., 2014).

Since infection with PRRSV increased the pool size for Lys, Met, Thr, Trp, Leu, Val, and Gln, it was claimed that 'PRRSV alters the utilization of dietary N and AA flux, as well as pool size, in growing pigs' (McGilvray et al., 2019b). It was theorized that the increase in Thr and Met flux in PRRSV+ pigs was associated with utilization of these AA for the synthesis of immune system metabolites. Therefore, dietary Met, Thr, and Val requirements may increase in pigs infected with PRRSV, relative to the requirements for other AA (McGilvray et al., 2019b).

An impact of cytokine production is a decrease in feed intake. Therefore, 'as feed intake decreases, the energy needed to drive protein synthesis is also decreased. This suggests that diets should still be formulated on a similar calorie:lysine ratio as those formulated for non-immune challenged pigs. While there may be transient changes in AA requirements for maintaining the immune system, decreased feed intake and poorer G:F suggest that highly digestible diets with ingredients that stimulate feed intake seem to be the best course of action in getting pigs through an immune system challenge' (Goodband et al., 2014).

5 Lipids and fatty acids

There are a variety of reasons for lipid supplementation of diets: increased energy density, improved feed efficiency and pellet quality, reduction of dust and improved palatability, a source of essential FA, and improved absorption of fat-soluble vitamins (NRC, 2012). Supplementation of up to 6% of the diet is not uncommon and apparently does not interfere with feed handling characteristics (NRC, 2012). Specific lipids, such as fish oil, have potential anti-inflammatory effects due to their FA composition, specifically ω-3 fatty acids, and their derivatives as eicosanoids (Simopoulos, 1991).

Fatty acids are involved in mucosal immune responses, epithelial barrier functions, oxidative stress, and inflammatory reactions (Lauridsen, 2020). Obviously, alterations in the milk FA composition require modifications of the sow diets. These alterations are necessary to enhance pig growth and gut health in pigs prior to weaning. After weaning, diets must include fat sources. Medium-chain fatty acids (MCFA) and short-chain fatty acids (SCFA) possess antibacterial effects and may thereby prevent overgrowth of pathogenic bacteria in the gastrointestinal tract (Lauridsen, 2020). The essential FA, linoleic (LA), and α-linolenic (ALA) FA are essential for the long-chain polyunsaturated n-3 and n-6 FA. According to Lauridsen (2020), 'the conversion of ALA and LA into n-3 and n-6 eicosanoids, respectively, influences the molecular structures of metabolites and inflammatory reactions and other immune responses upon bacterial challenges' (Lauridsen, 2020). As a result of oxidative stress associated with infectious diseases, the epithelial barrier function may be compromised from the development of lipid peroxides and other reactive oxygen products. Thus, 'manipulation of the dietary FA composition can influence the gut health and may support a normal immune system and modulate resistance to infectious diseases' (Lauridsen, 2020).

Porcine $CD8^+$ lymphocytes are involved in the cellular immune responses. It was postulated that nutritional modulation can also influence cell-mediated immunity (Bassaganya-Riera et al., 2001). Dietary supplementation with conjugated linoleic acid (CLA) for 42 days prior to a disease challenge may have prevented disease-associated growth suppression (Bassaganya-Riera et al., 2001). A subsequent study by Bassaganya-Riera and coworkers (2002) demonstrated that the effects of dietary CLA on immune cell phenotype persisted after the compound was withdrawn from the diet. Effector functions (i.e. antigen-stimulated proliferation and cytotoxicity) disappeared earlier (i.e. 25 days). Furthermore, 'prior dietary CLA supplementation interacted with viral immunization (i.e. modified-live pseudorabies virus (PRV) vaccine) by enhancing both pseudorabies-specific proliferative responses of CD8 αβ+ PBMC and granzyme activities of PBMC' (Bassaganya-Riera et al., 2002).

In contrast to the aforementioned reports, a recent study investigated 'whether the immunocompetence of piglets at weaning was modulated by including different sources of n-3 PUFA in the maternal diet from day 73 of gestation until weaning at 4 weeks. Sows were fed a palm oil-based diet (control group) or a diet including 1% linseed oil (C18:3n-3), 1% echium oil (C18:3n-3, C18:4n-3, C18:3n-6) or 1% fish oil (C20:5n-3, C22:6n-3)' (Tanghe et al., 2014). No significant dietary effects were observed on the thyroglobulin-specific IgG and IgA titers or on serum amyloid A (SAA) concentrations in the piglet serum. Interestingly, the inclusion of n-3 PUFA in sow diets failed to affect immunity in piglets after weaning. The researchers concluded that 'the immune response likely is influenced by animal- and environment-related factors, rather than by the FA composition of the maternal diet' (Tanghe et al., 2014). An earlier study compared fish oil to sunflower seed oil (2:1 and 10:1, n-6:n-3 ratios) in late gestation and lactation diets (Papadopoulos et al., 2009). The dietary treatments did not significantly change the TNFα nor IL-6 concentrations in sows; however, the oil supplementation did increase the SAA at days 3 and 8 after parturition. Evidently, the SAA levels can be influenced by nutritional factors (Papadopoulos et al., 2009).

A study by Duan and colleagues (2014) revealed that the 'growth performance of grow-finisher pigs fed a diet with an n-6:n-3 PUFA ratio of 5:1 was the best, but the group fed the diet with an n-6:n-3 PUFA ratio of 1:1 had the highest muscle mass and the lowest adipose tissue mass.' Pigs fed diets with lower the ratios of n-6:n-3 also had decreased concentrations of IL-6 and IL-1β. Additionally, the dietary ratios of n-6:n-3 PUFA of 1:1 and 5:1 markedly suppressed inflammatory cytokines IL-1β, TNF-α, and IL-6 (Duan et al., 2014). In recently weaned pigs, the inclusion of 2.5% fish oil (3:1, ω-6:ω-3) optimized growth performance; however, the ideal ω-6:ω-3 fatty acid ratio may differ when using immune responses as the major outcome (Huber et al., 2018).

Energy dense ingredients are commonly used in pig diets. These ingredients include rendered fat, restaurant grease, and vegetable oils, and oils from distiller's dried grains with solubles (DDGS) (Chang et al., 2019). However, the quality and peroxidative status of these sources are variable and could be detrimental to animal health (van Kempen and McComas, 2002; Song et al., 2014; Kerr et al., 2015). In a large-scale study with over 2000 weaned pigs, Chang and coworkers (2019) investigated the impact of lipid peroxidation on growth, health, and oxidative stress of nursery pigs. Their results indicated that the 'number of pigs removed for medical treatment, total number medically treated, pigs culled for low end weight, and mortality increased, and full-value pigs linearly decreased with increasing peroxidation.' Antibody titers to *Mycoplasma hyopneumoniae* and PCV2 were not affected by dietary treatment. It was not surprising to note that T-AOC and serum vitamin E concentrations decreased linearly with increasing peroxidation. Data showed a dose-dependent negative impact of lipid peroxidation on pig productivity when evaluated under commercial farm conditions (Chang et al., 2019). This study was one of the few conducted in real-world conditions.

6 Fiber, glucans and carbohydrates

Storage carbohydrates, such as starch, and non-starch polysaccharides represent the plant carbohydrates. Any carbohydrate that resists digestion in the upper gut of pigs is defined as fiber (Kerr and Shurson, 2013). Fiber can be soluble or insoluble, based on solubility in water. The characteristics of the two classifications of fiber were described elsewhere (Menegat et al., 2019). In general, the insoluble fiber often is used to reduce post-weaning diarrhea and *E. coli* colonization, and the vast majority of studies focused on the response in the gut.

The use of DDGS initially was examined to control *Lawsonia intracellularis* infections in growing pigs. The results were inconsistent; however, one study revealed that a diet with DDGS (10%) reduced the severity of lesions associated with ileitis (Whitney et al., 2006).

Recently, the focus of insoluble fiber or non-starch polysaccharides was on the use of degrading enzymes, such as xylanase, β-glucanase, mannanase, and galactosidase. These enzymes increase fiber digestibility and reportedly modified the concentrations of cytokines (IL-1β, IL-4, IL-17A, and IL-11) in the intestine and, specifically, the ileum (Ferrandis Vila et al., 2018). In fact, the inclusion of xylanase and carbohydrase blend in high fiber diets reduced plasma IL-8, TNFα, and IL-1β in weaned pigs (Li et al., 2019b). Furthermore, the role of oligosaccharides and polysaccharides of arabinoxylan and mannan received attention as immunomodulatory factors by preventing microbial attachment to the intestinal wall (Tiwari et al., 2020).

Four types of glucose-based polysaccharides are possible: 1,6- (starch), 1,4- (cellulose), 1,3- (laminarin), and 1,2-bonded glucans. Glucans (polymers) can initiate immune responses by binding to specific receptors. However, 'β-glucans from different sources also differ in their structure, conformation, physical properties, binding affinity to receptors, and thus biological functions' (Han et al., 2020). The β-glucans derived from fungi and yeast have immune modulating properties. Possibly, dietary β-glucans prime the host immune system and increase resistance against invading pathogens (Volman et al., 2008).

Glucans were considered to be promising vaccine adjuvants since they can stimulate various immune reactions including antibody production without any negative side effects (Vetvicka et al., 2020). Thus, it is not surprising that glucan inclusion in the diet could be used to modulate immunity in pigs. It was shown by Li et al. (2006) that diets supplemented with glucan partially downregulated the synthesis of TNF-α and IL-6 but also regulated metabolism of nutrients (Spurlock, 1997). Additional studies found that the addition of glucans to pig feed decreased Th-related cytokine production and protected against infection (Vetvicka et al., 2014). Supplementation with β(1-3)(1-6)-D-glucan (15 mg/kg/day) stimulated growth, phagocytic activity, and IL-2 production, while lowering cortisol and TNF-α levels after LPS challenge (Vetvicka and Oliveira, 2014).

Another study used β-glucans derived from seaweeds *Laminaria hyperborea* and *Laminaria digitata* and the yeast *Saccharomyces cerevisiae*. This study evaluated 'gene expression of cytokines, receptors, and signal transducing molecules relevant to the Th17 and TREG pathways in the porcine colon' (Ryan et al., 2012). All sources of β-glucans significantly decreased the expression of Th17-related cytokines, receptor IL23R, and IL-6; alterations to the TREG-related target, Foxp3, or to TGF-β were not observed (Ryan et al., 2012).

Some studies used *in vivo* models to examine the application of orally administered β-glucans to protect pigs from infectious agents. Dritz and coworkers (1995) suggested the 'existence of a complex interaction involving growth performance and resistance to *Streptococcus suis* in pigs fed.025% β-glucan.' A subsequent investigation examined the effects of orally administered β-glucans in protecting pigs against an ETEC infection after weaning (Stuyven et al., 2009). Pigs fed β-glucans (*S. cerevisiae* or *S. rolfsii*) for two weeks after weaning were less susceptible to an F4+ ETEC infection in comparison with the control group.

As mentioned previously, investigators evaluated the effects of dietary soluble and insoluble fiber without or with exogenous carbohydrases (xylanase, β-glucanase, and pectinase) on diarrhea incidence, selected immune responses, and growth performance in ETEC-challenged pigs (Li et al., 2019a). It was concluded that the soluble fiber improved growth, due in part to a reduction in inflammatory intermediates (Li et al., 2019a). In a follow-up study, these researchers showed that an enzyme blend (cellulase, ß-glucanase, and xylanase) reduced colonic *Lactobacillus* and cecal total volatile FA in weaned pigs. Also, the enzymes decreased IL-8, TNF-α, and IL-1β levels, suggesting decreased systemic immune activation (Li et al., 2019b).

7 Minerals

To reduce the use of antimicrobials in swine diets, some minerals have been added to provide bactericidal, bacteriostatic, or therapeutic properties, particularly in the nursery phase of production. These minerals often are given in levels that exceed the recommendations (NRC, 2012) and may be in an oxide or sulfate form. An excellent review of the minerals, other non-antibiotic feed additives, and their potential influence in the pig was published by Lui et al. (2018).

7.1 Copper

Copper (Cu) is critical in metabolism, production of hemoglobin and for the differentiation, maturation, and proliferation of neutrophils and macrophages (Osredkar and Sustar, 2011; Djoko et al., 2015). In addition, Cu has 'the potential to produce reactive oxygen species (ROS) that may cause severe oxidative stress and ultimately bacterial death' (Djoko et al., 2015).

Addition of pharmacological levels of Cu in pig diets was used to improve growth performance (Ma et al., 2015). Supplementing Cu (100–250 mg/kg) diets may reduce post-weaning diarrhea and improve ADG and ADFI. One study investigated the effects of cupric citrate (CuCit) on growth metabolites, serum immune indices, and tissue residues of Cu, zinc, and iron in weaned pigs (Peng et al., 2017a). The CuCit (30–60 mg/kg) enhanced growth performance and antioxidant status of weaned pigs (Peng et al., 2017a).

A large trial (Ding et al., 2021) with 2000 pigs evaluated the effects of different diets (gradient dosage of Cu and Zn additives) on growth performance and other parameters, including immune function, in piglets and growing pigs. Supplementation of the diet with Cu (120 mg Cu/kg feed) and Zn improved the antioxidant status and immune function of weaned piglets. This dose of Cu increased a number of enzymes, such as peroxidase (POD), glutathione peroxidase (GSH-PX), catalase (CAT), and copper/zinc superoxide dismutase (Cu/Zn-SOD). It should be noted that the addition of Cu and Zn had effect on neither serum IgG nor IgM concentrations (Ding et al., 2021). Nano-sized Cu particles were considered as appropriate substitutions for $CuSO_4$ in pig diets after weaning. Nano-size Cu (100 mg/kg/day) supplementation improved immunity by strengthening the antioxidant capacity of weaned pigs (Chang et al., 2018).

The inclusion of Cu at 75–250 mg/kg diet has growth-promoting actions in weaned and growing pigs. The Cu reduces post-weaning diarrhea. The effects of dietary Cu were attributed to its bacteriostatic and bactericidal properties as Cu may change bacterial populations in the intestine and thereby reduce inflammation (Espinosa and Stein, 2021).

Additional studies examined various aspects of Cu supplementation on metabolism. For example, its antioxidant capacity and role in intestinal health were described by Gonzales-Eguia et al. (2009). Diets with 3000 mg/kg of Zn and 250 mg/kg of Cu reduced plasma cytokine circulation after a LPS challenge (Namkung et al., 2006). Unfortunately, the immune response to common pathogens has rarely been assessed when growth-promoting levels of Cu were included in the diets.

7.2 Zinc

Dietary supplementation with Zn has been used for many years, particularly in the diets of recently weaned pigs. The inclusion of Zn oxide (ZnO) at 2000–3000 ppm/kg of the diet prevented post-weaning diarrhea. The high levels of ZnO conserved the intestinal flora during the critical period following the change of diet at weaning (Jensen-Waern et al., 1998). Recently, it was shown that pharmacological Zn modified the diversity of the intestinal microflora during the first two weeks post-weaning (Shannon and Hill, 2019).

Improved immunity in pigs, provided by high levels of Zn in the diets, was not consistently shown in some studies. For example, Kloubert and coworkers (2018) showed that innate (increased phagocytosis) and adaptive immunity were improved by including a basal level of 100 ppm Zn or supplementing with 2500 ppm Zn in the diets. With the higher level of Zn, the changes in adaptive immunity included increased numbers of CD3+ T cells, and CD3+CD4+Foxp3+ regulatory T cells, elevated IL-2 production, and decreased IL-10 production (Kloubert et al., 2018). In contrast, dietary Zn levels failed to influence the number of antiviral IFN-γ secreting cells or percentages of blood immune cell subsets after infection with PRRSV (Chai et al., 2014). In fact, higher levels (2500 ppm) of dietary ZnO did not alter systemic immune responses after vaccination and exposure to heterologous PRRSV infection (Chai et al., 2014).

It was reported that 2500 mg Zn/kg of diet immediately after weaning affected the immune responses of pigs infected with *Salmonella typhimurium* DT104. The beneficial effects of the Zn supplementation waned and after two weeks; there was higher shedding of salmonellae and lower T cell frequencies (Janczyk et al., 2013). In contrast, the use coated ZnO provided promising results as coated ZnO (380 mg Zn/kg or 570 mg Zn/kg) reduced the incidence of diarrhea 'by promoting intestinal development, protected the intestinal mucosal barrier from damage, stimulated the mucosal immune system and regulated the microbiota composition' (Shen et al., 2014). Furthermore, the use of nano-Zn appears to shows promise. Low doses of nano-ZnOs (150 mg/kg, 300 mg/kg, or 450 mg/kg) had beneficial effects on growth performance, intestinal morphology and microflora, and immunity in weaned pigs (Pei et al., 2019).

7.3 Selenium

Selenium (Se) is involved in several physiological processes, including functions of the immune system and often acts in synergy with vitamin E. It is beyond the scope of this chapter to review all the functions of Se, rather readers are encouraged to read the review by Hosnedlova et al. (2017).

Some disadvantages of supplementation with the inorganic form of Se include relative high toxicity, low transfer efficiency to milk, and inability to build and maintain Se reserves. Evidently, organic sources of Se are greater bioavailability than inorganic Se sources. Selenium was reported to inhibit the pathogenicity of certain viruses and, thus, the potential usefulness of dietary Se forms became a major interest in swine nutrition (Dalgaard et al., 2018). In relation to macrophage responses, selenoproteins act as major antioxidants to limit the cytotoxic effects of ROS (Vunta et al., 2008) and limit pathogen replication (Sumner et al., 2019). Recently, it was shown that the antioxidant status and Se level increased when sows were fed organic Se (Se-yeast) compared with sodium selenite (Chen et al., 2016a,b). In addition, these researchers demonstrated that Se and antioxidant status in the progeny including newborn and 21-d-old nursery piglets were improved when organic Se was added to the diet of sows (Chen et al., 2016b).

One study, using DL-selenomethionine (DL-SeMet) supplementation of weaned pigs, demonstrated that the immunity (IgA, IgM, IgG levels) among treatment groups (0.1 mg/kg, 0.3 mg/kg, 0.5 mg/kg, and 0.7 mg/kg of Se from DL-SeMet) was not affected (Cao et al., 2014). Also, 'different effects may be expected on antibody responses directed against T-dependent antigens versus T-independent antigens' (Dalgaard et al., 2018). Recently, an investigation elucidated the role of dietary selenium-enriched yeast (SeY) supplementation (250 mg/kg SeY) in weaned pigs' exposure to oxidative stress. Specifically, 'the expression levels of TNF-α, IL-6, IL-1β, Toll-like receptor 4 (TLR-4), and nuclear factor-κB (NF-κB) in the liver and thymus were downregulated by SeY upon diquat challenge' (Liu et al., 2021). These results suggested that SeY attenuates oxidative stress-induced growth retardation (Liu et al., 2021).

7.4 Iron

Iron (Fe) is important for enzyme metabolism and oxygen transport. It is typically bound to proteins, such as hemoglobin (Hb), with 65% in red blood cells and the rest stored as ferritin. If demands exceed intake, erythropoiesis and ferritin mobilization are initiated, and non-degenerative anemia or ferriprive anemia will occur when reserves are depleted. With few notable exceptions (Normand et al., 2012; Castevens et al., 2020), most studies focused on Fe and anemia in piglets. Both oral Fe and intramuscular injections of Fe dextran are used to prevent iron deficiency (ID). According to one study (Dong et al., 2020), 'oral Fe produced greater serum IgA levels, SOD, and GSH-Px activities, lower expressions of IL-1β and TNF-α in the liver, and lower expressions of IL-6 and TNF-α in the spleen compared to parenteral iron supplementation.' It was concluded that oral Fe was sufficient to sustain Fe homeostasis and oxidative and immune status in anemic neonatal pigs (Dong et al., 2020). In contrast, intramuscular administration of Fe-dextran-stimulated cytokine production of the intestine and enhanced the phagocytic capacity of monocytes (Pu et al., 2018).

In piglets, an additional injection of Fe 1 week prior to weaning failed to enhance humoral immunity to a T-cell-dependent antigen (Bruininx et al., 2000). Furthermore, an investigation was conducted to evaluate peripheral PRRSV viremia and inflammation when ID and respiratory infection occur simultaneously in neonatal piglets (Leyshon et al., 2019). It was not surprising that ID reduced hematocrit, hemoglobin, and serum iron. In contrast, PRRSV infection did not alter these red blood cell parameters; however, these investigators indicated that the 'ID either impaired immune function necessary for viral clearance or increased viral replication.' Also, the PBMC cytokine expression was generally decreased by ID. Collectively, the results suggested that ID decreases peripheral immune function leading to increased viremia (Leyshon et al., 2019).

As mentioned, the role of Fe in immunity is rarely considered, with the notable exception of limited studies in piglets prior to weaning. The overall influence of Fe in immunity has not been addressed for pigs at other stages of production. In contrast, the role of Fe in immunity received considerable attention in humans (Nairz and Weiss, 2020).

8 Vitamins

Dietary vitamin levels are often above the values suggested by nutritional tables (National Swine Nutrition Guide, 2010; NRC, 2012). This should not be surprising as a landmark study (Coehlo and Cousins, 1997) clearly demonstrated the benefits of vitamin supplementation in nursery and grow-finish diets. Using a 'stress model,' which included factors such as two pathogens (*E. coli* and *Salmonella*), aflatoxin, nutrient density, and pen space/pig density, these investigators showed that increasing the vitamin levels significantly above the industry averages resulted in decreased mortality, improvements in growth performance and carcass composition. These cost-effective improvements were particularly evident in high stress environments designed to mimic commercial production (Coehlo and Cousins, 1997). A survey in the US pork industry indicated that it is common, yet variable, to supplement vitamins and trace minerals at rates exceeding the NRC (2012) requirements (Flohr et al., 2016). This supplementation is intended to provide greater margins of safety.

One study assessed the effect of administering vitamin supplementation to sows in gestation and lactation and to their litters on the immune profile of the litters until slaughter (Santos et al., 2020). Vitamin supplementation positively impacted the performance of the litters in the early nursery stage; however, vitamin supplementation of sows failed to enhance the immune responses to vaccination against porcine circovirus or mycoplasma of nursery pigs (Santos et al., 2020). In contrast to the improved performance in the nursery phase, a previous report revealed that supplemental vitamins did not alter performance of pigs in high-density housing during the finishing phase (Zhang et al., 2013).

8.1 Vitamin A

Vitamin A is derived from 'precursors such as β-carotene, all-transretinol or retinyl esters which are converted to retinoids in the proximal portion of the gut'. The vitamin A metabolite, retinoic acid, is involved in the control of gene expression in immune function (Bono et al., 2016) and it influences both the adaptive and innate immunity, especially at the intestinal border (Larange and Cheroutre, 2016).

Based on the versatile role that vitamin A plays in the immune system, and a more critical role in mucosal integrity and intestinal lymphocyte trafficking, one study used oral vitamin A supplementation to boost lactogenic immunity and passive protection of nursing piglets against Porcine Epidemic Diarrhea Virus (PEDV) challenge (Langel et al., 2019). Gilts were given daily oral retinyl acetate (30000 IU) starting at gestation day 76 and throughout lactation. Compared to control gilts exposed to PEDV without supplemental retinyl acetate, the oral vitamin A supplementation reduced PEDV RNA shedding, reduced mortality of PEDV-challenged piglets, and increased PEDV IgA antibody-secreting cells and PEDV IgA antibodies in serum pre-partum and IgA+ β7+ (gut homing) cells in milk post piglet challenge. Thus, oral vitamin A supplementation enhanced maternal IgA and lactogenic immune protection in nursing piglets (Langel et al., 2019).

Vitamin A is easily degraded by environmental factors, and it is important to add antioxidants during vitamin A production. Ethoxyquin (EQ) and butylated hydroxytoluene were widely used; however, it was recently shown that vitamin C sodium appears to be a worthwhile antioxidant additive for vitamin A and the combination of vitamin C sodium and vitamin A improved the growth performance, antioxidant capacity, and immune function of weaned pigs (Zhou et al., 2021).

8.2 Vitamin D

Vitamin D is derived from 7-dehydrocholesterol in the skin or supplementation of the diet. It is then metabolized to its active form 1,25-dihydroxy vitamin D (1,25(OH)2D) (Mora et al., 2008). It is involved in calcium absorption in the gut and in the regulation of the intestinal barrier and immune functions (Yin and Agrawal, 2014). It was reported that 'vitamin D upregulates genes that encode proteins required for epithelial integrity and is critical for production of antimicrobial peptides such as cathelicidins' (Schwalfenberg, 2011) and prevents inflammatory damage and modulates adaptive immunity (Adams and Hewison, 2010).

The effects of vitamin D3 on the immune system remain somewhat elusive. One investigation (Konowalchuka et al., 2013) used a weaned pig model to examine the immune cellular changes resulting from vitamin supplementation. The control diet had 1500 IU/kg of vitamin D3 and the experimental diets had either 2000 additional IU/kg of vitamin D3 (ROVIMIX D3 500, DSM Nutritional Products Ltd) or 25(OH)D3 derived from a commercial ROVIMIX Hy·D 1.25% preparation. The vitamin 25(OH)D3 supplementation (3500 IU/kg) increased in leukocyte cell numbers results indicated that newly weaned pigs fed diets containing Hy·D® had significantly higher numbers of immune cells, as well as increased viability and phagocytic activity of those cells in serum, which also correlated to similar findings in bronchioalveolar lavage material. This *ex vivo* work shows that immune function and lung health can also potentially benefit from appropriate vitamin D levels circulating in serum. The investigators indicated that the study supported the positive immunomodulatory role of vitamin D3 and points to 25(OH)D3 as an appropriate dietary supplement for weanling pigs (Konowalchuka et al., 2013).

Combinations of vitamin A and vitamin D also are considered critical to immunity, particularly during lactation. The results from one study suggested that immune system development differed between low-birth-weight and high-birth-weight piglets and that early dietary supplementation with bovine colostrum and vitamins A and D (25-hydroxyvitamin D; 4000 IU) and copper ($CuSO_4$) had the potential to modulate development of immune functions (Verso et al., 2020).

8.3 Vitamin C

In humans, vitamin C or ascorbic acid is implicated in many biological processes, and ascorbic acid positively influenced lymphocyte development and function in humans (van Gorkom et al., 2018). In pigs, fewer studies addressed the value of supplemental vitamin C. One study concluded that weaned pig performance was not improved as a result of vitamin C supplementation; the results were inconclusive regarding the role of vitamin C in stimulating humoral immune function in pigs (Zhao et al., 2002). Another study demonstrated that supplementation with 1 g vitamin C/day from 10 d of age to 48 d of age was effective at preventing soybean (glycinin) allergies by increasing INFγ and decreasing IL-4 in cultured splenocytes of young pigs (Sun et al., 2009).

The inclusion of one vitamin without considering the potential influence of other vitamins may be misleading. For example, supplementation of sows' lactation diet with all-*rac*-α-tocopheryl acetate, and then supplementing 500 mg of vitamin C/kg to the piglet diet after weaning, increased the number of immune cells and IgM concentrations in weaned pigs (Lauridsen and Jensen, 2005). Conversely, another study used a combination of vitamin C (75 ppm) and β-glucan to assess the response to LPS in pigs. The authors concluded that the combination enhanced post-weaning growth and reduced TNF-α expression of the intestinal and liver tissues, suggesting an 'immunomodulatory role of the combination treatment' (Eicher et al., 2006). Recently, it was demonstrated the potential of 'rutin, α-tocopherol, and l-ascorbic acid in enhancing type I interferon-regulated genes and type I and II IFN expressions, and in reducing pro- and/or anti-inflammatory cytokine expressions in HP-PRRSV-inoculated monocyte-derived macrophages.' The findings suggested that rutin, α-tocopherol, and l-ascorbic acid may serve as 'effective immunomodulators for improving innate immune response to HP-PRRSV' (Suebsaard and Charerntantanakul, 2021).

8.4 Vitamin E

Vitamin E is involved in cell membrane integrity. It limits lipid peroxidation initiated by reactive oxygen species and free radicals in all cells. Vitamin E is important in limiting free radical damage in cells (Pinelli-Saavedra et al., 2008). The precise mechanism for immunomodulation by vitamin E in pigs has not been elucidated; however, one investigation revealed that vitamin E is able to suppress IL-10 production and to influence the production of IL-2 and IL-4 (Hernandez et al., 2009). Furthermore, the addition of vitamin E (250 IU/kg) to the maternal diet improved the weaning weights and enhanced humoral immune function and antioxidant activity in sows and piglets (Wang et al., 2017).

Feeding peroxidized lipids negatively impacted growth performance and antioxidant capacity of nursery pigs (Silva-Guillen et al., 2020). 'Supplementation of vitamin E and polyphenols improved total antioxidant capacity, especially in pigs fed control diets, but did not restore growth performance.' Vitamin E and plant-based polyphenols are antioxidants with high potential against oxidation (Silva-Guillen et al., 2020), but they do not serve as replacements for vitamin E.

9 Nonnutritive feed additives

In efforts to reduce the use of antimicrobials in pork production, numerous studies focused on nonnutritive feed additives and attempted to demonstrate the advantages of organic feeding systems. For example, one study examined the effect of organic or conventional feeding on the immune response of pigs in a fattening unit (Millet et al., 2005). Immunoglobulin 'responses against intramuscularly injected bovine thyroglobulin were determined as indicators of the antigen-specific immune responsiveness.' In addition, haptoglobin concentrations were measured at selected times. It was concluded that the immune responses following either a conventional or an organic diet are comparable (Millet et al., 2005).

Alternative feed sources were evaluated for their effects on immunity. One study supplemented the diet with Wakame seaweed stalks as 1% seaweed powder (Shimazu et al., 2019). The percentages of peripheral blood natural killer cells and CD3+CD4−CD8+ cells of the treatment groups were significantly higher than that of the control group in 20–30 kg and 70 kg body weight (BW) pigs. The authors indicated that Wakame seaweed can be used as a supplement to improve the gut health and immunity of pigs (Shimazu et al., 2019). Unfortunately, the pigs were not challenged with a pathogen or other ISS to assess immunity.

Finally, a recent study revealed that the inclusion of hydrolyzed yeast (1.0 or 1.5 g/kg) in the diet of weaned pigs improved BW, immunoglobulin secretion, and antioxidant enzyme activity. Also, the addition of hydrolyzed yeast lowered 'diarrhea occurrence, lipid peroxidation, and pathogenic bacteria in weaning pigs' (Boontiam et al., 2020).

Despite regulatory hurdles, approximately 40 papers are published each year on various probiotics and swine (Ma et al., 2018). Needless to say, it is a challenge to assess the validity of each publication and each product. The research and commercialization of probiotics expanded to include prebiotics, synbiotics, and proteobiotics. The majority of these products are aimed at gut health or enteric pathogens, with few directed at systemic immunity.

9.1 Probiotics

The FAO/WHO (2001) defined probiotics as 'live microorganisms which when administered in adequate amounts confer a health benefit on the host' (Schlundt, 2012); however, the European Food Safety Authority (EFSA) evidently has concerns regarding the health claims on probiotic products. The EFSA 'rejected all petitions by commercial manufacturers for health claims on probiotic products in Europe' (Turck et al., 2019). The rejections were 'due to insufficient evidence for a cause-and-effect mechanism for benefit' (Turck et al., 2019). In addition, the approval process for probiotics is complicated in the USA, and approval varies from country to country.

The gut microbiota interacts with the host immune functions. However, studies reporting the interactions between gut microbiota and immune function often are restricted to a single pathogen. Rodent models often were used to evaluate microbiota – innate immunity interactions and there is limited, albeit growing, understanding of these interactions in pigs. The gut microbial composition is affected by diet and hence, different feeding conditions and dietary interventions will influence the microbiota and presumably immunity in the gut (Ma et al., 2018). In addition, the influence of alterations in the microbiota on adaptive immunity in pigs often is extrapolated from studies in laboratory animals.

The mechanisms of action of probiotics previously were reviewed (Angelakis, 2017). There are direct interactions with host cells, inhibition of pathogens, and modulations of the host immune response (Ma et al., 2018). In swine, *Lactobacillus*, *Enterococcus*, *Bacillus*, *Bifdobacterium*, and *Brevibacterium* species were used as dietary supplements. Several studies demonstrated changes in adaptive immunity, including IL-10, TGFβ, serum IgG, fecal IgA, intestinal lymphocytes, and a decrease in IL-1β and IFNγ. The mechanism(s) responsible for these changes has yet to be clearly defined and, once again, the significance of alterations in immune factors without an appropriate immune challenge remains a cause for skepticism.

9.2 Prebiotics

Prebiotics are 'compounds that induce the growth or activity of beneficial microorganisms such as bacteria and fungi' (Hutkins et al., 2016). The prebiotics act to modify the composition or populations of organisms in the gut microbiome. Non-digestible fiber compounds often are used as prebiotics. Their mechanism of action is to modulate or stimulate the activity of the bacteria in the large intestine, while passing undigested through the small intestine (Hutkins et al., 2016).

9.3 Synbiotics

Synbiotics refer to 'food ingredients or dietary supplements combining probiotics and prebiotics in a form of synergism' (Pandey et al., 2015). The combination was designed to provide greater advantages to the intestinal microbiota than probiotics. Also, 'synbiotic preparations prevent the negative impact of weaning on piglets' microbial population in the gastrointestinal tract,' and thus, 'reduce the occurrence of diarrhea' (Chlebicz-Wójcik and Śliżewska, 2020).

One study determined 'if combining dietary synbiotics and 0.20% l-glutamine would improve growth performance and intestinal health following weaning and transport when compared with traditionally used dietary antibiotics' (McConn et al., 2020). These investigators supplemented diets with 0.20% l-glutamine (GLN) and synbiotics (SYN; 3 strains of *Lactobacillus* + β-glucan + fructooligosaccharide). Some metrics of pig health (i.e. haptoglobin and goblet cell count) were improved by the addition of GLN and SYN; however, the dietary treatments failed to differ (McConn et al., 2020). In contrast, the dietary supplementation with xylanase and probiotics had beneficial effects when nursery pigs were challenged with ETEC (Duarte et al., 2020). Another investigation evaluated 'supplementation of sows' and growing pigs' diets with three synbiotic and two probiotic products on immune parameters under field conditions' (Czyżewska-Dors et al., 2018). No significant differences between treatment groups and controls were found with regard to the immune parameters evaluated except for serum immunoglobulin concentrations. Serum Ig concentrations increased with synbiotic products which included (1) *S. cerevisiae, Lactobacillus rhamnosus*, inulin, *Lactobacillus reuteri, Lactobacillus plantarum*, and *Lactobacillus pentosus*; (2) *S. cerevisiae, Lactobacillus rhamnosus, Lactobacillus paracasei, Inulin, Bacillus subtilis*, and *Bacillus licheniformis*; and (3) calcium carbonate and *Enterococcus faecium* (Czyżewska-Dors et al., 2018).

By 2022, the use of medicinal ZnO must be phased out of pig production in the EU. Thus, one study assessed the impact on weaned pig performance, diarrhea incidence, and gut development, when (1) 'dietary ZnO supplementation was substituted by alternative commercial products based on macroalgae, specific probiotics or synbiotics, or (2) dietary ZnO inclusion was reduced from 2500 ppm to 1500 ppm.' The investigators concluded that 'none of the alternative dietary supplements were able to improve piglet performance or gut health, when ZnO was omitted from the diet' (Satessa et al., 2020). Despite the promising uses of synbiotics, it is evident that the 'ideal' combination remains elusive.

9.4 Proteobiotics

Metabolites derived from probiotics were identified and named as proteobiotics. The proteobiotics possess antimicrobial activity by interrupting cell-to-cell communication of bacteria (Tarsillo and Priefer, 2020), and by reducing the virulence of pathogenic bacteria. Molecules secreted by *Bifidobacterium bifidum* interfere with both attachment and invasion by bacteria and that probiotic *Bifidobacteria* strains provide an alternative approach to control food-borne enteric pathogens (Bayoumi and Griffiths, 2012). The first product, derived from *Lactobacillus acidophilus* and intended for use in piglets, is on the market by MicroSintesis Inc. It is claimed to support the growth and health of pigs specifically in the post-weaning phase (Nordeste et al., 2017).

Collectively, it is evident that there is considerable promise with the use of proteobiotics to control enteric pathogens; however, additional research is needed to demonstrate their effectiveness in large commercial farms.

10 Conclusion and future trends

For the vast majority of nutrients, researchers attempted to demonstrate the potential to improve immunity in pigs. Nutritional immunity is a concept that has been explored more extensively in other species, but it is well established that the gastrointestinal tract is the largest interface between the external and internal environments of the pig. While our knowledge of nutrient requirements has increased dramatically, techniques to assess immunity also have greatly improved and, to some degree, some of the nutrition research designed to evaluate immune function may be over simplified.

It is not surprising that many studies focused on nutrient interactions with intestinal immunity, and as microbiome research progresses, it is likely that our understanding of nutrition, immunity, and gut microbes will increase. However, further research is required to fully understand the influence of nutrition on systemic immunity. All pathogens are not created equal and there is considerable variation within a single pathogen, let alone within a disease category. Hence, we need to learn more about how each antigen interferes with metabolism, and disease challenges are likely multifactorial, which makes it challenging to plan for each disease-related scenario. Finally, one must consider the influence of the genetics of pigs. With intensive selection for lean growth, feed efficiency, and reproductive performance, it is likely that impact of nutritional modifications to improve immunity will differ among the various genetic lines of pigs. This aspect has rarely been addressed in the research studies.

Environmental regulations are becoming stricter and these regulations vary from country to country. Hence, some of the nutrients, such as zinc, which provided alternatives to antibiotic therapies, face more obstacles for practical, on-farm use. In addition, the global pork industry is facing additional challenges with various pathogens; however, there is a greater need for investigations to focus on nutrient strategies to minimize the effects of these pathogens on pig livability and growth performance. With few notable exceptions, most research was conducted in controlled experiments with relatively few animals in treatment groups. Often, the control animals receive a basal diet with the minimum requirements. Thus, it is not surprising that differences are detected between control and treatment groups. The results of the controlled studies need to be replicated in commercial farms, under typical production settings and within large populations of pigs.

It appears that more research and applications are being conducted in the private sector, rather than government-sponsored research within public institutions. Perhaps the private sector will generate more breakthroughs on nutrient strategies to boost immunity in pigs since the companies have financial incentives to produce cost-effective products. Unfortunately, results of research in the private sector may not be shared with the global pork industry. The negative results or studies that fail to demonstrate the efficacy of a product are rarely, if ever, published. This should not be surprising as the private sector has to consider financial returns and marketing strategies.

Despite the numerous published research papers, it is rare to note any cost:benefit ratios or independent financial analyses. Perhaps a nutrient strategy is effective to boost immunity, but if the cost exceeds the benefits, then the strategy will not be adopted by the pork industry. As stated by

Goodband et al. (2014), 'nutritionists and swine producers should resist the pressure to alter the diet, limit feed, or add expensive feed additives during an immune challenge.' This statement is valid; however, if the pork industry desires to continue to thrive, then cost-effective nutrient strategies need to be developed to deal with the variety of immune and pathogen challenges.

11 Where to look for further information

- Ding, H., Zhang, Q., Xu, H., Yu, X., Chen, L., Wang, Z. and Feng, J. (2021). Selection of copper and zinc dosages in pig diets based mutual benefit of animal growth and environmental protection. *Ecotoxicol. Environ. Saf.* 216: 112177. doi: 10.1016/j.ecoenv.2021.112177.
- Goodband, B., Tokach, M., Dritz, S., Derouchey, J. and Woodworth, J. (2014). Practical starter pig amino acid requirements in relation to immunity, gut health and growth performance. *J. Anim. Sci. Biotechnol.* 5:12 http://www.jasbsci.com/content/5/1/12.
- Liu, Y., Espinosa, C. D., Abelilla, J. J. Casas, G. A., Lagos, L. V., Lee, S. A., Kwon, W. B., Mathai, J. K., Navarro, D. M. D. L., Jaworski, N. W. and Stein, H. H (2018). Non-antibiotic feed additives in diets for pigs: A review. *Anim. Nutr.* 4(2):113–125. doi: 10.1016/j.aninu.2018.01.007.
- National Research Council (NRC). (2012). *Nutrient Requirements of Swine*. 11th Edition. National Academies Press, Washington DC.
- Pearce, E. L. and Pearce, E. J. (2013). Metabolic pathways in immune cell activation and quiescence. *Immunity* 38(4): 633–643. doi:10.1016/j.immuni.2013.04.005.
- Rodrigues, L. A., Felipe, N. A., Ferreira, M. O., Costa, M. O., Wellington, M. O. and Columbus, D. A. (2021). Factors affecting performance response of pigs exposed to different challenge models: A multivariate approach. *J. Anim. Sci.* 99: skab035. doi: 10.1093/jas/skab035.
- Saalmuller, A. and Gerner W. (2016). The immune system of swine. In: *Encyclopedia of Immunobiology*, Volume 1, pp. 538–548. doi: 10.1016/B978-0-12-374279-7.12019-3.
- Vitamin Nutrition Compendium. (2022). e-book. DSM Animal Nutrition & Health. https://www.dsm.com/anh/news/downloads/e-books-and-magazines/vitamin-compendium-e-book.html.

12 References

Adams, J. S. and Hewison, M. (2010). Update in vitamin D. *J. Clin. Endocrinol. Metab.* 95(2):471–478. DOI: 10.1210/jc.2009-1773.

Angelakis, E. (2017). Weight gain by gut microbiota manipulation in productive animals. *Microb. Pathog.* 106:162–170.

Bassaganya-Riera, J., Hontecillas-Magarzo, R., Bregendahl, K., Wannemuehler, M. J. and Zimmerman, D. R. (2001). Effects of dietary conjugated linoleic acid in nursery pigs of dirty and clean environments on growth, empty body composition, and immune competence. *J. Anim. Sci.* 79(3):714–721. DOI: 10.2527/2001.793714x.

Bassaganya-Riera, J., Hontecillas, R., Zimmerman, D. R. and Wannemuehler, M. J. (2002). Long-term influence of lipid nutrition on the induction of CD8 responses to viral or bacterial antigens. *Vaccine* 20:1435–1444.

Bayoumi, M. A. and Griffiths, M. W. (2012). In vitro inhibition of expression of virulence genes responsible for colonization and systemic spread of enteric pathogens using *Bifidobacterium bifidum* secreted molecules. *Int. J. Food Microbiol.* 156(3):255–263.

Bono, M. R., Tejon, G., Flores-Santibanez, F., Fernandez, D., Rosemblatt, M. and Sauma, D. (2016). Retinoic acid as a modulator of T cell immunity. *Nutrients* 8(6):349. DOI: 10.3390/nu8060349.

Boontiam, W., Wachirapakorn, C. and Phaengphairee, P. (2020). Effects of hydrolyzed yeast supplementation on growth performance, immunity, antioxidant capacity, and microbial shedding in weaning pigs. *Vet. World* 13(9):1902–1909.

Bruininx, E. M. A. M., Swinkels, J. W. G. M., Parmentier, H. K., Jetten, C. W. J., Gentry, J. L. and Schrama, J. W. (2000). Effects of an additional iron injection on growth and humoral immunity of weanling pigs. *Livest. Prod. Sci.* 67(1–2):31–39.

Cao, J., Guo, F., Zhang, L., Dong, B. and Gong, L. (2014). Effects of dietary selenomethionine supplementation on growth performance, antioxidant status, plasma selenium concentration, and immune function in weaning pigs. *J. Anim. Sci. Biotechnol.* 5(1):46. DOI: 10.1186/2049-1891-5-46.

Caputa, G., Castoldi, A. and Pearce, E. J. (2019). Metabolic adaptations of tissue-resident immune cells. *Nat. Immunol.* 20(7):793–801.

Castevens, K., Ferreira, J. B., Gillespie, T., Olsen, C., Nielsen, J. P. and Almond, G. (2020). Assessment of hemoglobin concentration in relation to sow reproductive stage and parity. *J. Swine Health Prod.* 28:254–257.

Center for Veterinary Medicine and Office of Regulatory Affairs. (1995). CPG Sec. 689.100. *Direct-Fed Microbial Products*. Available at: https://www.fda.gov/regulatory-information/search-fda-guidance-documents/cpg-sec-689100-direct-fed-microbial-products.

Chai, W., Wang, Z., Janczyk, P., Twardziok, S., Blohm, U., Osterrieder, N. and Burwinkel, M. (2014). Elevated dietary zinc oxide levels do not have a substantial effect on porcine reproductive and respiratory syndrome virus (PPRSV) vaccination and infection. *Virol. J.* 11:140. Available at: http://www.virologyj.com/content/11/1/140.

Chang, P. L., Boyd, R. D., Zier-Rush, C., Rosero, D. S. and van Heugten, E. (2019). Lipid peroxidation impairs growth and viability of nursery pigs reared under commercial conditions1. *J. Anim. Sci.* 97(8):3379–3389. DOI: 10.1093/jas/skz183.

Chang, Z., Zhang, H., Dong, H., Mehmood, K., Ijaz, M., Ahmad, H. I., Naeem, M. A., Wu, Q., Nabi, F. and Zhu, H. (2018). Effect of $CuSO_4$ and nano copper on serum antioxidant capacity in weaned piglets. *J. Biol. Regul. Homeost. Agents* 32(2):219–224.

Chase, C. and Lunney, J. K. (2019). Immune system. In: Zimmerman, J. J., Karriker, L. A. and Ramirez, A. (Eds) *Diseases of Swine* (11th edn.). Hobokon, NJ: Wiley Blackwell, pp. 264–291.

Chen, J., Chen, J., Zhang, Y., Lv, Y., Qiao, H., Tian, M., Cheng, L., Chen, F., Zhang, S. and Guan, W. (2021). Effects of maternal supplementation with fully oxidised β-carotene on the reproductive performance and immune response of sows, as well as the growth performance of nursing piglets. *Br. J. Nutr.* 125(1):62-70.

Chen, J., Han, J. H., Guan, W. T., Chen, F., Wang, C. X., Zhang, Y. Z., Lv, Y. T. and Lin, G. (2016a). Selenium and vitamin E in sow diets: I. Effect on antioxidant status and reproductive performance in multiparous sows. *Anim. Feed Sci. Technol.* 221:111-123. DOI: 10.1016/j.anifeedsci.2016.08.022.

Chen, J., Han, J. H., Guan, W. T., Chen, F., Wang, C. X., Zhang, Y. Z., Lv, Y. T. and Lin, G. (2016b). Selenium and vitamin E in sow diets: II. Effect on selenium status and antioxidant status of the progeny. *Anim. Feed Sci. Technol.* 221(A):101-110. DOI: 10.1016/j.anifeedsci.2016.08.021.

Chlebicz-Wójcik, A. and Śliżewska, K. (2020). The effect of recently developed synbiotic preparations on dominant fecal microbiota and organic acids concentrations in feces of piglets from nursing to fattening. *Animals (Basel)* 10(11):1999. DOI: 10.3390/ani10111999.

Coelho, M. B. and Cousins, B. (1997). Vitamin supplementation supports higher performance. *Feedstuffs* 27:10-21.

Czyżewska-Dors, E., Kwit, K., Stasiak, E., Rachubik, J., Śliżewska, K. and Pomorska-Mól, M. (2018). Effects of newly developed synbiotic and commercial probiotic products on the haematological indices, serum cytokines, acute phase proteins concentration, and serum immunoglobulins amount in sows and growing pigs - A pilot study. *J. Vet. Res.* 62(3):317-328. DOI: 10.2478/jvetres-2018-0046.

Dalgaard, T. S., Briens, M., Engberg, R. M. and Lauridsen, C. (2018). The influence of selenium and selenoproteins on immune responses of poultry and pigs. *Anim. Feed Sci. Technol.* 238:73-83.

Davanzo, G. G., Castro, G. and Moraes-Vieira, P. M. M. (2021). Immunometabolic regulation of adipose tissue resident immune cells. *Curr. Opin. Pharmacol.* 58:44-51. DOI: 10.1016/j.coph.2021.03.004.

de Ridder, K., Levesque, C. L., Htoo, J. K. and de Lange, C. F. M. (2012). Immune system efficiency of tryptophan utilization for body protein deposition in growing pigs. *J. Anim. Sci.* 90:3485-3491. DOI: 10.2527/jas.2011-4830.

Ding, H., Zhang, Q., Xu, H., Yu, X., Chen, L., Wang, Z. and Feng, J. (2021). Selection of copper and zinc dosages in pig diets based mutual benefit of animal growth and environmental protection. *Ecotoxicol. Environ. Saf.* 216:112-177. DOI: 10.1016/j.ecoenv.2021.112177.

Djoko, K. Y. Y., Ong, C. L., Walker, M. J. and McEwan, A. G. (2015). The role of copper and zinc toxicity in innate immune defense against bacterial pathogens. *J. Biol. Chem.* 290(31):18954-18961. DOI: 10.1074/jbc.R115.647099.

Dong, Z., Wan, D., Li, G., Zhang, Y., Yang, H., Wu, X. and Yin, Y. (2020). Comparison of oral and parenteral iron administration on iron homeostasis, oxidative and immune status in anemic neonatal pigs. *Biol. Trace Elem. Res.* 195(1):117-124. DOI: 10.1007/s12011-019-01846-9.

Dritz, S. S., Shi, J., Kielian, T. L., Goodband, R. D., Nelssen, J. L., Tokach, M. D., Chengappa, M. M., Smith, J. E. and Blecha, F. (1995). Influence of dietary beta-glucan on growth performance, nonspecific immunity, and resistance to Streptococcus suis infection in weanling pigs. *J. Anim. Sci.* 73(11):3341-3350. DOI: 10.2527/1995.73113341x.

Duan, Y., Li, F., Li, L., Fan, J., Sun, X. and Yin, Y. (2014). n-6:n-3 PUFA ratio is involved in regulating lipid metabolism and inflammation in pigs. *Br. J. Nutr.* 111(3):445-451. DOI: 10.1017/S0007114513002584.

Duarte, M. E., Tyus, J. and Kim, S. W. (2020). Synbiotic effects of enzyme and probiotics on intestinal health and growth of newly weaned pigs challenged with enterotoxigenic F18+Escherichia coli. *Front. Vet. Sci.* 7:573. DOI: 10.3389/fvets.2020.00573.

Eicher, S. D., McKee, C. A., Carroll, J. A. and Pajor, E. A. (2006). Supplemental vitamin C and yeast cell wall beta-glucan as growth enhancers in newborn pigs and as immunomodulators after an endotoxin challenge after weaning. *J. Anim. Sci.* 84(9):2352–2360. DOI: 10.2527/jas.2005-770.

Espinosa, C. D. and Stein, H. H. (2021). Digestibility and metabolism of copper in diets for pigs and influence of dietary copper on growth performance, intestinal health, and overall immune status: A review. *J. Anim. Sci. Biotechnol.* 12(1):13. DOI: 10.1186/s40104-020-00533-3.

Fan, P., Liu, P., Song, P., Chen, X. and Ma, X. (2017). Moderate dietary protein restriction alters the composition of gut microbiota and improves ileal barrier function in adult pig model. *Sci. Rep.* 7:43412.

FAO/WHO Joint Report. (2001). Health and nutritional properties of probiotics in food including powder milk with live lactic acid bacteria. report of a joint FAO/WHO expert consultation. Available at: http://www.who.int/foodsafety/publications/fs_management/en/probiotics.pdf.

Faure, M., Choné, F., Mettraux, C., Godin, J. P., Béchereau, F., Vuichoud, J., Papet, I., Breuillé, D. and Obled, C. (2007). Threonine utilization for synthesis of acute phase proteins, intestinal proteins, and mucins is increased during sepsis in rats. *J. Nutr.* 137(7):1802–1807. DOI: 10.1093/jn/137.7.1802.

Ferrandis Vila, M., Trudeau, M. P., Hung, Y. T., Zeng, Z., Urriola, P. E., Shurson, G. C. and Saqui-Salces, M. (2018). Dietary fiber sources and non-starch polysaccharide-degrading enzymes modify mucin expression and the immune profile of the swine ileum. *PLoS ONE* 13(11):e0207196. DOI: 10.1371/Journal.Pone.0207196.

Flohr, J. R., DeRouchey, J. M., Woodworth, J. C., Tokach, M. D., Goodband, R. D. and Dritz, S. S. (2016). A survey of current feeding regimens for vitamins and trace minerals in the US swine industry. *J. Swine Health Prod.* 24(6):290–303.

Garcia-Launay, F., van der Werf, H. M. G. and Nguyen, T. T. H. (2014). Evaluation of the environmental implications of the incorporation of feed use amino acids in pig production using life cycle assessment. *Livest. Sci.* 161:158–175. DOI: 10.1017/S1751731111001078.

Gloaguen, M., Le Floc'h, N., Corrent, E., Primot, Y. and van Milgen, J. (2014). The use of free amino acids allows formulating very low crude protein diets for piglets. *J. Anim. Sci.* 92(2):637–644. DOI: 10.2527/jas.2013-6514.

Gonzales-Eguia, A., Fu, C.-M., Lu, F.-Y. and Lien, T.-F. (2009). Effects of nanocopper on copper availability and nutrients digestibility, growth performance and serum traits of piglets. *Livest. Sci.* 126(1–3):122–129.

Goodband, B., Tokach, M., Dritz, S., Derouchey, J. and Woodworth, J. (2014). Practical starter pig amino acid requirements in relation to immunity, gut health and growth performance. *J. Anim. Sci. Biotechnol.* 5(1):12. Available at: http://www.jasbsci.com/content/5/1/12.

Han, B., Baruah, K., Cox, E., Vanrompay, D. and Bossier, P. (2020). Structure-functional activity relationship of β-glucans from the perspective of immunomodulation: A mini-review. *Front. Immunol.* 11:658. DOI: 10.3389/fimmu.2020.00658.

Hasan, S., Orro, T., Valros, A., Junnikkala, S., Peltoniemi, O. and Oliviero, C. (2019). Factors affecting sow colostrum yield and composition, and their impact on piglet growth and health. *Livest. Sci.* 227:60–67.

Hasan, S., Saha, S., Junnikkala, S., Orro, T., Peltoniemi, O. and Oliviero, C. (2018). Late gestation diet supplementation of resin acid-enriched composition increases sow colostrum immunoglobulin G content, piglet colostrum intake and improve sow gut microbiota. *Animal* 1-8. DOI: 10.1017/S1751731118003518.

Hernández, J., Soto-Canevett, E., Pinelli-Saavedra, A., Resendiz, M., Moya-Camarena, S. Y. and Klasing, K. C. (2009). In vitro effect of vitamin E on lectin-stimulated porcine peripheral blood mononuclear cells. *Vet. Immunol. Immunopathol.* 131(1-2):9-16. DOI: 10.1016/j.vetimm.2009.03.001.

Hosnedlova, B., Kepinska, M., Skalickova, S., Fernandez, C., Ruttkay-Nedecky, B., Malevu, T. D., Sochor, J., Baron, M., Melcova, M., Zidkova, J. and Kizek, R. (2017). A summary of new findings on the biological effects of selenium in selected animal species–A critical review. *Int. J. Mol. Sci.* 18(10):2209. DOI: 10.3390/ijms18102209.

Huber, L. A., Hooda, S., Fisher-Heffernan, R. E., Karrow, N. A. and de Lange, C. F. M. (2018). Effect of reducing the ratio of omega-6-to-omega-3 fatty acids in diets of low protein quality on nursery pig growth performance and immune response. *J. Anim. Sci.* 96(10):4348–4359. DOI: 10.1093/jas/sky296.

Hutkins, R. W., Krumbeck, J. A., Bindels, L. B., Cani, P. D., Fahey, G., Goh, Y. J., Hamaker, B., Martens, E. C., Mills, D. A., Rastal, R. A., Vaughan, E. and Sanders, M. E. (2016). Prebiotics: Why definitions matter. *Curr. Opin. Biotechnol.* 37:1–7. DOI: 10.1016/j.copbio.2015.09.001.

Igata, M., Islam, M. A., Tada, A., Takagi, M., Kober, A. K. M. H., Albarracin, L., Aso, H., Ikeda-Ohtsubo, W., Miyazawa, K., Yoda, K., He, F., Takahashi, H., Villena, J. and Kitazawa, H. (2019). Transcriptome modifications in porcine adipocytes via toll-like receptors activation. *Front. Immunol.* 10:1180. DOI: 10.3389/fimmu.2019.01180.

Janczyk, P., Kreuzer, S., Assmus, J., Nöckler, K. and Brockmann, G. A. (2013). No protective effects of high-dosage dietary zinc oxide on weaned pigs infected with Salmonella enterica serovar Typhimurium DT104. *Appl. Environ. Microbiol.* 79(9):2914–2921.

Jensen-Waern, M., Melin, L., Lindberg, R., Johannisson, A., Petersson, L. and Wallgren, P. (1998). Dietary zinc oxide in weaned pigs–Effects on performance, tissue concentrations, morphology, neutrophil functions and fecal microflora. *Res. Vet. Sci.* 64(3):225-231.

Jiao, N., Wu, Z., Ji, Y., Wang, B., Dai, Z. and Wu, G. (2015). L-glutamate enhances barrier and antioxidative functions in intestinal porcine epithelial cells. *J. Nutr.* 145(10):2258-2264. DOI: 10.3945/jn.115.217661.

Kampman-van de Hoek, E., Jansman, A. J., van den Borne, J. J., van der Peet-Schwering, C. M., van Beers-Schreurs, H. and Gerrits, W. J. (2016). Dietary amino acid deficiency reduces the utilization of amino acids for growth in growing pigs after a period of poor health. *J. Nutr.* 146(1):51–58. DOI: 10.3945/jn.115.216044.

Kampman-van de Hoek, E., Sakkas, P., Gerrits, W. J., van den Borne, J. J., van der Peet-Schwering, C. M. and Jansman, A. J. (2015). Induced lung inflammation and dietary protein supply affect nitrogen retention and amino acid metabolism in growing pigs. *Br. J. Nutr.* 113(3):414–425. DOI: 10.1017/S0007114514003821.

Kerr, B. J., Kellner, T. A. and Shurson, G. C. (2015). Characteristics of lipids and their feeding value in swine diets. *J. Anim. Sci. Biotechnol.* 6(1):30. DOI: 10.1186/s40104-015-0028-x.

Kerr, B. J. and Shurson, G. C. (2013). Strategies to improve fiber utilization in swine. *J. Anim. Sci. Biotechnol.* 4(1):11.

Kloubert, V., Blaabjerg, K., Dalgaard, T. S., Poulsen, H. D., Rink, L. and Wessels, I. (2018). Influence of zinc supplementation on immune parameters in weaned pigs. *J. Trace Elem.Med. Biol.* 49:231-240.

Konowalchuka, J. D., Riegera, A. M., Kiemeleb, M. D., Ayres, D. C. and Barreda, D. R. (2013). Modulation of weanling pig cellular immunity in response to diet supplementation with 25-hydroxyvitamin D3. *Vet. Immunol. Immunopathol.* 155(1-2):57-66.

Kvidera, S. K., Horst, E. A., Mayorga, E. J., Sanz-Fernandez, M. V., Abuajamieh, M. and Baumgard, L. H. (2017). Estimating glucose requirements of an activated immune system in growing pigs. *J. Anim. Sci.* 95(11):5020-5029. DOI: 10.2527/jas2017.1830.

Langel, S. N., Paim, F. C., Alhamo, M. A., Lager, K. M., Vlasova, A. N. and Saif, L. J. (2019). Oral vitamin A supplementation of porcine epidemic diarrhea virus infected gilts enhances IgA and lactogenic immune protection of nursing piglets. *Vet. Res.* 50(1):101. DOI: 10.1186/s13567-019-0719-y.

Larange, A. and Cheroutre, H. (2016). Retinoic acid and retinoic acid receptors as pleiotropic modulators of the immune system. *Annu. Rev. Immunol.* 34:369-394.

Lauridsen, C. (2020). Effects of dietary fatty acids on gut health and function of pigs pre- and post-weaning. *J. Anim. Sci.* 98(4). DOI: 10.1093/jas/skaa086.

Lauridsen, C. and Jensen, S. K. (2005). Influence of supplementation of all-rac-α-tocopheryl acetate preweaning and vitamin C postweaning on α-tocopherol and immune responses of piglets. *J. Anim. Sci.* 83(6):1274-1286.

Le Floc'h, N., Deblanc, C., Cariolet, R., Gautier-Bouchardon, A. V., Merlot, E. and Simon, G. (2014). Effect of feed restriction on performance and postprandial nutrient metabolism in pigs co-infected with Mycoplasma hyopneumoniae and swine influenza virus. *PLoS ONE* 9(8):e104605. DOI: 10.1371/journal.pone.0104605.

Le Floc'h, N., Lebellego, L., Matte, J. J., Melchior, D. and Sève, B. (2009)The effect of sanitary status degradation and dietary tryptophan content on growth rate and tryptophan metabolism in weaning pigs. *J. Anim. Sci.* 87(5):1686-1694. DOI: 10.2527/jas.2008-1348.

Le Floc'h, N., Wessels, A., Corrent, E., Wu, G. and Bosi, P. (2018). The relevance of functional amino acids to support the health of growing pigs. *Anim. Feed Sci. Tech*nol. 245:104-116. DOI: 10.1016/j.anifeedsci.2018.09.007.

Le Jan, C. (1994). A study by flow cytometry of lymphocytes in sow colostrum. *Res. Vet. Sci.* 57(3):300-304.

Leonard, S. G., Sweeney, T., Bahar, B., Lynch, B. P. and O'Doherty, J. V. (2010). Effect of maternal fish oil and seaweed extract supplementation on colostrum and milk composition, humoral immune response, and performance of suckled piglets. *J. Anim. Sci.* 88(9):2988-2997.

Leyshon, B. J., Ji, P., Caputo, M. P., Matt, S. M. and Johnson, R. W. (2019). Dietary iron deficiency impaired peripheral immunity but did not alter brain microglia in PRRSV-infected neonatal piglets. *Front. Immunol.* 9:Article 3150.

Li, J., Li, D. F., Xing, J. J., Cheng, Z. B. and Lai, C. H. (2006). Effects of β-glucan extracted from *Saccharomyces cerevisiae* on growth performance and immunological and somatotropic responses of pigs challenged with Escherichia coli lipopolysaccharide. *J. Anim. Sci.* 84(9):2374-2381.

Li, Q., Burrough, E. R., Gabler, N. K., Loving, C. L., Sahin, O., Gould, S. A. and Patience, J. F. (2019a). A soluble and highly fermentable dietary fiber with carbohydrases improved gut barrier integrity markers and growth performance in F18 ETEC challenged pigs1. *J. Anim. Sci.* 97(5):2139-2153. DOI: 10.1093/jas/skz093.

Li, Q., Schmitz-Esser, S., Loving, C. L., Gabler, N. K., Gould, S. A. and Patience, J. F. (2019b). Exogenous carbohydrases added to a starter diet reduced markers of systemic immune activation and decreased Lactobacillus in weaned pigs1. *J. Anim. Sci.* 97(3):1242–1253. DOI: 10.1093/jas/sky481.

Liu, L., Chen, D., Yu, B., Luo, Y., Huang, Z., Zheng, P., Mao, X., Yu, J., Luo, J., Yan, H. and He, J. (2021). Influences of selenium-enriched yeast on growth performance, immune function, and antioxidant capacity in weaned pigs exposure to oxidative stress. *BioMed Res. Int.* 2021:Article ID 5533210, 11 pages. DOI: 10.1155/2021/5533210.

Ma, T., Suzuki, Y. and Guan, L. L. (2018). Dissect the mode of action of probiotics in affecting host-microbial interactions and immunity in food producing animals. *Vet. Immunol. Immunopathol.* 205:35–48.

Ma, Y. L., Zanton, G. I., Zhao, J., Wedekind, K., Escobar, J. and Vazquez-Añón, M. (2015). Multitrial analysis of the effects of copper level and source on performance in nursery pigs. *J. Anim. Sci.* 93(2):606–614. DOI: 10.2527/jas.2014-7796.

Mao, X., Lai, X., Yu, B., He, J., Yu, J., Zheng, P., Tian, G., Zhang, K. and Chen, D. (2014). Effects of dietary threonine supplementation on immune challenge induced by swine Pseudorabies live vaccine in weaned pigs. *Arch. Anim. Nutr.* 68(1):1–15. DOI: 10.1080/1745039X.2013.869988.

McConn, B. R., Duttlinger, A. W., Kpodo, K. R., Eicher, S. D., Richert, B. T. and Johnson, J. S. (2020). Replacing dietary antibiotics with 0.20% l-glutamine and synbiotics following weaning and transport in pigs. *J. Anim. Sci.* 98(9):skaa272. DOI: 10.1093/jas/skaa272.

McGilvray, W. D., Johnson, B., Wooten, H., Rakhshandeh, A. R. and Rakhshandeh, A. (2019a). Immune system stimulation reduces the efficiency of whole-body protein deposition and alters muscle fiber characteristics in growing pigs. *Animals (Basel)* 9(6):323. DOI: 10.3390/ani9060323.

McGilvray, W. D., Klein, D., Wooten, H., Dawson, J. A., Hewitt, D., Rakhshandeh, A. R., De Lange, C. F. M. and Rakhshandeh, A. (2019b). Immune system stimulation induced by porcine reproductive and respiratory syndrome virus alters plasma free amino acid flux and dietary nitrogen utilization in starter pigs1. *J. Anim. Sci.* 97(6):2479-2492. DOI: 10.1093/jas/skz120.

Melchior, D., Sève, B. and Le Floc'h, N. (2004). Chronic lung inflammation affects plasma amino acid concentrations in pigs. *J. Anim. Sci.* 82(4):1091-1099. DOI: 10.2527/2004.8241091x.

Menegat, M. B., Goodband, R. D., DeRouchey, J. M., Woodworth, J. C., Bryte, J., Dritz, S. S. and Tokach, M. D. (2019). *Fiber in Nursery Diets*. Kansas State University Nutrition Guide.

Millet, S., Cox, E., Buyse, J., Goddeeris, B. M. and Janssens, G. P. J. (2005). Immunocompetence of fattening pigs fed organic versus conventional diets in organic versus conventional housing. *Vet. J.* 169(2):293–299.

Nairz, M. and Weiss, G. (2020). Iron in infection and immunity. *Mol. Aspects Med.* 75:100864. DOI: 10.1016/j.mam.2020.100864.

Namkung, H., Gong, J., Yu, H. and de Lange, C. F. M. (2006). Effect of pharmacological intakes of zinc and copper on growth performance, circulating cytokines and gut microbiota of newly weaned piglets challenged with coliform lipopolysaccharides. *Can. J. Anim. Sci.* 86(4):511-522.

National Research Council (NRC). (2012). *Nutrient Requirements of Swine* (11th edn.). Washington, DC: National Academies Press.

National Swine Nutrition Guide. (2010). *U.S. Pork Center of Excellence*. Ames, IA: Iowa State University.

Nguyen, K., Cassar, G., Friendship, R. M. and Hodgins, D. (2013). An investigation of the impacts of induced parturition, birth weight, birth order, litter size and sow parity on piglet serum concentrations of immunoglobulin G. *J. Swine Health Prod.* 21:139–143.

Nordeste, R., Tessema, A., Sharma, S., Kovač, Z., Wang, C., Morales, R. and Griffiths, M. W. (2017). Molecules produced by probiotics prevent enteric colibacillosis in pigs. *BMC Vet. Res.* 13(1):335. DOI: 10.1186/s12917-017-1246-6.

Normand, V., Perrin, H., Auvigne, V., Robert, N. and Laval, A. (2012). Anaemia in the sow: A cohort study to assess factors with an impact on haemoglobin concentration, and the influence of haemoglobin concentration on the reproductive performance. *Vet. Rec.* 171(14):350.

Ogawa, S., Tsukahara, T., Imaoka, T., Nakanishi, N., Ushida, K. and Inoue, R. (2016). The effect of colostrum ingestion during the first 24 hours of life on early postnatal development of piglet immune systems. *Anim. Sci. J.* 87(12):1511–1515.

Osredkar, J. and Sustar, N. (2011). Copper and zinc, biological role and significance of copper/zinc imbalance. *J. Clin. Tox. Suppl.* 3:1.

Pandey, K. R., Naik, S. R. and Vakil, B. V. (2015). Probiotics, prebiotics and synbiotics - A review. *J. Food Sci. Technol.* 52(12):7577–7587. DOI: 10.1007/s13197-015-1921-1.

Papadopoulos, G. A., Maes, D. G. D., Van Weyenber, S., van Kempen, T. A., Buyse, J. and Janssens, G. P. (2009). Perpartal feeding strategy with different n-6:n-3 ratios in sows: Effects on sows' performance, inflammatory and periparturient metabolic parameters. *Br. J. Nutr.* 101(3):348–357. DOI: 10.1017/S0007114508026160.

Pastorelli, H., van Milgen, J., Lovatto, P. and Montagne, L. (2012). Meta-analysis of feed intake and growth responses of growing pigs after a sanitary challenge. *Animal* 6(6):952–961. DOI: 10.1017/S175173111100228X.

Pearce, E. L. and Pearce, E. J. (2013). Metabolic pathways in immune cell activation and quiescence. *Immunity* 38(4):633–643. DOI: 10.1016/j.immuni.2013.04.005.

Pei, X., Xiao, Z., Liu, L., Wang, G., Tao, W., Wang, M., Zou, J. and Leng, D. (2019). Effects of dietary zinc oxide nanoparticles supplementation on growth performance, zinc status, intestinal morphology, microflora population, and immune response in weaned pigs. *J. Sci. Food Agric.* 99(3):1366–1374. DOI: 10.1002/jsfa.9312.

Peng, C. C., Yan, J. Y., Dong, B., Zhu, L., Tian, Y. Y. and Gong, L. M. (2017a). Effects of graded levels of cupric citrate on growth performance, antioxidant status, serum lipid metabolites and immunity, and tissue residues of trace elements in weaned pigs Asian-Australas. *J. Anim. Sci.* 30(4):538–545. DOI: 10.5713/ajas.16.0182.

Peng, Y., Yu, K., Mu, C., Hang, S., Che, L. and Zhu, W. (2017b). Progressive response of large intestinal bacterial community and fermentation to the stepwise decrease of dietary crude protein level in growing pigs. *Appl. Microbiol. Biotechnol.* 101(13):5415–5426. DOI: 10.1007/s00253-017-8285-6.

Pérez, V. G., Waguespack, A. M., Bidner, T. D., Southern, L. L., Fakler, T. M., Ward, T. L., Steidinger, M. and Pettigrew, J. E. (2011). Additivity of effects from dietary copper and zinc on growth performance and fecal microbiota of pigs after weaning. *J. Anim. Sci.* 89(2):414–425. DOI: 10.2527/jas.2010-2839.

Pinelli-Saavedra, A., Calderón de la Barca, A. M., Hernández, J., Valenzuela, R. and Scaife, J. R. (2008). [Alpha]-tocopherol to piglet tissues, colostrum, and milk: aspects of immune status of piglets. *Res. Vet. Sci.* 85(1):92–100.

Pu, Y., Li, S., Xiong, H., Zhang, X., Wang, Y. and Du, H. (2018). Iron promotes intestinal development in neonatal piglets. *Nutrients* 10(6):726. DOI: 10.3390/nu10060726.

Quiniou, N., Courboulay, V., Salaün, Y. and Chevillon, P. (2010). Impact of the non-castration of male pigs on growth performance and behavior - Comparison with barrows and gilts. *A. M. EAAP* 61:1-7.

Rakhshandeh, A., Dekkers, J. C. M., Kerr, B. J., Weber, T. E., English, J. and Gabler, N. K. (2012). Effect of immune system stimulation and divergent selection for residual feed intake on digestive capacity of the small intestine in growing pigs. *J. Anim. Sci.* 90(Suppl. 4):233–235. DOI: 10.2527/jas.53976.

Rakhshandeh, A., Htoo, J. K. and de Lange, C. F. M. (2010). Immune system stimulation of growing pigs does not alter apparent ileal amino acid digestibility but reduces the ratio between whole body nitrogen and sulfur retention. *Livest. Sci.* 134(1-3):21-23. DOI: 10.1016/j.livsci.2010.06.085.

Rémond, D., Buffière, C., Godin, J. P., Mirand, P. P., Obled, C., Papet, I., Dardevet, D., Williamson, G., Breuillé, D. and Faure, M. (2009). Intestinal inflammation increases gastrointestinal threonine uptake and mucin synthesis in enterally fed minipigs. *J. Nutr.* 139(4):720–726. 10.3945/jn.108.101675.

Rodrigues, L. A., Ferreira, F. N. A., Costa, M. O., Wellington, M. O. and Columbus, D. A. (2021). Factors affecting performance response of pigs exposed to different challenge models: A multivariate approach. *J. Anim. Sci.* 99(6):skab035. DOI: 10.1093/jas/skab035.

Rudar, M., Zhu, C. L. and de Lange, C. F. M. (2017). Dietary leucine supplementation decreases whole-body protein turnover before, but notvduring, immune system stimulation in pigs. *J. Nutr.* 147(1):45–51.

Ryan, M. T., O'Shea, C. J., Collins, C. B., O'Doherty, J. V. and Sweeney, T. (2012). Effects of dietary supplementation with Laminaria Hyperborea, Laminaria digitata, and Saccharomyces cerevisiae on the IL-17 pathway in the porcine colon. *J. Anim. Sci.* 90(Suppl. 4):263–265. DOI: 10.2527/jas.53802.

Saalmuller, A. and Gerner, W. (2016). The immune system of swine. In: *Encyclopedia of Immunobiology* (vol. 1), pp. 538–548. DOI: 10.1016/B978-0-12-374279-7.12019-3.

Salmon, H., Berri, M., Gerdts, V. and Meurens, F. (2009). Humoral and cellular factors of maternal immunity in swine. *Dev. Comp. Immunol.* 33(3):384–393.

Santos, R. K. S., Novais, A. K., Borges, D. S., Alves, J. B., Dario, J. G. N., Frederico, G., Pierozan, C. R., Batista, J. P., Pereira, M. and Silva, C. A. (2020). Increased vitamin supplement to sows, piglets and finishers and the effect in productivity. *Animal* 14(1):86-94. DOI: 10.1017/S1751731119001927.

Satessa, G. D., Kjeldsen, N. J., Mansouryar, M., Hansen, H. H., Bache, J. K. and Nielsen, M. O. (2020). Effects of alternative feed additives to medicinal zinc oxide on productivity, diarrhoea incidence and gut development in weaned piglets. *Animal* 26:1-9. DOI: 10.1017/S1751731120000154.

Schlundt, J. (2012). *Health and Nutritional Properties of Probiotics in Food including Powder Milk with Live Lactic Acid Bacteria" Report of a Joint FAO/WHO Expert Consultation on Evaluation of Health and Nutritional Properties of Probiotics in Food Including Powder Milk with Live Lactic Acid Bacteria.* Geneva: Food and Agriculture Organization/World Health Organization.

Schwalfenberg, G. K. (2011). A review of the critical role of vitamin D in the functioning of the immune system and the clinical implications of vitamin D deficiency. *Mol. Nutr. Food Res.* 55(1):96–108. DOI: 10.1002/mnfr.201000174.

Shannon, M. C. and Hill, G. M. (2019). Trace mineral supplementation for the intestinal health of young monogastric animals. *Front. Vet. Sci.* 6:73. DOI: 10.3389/fvets.2019.00073.

Shen, J., Chen, Y., Wang, Z., Zhou, A., He, M., Mao, L., Zou, H., Peng, Q., Xue, B., Wang, L., Zhang, X., Wu, S. and Lv, Y. (2014). Coated zinc oxide improves intestinal immunity function and regulates microbiota composition in weaned piglets. *Br. J. Nutr.* 111(12):2123–2134. DOI: 10.1017/S0007114514000300.

Shimazu, T., Borjigin, L., Katoh, K., Roh, S. G., Kitazawa, H., Abe, K., Suda, Y., Saito, H., Kunii, H., Nihei, K., Uemoto, Y., Aso, H. and Suzuki, K. (2019). Addition of Wakame seaweed (Undaria pinnatifida) stalk to animal feed enhances immune response and improves intestinal microflora in pigs. *Anim. Sci. J.* 90(9):1248–1260.

Silva-Guillen, Y. V., Arellano, C., Boyd, R. D., Martinez, G. and van Heugten, E. (2020). Growth performance, oxidative stress and immune status of newly weaned pigs fed peroxidized lipids with or without supplemental vitamin E or polyphenols. *J. Anim. Sci. Biotechnol.* 11(1):22. DOI: 10.1186/s40104-020-0431-9.

Simopoulos, A. P. (1991). Omega-3 fatty acids in health and disease and in growth and development. *Am. J. Clin. Nutr.* 54(3):438–463. DOI: 10.1093/ajcn/54.3.438.

Song, R., Chen, C., Johnston, L. J., Kerr, B. J., Weber, T. E. and Shurson, G. C. (2014). Effects of feeding diets containing highly peroxidized distillers dried grains with solubles and increasing vitamin E levels to wean-finish pigs on growth performance, carcass characteristics, and pork fat composition. *J. Anim. Sci.* 92(1):198–210. DOI: 10.2527/jas.2013-6334.

Spurlock, M. E. (1997). Regulation of metabolism and growth during immune challenge: An overview of cytokine function. *J. Anim. Sci.* 75(7):1773–1783.

Stuart, W. D., Burkey, T. E., Gabler, N. K., Schwartz, K. J., Klein, D., Dawson, J. A., Pendleton, A. R., de Lange, C. F. M. and Rakhshandeh, A. (2016). 110 Immune system stimulation (ISS) induced by *E. coli* lipopolysaccharide (LPS) alters amino acid metabolism in growing pigs. *J. Anim. Sci.* 94(Suppl 2):51–52.

Stuyven, E., Cox, E., Vancaeneghem, S., Arnouts, S., Deprez, P. and Goddeeris, B. M. (2009). Effect of β-glucans on an ETEC infection in piglets. *Vet. Immunol. Immunopathol.* 128(1-3):60–66.

Suebsaard, P. and Charerntantanakul, W. (2021). Rutin, α-tocopherol, and l-ascorbic acid up-regulate type I interferon-regulated gene and type I and II interferon expressions and reduce inflammatory cytokine expressions in monocyte-derived macrophages infected with highly pathogenic porcine reproductive and respiratory syndrome virus. *Vet. Immunol. Immunopathol.* 235:110231. DOI: 10.1016/j.vetimm.2021.110231.

Sumner, S. E., Markley, R. L. and Kirimanjeswara, G. S. (2019). Role of selenoproteins in bacterial pathogenesis. *Biol. Trace Elem. Res.* 192(1):69–82. DOI: 10.1007/s12011-019-01877-2.

Sun, P., Li, D., Dong, B., Qiao, S., Ma, X. and Chen, X. (2009). Vitamin C: An immunomodulator that attenuates anaphylactic reactions to soybean glycinin hypersensitivity in a swine model. *Food Chem.* 113(4):914–918. DOI: 10.1016/j.foodchem.2008.08.018.

Tan, L., Wei, T., Yuan, A., He, J., Liu, J., Xu, D. and Yang, Q. (2017). Dietary supplementation of Astragalus polysaccharides enhanced immune components and Growth Factors

EGF and IGF-1 in sow colostrum. *J. Immunol. Res.* 2017:Article ID 9253208. DOI: 10.1155/2017/9253208.
Tanghe, S., Cox, E., Melkebeek, V., De Smet, S. and Millet, S. (2014). Effect of fatty acid composition of the sow diet on the innate and adaptive immunity of the piglets after weaning. *Vet. J.* 200(2):287–293.
Tarsillo, B. and Priefer, R. (2020). Proteobiotics as a new antimicrobial therapy. *Microb. Pathog.* 142:104093. DOI: 10.1016/j.micpath.2020.104093.
Tiwari, U. P., Fleming, S. A., Rasheed, M. S. A., Jha, R. and Dilger, R. N. (2020). The role of oligosaccharides and polysaccharides of xylan and mannan in gut health of monogastric animals. *J. Nutr. Sci.* 9:e21. DOI: 10.1017/jns.2020.14.
Trevisi, P., Melchior, D., Mazzoni, M., Casini, L., De Filippi, S., Minieri, L., Lalatta-Costerbosa, G. and Bosi, P. (2009). A tryptophan-enriched diet improves feed intake and growth performance of susceptible weanling pigs orally challenged with Escherichia coli K88. *J. Anim. Sci.* 87(1):148–156. DOI: 10.2527/jas.2007-0732.
Turck, D., Castenmiller, J. and De Henauw, S. (2019). Nutrimune and immune defence against pathogens in the gastrointestinal and upper respiratory tracts: Evaluation of a health claim pursuant to Article 14 of Regulation (EC) No 1924/2006. *Eur. Food Saf. Authority Panel Nutr. Novel Foods Food Allergens* 17(4):e05656.
van Heugten, E., Spears, J. W. and Coffey, M. T. (1994). The effect of dietary protein on performance and immune response in weanling pigs subjected to an inflammatory challenge. *J. Anim. Sci.* 72(10):2661–2669.
van Kempen, T. A. and McComas, S. (2002). Infrared spectroscopy as a tool for assessing fat quality. *J. Appl. Poult. Res.* 11(2):191–201. DOI: 10.1093/japr/11.2.191.
Verso, L. L., Mattea, J. J., Lapointe, J., Talbot, G., Bissonnette, N., Blais, M., Guay, F. and Lessard, M. (2020). Impact of birth weight and neonatal nutritional interventions with micronutrients and bovine colostrum on the development of piglet immune response during the peri-weaning period. *Vet. Immunol. Immunopathol.* 226:110072. DOI: 10.1016/j.vetimm.2020.110072.
Vetvicka, V. and Oliveira, C. (2014). β(1-3)(1-6)-D-glucans modulate immune status in pigs: Potential importance for efficiency of commercial farming. *Ann. Transl. Med.* 2(2):16. DOI: 10.3978/j.issn.2305-5839.2014.01.04.
Vetvicka, V., Vannucci, L. and Sima, P. (2014). The effects of β-glucan on pig growth and immunity. *Open Biochem. J.* 8:89–93.
Vetvicka, V., Vannucci, L. and Sima, P. (2020). β-glucan as a new tool in vaccine development. *Scand. J. Immunol.* 91(2):e12833. DOI: 10.1111/sji.12833.
Volman, J. J., Ramakers, J. D. and Plat, J. (2008). Dietary modulation of immune function by β-glucans. *Physiol. Behav.* 94(2):276–284.
Vunta, H., Belda, B. J., Arner, R. J., Channa Reddy, C., Vanden Heuvel, J. P. and Sandeep Prabhu, K. (2008). Selenium attenuates pro-inflammatory gene expression in macrophages. *Mol. Nutr. Food Res.* 52(11):1316–1323.
Wang, L., Xiaodong Xu, X., Su, G., Shi, B. and Shan, A. (2017). High concentration of vitamin E supplementation in sow diet during the last week of gestation and lactation affects the immunological variables and antioxidative parameters in piglets. *J. Dairy Res.* 84(1):8–13. DOI: 10.1017/S0022029916000650.
Wang, X., Qiao, S. Y., Liu, M. and Ma, Y. X. (2006). Effects of graded levels of true ileal digestible threonine on performance, serum parameters and immune function of 10–25 kg pigs. *Anim. Feed Sci. Technol.* 129(3–4):264–278.

Wellington, M. O., Agyekum, A. K., Hamonic, K., Htoo, J. K., Van Kessel, A. G. and Columbus, D. A. (2019). Effect of supplemental threonine above requirement on growth performance of Salmonella typhimurium challenged pigs fed high-fiber diets. *J. Anim. Sci.* 97(9):3636–3647. DOI: 10.1093/jas/skz225.

Wellington, M. O., Htoo, J. K., Van Kessel, A. G. and Columbus, D. A. (2018). Impact of dietary fiber and immune system stimulation on threonine requirement for protein deposition in growing pigs. *J. Anim. Sci.* 96(12):5222–5232. DOI: 10.1093/jas/sky381.

Whitney, M. H., Shurson, G. C. and Guedes, R. C. (2006). Effect of including distillers dried grains with solubles in the diet, with or without antimicrobial regimen, on the ability of growing pigs to resist a Lawsonia intracellularis challenge. *J. Anim. Sci.* 84(7):1870–1879.

Williams, N. H., Stahly, T. S. and Zimmerman, D. R. (1997a). Effect of level of chronic immune system activation on the growth and dietary lysine needs of pigs fed from 6 to 112 kg. *J. Anim. Sci.* 75(9):2481–2496. DOI: 10.2527/1997.7592481x.

Williams, N. H., Stahly, T. S. and Zimmerman, D. R. (1997b). Effect of chronic immune system activation on body nitrogen retention, partial efficiency of lysine utilization, and lysine needs of pigs. *J. Anim. Sci.* 75(9):2472–2480. DOI: 10.2527/1997.7592472x.

Williams, P. N., Collier, C. T., Carroll, J. A., Welsh, T. H. and Laurenz, J. C. (2009). Temporal pattern and effect of sex on lipopolysaccharide-induced stress hormone and cytokine response in pigs. *Domest. Anim. Endocrinol.* 37(3):139–147. DOI: 10.1016/j.domaniend.2009.04.004.

Wu, G. (2014). Dietary requirements of synthesizable amino acids by animals: A paradigm shift in protein nutrition. *J. Anim. Sci. Biotechnol* 5(1):34. DOI: 10.86/2049-1891-5-34.

Wu, G., Bazer, F. W., Dai, Z., Li, D., Wang, J. and Wu, Z. (2014). Amino acid nutrition in animals: Protein synthesis and beyond. *Annu. Rev. Anim. Biosci.* 2:387–417. DOI: 10.1146/annurev-animal-022513-114113.

Yao, W., Li, J., Jun Wang, J. J., Zhou, W., Wang, Q., Zhu, R., Wang, F. and Thacker, P. (2012). Effects of dietary ratio of n-6 to n-3 polyunsaturated fatty acids on immunoglobulins, cytokines, fatty acid composition, and performance of lactating sows and suckling piglets. *J. Anim. Sci. Biotechnol.* 3(1):43. DOI: 10.1186/2049-1891-3-43.

Yin, K. and Agrawal, D. K. (2014). Vitamin D and inflammatory diseases. *J. Inflam. Res.* 7:69–87.

Yoo, S. S., Field, C. J. and McBurney, M. I. (1997). Glutamine supplementation maintains intramuscular glutamine concentrations and normalizes lymphocyte function in infected early weaned pigs. *J. Nutr.* 127(11):2253–2259.

Zhang, Z. F., Li, J., Park, J. C. and Kim, I. H. (2013). Effect of vitamin levels and different stocking densities on performance, nutrient digestibility, and blood characteristics of growing pigs. *Asian-Australas. J. Anim. Sci.* 26(2):241-246.

Zhao, J., Li, D., Piao, X., Yang, W. and Wang, F. (2002). Effects of vitamin C supplementation on performance, iron status and immune function of weaned piglets. *Arch. Tierernahr.* 56(1):33–40. DOI: 10.1080/00039420214176.

Zhou, H. B., Huanga, X. Y., Bi, Z., Hu, Y. H., Wang, F. Q., Wang, X. X., Wang, Y. Z. and Lu, Z. Q. (2021). Vitamin A with L-ascorbic acid sodium salt improves the growth performance, immune function and antioxidant capacity of weaned pigs. *Animal* 15(2):100133. DOI: 10.1016/j.animal.2020.100133.

Chapter 8

Host-pathogen interactions and genetic tools for resistance to porcine reproductive and respiratory syndrome virus

Raymond (Bob) Rowland, University of Illinois, USA

1 Introduction

Respiratory pathogens are a significant source of loss to the swine industry. They include porcine circovirus type 2 (PCV2), swine influenza virus (SIV), *Glaesserella parasuis*, and *Mycoplasma hyopneumoniae*. The impact of these

http://dx.doi.org/10.19103/AS.2022.0103.08

respiratory pathogens goes well beyond just affecting pulmonary function. Other syndromes associated with respiratory infections include (Brown and Davis, 2004):

(1) reduced feed intake;
(2) increased use of antibiotics and other medications;
(3) the diversion of energy toward the immune response and away from muscle growth;
(4) the physiological disruptions that arise from cytokine imbalances.

Among respiratory pathogens, porcine reproductive and respiratory syndrome virus (PRRSV) is the main cause of respiratory disease and reproductive failure, costing US producers alone about $550 million per year in disease-associated losses (Neumann et al., 2005; Holtkamp et al., 2013). PRRSV is characterized by respiratory disease in nursery-age pigs, slow growth performance in growing and finishing pigs, and late-term abortions in gilts and sows.

The disease syndrome was first described in 1987 as a mystery disease affecting pigs in the United States and then later in Europe (Keffaber, 1989). PRRSV was first isolated and sequenced in 1991 in the Netherlands and called Lelystad virus (Wensvoort et al., 1991). In 1992, the virus was isolated in the United States and called VR-2332 (VR; Benfield et al., 1992; Collins et al., 1992). The full genome-length sequence of VR was not published until 1999 (Allende et al., 1999). Even though North American and European strains are similar in disease-causing abilities, their genome identities are only 60–70%, and have been classified as PRRSV-2 and PRRSV-1, respectively.

As a respiratory pathogen, PRRSV has continued to evolve in the field. PRRSV-2 isolates appeared in the late 1990s, associated with severe disease. These isolates described as 'atypical' emerged in vaccinated herds (Botner et al., 1997). Outbreaks of PRRSV infection associated with severe disease continue to appear in the United States. In 2006, the emergence of porcine high fever disease in China was linked to a novel PRRSV strain known as highly pathogenic or HP-PRRSV (Tian et al., 2007). Affected herds experience a greatly enhanced respiratory disease accompanied by higher morbidity and increased mortality. A similar highly pathogenic strain, called Lena, was the source of outbreaks of severe PRRSV in Europe (Karniychuk et al., 2010). HP-PRRSV, Lena, and atypical PRRSV remain important disease challenges for Asia, Europe, and the United States, respectively.

PRRSV is a member of the genus *Arterivirus*, family *Arteriviridae*, order *Nidovirales*. As a group, the nidoviruses are characterized as possessing a nested set of subgenomic mRNAs with a common leader sequence and poly-A tail. This structure gives nidoviruses a remarkable genetic complexity among RNA viruses (Siddell and Snijder, 2008). Up until 2007, there were only four recognized members of the arterivirus group: equine arteritis virus (*Eqarterivirus*), simian hemorrhagic fever virus (*Simarterivirus*), lactate dehydrogenase-elevating virus of mice (*Porarterivirus*), and PRRSV (*Porarterivirus*). The original arterivirus group of four viruses shares several novel properties related to viral pathogenesis, including cytopathic replication in macrophages (a type of white blood cell involved in immune function), the capacity to establish a persistent infection, and cause severe disease (Snijder et al., 2013). The arteriviruses have been further expanded to include wobbly possum disease virus (*Diparterivirus*), African pouched rat virus (*Nesarterivirus*), and Oliver's shrew virus (*Crocarterivirus*; Vanmechelen et al., 2018).

The structure of the PRRSV genome helps to explain its pathogenicity. The genome possesses at least ten open reading frames (ORFs) flanked by 5′ and 3′ untranslated regions and a 3′ poly-A tail. The nonstructural proteins are derived from the proteolytic cleavage of two large polyproteins, pp1a and pp1ab, which are derived from ORF1. Translation is normally terminated early to yield only pp1A. However, a frame shift with ORF1 yields the larger pp1ab product. The proteolytic cleavage of pp1ab yields 14 nonstructural proteins, which possess protease, replicase, and functions that subvert host defenses (Snijder et al., 2013).

The 3′ end of the genome codes for at least eight structural proteins translated from a nested 3′-coterminal set of subgenomic mRNAs. The major structural proteins are GP5 (ORF5), matrix (M, ORF6), and nucleocapsid (N, ORF7). The minor structural proteins, GP2 (ORF2a), GP3 (ORF3), and GP4 (ORF4), form a heterotrimer on the cell surface (Snijder et al., 2013). Two small non-glycosylated proteins, 2b or E, and 5a are translated from ORF2b and ORF5a, respectively (Johnson et al., 2011; Wu et al., 2001). All surface glycoproteins and M of the virion interact with the viral receptor(s) and are potential targets for neutralizing antibodies (Chand et al., 2012; Popescu et al., 2017; Stoian and Rowland, 2019; Trible et al., 2015).

2 Porcine reproductive and respiratory syndrome and porcine respiratory disease complex

Respiratory pathogens are frequently categorized based on replication in either the upper or lower respiratory tract. Since CD163-positive cells of the monocyte-macrophage lineage are found in the lung and throughout the pig, respiratory disease following infection with PRRSV is independent of the route of infection. The same respiratory disease outcomes occur if the virus is inoculated intranasally, injected peripherally into the muscle, or applied to any exposed mucosal surface. Pulmonary macrophages are the principal site of replication during acute infection. After acute pulmonary infection, virus is found in macrophages of every organ and tissue with the greatest concentration of virus in primary and secondary immune organs and tissues (Rowland et al., 2003; Lunney et al., 2016). Even though PRRSV is considered a respiratory pathogen, the virus is shed by several non-respiratory routes, such as semen from infected boars, creating an efficient means of transmission (Christopher-Hennings et al., 1998). Therefore, any successful treatment of respiratory disease caused by PRRSV must incorporate a strategy for controlling systemic infection. PRRSV can establish a productive infection in the late gestation fetus, which lacks alveolar macrophages (AM). Within the infected fetus, virus replication is primarily localized to cells in the thymus (Rowland, 2010).

Acute PRRSV infection causes a mild to severe interstitial pneumonia. The standard scoring system for assessing clinical signs includes qualitative measures of dyspnea and/or tachypnea in pigs that are stressed or at rest (Halbur et al., 1995). Clinical scores range from 0 (normal) to 6 (severe dyspnea and/or tachypnea when the pig is at rest). At the gross anatomical level, the lungs appear mottled and firm. The degree of gross pathology is evaluated according to the percent of affected area on the individual lung lobes. At the microscopic level, PRRS is characterized by interstitial pneumonia along with the infiltration of lymphocytes. Proteinaceous debris accumulation is evident in the alveolar spaces along with type 2 pneumocyte hypertrophy and hyperplasia (Halbur et al., 1995). The advantages and limitations in the use of qualitative measurements on estimating the genetic association with disease are discussed in more detail.

During infection in the nursery, respiratory disease usually appears within a week after infection and is resolved within 28 days. Recovery from respiratory disease correlates with the appearance of neutralizing antibody and interferon-producing CD4 and CD8 cells (Lunney et al., 2011).

More than 20 PRRSV vaccines are commercially available worldwide. In general, inactivated virus or subunit vaccines are not effective. Modified live virus (MLV) vaccines, when used appropriately, are considered effective at reducing the impact of disease and improving pig performance in PRRSV-infected herds. Vaccines are frequently incorporated as a means to 'stabilize' a herd during an outbreak. Vaccine effectiveness is best achieved when vaccine and field strains are genetically similar or 'homologous'. Much less protection is afforded when the vaccine and field strain are genetically separated or heterologous (Huang and Meng, 2010; Meng, 2020). Homologous versus heterologous comparisons of field and vaccine strains are made based on the nucleotide sequence of ORF5, which codes for the major surface glycoprotein of the virus. Besides inconsistent protection, other limitations of MLV vaccines include the potential for reversion to virulence, persistent infection, and vaccine virus shedding.

Most studies focusing on the virulence properties of PRRSV are designed to evaluate differences in respiratory disease, virus replication, and immune response between different virus isolates (Halbur et al., 1995). Making quantitative measurements of virulence is difficult because disease outcomes and severity can be overridden by other factors, such as those listed in Table 1. One example is the disease syndrome, porcine respiratory disease complex (PRDC), which is usually caused by PRRSV along with coinfection by other 'low mortality' pathogens, such as swine influenza A virus (swIAV), *M. hyopneumoniae*, and PCV2. PRDV is a common source of severe clinical disease accompanied by increased mortality. It should be noted that several different combinations of virus and bacteria can produce PRDC even in the absence of PRRSV.

M. hyopneumoniae is an endemic infection of pigs and is primarily responsible for a chronic respiratory condition known as enzootic pneumonia. Disease is the result of bacteria colonizing the ciliated cells of the upper respiratory tract. One outcome is a reduction or complete loss of ciliary activity, which hampers the movement of mucous and clearance of small particles and pathogens from the lungs. Primary virulence factors include bacterial adhesions, which promote the colonization of the ciliated surface. The ultimate outcome of infection is chronic inflammation, which further hampers lung function and performance. The chronic nature of the disease is linked to the ability of *M. hyopneumoniae* to suppress B and T cells responses. The control of disease is through improved health management, including the judicious use of antibiotics and vaccines. Incomplete vaccine protection combined with antibiotic resistance remain important challenges to the control of *M. hyopneumoniae*.

As a component of PRDC, it is interesting to note that *M. hyopneumoniae* and PRRSV occupy different compartments within the pig lung. *M. hyopneumoniae* colonization primarily locates to the ciliated cells of the bronchi, whereas PRRSV infection locates to the AM. An *in vitro* model showed increased production of inflammatory cytokines, such as IL-8 and tumor necrosis factor-alpha (TNF-α) in coinfected cultures (Thanawongnuwech et al., 2001). An *in vivo* study by the same group showed increased interstitial pneumonia in PRRSV-infected pigs coinfected with *M. hyopneumoniae*. Increased pneumonia persisted up to 42 days, the length of the study and a time when PRRSV was no longer present

Table 1 Factors affecting the host response to PRRSV infection

1.	Virulence (genetic) properties of the virus
2.	The breed of pig and other host genetic factors
3.	The activation state of alveolar and interstitial macrophages in the lung
4.	Polymicrobial interactions with other bacteria and viruses
5.	Environmental factors (stressors)
6.	Management practices
7.	Vaccine status of the pig

in the lungs. Increased inflammatory cytokine production was observed in AM from the coinfected pigs. These experimental observations provide a good representation of the impact of *M. hyopneumoniae* on PRRSV infection in the absence of other complicating cofactors. Therefore, the mechanism for how *M. hyopneumoniae* enhances PRRSV pneumonia is likely a result of enhanced cytokine production.

swIAV is another common infection of pigs and is associated with PRDC. *M. hyopneumoniae* and PRRSV infect different cells within the lung. In the same manner, swIAV primarily targets pulmonary epithelial cells. Even though swIAV can infect AM, clinical disease is primarily restricted to the upper respiratory tract. Early experimental studies showed enhanced respiratory clinical signs in coinfected pigs. However, several studies indicate only small or no effect of swIAV infection on PRRSV respiratory disease (Bougon et al., 2021). The reason for this discrepancy is unclear but may be related to other confounding factors present in the experimental design.

One role of PRRSV is to enhance PCV2 replication, a likely the result of PRRSV-induced immunosuppression combined with PRRSV-stimulated hyperplasia, which increases the number of PCV2-permissive cells (Trible et al., 2011, 2012).

PCV2 and PRRSV are pathogens associated with interstitial pneumonia. At the anatomic level, PCV2- and PRRSV-affected lungs can be indistinguishable. Both viruses produce a systemic infection. In addition to respiratory disease, coinfection often results in a profound wasting disease. The assumption is that PCV2 is responsible for enhancing PRRSV infection, presumably through a mechanism involving PCV2-induced immune suppression. But, the reality is the opposite may be occurring; PRRSV infection is enhancing PCV2 pathogenesis. For example, pigs vaccinated with live modified PRRS vaccine followed 1 month later with PCV2 exhibit enhanced PCVAD, including severe respiratory disease (Niederwerder et al., 2015). Coinfected pigs showed increased PCV2 viremia. Again, the mechanistic basis for how coinfection enhances respiratory disease is unclear. The likely contribution of PRRSV to PCVAD is through PRRSV-induced lymphocyte hyperplasia, which results in increased numbers of PCV2-permissve cells (Trible et al., 2011, 2012).

After nearly 30 years of research, the mechanistic bases for PRRSV pathogenesis, including mechanism responsible for PRDC, remain largely unclear. Therefore, in the absence of any negative cofactors, a virulent PRRSV isolate may produce only mild clinical disease. Under the worst of circumstances, infection with an attenuated virus, such as a modified live PRRS vaccine along with a second pathogen can result in a negative impact on pig performance. Strategies for the control of PRRSV respiratory disease have placed a strong emphasis on good herd management, including effective biosecurity protocols, good nutrition, maintenance of a low stress environment, implementation of infectious disease control programs, and the judicious use of vaccines.

3 The lung macrophage as the target for porcine reproductive and respiratory virus infection

Acute respiratory disease during PRRSV infection is a product of the interactions between the virus and pulmonary macrophages. Two major macrophage populations in the lung are the AM and interstitial macrophages (IM). AM and IM can be further divided into two major functional groups:

1 classically activated (M1) macrophages;
2 alternate activated (M2) macrophages.

These groups have been reviewed in Mosser and Edwards (2008) and Wang et al. (2017). Within the lung architecture, M1 macrophages are enriched in the interstitium, whereas M2 macrophages primarily localize to the alveolar spaces (Liu et al., 2019; Schyns et al., 2018).

When stimulated with a pro-inflammatory mediator, such as LPS or IFNγ, M1 macrophages become activated and secrete pro-inflammatory cytokines, such as TNFα, IFNγ, IL-1, IL-6, along with the production of reactive oxygen species (Stein et al., 1992; Mosser and Edwards, 2008; Orecchioni et al., 2019). M1 macrophages are effective at the early detection and killing of pathogens and for initiating effective T cell responses. One hallmark of PRRS lung pathology is the accumulation of lymphocytes into the interstitial space, including the recruitment of activated T-helper (Th) 1 cells. Activated Th1 cells produce the same M1 pro-inflammatory cytokine pattern, further amplifying the inflammatory response.

In contrast, M2 macrophages exhibit anti-inflammatory properties. When stimulated, M2 cells secrete anti-inflammatory cytokines such as IL-10. The primary role of M2 macrophages is to dampen the inflammatory response, clean up the debris left from cell and tissue damage, and set the stage for the wound healing process. Activated Th2 cells produce anti-inflammatory cytokines and further support M2 cell function. Under normal circumstances, the macrophages occupying the alveolar spaces are maintained in an anti-inflammatory or M2-like activation state. This is reflected by increased numbers of M2 macrophages, which reside within the alveoli. Epithelial cells, which line the alveoli, secrete IL-10, and other anti-inflammatory cytokines, which further supports the anti-inflammatory environment (reviewed in Snelgrove et al., 2011). The purpose of the anti-inflammatory state is to support an environment favoring optimal alveolar function by blocking inflammation.

As this discussion suggests, the M1 versus M2 designations largely represent polarized extremes in macrophage function. Pulmonary macrophages likely exist in at least three major groups: host defense, wound healing, and immune regulation (Mosser and Edwards, 2008). The functional states of macrophages are not polarized but exist in various 'shades' of activation. For many viral pathogens, respiratory disease during infection is a consequence of the dysregulation of macrophage function. For example, an overactive M1 cytokine response and corresponding decrease in M2-related function can accelerate and magnifying pro-inflammatory cytokine production. The outcome is sometimes referred to as a 'cytokine storm' and can result in loss of lung function.

4 CD163: a functional marker for anti-inflammatory M2 macrophages and a receptor for porcine reproductive and respiratory virus

The CD163 protein is both a marker for M2 cells and the receptor for PRRSV (Sulahian et al., 2000; Calvert et al., 2007). Lungs of young pigs are enriched in CD163-positive macrophages which means that lung lavage of young pigs provides a convenient means to retrieve large numbers of permissive macrophages. AM can be used for growing virus or for investigation of virus infection during *in vitro* culture. A CD163-independent mechanism for infection is the direct cell-to-cell transfer of virus (Guo et al., 2016). Direct transfer of virus from an infected to a naïve cell is observed in cell lines and cultures of AM. The relevance of receptor-independent infection in respiratory disease remains unknown.

Porcine CD163 is member of the group B protein cluster scavenger receptor cysteine-rich superfamily (SRCR-SF; reviewed in Fabriek et al., 2005). The gene, located on chromosome 5, consists of 17 exons, which code for a peptide signal sequence, followed by 9 SRCR domains, labeled SRCR 1–9, and 2 proline serine threonine (PST) domains, located after SRCR 6 (PSTI), and SRCR 9 (PSTII). The protein is anchored to the cytoplasmic membrane by a transmembrane domain followed by a short cytoplasmic tail.

CD163 was first characterized based on its role as a scavenger protein involved in the removal of excess hemoglobin from the blood and damaged tissues. Free hemoglobin released from damaged red blood cells generates oxygen radicals, which can directly damage cells and tissues. The first step in the removal of hemoglobin is binding to haptoglobin to form hemoglobin-haptoglobin (HbHp) complexes. HbHp binds to CD163 on the surface of macrophages via the interaction of the CD163 SRCR3 domain (Kristiansen et al., 2001). After internalization, hemoglobin is further broken down into free iron, biliverdin, bilirubin, and carbon monoxide. The Hb breakdown products are potent anti-inflammatory compounds (Jeney et al., 2013).

In a separate pro-inflammatory capacity, the direct crosslinking of CD163 on the macrophage surface by anti-CD163 antibodies or bacteria results in the enzymatic cleavage of CD163 from the macrophage surface and the release of pro-inflammatory cytokines, including IL-6, GM-CSF, TNFα, and IL-1β (Polfliet et al., 2006; Fabriek et al., 2009; Sulahian et al., 2004). The removal of CD163 from the macrophage surface immediately downregulates any further pro-inflammatory activity. Circulating soluble CD163 (sCD163) is used as a marker for inflammation (Shmagel et al., 2016).

CD163 as a receptor for PRRSV was first described by Calvert et al. (2007) who showed that expression of CD163 nonpermissive cell lines was sufficient to render cells fully permissive to infection. In addition to CD163, there is a large body of work describing multiple coreceptors on macrophages and other cells that interact with PRRSV proteins (Popescu et al., 2017; Zhang and Yoo, 2015). These include sialoadhesin (CD169, siglec-1), heparin sulfate, CD209, vimentin, and others.

CD163 as a marker for M2 macrophages and as a target for PRRSV infection helps to explain the immune dysregulation associated with PRRSV infection of pulmonary macrophages and the onset of respiratory disease. Acute PRRSV infection results in the cytolytic removal of M2 macrophages. In the absence of the M2-mediated anti-inflammatory regulatory mechanism, the lung macrophage population shifts toward an M1, pro-inflammatory state, with the resulting release of pro-inflammatory cytokines.

5 The porcine reproductive and respiratory syndrome Host Genetics Consortium and the search for host genes affecting the response to porcine reproductive and respiratory syndrome infection

A genetic basis for improving the resistance of pigs to PRRS was first investigated by comparing the response of different pig breeds to PRRSV infection. Researchers evaluated viremia, lung pathology, and antibody responses following PRRSV infection of Duroc, Hampshire, and Meishan pigs (Halbur et al., 1998). Based on the assessment of anatomic pathology, Meishan pigs were described as more resistant to PRRSV, followed by Duroc and Hampshire breeds. These results raised the possibility that breeding could be used as a tool for improving resistance to respiratory disease during PRRSV infection. However, these types of comparison studies fail to reveal specific host factors, genetic markers, or genes responsible for the different outcomes.

Other studies relied on the responses of infected cultured macrophages derived from different pig breeds. For example, macrophages from different pig breeds were classified as supporting a high percentage of infected cells in culture versus low levels of infection (Vincent et al., 2005). The hypothesis was that low levels of *in vitro* infection would correlate with the magnitude of respiratory disease in the pig. The results showed that pig breeds with lower infection rates *in vitro* possessed lower viremia. However, the same pig breeds with low *in vitro* infection rates possessed higher clinical scores and increased lung pathology (Vincent et al., 2006).

The PRRS Host Genetics Consortium (PHGC) was formed in 2007 to find genomic markers and genes associated with the response of pigs to PRRSV, including the identification of markers/genes associated with disease progression, resistance, chronic infection (persistence), and response following vaccination (Lunney et al., 2011). This first-of-its-kind approach for understanding the genomic basis for disease resistance is similar to the application of genomic tools for improving meat quality and other desirable traits. The goal of the PHGC was to identify genomic markers linked to PRRS resistance, which could then be integrated into marker-assisted selection strategies for improved response to PRRS (Dekkers et al., 2017). In the search for disease-associated genomic markers, the PHGC incorporated a nursery pig infection model designed to replicate the impact of PRRSV during early stage of production when pigs are the most susceptible to respiratory disease. This is also the period when the magnitude of respiratory disease has its greatest impact on production.

One initial challenge to the study design was the identification of appropriate phenotypic measurements of respiratory disease for correlation with specific single-nucleotide polymorphism (SNP) markers on the pig genome. One solution was the application of established clinical and anatomical scoring methods used for assessment of respiratory disease outcomes in studies of breed differences (Halbur et al., 1995, 1998). The advantage to this approach is that the assignment of a clinical score, between 0 (no disease) and 6 (severe respiratory disease), is noninvasive, which means data can be collected multiple times during the study period. One important limitation to clinical scoring is the subjective nature of the measurement tool. Variations in score occurs by the subjective nature of the observation. Different observers may obtain different scores. Another limitation is that clinical respiratory signs are not strictly pathognomonic for PRRSV infection. Other respiratory pathogens may be present, which can mimic or alter PRRS clinical signs. Another option for the measurement of respiratory disease is the evaluation of lung pathology. The assessment of the percentage of lesions covering the surface of a lung combined with microhistopathology provide a useful characterization of respiratory disease. However, this methodology requires that pigs be removed from the study and necropsied.

As indicated earlier, respiratory disease contributes to other clinical syndromes that negatively impact pig health. One example is weight gain, which is decreased during acute PRRS infection and a good overall measure of the health status of the pig. The assumption is that the magnitude of respiratory clinical signs negatively correlates with weight gain. Therefore, virus load (VL) and weight gain were selected as the disease phenotypes. Both measurements yield a numerical value that is quantitative and reproducible between infection trials. VL for each pig was calculated as the area under the viremia curve during the first 21 days after infection (Boddicker et al., 2012). The first 21–28 days is the acute phase of infection, when virus can be easily recovered from the blood. The second phase of infection is persistence, which is characterized by the absence of circulating virus but the presence of virus in organs and tissues, such as tonsil (Hess et al., 2019; Lunney et al., 2016).

A single PHGC nursery pig trial consists of approximately 200 three-week old commercially bred and genotyped pigs. On entry to the infection facility, pigs are negative for common respiratory pathogens, including PRRSV, *M. hyopneumoniae,* and swine influenza. Pigs are genotyped using the >60K SNP chip (Ramos et al., 2009) and infected at about 28 days of age. Well-characterized PRRSV isolates are used in these studies.

An example of one trial showing results for viremia and weight are presented in Fig. 1a (Boddicker et al., 2012). The figure illustrates the typical pattern during the acute phase of PRRSV infection. In this example, virus level peaks at about 7 days and then begins to decline after 14 days. By 28–35 days after infection, circulating virus is cleared from the blood. In about 25% of pigs, the virus levels in serum show a rebound, returning to near peak levels. The mechanism for virus rebound remains unclear. One possibility is that virus escapes neutralizing antibody through a mechanism involving antigenic variation within important neutralizing epitopes (Popescu et al., 2017). Another possibility is that virus rebound is a product of the rebound in the number of permissive CD163-positive macrophages, which are initially depleted during the acute phase of infection and then recover. Virus rebound is an illustration of PRRSV persistence at the herd level.

Examples for weights during PRRSV infection are presented in Fig. 1bc. Figure 1b shows pig weights over time. In this example, pigs at the beginning of the study showed weights that ranged between 2.5 kg and 5.0 kg. By the end of the study, the weights ranged between 5 kg and 26 kg. This large variation in weight gain is reflected in the weight distribution at 42 days after infection (see Fig. 1c). By this time, pigs are about 10 weeks of age and have been infected for 42 days. Figure 1c is a good representation of how PRRSV infection results in a multimodal weight distribution, which includes a subpopulation of 'light' pigs. In a production system, these low-weight pigs can eventually catch up, but with the addition of extra time and feed.

When conducting studies on the relationship between host genes and infectious disease, there are important factors, which should be taken into consideration. The first is the complex nature of the different interactions between pathogen and host genomes. The participation of innate and adaptive immune response pathways suggests that multiple genes participate in the disease process. This makes it difficult, if not impossible, to breed pigs

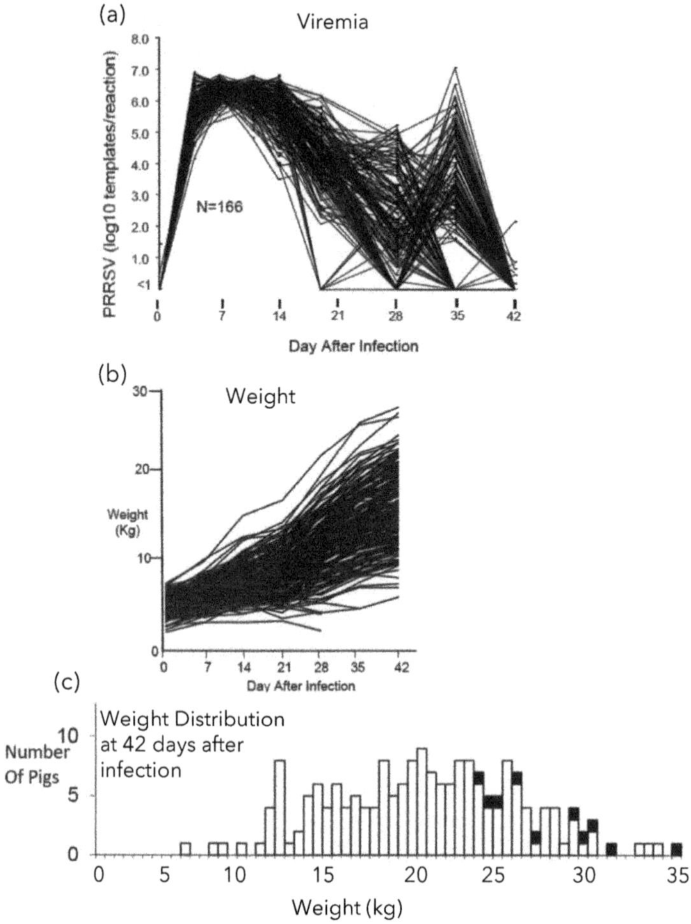

Figure 1 Viremia and weight gain responses of pigs infected with PRRSV. (a) RT-PCR measurement of PRRSV nucleic acid in serum. Results are shown for pigs that possessed a complete set of data. (b) Measurement of weight over time. (c) Weight distribution at 42 days after infection. Black boxes show the weights for uninfected control pigs. Data are shown for pigs infected with PRRSV isolate, NVSL-97-7895 (NVSL) infected at 3 weeks of age. Results are reproduced with permission from the author and according to *Frontiers in Genetics* (Rowland et al., 2012).

with improved responses to infection. A second consideration is the number of pigs that are needed to obtain a meaningful association between a specific genetic marker and a specific disease-associated phenotype. The selection of quantitative and reproducible disease traits, the recording of measurements over time, and the incorporation of well-characterized virus isolates decrease the number of pigs and improve the accuracy of the association between a disease trait and genomic marker.

The inheritance of a particular phenotypic trait is one means to determine if disease outcome is linked to the genome. The 'heritability estimate' is calculated based on the variation in genotype in combination with variation in phenotype. The result is a number between 0 and 1, with 1 being the greatest association between a disease phenotype and the genome. An example of heritability estimates for VL and weight gain at 42 days after infection with PRRSV are summarized in Table 2. The data are from five trials incorporating 959 pigs. The estimates of heritability were calculated as described in Boddicker et al. (2012). Mortality ranged between 1.0% and 8.5%, which illustrates the variable outcomes during PRRSV infection. The results in Table 2 show that VL and weight gain at 42 days (WG42d) phenotypes are considered moderately heritable with estimates ranging between 0.03 and 0.49 for VL (mean = 0.39) and 0.20–0.40 for WG42d (mean = 0.34).

A demonstration of heritability in the host response to PRRSV is the first step in identifying genomic markers or SNPs linked to disease genotypes. The standard approach is the use of a genome-wide association study (GWAS) to identify SNPs. The approach for GWAS during PRRSV infection is described in detail in Boddicker et al. (2012).

The GWAS results identified a single SNP, WUR10000125 (WUR), which was linked to both VL and weight gain. Unlike the previous assumption that multiple markers would be associated with a disease trait, this study showed multiple traits associated with a single marker. As summarized in Table 2, The WUR SNP explained 15% and 11% of the genetic variance for VL and weight gain at 42 days after infection, respectively. Four additional markers were weakly associated with the two phenotypes (Boddicker et al., 2012). Pigs possessing the favorable WUR genotype AB or BB showed an approximate 10% increase in weight gain over 42 days compared with pigs possessing the unfavorable genotype, AA. There was a corresponding decrease in viremia in the AA pigs. Importantly, only 15% of pigs currently in production possess the favorable WUR genotype, which creates the opportunity to improve pig production through marker-assisted selection of pigs possessing the B allele. In the context of PRRS, a 10% improvement in weight makes significant progress toward recovering some of the losses associated with PRRSV infection and respiratory disease. The 'WUR pig', a product of 10 years of research, is now commercially available (https://topigsnorsvin.us/news-us1/topigs-norsvin-implements-prrs-restance-in-breeding-value-estimation/).

6 The WUR marker and its association with guanylate binding protein 5

The WUR SNP is found within a 1 megabase region in the genome: *Sus scrofa* chromosome 4 occupied by a gene family of guanylate binding proteins (GBPs).

Table 2 Heritability estimates for PHGC trials 1-5 for pigs infected with NVSL isolate

				Heritability		WUR10000125[a]	
PHGC	Breed cross	Number of pigs	Mortality	VL[b]	WG42d[c]	VL	WG42d
1-3[d]	LR/LW	565	8.5%	0.49	0.37	0.16	0.15
4	D/(LW-LR-Y)	195	1.0%	0.03	0.40	0.14	0.18
5	D/(LR/Y)	199	7.0%	0.34	0.20	0.12	0.24
1-5		959	6.7%	0.39	0.34	0.15	0.11

LR, Landrace; LW, Large White; D, Duroc; Y, Yorkshire.
[a]The WUR SNP represents the 1 Mb region in chromosome 4. Results are shown for percent genetic variance explained by the 1 Mb region.
[b]VL calculated as area under the viremia curve of log-transformed during the first 21 days after infection.
[c]WG42d defined as the weight at 42 days after infection minus the initial weight.
[d]Results for first three trials are described in Boddicker et al. (2012).
Source: Boddicker et al. (2014).

GBPs are a group of genes that participates in the inflammatory response to bacteria, viruses, and cancer. GBP expression is regulated by interferon and other pro-inflammatory cytokines (reviewed in Tretina et al., 2019).

A more careful analysis of the WUR 1 megabase region identified *GBP5* as the gene most likely to contribute to the favorable AB, BB genotype (Koltes et al., 2015). The principal function associated with GPB5 is its participation in the activation of NOD-, LRR-, and pyrin domain-containing protein 3 (NLRP3)-mediated assembly of antibacterial inflammasomes. Mice lacking GPB5 show impaired responses to bacterial infections (Shenoy et al., 2012). A summary of GPB5-associated functions including its potential roles in controlling PRRSV replication is illustrated in Fig. 2.

Like other GBPs, GBP5 possesses a GTPase head followed by a tail on the C-terminal end. This structure is associated with virus inhibition. During the assembly of inflammasomes, the tail is responsible for the assembling of GBP multimers. Unique to GPB5 and a few other GBPs is the presence of a CaaX motif (C = cysteine, a = aliphatic amino acid, X = terminal residue) on the C-terminal end of the protein, which anchors the GBP within the Golgi membrane. The less favorable WUR A marker is linked to a mutant GBP5, which lacks the CaaX motif. The result is a lower amount of GBP5 combined with the inability of GBP5 to localize to the Golgi. In contrast, the WUR B genotype expresses the intact GBP5.

The localization of GBP5 in the Golgi places it near sights of virus protein synthesis and replication. For example, inhibition of respiratory syncytial virus (RSV) is dependent on the presence of an intact CaaX domain (Li et al., 2020). In another example, HIV-1 replication in macrophages is inhibited through the IFNγ-induced upregulation of GBP5 (Krapp et al., 2016). The proposed mechanism of inhibition is through the GBP5 inhibition of virus Env

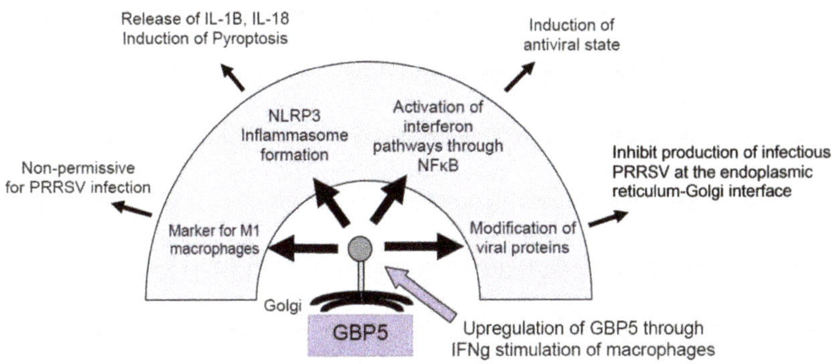

Figure 2 Roles for GBP5 in the macrophage response to PRRSV infection. Key: PRRSV, porcine reproductive and respiratory syndrome virus; NLRP3, NOD-, LRR- and pyrin domain-containing protein 3; IL, interleukin; NFkB, nuclear factor kappa B; GBP, guanylate binding protein; IFNg, interferon-gamma.

protein processing through the Golgi. The modified Env protein cannot be incorporated into virus assembly process (Hotter et al., 2007). The inhibition of influenza A virus (IAV) infection of A549 cells requires involves GBP5. GBP5 is also marker for the pro-inflammatory M1 macrophage group (Fujiwara et al., 2016).

7 The interplay between porcine reproductive and respiratory syndrome virus, M1 and M2 macrophages, CD163 and guanylate binding protein 5 in the onset and recovery from acute respiratory disease

During PRRSV infection, the CD163-positive M2 macrophages are the primary targets for PRRSV replication. Therefore, prior to infection, a defective GPB5 in AA genotype pigs would naturally skew macrophages away from M1 polarization, favoring macrophages with the M2 phenotype. During acute infection, cytolytic removal depletes the M2 cells and shifts the balance toward an M1 pro-inflammatory cytokine responses. Tissue damage and inflammation results in the accumulation of cytolytic cell debris in the alveoli, increased interstitial fluid accumulation, and onset of respiratory disease. After achieving peak viremia at around 14 days after infection, the cytolytic depletion of M2 macrophages, combined with the appearance of adaptive immunity, results in the disappearance of virus from the lungs and resolution of acute disease. At about 28 days after infection, the M2 population number rebounds and M2 macrophages are recruited into the lung to assist in tissue repair. However, the rebound in CD163-positive cells may result in the rebound of PRRSV replication.

For the WUR B genotype pigs, which express a wild-type GBP5 in macrophages, the outcome is likely to be different. Because of increased GBP5 prior to infection, the antiviral activation state in M1 macrophages is naturally higher resulting in a macrophage population naturally skewed toward an M1 response. Immediately after infection, the antiviral activity of GBP5 combined with decreased numbers of CD163-positive cells would immediately suppress PRRSV infection, resulting in lower viremia and a more rapid resolution of respiratory disease. Decreased VL along with a reduction in respiratory disease would result in increased weight gain.

As illustrated in Fig. 2, GBP5 may directly inhibit virus replication. At the cellular level, PRRSV virus replication occurs in double membrane vesicular structures in the perinuclear region of the cell. Virions bud at the endoplasmic reticulum-Golgi interface. In a similar way to HIV infection, GPB5 may directly inhibit the synthesis/processing of the virion-associated glycoproteins.

8 Porcine reproductive and respiratory syndrome resistance through modification of CD163

The gene editing technique CRISPR/Cas9 has emerged as a technology with broad applications in the control of infectious disease in humans and animals. CRISPR/Cas9 has been used to knockout (KO) CD163 expression in pigs (Whitworth et al., 2014). Macrophages from the CD163 KO pigs are negative for CD163 expression on the surface but retain the expression of other macrophage-associated markers, such as CD169. CD163 KO pigs are completely resistant to PRRSV, including the absence of circulating virus along with undetectable levels of PRRSV antibody (Whitworth et al., 2016). These types of studies confirm that CD163 is the primary receptor for PRRSV and that pigs lacking CD163 are completely resistant to virus infection. This result is the first demonstration of an effective preventive measure for PRRS.

Several questions remain regarding the potential negative consequences for pigs lacking a functional CD163 receptor. For example, CD163 KO pigs possess elevated levels of circulating haptoglobin, a likely result of the absence of a CD163 hemoglobin-clearing mechanism. Further improvements in CD163 modifications include the deletion of individual peptide domains in CD163 to create CD163 modifications that protect pigs from PRRSV infection while preserving important functions of CD163. The individual domains in CD163 involved in PRRSV infection were first investigated by Van Gorp et al. (2010). The approach was to express CD163 cDNAs in HEK293 (HEK) cells, a human cell line nonpermissive for PRRSV infection. The study identifies several CD163 domains participating in PRRSV infection. For example, the complete deletion of the (scavenger receptor cysteine-rich) SRCR5 can make cells nonpermissive for PRRSV infection.

Many of the experiments conducted by Van Gorp et al. (2010) were repeated in CD163-modified pigs. Macrophages from CD163-edited pigs possessing a deletion in the SD163 SRCR5 domain are resistant to PRRSV infection (Burkard et al., 2017). Pigs have also been constructed that possess a domain swap between porcine SRCR5 and its human homolog, hCD163L1 SRCR8. Pigs and AM were completely resistant to infection with PRRSV-1, confirming the results of Van Gorp et al. (2010). Unexpectedly, the domain-swap pigs retained the ability to support infection with PRRSV-2 isolates (Wells et al., 2017). When taken together, the results demonstrate that SRCR5 is required for PRRSV infection; however, PRRSV-1 and PRRSV-2 possess different recognition requirements. Further refinement of genetic modification of the methodology is the identification of the exact peptide sequences within the SRCR5 polypeptide involved in protection from PRRSV infection.

Reproductive PRRS originates from the infection of late-gestation gilts and sows at about 90 days of gestation. Virus crosses the placenta and infects

Table 3 Protection of fetuses by CD163 KO dams

CD163 genotype[a]		Number of fetuses		
Dam	Fetus	Total	Pathology[b]	Infected (range)[c]
WT/WT	WT/WT	16	8 (50%)	13 (<1-8.8)
WT/WT	WT/WT	12	8 (67%)	11 (1.0-5.0)
KO/KO	WT/KO	14	1 (7%)	0
KO/KO	WT/KO	17	1(6%)	0
KO/KO	WT/KO	11	0 (0%)	0

[a] Genotype of fetus determined based on crossing WT or KO dam with a WT/WT boar.
[b] Anatomical pathology included small fetus, detached or necrotic placenta, meconium staining.
[c] Infection measured by RT-PCR for PRRSV nucleic acid in serum. Results reported as log10 templates per PCR reaction.

individual fetuses. The outcomes of reproductive PRRS can be devastating, including abortions, mummified and dead fetuses, weak-born piglets, and delays in the return to estrus. During *in utero* development, thymic cells support PRRSV replication in the infected fetus. A unique application of the CD163 KO pig technology is the protection of fetuses and prevention of reproductive disease. CD163 KO gilts were bred with homozygous wild-type boars (Prather et al., 2017b). The pregnant *CD163+/+* and *CD163−/−* gilts were infected with PRRSV at 90 days of gestation. The fetuses were collected at around 109 days of the 114-day gestation period. An example of the results for two CD163 WT and three KO dams is shown in Table 3. As shown in the table, not all fetuses are infected, which is typical for reproductive PRRS. The table also lists the anatomic pathology associated with reproductive PRRS, including smaller than normal fetuses, meconium staining, and detached/necrotic placentas. In this example, no infected fetuses were found in the litters from the CD163 KO dams. Only a few fetuses showed pathology, which is typical for normal litters. By incorporating this breeding strategy, the fetuses are all wild-type heterozygous and expressed CD163. After farrowing the piglets would be normal, but susceptible to PRRSV. This application of CD163 KO technology creates an opportunity to eliminate PRRSV in the breeding herd.

9 Genetic association between the pig genome and porcine circovirus type 2 and swine influenza A virus

Similar to studies of PRRSV infection, a genetic association between PCV2 and the host genome was first established by comparing the responses of different pig breeds. In one study, immunological, virological, and histopathological responses of Landrace and Pietrain pigs were compared following infection with PCV2. Even though both breeds possessed the same concentration of virus in the blood and produced anti-PCV2 antibodies, the Landrace pigs showed more severe lymphoid lesions at 21 days after infection, including lymphoid depletion and replacement of lymphocytes with histiocytes. Histological changes were associated with increased quantities of PCV2 antigen in tissues (Opriessnig et al., 2009).

The association of PCV2 infection outcomes with specific genomic markers was investigated in several studies (Engle et al., 2014; McKnite et al., 2014; Walker et al., 2018). The experimental approach was similar to PRRSV infection studies, including locating pig genomic SNP markers linked with weight gain and VL in a large number of PCV2 experimentally infected pigs. The contribution of the pig genome to PCV2 infection was greatest for VL and much less for weight gain. SNP markers related to VL located to chromosomes 7 and 12. The chromosome 7 SNP marker was associated with a cluster of immune-related genes, which produce swine leukocyte antigens II (SLA2). The family of SLA2 proteins are expressed on the surface of antigen-presenting cells and are responsible for presenting foreign antigens to T cells; the first step in the induction of humoral and cellular immune responses. For several species, the genetic association with disease susceptibility or resistance frequently resides within this family of genes. The chromosome 12 marker was associated to a gene, *SYNGR2*, which codes for synaptogyrin-2 (SYNGR2), which belongs to family of genes expressed in neurons. Neuronal SYNGR proteins participate in the formation and recycling of synaptic vesicles. During PCV2 infection, SYNGR2 plays a nonneuronal role in blocking infection by interacting with PCV2 at the early stages of replication.

As discussed above, the variable responses of pigs following infection with PRRSV or PCV2 were the first clues for a genetic basis for disease resistance. For other viruses associated with PRDC, such as swIAV, the identification of host genes involved in disease resistance is based on studies in mice. The presumption is that discoveries from experiments involving IAV-infected mice are directly applicable to finding genetic solutions for swIAV infection of pigs. One example is the Mx1 protein, an interferon-inducible protein that inhibits IAV infection of cells. Transgenic mice overexpressing Mx1 of MxA (the human equivalent), show resistance to IAV infection. Pigs made transgenic for the overexpression of Mx1 showed increased levels of MX1 protein. Infection of fibroblast cells from the transgenic pigs with swIAV produced a 10-fold reduction in virus (Yan et al., 2014).

10 Conclusion and future trends

Increased knowledge of the livestock genome creates opportunities to decrease the impacts of respiratory disease and reduce the cost of food production. This chapter has described two approaches for the genetic improvement of livestock. The first approach incorporates heritability estimates and GWAS to identify the WUR SNP and GBP5 as having a role in protecting pulmonary macrophages from infection and promote the rapid clearance of virus. Genetic modification can achieve similar benefits, but through a one-step process. Continued improvements in genetic modification technologies along with the approval of genetically modified livestock for human consumption will pave the way for healthy animals and expand the availability of high-quality protein for a growing world.

In December 2020, the FDA approved a genetically modified pig for human consumption. The approved pig possesses a gene deletion that removes a protein responsible for causing allergic reactions in some people.

11 Where to look for further information

11.1 Further reading

- Good overview of pig diseases in general is *Diseases of Swine*, 10th edition.
- Good review on macrophage polarization is found in Mosser, D. M. and Edwards, J. P. (2008).
- Good review of CD163 and other scavenger receptor proteins is found in Fabriek et al. (2005).

11.2 Professional organizations and conferences

- An annual conference on PRRS is the North American PRRS Symposium.
- The American Association of Swine Veterinarians (AASV) sponsors an annual conference.
- A good conference on the effect of disease on pig production is the annual Leman Conference sponsored by the University of Minnesota.

12 References

Allende, R., Lewis, T. L., Lu, Z., Rock, D. L., Kutish, G. F., Ali, A., Doster, A. R. and Osorio, F. A. (1999). North American and European porcine reproductive and respiratory syndrome viruses differ in non-structural protein coding regions *J. Gen. Virol.* 80(2), 307-315.

Benfield, D. A., Nelson, E., Collins, J. E., Harris, L., Goyal, S. M., Robison, D., Christianson, W. T., Morrison, R. B., Gorcyca, D. and Chladek, D. (1992). Characterization of swine infertility and respiratory syndrome (SIRS) virus (Isolate ATCC VR-2332) *J. Vet. Diagn. Invest.* 4(2), 127-133.

Boddicker, N., Waide, E. H., Rowland, R. R., Lunney, J. K., Garrick, D. J., Reecy, J. M. and Dekkers, J. C. (2012). Evidence for a major QTL associated with host response to porcine reproductive and respiratory syndrome virus challenge *J. Anim. Sci.* 90(6), 1733–1746.

Botner, A., Strandbygaard, B., Sorensen, K. J., Have, P., Madsen, K. G., Madsen, E. S. and Alexandersen, S. (1997). Appearance of acute PRRS-like symptoms in sow herds after vaccination with a modified live PRRS vaccine *Vet. Rec.* 141(19), 497–499.

Bougon, J., Deblanc, C., Renson, P., Quéguiner, S., Gorin, S., Mahé, S., Le Dimna, M., Barbier, N., Paboeuf, F., Simon, G. and Bourry, O. (2021). Successive inoculations of pigs with porcine reproductive and respiratory syndrome virus 1 (PRRSV-1) and swine H1N2 influenza virus suggest a mutual interference between the two viral infections *Viruses* 13(11), 2169.

Brown, P. and Davis, A. (2004). Effects of disease on pig performance: a review Available at: https://www.thepigsite.com/articles/effects-of-disease-on-pig-performance-a-review.

Burkard, C., Lillico, S. G., Reid, E., Jackson, B., Mileham, A. J., Ait-Ali, T., Whitelaw, C. B. and Archibald, A. L. (2017). Precision engineering for PRRSV resistance in pigs: macrophages from genome edited pigs lacking CD163 SRCR5 domain are fully resistant to both PRRSV genotypes while maintaining biological function *PLOS Pathog.* 13(2), e1006206.

Calvert, J. G., Slade, D. E., Shields, S. L., Jolie, R., Mannan, R. M., Ankenbauer, R. G. and Welch, S. K. (2007). CD163 expression confers susceptibility to porcine reproductive and respiratory syndrome viruses *J. Virol.* 81(14), 7371-7379.

Chand, R. J., Trible, B. R. and Rowland, R. R. (2012). Pathogenesis of porcine reproductive and respiratory syndrome virus *Curr. Opin. Virol.* 2(3), 256–263.

Christopher-Hennings, J., Nelson, E. A., Nelson, J. K., Rossow, K. D., Shivers, J. L., Yaeger, M. J., Chase, C. C., Garduno, R. A., Collins, J. E. and Benfield, D. A. (1998). Identification of porcine reproductive and respiratory syndrome virus in semen and tissues from vasectomized and nonvasectomized boars *Vet. Pathol.* 35(4), 260–267.

Collins, J. E., Benfield, D. A., Christianson, W. T., Harris, L., Hennings, J. C., Shaw, D. P., Goyal, S. M., McCullough, S., Morrison, R. B. and Joo, H. S., Gorcyca, D., Chladek, D. (1992). Isolation of swine infertility and respiratory syndrome virus (isolate ATCC VR-2332) in North America and experimental reproduction of the disease in gnotobiotic pigs *J. Vet. Diagn. Invest.* 4(2), 117–126.

Dekkers, J., Rowland, R. R. R., Lunney, J. K. and Plastow, G. (2017). Host genetics of response to porcine reproductive and respiratory syndrome in nursery pigs *Vet. Microbiol.* 209, 107-113.

Engle, T. B., Jobman, E. E., Moural, T. W., McKnite, A. M., Bundy, J. W., Barnes, S. Y., Davis, E. H., Galeota, J. A., Burkey, T. E., Plastow, G. S., Kachman, S. D. and Ciobanu, D. C. (2014). Variation in time and magnitude of immune response and viremia in experimental challenges with porcine circovirus 2b *BMC Vet. Res.* 10, 286.

Fabriek, B. O., Dijkstra, C. D. and van den Berg, T. K. (2005). The macrophage scavenger receptor CD163 *Immunobiology* 210(2–4), 153–160.

Fabriek, B. O., van Bruggen, R., Deng, D. M., Ligtenberg, A. J., Nazmi, K., Schornagel, K., Vloet, R. P., Dijkstra, C. D. and van den Berg, T. K. (2009). The macrophage scavenger receptor CD163 functions as an innate immune sensor for bacteria *Blood* 113(4), 887–892.

Feng, J., Cao, Z., Wang, L., Wan, Y., Peng, N., Wang, Q., Chen, X., Zhou, Y. and Zhu, Y. (2017). Inducible GBP5 mediates the antiviral response via interferon-related pathways during influenza A virus infection *J. Innate Immun.* 9(4), 419–435.

Fujiwara, Y., Hizukuri, Y., Yamashiro, K., Makita, N., Ohnishi, K., Takeya, M., Komohara, Y. and Hayashi, Y. (2016). Guanylate-binding protein 5 is a marker of interferon-γ-induced classically activated macrophages *Clin. Transl. Immunology* 5(11), e111.

Guo, R., Katz, B. B., Tomich, J. M., Gallagher, T. and Fang, Y. (2016). Porcine reproductive and respiratory syndrome virus utilizes nanotubes for intercellular spread *J. Virol.* 90(10), 5163-5175.

Halbur, P. G., Paul, P. S., Frey, M. L., Landgraf, J., Eernisse, K., Meng, X. J., Lum, M. A., Andrews, J. J. and Rathje, J. A. (1995). Comparison of the pathogenicity of two U.S. porcine reproductive and respiratory syndrome virus isolates with that of the Lelystad virus *Vet. Pathol.* 32(6), 648-660.

Halbur, P. G., Paul, P. S., Meng, X. J., Lum, M. A., Andrews, J. J. and Rathje, J. A. (1996). Comparative pathogenicity of nine U.S. porcine reproductive and respiratory syndrome virus (PRRSV) isolates in a 5-week-old cesarean-derived-colostrum-derived pig model *J. Vet. Diagn. Invest.* 8(1), 11-20.

Halbur, P. G., Rothschild, M. F., Thacker, B. J., Meng, X. -J., Paul, P. S. and Bruna, J. D. (1998). Differences in susceptibility of Duroc, Hampshire, and Meishan pigs to infection with a high virulence strain (VR2385) of porcine reproductive and respiratory syndrome virus (PRRSV) *J. Anim. Breed. Genet.* 115(1-6), 181-189.

Hess, A. S., Lunney, J. K., Abrams, S., Choi, I., Trible, B. R., Hess, M. K., Rowland, R. R. R., Plastow, G. S. and Dekkers, J. C. M. (2019). Identification of factors associated with virus level in tonsils of pigs experimentally infected with porcine reproductive and respiratory syndrome virus *J. Anim. Sci.* 97(2), 536-547.

Holtkamp, D. J., Kliebenstein, J. B., Neumann, E., Wang, C., Zimmerman, J., Rotto, H., Yoder, T., Wang, C., Yeske, P., Mowrer, C. and Haley, C. (2013). Assessment of the economic impact of porcine reproductive and respiratory syndrome virus on United States pork producers *J. Swine Health Prod.* 21, 72-84.

Hotter, D., Sauter, D. and Kirchhoff, F. (2007). Guanylate binding protein 5: impairing virion infectivity by targeting retroviral envelope glycoproteins *Small GTPases* 8(1), 31-37.

Huang, Y. W. and Meng, X. J. (2010). Novel strategies and approaches to develop the next generation of vaccines against porcine reproductive and respiratory syndrome virus (PRRSV) *Virus Res.* 154(1-2), 141-149.

Jeney, V., Eaton, J. W., Balla, G. and Balla, J. (2013). Natural history of the bruise: formation, elimination, and biological effects of oxidized hemoglobin *Oxid. Med. Cell. Longev.* 2013, 703571.

Johnson, C. R., Griggs, T. F., Gnanandarajah, J. and Murtaugh, M. P. (2011). Novel structural protein in porcine reproductive and respiratory syndrome virus encoded by an alternative ORF5 present in all arteriviruses *J. Gen. Virol.* 92(5), 1107-1116.

Karniychuk, U. U., Geldhof, M., Vanhee, M., Van Doorsselaere, J., Saveleva, T. A. and Nauwynck, H. J. (2010). Pathogenesis and antigenic characterization of a new east European subtype 3 porcine reproductive and respiratory syndrome virus isolate *BMC Vet. Res.* 6, 30.

Keffaber, K. (1989). Reproductive failure of unknown etiology *Am. Assoc. Swine Pract. Newsl.* 1, 1-9.

Koltes, J. E., Fritz-Waters, E., Eisley, C. J., Choi, I., Bao, H., Kommadath, A., Serão, N. V., Boddicker, N. J., Abrams, S. M., Schroyen, M., Loyd, H., Tuggle, C. K., Plastow, G. S., Guan, L., Stothard, P., Lunney, J. K., Liu, P., Carpenter, S., Rowland, R. R., Dekkers, J. C. and Reecy, J. M. (2015). Identification of a putative quantitative trait nucleotide in guanylate binding protein 5 for host response to PRRS virus infection *BMC Genomics* 16, Article number: 412.

Krapp, C., Hotter, D., Gawanbacht, A., McLaren, P. J., Kluge, S. F., Stürzel, C. M., Mack, K., Reith, E., Engelhart, S., Ciuffi, A., Hornung, V., Sauter, D., Telenti, A. and Kirchhoff, F. (2016). Guanylate binding protein (GBP) 5 is an interferon-inducible inhibitor of HIV-1 infectivity *Cell Host Microbe* 19(4), 504–514.

Kristiansen, M., Graversen, J. H., Jacobsen, C., Sonne, O., Hoffman, H. J., Law, S. K. and Moestrup, S. K. (2001). Identification of the haemoglobin scavenger receptor *Nature* 409(6817), 198–201.

Li, Z., Qu, X., Liu, X., Huan, C., Wang, H., Zhao, Z., Yang, X., Hua, S. and Zhang, W. (2020). GBP5 is an interferon-induced inhibitor of respiratory syncytial virus *J. Virol.* 94(21), e01407-01420.

Liu, H., Liu, J., Huang, J., Bai, X. and Wang, Q. (2019). Heterogeneity and plasticity of porcine alveolar macrophage and pulmonary interstitial macrophage isolated from healthy pigs in vitro *Biol. Open* 8(10), bio046342.

Lunney, J. K., Fang, Y., Ladinig, A., Chen, N., Li, Y., Rowland, B. and Renukaradhya, G. J. (2016). Porcine reproductive and respiratory syndrome virus (PRRSV): pathogenesis and interaction with the immune system *Annu. Rev. Anim. Biosci.* 4, 129–154.

Lunney, J. K., Steibel, J. P., Reecy, J. M., Fritz, E., Rothschild, M. F., Kerrigan, M., Trible, B. and Rowland, R. R. (2011). Probing genetic control of swine responses to PRRSV infection: current progress of the PRRS host genetics consortium *BMC Proc., Suppl.* 5, S30.

McKnite, A. M., Bundy, J. W., Moural, T. W., Tart, J. K., Johnson, T. P., Jobman, E. E., Barnes, S. Y., Qiu, J. K., Peterson, D. A., Harris, S. P., Rothschild, M. F., Galeota, J. A., Johnson, R. K., Kachman, S. D. and Ciobanu, D. C. (2014). Genomic analysis of the differential response to experimental infection with porcine circovirus 2b *Anim. Genet.* 45(2), 205–214.

Meng, X. J. (2020). Heterogeneity of porcine reproductive and respiratory syndrome virus: implications for current vaccine efficacy and future vaccine development *Vet. Microbiol.* 74(4), 309–329.

Mosser, D. M. and Edwards, J. P. (2008). Exploring the full spectrum of macrophage activation *Nat. Rev. Immunol.* 8(12), 958–969.

Neumann, E. J., Kliebenstein, J. B., Johnson, C. D., Mabry, J. W., Bush, E. J., Seitzinger, A. H., Green, A. L. and Zimmerman, J. J. (2005). Assessment of the economic impact of porcine reproductive and respiratory syndrome on swine production in the United States *J. Am. Vet. Med. Assoc.* 227(3), 385–392.

Niederwerder, M. C., Bawa, B., Serão, N. V., Trible, B. R., Kerrigan, M. A., Lunney, J. K., Dekkers, J. C. and Rowland, R. R. (2015). Vaccination with a porcine reproductive and respiratory syndrome (PRRS) modified live virus vaccine followed by challenge with PRRS virus and porcine circovirus Type 2 (PCV2) protects against PRRS but enhances PCV2 replication and pathogenesis compared to results for nonvaccinated co-challenged controls *Clin. Vaccine Immunol.* 22(12), 1244–1254.

Opriessnig, T., Patterson, A. R., Madson, D. M., Pal, N., Rothschild, M., Kuhar, D., Lunney, J. K., Juhan, N. M., Meng, X. J. and Halbur, P. G. (2009). Difference in severity of porcine circovirus type two-induced pathological lesions between Landrace and Pietrain pigs *J. Anim. Sci.* 87(5), 1582–1590.

Orecchioni, M., Ghosheh, Y., Pramod, A. B., et al. (2019). Macrophage polarization: different gene signatures in M1(LPS+) vs. classically and M2(LPS-) vs. alternatively activated macrophages *Front. Immunol.* 10, 1084.

Polfliet, M. M., Fabriek, B. O., Daniels, W. P., Dijkstra, C. D. and van den Berg, T. K. (2006). The rat macrophage scavenger receptor CD163: expression, regulation and role in inflammatory mediator production *Immunobiology* 211(6-8), 419-425.

Popescu, L. N., Trible, B. R., Chen, N. and Rowland, R. R. R. (2017). GP5 of porcine reproductive and respiratory syndrome virus (PRRSV) as a target for homologous and broadly neutralizing antibodies *Vet. Microbiol.* 209, 90-96.

Prather, R. S., Wells, K. D., Whitworth, K. M., Kerrigan, M. A., Samuel, M. S., Mileham, A., Popescu, L. N. and Rowland, R. R. R. (2017b). Knockout of maternal CD163 protects fetuses from infection with porcine reproductive and respiratory syndrome virus (PRRSV) *Sci. Rep.* 7(1), 13371.

Rowland, R. R. R., Lunney, J. and Dekkers, J. (2012). Control of porcine reproductive and respiratory syndrome (PRRS) through genetic improvements in disease resistance and tolerance *Front. Genet.* 3, 260.

Rowland, R. R. R. (2010). The interaction between PRRSV and the late gestation pig fetus *Virus Res.* 154(1-2), 114-122.

Schyns, J., Bureau, F. and Marichal, T. (2018). Lung interstitial macrophages: past, present, and future *J. Immunol. Res.* 20(18), 5160794, Published online.

Shenoy, A. R., Wellington, D. A., Kumar, P., Kassa, H., Booth, C. J., Cresswell, P. and MacMicking, J. D. (2012). GBP5 promotes NLRP3 inflammasome assembly and immunity in mammals *Science* 336(6080), 481-485.

Shmagel, K. V., Saidakova, E. V., Shmagel, N. G., Korolevskaya, L. B., Chereshnev, V. A., Robinson, J., Grivel, J. C., Douek, D. C., Margolis, L., Anthony, D. D. and Lederman, M. M. (2016). Systemic inflammation and liver damage in HIV/hepatitis C virus coinfection *HIV Med.* 17(8), 581-589.

Siddell, S. and Snijder, E. J. (2008). An introduction to nidoviruses. In: Perlan, S., et al. (Eds), *Nidoviruses*, ASM Press, Washington, DC, pp. 1-13.

Snelgrove, R. J., Godlee, A. and Hussell, T. (2011). Airway immune homeostasis and implications for influenza-induced inflammation *Trends Immunol.* 32(7), 328-334.

Snijder, E. J., Kikkert, M. and Fang, Y. (2013). Arterivirus molecular biology and pathogenesis *J. Gen. Virol.* 94(10), 2141-2163.

Stein, M., Keshav, S., Harris, N. and Gordon, S. (1992). Interleukin 4 potently enhances murine macrophage mannose receptor activity: a marker of alternative immunologic macrophage activation *J. Exp. Med.* 176(1), 287-292.

Stoian, A. M. M. and Rowland, R. R. R. (2019). Challenges for porcine reproductive and respiratory syndrome (PRRS) vaccine design: reviewing virus glycoprotein Interactions with CD163 and targets of virus neutralization *Vet. Sci.* 6(1), 9.

Sulahian, T. H., Högger, P., Wahner, A. E., Wardwell, K., Goulding, N. J., Sorg, C., Droste, A., Stehling, M., Wallace, P. K., Morganelli, P. M. and Guyre, P. M. (2000). Human monocytes express CD163, which is upregulated by IL-10 and identical to p155 *Cytokine* 12(9), 1312-1321.

Sulahian, T. H., Pioli, P. A., Wardwell, K. and Guyre, P. M. (2004). Cross-linking of Fc gamma R triggers shedding of the hemoglobin-haptoglobin scavenger receptor CD163 *J. Leukoc. Biol.* 76(1), 271-277.

Thanawongnuwech, R., Young, T. F., Thacker, B. J. and Thacker, E. L. (2001). Differential production of proinflammatory cytokines: in vitro PRRSV and *Mycoplasma hyopneumoniae* co-infection model *Vet. Immunol. Immunopathol.* 79(1-2), 115-127.

Thanawongnuwech, R. and Thacker, E. L. (2003). Interleukin-10, interleukin-12, and interferon-gamma levels in the respiratory tract following *Mycoplasma hyopneumoniae* and PRRSV infection in pigs *Viral Immunol.* 16(3), 357-367.

Tian, K., Yu, X., Zhao, T., Feng, Y., Cao, Z., Wang, C., Hu, Y., Chen, X., Hu, D., Tian, X., Liu, D., Zhang, S., Deng, X., Ding, Y., Yang, L., Zhang, Y., Xiao, H., Qiao, M., Wang, B., Hou, L., Wang, X., Yang, X., Kang, L., Sun, M., Jin, P., Wang, S., Kitamura, Y., Yan, J. and Gao, G. F. (2007). Emergence of fatal PRRSV variants: unparalleled outbreaks of atypical PRRS in China and molecular dissection of the unique hallmark *PLoS ONE* 2(6), e526.

Tretina, K., Park, E. S., Maminska, A. and MacMicking, J. D. (2019). Interferon-induced guanylate-binding proteins: guardians of host defense in health and disease *J. Exp. Med.* 216(3), 482-500.

Trible, B. R., Kerrigan, M., Crossland, N., Potter, M., Faaberg, K., Hesse, R. and Rowland, R. R. (2011). Antibody recognition of porcine circovirus type 2 capsid protein epitopes associated with vaccination, infection and disease *Clin. Vaccine Immunol.* 18(5), 749-757.

Trible, B. R., Ramirez, A., Suddith, A., Fuller, A., Kerrigan, M., Hesse, R., Nietfeld, J., Guo, B., Thacker, E. and Rowland, R. R. (2012). Antibody responses following vaccination versus infection in a porcine circovirus-type 2 (PCV2) disease model show distinct differences in virus neutralization and epitope recognition *Vaccine* 30(27), 4079-4085.

Van Gorp, H., Van Breedam, W., Van Doorsselaere, J., Delputte, P. L. and Nauwynck, H. J. (2010). Identification of the CD163 protein domains involved in infection of the porcine reproductive and respiratory syndrome virus *J. Virol.* 84(6), 3101-3105.

Vanmechelen, B., Vergote, V., Laenen, L., Koundouno, F. R., Bore, J. A., Wada, J., Kuhn, J. H., Carroll, M. W. and Maes, P. (2018). Expanding the Arterivirus host spectrum: Olivier's shrew virus 1, a novel Arterivirus discovered in African giant shrews *Sci. Rep.* 8(1), 11171.

Vincent, A. L., Thacker, B. J., Halbur, P. G., Rothschild, M. F. and Thacker, E. L. (2005). In vitro susceptibility of macrophages to porcine reproductive and respiratory syndrome virus varies between genetically diverse lines of pigs *Viral Immunol.* 18(3), 506-512.

Vincent, A. L., Thacker, B. J., Halbur, P. G., Rothschild, M. F. and Thacker, E. L. (2006). An investigation of susceptibility to porcine reproductive and respiratory syndrome virus between two genetically diverse commercial lines of pigs *J. Anim. Sci.* 84(1), 49-57.

Walker, L. R., Engle, T. B., Vu, H., Tosky, E. R., Nonneman, D. J., Smith, T. P. L., Borza, T., Burkey, T. E., Plastow, G. S., Kachman, S. D. and Ciobanu, D. C. (2018). Synaptogyrin-2 influences replication of porcine circovirus 2 *PLoS Genet.* 14(10), e1007750.

Wang, L., Hu, S., Liu, Q., Li, Y., Xu, L., Zhang, Z., Cai, X. and He, X. (2017). Porcine alveolar macrophage polarization is involved in inhibition of porcine reproductive and respiratory syndrome virus (PRRSV) replication *J. Vet. Med. Sci.* 79(11), 1906-1915.

Wells, K. D., Bardot, R., Whitworth, K. M., Trible, B. R., Fang, Y., Mileham, A., Kerrigan, M. A., Samuel, M. S., Prather, R. S. and Rowland, R. R. R. (2017). Replacement of porcine CD163 scavenger receptor cysteine-rich domain 5 with a CD163-like homolog confers resistance of pigs to genotype 1 but not genotype 2 porcine reproductive and respiratory syndrome virus *J. Virol.* 91(2): EO1521-1516.

Wensvoort, G., Terpstra, C., Pol, J. M., ter Laak, E. A., Bloemraad, M., de Kluyver, E. P., Kragten, C., van Buiten, L., den Besten, A., Wagenaar, F., Broekhuijsen, J. M., Moonen, P. L. J. M., Zetstra, T., de Boer, E. A., Tibben, H. J., de Jong, M. F., van't Veld, P., Greenland, G. J. R., van Gennep, J. A., voets, M. Th., Verheijden, J. H. M. and

Braamskamp, J. (1991). Mystery swine disease in the Netherlands: the isolation of Lelystad virus *Vet. Q.* 13(3), 121–130.

Whitworth, K. M., Lee, K., Benne, J. A., Beaton, B. P., Spate, L. D., Murphy, S. L., Samuel, M. S., Mao, J., O'Gorman, C., Walters, E. M., Murphy, C. N., Driver, J., Mileham, A., McLaren, D., Wells, K. D. and Prather, R. S. (2014). Use of the CRISPR/Cas9 system to produce genetically engineered pigs from in vitro-derived oocytes and embryos *Biol. Reprod.* 91(3), 78.

Whitworth, K. M., Rowland, R. R. R., Ewen, C. L., Trible, B. R., Kerrigan, M. A., Cino-Ozuna, A. G., Samuel, M. S., Lightner, J. E., McLaren, D. G., Mileham, A. J., Wells, K. D. and Prather, R. S. (2016). Gene-edited pigs are protected from porcine reproductive and respiratory syndrome virus *Nat. Biotechnol.* 34(1), 20–22.

Wu, W. H., Fang, Y., Farwell, R., Steffen-Bien, M., Rowland, R. R., Christopher-Hennings, J. and Nelson, E. A. (2001). A 10-kDa structural protein of porcine reproductive and respiratory syndrome virus encoded by ORF2b *Virology* 287(1), 183–191.

Yan, Q., Yang, H., Yang, D., Zhao, B., Ouyang, Z., Liu, Z., Fan, N., Ouyang, H., Gu, W. and Lai, L. (2014). Production of transgenic pigs over-expressing the antiviral gene Mx1 *Cell Regen.* 3(1), 11.

Zhang, Q. and Yoo, D. (2015). PRRS virus receptors and their role for pathogenesis *Vet. Microbiol.* 177(3–4), 229–241.

Part 3

Disease prevention

Chapter 9

On-farm strategies for preventing pig diseases: improving biosecurity

Jeroen Dewulf and Dominiek Maes, Ghent University, Belgium

1 Introduction

Pigs are susceptible to a wide range of diseases, including zoonotic infections, which can affect health, welfare and productivity, and thereby have a major economic impact. The implementation of biosecurity measures along the production chain is one of the major solutions to minimize the risk of introduction of these diseases into a farm, as well as their spread within the farm (Dewulf and Van Immerseel, 2018).

Recently, several studies have demonstrated a positive association between biosecurity and production parameters (Laanen et al., 2013; Postma et al., 2016) and between biosecurity and farm profitability (Corrégé et al., 2012; Siekkinen et al., 2012; Rojo-Gimeno et al., 2016; Collineau et al., 2017). Biosecurity has also been shown to have a positive impact on reducing the amount of antimicrobials used in pig production (Laanen et al., 2013; Postma et al., 2016). This is important considering that antimicrobial use in pig production has been identified as one of the highest among livestock sectors in European Union (EU) countries (Filippitzi et al., 2014; Carmo et al., 2017; Sarrazin et al., 2019).

Despite these documented associations and the recognized importance of biosecurity measures, there are still major shortcomings in the implementation of these measures in pig farming (Laanen et al., 2013; Backhans et al., 2015; Filippitzi et al., 2017). There are several examples of spread of infections due

http://dx.doi.org/10.19103/AS.2022.0103.10

to insufficient implementation of biosecurity measures, such as the spread of porcine epidemic diarrhea (PED) (Scott et al., 2016), the spread of highly pathogenic strains of porcine reproductive and respiratory syndrome virus (HP-PRRSV) (Brookes et al., 2015), the foot-and-mouth disease (FMD) epidemic in the United Kingdom in 2001 (Ellis-Iversen et al., 2011) and, most recently, the African swine fever (ASF) epidemic in Europe and Asia (Blome et al., 2020).

2 What is biosecurity?

Biosecurity consists of the combination of all measures implemented to reduce the risk of introduction and spread of disease agents (Dewulf and Van Immerseel, 2018). The measures to be established should not be seen as constraints but rather as part of a process aimed at improving the health of animals, people and the environment. Biosecurity can be subdivided into two main components:

1 External biosecurity, which is focused on keeping pathogens out of the herd; and
2 Internal biosecurity or bio-management, which is focused on preventing the spread of pathogens within the herd.

Implementing biosecurity requires the adoption of a set of attitudes and behaviors to reduce the risk in all activities involving animal production or animal care.

Infectious diseases may spread through many different transmission routes. Some are spread by airborne transmission whereas others are not; some pathogens may be transmitted through vectors while some may be spread through semen. Biosecurity measures generally aim at preventing these different transmission routes in an attempt to break the infection cycle. When designing biosecurity measures, one can approach the topic from the point of view of one specific pathogen and design measures that are specifically adapted to the epidemiology of that pathogen. Alternatively, a biosecurity plan can be made more generic and include the majority of the transmission routes with a focus on those that are more important, either because they are important in the transmission routes of many different pathogens or because they are of importance in the transmission pathways of the most prevalent or damaging diseases.

When designing biosecurity programs, there are some general principles that are of value in all settings:

1 Separation of high- and low-risk animals and environments;
2 Reduction of general infection pressure;

3 Recognition that not every transmission route is equally important;
4 Assessing risk as a combination of probability of transmission and frequency of occurrence of transmission routes; and
5 Recognizing that larger animal groups pose higher risks.

These principles are discussed in the following sections.

2.1 Separation of high- and low-risk animals and environments

To avoid pathogen transmission, it is important to try to keep the sources of infection (e.g. animals, persons, vermin), i.e. small animals or insects that may transmit pathogens, separated from susceptible contacts as much as possible. This can be achieved through the prevention of direct contacts between high-risk (infectious) and low-risk (susceptible) animals as well as the prevention of indirect contacts. Whenever contacts between high- and low-risk animals or compartments cannot be avoided (e.g. buying new breeding stock, visitors to the farm), precautionary measures should be implemented (e.g. quarantine period, changing of clothing and footwear, etc. – discussed later in the chapter).

2.2 Reduction of general infection pressure

Even with the best possible biosecurity, it is not possible, and probably not even desirable, to keep animals under sterile conditions. It is therefore very hard to prevent contact with potentially harmful infectious agents altogether. Biosecurity measures aim at reducing the infection pressure to below a level that allows the natural immunity of the animals to cope with infections. As a result, biosecurity is not about 'all or nothing', but it is about controlling the infection pressure as much as possible. For some specific pathogens, however (e.g. notifiable diseases or other pathogens such as *Brachyspira hyodysenteriae*), the aim is to fully prevent contact with the pathogen and its introduction into the farm.

2.3 Recognition that not every transmission route is equally important

Some transmission routes may transmit many different pathogens with a high efficiency (e.g. direct animal contact), whereas other transmission routes (e.g. feed) are of less importance (Dewulf et al., 2018) (Fig. 1). Therefore, when designing biosecurity control programs, it is important to focus first on high-risk transmission routes and only subsequently on lower-risk transmission routes. A frequently observed mistake in biosecurity is that emphasis is put on the wrong measures and much energy is put into the prevention of low-risk

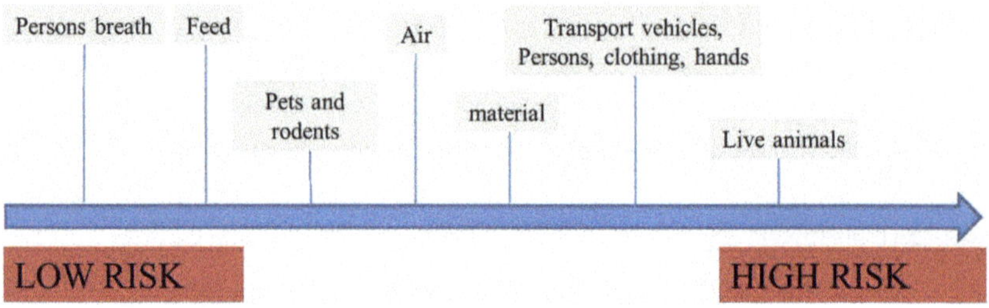

Figure 1 Theoretical ranking of different routes for infection transmission from low to high risk (after Boklund et al., 2008).

transmission routes whereas high-risk routes are ignored. This, of course, can depend on the pathogen involved, because some are specifically associated with so-called low transmission routes (e.g. some *Salmonella* serotypes with certain feed components).

2.4 Assessing risk as a combination of probability of transmission and frequency of occurrence of transmission routes

Besides the probability of transmission of a pathogen through different transmission routes, the frequency of occurrence of the transmission route is also important. If a certain transmission route (e.g. transmission via hands of animal care takers) has only a low probability of occurrence but handling is repeated very frequently (e.g. the animal care taker is touching the animals several times a day without precautionary measures), the risk of infection transmission ends up being very substantial. Therefore, when designing biosecurity programs, one also has to focus on those acts that are repeated very regularly as they, even if they carry only low risks, will still become important due to the frequency of occurrence.

2.5 Recognizing that larger animal groups pose higher risks

The larger the herds become, the more important biosecurity measures will be. This is firstly because large herds have more frequent contacts. There are more feed-delivery trucks coming to the herd, there are more frequent movements of animals, there are more professional visitors to the farm, etc. All of these carry a certain risk of infection transmission. Secondly, in large groups of animals, there are more animals that can become infected, maintain an infection cycle and build up infection pressure above the limit that the animals can cope with. It is often observed that farms gradually increase in size without fundamentally adapting their biosecurity measures. In these cases, the managers continue working according to their standards of the past and do not understand why they should adapt. The risks related to introduction and spread of an infection are, however, much more important in large herds.

3 External biosecurity measures

Based upon transmission routes, biosecurity measures have been developed, which aim to prevent either the introduction or spread of these pathogens in a pig herd. These measures can be subdivided into:

1. external biosecurity measures, aiming at prevention of pathogen introduction; and
2. internal biosecurity measures, aiming at prevention of within-herd spread of infections.

These are discussed in this and the following section.

In general terms, it can be stated that external biosecurity measures are mainly linked either to infrastructural aspects, such as organization of the buildings on the farm, measures to restrict entrance for animals and persons (e.g. hygiene lock, quarantine pen) or to measures affecting those outside the farm (e.g. entrance restrictions for visitors, hygiene of transport vehicles, safety of feed). External biosecurity measures include:

- purchasing policy;
- transport of animals, removal of manure and carcasses;
- supply of feed, water and equipment;
- access of personnel and visitors;
- vermin and bird control; and
- Location and environment.

In many cases, external biosecurity is better understood and implemented by farmers as compared to internal biosecurity. This is probably because external biosecurity measures have received more attention in the past in control of epidemic diseases.

3.1 Purchasing policy

The introduction of non-proprietary animals or genetic material (e.g. semen) might lead to the introduction of pathogens for which there is no farm immunity. Pathogen transmission occurs very effectively via direct contact between infected and susceptible animals (Filippitzi et al., 2017). Therefore, the importance of biosecurity in purchasing policy is high in protecting a farm from many pathogens. From a biosecurity viewpoint, the primary aim should be to avoid the purchase of animals or genetic material as much as possible (Amass and Baysinger, 2006; Dewulf, 2014; Filippitzi et al., 2017). A fully closed herd or production system, that is, a system where no other animals are purchased, has a substantially lower risk of pathogen introduction. Moreover, besides the risk of pathogen introduction, the frequent addition of 'naïve' animals may also favor the continuous circulation of herd-specific pathogens. This may hamper the control and eradication of certain pathogens in a herd.

However, avoiding the introduction of new animals in modern pig production is often very difficult or unwelcome because farms want to keep up with the genetic progress made by breeding companies. In addition, some farmers do not have the facilities, management skills or time to raise their own breeding gilts. The male offspring of breeding sows (production of new breeding gilts) on the farm generally also have a lower value compared to the offspring from the other sows (production of fattening pigs). Bernaerdt et al. (2021) showed that 57% of the farms were purchasing breeding gilts in Belgium. Garza-Moreno et al. (2017) described, based on 321 questionnaires completed by 118 pig veterinarians from 18 countries, that approximately half of the surveyed farms purchased replacement gilts. Therefore, whenever new animals are introduced, a number of precautionary measures should be taken.

3.1.1 Limit the frequency of introduction

Both the frequency of introduction and the number of purchased animals will influence the risk of pathogen introduction (Fèvre et al., 2006; Laanen et al., 2013). In both cases, the key principle from a biosecurity viewpoint is 'the less, the better'. However, sometimes it can be advisable to increase the size of the group of purchased animals (e.g. new gilts) as this may reduce the frequency of new introductions. It is generally less risky to buy 20 gilts five times a year rather than 10 gilts ten times a year.

3.1.2 Limit the number of sources

It is very important to limit the number of source herds for animals and AI centers for semen as much as possible (Dewulf, 2014). Several studies have shown that introducing animals from different source herds increases the risk of disease introduction (Hege et al., 2002; Lo Fo Wong et al., 2004). Source herds should have a documented high health status (Pritchard et al., 2005; Kirwan, 2008; Dewulf, 2014). This status may include the certified absence of a number of infectious diseases (e.g. specific pathogen-free status), which avoids the unintended introduction of new pathogens in the receptor herd (Laanen et al., 2010; Filippitzi et al., 2017).

3.1.3 Respect good quarantine

Newly purchased animals should always be introduced first in a quarantine stable. A good quarantine stable is fully isolated from the other animal facilities and should be accessed through a separate entry with its own hygiene lock (Fig. 2). During the quarantine period, animals should be clinically inspected to confirm that no signs of any diseases are present. Animals can also be sampled to detect any current infections and to assess the immune status of animals. The quarantine period also allows implementation of acclimatization procedures, for example, vaccinating the newly introduced animals to assure a sufficient level of immunity when they are brought into contact with the resident animals (Barceló and Marco, 1998; Corrégé, 2002; Pritchard et al., 2005; Calvar et al., 2012; Dewulf, 2014). A quarantine period should last at least 4 weeks; however, for some diseases longer periods are recommended (PRRSV and PCV2, 6–8 weeks; *Mycoplasma hyopneumoniae*, 8–10 weeks) (Eijck, 2003; Pritchard et al., 2005). Besides quarantine, vaccination and/or acclimatization may be required even earlier in the life of gilts (Pieters and Fano, 2016). In a study by Bernaerdt et al. (2021), 95% of the farms that purchased breeding gilts practiced a quarantine period. Comparable results were obtained in Sweden (Backhans et al., 2015). On most of these farms, the quarantine was located on the farm site. The median duration of the quarantine period was 42 days. The most common acclimatization practice was vaccination, although in some farms exposure of gilts to farm-specific microorganisms was also undertaken.

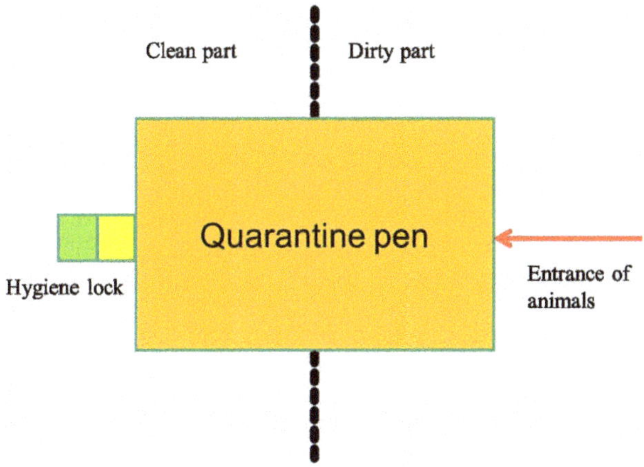

Figure 2 Design of a good quarantine stable.

3.2 Transport of animals, removal of manure and carcasses

Pathogens can spread through the transport of live animals and/or the removal of carcasses or manure. This spreading of infectious germs can be achieved directly (via secreta and excreta of diseased animals and carcasses) or indirectly (via fomites or carcasses, the rendering truck, people and their material, rodents, domestic animals and manure).

3.2.1 Use clean animal-transport vehicles

Epidemiological field studies have identified contagious livestock lorries as a main source of contamination for many disease-causing agents, including Classical swine fever (CSF) virus (Fritzemeier et al., 2000), *M. hyopneumoniae* (Hege et al., 2002), *Actinobacillus pleuropneumoniae* (Fussing et al., 1998 and Hege et al., 2002), *Brachyspira hyodysenteriae* (Windsor and Simmons, 1981) and *Salmonella* (Rajkowski et al., 1998). Lorries collecting livestock should always be empty, cleaned and disinfected and dry before entering the premises (Pritchard et al., 2005; Dewulf, 2014). Although this is a well-known principle, this is not sufficiently respected in practice. Very cold weather conditions may, for example, hamper thorough cleaning and disinfection. It is also the case that trucks that collect culled breeding sows are not always empty upon arrival on the farm. It is sometimes required that a lorry for the transport of livestock should be empty for at least a couple of hours or days before it can enter the farm. This might provide additional risk reduction. However, it is clear that thorough cleaning, disinfection and drying are the principal biosecurity measures. These cannot be replaced in favor of 'downtime', when lorries are empty. Pigs that have been in contact with the lorry during loading may not be moved back to the stable in order to minimize the chance of introducing pathogens through an insufficiently cleaned lorry. For the same reason, the lorry-driver should not be allowed to enter the stables. A loading bay is advisable, and, if available, it should also be cleaned and disinfected after each movement of animals (Pritchard et al., 2005; Backhans et al., 2015).

3.2.2 Ensure separation between the clean and the dirty area

The principle of the clean and dirty 'road' in a pig farm means that there is a clear separation between the clean and the dirty (risky) sections of the premises (Hémonic et al., 2010; Anonymous, 2010; Neumann, 2012; Filippitzi et al., 2017). All inbound and outbound traffic involving external parties (e.g. feed supply, movement of liquid manure, external transport of animals, etc.) are always directed via the 'dirty road'. The 'clean road' is restricted to within-farm movements of animals and goods and potentially the supply of safe products, but only in fully cleaned and disinfected lorries.

Only the 'dirty road' is relatively easily accessible to visitors, suppliers and consumers. The collection of carcasses is for obvious reasons part of the dirty section. Barrels, wheelbarrows and other tools used for this purpose may only be returned to the clean section after they have been thoroughly cleaned and disinfected.

Liquid manure should always be conveyed via the dirty road. Furthermore, it is advisable to the farmer to use farm-specific discharge pipes in order to prevent pipes from a manure removal company, which recently have been in contact with manure on other farms, from also being used on the farm. Some farms also have a manure processing plant on or nearby the pig stables. It is obvious that all this movements relating to manure should take place via the dirty road.

Recent studies have indicated that the clean-dirty area principle is not completely respected by manure removal and supply companies in a number of EU countries (Filippitzi et al., 2017). This means that farmers should ensure that clean-dirty areas are clearly defined with signs indicating how to adhere to these areas. The use of 'nudging' might also help promote the right behavior in these circumstances (e.g. Reddy et al., 2017).

3.2.3 Management of carcasses

Carcasses should always be considered as a major source of infectious material. Animals often die due to an infection and will thereby potentially spread a lot of infectious material. Carcasses should therefore be removed from pens and the stables as soon as possible. They should be stored in a well-insulated place (Meroz and Samberg, 1995; Pritchard et al., 2005) that cannot be accessed by vermin (rodents, insects, etc.) as they could spread the infectious material. After the collection of carcasses, the cadaver storage room should be thoroughly cleaned and disinfected. The persons handling carcasses should always wear disposable gloves for their own safety as well as to avoid further spread of pathogens (Pritchard et al., 2005; Filippitzi et al., 2017).

The cadaver storage room should be located in a place where the rendering company can collect the carcasses without entering the farm to avoid pathogen introduction through these potentially risky channels (Evans and Sayer, 2000; McQuiston et al., 2005; Pritchard et al., 2005; Anonymous, 2010). It is also recommended to have a cooled cadaver storage room for hygienic reasons (to reduce the growth of pathogenic or spoilage bacteria as well as reduce odors). Such cooled rooms generally have a higher storage capacity, making it possible to reduce the frequency of visits by the rendering company (Fig. 3).

3.3 Supply of feed, water and equipment

Feed itself should generally not pose a risk due to the strict hygienic conditions of its production. However, swill feeding (banned for decades under EU law)

Figure 3 Cooled cadaver storage close to the public road.

is a practice that has previously been associated with large outbreaks of infectious diseases, including CSF (Horst et al., 1997; Fritzemeier et al., 2000). In the recent ASF outbreak in Asia, swill feeding was also considered to be one of the major drivers of virus spread (Blome et al., 2020).

The quality of drinking water for pigs often leaves much to be desired. The water may originate from different sources (rainwater, wells, etc.) after which it is typically stored in a reservoir and then distributed to the animals. The water can be contaminated in the reservoir or in pipes leading to the drinking nipples used by pigs, and biofilms may be formed. Regular examination of the drinking water quality (at least once a year) both at the source (prior to entering the stables) and at the drinking nipples in the stables is therefore advisable. Proper treatment procedures for water and/or the pipes should be implemented in case of microbiological contamination.

The introduction of all sorts of equipment, such as floating panels or shovels, which come into contact with animals and their manure may also introduce pathogens. Therefore, it is advisable to avoid the introduction of new equipment as much as possible or, if introduced, it should be disinfected first.

3.4 Access of personnel and visitors

Humans can act as a vector for pathogens if they have been in contact with infected animals and subsequently come in to contact with susceptible animals without taking any preventive measures. This type of transmission has been proven through experiments with several pathogens, including the transmissible gastroenteritis (TGE) virus (Alvarez et al., 2001), *Escherichia coli* (Amass et al., 2003) and CSFV (Ribbens et al., 2007). The transmission occurs mainly through particles of excreta from infected animals on footwear and clothing. There is also the possibility of biological transmission of pathogens between persons and pigs that can infect both, such as the H1N1 influenza virus (Wentworth et al., 1997) or *methicillin-resistant Staphylococcus aureus* (MRSA) ST398 (Huijsdens et al., 2006). The first measure to be taken is therefore to limit the number of persons that can access the stables to the absolute minimum. This requires measures to prevent visitors entering accidentally (Fig. 4).

When visitors and personnel enter the stables, they should always wear clean, herd-specific clothes and footwear. They should also wash their hands properly (Pritchard et al., 2005; Hémonic et al., 2010; Dewulf, 2014) (Fig. 5). This latter is a simple and very useful measure, but one which is often neglected. Pathogens are efficiently transferred via the hands of animal care takers through direct contact with the animals (Vangroenweghe et al., 2009; Hémonic et al., 2010; Backhans et al., 2015). A study by Lo Fo Wong et al. (2004) has shown that the chance of testing positive for *Salmonella spp.* is reduced as a result of consistent hand washing before entering a section with pigs.

To ensure a change of clothing and washing of hands by personnel and visitors, a good hygiene lock should be in place (Vangroenweghe et al., 2009)

Figure 4 Clear entrance control at the farm to avoid unwanted visitors.

Figure 5 Proper washing of hands.

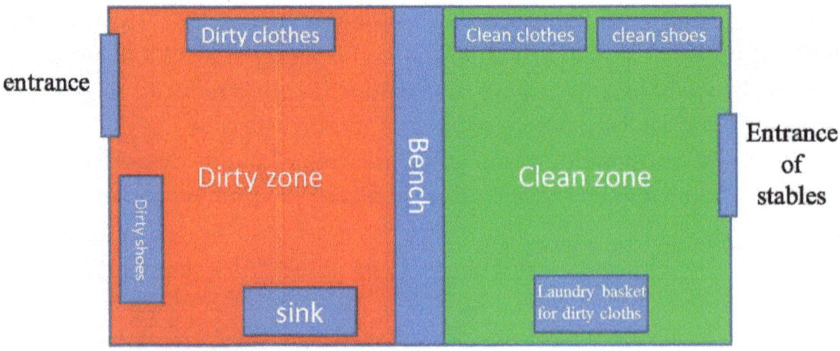

Figure 6 Design of a hygiene lock.

(Fig. 6). In this hygiene lock, a clear physical separation (e.g. bench) between the dirty and the clean area should be provided.

When entering the hygiene lock, the following steps should be followed:

1. Take off your jacket and shoes.
2. Wash your hands with disinfecting soap.
3. Step over the bench and put on a clean overall and boots.
4. Disinfect the boots with the boot washer before entering the stable.
5. When returning to the hygiene lock, clean and disinfect the boots with the boot washer.
6. Put the boots on the appropriate shelf or in a disinfectant bath.
7. Take off the dirty overall and put it in the laundry basket.
8. Step over the bench and wash your hands before you put on your own jacket and shoes again.

9 In farms with high health standards, visitors and personnel are often obliged to shower before entering the farm. The main benefit of this requirement is to ensure that all potentially contaminated clothing will be replaced by farm-specific clothing and that the hands are washed thoroughly. In addition, it discourages unnecessary visits and increases awareness of the need to implement proper biosecurity measures (Moore, 1992; Amass and Clark, 1999).

Often, a quarantine period of 24 or even 48 h, during which no contact with pigs can occur, is also required for visitors before they can have access to the farm. This is based on the argument that germs excreted by pigs can survive on humans for a specific period. During this period, persons could passively excrete germs and transfer them to susceptible animals. There is little evidence in the scientific literature that this is a significant risk. The authors are aware of only one study from 1970, which found that FMDV could be isolated from the nose and mouth of people who had been in contact with animals infected with the FMDV. On one person, the virus could still be isolated 28 h after contact with the infected pigs. After 48 h, this was no longer possible (Sellers et al., 1970). If all required precautionary measures are taken as described above, this downtime period probably has little additional value. It is clear that, whatever downtime is used, it is no substitute for the hygiene measures on entering a farm described here.

3.5 Vermin and bird control

A number of pathogens can be transmitted from outside the farm or between different areas within the farm directly or indirectly by rodents, birds, dogs and cats. They may act as reservoirs for herd-specific pathogens that will continue to circulate in the farm (Andres and Davies, 2015). Rodents and birds can also cause damage to equipment and farm buildings or contaminate feed if they can access it (Backhans and Fellstrom, 2012).

To control vermin, an efficient control program is required. This is often developed in collaboration with specialized companies (Lister, 2008; Hémonic et al., 2010; Dewulf, 2014; Backhans et al., 2015; Filippitzi et al., 2017). It is important to prevent vermin living in close proximity to stables. This can be achieved by removing potential hiding places near to the stables (e.g. plants, piles of dirt, etc.). Feed should be stored in closed containers with no access for rodents or birds (Lister, 2008; Anonymous, 2010). The entrance of birds into stables can be prevented by covering all air inlets with nets (Loncke and Dewulf, 2018). Pets should also be kept out of the stables as they can be vectors for pathogens. The use of cats or dogs to control rats and mice is therefore not advisable (Vangroenweghe et al., 2009).

3.6 Location and environment

The location of the farm and the density of pig farming in the immediate area are important factors for airborne and vector-borne disease transmission. The presence of wildlife in the neighborhood of pig farms may also pose certain risks.

3.6.1 Airborne transmitted diseases

Bacterial transmission through the air is particularly important over short distances (<2 km), hence the importance of distance to the nearest neighbor. Rose and Madec (2002) concluded that the number of farms within a range of 2 km significantly increased the frequency of respiratory disorders in a farm. In the case of airborne transmission of *M. hyopneumoniae*, the distance to neighboring farms proved to be the most important factor (Goodwin, 1985; Dee et al., 2009). Maes et al. (2000a) reported increased prevalence of *M. hyopneumoniae*, swine influenza viruses and Aujeszky's disease virus in herds located in areas with a high density of pig farming. Mintiens et al. (2003) showed that the combination of the distance to a neighboring farm and pig herd density in an area are major risk factors for the spread of CSFV. When the building of a new pig farm is being planned, the distance to the nearest neighboring pig farm can be a determining factor in choosing a location. The predominant direction of the wind could be taken into account as well. Knowledge of the presence of diseases in neighboring farms is equally important. The spread of liquid manure originating from other farms should also be avoided in the neighborhood of the farm.

It has become possible to equip stables with high-performance air filtration systems that can block airborne pathogens. This might be worth considering in densely-populated livestock areas, especially if animals in the farm are free of endemic diseases in the area. Filtration of incoming air, in combination with standard biosecurity procedures, has been demonstrated to prevent transmission of PRRSV in susceptible herds (Alonso et al., 2013). The authors showed that air filtration reduced the risk of introduction of novel PRRSV strains by approximately 80%, indicating that on large sow farms with good biosecurity in pig-dense regions, approximately four out of five PRRSV outbreaks may be attributable to aerosol transmission. Air filtration may also be helpful to prevent airborne transmission of other pathogens, such as swine influenza virus and *M. hyopneumoniae*.

3.6.2 Wild animals

Direct or indirect contact with wild boars may cause transmission of pathogens (e.g. CSFV (Fritzemeier et al., 2000), Aujeszky's disease virus (Artois et al., 2002) and ASFV (Blome et al., 2020)). In regions where free ranging wild boars are present, it is therefore important to keep wild animals out of the farm with a solid fence (Amass and Clark, 1999) with an underground depth of 30 to 40 cm (Hartung, 2005). Even if pigs are kept indoors, wild boars should not be able to come near the farm in order to avoid the indirect transmission of pathogens (e.g. airborne transmission, through vectors, through contact with stored feed, etc.).

4 Internal biosecurity measures

Internal biosecurity measures include:

- Compartmentalization, working lines and equipment;
- Management of diseases;
- Measures during the farrowing and suckling period;
- Management of the nursery and fattening unit; and
- Cleaning and disinfection.

These are discussed in the following sections.

4.1 Compartmentalization, working lines and equipment

Animals of different ages may have different levels of susceptibility to specific pathogens. It is therefore crucial to keep different age groups separate and to work in a well-defined sequence. Equipment and materials (e.g. bedding material, feeders, drinking troughs, boots, spades, syringes, needles, etc.) may also play an important role in the transmission of a large number of diseases.

4.1.1 Working lines and separate hygiene locks

An important basic rule to prevent the spread of infections between different age groups is establishing and maintaining working lines within the farm. A fixed routine is created that is always used to visit and work in the stables. During rounds to the stables, the youngest animals are visited first, followed by the pregnant sows, the older age groups, the quarantine and sick animals and, finally, the cadaver storage unit. For each age category, and especially for risk-bearing groups (e.g. quarantine stables, sickbay), it is recommended to provide an additional hygiene lock allowing proper changing of clothing, footwear and washing of hands.

4.1.2 Equipment in the various compartments

Designated equipment should be provided for each working line. A brush, a shovel or floating panels can easily be contaminated with feces that contain a large number of germs. It is therefore recommended to use dedicated equipment in each section. Equipment has to be clearly recognizable (e.g. using different colors) (Fig. 7) to avoid moving it from one section to another (Vangroenweghe et al., 2009; Laanen, 2011; Gelaude et al., 2014). The same rule applies to clothing and footwear for the same reason.

4.1.3 Boot washers and disinfection baths

To avoid germs on footwear, boot washers and disinfection baths can be placed between the different production units. Effective disinfection can only be achieved if dirt and feces are removed from boots in advance. This can be done with a boot washer and water (preferably adding a detergent). The boots then have to be placed in a visually clean solution with a disinfectant. This protocol requires that the concentration of the disinfectant and the duration of contact must follow manufacturer instructions (Amass et al., 2000). Disinfection baths that are not used properly will inadvertently increase the number of germs on the boots. This wastes both time and money and can even increase the risk of disease spread. However, it is not practical to stand for many minutes in a disinfection bath before being able to go to another section. This problem can be avoided by providing an extra pair of boots at each disinfection bath, so there's always a pair of boots ready to be used at each bath while the other boots are placed in the disinfecting solution. The presence of foot baths also draws the attention of staff and visitors to the importance of biosecurity on the farm (Amass et al., 2000).

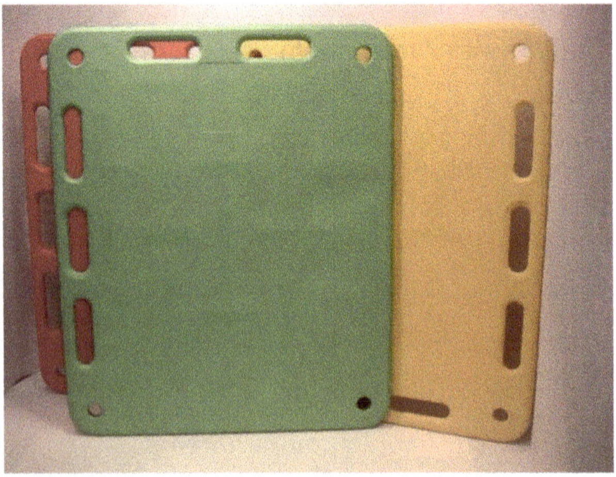

Figure 7 Floating panels in different colors to be used in different compartments and to avoid using the same material in different age groups.

4.2 Management of diseases

Disease management concerns all actions related to the correct handling and treatment of diseased animals. This includes proper diagnostics, isolation of sick animals and disease registration as well as improvement of the immunity status of susceptible animals, in particular through vaccination. Good herd heath management should result in a good understanding of the specific health status of the herd and in use of appropriate preventive treatments to avoid disease and its subsequent losses.

4.2.1 Sickbay

Diseased pigs should be isolated in a sick bay in order to protect other animals from exposure to pathogens through infected excretions and secretions. A good sickbay is fully separated from the other animals, preferably in a separate building (Hémonic et al., 2010; Dewulf, 2014). Once an animal has been in the sick bay, it should not return to the regular stables as it is very likely that it will transmit any remaining pathogens to healthy animals. The sick bay should also be accessed separately by farm workers and the necessary hygienic measures (e.g. changing of coverall, footwear, washing hands, etc.) should be implemented when entering and leaving. The sick bay should ideally only be visited at the end of the working round (Vangroenweghe et al., 2009; Backhans et al., 2015).

4.2.2 Use of needles and medicines

There is extensive literature on the spread of germs via injection equipment (needles and syringes) (Hémonic et al., 2010; Filippitzi et al., 2017). In practice, the same needles are often used on individual pig farms and only replaced when they become blunt! These needles may get contaminated with bacteria through use and storage (Fig. 8). In addition, when injecting sick animals, the needle (and consequently the bottle) can become contaminated with a pathogen. Injecting multiple animals with the same needle thus increases the risk of spreading germs.

Ideally, single use needles should be used (Hémonic et al., 2010). If this is not feasible, it is highly advisable to use one needle for each group (e.g. litter or pen). Using the same needles for different age groups should be avoided and needles should be replaced before they become blunt, both for hygiene and animal welfare reasons. Previously opened bottles should be stored in a hygienic environment at the recommended environmental conditions such as temperature and light. Needle-free injection devices, often used for vaccination, may also help to decrease the risk of iatrogenic infection transmission.

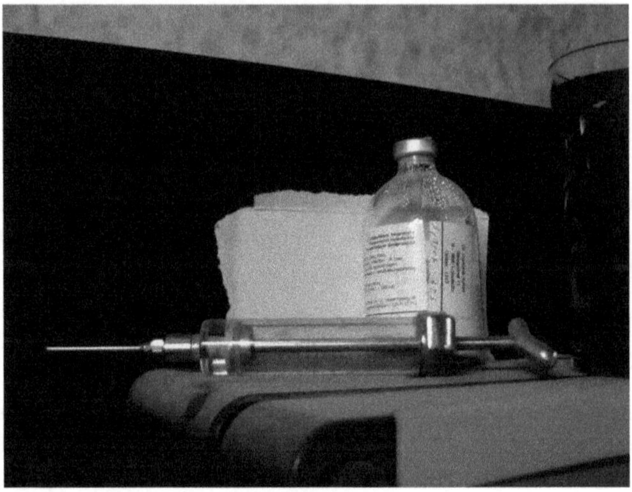

Figure 8 Improper stored needles, syringes and medicines may be a source of infection.

4.2.3 Returning to younger age group

In case of slower growth rate of some piglets compared to the remainder of the group, it is important to avoid moving more slowly growing piglets to a batch of younger piglets. These slower growing piglets may be suffering from one or more infectious diseases. By transferring these pigs to a younger age group, possible carriers of germs are brought into a susceptible population (Vangroenweghe et al., 2009; Dewulf, 2014; Filippitzi et al., 2017). When it is expected that a piglet has a low probability of reaching slaughter age, euthanasia should be considered. This might be a better option both for the piglet, for welfare reasons, and for the whole litter, because their health is protected. If euthanasia is not believed to be the right option, then these animals should be isolated in the sickbay until they can be transported to the slaughterhouse.

4.3 Measures during the farrowing and suckling period

Pathogens can be transmitted from sows to offspring via the placenta or via contaminated colostrum and milk. They can also be transmitted by direct contact, for example, through the nose, skin, nipples and the udder. Cross-fostering in particular increases the risk of pathogen transmission from infected or carrier sows to susceptible piglets. This is particularly the case if cross-fostered piglets do not have (sufficient) maternal antibodies, for example, of PRRSV, as indicated by Zimmerman et al. (2012). Another route of pathogen transmission in the farrowing unit is the repeated use of equipment or materials (e.g. castration blades, elastrator for tail docking, ear-tagger, iron injection needles) from one piglet to another without intermediate cleaning and disinfection.

4.3.1 Washing the sows

Before sows are placed in the farrowing pen, they should be dewormed and washed in order to prevent the transmission of germs from the sow gestation unit to the farrowing unit. Washing of sows should be done before they enter the farrowing pen. If it is done in the farrowing unit, then contaminated aerosols are generated and spread over the entire farrowing unit. This significantly decreases the effect of washing and is contrary to the basic principle of using separate hygiene measures for different age groups or stables in a farm.

4.3.2 Cross-fostering

Mixing litters in the farrowing pen is a significant way for spread of infection to different animal groups. Sows that carry *Streptococcus suis* can infect their piglets during or immediately after parturition (Amass et al., 1996). Further spread of *S. suis* may take place if piglets are moved to other litters. This principle applies to other germs as well. If 5% of piglets are moved between pens more than 48 hours after birth, there is an increased probability of problems with PRRS (Duinhof et al., 2006). From a biosecurity viewpoint, it is advisable to avoid cross-fostering as much as possible. If this is not possible, it is recommended to limit cross-fostering to one occasion and to perform it shortly after birth.

4.3.3 Equipment for treatment of the piglets

Equipment that is used in the farrowing pen, such as blades for castration, is exposed to secretions and excretions from piglets. Germs can then be transferred from one piglet to another. It is necessary to clean and disinfect equipment when using it for a different piglet (immerse in disinfectant). This limits the chance of disease transmission (Vangroenweghe et al., 2009; Filippitzi et al., 2017).

4.4 Management of the nursery and fattening unit

4.4.1 All-in/all-out

The all-in/all-out (AI/AO) principle helps to prevent cross-contamination between consecutive production batches and makes it possible to clean and disinfect stables thoroughly between different production batches. The strict application of this AI/AO principle is a very important measure to break the infection cycle from one production batch to the next (Clark et al., 1991). It is fundamentally important that the stable is empty at the end of a production batch. Sometimes only 95% of a batch is removed with a few (light weight) animals kept in the stables and mixed with animals in the next batch. Although few, these animals are very likely a source of infection for the next groups (as discussed earlier). Moving piglets back to younger age groups is also a risky practice since it can also bring pathogens to a susceptible population. Weaners are a vulnerable age group due to their temporary lower immune status, the higher presence of diverse pathogens in this life period (Johnson et al., 2012) and the risk of fighting when pigs are mixed (Cameron, 2012). When moving animals from one production stage to the next (e.g. from the farrowing house to the nursery pen), it is advisable to keep the groups together as much as possible rather than sorting all animals according to weight. The latter will result in a lot of mixing which substantially increases the likelihood of spread of infections (Maes et al., 2008; Hémonic et al., 2010).

4.4.2 Stocking density

A high stocking density induces stress, which results in an increased susceptibility to infections and an increased excretion of germs. Many infected pigs in a small area can cause a sharp rise in infection pressure. Various studies have shown that a higher stocking density in different production phases increases the occurrence of respiratory as well as digestive tract disorders (Pointon et al., 1985; Maes et al., 2000a,b; Stärk, 2000; Laanen, 2011). In addition, it has been shown repeatedly that there is a positive connection between available space per animal and its daily growth (Dewulf et al., 2007). In many cases recommended stocking densities, as prescribed in EU legislation, are based on outdated research and have not evolved with recent advances in the industry. It is therefore recommended to consider these norms as the absolute minimum requirements and not as target values (Dewulf et al., 2007). Studies have shown that the optimal values for stocking density are on average 20 to 24 % above EU requirements (Hamilton et al., 2003; Laanen, 2011).

4.5 Cleaning and disinfection

Pens, feeding troughs and equipment contaminated by feces can maintain an infection cycle because new animals continue to become infected. These animals will consequently secrete germs and will re-infect the environment. To break the infection cycle between consecutive litters, a thorough cleaning and disinfection (C&D) of pens is required.

A complete C&D protocol consists of seven steps (Vangroenweghe et al., 2009a; Hémonic et al., 2010; Laanen, 2011; Dewulf, 2014; Luyckx, 2016; Van Immerseel et al., 2018):

1 Dry cleaning to remove all organic material;
2 Soaking of all surfaces preferably with detergent;
3 High pressure cleaning with water to remove all dirt. This step will be much easier, faster and more effective if a good soaking step is performed beforehand;
4 Drying of the stable to avoid dilution of the disinfectant applied in the next step;
5 Disinfection of the stable to achieve further reduction of the concentration of germs;
6 Drying of the stable to assure that animals afterwards cannot come into contact with pools of remaining disinfectant; and
7 Testing the efficiency of the procedure through sampling of the surface.

With the aid of pressure plates, it is possible to simply and rapidly check all surfaces for microbial contamination. These plates measure and quantify the presence of bacterial contamination. The results are expressed in colony-forming units (CFU) per plate. When all these steps are performed correctly, an additional empty period to further reduce the infection load is not required (Luycks, 2016).

5 Measuring biosecurity

'You need to be able to measure, to be able to improve' is one of the most famous quotes of William Thomson (better known as Lord Kelvin), a famous British scientist of the nineteenth century. This is certainly true for biosecurity and hygiene. The inability to measure accurately and reproducibly the biosecurity and hygiene status of a farm has long been one of the main obstacles to improving safety. If farm managers are to enhance the biosecurity or hygiene status of their farm, it is essential to provide them with quantitative goals and benchmarks relating to the biosecurity and hygiene status of a farm. This makes it possible to identify measures for improvement and assess their effectiveness, if possible quantitatively.

Several systems have been developed to provide an inventory of biosecurity measures for animal production. They are often developed as checklists or as manuals either by independent advisory organizations or as supporting material in the marketing of specific disease prevention products (e.g. vaccines). An example of the latter is the COMBAT system developed by the pharmaceutical company Boehringer Ingelheim that helps to identify

Table 1 Overview of different biosecurity scoring tools or checklists with some of their characteristics

System	Developer/owner	Disease focus	Language	Risk-based scoring	Benchmarking	Availability	Weblink
COMBAT	Boehringer Ingelheim	PRRS	Multiple languages	Yes	Yes	Free	https://www.prrs.com/disease-control/control/combat
Biocheck.Ugent	Ghent University	Generic	Multiple languages	Yes	Yes	Free	https://biocheck.ugent.be/en
CFSPH Biosecurity	Iowa State University	Generic	English	No	No	Free	https://www.cfsph.iastate.edu/biosecurity/
APIQ√	Australian Pork	Generic	English	No	No	Free	https://www.apiq.com.au/

biosecurity hazards in relation to the PRRSV infections in pig production. Like the COMBAT PRRS system, many of these systems were developed from the point of view of the control of a specific disease. In the United States, there is also the more generic biosecurity evaluation tool developed by the Center for Food Security and Public Health at Iowa State University. This system consists of a series of checklists for different components of biosecurity. Other countries have published fact sheets on the principles of biosecurity, for example to reduce the risk of introduction and spread of CSFV. In Australia, information on biosecurity is also available for pig producers via the Australian Pork Industry (2003). More recently, they have also developed manuals for pig producers, including an internal audit system called APIQ√® which stands for Australian Pork Industry Quality Assurance Program. This system enables producers to demonstrate that their on-farm practices reflect good farming practice for management, animal welfare, food safety, biosecurity and traceability.

At Ghent University, a risk-based biosecurity scoring system (Biocheck. UGent™) has been developed to quantify on-farm biosecurity (Laanen et al., 2010). It does not focus on a specific disease but rather approaches biosecurity in general and focuses on those aspects that are common for the transmission of many different types of infectious diseases. The Biocheck.UGent™ system consists of a number of questions divided into several subcategories for internal and external biosecurity. Depending on the importance of a particular biosecurity measure, the score per question is multiplied by a weight factor (Laanen et al., 2013; Gelaude et al., 2014). The subcategories also have a specific weight factor corresponding to their relative importance in disease transmission. The Biocheck.UGent™ scoring system thus provides a risk-based score that takes into account the relative importance of each biosecurity measure. The Biocheck.UGent™ scoring tool is accessible to everybody and is free of charge (www.biocheck.ugent.be). After filling in the questionnaire, the results allow evaluation of strong and weak points in the biosecurity on a farm and provide a basis for improvement. Besides the systems discussed here, there are likely to be others, with a focus on a specific country or region, which are unknown to the authors or inaccessible due to language limitations. Table 1 provides an overview of the systems described in this section.

6 Conclusion

As discussed in this chapter, biosecurity in pig production is a combination of many different measures aimed at preventing the introduction and spread of pathogens in a farm. It forms the basis of any disease control program. Both external and internal biosecurity measures are of utmost importance both to prevent introduction of pathogens on a farm and to avoid infection spread between animal populations in a farm. Although most of the measures to be implemented are logical and generally easy to apply, it requires strong discipline to adhere to the measures in daily farm practice. Yet, those who do surely will see the benefits.

7 Where to look for further information

More information can be found in the book: Biosecurity in Animal Production and Veterinary Medicine, from principles to practice. Edited by: Jeroen Dewulf, Ghent University, Belgium, Filip Van Immerseel, Ghent University, Belgium

Also the website www.biocheck.ugent.be contains tons of tips and tricks on biosecurity.

8 References

Alonso, C., Murtaugh, M. P., Dee, S. A. and Davies, P. R. (2013). Epidemiological study of air filtration systems for preventing PRRSV infection in large sow herds. *Preventive Veterinary Medicine* 112(1–2), 109–117.

Alvarez, R. M., Amass, S. F., Stevenson, G. W., Spicer, P. M., Anderson, C., Ragland, D., Grote, L., Dowell, C. and Clark, L. K. (2001). Evaluation of biosecurity protocols to prevent mechanical transmission of transmissible gastro-enteritis virus of swine by pork production unit personnel. *Pig Journal* 48, 22–33.

Amass, S. F. and Baysinger, A. (2006). Swine disease transmission and prevention. In: Straw, B. E., Zimmerman, J. J., D'Allaire, S. and Taylor, D. J. (Eds), *Diseases of Swine* (9th edn). Blackwell Publishing Ltd., Oxford, UK, 1075–1098.

Amass, S. F. and Clark, L. K. (1999). Biosecurity considerations for pork production units. *Journal of Swine Health and Production* 7, 217–228.

Amass, S. F., Clark, L. K., Knox, K., Wu, C. C. and Hill, M. A. (1996). *Streptococcus suis* colonization of piglets during parturition. *Journal of Swine Health and Production* 4, 269–272.

Amass, S. F., Halbur, P. G., Byrne, B. A., Schneider, J. L., Koons, C. W., Cornick, N. and Ragland, D. (2003). Mechanical transmission of enterotoxigenic *Escherichia coli* to weaned pigs by people, and biosecurity procedures that prevented such transmission. *Journal of Swine Health and Production* 11, 61–68.

Amass, S. F., Vyverberg, B. D., Ragland, D., Dowell, C. A., Anderson, C. D., Stover, J. H. and Beaudry, D. J. (2000). Evaluating the efficacy of boot baths in biosecurity protocols. *Journal of Swine Health and Production* 8, 169–173.

Andres, V. M. and Davies, R. H. (2015). Biosecurity measures to control *Salmonella* and other infectious agents in pig farms: a review. *Comprehensive Reviews in Food Science and Food Safety* 14(4), 317–335.

Anonymous (2010). Food and agriculture organization of the United Nations/world organisation for animal health/world bank. Good practices for biosecurity in the pig sector – issues and options in developing and transition countries. *FAO Animal Production and Health Paper No. 169*.

Artois, M., Depner, K. R., Guberti, V., Hars, J., Rossi, S. and Rutili, D. (2002). Classical swine fever (hog cholera) in wild boar in Europe. *Revue Scientifique et Technique* 21(2), 287–303.

Backhans, A. and Fellstrom, C. (2012). Rodents on pig and chicken farms - a potential treath to human and animal health. *Infection Ecology and Epidemiology* 2, 17093.

Backhans, A., Sjölund, M., Lindberg, A. and Emanuelson, U. (2015). Biosecurity level and health management practices in 60 Swedish farrow-to-finish herds. *Acta Veterinaria Scandinavica* 57, 14.

Barceló, J. and Marco, E. (1998). On farm biosecurity. Proceedings of the 15th IPVS Congress, Birmingham, England, 5-9 July, 129-133.

Bernaerdt, E., Dewulf, J., Verhulst, R., Bonckaert, C. and Maes, D. (2021). Purchasing policy, quarantine and acclimation practices of breeding gilts in Belgian pig farms. *Porcine Health Management* 7(1), 25.

Blome, S., Franzke, K. and Beer, M. (2020). African swine fever - a review of current knowledge. *Virus Research* 287, 198099.

Boklund, A., Barfod, K., Mortensen, S., Houe, H. and Uttenthal, Å. (2008). Exotic diseases in swine: evaluation of biosecurity and control strategies for classical swine fever.

Brookes, V. J., Hernández-Jover, M., Holyoake, P. and Ward, M. P. (2015). Industry opinion on the likely routes of introduction of highly pathogenic porcine reproductive and respiratory syndrome into Australia from South-East Asia. *Australian Veterinary Journal* 93(1-2), 13-19.

Calvar, C., Heugebaert, S., Caille, M. E. and Roy, H. (2012). La quarantaine. Des préconisations de techniciens diversifies. Des conduits multiples chez de très bons éleveurs. *Rapport d'étude, Chambres d'agriculture de Bretagne*.

Cameron, R. (2012). Integumentary system: skin, hoof, and claw. In: Zimmermann, J. J., Karriker, L. A., Ramirez, A., Schwartz, K. J. and Stevenson, G. W. (Eds), *Diseases of Swine*. Wiley-Blackwell, Chichester, West Sussex, UK, 926-991.

Carmo, L. P., Schuepbach-Regula, G., Muentener, C., Chevance, A., Moulin, G. and Magouras, I. (2017). Approaches for quantifying antimicrobial consumption per animal species based on national sales data: a Swiss example (2006–2013). *Eurosurveillance* 22(6), 30458.

Clark, L., Freeman, M., Scheidt, A. and Knox, K. (1991). Investigating the transmission of Mycoplasma hyopneumoniae in a swine herd with enzootic pneumonia. *Veterinary Medicine* 86, 543-550.

Collineau, L., Rojo-Gimeno, C., Léger, A., Backhans, A., Loesken, S., Nielsen, E. O., Postma, M., Emanuelson, U., Beilage, E. G., Sjölund, M., Wauters, E., Stärk, K. D. C., Dewulf, J., Belloc, C. and Krebs, S. (2017). Herd-specific interventions to reduce antimicrobial usage in pig production without jeopardising technical and economic performance. *Preventive Veterinary Medicine* 144, 167-178.

Corrégé, I. (2002). Problématique de l'introduction des reproducteurs. *Techniporc* 25, 27-30.

Corrégé, I., Fourchon, P., Le Brun, T. and Berthelot, N. (2012). Biosécurité et hygiène en élevage de porcs: état des lieux et impact sur les performances technico-économiques. *Journées Recherche Porcine* 44, 101-102.

Dee, S., Otake, S., Oliveira, S. and Deen, J. (2009). Evidence of long distance airborne transport of porcine reproductive and respiratoy syndrome virus and Mycoplasma hyopneumoniae. *Veterinary Research* 40, 1.

Dewulf, J. (2014). An online risk-based biosecurity scoring system for pig farms. *Veterinary Ireland Journal* 4, 426-429.

Dewulf, J., Tuyttens, F., Lauwers, L., Van Huylenbroeck, G. and Maes, D. (2007). De invloed van de hokbezettingsdichtheid bij vleesvarkens op productie, gezondheid en welzijn. *Vlaams Diergeneeskundig Tijdschrift* 76, 410–416.

Dewulf, J. and Van Immerseel, F. (2018). General principles of biosecurity in animal production and veterinary medicine. In: Dewulf, J. and Van Immerseel, F. (Eds), *Biosecurity in Animal Production and Veterinary Medicine, from Principle to Practice*. ACCO (vol. 2018), 63–76.

Dewulf, J., Postma, M., Vanbeselaere, B., Maes, D. and Filippitzi, M. E. (2018). Transmission of pig diseases and biosecurity in pig production. In: Dewulf, J. and Van Immerseel, F. (Eds), *Biosecurity in Animal Production and Veterinary Medicine, from Principle to Practice*. ACCO (vol. 2018), 295–328.

Duinhof, T., Van de Ven, S. C. G. and Van Schaik, G. (2006). A survey among veterinarians and pig farmers in the Netherlands: more focus needed on diagnostic approach and on-farm contact structures in the control of PRRS. Proceedings of the 19th IPVS Congress, Copenhagen, Denmark (vol. 1), 234.

Eijck, I. A. J. M. (2003). *Gezond starten, gezond blijven. Praktijkonderzoek veehouderij*. Animal Science Group, Wageningen.

Ellis-Iversen, J., Smith, R. P., Gibbens, J. C., Sharpe, C. E., Dominguez, M. and Cook, A. J. (2011). Risk factors for transmission of foot-and-mouth disease during an outbreak in southern England in 2007. *Veterinary Record* 168(5), 128.

Evans, S. J. and Sayer, A. R. (2000). A longitudinal study of Campylobacter infection of broiler flock in Great Britain. *Preventive Veterinary Medicine* 46(3), 209–223.

Fèvre, E. M., Bronsvoort, B. M. C., Hamilton, K. A. and Cleaveland, S. (2006). Animal movements and the spread of infectious diseases. *Trends in Microbiology* 14(3), 125–131.

Filippitzi, M. E., Brinch Kruse, A., Postma, M., Sarrazin, S., Maes, D., Alban, L., Nielsen, L. R. and Dewulf, J. (2017). Review of transmission routes of 24 infectious diseases preventable by biosecurity measures and comparison of the implementation of these measures in pig herds in six European countries. *Transboundary and Emerging Diseases* 65(2), 381–398. doi: 10.1111/tbed.12758.

Filippitzi, M. E., Callens, B., Pardon, B., Persoons, D. and Dewulf, J. (2014). Antimicrobial use in pigs, broilers and veal calves in Belgium. *Vlaams Diergeneeskundig Tijdschrift* 83(5), 215–224.

Fritzemeier, J., Teuffert, J., Greiser-Wilke, I., Staubach, Ch., Schlüter, H. and Moennig, V. (2000). Epidemiology of classical swine fever in Germany in the 1990s. *Veterinary Microbiology* 77(1–2), 29–41.

Fussing, V., Barfod, K., Nielsen, R., Møller, K., Nielsen, J. P., Wegener, H. C. and Bisgaard, M. (1998). Evaluation and application of ribotyping for epidemiological studies of Actinobacillus pleuropneumoniae in Denmark. *Veterinary Microbiology* 62(2), 145–162.

Garza-Moreno, L., Segalés, J., Pieters, M., Romagosa, A. and Sibila, M. (2017). Survey on *Mycoplasma hyopneumoniae* gilt acclimation practices in Europe. *Porcine Health Management* 3, 21.

Gelaude, P., Schlepers, M., Verlinden, M., Laanen, M. and Dewulf, J. (2014). Biocheck.UGent: a quantitative tool to measure biosecurity at broiler farms and the relationship with technical performances and antimicrobial use. *Poultry Science* 93(11), 2740–2751.

Goodwin, R. F. W. (1985). Apparent reinfection of enzootic-pneumonia-free pig herds: search for possible causes. *The Veterinary Record* 116(26), 690–694.

Hamilton, D. N., Ellis, M., Wolters, B. F., Schninckel, A. P. and Wilson, E. R. (2003). The growth performance of the progeny of two swine sire lines reared under different floor space allowances. *Journal of Animal Science* 81(5), 1126–1135.

Hartung, J. (2005). Zur Abschirmung von Beständen aus tierhygienischer Sicht. *Deutsche Tierärztliche Wochenschrift* 112, 313–316.

Hege, R., Zimmermann, W., Scheidegger, R. and Stärk, K. D. C. (2002). Incidence of reinfections with *Mycoplasma hyopneumoniae* and *Actinobaillus pleuropneumoniae* in pig farms located in respiratory-disease-free regions of Switzerland - identification and quantification of risk factors. *Acta Veterinaria Scandinavica* 43(3), 145–156.

Hémonic, A., Corrégé, I. and Lanneshoa, M. (2010). Quelles sont les pratiques de bioséccurité et d'hygiène en élevages de porcs? *Techni-Porc.* 33, 7–13.

Horst, H. S., Huirne, R. B. and Dijkhuizen, A. A. (1997). Risks and economic consequences of introducing classical swine fever into the Netherlands by feeding swill to swine. *Revue Scientifique et Technique* 16(1), 207–214.

Huijsdens, X. W., van Dijke, B. J., Spalburg, E., van Santen-Verheuvel, M. G., Heck, M. E., Pluister, G. N., Voss, A., Wannet, W. J. and de Neeling, A. J. (2006). Community-acquired MRSA and pig-farming. *Annals of Clinical Microbiology and Antimicrobials*. A.K. Press, Johnson 5, 26.

Johnson, A. K., Edwards, L. N., Niekamp, S. R., Philips, C. E., Sutherland, M. A., Torrey, S., Casey-Trott, T., Tucker, A. L., Widowski, T. (2012). Behavior and welfare. In: Zimmermann, J. J., Karriker, L. A., Ramirez, A., Schwartz, K. J. and Stevenson, G. W. (Eds), *Diseases of Swine*. Wiley-Blackwell, Chichester, West Sussex, UK, 180–249.

Kirwan, P. (2008). Biosecurity in the pig industry - an overview. *Cattle Practice* 16, 147–154.

Laanen, M. (2011). Waarop letten voor meer bioveiligheid? *Varkensbedrijf* 11, 30–33.

Laanen, M., Beek, J., Ribbens, S., Vangroenweghe, F., Maes, D. and Dewulf, J. (2010). Bioveiligheid op varkensbedrijven: ontwikkeling van een online scoresysteem en de resultaten van de eerste 99 deelnemende bedrijven. *Vlaams Diergeneeskundig Tijdsschrift* 79, 302–306.

Laanen, M., Persoons, D., Ribbens, S., de Jong, E., Callens, B., Strubbe, M., Maes, D. and Dewulf, J. (2013). Relationship between biosecurity and production/antimicrobial treatment characteristics in pig herds. *Veterinary Journal* 198(2), 508–512.

Lister, S. A. (2008). Biosecurity in poultry management. In: Patisson, M., McMullin, P. F., Bradburry, J. M. and Alexander, D. J. (Eds), *Poultry Diseases* (6th edn.). Saunders Elsevier, Beijing, China, 48–65,

Lo Fo Wong, D. M. A., Dahl, J., Stege, H., van der Wolf, P. J., Leontides, L., von Altrock, A. and Thorberg, B. M. (2004). Herd-level risk factors for subclinical *Salmonella* infection in European finishing-pig herds. *Preventive Veterinary Medicine* 62(4), 253–266.

Loncke, T. and Dewulf, J. (2018). Rodent control in animal production. In: Dewulf, J. and Van Immerseel, F. (Eds), *Biosecurity in Animal Production and Veterinary Medicine, from Principle to Practice*. ACCO (vol. 2018), 283–294.

Luyckx, K. (2016). Evaluation and implication of cleaning and disinfection of broiler houses and pig nursery units. Phd thesis. Ghent University. Faculty of Veterinary Medicine, Merelbeke, Belgium. Available at: https://biblio.ugent.be/publication/8081692.

Maes, D., Deluyker, H., Verdonck, M., Castryck, F., Miry, C., Vrijens, B. and de Kruif, A. (2000a). Herd factors associated with the seroprevalences of four major respiratory pathogens in slaughter pigs from farrow-to-finish pig herds. *Veterinary Research* 31(3), 313–327.

Maes, D., Deluyker, H., Verdonck, M., Castryck, F., Miry, C., Vrijens, B., Ducatelle, R. and de Kruif, A. (2000b). Noninfectious herd factors associated with macroscopic and microscopic lung lesions in slaughter pigs from farrow to-finish pig herds. *The Veterinary Record* 148, 41–46.

Maes, D., Segales, J., Meyns, T., Sibila, M., Pieters, M. and Haesebrouck, F. (2008). Control of Mycoplasma hyopneumoniae infections in pigs. *Veterinary Microbiology* 126(4), 297–309.

McQuiston, J. H., Garber, L. P., Porter-Spalding, B. A., Hahn, J. W., Pierson, F. W., Wainwright, S. H., Senne, D. A., Brignole, T. J., Akey, B. L. and Holt, T. J. (2005). Evaluation of risk factors for the spread of low pathogenicity H7N2 avian influenza virus among commercial poultry farms. *Journal of the American Veterinary Medical Association* 226(5), 767–772.

Meroz, M. and Samberg, Y. (1995). Disinfecting poultry production premises. Revue scientifique et technique. *Revue Scientifique et Technique*. Office International des Epizooties 14(2), 273–291.

Mintiens, K., Laevens, H., Dewulf, J., Boelaert, F., Verloo, D. and Koenen, F. (2003). Risk analysis of the spread of classical swine fever virus through 'neighbourhood infections' for different regions in Belgium. *Preventive Veterinary Medicine* 60(1), 27–36.

Moore, C. (1992). Biosecurity and minimal disease herds. *Veterinary Clinics of North America. Food Animal Practice* 8(3), 461–474.

Neumann, E. J. (2012). Disease transmission and biosecurity. In: Zimmerman, J. J., Karriker, L. A., Ramirez, A., Schwartz, K. J. and Stevenson, G. W. (Eds), *Diseases of Swine* (10th edn.). John Wiley & Sons Inc., Iowa, 141–164.

Pieters, M. and Fano, E. (2016). *Mycoplasma hyopneumoniae* management in gilts. *Veterinary Record* 178(5), 122–123.

Pointon, A. M., Heap, P. and McCloud, P. (1985). Enzootic pneumonia of pigs in South Australia–factors relating to incidence of disease. *Australian Veterinary Journal* 62(3), 98–101.

Postma, M., Backhans, A., Collineau, L., Loesken, S., Sjölund, M., Belloc, C., Emanuelson, U., Grosse Beilage, E., Stärk, K. D. C. and Dewulf, J. (2016). The biosecurity status and its associations with production and management characteristics in farrow-to-finish pig herds. *Animal* 10(3), 478–489.

Pritchard, G., Dennis, I. and Waddilove, J. (2005). Biosecurity: reducing disease risks to pig breeding herds. *In Practice* 27(5), 230–237.

Rajkowski, K. T., Eblen, S. and Laubauch, C. (1998). Efficacy of washing and sanitizing trailers used for swine transport in reduction of *Salmonella* and *Escherichia coli*. *Journal of Food Protection* 61(1), 31–35.

Reddy, S. M. W., Montambault, J., Masuda, Y. J., Keenan, E., Butler, W., Fisher, J. R. B., Asah, S. T. and Gneezy, A. (2017). Advancing conservation by understanding and influencing human behavior. *Conservation Letters* 10(2), 248–256.

Ribbens, S., Dewulf, J., Koenen, F., Maes, D. and De Kruif, A. (2007). Evidence of indirect transmission of classical swine fever virus through contacts with people. *The Veterinary Record* 160(20), 687–690.

Rojo-Gimeno, C., Postma, M., Dewulf, J., Hogeveen, H., Lauwers, L. and Wauters, E. (2016). Farm-economic analysis of reducing antimicrobial use whilst adopting improved management strategies on farrow-to-finish pig farms. *Preventive Veterinary Medicine* 129, 74–87.

Rose, N. and Madec, F. (2002). Occurrence of respiratory disease outbreaks in fattening pigs: relation with the features of a densely and a sparsely populated pig area in France. *Veterinary Research* 33(2), 179–190.

Sarrazin, S., Joosten, P., Van Gompel, L., Luiken, R. E. C., Mevius, D. J., Wagenaar, J. A., Heederik, D. J. J., Dewulf, J. and EFFORT Consortium (2019). Quantitative and qualitative analysis of antimicrobial usage patterns in 180 selected farrow-to-finish pig farms from nine European countries based on single batch and purchase data. *Journal of Antimicrobial Chemotherapy* 74(3), 807–816.

Scott, A., McCluskey, B., Brown-Reid, M., Grear, D., Pitcher, P., Ramos, G., Spencer, D. and Singrey, A. (2016). Porcine epidemic diarrhea virus introduction into the United States: root cause investigation. *Preventive Veterinary Medicine* 123, 192–201.

Sellers, R. F., Donaldson, A. I. and Herniman, K. A. J. (1970). Inhalation, persistence and dispersal of foot-and-mouth disease virus by man. *Journal of Hygiene* 68(4), 565–573.

Siekkinen, K. M., Heikkilä, J., Tammiranta, N. and Rosengren, H. (2012). Measuring the costs of biosecurity on poultry farms: a case study in broiler production in Finland. *Acta Veterinaria Scandinavica* 54, 12.

Stärk, K. D. (2000). Epidemiological investigation of the influence of environmental risk factors on respiratory disease in swine. *Veterinary Journal* 159(1), 37–56.

Vangroenweghe, F., Ribbens, S., Vandersmissen, T., Beek, J., Dewulf, J., Maes, D. and Castryck, F. (2009). Keeping Pigs Healthy (in Dutch). In: Vangroenweghe, F., Print, D. C. L. and Signs (Eds), (1st edn). Zelzate, Belgium.

Van Immerseel, F., Luyckx, K., De Reu, K. and Dewulf, J. (2018). Cleaning and disinfection. In: Dewulf, J. and Van Immerseel, F. (Eds), *Biosecurity in Animal Production and Veterinary Medicine, from Principle to Practice*. ACCO (vol. 2018), 133–158.

Wentworth, D. E., McGregor, M. W., Macklin, M. D., Neumann, V. and Hinshaw, V. S. (1997). Transmission of swine influenza virus to humans after exposure to experimentally infected pigs. *Journal of Infectious Diseases* 175(1), 7–15.

Windsor, R. S. and Simmons, J. R. (1981). Investigation into the spread of swine dysentery in 25 herds in East Anglia and assessment of its economic significance in five herds. *The Veterinary Record* 109(22), 482–484.

Zimmerman, J. J., Benfield, D. A., Dee, S. A., Murtaugh, M. P., Stadejek, T., Stevenson, G. W. and Torremorell, M. (2012). Porcine reproductive and respiratory syndrome virus. In: Zimmermann, J. J., Karriker, L. A., Ramirez, A., Schwartz, K. J. and Stevenson, G. W. (Eds), *Diseases of Swine*. Wiley-Blackwell, Chichester, West Sussex, UK, 1675–1777.

Chapter 10

Managing feed to optimize pig health

Sam Millet, Flanders Research Institute for Agriculture, Fisheries and Food (ILVO), Belgium; and Nadia Everaert, TERRA Teaching and Research Centre, Gembloux Agro-Bio Tech, Liège University, Belgium

1 Introduction

Feed is the most important cost factor in pig production. Optimal feeding goes far beyond merely providing the right set of nutrients to optimize the animal's growth: feeding plays a role in gastrointestinal diseases, such as weaning diarrhea and swine dysentery, but it also affects body condition, lameness, and immunity. Proper nutrition may increase resilience and prevent disease. But what comprises proper nutrition? There is no simple answer, as it often depends on the farm situation. While optimal performance has economic and ecological benefits and should be the main goal on farms with optimal health conditions, most farms need to balance trade-offs between feeding for optimal performance and minimizing the risk of health problems.

Feed management means choosing the right set of ingredients to meet the pigs' nutrient requirements. Defining those requirements may depend not only on the physiological status of the pig but also on its health status. At certain time points, pigs may require feeding below their requirements to improve their health. This chapter begins with a discussion of how balanced energy and protein levels are required for optimal health and performance (Section 2). A sufficient quantity and good quality of drinking water are also important in

http://dx.doi.org/10.19103/AS.2022.0103.11

this respect, but it is not discussed in this chapter. Vitamins and minerals are essential for physiological functions and metabolism, but the animal's health status may be affected when too much or too little of these micronutrients are provided relative to the animal's physiological requirements (Section 3). Fiber should not only be seen as a bulking agent but may also be considered in the context of intestinal health (Section 4). The choice of ingredients for minimal feed cost may differ from the choice for optimal health; certain ingredients are discussed in the health context of piglets (Section 5). In addition, the form and structure of the feed are important characteristics as they can affect both health and performance (Section 6).

2 Healthy energy and protein intake

Energy and amino acid levels are major determinants of pig diets, especially in relation to optimal performance. It is generally assumed that pigs eat to meet their energy needs. When energy level decreases, feed intake (FI) is likely to increase (Henry, 1985). For this reason, the requirements of other nutrients are generally expressed relative to energy. A balanced amino acid:energy ratio leads to optimal muscle deposition and efficient growth, with the health status of the animal possibly affecting the amino acid pattern. If one or more amino acids are provided below the requirement, muscle deposition will decrease without any evident health risks. In growing-finishing pigs, the dietary amino acid:energy ratio mainly affects feed efficiency, with a lesser effect on health. In growing-finishing pigs under heat stress, increasing the dietary nutrient density and lowering the crude protein (CP) level (without affecting the ileal digestible amino acid level) may decrease heat production, as dietary protein leads to higher heat increment (the heat production following feed consumption) than fat or starch do (Noblet et al., 1994; Renaudeau et al., 2008).

In piglets, the dietary CP level is of main concern, but FI itself is also crucial to overcome post-weaning diarrhea (PWD). In sows, energy intake and body condition may be determinants of longevity.

The importance of gastric pH in piglets

A low gastric pH serves as a barrier to pathogens attempting to enter the small intestine (Hansen et al., 2007). Low gastric pH is important for protein digestion, as it is key for denaturizing dietary proteins as the first step in the digestive process. The transformation of pepsinogen into pepsin occurs most efficiently at low pH. Moreover, pepsin is most effective at pH 2.0-3.5 (Lawlor et al., 2005). In newly weaned piglets, HCl secretion is limited and only starts after consumption of solid feed (Cranwell, 1985). As a result, pathogens may better survive the stomach environment and protein digestion may be impaired, leading to excess levels of protein escaping the small intestine where it becomes a substrate for potentially harmful bacteria in the large intestine. The dietary acid-binding capacity of piglet feed should therefore be kept as low as possible, which can be achieved by careful ingredient selection (Lawlor et al., 2005). Weaned pigs may also benefit from whey or lactose in starter diets, as the ongoing lactic acid production lowers gastric pH.

2.1 The protein level in piglet diets

Prior to weaning and in the first weeks afterward, piglets eat little solid feed, which hampers their growth potential. To maximize performance, they need high dietary amino acid levels. However, if the protein level is too high, gastric functioning can be negatively affected and can result in an increased risk for intestinal problems. In piglets, high CP levels have been shown to increase the risk for PWD (Nyachoti et al., 2006; Wellock et al., 2007, 2008a; Heo et al., 2009; Rist et al., 2013). Different mechanisms may be at play here: first, protein increases the buffering capacity in the stomach, with higher pH leading to impaired protein digestion and greater passage of potential pathogens, leading to the growth of opportunistic pathogens like *E. coli* in the small intestine. The toxins produced by pathogenic *E. coli* cause elevated secretion of water and chloride from small intestinal crypt cells, leading to secretory diarrhea (Sun and Kim, 2017). Second, higher protein levels may lead to higher amounts of undigested protein entering the large intestine. Protein fermentation, accompanied by the formation of potentially harmful compounds, is regarded as a factor in the development of enteric diseases, especially under stressful conditions (Rist et al., 2013). Protein fermentation in the large intestine, with its potentially toxic metabolites, is reduced along with CP level (Sun and Kim, 2017). Lowering the CP level while maintaining essential amino acid levels is the first way to reduce the risk of diarrhea without affecting the performance (Htoo et al., 2007).

When reducing the protein level, good knowledge of essential amino acid requirements is essential for maintaining performance. Even then, protein reduction without affecting performance is only possible within limits: Jansman et al. (2016) could not reduce CP level below 160 g/kg. Gloaguen et al. (2014) achieved a reduction in CP for piglets weighing 10–20 kg from 176 g/ kg to 135 g/ kg without significantly affecting performance, but in that trial, the lysine level was below requirement. At lower CP levels, semi-essential or even nonessential amino acids may become limiting (Wu, 2014). At some point, nitrogen itself may limit growth. Lysine makes up 7.6 g/100 g of protein in the carcass (Mahan and Shields, 1998). As a practical guideline, a maximal

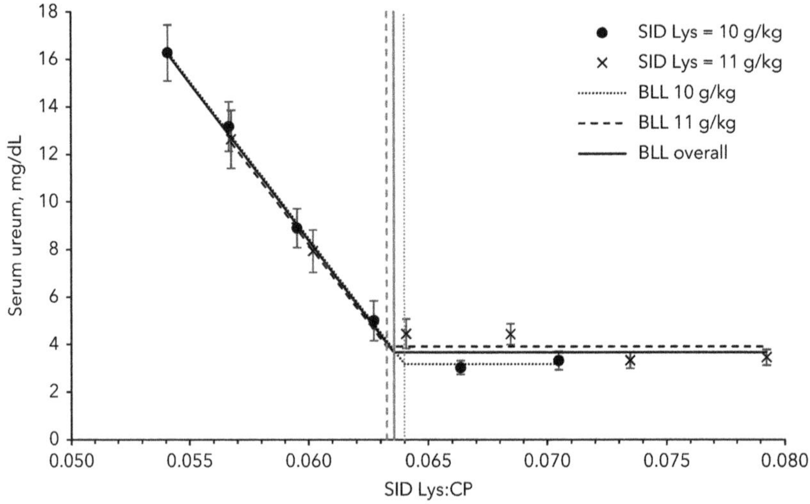

Figure 1 Effect of dietary standardized ileal digestible (SID) Lys to CP ratio on serum urea levels in piglets at 7 weeks of age. Piglets receiving 10 g SID Lys/kg or 11 g SID Lys/kg diet are presented, together with their corresponding and overall broken line models with linear ascending portions. Vertical gray lines represent breakpoints. Urea level decreases with decreasing CP level (=increasing SID Lys:CP ratio) until a minimum is reached. This indicates the point of maximal nitrogen retention and minimal protein breakdown. A logical consequence is that low CP diets for piglets may contain amino acid levels below the requirement for maximal growth and efficiency. In this period, in our opinion, it is a valid choice to feed below the requirements for optimal growth and as such decrease the risk for gastrointestinal disturbance. Source: Figure from Millet et al. (2018), with permission from Oxford University Press.

standardised ileal digestible (SID) lysine:CP ratio of 0.064 (SID Lys:ATTD CP = 0.079) can be used in piglet diets (Millet et al., 2018). At this point, serum urea level (as a measure of protein breakdown) is the lowest (Fig. 1). In our opinion, the optimal CP level in the diet may depend on the farm situation. In good health conditions, maximal performance can be achieved by using high dietary amino acid levels and possibly also higher CP levels. For most herds, however, nutritionists will lower the CP level to decrease the risk of diarrhea. This implies that suboptimal growth performance may be accepted to decrease the risk for disease and mortality. Furthermore, feed costs may be reduced by adapting the dietary SID lysine level to this lower CP level.

2.2 Amino acid levels for optimal health

In addition to protein biosynthesis, amino acids exert functions in regulating immunity, anti-oxidation, and redox regulation (Liu et al., 2017). According to Goodband et al. (2014), increase in amino acid requirements for immune function in times of disease are more than offset by the associated decrease in protein accretion and increased muscle protein degradation, which implies there would be no need for increased dietary amino acid levels. Still, the balance between amino acids may be altered. A study by van der Meer et al. (2016) suggested that methionine, threonine, and tryptophan supplementation improved the performance of pigs especially under poor sanitary conditions, which indicates again that the ideal amino acid pattern may depend on the health status of the animals. Similarly, supplementation with threonine, methionine, and tryptophan improved immune status in weaned pigs challenged with *Salmonella* (Rodrigues et al., 2021). The disadvantage of these studies is that it is impossible to separate the effect of a single amino acid. Particular amino acids may exert beneficial effects apart from their function as a building block for protein synthesis. Some reviews have been written on the role of individual amino acids in immune functioning and pig health (Li et al., 2007; Liu et al., 2017; Le Floc'h et al., 2018). Here, in the following sections, we describe three amino acids that have been shown to be important for maintaining health.

2.2.1 Glutamine

Glutamine, a unique amino acid, serves among others as a preferred respiratory fuel for cells that proliferate rapidly, like enterocytes and lymphocytes (Lacey and Wilmore, 1990; Wu et al., 1996). Even though it is abundantly present in the body, during stress the individual's body may not be able to produce sufficient amounts of this amino acid. Glutamine may therefore become a 'conditionally essential' amino acid in critically ill animals and during periods of stress such as the weaning period (Lacey and Wilmore, 1990). Sufficient glutamine intake may be needed for intestinal integrity. In research on piglets, glutamine supplementation beneficially affected the performance of early weaning piglets in the first 2 weeks after weaning, especially when fed a low protein diet (Millet et al., 2020a). L-glutamine in creep feed and nursery diet was shown to improve feed conversion in the first 3 weeks post-weaning, possibly due to improved gut health (Cabrera et al., 2013). Johnson and Lay (2017) observed that intestinal morphology was improved by L-glutamine supplementation and therefore stated that L-glutamine in piglet diets can be a cost-effective alternative to antibiotics following weaning and transport. Recently Ji et al. (2019) reviewed the studies of glutamine in weaning pigs.

2.2.2 Threonine

Threonine is considered an important amino acid for intestinal health. During acute intestinal inflammation, threonine fluxes across the portal-drained viscera (PDV) increase. Threonine is needed for the synthesis of mucin and plasma immunoglobulin production in animals (Li et al., 2007; Trevisi et al., 2015). Hamard et al. (2010) found that moderate threonine deficiency affected the functionality of the small intestine of piglets. Similarly, dietary L-threonine supplementation attenuated inflammatory responses, facilitated Muc2 synthesis, and promoted goblet cell differentiation in the ileum of intrauterine-growth-retarded weanling piglets (Zhang et al., 2019). Furthermore, Trevisi et al. (2015) showed an altered microbiota composition as evidenced by a reduced *E. coli* shedding in pigs fed at higher threonine levels.

2.2.3 Arginine

Arginine plays an important role in intestinal health: it protects against inflammatory stress and inflammation, modulates mTOR in intestinal tissue and the intestinal inflammatory response, and attenuates villus atrophy (Le Floc'h et al., 2018). Part of its effect is due to its role in the formation of nitric oxide, which acts as a vasodilator, signaling molecule, and mediator of immune response (Le Floc'h et al., 2018). It is considered a conditionally essential amino acid (NRC, 2012). Liu et al. (2008) observed that arginine supplemented in the piglet diet attenuated the negative effects of *E. coli* lipopolysaacharide on the intestinal inflammatory response and intestinal morphology. Similarly, Zheng et al. (2017) observed a positive effect in an experimental model of oxidative stress in piglets by decreasing inflammatory cytokine expression.

Recently, additional free amino acids have become commercially available for use in feeds. This allows for more targeted amino acid nutrition. This evolution is expected to lead to more research into their functions beyond muscle formation.

2.3 Feed intake as a driver for healthy weaning

The stressful period shortly after weaning is characterized by reduced FI. Weaning age, weaning stress, FI, and diet composition all affect intestinal integrity after weaning (Wijtten et al., 2011). FI affects the flow of nutrients through the intestinal tract and intestinal morphology. Ensuring sufficient nutrient intake after weaning is a challenge, but it remains key to a proper gut structure and function and to the prevention of several problems associated with weaning. This process starts before weaning. Bruininx et al. (2002) observed that piglets that ate creep feed before weaning showed more favorable FI behavior and thus a higher daily gain in the first 8 days post-weaning than the non-eaters. Several management strategies in the pre- and post-weaning phases have been investigated to ensure sufficient nutrient intake after weaning. For instance, restricted feeder space increases competition and reduces FI, and thus negatively influences gut development and health as well as performance

(He et al., 2016). Other factors include creep feed provision, the composition of creep feed and weaning diet, mat feeding, sow management, barn climate, pellet size, nursery placement strategies, water access points and location, and diet form. Flavor imprinting is another strategy that may stimulate FI of weaned piglets and reduce diarrhea incidence (Oostindjer et al., 2010). An overview of these strategies was published recently by Wensley et al. (2021a,b).

The amino acid profile may also have a profound effect on FI. Especially when the ratio between amino acids is unbalanced, the animals react by decreasing their FI. Especially excess leucine depresses FI and the growth of piglets (Goodband et al., 2019). High brain leucine levels reduce FI by hypothalamic mTOR signaling (Cota et al., 2006). This negative effect can be (partially) counteracted by adding more valine to the diet (Millet et al., 2015). Valine has been shown to increase FI (Gloaguen et al., 2011; Millet et al., 2020b). It is important to note that the ratios between amino acids may be more important than their absolute values. In recent research, FI was largely affected by decreasing lysine levels when piglets were fed a high protein diet and thus an excessive leucine:lysine ratio, while this effect was less pronounced when they were fed a diet with a lower CP level and more balanced amino acid profile (Millet et al., 2020b) (Fig. 2). In addition, the sanitary conditions may affect the

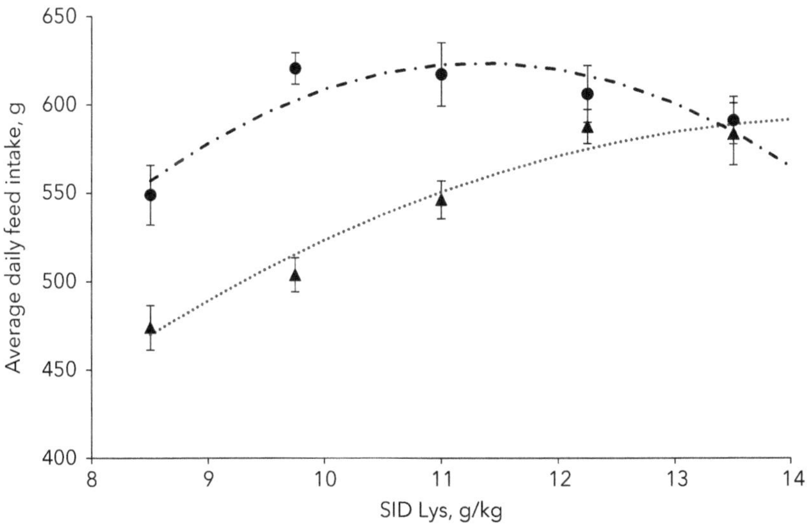

Figure 2 Measured means (▲ for Exp. 1, ● for Exp. 2) and standard error bars for each treatment are presented, with their fitted quadratic polynomial models. With decreasing lysine levels, feed intake decreased especially in experiment 1. In this experiment, a high-protein diet was used and with decreasing lysine levels the amino acid ratios were less and less balanced. In experiment 2, the balance between amino acids was maintained better by using a low CP diet and free amino acids. Source: Reprinted from Millet et al. (2020b), with permission from Elsevier.

optimal AA ratio. For example, a higher optimal ratio of SID threonine:lysine for weaned piglets was shown under unclean sanitary conditions compared to clean sanitary conditions. Poor air quality may have affected FI and growth performance (Jayaraman et al., 2015).

2.4 Feed and energy intake and body condition of sows

In sows, optimal body condition is necessary to maintain health and longevity. While sufficient amino acid provision is important, daily energy intake may be even more important: 'thin sow syndrome,' 'fat sow syndrome,' and 'second parity syndrome' have all been related to the dynamics of body condition management (Charette et al., 1996). Body condition is managed by controlling FI and the energy level in the diet. During gestation, energy and amino acids are needed to grow the fetuses and for the sow's own growth, but a good diet is also needed to build sufficient body reserves to compensate for a possible nutritional deficit that may occur in the following lactation (Dourmad et al., 1994). During lactation, sows mobilize fat and muscle mass to compensate for the negative energy balance, as FI during lactation is often not high enough to sustain the milk production needed for large litters (Eissen et al., 2000; Bergsma et al., 2009). Body condition at the end of lactation thus determines the energy requirements during gestation. The speed of body condition recovery is still a matter of research. In general, studies suggest it may be good to increase the FI shortly after weaning (Leal et al., 2019), but others claim this may negatively affect embryo survival and litter size (Mallmann et al., 2020). The importance of increasing the total FI of sows during late gestation has been questioned by Goodband et al. (2013). They suggest increasing the amino acid intake, without the sows becoming too fat, as overfeeding in gestation often leads to lower FI during lactation (Goodband et al., 2013). A logical conclusion is that two-phase feeding during gestation may be optimal.

Sows that are too thin at the start of lactation have been associated with a higher number of stillborn piglets (Maes et al., 2004b), lower colostrum production (Decaluwe et al., 2014), and more shoulder lesions (Bonde et al., 2004; Zurbrigg, 2006; KilBride et al., 2009; Rioja-Lang et al., 2018). In contrast, fat sows or sows overfed during gestation have lower FI and less milk production during lactation (Weldon et al., 1994). Moreover, excessive fattening during late pregnancy is related to a longer farrowing duration (Oliviero et al., 2010), with a higher risk for perinatal mortality. Decaluwé et al. (2013) observed that colostrum yield was negatively associated with the use of body protein before farrowing, whereas it was positively associated with the use of body protein on the first day of lactation. The restriction of feeding prior to parturition significantly reduces the risk of postpartum dysgalactia syndrome (PDS) (Peltoniemi et al., 2007). In conclusion, sows need optimal body condition management during gestation and lactation. Both overfeeding and underfeeding pose risks for sow health and longevity. For further reading on this topic, we suggest the review paper of Solà-Oriol and Gasa (2017).

3 Mineral and vitamin nutrition for resilient animals

Minerals and vitamins both play an important role in animal metabolism. Deficiencies as well as toxicities are problematic and lead to different types of symptoms. In contemporary pig production, such symptoms are rare, as premixes meant to supplement the diet contain a balanced mixture of vitamins and minerals. Therefore, we focus only on a few important macro- and micro minerals that can still be linked to a higher risk for production diseases. For advised levels, we refer to NRC (2012), CVB (2020), GfE (German Society for Nutrition Physiology, 2006), and local recommendations.

3.1 Calcium and phosphorus

Calcium (Ca) and phosphorus (P) are essential minerals for optimal bone mineralization. Both minerals have to be considered together when looking at the requirements as they are homeostatically regulated. The Ca:P ratio should remain constant, with a minimum of 1:1 (van Riet et al., 2013). Since phosphorus excreted in the manure is considered a pollutant, the use of low phosphorus diets in growing-finishing pigs is encouraged. The use of phytase can be particularly helpful for this. More than 95% of body calcium is stored in the bones, while phosphorus is mainly stored in the bones as well as in lean tissue and organs (Bikker and Blok, 2017). Excessive calcium, in comparison to phosphorus, is excreted through the urine. An unbalanced calcium:phosphorus ratio may lead to urolithiasis (Lorenzett et al., 2019). Uroliths may be present in a larger proportion of fattening pigs than usually expected (Vrielinck et al., 2019), despite the small number of reports of obstructive urolithiasis in pigs (Maes et al., 2004a). The mineral composition of the diet, together with the amount of drinking water and the urinary pH, is considered the major risk factors for urolithiasis (Maes et al., 2004a). When pigs eat diets deficient in phosphorus, priority is given to phosphorus retention in soft tissue instead of phosphorus retention in bones (Misiura et al., 2020). Insufficient calcium and phosphorus levels thus lead to poor bone mineralization and eventually osteomalacia or hind leg paralysis (NRC, 2012). The exact role of calcium or phosphorus shortages is not entirely clear, as the reason for lameness is mostly multifactorial (Wegner et al., 2020). Apart from its impacts on animal welfare, lameness is an important reason for the early culling of sows (Pluym et al., 2013).

Recommended levels for growing pigs are based on a total phosphorus level of 5.3 g/kg empty body weight gain (Bikker and Blok, 2017). Translating this to dietary levels may not be easy. Especially in highly efficient animals, the ingested calcium and phosphorus level may be lower than optimal because of their low FI in comparison to their growth rate. The advice is therefore to compensate for the lower FI per kilogram gain in highly efficient animals through increased calcium and phosphorus levels per kilogram of feed (Bikker and Blok, 2017). In sows, mineral requirements vary throughout a sows' cycle, with the highest requirements at the end of gestation and during lactation. The bone acts as a reservoir with bone mineralization during gestation and mobilization in lactation, as reflected in the concentration of bone mineralization markers (van Riet et al., 2016b). Because bone mineralization is dynamic, assessment of 'optimal' mineralization is difficult.

Feeding below calcium and phosphorus requirements has been advised for young piglets, as calcium sources such as limestone are characterized by a high buffering capacity. Jongbloed et al. (2003) recommended the use of a maximum calcium content below the optimum, while accepting a transient reduction in bone mineralization. This was adopted in feeding recommendations of Bikker and Blok: instead of the calculated requirements of 5.9 g/kg and 3.4 g/kg for calcium and phosphorus, respectively, they suggest using 5.0 g/kg and 3.0 g/kg during the first weeks after weaning to reduce the calcium content and buffer capacity of the weaning diet (Bikker and Blok, 2017).

3.2 Zinc and copper

The microminerals zinc and copper have received considerable attention in recent years. They are involved in many metabolic functions as cofactors for a variety of enzymes (Jondreville et al., 2003) and play a crucial role in bone formation and claw integrity (van Riet et al., 2013, 2016a), skin development (Tucker and Salmon, 1955), and immune function (Kloubert et al., 2018). To meet the copper requirements, Bikker and Jongbloed (2013) calculated copper levels of 18 mg/kg for lactating sows and 12 mg/kg for other pigs, while maximum allowed levels in the EU are 150 mg/kg for piglets during the first 4 weeks after weaning, 100 mg/kg until 8 weeks after weaning, and 25 mg/kg for other animals (European Commission, 2018). For zinc, the recommended dietary levels to meet the requirements vary from 65 mg/kg in growing-finishing pigs to 85 mg/kg in lactating sows, with maximum allowed levels of 150 mg/kg for piglets and sows and 120 mg/kg in other animals.

However, most of these ingested microminerals are excreted through the feces. In the soil, they may be toxic to plants and microorganisms and have thus become an environmental concern (Jondreville et al., 2003). Dietary levels are therefore restricted by legislation.

Although the use of microminerals as growth promoters has been criticized, especially zinc oxide (ZnO) at 'pharmacological' concentrations (2,500 mg/kg) has been shown to be effective in fighting PWD. The mechanism of action is not well understood (Debski, 2016), but it seems to be multifactorial and goes beyond the fulfilment of daily nutritional requirements (Bonetti et al., 2021). It has been shown that ZnO exerts anti-inflammatory effects (Lallès and Montoya, 2021) and improves intestinal morphology and intestinal barrier function (Hu et al., 2013; Song et al., 2015), thereby probably preventing translocation of bacteria through the intestinal barrier and increasing the activity of digestive enzymes (Hedemann et al., 2006). While ZnO does have antibacterial properties, its ability to reduce *E. coli*, even at high doses, is questioned (Liedtke and Vahjen, 2012). Apart from inorganic forms, such as zinc sulfate and zinc chloride, organic microminerals, such as Zn-methionine or Zn-lysine, are now commonly available in commercial pig feeding. They have been claimed to be more effective than inorganic mineral sources in improving intestinal barrier function (Pearce et al., 2015), immune function (Richards et al., 2010), or claw health (Lisgara et al., 2016), although many studies showed little or no difference in the bioavailability of organic versus inorganic mineral sources (Richards et al., 2010; Schlegel et al., 2013). Similarly, several researchers did not see differences between zinc source on immune function parameters (van Heugten et al., 2003) or performance (Hollis et al., 2005; Holen et al., 2018).

The common usage of ZnO at pharmacological levels in pig farming has raised several concerns. Apart from the resistance of microorganisms to ZnO itself, the use of ZnO in animal diets seems to promote the spread of antibiotic resistance in *E. coli* (Bednorz et al., 2013). Together with environmental concerns due to the excretion of excess zinc (Bonetti et al., 2021), it makes ZnO a less sustainable alternative for antibiotics. This has led to a phase-out of therapeutic ZnO use in Europe by June 2022. To cope with reduced copper and zinc levels in European legislation, new forms such as nanoparticles have been developed and marketed. Positive effects on performance and health parameters have been reported (Wang et al., 2018; Bai et al., 2019; Pei et al., 2019), although most of these studies were carried out with concentrations higher than 150 mg/kg. In a study by Milani et al. (2017), doses up to 60 mg/kg were not able to improve diarrhea scores (in contrast to therapeutic levels of ZnO). In summary, there is at this moment no sustainable mineral-based alternative for the pharmacological use of ZnO, although research on alternative forms and doses is still ongoing.

3.3 Optimal vitamin levels

The fat-soluble vitamins (A, D, E, K) can be accumulated in the body (mainly the liver), with a possible risk of toxicity, while this is not the case for the water-soluble vitamins (all Bs and C). Although little is known regarding the quantitative need for vitamins for supporting the optimal development of innate and adaptive immune responses, the use of a premix rich in vitamins has largely eliminated the risk of vitamin deficiency in modern intensive pig production (Lauridsen et al., 2021). However, the rapid genetic improvement and high performance of modern breeds may lead to disrupted redox balance and uncontrolled inflammation. This increases the need to control reactive oxygen species and peroxides. Antioxidant vitamins (vitamin E and vitamin C) may therefore be considered as a strategy for enhanced enteric immunity and inflammation control (Lauridsen, 2019). Critical periods are soon after birth, especially under limited colostrum intake, and post-weaning under conditions of limited FI and impaired digestive functioning. For larger pigs, heat stress increases micronutrient needs due to increased utilization coupled with reduced FI (Cottrell et al., 2015). This may trigger oxidative stress, including in the gut (Liu et al., 2016). Besides the abovementioned vitamins, selenium is also a well-known biological antioxidant. Results of supplementing sows with either vitamin E or selenium have included a higher number of piglets born and weaned, increased piglet weight at weaning, enhanced milk fat content, better humoral immune function (IgG and IgA), and improved oxidant activity of sows and piglets (Mavromatis et al., 1999; Pinelli-Saavedra et al., 2008; Wang et al., 2017). Selenium and vitamin E supplementation in growing pigs has been shown to reduce intestinal leakiness caused by heat stress (Liu et al., 2016). Vitamins A and D can also regulate intestinal immune functions (Lauridsen et al., 2021). Indeed, in a weanling pig model, supplementation with vitamin D_3 or its metabolite $25OHD_3$ increased leukocyte cell numbers and had a potential impact on systemic and mucosal antimicrobial responses (Konowalchuk et al., 2013). For an extensive review of the role of vitamins on gastrointestinal functionality and health of pigs, we refer to the review of Lauridsen et al. (2021). Optimal levels of each of the vitamins are discussed by Isabel et al. (2012).

4 Fiber, an overlooked nutrient?

Dietary fiber (DF) is defined as 'carbohydrate polymers and lignin that are not hydrolysed by endogenous enzymes and are substrates for bacterial fermentation in the hindgut' (Lupton et al., 2009). It includes plant cell walls, as well as resistant starch and non-digestible oligosaccharides, which include plant cell contents and include fructo-oligosaccharides (Agyekum and Nyachoti, 2017). The main cell wall polysaccharides in cereal grains are cellulose, arabinoxylans, and mixed linked β-(1,3) (1,4)-*D*-glucans, while cell walls of oilseeds, oilseed meal, and leguminous crops contain cellulose, pectic polysaccharides, lignin, and xyloglucans. Legumes also contain significant quantities of galacto-oligosaccharides including raffinose, stachyose, and verbascose (Navarro et al., 2019).

DF is classified based on physicochemical properties, that is viscosity, hydration, and fermentability, in order to provide more information on its metabolic and physiological activities (Agyekum and Nyachoti, 2017). Solubility is a major factor: non-starch polysaccharides (NSP) can be classified as insoluble (cellulose and some hemicelluloses) or soluble (pectins, gums, and β-glucans) based on their solubility in water or weak alkali (Bach Knudsen, 2001).

Contradictory results on the use of DF have been reported, which can be explained by the influence of different factors such as variation in ingredient composition, the inclusion of different fiber types with different physicochemical properties, inclusion levels, and the matrix provided by the other ingredients. The review of Agyekum and Nyachoti (2017) provides more information on the physicochemical properties of fibers and the mode of action on digestive physiology, health, and performance of pigs. For an extensive review of the structure and characteristics of the carbohydrates that compose DF, see Navarro et al. (2019).

4.1 Role of fibers in the digestive tract

Soluble fibers may increase digesta viscosity, thus hampering nutrient digestion and absorption due to increased cell exfoliation causing villus atrophy (Jha and Berrocoso, 2016). In addition, they may hinder contact between digestive enzymes and nutrients. In general, soluble fiber appears to have a greater impact on starch digestibility than insoluble fiber due to the increase in luminal viscosity and the nutrient-encapsulation effect of intact cell wall polysaccharides (Agyekum and Nyachoti, 2017). Insoluble fiber increases digesta passage rate, allowing less time for mixing of digestive enzymes and dietary components, which also leads to decreased nutrient digestibility. Fiber may reduce apparent ileal protein digestibility through increased endogenous and exogenous nitrogen and amino acid losses. In the large intestine, fiber may stimulate bacterial protein synthesis, leading to decreased apparent total tract protein digestibility. Inclusion of fibrous feedstuffs in the diet, therefore, shifts nitrogen excretion from urea in urine to bacterial protein in feces (Canh et al., 1997). Furthermore, due to the increased workload of the secretary organs, DF may increase hypertrophy, which increases visceral organ mass and cell exfoliation (Agyekum and Nyachoti, 2017). However, DF can contribute to some extent to the energy supply of the pigs due to fermentation by the microbiota in the distal small intestine and large intestine. This produces short-chain fatty acids (SCFA; mainly acetate, propionate, and butyrate) although energy losses also occur due to the production of methane and hydrogen, as well as heat from the fermentation process. While butyrate is used as an energy source for colonic cell proliferation, propionate and a certain amount of butyrate are both used for hepatic gluconeogenesis and nearly two-thirds of acetate is metabolized in muscle cells (Slavin, 2013).

Another physiological effect of fibers is their effect on FI and satiety. FI is typically influenced by dietary energy content (Henry, 1985), but the compensation for reduced energy levels is often not complete. A prolonged satiety effect has been observed by including DF (Danielsen and Vestergaard, 2001), which has been ascribed to their ability to delay gastric emptying, the increased swelling of the stomach content, and the release of satiety-related gut hormones upon the increased abundance of SCFA in the hindgut (Agyekum and Nyachoti, 2017). Therefore, DF is often incorporated at higher levels in the diet of gestating sows to limit a high energy intake, to reduce the sensation of hunger, and to overcome the aggression and behavioral problems that are associated with restricted feeding (de Leeuw et al., 2008).

4.2 Fibers and gut health

The microbial fermentation by bacteria, fungi, protozoa, and archaea results in the production of SCFA, branched-chain fatty acids (BCFA), lactate, amines, indoles, phenols, and various gases like hydrogen, carbon dioxide, and methane (Jha and Berrocoso, 2016). Soluble fiber generally is more extensively and more quickly fermented than insoluble fiber. Fiber inclusion modulates the gut microbiota. For example, resistant starch selectively promotes bifidobacteria and lactobacilli in weaned and growing pigs (Metzler-Zebeli et al., 2019). The fiber fermentation and subsequent production of SCFA contribute to the maintenance of an anaerobic environment, preventing the proliferation of facultative anaerobic pathogens such as *E. coli* or *Salmonella* (Jha et al., 2019). Moreover, the microbial metabolites can promote the proliferation of the mucosal epithelium by stimulating goblet cell proliferation, increasing villus height and hence the absorptive surface area, and in some cases decreased intestinal permeability. Indeed, supplementation with insoluble fiber sources improves gut morphology, with a rise of the villus height to crypt depth ratio as well as an increased number of colonic goblet cells and mucin areas (Hedemann et al., 2006; Chen et al., 2013). The SCFA are also potential immune mediators for the host due to their immunomodulatory and anti-inflammatory properties (Correa-Oliveira et al., 2016). For example, the inclusion of oat bran in the diet of growing pigs modulated the colonic microbiota and decreased mRNA expression of IL-8 in the caecum and reduced IL-8, NF-κB, and TNF-α gene levels in the colon, ameliorating inflammatory responses in the hindgut of growing pigs (He et al., 2018). However, the understanding of how microbiota improve immune function is in its infancy (Jha et al., 2019), especially in pigs, as most knowledge comes from rodent studies as a model for humans.

It is important to note that not all studies support the ability of DF to promote intestinal health (Agyekum and Nyachoti, 2017). Conflicting results have been observed on the inclusion of soluble fiber and *E. coli* proliferation (Hopwood et al., 2004) and occurrence of PWD (Wellock et al., 2008b; Halas et al., 2009). Moreover, the inclusion of resistant starch increased the pro-inflammatory cytokine IL-1β, indicating a negative impact on the immune system (Sun et al., 2015). Pectin was associated with a lower number of goblet cells (Święch et al., 2012) and consequently a reduction of mucin areas in the small intestine crypts (Hedemann et al., 2006).

4.3 Optimizing the use of fibers

To improve the nutritive value of high-fiber diets, several strategies can be applied such as the formulation of diets based on net energy or available nutrient basis, particle size reduction, and pelleting. However, these may counteract the health effects (see below). Several enzyme preparations are commercially available to degrade not only the fiber in pig feeds, such as xylanases, β-glucanases, but also combinations with β-mannanases, α-galactosidases, and pectinase. Also in this field – contrary to poultry studies – inconsistent results have been reported on nutrient utilization and growth performance (Agyekum and Nyachoti, 2017). We hypothesize that one of the reasons may be that these enzymes only speed up digestion by digesting components in the small intestine that would have normally broken down in the large intestine through fermentation and bacterial enzymes.

5 Ingredient composition

The pig diet is comprised of a mix of ingredients. Especially in piglet diets, the choice of ingredients in relation to weaning issues has been studied, although it is often difficult to disconnect the effect of the nutrients it brings from other properties. For example, cooked white rice has consistently been shown to protect against weaning diarrhea (McDonald et al., 1999; Hopwood et al., 2004; Montagne et al., 2004) and *Brachyspira hyodysenteriae* infection (Pluske et al., 1996; Siba et al., 1996). Still, it is difficult to conclude whether this is the result of the low fiber content of white rice or other factors inherent in this ingredient.

In particular, the search for optimal protein sources is ongoing. Animal-based products are of value, such as dairy products, spray-dried plasma (SDP), and egg powder. Spray-dried plasma, a protein-rich product obtained from the industrial fractionation of blood from healthy animals, has been shown to have positive effects on the health and performance of piglets (reviewed by Pérez-Bosque et al. [2016]). Skim milk powder improved intestinal morphology compared to hydrolyzed feather meal (Vente-Spreeuwenberg et al., 2004).

Immunized egg powder has been shown to protect against *E. coli* diarrhea (Yokoyama et al., 1992; Marquardt et al., 1999; Wang et al., 2019) although not in all studies (Aluko et al., 2017).

In comparison with animal protein sources, plant protein sources have higher levels of DF and might contain anti-nutritional factors (Montagne et al., 2004). Soybean meal – although easily digested and an interesting ingredient in growing pig diets – has been linked to PWD and its use in piglet diets is therefore limited (Friesen et al., 1993). The link between soy and PWD may be an aberrant immune response to soy proteins, leading to hypersensitivity (Engle, 1994). Moreover, the use of soybean products in pig diets is criticized because of its environmental impact. The soybean meal used in Europe is predominantly produced overseas. Especially when soy production leads to deforestation, it may be a threat to tropical biodiversity and may increase the global warming potential, energy, and land use of the pig feed (Fearnside, 2002; van Zanten et al., 2015). Ingredients that have shown potential to replace soybean meal include fermented soybean meal (Kiers et al., 2003; Feng et al., 2007; Xie et al., 2017; Wang et al., 2020), (hydrolyzed) wheat gluten (Blasco et al., 2005; Wang et al., 2011), and potato protein (Pedersen and Lindberg, 2004; de Souza et al., 2019). Recently, Cruz et al. (2019) showed that also yeast, even up to 40% of dietary CP, may be a valuable protein source that positively influences intestinal morphology and fecal consistency. It should be taken into account that distribution of these specialty ingredients has commercial value and research bias can therefore be anticipated (Resnik, 2000). Apart from the major ingredients, a great deal of research has been performed on the development and evaluation of feed additives in pig production. Especially for piglets, the search for alternatives to antibiotics is ongoing. They include pre-, pro-, or synbiotics, organic acids, exogenous enzymes, essential oils, and phytochemicals. The mechanisms associated with the beneficial effects of pre-, pro-, and synbiotics include manipulation of intestinal microbial communities (Guevarra et al., 2019), the neutralization of toxins, restriction of colonization and adhesion of opportunistic pathogens (Barba-Vidal et al., 2018), enhancement of the mucosal barrier (Trevisi et al., 2017), and stimulation of the immune system (Lessard et al., 2009). The mode of action of plant-derived substances, that is essential oils, organic acids, or phytochemicals, is mostly attributed to their antimicrobial and antioxidant properties, with the former stabilizing the intestinal microbiota (Lillehoj et al., 2018). In addition, organic acids may lower both the dietary and the digesta pH and promote intestinal morphology in weaned pigs (Tugnoli et al., 2020). Feed enzymes are also considered to play a positive role in gut health in pigs by improving nutrient digestibility, producing possible prebiotic oligosaccharides, or modulating gut microbiota, as reviewed by Kiarie et al. (2013). Pre-, pro-, and synbiotics in animal nutrition have been reviewed by Markowiak and Slizewska (2018). Although several positive effects of these supplements have been observed, not every study confirms this, as for example shown in a probiotic study that failed to reduce *Salmonella* Typhimurium prevalence (Peeters et al., 2019). Essential oils, organic acids, and phytochemicals were reviewed by Lillehoj et al. (2018), Omonijo et al. (2018), and Tugnoli et al. (2020).

6 Case study: the importance of feed structure for health and performance

Throughout this chapter, we have shown that optimal nutrition may depend on the situation. Under stressful conditions such as the weaning period, diets may be formulated either with more expensive ingredients or below the nutrient requirements for optimal growth performance. The search for the optimal diet form and feed structure may affect both performance and gastrointestinal health, as was observed in trials with finishing pigs at Flanders Research Institute for Agriculture, Fisheries and Food (ILVO) and Ghent University. In our first trial (Millet et al., 2012b), we evaluated the effect of particle size distribution and fiber content on gastric mucosa integrity. To do this, we formulated a high- and low-fiber diet with similar net energy and amino acid levels. The diets were ground with a hammer mill provided with a 1.5 mm or 6 mm screen. This resulted in four treatment groups: low-fiber finely ground, low-fiber coarsely ground, high-fiber finely ground, and high-fiber coarsely ground. Only one diet, the coarsely ground high-fiber diet, was successful in limiting the number of gastric lesions (Table 1). Unfortunately, the feed efficiency, expressed as kilogram of feed per kilogram of carcass gain, was significantly worse on this coarsely ground high-fiber diet compared to the other diets. This trial showed that feed structure, as a combination of physical and chemical characteristics, can affect gastric health. It also showed that a diet for optimal health does not always lead to optimal performance. In a similar trial (Millet et al., 2012a), we compared a mash diet with an expanded diet and a crumbled pelleted diet. Expanding and pelleting the diet led to better feed efficiency. However, these technological treatments had a negative impact on gastric mucosa integrity (worse gastric lesion scores) and were associated with increased *Helicobacter suis* colonization. Thus, in both trials, the diets with the best performance negatively affected gastric health. Both were probably the result of the dietary particle size, as pelleting and crumbling can reduce the particle size of the diet (Wolf et al., 2010; Vukmirović et al., 2017). Smaller particles lead to a higher surface area, allowing better contact with digestive enzymes (Vukmirović et al., 2017). Conversely, they also result in higher fluidity in the stomach, eventually leading to a less distinct pH gradient in the stomach and the formation of gastric ulcers in the esophageal part (Maxwell et al., 1970). Reduced particle size led to similar results with respect to *Salmonella* infections. Pigs fed a non-pelleted diet were less susceptible to *Salmonella enterica* serovar Typhimurium colonization,

Table 1 Frequency of macroscopic lesion scores in the stomach in pigs consuming diets with a low- or a high-fiber content that were finely or coarsely ground (n = 47 per treatment)

Fiber content	Low		High		*P*	
Particle size	Fine	Coarse	Fine	Coarse	Fibre content	Particle size
Lesion score 0	4.3	2.1	0.0	46.8	<0.001	<0.001
Lesion score 1	8.5	12.8	25.5	27.7		
Lesion score 2	55.3	53.2	59.6	19.1		
Lesion score 3	21.3	17.0	6.4	2.1		
Lesion score 4	4.3	8.5	0.0	2.1		
Lesion score 5	6.4	6.4	8.5	2.1		

Note: Only pig receiving the diet with high-fiber level that was coarsely ground showed better scores for the gastric mucosa.
Source: Adapted from: Millet et al. (2012b), with permission from Elsevier.

especially when feed was coarsely ground (Mikkelsen et al., 2004; Hedemann et al., 2005). Similarly, in a study with an experimental *E. coli* infection, coarsely (but not finely) ground wheat bran reduced the incidence of diarrhea in piglets (Molist et al., 2012).

In feed mills, different procedures for milling, processing, and pelleting are used and will continue to be used. The current awareness of the importance of feed structure is high, and further study of optimal diet form for performance and health is necessary.

7 Conclusion

This chapter has addressed several aspects of feeding that may affect pig health and well-being. It is clear that the diet has a large impact on an animal's functioning. The optimal diet on any given farm may depend on the farm health status: with a good health status, nutrient levels for maximal performance may be used. In contrast, farms with poorer sanitary conditions can use feeding management to decrease the risk of disease.

7.1 Key messages

- Piglets may be fed below their amino acid, calcium, and phosphorus requirements to decrease dietary buffering capacity and the risk for PWD.
- FI stimulation can be attained via both feeding management and dietary measures, affecting growth, body condition, and the health status of animals.
- Amino acids have physiological functions besides their role as a building block for protein synthesis.

- Fiber can be used to prevent disease, but more research is needed on the physiological and health-promoting role of different fiber fractions.
- Dietary energy and protein concentration, together with antioxidant vitamins and minerals may affect susceptibility to heat stress.
- Feed structure, comprising both chemical composition and physical characteristics, is important for gastric and intestinal health.
- Nutritionists face trade-offs between diets for optimal health and optimal performance.

8 Future trends

After the ban on growth-promoting antibiotics (Millet and Maertens, 2011), the pressure to diminish the use of prophylactic and therapeutic use of antibiotics is also growing. Hence, the search for feeding strategies to replace antibiotics and increase the resilience of animals, especially shortly after weaning, is ongoing and is expected to increase. We expect that feed and premix companies will increase their investments in research and we anticipate an increase in publications on dietary antibiotic replacers. Publication bias is a potential concern, as industry-funded papers may be an important marketing tool. With antibiotic resistance being one of the greatest challenges for humans in the decades to come, commercially sponsored research must be counterbalanced by adequate public funding for research. Besides strategies improving intestinal health, novel molecules targeting pathogens or toxins will be further developed; thanks to the use of advanced biotechnology, such as engineered peptides, nanoantibiotics, chicken- or plant-derived immunoglobulines, bacteriophages, eubiotics, or the use of new gene-editing technology (CRISPR/Cas9) (Marquardt and Li, 2018).

From the point of view of the environment as well as health, a decrease in dietary CP level and increase in the use of free amino acids can be expected, especially as the number of commercially available feed grade amino acids increases (see above). In the next decade, we expect to see additional research in the requirements and especially the health-promoting functions of specific amino acids.

As stated above, the role of fiber and feed structure in pig health needs further study. The rising environmental concern will likely lead to the increased use of byproducts - often containing higher fiber levels - in pig diets. Analytical methods are being developed to discriminate between the different fiber fractions. Therefore, we expect more research on the physiological and health-promoting effects of different fiber fractions and how to apply this in pig health management. The interactions between these fiber fractions and feed processing deserves also further attention.

The past decades have seen a great deal of work on the nutritional requirements of pigs. Most research is carried out in optimal conditions, while the on-farm health situation may be more variable. It can be anticipated that in the future, more emphasis will be placed on the search for precision feeding depending on the animal's characteristics, including its health status. In lactating sows and in gestating sows fed with electronic sow feeders, this is already partially done by adapting the FI to the energy requirements. Through the use of data and decision support tools, dietary strategies may be adapted to the animals' energy and amino acid needs to maximize both health and performance. The first step, albeit difficult, is to adapt the diet to the group of animals on the farm that shares similar conditions. The second step is to adapt to the individual animals' needs. To cope with the specific challenges, will require both accurate knowledge of the health status of the farm and knowledge about dietary characteristics. Tools will be needed to accurately determine the health status and the genetic potential of the animals on the farm and to translate this to farm-specific dietary advice in combination with adapted management practices. Scientific knowledge will be key in this development, although good feeding and feeding advice may go beyond the knowledge documented in scientific papers. It may require a transdisciplinary approach involving animal scientists, veterinarians, data scientists, farmers, and farm advisors. Developments in artificial intelligence may speed up this process.

9 References

Agyekum, A. K. and Nyachoti, C. M. 2017. Nutritional and metabolic consequences of feeding high-fiber diets to swine: A review. *Engineering* 3(5):716–725. doi: 10.1016/J.ENG.2017.03.010.

Aluko, K., Velayudhan, D. E., Khafipour, E., Fang, L. and Nyachoti, M. 2017. Effect of chicken egg anti-F4 antibodies on performance and diarrhea incidences in enterotoxigenic Escherichia coli K88(+)-challenged piglets. *Anim. Nutr.* 3(4):353–358. doi: 10.1016/j.aninu.2017.06.006.

Bach Knudsen, K. E. 2001. The nutritional significance of "dietary fibre" analysis. *Anim. Feed Sci. Technol.* 90(1–2):3–20. doi: 10.1016/S0377-8401(01)00193-6.

Bai, M. M., Liu, H. N., Xu, K., Wen, C. Y., Yu, R., Deng, J. P. and Yin, Y. L. 2019. Use of coated Nano zinc oxide as an additive to improve the zinc excretion and intestinal morphology of growing pigs1. *J. Anim. Sci.* 97(4):1772–1783. doi: 10.1093/jas/skz065.

Barba-Vidal, E., Martin-Orue, S. M. and Castillejos, L. 2018. Review: are we using probiotics correctly in post-weaning piglets? *Animal* 12(12):2489–2498. doi: 10.1017/S1751731118000873.

Bednorz, C., Oelgeschlager, K., Kinnemann, B., Hartmann, S., Neumann, K., Pieper, R., Bethe, A., Semmler, T., Tedin, K., Schierack, P., Wieler, L. H. and Guenther, S. 2013. The broader context of antibiotic resistance: Zinc feed supplementation of piglets increases the proportion of multi-resistant Escherichia coli in vivo. *Int. J. Med. Microbiol.* 303(6–7):396–403. doi: 10.1016/j.ijmm.2013.06.004.

Bergsma, R., Kanis, E., Verstegen, M. W. A., van der Peet-Schwering, C. M. C. and Knol, E. F. 2009. Lactation efficiency as a result of body composition dynamics and feed intake in sows. *Livest. Sci.* 125(2–3):208–222. doi: 10.1016/j.livsci.2009.04.011.

Bikker, P. and Blok, M. C. 2017. *Phophorus and Calcium Requirements of Growing Pigs and Sows.*

Bikker, P. and Jongbloed, A. W. 2013. *Koper-en zinknormen voor varkens.* 1570-8616, Wageningen UR Livestock Research.

Blasco, M., Fondevila, M. and Guada, J. A. 2005. Inclusion of wheat gluten as a protein source in diets for weaned pigs. *Anim. Res.* 54(4):297–306. doi: 10.1051/animres:2005026.

Bonde, M., Rousing, T., Badsberg, J. H. and Sørensen, J. T. 2004. Associations between lying-down behaviour problems and body condition, limb disorders and skin lesions of lactating sows housed in farrowing crates in commercial sow herds. *Livest. Prod. Sci.* 87(2–3):179–187. doi: 10.1016/j.livprodsci.2003.08.005.

Bonetti, A., Tugnoli, B., Piva, A. and Grilli, E. 2021. Towards zero zinc oxide: feedingstrategies to manage post-weaning diarrhea in piglets. *Animals (Basel)* 11(3):642.

Bruininx, E. M., Binnendijk, G. P., van der Peet-Schwering, C. M., Schrama, J. W., den Hartog, L. A., Everts, H. and Beynen, A. C. 2002. Effect of creep feed consumption on individual feed intake characteristics and performance of group-housed weanling pigs. *J. Anim. Sci.* 80(6):1413–1418. doi: 10.2527/2002.8061413x.

Cabrera, R. A., Usry, J. L., Arrellano, C., Nogueira, E. T., Kutschenko, M., Moeser, A. J. and Odle, J. 2013. Effects of creep feeding and supplemental glutamine or glutamine plus glutamate (Aminogut) on pre- and post-weaning growth performance and intestinal health of piglets. *J. Anim. Sci. Biotechnol.* 4(1):29. doi: 10.1186/2049-1891-4-29.

Canh, T. T., Verstegen, M. W., Aarnink, A. J. and Schrama, J. W. 1997. Influence of dietary factors on nitrogen partitioning and composition of urine and feces of fattening pigs. *J. Anim. Sci.* 75(3):700–706. doi: 10.2527/1997.753700x.

Charette, R., Bigras-Poulin, M. and Martineau, G.-P. 1996. Body condition evaluation in sows. *Livest. Prod. Sci.* 46(2):107–115. doi: 10.1016/0301-6226(96)00022-X.

Chen, H., Mao, X., He, J., Yu, B., Huang, Z., Yu, J., Zheng, P. and Chen, D. 2013. Dietary fibre affects intestinal mucosal barrier function and regulates intestinal bacteria in weaning piglets. *Br. J. Nutr.* 110(10):1837–1848. doi: 10.1017/S0007114513001293.

Correa-Oliveira, R., Fachi, J. L., Vieira, A., Sato, F. T. and Vinolo, M. A. 2016. Regulation of immune cell function by short-chain fatty acids. *Clin. Transl. Immunol.* 5(4):e73. doi: 10.1038/cti.2016.17.

Cota, D., Proulx, K., Smith, K. A., Kozma, S. C., Thomas, G., Woods, S. C. and Seeley, R. J. 2006. Hypothalamic mTOR signaling regulates food intake. *Science* 312(5775):927–930. doi: 10.1126/science.1124147.

Cottrell, J. J., Liu, F., Hung, A. T., DiGiacomo, K., Chauhan, S. S., Leury, B. J., Furness, J. B., Celi, P. and Dunshea, F. R. 2015. Nutritional strategies to alleviate heat stress in pigs. *Anim. Prod. Sci.* 55(12):1391–1402.

Cranwell, P. D. 1985. The development of acid and pepsin (EC 3.4.23.1) secretory capacity in the pig; the effects of age and weaning. 1. Studies in anaesthetized pigs. *Br. J. Nutr.* 54(1):305–320. doi: 10.1079/bjn19850113.

Cruz, A., Håkenåsen, I. M., Skugor, A., Mydland, L. T., Åkesson, C. P., Hellestveit, S. S., Sørby, R., Press, C. M. and Øverland, M. 2019. Candida utilis yeast as a protein source for weaned piglets: effects on growth performance and digestive function. *Livest. Sci.* 226:31–39. doi: 10.1016/j.livsci.2019.06.003.

CVB. 2020. *Tabellenboek Voeding Varkens 2020*.

Danielsen, V. and Vestergaard, E.-M. 2001. Dietary fibre for pregnant sows: Effect on performance and behaviour. *Anim. Feed Sci.Technol.* 90(1-2):71–80. doi: 10.1016/S0377-8401(01)00197-3.

de Leeuw, J. A., Bolhuis, J. E., Bosch, G. and Gerrits, W. J. 2008. Effects of dietary fibre on behaviour and satiety in pigs. *Proc. Nutr. Soc.* 67(4):334–342. doi: 10.1017/S002966510800863X.

de Souza, T. C. R., Barreyro, A. A., Rubio, S. R., González, Y. M., García, K. E., Soto, J. G. G. and Mariscal-Landín, G. 2019. Growth performance, diarrhoea incidence, and nutrient digestibility in weaned piglets fed an antibiotic-free diet with dehydrated porcine plasma or potato protein concentrate. *Ann. Anim. Sci.* 19(1):159–172.

Debski, B. 2016. Supplementation of pigs diet with zinc and copper as alternative to conventional antimicrobials. *Pol. J. Vet. Sci.* 19(4):917–924. doi: 10.1515/pjvs-2016-0113.

Decaluwe, R., Maes, D., Cools, A., Wuyts, B., De Smet, S., Marescau, B., De Deyn, P. P. and Janssens, G. P. 2014. Effect of peripartal feeding strategy on colostrum yield and composition in sows. *J. Anim. Sci.* 92(8):3557–3567. doi: 10.2527/jas.2014-7612.

Decaluwe, R., Maes, D., Declerck, I., Cools, A., Wuyts, B., De Smet, S. and Janssens, G. P. 2013. Changes in back fat thickness during late gestation predict colostrum yield in sows. *Animal* 7(12):1999–2007. doi: 10.1017/S1751731113001791.

Dourmad, J. Y., Etienne, M., Prunier, A. and Noblet, J. 1994. The effect of energy and protein intake of sows on their longevity: A review. *Livest. Prod. Sci.* 40(2):87–97. doi: 10.1016/0301-6226(94)90039-6.

Eissen, J. J., Kanis, E. and Kemp, B. 2000. Sow factors affecting voluntary feed intake during lactation. *Livest. Prod. Sci.* 64(2–3):147–165. doi: 10.1016/S0301-6226(99)00153-0.

Engle, M. J. 1994. The role of soybean meal hypersensitivity in postweaning lag and diarrhea in piglets. *J. Swine Health Prod.* 2:7–10.

European Commission. 2018. Commission implementing regulation (EU) 2018/1039 of 23 July 2018 concerning the authorisation of copper(II) diacetate monohydrate, copper(II) carbonate dihydroxy monohydrate, copper(II) chloride dihydrate, copper(II) oxide, copper(II) sulphate pentahydrate, copper(II) chelate of amino acids hydrate, copper(II) chelate of protein hydrolysates, copper(II) chelate of glycine hydrate (solid) and copper(II) chelate of glycine hydrate (liquid) as feed additives for all animal species and amending Regulations (EC) no 1334/2003, (EC). *Off. J. Eur. Union*. European Union No 1230/2014 and (EU) 2016/2261479/2006 and (EU) No 349/2010 and Implementing Regulations (EU) No 269/2012:L186/183-L186/124.

Fearnside, P. M. 2002. Soybean cultivation as a threat to the environment in Brazil. *Environ. Conserv.* 28(1):23–38. doi: 10.1017/S0376892901000030.

Feng, J., Liu, X., Xu, Z. R., Lu, Y. P. and Liu, Y. Y. 2007. Effect of fermented soybean meal on intestinal morphology and digestive enzyme activities in weaned piglets. *Dig. Dis. Sci.* 52(8):1845–1850. doi: 10.1007/s10620-006-9705-0.

Friesen, K. G., Goodband, R. D., Nelssen, J. L., Blecha, F., Reddy, D. N., Reddy, P. G. and Kats, L. J. 1993. The effect of pre- and postweaning exposure to soybean meal on growth performance and on the immune response in the early-weaned pig. *J. Anim. Sci.* 71(8):2089–2098. doi: 10.2527/1993.7182089x.

German Society for Nutrition Physiology. 2006. *Empfehlungen zur Energie-und Nährstoffversorgung von Schweinen*.

Gloaguen, M., Le Floc, N., Brossard, L., Barea, R., Primot, Y., Corrent, E. and van Milgen, J. 2011. Response of piglets to the valine content in diet in combination with the supply of other branched-chain amino acids. *Animal* 5(11):1734–1742. doi: 10.1017/S1751731111000760.

Gloaguen, M., Le Floc, N., Corrent, E., Primot, Y. and van Milgen, J. 2014. The use of free amino acids allows formulating very low crude protein diets for piglets. *J. Anim. Sci.* 92(2):637–644. doi: 10.2527/jas.2013-6514.

Goodband, B., Tokach, M., Dritz, S., Derouchey, J. and Woodworth, J. 2014. Practical starter pig amino acid requirements in relation to immunity, gut health and growth performance. *J. Anim. Sci. Biotechnol.* 5(1):12. doi: 10.1186/2049-1891-5-12.

Goodband, R. D., DeRouchey, J. M., Dritz, S. S., Woodworth, J. C., Tokach, M. D., and Cemin, H. S. 2019. Branched-chain amino acid interactions in growing pig diets1. *Translational. Animal. Science* 3(4):1246--1253. doi: 10.1093/tas/txz087.

Goodband, R. D., Tokach, M. D., Goncalves, M. A. D., Woodworth, J. C., Dritz, S. S. and DeRouchey, J. M. 2013. Nutritional enhancement during pregnancy and its effects on reproduction in swine. *Anim. Front.* 3(4):68–75. doi: 10.2527/af.2013-0036.

Guevarra, R. B., Lee, J. H., Lee, S. H., Seok, M. J., Kim, D. W., Kang, B. N., Johnson, T. J., Isaacson, R. E. and Kim, H. B. 2019. Piglet gut microbial shifts early in life: Causes and effects. *J. Anim. Sci. Biotechnol.* 10:1. doi: 10.1186/s40104-018-0308-3.

Halas, D., Hansen, C. F., Hampson, D. J., Mullan, B. P., Wilson, R. H. and Pluske, J. R. 2009. Effect of dietary supplementation with inulin and/or benzoic acid on the incidence and severity of post-weaning diarrhoea in weaner pigs after experimental challenge with enterotoxigenic Escherichia coli. *Arch. Anim. Nutr.* 63(4):267–280. doi: 10.1080/17450390903020414.

Hamard, A., Mazurais, D., Boudry, G., Le Huerou-Luron, I., Seve, B. and Le Floc, N. 2010. A moderate threonine deficiency affects gene expression profile, paracellular permeability and glucose absorption capacity in the ileum of piglets. *J. Nutr. Biochem.* 21(10):914-921. doi: 10.1016/j.jnutbio.2009.07.004.

Hansen, C. F., Riis, A. L., Bresson, S., Højbjerg, O. and Jensen, B. B. 2007. Feeding organic acids enhances the barrier function against pathogenic bacteria of the piglet stomach. *Livest. Sci.* 108(1–3):206–209. doi: 10.1016/j.livsci.2007.01.059.

He, B., Bai, Y., Jiang, L., Wang, W., Li, T., Liu, P., Tao, S., Zhao, J., Han, D. and Wang, J. 2018. Effects of oat bran on nutrient digestibility, intestinal microbiota, and inflammatory responses in the hindgut of growing pigs. *Int. J. Mol. Sci.* 19(8):2407.

He, Y., Deen, J., Shurson, G. C., Wang, L., Chen, C., Keisler, D. H. and Li, Y. Z. 2016. Identifying factors contributing to slow growth in pigs. *J. Anim. Sci.* 94(5):2103–2116. doi: 10.2527/jas.2015-0005.

Hedemann, M. S., Eskildsen, M., Laerke, H. N., Pedersen, C., Lindberg, J. E., Laurinen, P. and Knudsen, K. E. 2006. Intestinal morphology and enzymatic activity in newly weaned pigs fed contrasting fiber concentrations and fiber properties. *J. Anim. Sci.* 84(6):1375–1386. doi: 10.2527/2006.8461375x.

Hedemann, M. S., Mikkelsen, L. L., Naughton, P. J. and Jensen, B. B. 2005. Effect of feed particle size and feed processing on morphological characteristics in the small and large intestine of pigs and on adhesion of *Salmonella enterica* serovar

Typhimurium DT12 in the ileum in vitro. *J. Anim. Sci.* 83(7):1554–1562. doi: 10.2527/2005.8371554x.

Henry, Y. 1985. Dietary factors involved in feed intake regulation in growing pigs: A review. *Livest. Prod. Sci.* 12(4):339–354. doi: 10.1016/0301-6226(85)90133-2.

Heo, J. M., Kim, J. C., Hansen, C. F., Mullan, B. P., Hampson, D. J. and Pluske, J. R. 2009. Feeding a diet with decreased protein content reduces indices of protein fermentation and the incidence of postweaning diarrhea in weaned pigs challenged with an enterotoxigenic strain of Escherichia coli. *J. Anim. Sci.* 87(9):2833–2843. doi: 10.2527/jas.2008-1274.

Holen, J. P., Rambo, Z., Hilbrands, A. M. and Johnston, L. J. 2018. Effects of dietary zinc source and concentration on performance of growing-finishing pigs reared with reduced floor space. *Prof. Anim. Sci.* 34(2):133–143.

Hollis, G. R., Carter, S. D., Cline, T. R., Crenshaw, T. D., Cromwell, G. L., Hill, G. M., Kim, S. W., Lewis, A. J., Mahan, D. C., Miller, P. S., Stein, H. H., Veum, T. L. and NCRCoS. 2005. Effects of replacing pharmacological levels of dietary zinc oxide with lower dietary levels of various organic zinc sources for weanling pigs. *J. Anim. Sci.* 83(9):2123–2129. doi: 10.2527/2005.8392123x.

Hopwood, D. E., Pethick, D. W., Pluske, J. R. and Hampson, D. J. 2004. Addition of pearl barley to a rice-based diet for newly weaned piglets increases the viscosity of the intestinal contents, reduces starch digestibility and exacerbates post-weaning colibacillosis. *Br. J. Nutr.* 92(3):419–427. doi: 10.1079/bjn20041206.

Htoo, J. K., Araiza, B. A., Sauer, W. C., Rademacher, M., Zhang, Y., Cervantes, M. and Zijlstra, R. T. 2007. Effect of dietary protein content on ileal amino acid digestibility, growth performance, and formation of microbial metabolites in ileal and cecal digesta of early-weaned pigs. *J. Anim. Sci.* 85(12):3303–3312. doi: 10.2527/jas.2007-0105.

Hu, C., Song, J., Li, Y., Luan, Z. and Zhu, K. 2013. Diosmectite-zinc oxide composite improves intestinal barrier function, modulates expression of pro-inflammatory cytokines and tight junction protein in early weaned pigs. *Br. J. Nutr.* 110(4):681–688. doi: 10.1017/S0007114512005508.

Isabel, B., Rey, A. I. and López Bote, C. 2012. *Optimum Vitamin Nutrition in Pigs, Optimum Vitamin Nutrition in the Production of Quality Animal Foods. 5M Publishing*. Sheffield, pp. 243–308.

Jansman, A., van Diepen, J., Rovers, M. and Corrent, E. 2016. *Lowering the Dietary Protein Content in Piglets: How Far Can We Go?, Energy and Protein Metabolism and Nutrition*, pp. 161–162.

Jayaraman, B., Htoo, J. and Nyachoti, C. M. 2015. Effects of dietary threonine:lysine ratioes and sanitary conditions on performance, plasma urea nitrogen, plasma-free threonine and lysine of weaned pigs. *Anim. Nutr.* 1(4):283–288. doi: 10.1016/j.aninu.2015.09.003.

Jha, R. and Berrocoso, J. F. D. 2016. Dietary fiber and protein fermentation in the intestine of swine and their interactive effects on gut health and on the environment: A review. *Anim. Feed Sci. Technol.* 212:18–26. doi: 10.1016/j.anifeedsci.2015.12.002.

Jha, R., Fouhse, J. M., Tiwari, U. P., Li, L. and Willing, B. P. 2019. Dietary fiber and intestinal health of monogastric animals. *Front. Vet. Sci.* 6:48. doi: 10.3389/fvets.2019.00048.

Ji, F. J., Wang, L. X., Yang, H. S., Hu, A. and Yin, Y. L. 2019. Review: the roles and functions of glutamine on intestinal health and performance of weaning pigs. *Animal* 13(11):2727–2735. doi: 10.1017/S1751731119001800.

Johnson, J. S. and Lay, D. C. 2017. Evaluating the behavior, growth performance, immune parameters, and intestinal morphology of weaned piglets after simulated transport and heat stress when antibiotics are eliminated from the diet or replaced with L-glutamine. *J. Anim. Sci.* 95(1):91–102.

Jondreville, C., Revy, P. S. and Dourmad, J. Y. 2003. Dietary means to better control the environmental impact of copper and zinc by pigs from weaning to slaughter. *Livest. Prod. Sci.* 84(2):147–156. doi: 10.1016/j.livprodsci.2003.09.011.

Jongbloed, A., van Diepen, J. T. M., Kemme, P. and Veevoederbureau, C. 2003. *Fosfornormen voor varkens: Herziening 2003. Centraal Veevoederbureau.*

Kiarie, E., Romero, L. F. and Nyachoti, C. M. 2013. The role of added feed enzymes in promoting gut health in swine and poultry. *Nutr. Res. Rev.* 26(1):71–88. doi: 10.1017/S0954422413000048.

Kiers, J. L., Meijer, J. C., Nout, M. J., Rombouts, F. M., Nabuurs, M. J. and van der Meulen, J. 2003. Effect of fermented soya beans on diarrhoea and feed efficiency in weaned piglets. *J. Appl. Microbiol.* 95(3):545–552. doi: 10.1046/j.1365-2672.2003.02011.x.

KilBride, A. L., Gillman, C. E. and Green, L. E. 2009. A cross sectional study of the prevalence, risk factors and population attributable fractions for limb and body lesions in lactating sows on commercial farms in England. *BMC Vet. Res.* 5:30. doi: 10.1186/1746-6148-5-30.

Kloubert, V., Blaabjerg, K., Dalgaard, T. S., Poulsen, H. D., Rink, L. and Wessels, I. 2018. Influence of zinc supplementation on immune parameters in weaned pigs. *J. Trace Elem. Med. Biol.* 49:231–240. doi: 10.1016/j.jtemb.2018.01.006.

Konowalchuk, J. D., Rieger, A. M., Kiemele, M. D., Ayres, D. C. and Barreda, D. R. 2013. Modulation of weanling pig cellular immunity in response to diet supplementation with 25-hydroxyvitamin D(3). *Vet. Immunol. Immunopathol.* 155(1–2):57–66. doi: 10.1016/j.vetimm.2013.06.002.

Lacey, J. M. and Wilmore, D. W. 1990. Is glutamine a conditionally essential amino acid? *Nutr. Rev.* 48(8):297–309. doi: 10.1111/j.1753-4887.1990.tb02967.x.

Lallès, J.-P. and Montoya, C. A. 2021. Dietary alternatives to in-feed antibiotics, gut barrier function and inflammation in piglets post-weaning: Where are we now? *Anim. Feed Sci. Technol.* 274:274. doi: 10.1016/j.anifeedsci.2021.114836.

Lauridsen, C. 2019. From oxidative stress to inflammation: Redox balance and immune system. *Poult. Sci.* 98(10):4240–4246. doi: 10.3382/ps/pey407.

Lauridsen, C., Matte, J. J., Lessard, M., Celi, P. and Litta, G. 2021. Role of vitamins for gastro-intestinal functionality and health of pigs. *Anim. Feed Sci. Technol.* 273:273. doi: 10.1016/j.anifeedsci.2021.114823.

Lawlor, P. G., Lynch, P. B., Caffrey, P. J., O'Reilly, J. J. and O'Connell, M. K. 2005. Measurements of the acid-binding capacity of ingredients used in pig diets. *Ir. Vet. J.* 58(8):447–452. doi: 10.1186/2046-0481-58-8-447.

Le Floc'h, N., Wessels, A., Corrent, E., Wu, G. and Bosi, P. 2018. The relevance of functional amino acids to support the health of growing pigs. *Anim. Feed Sci. Technol.* 245:104–116. doi: 10.1016/j.anifeedsci.2018.09.007.

Leal, D. F., Muro, B. B. D., Nichi, M., Almond, G. W., Viana, C. H. C., Vioti, G., Carnevale, R. F. and Garbossa, C. A. P. 2019. Effects of post-insemination energy content of feed on embryonic survival in pigs: A systematic review. *Anim. Reprod. Sci.* 205:70–77. doi: 10.1016/j.anireprosci.2019.04.005.

Lessard, M., Dupuis, M., Gagnon, N., Nadeau, E., Matte, J. J., Goulet, J. and Fairbrother, J. M. 2009. Administration of Pediococcus acidilactici or Saccharomyces cerevisiae boulardii modulates development of porcine mucosal immunity and reduces intestinal bacterial translocation after Escherichia coli challenge. *J. Anim. Sci.* 87(3):922–934. doi: 10.2527/jas.2008-0919.

Li, P., Yin, Y. L., Li, D., Kim, S. W. and Wu, G. 2007. Amino acids and immune function. *Br. J. Nutr.* 98(2):237–252. doi: 10.1017/S000711450769936X.

Liedtke, J. and Vahjen, W. 2012. In vitro antibacterial activity of zinc oxide on a broad range of reference strains of intestinal origin. *Vet. Microbiol.* 160(1–2):251–255. doi: 10.1016/j.vetmic.2012.05.013.

Lillehoj, H., Liu, Y., Calsamiglia, S., Fernandez-Miyakawa, M. E., Chi, F., Cravens, R. L., Oh, S. and Gay, C. G. 2018. Phytochemicals as antibiotic alternatives to promote growth and enhance host health. *Vet. Res.* 49(1):76. doi: 10.1186/s13567-018-0562-6.

Lisgara, M, Skampardonis, V. and Leontides, L. 2016. Effect of diet supplementation with chelated zinc, copper and manganese on hoof lesions of loose housed sows. *Porcine Health Manag.* 2:6. doi: 10.1186/s40813-016-0025-2.

Liu, F., Cottrell, J. J., Furness, J. B., Rivera, L. R., Kelly, F. W., Wijesiriwardana, U., Pustovit, R. V., Fothergill, L. J., Bravo, D. M., Celi, P., Leury, B. J., Gabler, N. K. and Dunshea, F. R. 2016. Selenium and vitamin E together improve intestinal epithelial barrier function and alleviate oxidative stress in heat-stressed pigs. *Exp. Physiol.* 101(7):801–810. doi: 10.1113/EP085746.

Liu, Y., Huang, J., Hou, Y., Zhu, H., Zhao, S., Ding, B., Yin, Y., Yi, G., Shi, J. and Fan, W. 2008. Dietary arginine supplementation alleviates intestinal mucosal disruption induced by Escherichia coli lipopolysaccharide in weaned pigs. *Br. J. Nutr.* 100(3):552–560. doi: 10.1017/S0007114508911612.

Liu, Y., Wang, X., Hou, Y., Yin, Y., Qiu, Y., Wu, G. and Hu, C. A. 2017. Roles of amino acids in preventing and treating intestinal diseases: Recent studies with pig models. *Amino Acids* 49(8):1277–1291. doi: 10.1007/s00726-017-2450-1.

Lorenzett, M. P., Cruz, R. A. S., Cecco, B. S., Schwertz, C. I., Hammerschmitt, M. E., Schu, D. T., Driemeier, D. and Pavarini, S. P. 2019. Obstructive urolithiasis in growing-finishing pigs. *Pesq. Vet. Bras.* 39(6):382–387. doi: 10.1590/1678-5150-pvb-6229.

Lupton, J. R., Betteridge, V. A. and Pijls, L. T. J. 2009. Codex final definition of dietary fibre: Issues of implementation. *Qual. Assur. Saf. Crops Foods* 1(4):206–212. doi: 10.1111/j.1757-837X.2009.00043.x.

Maes, D. G., Vrielinck, J., Millet, S., Janssens, G. P. and Deprez, P. 2004a. Urolithiasis in finishing pigs. *Vet. J.* 168(3):317–322. doi: 10.1016/j.tvjl.2003.09.006.

Maes, D. G. D., Janssens, G. P. J., Delputte, P., Lammertyn, A. and de Kruif, A. 2004b. Back fat measurements in sows from three commercial pig herds: Relationship with reproductive efficiency and correlation with visual body condition scores. *Livest. Prod. Sci.* 91(1–2):57–67. doi: 10.1016/j.livprodsci.2004.06.015.

Mahan, D. C. and Shields, R. G., Jr. 1998. Essential and nonessential amino acid composition of pigs from birth to 145 kilograms of body weight, and comparison to other studies. *J. Anim. Sci.* 76(2):513–521. doi: 10.2527/1998.762513x.

Mallmann, A. L., Oliveira, G. S., Ulguim, R. R., Mellagi, A. P. G., Bernardi, M. L., Orlando, U. A. D., Goncalves, M. A. D., Cogo, R. J. and Bortolozzo, F. P. 2020. Impact of feed intake in early gestation on maternal growth and litter size according to body reserves at weaning of young parity sows. *J. Anim. Sci.* 98(3).

Markowiak, P. and Slizewska, K. 2018. The role of probiotics, prebiotics and Synbiotics in animal nutrition. *Gut Pathog.* 10:21. doi: 10.1186/s13099-018-0250-0.

Marquardt, R. R., Jin, L. Z., Kim, J. W., Fang, L., Frohlich, A. A. and Baidoo, S. K. 1999. Passive protective effect of egg-yolk antibodies against enterotoxigenic Escherichia coli K88+ infection in neonatal and early-weaned piglets. *FEMS Immunol. Med. Microbiol.* 23(4):283–288. doi: 10.1111/j.1574-695X.1999.tb01249.x.

Marquardt, R. R. and Li, S. 2018. Antimicrobial resistance in livestock: Advances and alternatives to antibiotics. *Anim. Front.* 8(2):30–37. doi: 10.1093/af/vfy001.

Mavromatis, J., Koptopoulos, G., Kyriakis, S. C., Papasteriadis, A. and Saoulidis, K. 1999. Effects of alpha-tocopherol and selenium on pregnant sows and their piglets' immunity and performance. *Zentralbl. Veterinarmed. A* 46(9):545–553. doi: 10.1046/j.1439-0442.1999.00244.x.

Maxwell, C. V., Reimann, E. M., Hoekstra, W. G., Kowalczyk, T., Benevenga, N. J. and Grummer, R. H. 1970. Effect of dietary particle size on lesion development and on the contents of various regions of the swine stomach. *J. Anim. Sci.* 30(6):911–922. doi: 10.2527/jas1970.306911x.

McDonald, D. E., Pethick, D. W., Pluske, J. R. and Hampson, D. J. 1999. Adverse effects of soluble non-starch polysaccharide (guar gum) on piglet growth and experimental colibacillosis immediately after weaning. *Res. Vet. Sci.* 67(3):245–250. doi: 10.1053/rvsc.1999.0315.

Metzler-Zebeli, B. U., Canibe, N., Montagne, L., Freire, J., Bosi, P., Prates, J. A. M., Tanghe, S. and Trevisi, P. 2019. Resistant starch reduces large intestinal pH and promotes fecal lactobacilli and bifidobacteria in pigs. *Animal* 13(1):64–73. doi: 10.1017/S1751731118001003.

Mikkelsen, L. L., Naughton, P. J., Hedemann, M. S. and Jensen, B. B. 2004. Effects of physical properties of feed on microbial ecology and survival of *Salmonella enterica* serovar Typhimurium in the pig gastrointestinal tract. *Appl. Environ. Microbiol.* 70(6):3485–3492. doi: 10.1128/AEM.70.6.3485-3492.2004.

Milani, N. C., Sbardella, M., Ikeda, N. Y., Arno, A., Mascarenhas, B. C. and Miyada, V. S. 2017. Dietary zinc oxide nanoparticles as growth promoter for weanling pigs. *Anim. Feed Sci. Technol.* 227:13–23. doi: 10.1016/j.anifeedsci.2017.03.001.

Millet, S., Aluwe, M., Ampe, B. and De Campeneere, S. 2015. Interaction between amino acids on the performances of individually housed piglets. *J. Anim. Physiol. Anim. Nutr. (Berl)* 99(2):230–236. doi: 10.1111/jpn.12227.

Millet, S., Aluwe, M., De Boever, J., De Witte, B., Douidah, L., Van den Broeke, A., Leen, F., De Cuyper, C., Ampe, B. and De Campeneere, S. 2018. The effect of crude protein reduction on performance and nitrogen metabolism in piglets (four to nine weeks of age) fed two dietary lysine levels1. *J. Anim. Sci.* 96(9):3824–3836. doi: 10.1093/jas/sky254.

Millet, S., Kumar, S., De Boever, J., Ducatelle, R. and De Brabander, D. 2012a. Effect of feed processing on growth performance and gastric mucosa integrity in pigs from weaning until slaughter. *Anim. Feed Sci. Technol.* 175(3–4):175–181. doi: 10.1016/j.anifeedsci.2012.05.010.

Millet, S., Kumar, S., De Boever, J., Meyns, T., Aluwe, M., De Brabander, D. and Ducatelle, R. 2012b. Effect of particle size distribution and dietary crude fibre content on growth performance and gastric mucosa integrity of growing-finishing pigs. *Vet. J.* 192(3):316–321. doi: 10.1016/j.tvjl.2011.06.037.

Millet, S. and Maertens, L. 2011. The European ban on antibiotic growth promoters in animal feed: From challenges to opportunities. *Vet. J.* 187(2):143–144. doi: 10.1016/j.tvjl.2010.05.001.

Millet, S., Chalvon-Demersay, T., Schroyen, M., De Cuyper, C., Aluwé, M., Ampe, B., Lambert, W., Bannai, M., Yasuhiro, O. and Simongiovanni, A. 2020a. La performance des porcelets au sevrage répond à une supplémentation en glutamine, en particulier avec un régime faible en protéines. *J. Recherche porcine* 52:167–168.

Millet, S., Minussi, I., Lambert, W., Aluwé, M., Ampe, B., De Sutter, J. and De Campeneere, S. 2020b. Standardized ileal digestible lysine and valine-to-lysine requirements for optimal performance of 4 to 9-week-old Piétrain cross piglets. *Livest. Sci.* 241:241. doi: 10.1016/j.livsci.2020.104263.

Misiura, M. M., Filipe, J. A. N., Walk, C. L. and Kyriazakis, I. 2020. How do pigs deal with dietary phosphorus deficiency? *Br. J. Nutr.* 124(3):256–272. doi: 10.1017/S0007114520000975.

Molist, F., Manzanilla, E. G., Perez, J. F. and Nyachoti, C. M. 2012. Coarse, but not finely ground, dietary fibre increases intestinal Firmicutes:Bacteroidetes ratio and reduces diarrhoea induced by experimental infection in piglets. *Br. J. Nutr.* 108(1):9–15. doi: 10.1017/S0007114511005216.

Montagne, L., Cavaney, F. S., Hampson, D. J., Lalles, J. P. and Pluske, J. R. 2004. Effect of diet composition on postweaning colibacillosis in piglets. *J. Anim. Sci.* 82(8):2364–2374. doi: 10.2527/2004.8282364x.

Navarro, D. M. D. L., Abelilla, J. J. and Stein, H. H. 2019. Structures and characteristics of carbohydrates in diets fed to pigs: A review. *J. Anim. Sci. Biotechnol.* 10:39. doi: 10.1186/s40104-019-0345-6.

Noblet, J., Fortune, H., Shi, X. S. and Dubois, S. 1994. Prediction of net energy value of feeds for growing pigs. *J. Anim. Sci.* 72(2):344–354. doi: 10.2527/1994.722344x.

NRC. 2012. *National Research Council. Nutrient Requirements of Swine* (11th Rev. Edn.). Washington, DC: National Academy Press.

Nyachoti, C. M., Omogbenigun, F. O., Rademacher, M. and Blank, G. 2006. Performance responses and indicators of gastrointestinal health in early-weaned pigs fed low-protein amino acid-supplemented diets. *J. Anim. Sci.* 84(1):125–134. doi: 10.2527/2006.841125x.

Oliviero, C., Heinonen, M., Valros, A. and Peltoniemi, O. 2010. Environmental and sow-related factors affecting the duration of farrowing. *Anim. Reprod. Sci.* 119(1–2):85–91. doi: 10.1016/j.anireprosci.2009.12.009.

Omonijo, F. A., Ni, L., Gong, J., Wang, Q., Lahaye, L. and Yang, C. 2018. Essential oils as alternatives to antibiotics in swine production. *Anim. Nutr.* 4(2):126–136. doi: 10.1016/j.aninu.2017.09.001.

Oostindjer, M., Bolhuis, J. E., van den Brand, H., Roura, E. and Kemp, B. 2010. Prenatal flavor exposure affects growth, health and behavior of newly weaned piglets. *Physiol. Behav.* 99(5):579–586. doi: 10.1016/j.physbeh.2010.01.031.

Pearce, S. C., Sanz Fernandez, M. V., Torrison, J., Wilson, M. E., Baumgard, L. H. and Gabler, N. K. 2015. Dietary organic zinc attenuates heat stress-induced changes in pig intestinal integrity and metabolism. *J. Anim. Sci.* 93(10):4702–4713. doi: 10.2527/jas.2015-9018.

Pedersen, C. and Lindberg, J. E. 2004. Comparison of low-glycoalkaloid potato protein and fish meal as protein sources for weaner piglets. *Acta Agric. Scand. A* 54(2):75–80.

Peeters, L., Mostin, L., Wattiau, P., Boyen, F., Dewulf, J. and Maes, D. 2019. Efficacy of Clostridium butyricum as probiotic feed additive against experimental Salmonella Typhimurium infection in pigs. *Livest. Sci.* 221:82–85. doi: 10.1016/j.livsci.2018.12.019.

Pei, X., Xiao, Z., Liu, L., Wang, G., Tao, W., Wang, M., Zou, J. and Leng, D. 2019. Effects of dietary zinc oxide nanoparticles supplementation on growth performance, zinc status, intestinal morphology, microflora population, and immune response in weaned pigs. *J. Sci. Food Agric.* 99(3):1366–1374. doi: 10.1002/jsfa.9312.

Peltoniemi, O. A. T., Oliviero, C., Hälli, O. and Heinonen, M. 2007. Feeding affects reproductive performance and reproductive endocrinology in the gilt and sow. *Acta Vet. Scand.* 49(Suppl 1).

Perez-Bosque, A., Polo, J. and Torrallardona, D. 2016. Spray dried plasma as an alternative to antibiotics in piglet feeds, mode of action and biosafety. *Porcine Health Manag.* 2:16. doi: 10.1186/s40813-016-0034-1.

Pinelli-Saavedra, A., Calderon de la Barca, A. M., Hernandez, J., Valenzuela, R. and Scaife, J. R. 2008. Effect of supplementing sows' feed with alpha-tocopherol acetate and vitamin C on transfer of alpha-tocopherol to piglet tissues, colostrum, and milk: Aspects of immune status of piglets. *Res. Vet. Sci.* 85(1):92–100. doi: 10.1016/j.rvsc.2007.08.007.

Pluske, J. R., Siba, P. M., Pethick, D. W., Durmic, Z., Mullan, B. P. and Hampson, D. J. 1996. The incidence of swine dysentery in pigs can be reduced by feeding diets that limit the amount of fermentable substrate entering the large intestine. *J. Nutr.* 126(11):2920–2933. doi: 10.1093/jn/126.11.2920.

Pluym, L. M., Van Nuffel, A., Van Weyenberg, S. and Maes, D. 2013. Prevalence of lameness and claw lesions during different stages in the reproductive cycle of sows and the impact on reproduction results. *Animal* 7(7):1174–1181. doi: 10.1017/S1751731113000232.

Renaudeau, D., Gourdine, J., Silva, B. and Noblet, J. 2008. *Nutritional Routes to Attenuate Heat Stress in Pigs, Livestock and Global Climate Change*. Cambridge: Cambridge University Press, pp. 134–138.

Resnik, D. B. 2000. Financial interests and research bias. *Perspect. Sci.* 8(3):255–285. doi: 10.1162/106361400750340497.

Richards, J. D., Zhao, J., Harrell, R. J., Atwell, C. A. and Dibner, J. J. 2010. Trace mineral nutrition in poultry and swine. *Asian-Australas. J. Anim. Sci.* 23(11):1527–1534.

Rioja-Lang, F. C., Seddon, Y. M. and Brown, J. A. 2018. Shoulder lesions in sows: A review of their causes, prevention, and treatment. *J. Swine Health Prod.* 26(2):101–107.

Rist, V. T., Weiss, E., Eklund, M. and Mosenthin, R. 2013. Impact of dietary protein on microbiota composition and activity in the gastrointestinal tract of piglets in relation to gut health: A review. *Animal* 7(7):1067–1078. doi: 10.1017/S1751731113000062.

Rodrigues, L. A., Wellington, M. O., Gonzalez-Vega, J. C., Htoo, J. K., Van Kessel, A. G. and Columbus, D. A. 2021. Functional amino acid supplementation, regardless of dietary protein content, improves growth performance and immune status of weaned pigs challenged with Salmonella Typhimurium. *J. Anim. Sci.* 99(2):skaa365.

Schlegel, P., Sauvant, D. and Jondreville, C. 2013. Bioavailability of zinc sources and their interaction with phytates in broilers and piglets. *Animal* 7(1):47–59. doi: 10.1017/S1751731112001000.

Siba, P. M., Pethick, D. W. and Hampson, D. J. 1996. Pigs experimentally infected with *Serpulina hyodysenteriae* can be protected from developing swine dysentery by feeding them a highly digestible diet. *Epidemiol. Infect.* 116(2):207–216. doi: 10.1017/s0950268800052456.

Slavin, J. 2013. Fiber and prebiotics: Mechanisms and health benefits. *Nutrients* 5(4):1417–1435. doi: 10.3390/nu5041417.

Solà-Oriol, D. and Gasa, J. 2017. Feeding strategies in pig production: sows and their piglets. *Anim. Feed Sci. Technol.* 233:34–52. doi: 10.1016/j.anifeedsci.2016.07.018.

Song, Z. H., Xiao, K., Ke, Y. L., Jiao, le F. and Hu, C. H. 2015. Zinc oxide influences mitogen-activated protein kinase and TGF-beta1 signaling pathways, and enhances intestinal barrier integrity in weaned pigs. *Innate Immun.* 21(4):341-348. doi: 10.1177/1753425914536450.

Sun, Y. and Kim, S. W. 2017. Intestinal challenge with enterotoxigenic Escherichia coli in pigs, and nutritional intervention to prevent postweaning diarrhea. *Anim. Nutr.* 3(4):322–330. doi: 10.1016/j.aninu.2017.10.001.

Sun, Y., Zhou, L., Fang, L., Su, Y. and Zhu, W. 2015. Responses in colonic microbial community and gene expression of pigs to a long-term high resistant starch diet. *Front. Microbiol.* 6:877. doi: 10.3389/fmicb.2015.00877.

Święch, E., Tuśnio, A., Taciak, M., Boryczka, M. and Buraczewska, L. 2012. The effects of pectin and rye on amino acid ileal digestibility, threonine metabolism, nitrogen retention, and morphology of the small intestine in young pigs. *J. Anim. Feed Sci.* 21(1):89–106. doi: 10.22358/jafs/66055/2012.

Trevisi, P., Corrent, E., Mazzoni, M., Messori, S., Priori, D., Gherpelli, Y., Simongiovanni, A. and Bosi, P. 2015. Effect of added dietary threonine on growth performance, health, immunity and gastrointestinal function of weaning pigs with differing genetic susceptibility to Escherichia coli infection and challenged with E. coli K88ac. *J. Anim. Physiol. Anim. Nutr. (Berl)* 99(3):511–520. doi: 10.1111/jpn.12216.

Trevisi, P., Latorre, R., Priori, D., Luise, D., Archetti, I., Mazzoni, M., D'Inca, R. and Bosi, P. 2017. Effect of feed supplementation with live yeast on the intestinal transcriptome profile of weaning pigs orally challenged with Escherichia coli F4. *Animal* 11(1):33–44. doi: 10.1017/S1751731116001178.

Tucker, H. F. and Salmon, W. D. 1955. Parakeratosis or zinc deficiency disease in the pig. *Proc. Soc. Exp. Biol. Med.* 88(4):613-616.

Tugnoli, B., Giovagnoni, G., Piva, A. and Grilli, E. 2020. From acidifiers to intestinal health enhancers: How organic acids can improve growth efficiency of pigs. *Animals (Basel)* 10(1), 134. doi: 10.3390/ani10010134.

van der Meer, Y., Lammers, A., Jansman, A. J., Rijnen, M. M., Hendriks, W. H. and Gerrits, W. J. 2016. Performance of pigs kept under different sanitary conditions affected by protein intake and amino acid supplementation. *J. Anim. Sci.* 94(11):4704–4719. doi: 10.2527/jas.2016-0787.

van Heugten, E., Spears, J. W., Kegley, E. B., Ward, J. D. and Qureshi, M. A. 2003. Effects of organic forms of zinc on growth performance, tissue zinc distribution, and immune response of weanling pigs. *J. Anim. Sci.* 81(8):2063-2071. doi: 10.2527/2003.8182063x.

van Riet, M. M., Janssens, G. P., Cornillie, P., Van Den Broeck, W., Nalon, E., Ampe, B., Tuyttens, F. A., Maes, D., Du Laing, G. and Millet, S. 2016a. Marginal dietary zinc concentration affects claw conformation measurements but not histological claw characteristics in weaned pigs. *Vet. J.* 209:98–107. doi: 10.1016/j.tvjl.2016.01.007.

van Riet, M. M., Millet, S., Liesegang, A., Nalon, E., Ampe, B., Tuyttens, F. A., Maes, D. and Janssens, G. P. 2016b. Impact of parity on bone metabolism throughout the reproductive cycle in sows. *Animal* 10(10):1714–1721. doi: 10.1017/S1751731116000471.

van Riet, M. M. J., Millet, S., Aluwé, M. and Janssens, G. P. J. 2013. Impact of nutrition on lameness and claw health in sows. *Livest. Sci.* 156(1-3):24–35. doi: 10.1016/j.livsci.2013.06.005.

van Zanten, H. H., Bikker, P., Mollenhorst, H., Meerburg, B. G. and de Boer, I. J. 2015. Environmental impact of replacing soybean meal with rapeseed meal in diets of finishing pigs. *Animal* 9(11):1866–1874. doi: 10.1017/S1751731115001469.

Vente-Spreeuwenberg, M. A. M., Verdonk, J. M. A. J., Bakker, G. C. M., Beynen, A. C. and Verstegen, M. W. A. 2004. Effect of dietary protein source on feed intake and small intestinal morphology in newly weaned piglets. *Livest. Prod. Sci.* 86(1–3):169–177. doi: 10.1016/S0301-6226(03)00166-0.

Vrielinck, J., Sarrazin, S., Schoos, A., Janssens, G. P. J. and Maes, D. 2019. Prevalence and chemical composition of uroliths in fattening pigs in Belgium. *J. Anim. Physiol. Anim. Nutr. (Berl)* 103(6):1828–1836. doi: 10.1111/jpn.13169.

Vukmirović, Đ., Čolović, R., Rakita, S., Brlek, T., Đuragić, O. and Solà-Oriol, D. 2017. Importance of feed structure (particle size) and feed form (mash vs. pellets) in pig nutrition – A review. *Anim. Feed Sci.Technol.* 233:133–144. doi: 10.1016/j.anifeedsci.2017.06.016.

Wang, C., Zhang, L., Ying, Z., He, J., Zhou, L., Zhang, L., Zhong, X. and Wang, T. 2018. Effects of dietary zinc oxide nanoparticles on growth, diarrhea, mineral deposition, intestinal morphology, and barrier of weaned piglets. *Biol. Trace Elem. Res.* 185(2):364–374. doi: 10.1007/s12011-018-1266-5.

Wang, L., Xu, X., Su, G., Shi, B. and Shan, A. 2017. High concentration of vitamin E supplementation in sow diet during the last week of gestation and lactation affects the immunological variables and antioxidative parameters in piglets. *J. Dairy Res.* 84(1):8–13. doi: 10.1017/S0022029916000650.

Wang, W., Wang, Y., Hao, X., Duan, Y., Meng, Z., An, X. and Qi, J. 2020. Dietary fermented soybean meal replacement alleviates diarrhea in weaned piglets challenged with enterotoxigenic Escherichia coli K88 by modulating inflammatory cytokine levels and cecal microbiota composition. *BMC Vet. Res.* 16(1):245. doi: 10.1186/s12917-020-02466-5.

Wang, X., Feng, Y., Shu, G., Jiang, Q., Yang, J. and Zhang, Z. 2011. Effect of dietary supplementation with hydrolyzed wheat gluten on growth performance, cell immunity and serum biochemical indices of weaned piglets (Sus scrofa). *Agric. Sci. China* 10(6):938–945.

Wang, Z., Li, J., Li, J., Li, Y., Wang, L., Wang, Q., Fang, L., Ding, X., Huang, P., Yin, J., Yin, Y. and Yang, H. 2019. Protective effect of chicken egg yolk immunoglobulins (IgY) against enterotoxigenic Escherichia coli K88 adhesion in weaned piglets. *BMC Vet. Res.* 15(1):234. doi: 10.1186/s12917-019-1958-x.

Wegner, B., Tenhündfeld, J., Vogels, J., Beumer, M., Kamphues, J., Hansmann, F., Rieger, H., Grosse Beilage, E. and Hennig-Pauka, I. 2020. Lameness in fattening pigs-Mycoplasma hyosynoviae, osteochondropathy and reduced dietary phosphorus level as three influencing factors: A case report. *Porcine Health Manag.* 6(1):41.

Weldon, W. C., Lewis, A. J., Louis, G. F., Kovar, J. L., Giesemann, M. A. and Miller, P. S. 1994. Postpartum hypophagia in primiparous sows: I. Effects of gestation feeding level on feed intake, feeding behavior, and plasma metabolite concentrations during lactation. *J. Anim. Sci.* 72(2):387–394. doi: 10.2527/1994.722387x.

Wellock, I. J., Fortomaris, P. D., Houdijk, J. G. and Kyriazakis, I. 2008a. Effects of dietary protein supply, weaning age and experimental enterotoxigenic Escherichia coli infection on newly weaned pigs: Performance. *Animal* 2(6):825–833. doi: 10.1017/S1751731108001559.

Wellock, I. J., Fortomaris, P. D., Houdijk, J. G., Wiseman, J. and Kyriazakis, I. 2008b. The consequences of non-starch polysaccharide solubility and inclusion level on the health and performance of weaned pigs challenged with enterotoxigenic Escherichia coli. *Br. J. Nutr.* 99(3):520–530. doi: 10.1017/S0007114507819167.

Wellock, I. J., Fortomaris, P. D., Houdijk, J. G. M. and Kyriazakis, I. 2007. The effect of dietary protein supply on the performance and risk of post-weaning enteric disorders in newly weaned pigs. *Anim. Sci.* 82(3):327–335. doi: 10.1079/ASC200643.

Wensley, M. R., Tokach, M. D., Woodworth, J. C., Goodband, R. D., Gebhardt, J. T., DeRouchey, J. M. and McKilligan, D. 2021a. Maintaining continuity of nutrient intake after weaning I: Review of pre-weaning strategies. *Transl. Anim. Sci.* 5(1):txab021. doi: 10.1093/tas/txab021.

Wensley, M. R., Tokach, M. D., Woodworth, J. C., Goodband, R. D., Gebhardt, J. T., DeRouchey, J. M. and McKilligan, D. 2021b. Maintaining continuity of nutrient intake after weaning II: Review of post-weaning strategies. *Transl. Anim. Sci.* 5(1):txab022. doi: 10.1093/tas/txab022.

Wijtten, P. J., van der Meulen, J. and Verstegen, M. W. 2011. Intestinal barrier function and absorption in pigs after weaning: A review. *Br. J. Nutr.* 105(7):967–981. doi: 10.1017/S0007114510005660.

Wolf, P., Rust, P. and Kamphues, J. 2010. How to assess particle size distribution in diets for pigs? *Livest. Sci.* 133(1–3):78–80. doi: 10.1016/j.livsci.2010.06.030.

Wu, G. 2014. Dietary requirements of synthesizable amino acids by animals: A paradigm shift in protein nutrition. *J. Anim. Sci. Biotechnol.* 5(1):34. doi: 10.1186/2049-1891-5-34.

Wu, G., Meier, S. A. and Knabe, D. A. 1996. Dietary glutamine supplementation prevents jejunal atrophy in weaned pigs. *J. Nutr.* 126(10):2578–2584. doi: 10.1093/jn/126.10.2578.

Xie, Z., Hu, L., Li, Y., Geng, S., Cheng, S., Fu, X., Zhao, S. and Han, X. 2017. Changes of gut microbiota structure and morphology in weaned piglets treated with fresh fermented soybean meal. *World J. Microbiol. Biotechnol.* 33(12):213. doi: 10.1007/s11274-017-2374-7.

Yokoyama, H., Peralta, R. C., Diaz, R., Sendo, S., Ikemori, Y. and Kodama, Y. 1992. Passive protective effect of chicken egg yolk immunoglobulins against experimental enterotoxigenic Escherichia coli infection in neonatal piglets. *Infect. Immun.* 60(3):998–1007.

Zhang, H., Chen, Y., Li, Y., Zhang, T., Ying, Z., Su, W., Zhang, L. and Wang, T. 2019. l-threonine improves intestinal mucin synthesis and immune function of intrauterine growth-retarded weanling piglets. *Nutrition* 59:182–187. doi: 10.1016/j.nut.2018.07.114.

Zheng, P., Yu, B., He, J., Yu, J., Mao, X., Luo, Y., Luo, J., Huang, Z., Tian, G., Zeng, Q., Che, L. and Chen, D. 2017. Arginine metabolism and its protective effects on intestinal health and functions in weaned piglets under oxidative stress induced by diquat. *Br. J. Nutr.* 117(11):1495–1502. doi: 10.1017/S0007114517001519.

Zurbrigg, K. 2006. Sow shoulder lesions: Risk factors and treatment effects on an Ontario farm. *J. Anim. Sci.* 84(9):2509–2514. doi: 10.2527/jas.2005-713.

Chapter 11

Managing housing and stocking density to optimize health, welfare and production in pig herds

Isabel Hennig-Pauka and Alexandra von Altrock, University of Veterinary Medicine Hannover, Germany

1 Introduction

Pig meat production in the European Union (EU) is subject to Regulation (EC) No 178/2002 (the General Food Law Regulation), which provides an overarching guideline for food and feed legislation in EU member states. In the EU, the three main pork production countries (Spain, Germany and France) account for more than half of pork production. In general, a high density of farms is found with easy access to feed. Traditional pig farming using an agro-sylvo-pastural system is still followed in Spain. Among EU member states, pig farms are very different in terms of size and husbandry systems. In Romania, the country with

http://dx.doi.org/10.19103/AS.2022.0103.12

most individual farms, there are both large farms, characterized by intensive production, and traditional backyard farms. Integrated production is common practice in Denmark and Spain, while it is not yet the case in other countries, such as Germany.

Modern pig husbandry systems are currently the focus of public and political attention in some European countries. There is focus on improving production of safe and high-quality pork using climate- and resource-friendly production systems that maintain high animal-welfare standards. Important recent drivers are EU regulations for sustainable production systems (e.g. Farm-to-Fork Strategy, New Green Deal, One-Health-Initiative) and higher animal welfare standards. Integral to these requirements is the transparency of the entire production chain backed up by appropriate documentation for both producers and regulators.

This chapter describes major husbandry- and management-related factors affecting health, welfare and production efficiency in different swine husbandry systems. Given the current transformation in pig production, there are several conflicting goals that need to be addressed during design and adaptation of husbandry systems in the future. After an overview of welfare and sustainability issues, the following sections deal with the most important areas impacting pig health and welfare such as space, climate, flooring, water and feed supply. Since it is not possible to give a comprehensive overview of all systems, the authors summarize fundamental housing requirements that must be achieved in all systems. They also review basic principles in assessing the quality of housing systems. The very important human factor is addressed at the end of the chapter.

2 Welfare and sustainability issues affecting pig production

There are different perspectives on the definition of good animal welfare. The World Organisation for Animal Health (OIE) and the European Food Safety Authority (EFSA, Panel on Animal Health and Welfare (AHAW)) define welfare as the 'ability of an animal to cope with its environment' (Broom, 1996). The five freedoms (freedom from hunger and thirst, freedom from discomfort, freedom from pain, injury and disease, freedom from pain and distress, freedom to behave in a normal and natural way) are defined as animal welfare prerequisites by the UK Farm Animal Welfare Council (FAWC, 2009) (https://assets.publishing.service.gov.uk/government/uploads/system/uploads/attachment_data/file/319292/Farm_Animal_Welfare_in_Great_Britain_-_Past__Present_and_Future.pdf; accessed: 14.07.21).

Animal welfare is mainly impacted by husbandry conditions, management practices and the health status of animals. Welfare in farm animals is regulated by the Council Directive 98/58/EC and Council Directive 2008/120/EC, which lay down minimum requirements for husbandry conditions for pigs. The Commission had additionally adopted a Recommendation (EU) 2016/336 on the Council Directive 2008/120/EC, laying down minimum standards for the protection of pigs as regards measures to reduce the need for tail-docking. The EU-wide ban 'on the routine shortening of piglets' tails' (according to Annex I, Chap.I No.8 of Directive 2008/120/EC) requires pig farmers to implement an action plan to prevent tail-biting. Tail-biting has a multi-factorial origin. This requires significantly increased monitoring of abnormal behaviour to identify risk factors and keeping detailed records of husbandry and environmental conditions to identify potential triggers. However, conventional husbandry typically provides limited opportunities for pigs to express their natural exploratory and foraging behaviour. Boredom and frustration result in pathological behaviour patterns, such as tail, flank and ear biting or even vulva or penis biting, which are of great importance not only for animal health and animal welfare but also have important negative economic impacts.

The welfare of food animals is a growing concern. The ban of tail-docking in the EU is one important driver in re-thinking swine husbandry systems. An additional example is the requirements on the confinement of breeding sows, especially increasing space allowance, bedding and nesting material. These requirements have to be met by the farmers which, in turn, mean stall conversations and related investments with a high economic cost for many holdings.

The Directive 2010/75/EU sets out measures for environmental protection from emissions from intensive pig farms with capacity for more than 750 sows and 2000 fattening animals. To meet these requirements, the Commission Implementing Decision (EU) 2017/302 established best available techniques in the fields of nutrition, manure management and removal of dead animals. Water and air pollution are important environmental issues in pig production, mainly associated with manure processing. Important aspects are the introduction of nitrate (NO_3^-), ammonium (NH_4+), nitrogen (N), and phosphor (P) in surface and groundwater, emission of greenhouse gases (nitrous oxide (N_2O), carbon dioxide (CO_2), methane (CH_4)), acidification and air pollution by ammonia and nitrous oxide. Minimizing emissions requires measures such as adapted feeding strategies, no long-term storage of manure within the building, separate collection of solid feces and urine, emission-free storage of manure with solid flooring and a liquid manure collection tank, and covering of manure during storage. The vast majority of greenhouse gases result from feed production, while direct energy consumption is due to heating and ventilation. The European Green Deal Farm to Fork Strategy addresses sustainability of food systems with a particular focus on reduction of use of fertilizers, pesticides, antimicrobials and reducing biodiversity loss. It promotes a parallel increase in organic farming and animal welfare.

3 Fundamental housing requirements for pigs

Pig production systems differ in various ways, e.g. biosecurity, genetics, feeding system, bedding, flooring, ventilation, manure removal, group size, conventional versus alternative production, etc. The combination of those characteristics depends on different aspects, such as number of animals and level of production, geographic location (e.g. for outdoor farming), or if production must fulfil certain requirements (ecological farming, welfare labelling requirements). The housing system is the foundation for all measures and activities to guarantee the health, welfare and productivity of pigs, although the type of housing seems to be of less importance than the physical layout and how animals are managed by stock people (Hemsworth, 2018).

Pigs are usually kept in buildings with or without outdoor access. Minimal environmental requirements to guarantee welfare and health of pigs are set out by legislation (Table 1). The requirements for buildings address the:

i optimization of climatic conditions, such as air temperature, humidity, velocity and direction of air flow, air contaminants such as dust and gases;
ii the provision of adequate space for individual animals;
iii guaranteed access to feed and water;
iv the design of pens including flooring; and
v technical and logistical solutions to remove manure.

Key requirements also cover the optimization of light and maximum noise levels (Zulovich, 2012).

Factors that influence the choice of housing systems include the provision and distribution of feed, space allocation, manure removal, as well as the distance from neighboring farms and frequently used roads. The location of a building within a particular geographic area can have a significant influence on its design and operation. An illustration of this is eradication of airborne diseases, which has a high risk of failure if the building is located in an area with high pig density, where infection in the pig population is endemic and filtration of incoming air is not practiced. Areas with high pig density are also a risk for other diseases, e.g. those that can be transmitted by rodents or other vectors able to move between neighbouring farms (e.g. transmission of *Brachyspira* spp, *Salmonella* spp.) (Desrosiers, 2011; Wang et al., 2011).

The location and design of a building has a major impact on welfare in the context of a future shift towards warmer climates. During hot weather, outdoor

Table 1 Summary of the minimum standards for the protection of pigs in general described in the Council Directive 2008/120/EC

	General Conditions
Accommodation	• Floors must be smooth but not slippery and suitable for size and weight of the pigs. • The lying area must be comfortable, clean and dry and allows all animals to lie at the same time. • Pigs must see other pigs.
Environment	• Continuous noise levels as loud as 85 dBA shall be avoided. • Light intensity of at least 40 lux for a minimum period of 8 h per day. • Environment must correspond to the needs for exercise and investigatory behaviour.
Feed	• All pigs must be fed at least once a day and pigs over 2 weeks of age must have permanent access to a sufficient quantity of fresh water. • When fed in groups, all pigs must have access to the food at the same time (exception: ad libitum feeding or feeding by an automatic system).
Health	• Aggressive, attacked pigs, sick or injured pigs are to be placed in individual enclosures.
Behaviour	• Pigs should benefit to their needs for exercise and investigatory behaviour. • Pigs must have permanent access to a sufficient quantity of material to enable proper investigation and manipulative activities.
Allowed painful interventions	• Reduction of piglets' corner teeth and boars' tusks. • Docking of a part of the tails and castration of males (before the seventh day of life or after this age if carried out by a veterinarian and under anaesthesia and with additional prolonged analgesia). • Nose-ringing in outdoor husbandry systems. • The docking of tails or the reduction of corner teeth must not be carried out routinely.

and indoor climates are linked with solar irradiance on building surfaces having a significant impact on indoor temperature. Long, narrow buildings are cooler in summer and warmer in winter if the long axis runs from east to west. This is because the extent of solar radiation is reduced as a result of the large non-sunlit building surface on the north side of the building (northern hemisphere) or the south side (southern hemisphere) (Angrecka and Herbut, 2016). The pig shed should be situated to take advantage of prevailing winds for coolness in summer. Conversely, ventilation openings should be protected from prevailing winds in winter. Shelterbelts can affect the physical environment by effectively increasing the surrounding temperature in winter and reducing it in summer. Additionally, shelterbelts mitigate livestock odour and provide an aesthetic improvement to the landscape (Tyndall and Colletti, 2007). The eave width should be adapted in size so that the sun does not shine on the wall or into the interior of the piggery in summer, while the right width may also improve warming in winter. In confined livestock housing with forced ventilation systems (i.e. those relying on fans or blowers to circulate air in the pig barn), the exhaust air should be controlled to allow implementation of regenerative heat recovery as well as air purification.

Whilst some authors suggest positive effects of outdoor access for swine health and welfare compared to confined, indoor systems, there is a general recognition that both outdoor and indoor systems can have potential negative effects (Ludwiczak et al., 2021). There is also now an extensive literature on how far suitable enrichment in well-designed indoor settings can meet pig health and welfare needs (Düpjan et al., 2021).

Outdoor farming is only possible if regional soil and topographical requirements are fulfilled (e.g. sandy, well-drained soils). Outdoor farming systems are assessed to be beneficial in allowing pigs to display species-specific foraging behaviour typically with sufficient space to express this behaviour. Investment costs are also typically lower. Disadvantages include uncontrolled airborne emissions and seeping of manure components into the soil, high labor requirements and higher disease risk, particularly the lack of control of internal parasites. In outdoor farming systems, welfare and health issues differ significantly from closed systems. Sunburn in less-pigmented pig breeds is a common problem, other than mortality, due to predators such as owls and foxes. Biosecurity in outdoor settings is incomplete because birds, rodents and wild boars cannot be entirely prevented from entering the holding. Due to more limited opportunities for cleaning and disinfection, and the risk of soil contamination with pathogens, biosecurity can only be implemented to a limited degree (EFSA AHAW et al., 2021; Honeyman et al., 2003).

Organic pig farming in the EU represents about 1% of overall pig production. Based on the availability of land, soil characteristics, climate, tradition and national organic certification schemes, different housing systems have been developed, characterized by some access to a free-range area (Früh and Holinger, 2019). Organic pig husbandry systems can be divided into three types:

i indoor rearing with access to a concrete outside run;
ii outdoor farming with access to shelter; and
iii a mixed housing system with outdoor and indoor access during different production stages.

According to EU legislation, sows must have access to free-farrowing areas, farrowing units must provide a minimum area of 10m^2, including an outdoor area of at least 2.3 m^2, whilst piglets must be suckled for up to 40 days before weaning.

In all housing systems, pigs must, at a minimum, be protected from extreme weather conditions and sunburn, always have access to a dry, comfortable resting area and always have access to water. Overall housing conditions, including pen design and stable equipment, should also not increase the risk for development of disease or trauma. Legal requirements for husbandry (e.g. in terms of feeding, animal handling or monitoring) must also be fulfilled. As well as minimal space requirements, EU and national rules also include regulations criteria for air quality. In addition, the needs of diseased pigs, which can be bullied by pen mates, must be addressed by provision of separate hospital pens allowing resting in a warm environment. The optimal ratio of hospital pens needed in a piggery is still under discussion (Millman, 2007, 2015). Animal needs can best be investigated by preference tests to determine those housing conditions which are most beneficial to animals (Elmore et al., 2010).

4 Behavioural traits to be considered in planning housing

Pigs were domesticated about 9000 years ago, but their natural behavioural patterns are still nearly identical to those of their wild ancestors. In wild boars, social groups are formed, which explore, eat, and rest synchronously and spend half of their active time foraging. Peak activity times are during the morning and afternoon hours, about 70% of which are filled with exploratory behaviour for the purpose of feeding. The results of choice experiments show that pigs can independently, relatively, and accurately select a balanced diet for themselves. They can also adapt the composition of their diet to meet changing requirements (Kyriazakis and Emmans, 1992; Kyriazakis et al., 1990). The pig thus has an intrinsic need for foraging and synchronous feed intake, which influences eating behaviour as well as the social structure in the group. This suggests that feed intake behaviour may be also influenced by external factors such as type of feed, feeding systems and housing facilities, health, breed and environment (Maselyne et al., 2015).

Conventional housing systems often do not meet those behavioural needs since they typically provide a complete, readily available high-energy feed at defined times. Intensive housing conditions are also often relatively sterile environments devoid of stimulation, thus depriving the intelligent pig species the possibility to exhibit their natural exploratory behaviour. The result can be manipulative or aggressive behaviour towards pen mates, which poses significant welfare, health and production challenges for pig farmers. Housing conditions are in interaction with the social hierarchy in a group and need to take account of social dynamics. It should be kept in mind that lower-ranking animals have better chance for adequate and stress-free uptake of feed and water, and also for a comfortable resting place, when a sufficient number of drinkers and feeders are installed in relation to the number of pigs and the pen provides enough space for resting behaviour.

Issues related to the visual, auditory, and olfactory abilities of pigs are increasingly the subject of research. It is known that there are large differences between individuals in terms of sensory perception and learning ability. Olfactory differences are generally learned faster than visual differences. In addition to operant conditioning in the sense of associative learning, pigs can learn olfactory and visual discrimination tasks (Croney et al., 2003; Gieling et al., 2011; Lind et al., 2007). Pigs, for example, can remember food locations and discriminate between different attributes of food (Mendl et al., 2010; Mendl and Paul, 2004). The pig´s hearing range exceeds the human hearing range. This means that loud and sudden noises can stress pigs. Noise can impact the ability of sows to react to their piglets and can interfere with the farrowing process (Chapel et al., 2019).

Pigs are able to make decisions based on visual information from a very early age. Curiosity and intrinsically high motivation supports the use of indirect visual information for novel foraging situations (Nawroth and von Borell, 2015). Conflicting data exist regarding visual acuity at low light intensities. At 12 lux, visual acuity was not impaired with respect to black and white discrimination (Zonderland et al., 2008). With respect to color vision, blue, in particular, is discriminated from other colors (Tanida et al., 1991). Choice experiments show that pigs prefer brightness over darkness, and infrared light is preferred in combination with heat (Baldwin and Meese, 1977). For pig housing, EU legislation requires 40 lux and German legislation 80 lux for 8 h per day. The natural circardian rhythm of the pig must be the basis for the light-controlled structuring of the daily routine (Tilger, 2005).

Pig behavioural traits can be assigned to ten functional groups:

1 Locomotion;
2 Feed intake (feeding behaviour);
3 Thermoregulation;
4 Explorative behaviour;
5 Reproductive behaviour;
6 Resting behaviour;
7 Excretal behaviour;
8 Body care behaviour;
9 Social behaviour; and
10 Birth-giving behaviour.

All these functional groups are influenced by factors such as space, group size and pen structure. Housing and management conditions should, as far as possible, facilitate natural behaviour of pigs in all these functional groups. The degree to which this is prioritised depends on traditions and priorities in individual countries, which may change over time (e.g. due to economic pressures, disease outbreaks or changing political priorities). It is impossible to enable the full expression of natural pig behaviour in husbandry conditions other than for free-range pigs. Wild pigs with domesticated ancestors can still be found in some parts of Europe (e.g. Corsica, Sardinia) (Albarella et al., 2006; Jori et al., 2017).

The focus on enabling natural behaviour of pigs under farm conditions, as well as the need to comply with the ban on tail-docking (which increases the risk of tail biting resulting from aggressive behaviour), has led to an emphasis on enrichment to meet the basic behavioural needs of pigs. Enriching the environment for pigs with objects or substrates that 'enable proper investigation and manipulation activities' and which satisfy the pig's exploratory behaviour is required by law in the EU (Table 1). Specific requirements for enrichment are also defined in the Council Directive 2008/120/EC (Table 2). In principle, environmental enrichment should result in an increase in positive behaviour and a decrease in aggressive behaviour against pen mates and from engaging in stereotypic behaviours such as belly nosing and sham chewing, which are

Table 2 Sections of the Council Directive 2008/120/EC referring to enrichment materials for different age categories of swine (modified from van de Weerd and Ison (2019))

Directive Section	Age category	Text
Article 3 (5)	Sows and gilts	Member states shall ensure that, without prejudice to the requirements laid down in Annex I, sows and gilts have permanent access to manipulable material at least complying with the relevant requirements of that Annex.
Annex I, Chapter II: B3	Sows and gilts	In the week before the expected farrowing time sows and gilts must be given suitable nesting material in sufficient quantity unless it is not technically feasible for the slurry system used in the establishment.
Annex I, Chapter II: C.1	Piglets	A part of the total floor, sufficient to allow the animals to rest together at the same time, must be solid or covered with a mat, or be littered with straw or any other suitable material.
Annex I, Chapter II: D.3	Weaners and rearing pigs	When signs of severe fighting appear, the causes shall be immediately investigated and appropriate measures taken, such as providing plentiful straw to the animals, if possible, or other materials for investigation. Animals at risk or particularly aggressive animals shall be kept separate from the group.

ndicators of boredom or stress. However, since objects quickly lose their attractiveness through habituation, new and unfamiliar objects should be introduced regularly or a system of rotating enrichment objects should be implemented (Mkwanazi et al., 2019). The pig's natural curiosity is addressed by an unfamiliar object, thus stimulating the natural behaviour of investigating, manipulating it, and, potentially, concluding the interaction with reward feeding (van de Weerd et al., 2003). Straw is recognized as one of the most suitable materials for rooting and foraging behaviour, but it poses challenges for effective manure management (Pedersen et al., 2014).

It has been shown that stress affects the immune system via key hormones, such as corticosteroids and catecholamines. Corticosteroids lead to decreased proliferation of lymphocytes and antibody production, while catecholamines can affect the virulence of microorganisms, disturbing the balance between pathogen exposure and immune defence (Dalin et al., 1993; Lyte and Lyte, 2019; Lyte, 2016; Reiske et al., 2020). By reducing stress, enrichment can support immune function, so that, in addition to improved welfare by enabling rooting and avoiding aggressive behaviour, enrichment can play a role in preventing infectious diseases (Luo et al., 2020; van de Weerd and Ison, 2019; van Dixhoorn et al., 2016). Enrichment can reduce behaviour such as tail-biting and lead to higher overall growth rates due to better health (Carroll et al., 2018; Faucitano et al., 2020).

However, it is important to note that, for all their benefits, enrichment activities are not as important as overall management by farmers (Carroll and Groarke, 2019; Henningsen et al., 2018; Peden et al., 2019; Sørensen and Schrader, 2019). Enrichment activities can also be costly and time-consuming for the farmer to implement and maintain. Godyń et al. (2019) observed increased production costs, and lower farm output in farms providing enrichment. Most effective enrichment measures are expensive to manage (such as providing a regular supply of fresh straw), so that a significant reduction in injured pigs would be required to be profitable (Niemi et al., 2021). An animal-welfare program should financially support farmers to offset lower productivity. Cost-relevant health indicators are mostly not sufficiently improved by additional welfare measures, so that additional costs are not justified from an economic perspective (Tonsor and Wolf, 2019; Uehleke et al., 2021).

A legal obligation to provide environmental enrichment for pigs is not in force in all pork-producing regions. It is required by legislation in the EU, Switzerland and Norway, but, e.g. not in the USA or China (van de Weerd and Ison, 2019). In practice, enrichment is often not provided or is not adequate. Problems often observed on farms include the provision of hazardous material or objects or placing it incorrectly, in terms of reachability. In the future more emphasis should be laid on the economic benefits of effective enrichment (return on investment) as well as drivers for improvement, such as, e.g. benchmarking (van de Weerd and Ison, 2019).

5 Potential indicators for assessment of housing systems

The assessment and comparison of different systems and farms is hampered by the lack of robust and efficiently measurable indicators covering requirements in different areas (health, welfare, antibiotic usage, environmental impact and sustainability). The relationship between economic performance and animal welfare indicators has been found to be weak. Good farm managers can achieve high performance in both areas (Henningsen et al., 2018). Precise data collection on farms following the Welfare Quality® assessment protocol is expensive and time-consuming and therefore not used by many farms on a regular basis (Botreau et al., 2009; Temple et al., 2011). As a result, alternative ways of assessing welfare have been developed.

Roughly, two groups of welfare indicators can be distinguished:

- resource based; and
- animal based.

Indirect resource-based indicators (e.g. access to and consumption of feed) can help identify poor animal welfare conditions before animals are negatively affected (e.g. with reduced feed intake as a possible indicator of illness). Animal-based indicators are outcome-based indicators (e.g. relating to an animal's health status), which can be used to assess the interplay between housing environment, human–animal interaction and management as well as exposure to harmful microorganisms and abiotic stressors (EFSA, 2012). Researchers address animal-based welfare indicators in three different ways (Fraser et al., 1997):

- adequate biological functions (growth, reproduction, health);
- emotional state (no distress, fear, pain); and
- the ability to behave in a natural way.

An example of the two latter types of indicator, e.g. is stereotypical behaviour like bar-biting, which may suggest stress or boredom.

Different animal-based welfare indicators have been described for the collection of data (Table 3). An alternative approach to assess welfare is the use of data already available from routine inspections along the food chain (Nienhaus et al., 2020). Abattoir-based monitoring systems to assess the welfare status of pigs originating from specific farms are of growing interest (De Luca et al., 2021; Friedrich et al., 2020; Wadepohl et al., 2020).

Table 3 Indicators for assessing animal welfare on farms

Stages of production	Indicators	References
Sows	Ectoparasites, udder and teat lesions, claw alterations, frothy saliva	(Friedrich et al., 2020)
Piglets	Face lesions, carpal joint lesions, undersized animals, stereotypes	
Fattening pigs	Average daily gain, feed conversion ratio, treatment frequency, respiratory lesions, exterior lesions, animal management, and mortality	(Grosse-Kleimann et al., 2021)
Fattening pigs	Respiratory health: pneumonia, pericarditis, pleurisy animal management: milkspots (liver), intestinal alterations, whole carcass condemnation, dermal damage (handling) external injuries/alterations: tail lesions, bursitis, ear lesions, dermal alterations Salmonella status, antibiotic usage, mortality	(Nienhaus et al., 2020)
Fattening pigs	Bursa alterations, lameness, dirty animals, runts, and tail, flank and ear biting	(Wadepohl et al., 2020)

A key problem in using data across the food chain to assess welfare is data availability and access to databases which differ along supply chains as well as between countries and regions. Poor availability, quality and interconnectivity of data in many areas of livestock production may be due to the relatively late adoption of information technologies in this field (Faverjon et al., 2019). Large quantities of data are generated at all stages in the pork-production chain, but combining it using big data approaches is still largely restricted to the occasional pilot project (Faverjon et al., 2019). Sterchi et al. (2019) highlighted the role of pooling data to assess the effect of shared transport and truck contamination on disease transmission between farms. In a German pilot study, the lack of interconnectivity between data bases and the incompatibility of data formats were overcome using a relative scoring approach. Z-transformation of different variables from different data sources was used to generate an animal health score that allows comparison and benchmarking of farms (Table 3) (Nienhaus et al., 2020). Grosse-Kleimann et al. (2021) conducted a comparable study based on production data for German finishing pigs (Table 3), indicating that the health indicators examined are usable for welfare monitoring.

In the EU, animal carcasses are sent to rendering plants, providing a continuous flow of mortality data, which could be used for farm monitoring. Mortality can be considered as one of the most important animal health indicators, making it particularly important to record this parameter consistently in a database (Lopes Antunes et al., 2017). Monitoring of mortality rates in combination with antibiotic usage data was found to have a predictive value in forecasting infectious disease on farms (Lopes Antunes et al., 2019). So-called 'iceberg indicators' might be helpful to assess welfare, health, management and productivity on farms to evaluate and improve housing and management systems. Ice-berg indicators need also to be developed to include the carbon footprint as well as emission and resource use data.

6 Housing requirements: climate, ventilation, temperature and air quality

6.1 Climate and ventilation

Indoor climate is influenced by a wide range of factors, such as the external environment, building type and age, building materials, insulation, ventilation system as well as stocking density (Banhazi et al., 2008). Concrete as a building material can retain heat and be warmed up if dry, though damp concrete flooring accentuates the negative effects of low temperatures. Indoor climate quality can be measured by variables such as air temperature, humidity, air velocity and pollutant gas concentrations (CO_2, NH_3, H_2S). Measuring and managing climate requires a fundamental understanding of the relationship between humidity and temperature as well as algorithms to measure different aspects of air flow. A particular problem is rapidly changing weather conditions, which cannot be quickly compensated by indoor ventilation equipment.

The most important climatic stressors affecting pigs are draught and incorrect temperature. Intermittent draught has been shown to initiate sneezing and coughing. Air speed should be below 0.15 m/s at pig level (Scheepens, 1996). In general, a relatively hot and humid environment is considered beneficial, resulting in less coughing (Geers et al., 1989; Gordon, 1963). As well as temperature, air quality is important. Several air contaminants damaging the lungs have been detected at the slaughterhouse stage (Cargill et al., 2002; Donham, 1991). Under the so-called 'sweet house' conditions, harmful airborne particles in the range of <1–3 μm are reduced (Thomas, 2013). Good cleaning regimes influence air quality, e.g. the removal of sediment dust from new pigs before they enter the barn. In case of infectious respiratory diseases, the ventilation system should be monitored using sensors for variables such as temperature, air velocity, circulation and impurities (Gonyou et al., 2006). Slurry levels in pits below the slatted floor of pig barns must also be controlled since surfaces closer than 40 cm to the slat bottom increase the risk of releasing bacteria-containing aerosols.

Independent of the environmental conditions, a minimum amount of fresh air (depending on the number and age class of pigs) must be introduced regularly into a building to remove bacteria, water vapour, carbon dioxide, ammonia, airborne dust and odour. Uncontrolled air exchange via damage to building walls must be avoided. Cold air must be warmed to avoid draughts. In this respect, natural ventilation has been found to be inferior to forced ventilation systems with regards to the levels of NH_3 and CO_2 as well as thermal comfort, resulting in higher risk of respiratory disease (Chantziaras et al., 2020). This is partly because animal density (measured in kilograms of swine per cubic meter) is correlated with the degree of airborne contamination, with forced ventilation more efficient in adjusting for higher densities (Donham, 1991). Practitioners recommend an air volume of more than 3 cubic meters per finishing pig not heavier than 100 kg (Marco et al., 2020). High animal density in confined livestock buildings requires mechanical ventilation systems to ensure adequate air quality and temperature both in summer (at high ventilation rates) and in winter. Ventilation in general will reduce temperature so that insulation of roof and walls is important to minimize heat loss and draught. Insulation combined with a vapour barrier can reduce condensation and therefore the need for ventilation. Failure of a ventilation system in a confined system can be fatal with rapid changes in temperature potentially leading to heat shock and dehydration or hyperthermia. Airspace stocking densities of 2.5 m^3/grower and 3 m^3/finisher are recommended (Gonyou et al., 2006).

6.2 Temperature

The optimal temperature range for pigs is defined as the thermoneutral zone (optimising productivity, health and welfare) (Zulovich, 2012). Within this temperature range, heat production by the pig itself is not influenced by air temperature. Lower and higher temperatures beyond body-weight-specific thresholds can have increasingly negative effects. Below the lower critical temperature, homoeothermic conditions will be maintained by increased metabolic activity. Above the upper critical temperature, pigs will try to dissipate heat but are at serious risk from heat stress. Absolute critical temperatures depend on feed intake, age, weight and environment. To assess the effects of a housing system, the temperature humidity index (THI) can be used to quantify potential heat stress in pigs (Wegner et al., 2016). The THI combines air temperature and humidity whilst taking heat release into account. In summer, high temperatures and a high THI around insemination reduce litter sizes. These conditions around farrowing reduce numbers of liveborn piglets, while numbers of weaned piglets are not affected (Wegner et al., 2016). Older pigs can tolerate low temperatures when they are not exposed to draught.

Observing pig behaviour will identify inadequate temperatures. Shivering and huddling behaviour indicates that the temperature is too low. Avoidance of skin contact between pen mates and fouling additional areas, which are normally clean, indicate that temperatures may be too high. Since swine cannot sweat and heat dissipation under natural conditions is mainly achieved by changes in behaviour (e.g. wallowing), pigs are particularly vulnerable to heat stress. Heat transfer is influenced by air-vapor mixture. Sensible heat transfer needs a temperature difference between two surfaces whilst latent heat transfer is mediated by differences in moisture concentrations (Deshazer et al., 2009). Where there is direct contact between surfaces of different temperature, conduction of heat from the warmer to the colder surface will occur, mediated by velocity of air circulation. At temperatures lower than 18–20°C pigs prefer a solid floor while, above 20–25°C, a preference for lying on slatted floor (for thermoregulatory reasons) is observed (European Food Safety, 2005). It is important to be aware of different sources of heat and mechanisms of heat transfer which can counteract each other e.g. cold flooring leads to heat conduction, while heating lamps provide radiated heat.

Extreme weather events involving higher temperatures will become more frequent due to the effects of climate change and will be a significant challenge because most housing systems have insufficient cooling mechanisms in place. Initial measures to deal with this problem include removal of heat sources, avoiding stressful handling of stock and moving feeding to cooler periods (night). Buildings can be cooled by spraying water on the roof surface. A very effective means to reduce temperature is reducing stocking density. In the long run, the housing system should be equipped with additional cooling systems together with increased ventilation. Evaporation through spraying water can reduce temperature. However, installing such equipment means high investment and energy costs, which may not be economically or environmentally sustainable over time.

There are various ways of optimizing the efficiency of cooling systems. The thermodynamic properties of the inlet air can be influenced by evaporative cooling pads (Xuan et al., 2012), earth-air heat exchanger (Bisoniya et al., 2014) or an indirect evaporative cooling system combining both and avoiding air humidification (Heidarinejad et al., 2009). This last system cools inlet air by evaporation, which is subsequently used in an air-to-air heat exchanger (ASHRAE, 2008). All three air treatment devices are valuable adaptation measures but need significant capital investment (Olesen and Bindi, 2002). Heat exchangers are effective for cooling in summer as well as higher and more stable temperatures in winter, allowing a higher ventilation rate, which improves air quality (Vitt et al., 2017).

6.3 Air quality

As noted earlier, natural ventilation has been found to be less effective than forced ventilation in reducing levels of NH_3 and CO_2 (Chantziaras et al., 2020).

Indoor air quality, as defined by gaseous and particle components, is mainly influenced by ventilation and the system for handling manure, which is itself influenced by feed composition and barn and pen design. Particulate matter (PM) in stable air is a mixture of particles of different size and biological and chemical properties (Cambra-López et al., 2010). PM10, PM2.5 and PM1 define particles of which 50% pass a size-selective filter of 10 μm, 2.5 μm and 1 μm aerodynamic equivalent diameter (Ulens, 2015).

PM stay in the air when interaction with air molecules is stronger than gravity. PM can bind and carry noxious gases like ammonia and, after inhalation, will deposit (PM smaller than 100 μm) in the upper airways (<10 μm), or in the lung (<4 μm). In general a concentration of 2.4mg dust per m^3 should not be exceeded in confined conditions due to health hazards for human and swine (Donham, 1991). Humidity has been found to affect PM (Kim et al., 2008). This means absolute humidity should be as high as possible to reduce air-borne particles but relative humidity must not exceed 80%. A promising method for reduction of airborne particles is spraying oil or an oil-water mixture in buildings (Pedersen et al., 2000).

In general, carbon dioxide is a good indicator for the effectiveness of ventilation and air quality (Donham, 1991). High levels of CO_2 (>2000 ppm) indicate inadequate ventilation rates, which should guarantee an air exchange of 60 m^3 per hour and pig (Flesjå and Solberg, 1981). Ammonia concentration is another measure of air quality. Legal thresholds are usually defined at 20 ppm but 10 ppm is considered to be the maximum gas concentration to ensure no adverse effects from ammonia. Several factors affect ammonia concentration in swine buildings, including manure and feeding management, hygiene, animal age, stocking density, season and ventilation. A decrease in ventilation rate at night causes an increase in ammonia concentration (Rodriguez et al., 2020).

7 Housing requirements: pen design, group size, feed systems and flooring

7.1 Pen design

Pen design and arrangement varies between individual farms and should be based on good animal observation, so each behaviour has a functional area in the pen in which it can occur (feeding, activity, lying, defecation), taking into account social interaction in each functional area. Synchronous access to water and feed would meet pig needs best but is hard to achieve with no 1:1 pig-feeding place ratio in conventional systems. The floor space required depends in part on body weight. Space requirements for weaner or rearing pigs according to weight are given by Council Directive 2008/120/EC (Table 4).

Table 4 Recommended minimum unobstructed floor area available for weaner or rearing pigs kept in a group and the calculated constant *k* based on the maximum weight according to Council Directive 2008/120/EC

Live weight (kg)	m^3	Constant k $k = \frac{A}{BW^{0.67}}$
Not more than 10	0.15	0.032
More than 10 but not more than 20	0.20	0.027
More than 20 but not more than 30	0.30	0.030
More than 30 but not more than 50	0.40	0.029
More than 50 but not more than 85	0.55	0.028
More than 85 but not more than 110	0.65	0.027
More than 110	1.00	

The floor size specified by EU regulations cannot be considered as optimal. For better health and productivity outcomes, approximately 20% more space per pig has been recommended (Hamilton et al., 2003). In general, more space, and thus a lower animal density, can reduce the risk of transmission of pathogens and disease. A compromise must be found between stocking densities that do not impair animal health but are adequate for a sufficient return on space investment costs.

The resting area should be designed so that pigs passing to activity areas do not disturb their pen mates. A ratio of 1.5:1 to 2.5:1 is recommended for nursery and fattening pigs. Different flooring areas can either reduce heat conduction (e.g. plastic, solid floor) as well as increase heat conduction (e.g. concrete slatted floor, metal or wet floor). In high air temperatures sufficient space for pigs to lie in a fully recumbent position to dissipate heat is crucial. Space requirements should account for those demands (Carol Petherick, 1983; Corino et al., 2003; Ekkel et al., 2003; Pastorelli et al., 2006). Lying behaviour is not only affected by temperature but also by the floor type. Ekkel et al. (2003) suggested reducing space allowance in slatted systems by factor of 0.003 because pigs typically lie closer together in these conditions. In deep-litter housing systems, at least 30% more floor space per pig is recommended (Ekkel et al., 2003).

The lying area should preferably be quiet and dark, avoiding travel routes to other locations, but ideally with enclosed corners. Space for defecation should be at least 0.08 m^2 per animal independent of the whole space. Ideally 20–30% of the overall space should be provided for defecation in a pen. Additional space next to defecation and resting areas can serve as an activity area. Efficient space utilisation without adversely affecting welfare should be the goal of successful pork production (Anil et al., 2007).

7.2 Management of group size

Deciding on group size will be mostly influenced by the overall herd size and the feeding system. In smaller groups, locating individual animals and performing routine management tasks like pregnancy and estrus management are easier because of smaller pen space and the reduced number of animals. In small groups, synchronous feeding of all swine is feasible while meeting the nutritional needs of the individual after sorting by size and age. In sow herds, static grouping mostly results in small groups. Housing in small units facilitates use of the 'All In All Out' (AIAO) system.

'Large group size' is a relative term that can range from 50 to several hundred pigs per group. The advantages of large groups are the more efficient use of labor, space and pen divisions. In larger groups, observation and treatment of the individual pig is more challenging. New technology can make a significant contribution to managing large groups effectively, e.g. the electronic sow feeder (ESF) system where individual sows are identified by transponders in ear tags. These allow a computer-controlled feeder to allocate a specific amount of feed to each sow and can be used to monitor animal movement and feed consumption (e.g. by identifying sows that have not accessed the feeder).

Group size related to space is a relevant factor for health. Large numbers of animals in a group can lead to a higher frequency of fights to establish ranking order, but may also decrease aggression due to more opportunities to avoid dominant animals (Samarakone and Gonyou, 2009; Turner et al., 2003). To maintain social stability, it has been suggested that each pig within a group should be able to recognise every other pig and their relative social status. It has been estimated that pigs can recognise between 20 and 30 individuals (Fraser and Broom, 1990; Broom and Fraser, 2015). Turner et al. (2001) combined unfamiliar pigs from small and large groups and found that pigs from large groups showed a marked reduction in aggression towards unfamiliar pigs (Turner et al., 2001). Large groups facilitate avoidance or escape as well as formation of subgroups, reducing chances of aggressive interactions. The overall goal of an optimal pen structure is to allow every pig to fulfil its needs at all times without hindrance. Space and structure should guarantee resting times with deep sleep as well as opportunities for play and exploration. Contact with feces and urine should be avoidable, suggesting use of areas of slatted flooring in most systems. Devices or measures to avoid heat stress should be implemented.

7.3 Feeding systems

Pigs' feeding behaviour is controlled by hunger and satiety, but can be influenced by a range of other factors, such as type of diet, housing system, breed and health (Maselyne et al., 2015). Feed management must meet varying nutritional needs throughout the production process (lactation, reproduction and growth). Poor access to feed results in uneven and reduced growth and body condition, a poor feed conversion rate and an increased percentage of downgraded pigs. Trough feeding is generally considered to be better than feeding from the floor. Trough construction should prevent pigs from getting inside the trough, lying or defecating in it.

Pigs want to eat synchronously, which is attainable with an animal to trough space ratio of 1:1. But the available number of feeding places varies according to feeding system (ad libitum or rationed). Single-space or multi-space feeding systems can be combined with wet and dry feeders. Group size is oriented to feeding systems, with 25 pigs per mush feeder, at least 25 pigs per liquid feeding system and a maximum 25 piglets in a dry feeding system. The larger the group the higher the animal-feeding-place ratio (1:3 to 1:4). Feeder space sizes ranging from 0.15 m per pig (8 weeks of age) to 0.3 m per finishing pig and 0.45 m per sow are often too small and do not account for pigs' competitive eating behaviour. Slow growing pigs in particular seem to benefit from more feeder space (He et al., 2018).

The provision of safe and palatable water in sufficient amounts at the right temperature is essential to pig health and welfare. The number of drinkers depends on group size, but at least two drinkers must be installed in a pen so that they cannot be blocked by dominate pigs. Wet flooring risks pigs choosing the drinking area for defecation. More water is wasted by push-type nipple drinkers but they can be used to wet skin for cooling during hot weather. Waste water can be reduced by correct mounting of drinkers. While straight-out-pointing drinkers should be in the height of the smallest pig´s shoulder, drinkers at an 45° angle downwards should be 5 cm above the pig's back (http://www.omafra.gov.on.ca/english/livestock/swine/news/mayjun12a1.htm, accessed: 24.09.2021). Within the thermal comfort zone, drinking behaviour usually occurs within 10 minutes of eating (Gonyou, 2001). Diseased pigs show a different drinking behaviour from healthy pigs even in the subclinical stage of infection. If water usage drops about 30% on 1 day or is obviously decreased on three consecutive days, a health problem can be expected in the herd. Consequently, automatic monitoring of feed and water intake can provide early warning of disease or stress (Maselyne et al., 2016).

7.4 Flooring

In the case of slatted floors, flooring material as well as slat and gap width must be adapted to the pigs' claws size to avoid injuries. The maximum width of the openings and minimum slat width is set out by EU Directive 2008/20/EC (Table 5). In general, a pig's claw should not fit into gaps. The slat opening

Table 5 Recommended maximum width of the openings of concrete slatted floors and minimum slat width for different age categories of swine kept in groups according to Council Directive 2008/120/EC

Age category	Maximum width of the openings (mm)	Minimum slat width (mm)
Piglets	11	50
Weaners	14	50
Rearing pigs	18	80
Gilts after service and sows	20	80

should be uniform along its whole length to prevent claws becoming trapped if the gap narrows. The width should not be larger than half the width of the claw-floor-contact area (European Food Safety, 2005). A maximum 60% of gaps (in relation to the total contact area) for weaners with 8 kg body weight (bw), 51% for finishers of 100 kg bw and 40% for heavier pigs are considered beneficial for welfare and health (European Food Safety, 2005). Claw lesions and lameness are caused by inappropriate slat and gap width, wetness and dirtiness of the floor. Good flooring must be smooth and non-slippery, dry due to quick drainage and easy to clean. Concrete slats should have rounded edges and no sharp cuts or cracks, which might cause claw injuries. Deterioration of concrete floor, e.g. by cleaning procedures, which results in a rough surface, can cause injuries to claws and legs as well as infection. The roughness and hardness of flooring affects hoof abrasion. In general, abrasion should outweigh horn growth, avoiding both overgrown hooves and abrasion of soft hind claws. Other materials, like plastic and cast iron, are more slippery, but are warmer for the pigs to lie on.

Partly or fully slatted floors can be kept clean more easily than solid concrete floors. Where solid floors are used, efficient drainage is essential. A solid floor can become very contaminated, especially if ventilation and airflow are poor. In general, to prevent fouling it is important to ensure the right climatic conditions in lying areas (Larsen et al., 2018). Dust, gaseous emissions and airborne bacteria can be reduced using effluent channels in which manure is submerged in water. In partly slatted flooring systems, the minimal slatted area must be increased where the temperature is above 25°C and with higher stocking density. An appropriate proportion can be 40% or more, depending on both parameters (European Food Safety, 2005). Reducing the surface area, e.g. by lower width of slats, can reduce fouling and NH_3 emissions (Aarnink et al., 1997).

Deep bedding with material such as straw, sawdust, wood chips or peat usually requires a solid concrete floor. Bedding material provides the pig with physical and thermal comfort. Since it encourages rooting and foraging behaviour, it increases animal welfare. In managing bedding material, it is crucial to avoid high humidity and ammonia emissions, exposure to dust, as well as risks from mycotoxins and pathogens in the material. For any system providing bedding, costs will be increased by the need for composting or disposal.

Fresh straw as bedding material should be provided every 2–3 days. Damp straw, soaked with urine and faeces, soften the claws and increases the risk of claw infections (Maes et al., 2016). In fattening units, manure is usually not removed until slaughter. In cold conditions deep-litter bedding with straw or compost will provide thermal comfort. However, in warm conditions bedding produces a large amount of heat, increasing the risk of thermoregulatory problems. In so-called 'sloped manure confinements,' an inclined solid floor area with an approx. 6% slope is combined with access to bedding material in the upper side, which can be used by the pigs. Manure is collected on the lower side of the pen and can be removed automatically. This combination results in low emissions and a good microclimate. Other systems have a separate bedded area for lying and a bare area for defecation, which can have slatted or solid floor. Bedding in general requires more space and additional work to deal with dung.

8 Specific housing requirements for sows and suckling piglets

8.1 Sows

Three housing requirements are considered important for the health and welfare of sows:

1. Environmental enrichment for gestating sows to extend feeding and foraging times as well as to improve satiety.
2. Escape opportunities in group-housed sows to reduce aggressive behaviour.
3. Opportunities for nest-building behaviour and movement for farrowing and lactating sows

For groups of gilts, there needs to be sufficient space to support attaining puberty. Levis (1997) published the formula: $A\ (m^2) = 0.036 \times BW^{0.66}$ to assess minimum floor space for gilts.

8.2 Pregnant sows

About 75–85% of breeder sows in an intensive farrowing farm are kept as gestating sows, with the percentage depending on the length of lactation.

According to the Directive 2008/120/EC, gestating sows should be kept in groups up to 4 weeks after service to 1 week before the expected time of farrowing. Changes in farrowing crate regulation are expected in the near future and the time period for sows kept in crates will be reduced to a minimal number of days around farrowing and insemination.

One advantage of using crates is protection from aggressive behaviour, which is a particular issue during the first weeks after insemination. Individual housing also makes it easier to monitor health and tailor feeding to ensure optimal body condition. However, crated sows can suffer from problems such as urinary tract infections, overgrown hooves, and stereotypical behaviours resulting from confinement, including bar-biting, head-weaving, and sham-chewing. On the other hand, loose housing is often associated with problems such as lameness, foot and other diseases such as cystitis and pyelonephritis, particularly if floors are not kept clean and dry (Calderón Díaz et al., 2014). Slipping or foot disorders in sows can increase crushing of piglets (Calderón Díaz et al., 2014; Fitzgerald et al., 2012).

Group housing allows freedom of movement and social interactions. The number of sows in each group can vary greatly in size from less than 10 sows to hundreds. The optimum group size for pregnant sows has yet to be determined (Anil et al., 2006). Space requirements in group housing of sows are determined in the Council Directive 2008/120/EC (Table 6). The main source of stress is competition for resources so groups should ideally have synchronous access to feed and a comfortable resting area. Electric sow feeder systems (ESF) are frequently used in group housing systems for pregnant sows. Whilst they allow individually-tailored feeding, they can cause competition between sows (Jang et al., 2017). ESF should protect the sow inside the automated feeder box to prevent potential aggression. Low-ranking sows were found to be at a disadvantage in both static and dynamic groups, being subject to more aggression and injuries and poorer productivity compared with high-ranking sows (Bench et al., 2013). Sows should be provided with fermentable fibers in the diet to improve satiety.

Sow groups can either be stable, which means the group does not change, or may be dynamic, with sows leaving and entering the established group

Table 6 Space requirements for sows and gilts kept in groups according to Council Directive 2008/120/EC

	Area in square meters		
	Group size: up to 5 animals	Group size: 6–40 animals	Group size more than 40 animals
Gilts	≥1.80 m^2	≥1.64 m^2	≥1.48 m^2
Sows	≥2.48 m^2	≥2.25 m^2	≥2.03 m^2

multiple times. When sows are re-grouped after weaning or after insemination, increased aggression occurs due to establishing hierarchies. Full integration of small groups of gilts or sows in large dynamic groups may last 3–4 weeks, with the day of introduction producing the highest level of aggression (Spoolder, 1998). Pre-exposure in a separated area in the same room allows visual, olfactory and auditory perception of future pen-mates, which can reduce later fighting. Sustained aggression may be caused by competition for resources. Aggression can be more severe in circular compared to rectangular pens, as pigs prefer corners to hide from other pigs (Wiegand et al., 1994). In general, solid pen barriers can reduce aggression in group-housed sows. In large, dynamic sow groups, solid barriers can be used to create lying pens for sow subgroups. In general, escape opportunities must be provided to reduce aggression and minimise risks of injuries and stress.

8.3 Farrowing and lactating sows and suckling piglets

While there is no specific space allowance for farrowing pens in current EU regulations, 2.79 m^2 is recommended during farrowing and lactation by Baxter et al. (2011b). Twenty-four hours prior to farrowing sows exhibit nest-building behaviour with a peak 6–12 h before farrowing (Andersen et al., 2014). Sows showing intense nest-building activity are thought to be more maternal with less risk of crushing piglets (Iempeng et al., 2019). Maternal traits in combination with housing conditions are major factors in pre-weaning mortality. Pen design can reduce the number and type of crushing events, particularly when sows lie down or roll over (Baxter et al., 2011a). Providing sows with bowls will facilitate water uptake in comparison of nipple or press level drinkers, reducing the risk of conditions such as cystitis-pyelonephritis (Stalder et al., 2010).

Currently the vast majority of farrowing sows in the EU are housed in farrowing crates to allow safe handling of sows and piglets and minimize the risk of crushing. In future the number of days when a sow is confined around farrowing will be kept to a minimum of approximately five days. Research suggests a crate size of approximately 200 cm width × 280 cm length (Pedersen et al., 2013). Many crates are too narrow to allow adequate space for getting up and lying down. Crates must also be adjustable to differing sizes of sow to allow sufficient but mot excessive movement. Space for piglets requires approximately 56 cm to both sides of the sow for suckling as well as a lying area large enough to allow synchronized resting.

Research suggests that free-farrowing pens that do not confine the sow may be superior in terms of lower stillbirths, increased colostrum intake by piglets as well as fewer injuries, though devices such as farrowing rails are needed to protect piglets (Verhovsek et al., 2007; Kilbride et al., 2009a). Pens for free farrowing should be divided into a dedicated lying area separated from an area for defecation. About 50% of the floor should be solid to allow nest-building behaviour with adequate material. The slatted area at the rear of the pen should be long and wide enough (longer than the sow) to allow the sow to distance herself from the feeding trough. Floor plans can be found in various publications or on the website of the FreeFarrowing Project (www.freefarrowing.org, access: 22.9.2021). Pens are often equipped with sloped walls to assist the sow in lying down. Pens should include creeps (areas accessible only to piglets), which should be near to the udder (mostly at the side or in a corner). Recommendations for size of free farrowing pens range from 6.5 m^2 to 7.5 m^2 (Beilage, 2020). It is important to keep floors dry and warm to minimize injuries such as shoulder lesions, leg injuries and teat damage (Kilbride et al., 2009a,b).

Further research on housing systems for farrowing and lactating sows is needed to address the different requirements for piglets and sows in a limited space with respect to floor conditions and temperature. For lactating sows the upper critical temperature is approximately 21°C given the high risk of heat stress in sows due to increased metabolic rate, especially during birth and lactation (Ross et al., 2017). While normal breathing frequency is approximately 20 breaths/min, breathing frequency is increasing by an additional 30 breaths/min with every 2°C. This results in decreased feed intake followed by decreased milk production in lactating sows with a negative impact on piglet performance (Muns et al., 2016).

Sows prefer a room temperature of 18–20°C while the optimum temperature for new-born piglets is 32–34°C. A typical room temperature in farrowing units is 20–22°C, which is relatively warm for the sow but cold for piglets (Pedersen et al., 2013). During and after birth piglets are vulnerable to cold stress due to a lack of fatty tissue. Within 30 min, the body temperature can decrease by 5°C to a critical temperature of 33°C. Temperatures below 16°C result in rapid loss in body heat, which are potentially fatal. Readily accessible, heated creep areas for piglets are therefore indispensable (Villanueva-García et al., 2020). To ensure an air temperature of 32°C in the nest, the floor should be heated up to approximately 38–40°C with an even distribution of heat (Strauch, 2013). Heat lamps differ in wattage and light wavelength, which affects their use by piglets. Adequate heating can be assessed by the lying behaviour of piglets. The majority of piglets lie laterally with 40–50% touching each other (Lane, 2019). Recommendations for space of the nesting area range from 0.06 m^2 per piglet to 0.166 m^2 per piglet depending on lying recumbancy (www.freefarrowing.org, access 22.9.2021). The piglet nest should allowing relaxed lying by all piglets simultaneously (1.1 m^2 for 10 piglets at the end of lactation), with a roof, solid floor and ideally surrounded by three solid walls.

Suckling piglets should already have immediate access to water to avoid dehydration especially in hot weather conditions. Piglets drink about 50–65 mL/kg body weight. Larger water dispensers facilitate socialisation. Spillage from drinkers should not wet the floor. In well-designed farrowing pens piglets faeces and urine will be collected below the slatted area away from the warm and dry nesting area.

Group-farrowing pens allow early socialization of piglets who form social hierarchies rapidly with less aggression and injuries (D'Eath, 2004, 2005). However, group-farrowing pens are not widely used given problems such as cross-suckling of piglets and disturbing lactating sows. These disadvantages might not be outweighed by reduced later aggression between piglets (Pitts et al., 2000).

9 Specific housing requirements for nursery and fattening pigs

9.1 Nursery pigs

In nature, weaning is a gradual process that begins approximately 60–100 days after birth (Newberry and Wood-Gush, 1985). In accordance with Directive 2008/120/EC, no piglet should be weaned at less than 28 days of age, but they may be weaned up to 7 days earlier if they are moved into specialized housing separate from sows. In organic farms, piglets must be nursed for a minimum of 40 days. Stereotypic behaviour such as belly-nosing may indicate welfare issues during weaning and a slower growth rate (Straw and Bartlett, 2001). A comparison between pigs reared in confined or outdoor environments suggested belly-nosing and aggressive behaviour in outdoor piglets (Hötzel et al., 2004).

After weaning a transition period begins, in which gut maturation, odontiasis and behavioural changes take place; feed intake becomes biphasic with peaks in the morning and afternoon. The process of weaning is one of the most stressful events in the life of the pig. An additional challenge for the piglet is that, at the same time, an abrupt change of milk to solid feed occurs, during which passive immunity from sow milk declines while the piglet`s gut is still immature (Wei et al., 2021). It is therefore essential that transition of feeding should be as smooth as possible providing fresh and palatable creep feed from 7 days of age and feeding the same pre-starter in the first days after weaning. Higher intake of solid creep feed prior to weaning results in higher intake post-weaning although there are significant individual variations (Aherne et al., 1982; Muns and Magowan, 2018). For weaners bite-type nipple drinkers or pressure plate-operated bowl drinkers are used (1 drinker per 12 pigs), which should be installed above slats to avoid wet floors (Ocepek et al., 2018) (https://www.daf.qld.gov.au/business-priorities/agriculture/ animals/pigs/piggery-management/housing/basic-housing, accessed: 24.09.2021).

Weaners are often kept in specially-designed pens until approximately 10–12 weeks of age. Environmental parameters such as temperature and hygiene must be carefully controlled. Good pen design provides feeders and drinkers in the centre and lying areas next to walls to allow undisturbed resting. In some husbandry systems nursery pigs are kept in small groups, particularly where litters are kept together from weaning to slaughter. Leaving litters in the farrowing pen for some days after moving the sows reduces additional stress from moving and mixing with other animals whilst reducing pathogen transmission (Tobias et al., 2014). Leaving piglets from one litter together in a pen after moving to the nursery is preferable but not always economic (Faccin et al., 2020; Thomsson, 2008). 15m^2 is recommended as a minimum nursery pen size (Becker et al., 2020). The correct arrangement of feeding places, drinkers, enrichment and resting zones is critical for welfare. It is easier to monitor individuals in small groups of up to 20 pigs but pens are typically too small for an optimal layout.

Pen space can be used more efficiently when $\geq$5 litters of similar age are combined into larger groups. In large groups of more than 70 individuals, management is more challenging but an improved layout is easier to achieve given the larger space available (Becker et al., 2020). Growth performance of piglets can be impacted by large groups of 25–100 animals, especially if floor space is limited (Wolter et al., 2000, 2001). Group sizes of 20–50 pigs are considered a good compromise (Becker et al., 2020). High pig density is a significant factor in increasing the risk of tail biting (Bolhuis et al., 2006). Minimizing tail biting means adding more space which then allows more zoning and installation of toys or other enrichment materials (Moinard et al., 2003; Randolph et al., 1981). Nursery pens are often fully or partly slatted. A preferred maximum depth is 4 m with a width of 1.8 m which allows space for a shed width of 10 m (https://www.daf.qld.gov.au/business-priorities/agriculture/animals/pigs/piggery-management/housing/basic-housing, accessed: 23.9.2021). A common recommendation is, that the total depth of the pen should be at least double the width, because narrow and deep pens enable a separate area for defecation (EFSA), 2005). In pens with a partly slatted floor, one-third of the floor can be slatted with a slope of 1:25 and a clear demarcation of the solid floor. Fronts and gates of pens facing each other should be solid, while pigs of neighbouring pens should be able to see each other in the defecation areas (https://www.daf.qld.gov.au/business-priorities/agriculture/animals/pigs/piggery-management/housing/basic-housing, accessed: 23.9.2021). An overview of research in pen size recommendation and structuring of pens in different husbandry conditions can be found e.g. in recommendations by the European Food Safety Authority (EFSA), 2005).

9.2 Fattening pigs

In most farms pigs with an average weight of approximately 30 kg are moved to separate sections to be grown and finished for slaughter. Sorting pigs by weight when entering the fattening unit is also a common strategy to facilitate feeding management. Housing of fattening pigs has traditionally been characterized by limited space and a clean but stimulus-poor environment. This intensive rearing was justified on economical and health grounds (Millet et al., 2005). However, mixing of larger groups of fatteners in more confined conditions is associated with a higher incidence of stress and aggression (Meyer-Hamme et al., 2016). Alternative housing systems are characterized by more space and enrichment stimuli which can then result in improved weight gain and meat quality (Millet et al., 2005). Adding 25–50% more floor space for nearly-to-finishing pigs in the final 3 weeks of fattening was shown to significantly increase average daily weight gain and improve feed conversion rates, though more space can cause social tensions (DeDecker et al., 2005). Comparisons between outdoor and indoor systems reveal different issues. Slaughter data indicate that organic pigs reared in extensive systems had fewer respiratory problems, skin lesions and tail wounds compared to conventional pigs. On the other hand, joint lesions and white spot livers were more common with, in particular, a high level of exposure to endoparasites present in pasture conditions (Lindgren et al., 2014).

Minimal space allowance for fattening pigs up to 110 kg should be calculated using the formula: Area (m^2) = 0.047 × $BW^{0.67}$ in case, that heat stress can not be precluded, resulting in approximately 1.1 m2/pig. Those space requirements are significantly higher than those stated by the EU Council Directive 2001/88/CE (Table 4) and also higher than those calculated by other published formula: e.g. A (m^2) = 0.041 × $BW^{0.67}$ (Pastorelli et al., 2006). Based on the number of body lesions, Camp Montoro et al. (2021) found that, although compliant with EU legislation, space allowances for fattening pigs were detrimental to the welfare (Camp Montoro et al., 2021). The influence of stocking density can be seen in weight gain by fatteners. Pigs housed in a stocking density of 1.4 m^2/pig showed a 24 g higher average daily weight gain compared to those housed in a density of 1 m^2/pig (Corino et al., 2003). Other studies demonstrate the influence of stocking density on metabolism, intestinal morphology and immunity of growing pigs (Li et al., 2020).

Boar fattening is a new challenge in some European countries such as Germany, due to the ban on piglet castration without anaesthesia. It is common to leave boars intact, especially in Ireland, the UK and Portugal. Boars are more aggressive and show more mounting behaviours than castrated males. Risk factors for aggressive behaviour are too few eating places, restricted feeding, a low level of amino acids in the diet, insufficient water supply, disease, a suboptimal climate in the barn and fear of humans (Backus et al., 2016). Raising in sibling groups seems to reduce aggressive behaviour (Fredriksen et al., 2007). Maintaining pig health, stable social groups and sufficient physical resources are necessary to raise entire male pigs in accordance with good welfare standards (Borell et al., 2020).

The flooring system in fattening units is closely connected to handling of manure on a farm. Slatted floors are less labor intensive than solid flooring because feces fall through the slats into the slurry system, making it easier to keep floors clean. Most fattening systems are equipped with a fully slatted floor with no separation between lying and defecation/dunging areas. Gap width should be not more than 18 mm for fatteners. In general flooring and space affects the prevalence of disorders such as claw lesions, which can be assessed by abbatoir surveys (Alban et al., 2015). Claw lesions were found less often in finishing pigs kept on bedded floors compared to pigs on solid slatted floors. Pigs kept on slatted floors (partially and fully) show often specific heel, sole, white line and wall lesions (Mouttotou et al., 1999). This suggests that types of lesion are associated with different floor types (EFSA, 2005). Some authors have suggested that slat material (determining e.g. traction level, abrasiveness, hardness) is more important than the slat width for pig health and welfare.

10 Building-related biosecurity measures

The implementation of biosecurity measures in intensive pig production is easier than in extensive production which is more vulnerable to epidemic diseases such as African swine fever (Maes et al., 2019). External biosecurity measures are influenced by the location of the farm, the environment and economic aspects. Measures include fencing and hygiene locks to control access. A perimeter fence should demarcate the farm boundary and keep out wild animals such as wild boars. All visitors should stay at the perimeter. Containers for dead animals must be located outside the perimeter fence. Dead animals must be stored and disposed appropriately to avoid environmental contamination and infection of animals and humans. Birds are a potential vector of pathogens like Salmonella. This requires biosecurity measures to keep birds out by using bird-proof nets and closing doors and windows. Feed silos must be closed to avoid contamination by bird faeces. Where farmers do not have sufficient land to accommodate manure, it must be moved to other farms though this might lead to a spread of pathogens. A quarantine period for purchased gilts is mandatory. Contact between a quarantine unit and main farm must be prevented (Alarcón et al., 2021). A minimum safe distance to prevent transmission of diseases by air or vectors such as flies is 1000 m, although there are pathogens which can be transmitted over longer distances (e.g. foot and mouth disease virus) (Desrosiers, 2011).

Internal biosecurity measures must be implemented to prevent transmission of pathogens within the farm. Good hygiene practices include wearing personal protective equipment and preventing ill employees from entering the farm. Airborne transmission of pathogens between different units is a particular challenge. In herds located in pig-dense areas, airborne transmission between farms has been shown e.g. for PRRSV, *Mycoplasma hyopneumoniae,* swine influenza viruses and Aujeszky's disease virus (Maes et al., 2000; Desrosiers, 2011; Otake et al., 2010). Air filtration can help prevent infection (Alonso et al., 2013). Airborne transmission is less of an issue when negative air pressure caused by high ventilation rates in pens for fattening pigs (with a higher risk of infection) sucks in air from units with young animals with a lower risk. However, where a low number of animals is left in a fattening unit (e.g. because they are divided for sale by weight), air flow may go in the wrong direction, increasing the risk of infection of young pigs by pathogens shed from older pigs.

The movement of pigs through the farm is a critical factor in disease transmission. Diseased animals must be separated from healthy pigs. To efficiently remove an animal from a group for isolation or treatment, a small capture or hospital pen should be included in the pen design. There need to be hygiene steps implemented between different age groups to prevent passing disease on from one group to another. The AIAO system allows separation of age groups to minimize pathogen transmission between groups. An AIAO policy in the farrowing unit is e.g. particularly important to reduce transmission of *M. hyopneumoniae* (Nathues et al., 2013). Good hygiene practices such as routine cleaning and disinfection are essential in farrowing units, nursery and fattening units, including cleaning and disinfection of pens before a new group arrives. There is a temptation to hold back pigs of insufficient weight and integrate them into a younger batch but this should be avoided (Calderon Diaz et al., 2017). Batch farrowing of sows ideally with no more than 4–7 days between litters will increase the time available for cleaning floors, ceilings, walls, feeders and drinkers to reduce risk of pathogen spread between batches. A 5-week batch management system was found to achieve better disease control than a 4-week batch management system (Vangroenweghe et al., 2012).

Important cleaning steps are (https://www.thepigsite.com/articles/cleaning-and-disinfection; accessed: 15.07.21):

- Removal of movable pen equipment and separate cleaning in parallel with cleaning of walls and floors.
- Removal of dust by soaking the pen with water and detergent for approximately 24 h to dissolve proteins and fat.
- Pressure-cleaning with hot water using approximately 6900 kPa.
- Drying of all surfaces followed by disinfection of all surfaces
- The water dispensing system should be cleaned and disinfected and potential intruders such as insects or mice should be prevented from getting onto cleaned and disinfected surfaces.

Re-stocking 24 h after disinfection resulted in partially wet slats under natural ventilation, increasing soiling of floors which resulted in poor air quality and reduced growth rates (Murphy, 2011). More time is required (e.g. 2–3 days) to ensure pens are dry before they are restocked.

11 The importance of stockmanship

Stock people require a range of well-developed husbandry skills and knowledge to care for and manage farm animals. Negative behaviour like hitting and pushing pigs is associated with high fear levels in pigs, which is associated with reduced productivity in terms of reproductive performance of commercial sows (Hemsworth et al., 1994). Training to target technical skills and knowledge, as well as attitudes and behaviours, of stockpeople should be a primary component of human resource management practices on a farm.

Analogous to the 'Five Freedoms' three essential requirements of stockmanship are (Council, 2007):

- Knowledge of animal husbandry.
- Practical skills in animal husbandry.
- Personal qualities and attitudes.

Stock people need to understand the reasons for and have the training to implement appropriate practices to ensure the safety and welfare of pigs at all stages in their lifecycle, including skills in observing the animals in their care to identify potential health or welfare issues. In general, farms with large herds are able to offset the negative effects of herd size by a higher degree of awareness, better risk assessment processes, training, resources and procedures (Gardner et al., 2002; Laanen et al., 2013). The Report on Stockmanship and Farm Animal Welfare of the Farm Animal Welfare Council in the UK has highlighted issues in recruiting staff due to issues such as the negative image of livestock farming, long working hours and sometimes poor pay and conditions. In the long run there is a development towards automated monitoring, data analysis and process control (through precision livestock farming technologies) which will require a different type of stockperson (Benjamin and Yik, 2019).

12 Conclusion and future trends

To achieve sustainable livestock production, trade-offs between animal health, production efficiency, animal welfare standards and consumer protection must be avoided. There is a complex relationship between economic performance on the one hand and welfare on the other (Henningsen et al., 2018). Further complication comes from challenges such as antibiotic usage and environmental impact. However, a good farm manager can reconcile these requirements. Whilst there are many ways of measuring individual parameters, many are not practical on a large scale in practical farm settings. Selecting effective but workable criteria for the assessment of production systems in the face of conflicting goals will be the task.

Current and future challenges for pork production especially in Europe are:

i Implementation of new and more sustainable husbandry standards and stricter rules to meet environmental and animal welfare requirements. New regulations will increase costs for livestock production in the EU. Differences between EU countries with respect to national standards may lead to distortion of competition within the European Free Trade Association.

ii Developing new extensive husbandry systems. Whilst this may allow better expression of natural behaviours, the risk of (zoonotic) pathogens in the production chain will increase and external and internal biosecurity will be harder to manage. In addition, extensive systems make it hard to control emissions. The generally lower efficiency of extensive systems (higher feed conversion rates, lower average daily weight gain and higher animal losses) also raises issues of competitiveness as well as increased use of scarce land and feed resources.

iii Increased frequency of extreme weather events as a consequence of climate change. Periods of extreme heat in particular could lead to high production losses from heat stress, unless housing systems are suitably adapted.

iv Lower future availability of water. Water supply in Europe is a potential problem for almost half of its population, which is reflected by the Water Exploitation Index (how much water is consumed each year in relation to the total long-term freshwater resources in region). In the last 50 years renewable water resources per capita decreased by 24% in the European Union with about one third of the European countries being threatened by water shortage (https://www.eea.europa.eu/data-and-maps/indicators/use-of-freshwater-resources-3/assessment-4, accessed: 14.07.21).

v Lower future availability of suitably trained and knowledgeable personnel. Farmers and stock people must now not only be experts in animal husbandry. Their knowledge must cover disease prevention, disease recognition, reproduction management, neonatal care, nutrition, and understanding behaviour. They must also manage and record a large quantity of data to fulfil legal requirements. The implementation of a hazard analysis and critical control point (HACCP) system would help to manage biosecurity, health and welfare as well as consumer protection. A HACCP-type standard has not been developed so far in swine production.

Digitalization of data streams, data bundling and management have great potential in meeting increasing requirements for higher standards and greater transparency within the pork production chain. Several systems are under development for data collection and digitalized data flow. Sensors for continuous measurement of various environmental variables are also available on the market and are used for automatic adjustment to ensure a stable micro-climate in barns.

Smart farming systems help to digitalize, connect and evaluate any data gathered in the farm environment for tracing, tracking and managing individual activities. Mobile applications (Apps) as software installed on electronic devices such as a smartphone, tablet, computer or other electronic devices facilitate data recording and are provided by various companies. In principle, all data can be bundled and controlled in one complex system. An integrated sow management system can, e.g. be based on scanning of electronic ear tags and communication with the smartphone, so that data relating to all activities, e.g. insemination, farrowing or weaning, are digitized. Such technological advances will support livestock farming by providing standardized measures to predict and better manage both productivity and animal welfare.

13 Where to look for further information

Modern swine husbandry is undergoing a major change: In addition to high animal-welfare standards, the focus is on improving production of safe and high-quality pork using climate- and resource-friendly production systems. In addition, the implementation of new digital technologies provide stockman with opportunities that have not yet been fully exploited.

Efforts to push forward developments to make food systems fair, healthy and environmentally-friendly are supported by the European Commission. Details on the farm-to-fork strategy can be read on the following site: https://ec.europa.eu/food/horizontal-topics/farm-fork-strategy_en#documents (accessed: 28.04.2022).

While human and animal health are interconnected significantly influenced by the environment, the EU One Health Action Plan against AMR (antimicrobial resistance) also focus on the reduction of antimicrobials for farmed animals. Based on the knowledge that better animal welfare improves animal health, so that less medication is needed, the Commission will revise the animal welfare legislation. The actual development of EU Action on Antimicrobial Resistance can be read here:

https://ec.europa.eu/health/antimicrobial-resistance/eu-action-antimicrobial-resistance_en (accessed: 28.04.2022).

There are several approaches to collect innovative projects by the EU. For the pig sector the thematic network 'EU Pig Innovation Group' is supporting innovative practices into pig production. The initiative belongs to the European activity within EIP-AGRI (agriculture and innovation). Search for projects is possible on the EIP-Agri site (https://ec.europa.eu/eip/agriculture/en/find-connect/projects/eu-pig-eu-pig-innovation-group-thematic-network, accessed: 28.04.2022). Good agricultural practice is also continuously followed by the Food and Agricultural Organisation of the United Nations, because pig production systems differ worldwide und should be developed towards sustainable production guaranteeing a high food safety (https://www.ciwf.org.uk/media/5492194/gap_pig_book_full.pdf, accessed: 28.04.2022). Several national information platforms for research projects but also for pig husbandry and pig welfare standards exist. One example is the Farm Animals-Pig Welfare site in UK (https://www.ciwf.org.uk/farm-animals/pigs/pig-welfare/ accessed: 28.04.2022).

14 References

Aarnink, A. J. A., Swierstra, D., van den Berg, A. J. and Speelman, L. 1997. Effect of type of slatted floor and degree of fouling of solid floor on ammonia emission rates from fattening piggeries. *Journal of Agricultural Engineering Research* 66(2), 93-102.

Aherne, F. X., Danielsen, V. and Nielsen, H. E. 1982. The effects of creep feeding on pre- and post-weaning pig performance. *Acta Agriculturae Scandinavica* 32(2), 155-160.

Alarcón, L. V., Allepuz, A. and Mateu, E. 2021. Biosecurity in pig farms: a review. *Porcine Health Management* 7(1), 5.

Alban, L., Petersen, J. V. and Busch, M. E. 2015. A comparison between lesions found during meat inspection of finishing pigs raised under organic/freerange conditions and conventional, indoor conditions. *Porcine Health Management* 1, 4.

Albarella, U., Manconi, F., Rowley-Conwy, P. and Vigne, J.-D. 2006. Pigs of Sardinia and Corsica: a biometrical re-evaluation of their status and history. In: Tecchiati, U. and Sala, B. (Eds), Archaeozoological Studies in honour of Alfredo Riedel. Ripartizione beni culturali Ufficio beni archeologici, Bolzano , pp. 285-302.

Alonso, C., Murtaugh, M. P., Dee, S. A. and Davies, P. R. 2013. Epidemiological study of air filtration systems for preventing PRRSV infection in large sow herds. *Preventive Veterinary Medicine* 112(1-2), 109-117.

American Society of Heating, Refrigerating and Air-Conditioning Engineers, Inc. (ASHRAE). 2012. ASHRAE handbook–Heating, Ventilating, and Air-Conditioning Systems and Equipment. Altanta, GA.

Andersen, I. L., Vasdal, G. and Pedersen, L. J. 2014. Nest building and posture changes and activity budget of gilts housed in pens and crates. *Applied Animal Behaviour Science* 159, 29-33.

Angrecka, S. and Herbut, P. 2016. Impact of barn orientation on insolation and temperature of stalls surface. *Annals of Animal Science* 16(3), 887-896.

Anil, L., Anil, S. S. and Deen, J. 2007. Effects of allometric space allowance and weight group composition on grower-finisher pigs. *Canadian Journal of Animal Science* 87(2), 139–151.

Anil, L., Anil, S. S., Deen, J., Baidoo, S. K. and Walker, R. D. 2006. Effect of group size and structure on the welfare and performance of pregnant sows in pens with electronic sow feeders. *Canadian Journal of Veterinary Research = Revue Canadienne de Recherche Veterinaire* 70(2), 128–136.

Backus, G. B. C., van den Broek, E., van der Fels, B., Heres, L., Immink, V. M., Knol, E. F., Kornelis, M., Mathur, P. K., van der Peet-Schwering, C., van Riel, J. W., Snoek, H. M., de Smet, A., Tacken, G. M. L., Valeeva, N. I. and van Wagenberg, C. P. A. 2016. Evaluation of producing and marketing entire male pigs. *NJAS: Wageningen Journal of Life Sciences* 76(1), 29–41.

Baldwin, B. A. and Meese, G. B. 1977. Sensory reinforcement and illumination preference in the domesticated pig. *Animal Behaviour* 25, 497–507.

Banhazi, T. M., Seedorf, J., Rutley, D. L. and Pitchford, W. S. 2008. Identification of risk factors for sub-optimal housing conditions in Australian piggeries: part 3. Environmental parameters. *Journal of Agricultural Safety and Health* 14(1), 41–52.

Baxter, E. M., Jarvis, S., Sherwood, L., Farish, M., Roehe, R., Lawrence, A. B. and Edwards, S. A. 2011a. Genetic and environmental effects on piglet survival and maternal behaviour of the farrowing sow. *Applied Animal Behaviour Science* 130(1–2), 28–41.

Baxter, E. M., Lawrence, A. B. and Edwards, S. A. 2011b. Alternative farrowing systems: design criteria for farrowing systems based on the biological needs of sows and piglets. *Animal: An International Journal of Animal Bioscience* 5(4), 580–600.

Becker, C., Böck, N., Drexl, V., Elkmann, A., Freisfeld, F. and Häuser, S. 2020. DLG-Merkblatt 458 Strukturiereung von Buchten in Ferkelaufzucht und Schweinemast. In *DLG-Merkblätter (Deutsche Landwirtschaftliche Gesellschaft DLG e.V.).*

Beilage, E. G. 2020. *Literaturübersicht zur Unterbringung von Sauen währned Geburtsvorbereitung, Geburt und Säugezeit (Hessisches Ministerium für umwelt, Klimaschutz, Landwirtschaft und Verbraucherschutz).*

Bench, C. J., Rioja-Lang, F. C., Hayne, S. M. and Gonyou, H. W. 2013. Group gestation sow housing with individual feeding–II: how space allowance, group size and composition, and flooring affect sow welfare. *Livestock Science* 152(2–3), 218–227.

Benjamin, M. and Yik, S. 2019. Precision livestock farming in swine welfare: a review for swine practitioners. *Animals: An Open Access Journal from MDPI* 9(4), 133.

Bisoniya, T. S., Kumar, A. and Baredar, P. 2014. Study on calculation models of earth-air heat exchanger systems. *Journal of Energy* 2014, 1–15, http://dx.doi.org/10.1155/2014/859286.

Bolhuis, J. E., Schouten, W. G. P., Schrama, J. W. and Wiegant, V. M. 2006. Effects of rearing and housing environment on behaviour and performance of pigs with different coping characteristics. *Applied Animal Behaviour Science* 101(1–2), 68–85.

Borell, E. V., Bonneau, M., Holinger, M., Prunier, A., Stefanski, V., Zöls, S. and Weiler, U. 2020. Welfare aspects of raising entire male pigs and Immunocastrates. *Animals* 10(11), 2140, https://doi.org/10.3390/ani10112140.

Botreau, R., Veissier, I. and Perny, P. 2009. Overall assessment of animal welfare: strategy adopted in Welfare Quality®. *Animal Welfare* 18, 363–370.

Broom, D. 1996. Animal welfare defined in terms of attempts to cope with the environment. *Acta Agriculturae Scandinavica, Section A – Animal Science Suppl.* 27, 22–28.

Broom, D. M. and Fraser, A. F. 2015. *Domestic Animal Behaviour and Welfare* (5th edn.). CABI Publishing, Cambridge, MA, USA.

Calderon Diaz, J. A., Diana, A., Boyle, L. A., Leonard, F. C., McElroy, M., McGettrick, S., Moriarty, J. and Garcia Manzanilla, E. 2017. Delaying pigs from the normal production flow is associated with health problems and poorer performance. *Porcine Health Management* 3, 13.

Calderón Díaz, J. A., Fahey, A. G. and Boyle, L. A. 2014. Effects of gestation housing system and floor type during lactation on locomotory ability; body, limb, and claw lesions; and lying-down behavior of lactating sows. *Journal of Animal Science* 92(4), 1675–1685.

Cambra-López, M., Aarnink, A. J. A., Zhao, Y., Calvet, S. and Torres, A. G. 2010. Airborne particulate matter from livestock production systems: a review of an air pollution problem. *Environmental Pollution* 158(1), 1–17.

Camp Montoro, J., Boyle, L. A., Solà-Oriol, D., Muns, R., Gasa, J. and Garcia Manzanilla, E. 2021. Effect of space allowance and mixing on growth performance and body lesions of grower-finisher pigs in pens with a single wet-dry feeder. *Porcine Health Management* 7(1), 7.

Cargill, C., Murphy, T. and Banhazi, T. 2002. Hygiene and air quality in intensive housing facilities in Australia. *Animal Production in Australia* 24, 387-393.

Carol Petherick, J. 1983. A note on the space use for excretory behaviour of suckling piglets. *Applied Animal Ethology* 9(3–4), 367-371.

Carroll, G. A., Boyle, L. A., Hanlon, A., Collins, L., Griffin, K., Friel, M., Armstrong, D. and O'Connell, N. E. 2018. What can carcass-based assessments tell us about the lifetime welfare status of pigs? *Livestock Science* 214, 98–105.

Carroll, G. A. and Groarke, J. M. 2019. The importance of the social sciences in reducing tail biting prevalence in pigs. *Animals: An Open Access Journal from MDPI* 9(9), 591.

Chantziaras, I., De Meyer, D., Vrielinck, L., Van Limbergen, T., Pineiro, C., Dewulf, J., Kyriazakis, I. and Maes, D. 2020. Environment-, health-, performance- and welfare-related parameters in pig barns with natural and mechanical ventilation. *Preventive Veterinary Medicine* 183, 105150.

Chapel, N. M., Radcliffe, J. S., Stewart, K. R., Lucas, J. R. and Lay, D. C., Jr. 2019. The impact of farrowing room noise on sows' reactivity to piglets. *Translational Animal Science* 3(1), 175–184.

Corino, C., Rossi, R. and Pastorelli, G. 2003. Different types of flooring and space allowance in heavy pigs: effects on growth performances. *Italian Journal of Animal Science* 2, 388–390.

Croney, C. C., Adams, K. M., Washington, C. G. and Stricklin, W. R. 2003. A note on visual, olfactory and spatial cue use in foraging behavior of pigs: indirectly assessing cognitive abilities. *Applied Animal Behaviour Science* 83(4), 303–308.

D'Eath, R. B. 2004. Consistency of aggressive temperament in domestic pigs: the effects of social experience and social disruption. *Aggressive Behavior* 30(5), 435–448.

D'Eath, R. B. 2005. Socialising piglets before weaning improves social hierarchy formation when pigs are mixed post-weaning. *Applied Animal Behaviour Science* 93(3–4), 199–211.

Dalin, A. M., Magnusson, U., Häggendal, J. and Nyberg, L. 1993. The effect of transport stress on plasma levels of catecholamines, cortisol, corticosteroid-binding globulin, blood cell count, and lymphocyte proliferation in pigs. *Acta Veterinaria Scandinavica* 34(1), 59-68.

De Luca, S., Zanardi, E., Alborali, G. L., Ianieri, A. and Ghidini, S. 2021. Abattoir-based measures to assess swine welfare: analysis of the methods adopted in European slaughterhouses. *Animals: An Open Access Journal from MDPI* 11(1), 226.

DeDecker, J. M., Ellis, M., Wolter, B. F., Corrigan, B. P., Curtis, S. E., Parr, E. N. and Webel, D. M. 2005. Effects of proportion of pigs removed from a group and subsequent floor space on growth performance of finishing pigs. *Journal of Animal Science* 83(2), 449–454.

Deshazer, J., Hahn, G. and Xin, H. 2009. Chapter 1. Basic principles of the thermal environment and livestock energetics. In: Livestock Energetics and Thermal Environment Management. American Society of Agricultural and Biological Engineers, St. Joseph, MI, 2009. doi:10.13031/2013.28294.

Desrosiers, R. 2011. Transmission of swine pathogens: different means, different needs. *Animal Health Research Reviews* 12(1), 1–13.

Donham, K. J. 1991. Association of environmental air contaminants with disease and productivity in swine. *American Journal of Veterinary Research* 52(10), 1723–1730.

Düpjan, S., Düpjan, S., Moscovice, L.R., Puppe, B. 2021. The role of enrichment in optimizing pig behaviour and welfare. In: Edwards, S. (Ed.), *Understanding the Behaviour and Improving the Welfare of Pigs*. Part 3 Current welfare issues, Chapter 11, Burleigh Dodds Science Publishing, Cambridge, UK.

EFSA AHAW Panel. 2005. Opinion of the scientific panel on Animal Health and Welfare (AHAW) on a request from the commission related to welfare of weaners and rearing pigs: effects of different space allowances and floor. *EFSA Journal* 268, 1–19.

EFSA AHAW Panel 2012. Scientific Opinion on the use of animal-based measures to assess welfare in pigs. *EFSA Journal* 10(1), 2512.

EFSA Panel on Animal Health and Welfare (AHAW), Nielsen, S. S., Alvarez, J., Bicout, D. J., Calistri, P., Canali, E., Drewe, J. A., Garin-Bastuji, B., Gonzales Rojas, J. L., Herskin, M., Miranda Chueca, M. Á., Michel, V., Padalino, B., Pasquali, P., Roberts, H. C., Sihvonen, L. H., Spoolder, H., Stahl, K., Velarde, A., Viltrop, A., Winckler, C., Blome, S., More, S., Gervelmeyer, A., Antoniou, S. E. and Gortázar Schmidt, C. 2021. African swine fever and outdoor farming of pigs. *EFSA Journal* 19(6), e06639.

Ekkel, E. D., Spoolder, H. A. M., Hulsegge, I. and Hopster, H. 2003. Lying characteristics as determinants for space requirements in pigs. *Applied Animal Behaviour Science* 80(1), 19–30.

Elmore, M. R. P., Garner, J. P., Johnson, A. K., Richert, B. T. and Pajor, E. A. 2010. A flooring comparison: the impact of rubber mats on the health, behavior, and welfare of group-housed sows at breeding. *Applied Animal Behaviour Science* 123(1–2), 7–15.

F.A.W.C. (FAWC). 2009. *Farm Animal Welfare in Great Britain: Past, Present and Future*. London, UK.

Faccin, J. E. G., Laskoski, F., Quirino, M., Gonçalves, M. A. D., Mallmann, A. L., Orlando, U. A. D., Mellagi, A. P. G., Bernardi, M. L., Ulguim, R. R. and Bortolozzo, F. P. 2020. Impact of housing nursery pigs according to body weight on the onset of feed intake, aggressive behavior, and growth performance. *Tropical Animal Health and Production* 52(3), 1073–1079.

Farm Animal Welfare Council 2007. *FAWC Report on Stockmanship and Farm Animal Welfare*. ed. Farm Animal Welfare Council, London, UK .

Faucitano, L., Conte, S., Pomar, C., Paiano, D., Duan, Y., Zhang, P., Drouin, G., Rina, S., Guay, F. and Devillers, N. 2020. Application of extended feed withdrawal time preslaughter and its effects on animal welfare and carcass and meat quality of enriched-housed pigs. *Meat Science* 167, 108163.

Faverjon, C., Bernstein, A., Grütter, R., Nathues, C., Nathues, H., Sarasua, C., Sterchi, M., Vargas, M. E. and Berezowski, J. 2019. A transdisciplinary approach supporting the implementation of a big data project in livestock production: an example From the Swiss pig production industry. *Frontiers in Veterinary Science* 6, 215.

Fitzgerald, R. F., Stalder, K. J., Karriker, L. A., Sadler, L. J., Hill, H. T., Kaisand, J. and Johnson, A. K. 2012. The effect of hoof abnormalities on sow behavior and performance. *Livestock Science* 145(1-3), 230-238.

Flesjå, K. I. and Solberg, I. 1981. Pathological lesions in swine at slaughter. IV. Pathological lesions in relation to rearing system and herd size. *Acta Veterinaria Scandinavica* 22(2), 272-282.

Fraser, A. and Broom, D. 1990. *Farm Animal Behaviour and Welfare*. 3rd edn. Baillierre Tindall, London, UK.

Fraser, D., Weary, D., Pajor, E. and Milligan, B. 1997. A scientific conception of animal welfare that reflects ethical concerns. *Animal Welfare* 6, 187-205.

Fredriksen, B., Bjørn, M., Lium, B., Hexeberg, C., Mosveen, B. and Nafstad, O. 2007. Entire male pigs in farrow-to-finish pens--Effects on animal welfare. *Applied Animal Behaviour Science* 110, 258-268.

Friedrich, L., Krieter, J., Kemper, N. and Czycholl, I. 2020. Iceberg indicators for sow and piglet welfare. *Sustainability* 12(21), 8967, https://doi.org/10.3390/su12218967.

Früh, B. and Holinger, M. 2019. Organic pig farming: key characteristics, opportunities, advantages and challenges. In: Vaarst, M. and Roderick, S. (Eds), *Improving Organic Animal Farming*. Burleigh Dodds Science Publishing, Cambridge, UK, 287-306.

Gardner, I. A., Willeberg, P. and Mousing, J. 2002. Empirical and theoretical evidence for herd size as a risk factor for swine diseases. *Animal Health Research Reviews* 3(1), 43-55.

Geers, R., Dellaert, B., Goedseels, V., Hoogerbrugge, A., Vranken, E., Maes, F. and Berckmans, D. 1989. An assessment of optimal air temperatures in pig houses by the quantification of behavioural and health-related problems. *Animal Production* 48(3), 571-578.

Gieling, E. T., Nordquist, R. E. and van der Staay, F. J. 2011. Assessing learning and memory in pigs. *Animal Cognition* 14(2), 151-173.

Godyń, D., Nowicki, J. and Herbut, P. 2019. Effects of environmental enrichment on pig welfare-A review. *Animals (Basel)* 9(6), 383, doi: 10.3390/ani9060383.

Gonyou, H., Lemay, S. and Zhang, Y. 2006. Effects of the environment on productivity and disease. In: Straw, B., Zimmerman, J., D'Allaire, S. and Taylor, D. J. (Eds), *Diseases of Swine*. Blackwell Publishing Ltd., Oxford, UK, 1027-1038.

Gonyou, H. W. 2001. The social behaviours of pigs. In: Heeling, L. J. and Gonyou, H. W. (Eds), *Social Behaviour in Farm Animals*. CABI Publishing, Wallingford, UK, 147-176.

Gordon, W. A. M. 1963. Environmental studies in pig housing: V. The effects of housing on the degree and incidence of pneumonia in bacon pigs**. *British Veterinary Journal* 119(I), 307-314.

Grosse-Kleimann, J., Plate, H., Meyer, H., Gerhardy, H., Heucke, C. E. and Kreienbrock, L. 2021. Health monitoring of finishing pigs by secondary data use - a longitudinal analysis. *Porcine Health Management* 7(1), 20.

Hamilton, D. N., Ellis, M., Wolter, B. F., Schinckel, A. P. and Wilson, E. R. 2003. The growth performance of the progeny of two swine sire lines reared under different floor space allowances. *Journal of Animal Science* 81(5), 1126-1135.

He, Y., Deen, J., Shurson, G. C. and Li, Y. Z. 2018. Behavioral indicators of slow growth in nursery pigs. *Journal of Applied Animal Welfare Science: JAAWS* 21(4), 389-399.

Heidarinejad, G., Bozorgmehr, M., Delfani, S. and Esmaeelian, J. 2009. Experimental investigation of two-stage indirect/direct evaporative cooling system in various climatic conditions. *Building and Environment* 44(10), 2073-2079.

Hemsworth, P. H. 2018. Key determinants of pig welfare: implications of animal management and housing design on livestock welfare. *Animal Production Science* 58(8), 1375-1386.

Hemsworth, P. H., Coleman, G. J. and Barnett, J. L. 1994. Improving the attitude and behaviour of stockpersons towards pigs and the consequences on the behaviour and reproductive performance of commercial pigs. *Applied Animal Behaviour Science* 39(3-4), 349-362.

Henningsen, A., Czekaj, T. G., Forkman, B., Lund, M. and Nielsen, A. S. 2018. The relationship between animal welfare and economic performance at farm level: a quantitative study of Danish pig producers. *Journal of Agricultural Economics* 69(1), 142-162.

Honeyman, M., McGlone, J., Kliebenstein, J. and Larson, B. 2003. *Outdoor Pig Production*. Pork Industry Handbook, University Extension, ed. Iowa State University.

Hötzel, M. J., Pinheiro Machado F°, L. C., Wolf, F. M. and Costa, O. A. D. 2004. Behaviour of sows and piglets reared in intensive outdoor or indoor systems. *Applied Animal Behaviour Science* 86(1-2), 27-39.

Iempeng, P., Tartrakoon, W., Charoensook, R. and Somparn, P. 2019. Relationship of nest-building and nursing behaviour of sows in the farrowing crate and piglet crushing. *Khon Kaen Agriculture Journal* 47, 355-360.

Jang, J. C., Hong, J. S., Jin, S. S. and Kim, Y. Y. 2017. Comparing gestating sows housing between electronic sow feeding system and a conventional stall over three consecutive parities. *Livestock Science* 199, 37-45.

Jori, F., Relun, A., Trabucco, B., Charrier, F., Maestrini, O., Chavernac, D., Cornelis, D., Casabianca, F. and Etter, E. M. C. 2017. Questionnaire-based assessment of wild boar/domestic pig interactions and implications for disease risk management in Corsica. *Frontiers in Veterinary Science* 4, 198.

Kilbride, A., Gillman, C., Ossent, P. and Green, L. 2009a. Impact of flooring on the health and welfare of pigs. *In Practice* 31(8), 390-395.

Kilbride, A. L., Gillman, C. E. and Green, L. E. 2009b. A cross sectional study of the prevalence, risk factors and population attributable fractions for limb and body lesions in lactating sows on commercial farms in England. *BMC Veterinary Research* 5, 30.

Kim, K. Y., Jong Ko, H. J., Kim, Y. S. K. and Kim, C. N. 2008. Assessment of Korean Farmer's exposure level to dust in pig buildings. *Annals of Agricultural and Environmental Medicine: AAEM* 15(1), 51-58.

Kyriazakis, I. and Emmans, G. C. 1992. The effects of varying protein and energy intakes on the growth and body composition of pigs. 2. The effects of varying both energy and protein intake. *British Journal of Nutrition* 68(3), 615-625.

Kyriazakis, I., Emmans, G. C. and Whittemore, C. T. 1990. Diet selection in pigs: choices made by growing pigs given foods of different protein concentrations. *Animal Science* 51(1), 189-199.

Laanen, M., Persoons, D., Ribbens, S., de Jong, E., Callens, B., Strubbe, M., Maes, D. and Dewulf, J. 2013. Relationship between biosecurity and production/antimicrobial treatment characteristics in pig herds. *Veterinary Journal* 198(2), 508-512.

Lane, K. J. 2019. Heat lamps and heat mats in the farrowing house: effect on piglet production, piglet and sow behavior and energy usage. Thesis, Iowa State University.
Larsen, M. L. V., Bertelsen, M. and Pedersen, L. J. 2018. Review: factors affecting fouling in conventional pens for slaughter pigs. *Animal: An International Journal of Animal Bioscience* 12(2), 322-328.
Levis, D. G. 1997. Management of replacement gilts for efficient reproduction, EC97-274, B. *New Education* (Nebraska-Lincoln, Historical Materials from University of Nebraska-Lincoln Extension).
Li, X., Xiong, X., Wu, X., Liu, G., Zhou, K. and Yin, Y. 2020. Effects of stocking density on growth performance, blood parameters and immunity of growing pigs. *Animal Nutrition* 6(4), 529-534.
Lind, N. M., Moustgaard, A., Jelsing, J., Vajta, G., Cumming, P. and Hansen, A. K. 2007. The use of pigs in neuroscience: modeling brain disorders. *Neuroscience and Biobehavioral Reviews* 31(5), 728-751.
Lindgren, K., Bochicchio, D., Hegelund, L., Leeb, C., Mejer, H., Roepstorff, A. and Sundrum, A. 2014. Animal health and welfare in production systems for organic fattening pigs. *Organic Agriculture* 4(2), 135-147.
Lopes Antunes, A. C., Ersbøll, A. K., Bihrmann, K. and Toft, N. 2017. Mortality in Danish Swine herds: spatio-temporal clusters and risk factors. *Preventive Veterinary Medicine* 145, 41-48.
Lopes Antunes, A. C., Jensen, V. F. and Jensen, D. 2019. Unweaving tangled mortality and antibiotic consumption data to detect disease outbreaks - Peaks, growths, and foresight in swine production. *PLoS ONE* 14(10), e0223250.
Ludwiczak, A., Skrzypczak, E., Składanowska-Baryza, J., Stanisz, M., Ślósarz, P. and Racewicz, P. 2021. How housing conditions determine the welfare of pigs. *Animals: An Open Access Journal from MDPI* 11(12), 3484.
Luo, L., Jansen, C. A., Bolhuis, J. E., Arts, J. A. J., Kemp, B. and Parmentier, H. K. 2020. Early and later life environmental enrichment affect specific antibody responses and blood leukocyte subpopulations in pigs. *Physiology and Behavior* 217, 112799.
Lyte, J. M. and Lyte, M. 2019. Review: microbial endocrinology: intersection of microbiology and neurobiology matters to swine health from infection to behavior. *Animal* 13(11), 2689-2698.
Lyte, M. 2016. Microbial endocrinology in the pathogenesis of infectious disease. *Microbiology Spectrum* 4(2). doi 10.1128/microbiolspec.VMBF-0021-2015.
Maes, D., Deluyker, H., Verdonck, M., Castryck, F., Miry, C., Vrijens, B. and de Kruif, A. 2000. Herd factors associated with the seroprevalences of four major respiratory pathogens in slaughter pigs from farrow-to-finish pig herds. *Veterinary Research* 31(3), 313-327.
Maes, D. G. D., Dewulf, J., Piñeiro, C., Edwards, S. and Kyriazakis, I. 2019. A critical reflection on intensive pork production with an emphasis on animal health and welfare. *Journal of Animal Science* 98(Suppl 1), S15-S26.
Maes, D., Pluym, L. and Peltoniemi, O. 2016. Impact of group housing of pregnant sows on health. *Porcine Health Management* 2, 17.
Marco, M., Yeske, P. and Pieters, M. 2020. General control measures against Mycoplasma hyopneumoniae infections. In: Maes, D., Siblia, S. and Pieters, M. (Eds), *Mycoplasmas in Swine*. Uitgeverij Acco, Leuven, Belgium, Den Haag, The Netherlands, 163-180.
Maselyne, J., Adriaens, I., Huybrechts, T., De Ketelaere, B., Millet, S., Vangeyte, J., Van Nuffel, A. and Saeys, W. 2016. Measuring the drinking behaviour of individual pigs

housed in group using radio frequency identification (RFID). *Animal: An International Journal of Animal Bioscience* 10(9), 1557–1566.

Maselyne, J., Saeys, W. and Van Nuffel, A. 2015. Review: quantifying animal feeding behaviour with a focus on pigs. *Physiology and Behavior* 138, 37–51.

Mendl, M., Held, S. and Byrne, R. W. 2010. Pig cognition. *Current Biology: CB* 20(18), R796–R798.

Mendl, M. and Paul, E. S. 2004. Consciousness, emotion and animal welfare: insights from cognitive science. *Animal Welfare* 13, 17–25.

Meyer-Hamme, S. E., Lambertz, C. and Gauly, M. 2016. Does group size have an impact on welfare indicators in fattening pigs? *Animal: An International Journal of Animal Bioscience* 10(1), 142–149.

Millet, S., Moons, C., Oeckel, M. and Janssens, G. 2005. Welfare, performance and meat qualify of fattening pigs in alternative housing and management systems: a review. *Journal of the Science of Food and Agriculture* 85, 709–719

Millman, S. 2007. Sickness behaviour and its relevance to animal welfare assessment at the group level. *Animal Welfare* 16, 123–125.

Millman, S. 2015. *Caring for Pigs in Hospital Pens*. Proceedings of the American Association of Swine Veterinarians, Orlando, FL.

Mkwanazi, M. V., Ncobela, C. N., Kanengoni, A. T. and Chimonyo, M. 2019. Effects of environmental enrichment on behaviour, physiology and performance of pigs - a review. *Asian-Australasian Journal of Animal Sciences* 32(1), 1–13.

Moinard, C., Mendl, M., Nicol, C. J. and Green, L. E. 2003. A case control study of on-farm risk factors for tail biting in pigs. *Applied Animal Behaviour Science* 81(4), 333–355.

Mouttotou, N., Hatchell, F. M. and Green, L. E. 1999. Foot lesions in finishing pigs and their associations with the type of floor. *Veterinary Record* 144(23), 629–632.

Muns, R. and Magowan, E. 2018. The effect of creep feed intake and starter diet allowance on piglets' gut structure and growth performance after weaning. *Journal of Animal Science* 96(9), 3815–3823.

Muns, R., Malmkvist, J., Larsen, M. L., Sørensen, D. and Pedersen, L. J. 2016. High environmental temperature around farrowing induced heat stress in crated sows. *Journal of Animal Science* 94(1), 377–384.

Murphy, T. W. 2011. *The Effects of Individual and Combinations of Airborne Pollutants on Feed Intake, Immune Function and Physiology of the Pig*. Thesis, University of Adelaide, Australia.

Nathues, H., Woeste, H., Doehring, S., Fahrion, A. S., Doherr, M. G. and Beilage, Eg. 2013. Herd specific risk factors for Mycoplasma hyopneumoniae infections in suckling pigs at the age of weaning. *Acta Veterinaria Scandinavica* 55, 30.

Nawroth, C. and von Borell, E. 2015. Domestic pigs' (Sus scrofa domestica) use of direct and indirect visual and auditory cues in an object choice task. *Animal Cognition* 18(3), 757–766.

Newberry, R. and Wood-Gush, D. 1985. The suckling behaviour of domestic pigs in a semi-natural. *Environment and Behaviour* 95, 11–25.

Niemi, J. K., Edwards, S. A., Papanastasiou, D. K., Piette, D., Stygar, A. H., Wallenbeck, A. and Valros, A. 2021. Cost-effectiveness analysis of seven measures to reduce tail biting lesions in fattening pigs. *Frontiers in Veterinary Science* 8, 682330.

Nienhaus, F., Meemken, D., Schoneberg, C., Hartmann, M., Kornhoff, T., May, T., Heß, S., Kreienbrock, L. and Wendt, A. 2020. Health scores for farmed animals: screening pig health with register data from public and private databases. *PLoS ONE* 15(2), e0228497.

Ocepek, M., Goold, C. M., Busancic, M. and Aarnink, A. J. A. 2018. Drinker position influences the cleanness of the lying area of pigs in a welfare-friendly housing facility. *Applied Animal Behaviour Science* 198, 44-51.

Olesen, J. E. and Bindi, M. 2002. Consequences of climate change for European agricultural productivity, land use and policy. *European Journal of Agronomy* 16(4), 239-262.

Otake, S., Dee, S., Corzo, C., Oliveira, S. and Deen, J. 2010. Long-distance airborne transport of infectious PRRSV and Mycoplasma hyopneumoniae from a swine population infected with multiple viral variants. *Veterinary Microbiology* 145(3-4), 198-208.

Pastorelli, G., Musella, M., Zaninelli, M., Tangorra, F. and Corino, C. 2006. Static spatial requirements of growing-finishing and heavy pigs. *Livestock Science* 105(1-3), 260-264.

Peden, R. S. E., Akaichi, F., Camerlink, I., Boyle, L. A. and Turner, S. P. 2019. Pig farmers' willingness to pay for management strategies to reduce aggression between pigs. *PLoS ONE* 14(11), e0224924.

Pedersen, L., Malmkvist, J. and Andersen, H. 2013. Housing of sows during farrowing: a review on pen design, welfare and productivity. *Livestock Housing: Modern Management to Ensure Optimal Health and Welfare of Farm Animals*, 93-111.

Pedersen, L. J., Herskin, M. S., Forkman, B., Halekoh, U., Kristensen, K. M. and Jensen, M. B. 2014. How much is enough? The amount of straw necessary to satisfy pigs' need to perform exploratory behaviour. *Applied Animal Behaviour Science* 160, 46-55.

Pedersen, S., Nonnenmann, M., Rautiainen, R., Demmers, T. G., Banhazi, T. and Lyngbye, M. 2000. Dust in pig buildings. *Journal of Agricultural Safety and Health* 6(4), 261-274.

Pitts, A. D., Weary, D. M., Pajor, E. A. and Fraser, D. 2000. Mixing at young ages reduces fighting in unacquainted domestic pigs. *Applied Animal Behaviour Science* 68(3), 191-197.

Randolph, J. H., Cromwell, G. L., Stahly, T. S. and Kratzer, D. D. 1981. Effects of group size and space allowance on performance and behavior of swine. *Journal of Animal Science* 53(4), 922-927.

Reiske, L., Schmucker, S., Pfaffinger, B., Weiler, U., Steuber, J. and Stefanski, V. 2020. Intravenous infusion of cortisol, adrenaline, or noradrenaline alters porcine immune cell numbers and promotes innate over adaptive immune functionality. *Journal of Immunology* 204(12), 3205-3216.

Rodriguez, M. R., Losada, E., Besteiro, R., Arango, T., Velo, R., Ortega, J. A. and Fernandez, M. D. 2020. Evolution of NH3 concentrations in Weaner pig buildings based on setpoint temperature. *Agronomy* 10(1), 107. https://doi.org/10.3390/agronomy10010107.

Ross, J. W., Hale, B. J., Seibert, J. T., Romoser, M. R., Adur, M. K., Keating, A. F. and Baumgard, L. H. 2017. Physiological mechanisms through which heat stress compromises reproduction in pigs. *Molecular Reproduction and Development* 84(9), 934-945.

Samarakone, T. S. and Gonyou, H. W. 2009. Domestic pigs alter their social strategy in response to social group size. *Applied Animal Behaviour Science* 121(1), 8-15.

Scheepens, C. J. M. 1996. Climatic stress in swine: hazards for health. *The Pig Journal* 37, 130-136.

Sørensen, J. T. and Schrader, L. 2019. Labelling as a tool for improving animal welfare–the pig case. *Agriculture* 9(6), 123. https://doi.org/10.3390/agriculture9060123.

Spoolder, H. A. M. 1998. Effects of food motivation on stereotypies and aggression in group housed sows. Ph.D. Thesis. Wageningen University, Wageningen, The Netherlands.

Stalder, K., DÀllaire, S., Drolet, R. and Abell, C. 2010. Longevity in breeding animals, In: Zimmerman, J. J., Karriker, L. A., Ramirez, A., Schwartz, K. J. and Stevenson, G. W. (Eds), *Disases of Swine*. Wiley-Blackwell, Ames, IA, 50–59.

Sterchi, M., Faverjon, C., Sarasua, C., Vargas, M. E., Berezowski, J., Bernstein, A., Grütter, R. and Nathues, H. 2019. The pig transport network in Switzerland: structure, patterns, and implications for the transmission of infectious diseases between animal holdings. *PLoS ONE* 14(5), e0217974.

Strauch, L. 2013. *Untersuchungen zu Infrarotwärmeplatten als Ausrüstung für Ferkelnester im Vergleich mit konventionellen Warmwasserplatten*. Doctoral Thesis, University of Veterinary Medicine, Hannover.

Straw, B. and Bartlett, P. 2001. Flank or belly nosing in weaned pigs. *Journal of Swine Health and Production* 9, 19–23.

Tanida, H., Senda, K., Suzuki, S., Tanaka, T. and Yoshimoto, T. 1991. Color discrimination in weanling pigs. *Nihon Chikusan Gakkaiho* 62(11), 1029–1034.

Temple, D., Dalmau, A., Ruiz de la Torre, J. L., Manteca, X. and Velarde, A. 2011. Application of the Welfare Quality® protocol to assess growing pigs kept under intensive conditions in Spain. *Journal of Veterinary Behavior* 6(2), 138–149.

Thomas, R. J. 2013. Particle size and pathogenicity in the respiratory tract. *Virulence* 4(8), 847–858.

Thomsson, A. 2008. Weaning of pigs - effects of lectin exposure and weaning strategies on feeding behaviour, performance and health. Doctoral Thesis. Swedish University of Agricultural Sciences, Alnarp.

Tilger, M. 2005. Biologische Rhythmen bei Nutztieren- eine Literaturstudie. Doctoral Thesis. Ludwig-Maximilian-Universität München, Ludwig-Maximilian-Universität München.

Tobias, T. J., Bouma, A., van den Broek, J., van Nes, A., Daemen, A. J., Wagenaar, J. A., Stegeman, J. A. and Klinkenberg, D. 2014. Transmission of Actinobacillus pleuropneumoniae among weaned piglets on endemically infected farms. *Preventive Veterinary Medicine* 117(1), 207–214.

Tonsor, G. T. and Wolf, C. A. 2019. US farm animal welfare: an economic perspective. *Animals: An Open Access Journal from MDPI* 9(6).

Turner, S. P., Allcroft, D. J. and Edwards, S. A. 2003. Housing pigs in large social groups: a review of implications for performance and other economic traits. *Livestock Production Science* 82(1), 39–51.

Turner, S. P., Horgan, G. W. and Edwards, S. A. 2001. Effect of social group size on aggressive behaviour between unacquainted domestic pigs. *Applied Animal Behaviour Science* 74(3), 203–215.

Tyndall, J. and Colletti, J. 2007. Mitigating swine odor with strategically designed shelterbelt systems: a review. *Agroforestry Systems* 69(1), 45–65.

Uehleke, R., Seifert, S. and Hüttel, S. 2021. Do animal welfare schemes promote better animal health? An empirical investigation of German pork production. *Livestock Science* 247, 104481.

Ulens, T. 2015. Source-oriented techniques for improving indoor air quality and assessment of air emissions from pig husbandry. PhD Thesis, Ghent University, Belgium.

van de Weerd, H. and Ison, S. 2019. Providing effective environmental enrichment to pigs: how far have we come? *Animals (Basel)* 9(5), 254. doi: 10.3390/ani9050254.

van de Weerd, H. A., Docking, C. M., Day, J. E. L., Avery, P. J. and Edwards, S. A. 2003. A systematic approach towards developing environmental enrichment for pigs. *Applied Animal Behaviour Science* 84(2), 101–118.

van Dixhoorn, I. D., Reimert, I., Middelkoop, J., Bolhuis, J. E., Wisselink, H. J., Groot Koerkamp, P. W., Kemp, B. and Stockhofe-Zurwieden, N. 2016. Enriched housing reduces disease susceptibility to co-infection with porcine reproductive and respiratory virus (PRRSV) and Actinobacillus pleuropneumoniae (A. pleuropneumoniae) in young pigs. *PLoS ONE* 11(9), e0161832.

Vangroenweghe, F., Suls, L., Van Driessche, E., Maes, D. and De Graef, E. 2012. Health advantages of transition to batch management system in farrow-to-finish pig herds. *Veterinární Medicína* 57(2), 83–91.

Verhovsek, D., Troxler, J. and Baumgartner, J. 2007. Peripartal behaviour and teat lesions of sows in farrowing crates and in a loose-housing system. *Animal Welfare* 16, 273–276.

Villanueva-García, D., Mota-Rojas, D., Martinez-Burnes, J., Olmos-Hernandez, A., Mora-Medina, P., Salmerón, C., Gómez-Prado, J., Boscato, L., Gutiérrez-Pérez, O., Cruz, V., Reyes, B. and González-Lozano, M. 2020. Hypothermia in newly born piglets: mechanisms of thermoregulation and pathophysiology of death. *Journal of Animal Behaviour and Biometeorology* 9(9), 2101–2109:2101.

Vitt, R., Weber, L., Zollitsch, W., Hörtenhuber, S. J., Baumgartner, J., Niebuhr, K., Piringer, M., Anders, I., Andre, K., Hennig-Pauka, I., Schönhart, M. and Schauberger, G. 2017. Modelled performance of energy saving air treatment devices to mitigate heat stress for confined livestock buildings in Central Europe. *Biosystems Engineering* 164, 85–97.

Wadepohl, K., Blaha, T. and Meemken, D. 2020. Evaluation of a simplified herd health and welfare index for benchmarking in pig herds. *Tierarztl. Prax. Ausg. G Grosstiere Nutztiere* 48(3), 156–164.

Wang, Y. C., Chang, Y. C., Chuang, H. L., Chiu, C. C., Yeh, K. S., Chang, C. C., Hsuan, S. L., Lin, W. H. and Chen, T. H. 2011. Transmission of Salmonella between swine farms by the housefly (Musca domestica). *Journal of Food Protection* 74(6), 1012–1016.

Wegner, K., Lambertz, C., Das, G., Reiner, G. and Gauly, M. 2016. Effects of temperature and temperature-humidity index on the reproductive performance of sows during summer months under a temperate climate. *Animal Science Journal* 87(11), 1334–1339.

Wei, X., Tsai, T., Howe, S. and Zhao, J. 2021. Weaning induced gut dysfunction and nutritional interventions in nursery pigs: a partial review. *Animals: An Open Access Journal from MDPI* 11(5), 1279.

Wiegand, R. M., Gonyou, H. W. and Curtis, S. E. 1994. Pen shape and size: effects on pig behavior and performance. *Applied Animal Behaviour Science* 39(1), 49–61.

Wolter, B. F., Ellis, M., Curtis, S. E., Augspurger, N. R., Hamilton, D. N., Parr, E. N. and Webel, D. M. 2001. Effect of group size on pig performance in a wean-to-finish production system. *Journal of Animal Science* 79(5), 1067–1073.

Wolter, B. F., Ellis, M., Curtis, S. E., Parr, E. N. and Webel, D. M. 2000. Group size and floor-space allowance can affect weanling-pig performance. *Journal of Animal Science* 78(8), 2062–2067.

Xuan, Y. M., Xiao, F., Niu, X. F., Huang, X. and Wang, S. W. 2012. Research and application of evaporative cooling in China: a review (I) – Research. *Renewable and Sustainable Energy Reviews* 16(5), 3535–3546.

Zonderland, J. J., Cornelissen, L., Wolthuis-Fillerup, M. and Spoolder, H. A. M. 2008. Visual acuity of pigs at different light intensities. *Applied Animal Behaviour Science* 111(1-2), 28–37.

Zulovich, J. M. 2012. Effect of the environment on health. In: Zimmerman, J. J., Karriker, L. A., Ramirez, A., Schwartz, K. J. and Stevenson, G. W. (Eds), *Disease of Swine*. Wiley-Blackwell, Iowa, 60–66.

Published by Burleigh Dodds Science Publishing Limited, 2023.

Chapter 12

Advances in the development and use of vaccines for prevention of endemic diseases in pigs

Enric Mateu, Universitat Autònoma de Barcelona, Spain; Ivan Díaz, Centre de Recerca en Sanitat Animal (CReSA, IRTA-UAB) and OIE Collaborating Centre for the Research and Control of Emerging and Re-Emerging Swine Diseases in Europe (IRTA-CReSA), Spain; and Gerard Martín-Valls, Universitat Autònoma de Barcelona, Spain

1 Introduction

In the last 40 years, industrial pig production has undergone an important process of transformation. On one hand, the number of sows per farm and the reproductive performance of sow breeds have been steadily increasing. On the other hand, optimisation of herd performance has led to higher replacement rates and the consequent need for more gilts, often coming from specialist external sources. These high-output production systems can only be operated with careful planning and organisation of farrowing batches at regular intervals (every week, every third week, etc.). There also is a trend towards the separation of the different production phases (nurseries, fattening units, breeders) on different sites. All these developments suggest more pig movements within

http://dx.doi.org/10.19103/AS.2022.0103.13

larger groups, thus increasing the chances for transmission of diseases between farms. Figure 1 summarises some of these changes.

As a result of these trends in production, there is a constant influx of susceptible animals (gilts) into breeding herds and a constant flow of susceptible animals (piglets) into nurseries. These are ideal situations for the transmission of infectious agents within herds, but also provide good opportunities for limiting their impact through the implementation of appropriate biosecurity, management and vaccination protocols.

Prevention of diseases is crucial in the modern large-scale pig industry. Disease increases general costs and interrupts production. For example, abortions and infertility associated with disease require returning sows back to the insemination phase, thus disrupting the organisation of breeding batches. If infected sows need to be discarded, replacement rates increase, and planning and management of quarantines and acclimatisation areas are compromised. Similarly, disease in the nurseries may not only increase mortality and veterinary costs but also delay piglet growth and produce batches of less homogeneous weights for transfer to fattening units (reviewed by Muirhead et al., 2013).

It is clear then that management of disease in pig farms has a range of objectives. The first objective is limitation of health impacts, namely, mortality or development of clinical signs of disease. The second objective is limiting impact on production and on the secondary economic losses associated with disease such as weight losses, secondary infections, etc. The third objective is limiting, and ideally stopping, the transmission of infection within the farm. In addition, in the case of zoonotic agents, there is also the need to prevent transmission of pathogens within the farm to minimise the risk to humans as well.

Vaccination is a convenient way to accomplish these objectives. It can be planned, is safe, is easy to administer and is relatively cheap. However, the benefits of vaccination may differ substantially for different pathogens and vaccines. In some cases, vaccination produces full sterilising immunity. Vaccinated animals are not susceptible to the infection and will no longer contribute to transmission. For example, in a study testing different porcine parvovirus (PPV) vaccines, administration of a subunit vaccine based on the 27a strain (cluster D) prevented infection in gilts challenged with a cluster C strain (Noguera et al., 2021).

However, in most cases, vaccination produces non-sterilising immunity. Induced immunity is powerful enough to protect against the development of clinical disease and to substantially reduce (but not entirely eliminate) transmission of infection or the associated losses. Finally, in other cases, vaccination helps to reduce clinical signs and lesions but does not fully eliminate disease or stop transmission (reviewed by Opriessnig et al., 2021).

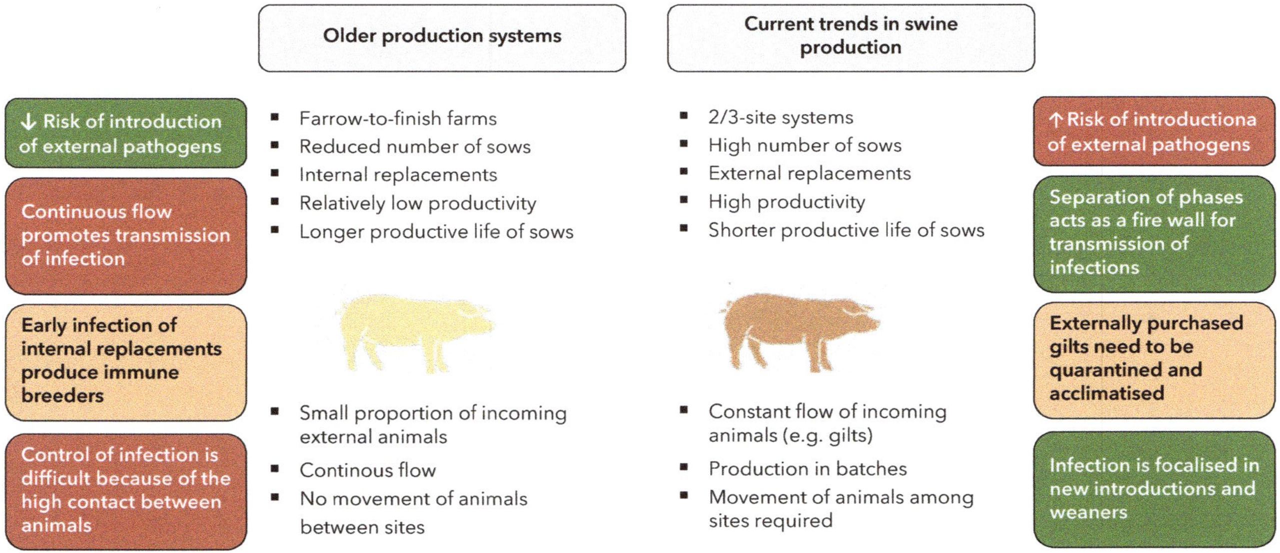

Figure 1 Characteristics of traditional production systems versus highly industrialised farms. The figure shows some of the characteristics of traditional production systems versus those of more industrialised farms and the implications for the entrance, transmission, and maintenance of pathogens. Knowledge of these elements is crucial to understand the epidemiology of diseases in different systems and to plan adequate vaccination strategies.

Even in these cases, vaccination may still help to control disease and have substantial benefits.

One important element to consider as far as vaccination is concerned is that we are on the verge of a revolution in the development of vaccines for humans and animals. For more than a 100 years, vaccines have been based on the use of inactivated or attenuated microorganisms, with or without the addition of adjuvants to boost immune response. This relatively simple approach has been one of the major breakthroughs in human history. However, in recent decades, our understanding of immune response and the technological means at our disposal have changed substantially (reviewed by Pollard and Bijker, 2021).

At present, the knowledge and the technologies available for the development of vaccines has no parallel in history. We know the basics on how pathogens are sensed by cells of the innate immune system, how antigens are processed and presented, how antigen-presenting cells drive the polarisation of T-cells and what effector actions are most suited to fight different pathogens. We also know how to manipulate the immune response to our advantage. Similarly, we now have technologies for producing DNA and RNA vaccines, for vectoring antigens in viruses or bacteria, for directing antigens to dendritic cells etc. The current Coronavirus Disease 2019 (COVID-19) pandemic is boosting the practical translation of all this knowledge into actual vaccines (Mukherjee, 2020). Sooner rather than later, this change will occur in veterinary vaccines.

This chapter reviews the current use of vaccines and vaccination programmes in controlling endemic diseases in pigs. It also discusses the future of vaccines and vaccination in pigs.

2 How infectious diseases become endemic in pig farms

The initial introduction of a new pathogen in a naïve population often causes an epidemic episode. Its intensity and duration are the result of several variables. These include, but are not limited to, the modes of transmission of the pathogen, the number and density of susceptible animals, the network of contacts (direct and indirect) within the farm or the likelihood of indirect contact through fomites (e.g. floors or walls). In highly connected populations infected with highly transmissible but labile pathogens that induce a sterilising and durable immune response, it is possible that the epidemic simply fades out through lack of susceptible individuals after rapidly infecting most of the population. This is an infrequent scenario. Most often, pathogens are not that easily transmitted, may persist in the environment, do not induce full sterilising immunity and may establish permanently in the population. They therefore become endemic, showing a more or less constant epidemiological pattern.

In many cases, the pattern of disease is determined by the inflow of susceptible gilts, the infection of a small proportion of piglets in farrowing crates and the outflow of infected piglets to nurseries. The introduction of susceptible gilts into the farm without an adequate acclimation period results in infection during the acclimation period or the first gestation. Depending on the nature of the pathogen, transmission to offspring can happen in various ways:

- *in utero* (e.g. in the case of porcine reproductive and respiratory syndrome virus (PRRSV), revised by Pileri and Mateu, 2016);
- during birth or immediately after birth; and
- Due to close and prolonged contact between the sow and her piglets (e.g. with *Mycoplasma hyopneumoniae*, revised by Sibila et al., 2009).

Transmission to the piglet can be accompanied by transfer of colostral antibodies. For pathogens such as *Streptococcus suis, Actinobacillus pleuropneumoniae, Glaesserella parasuis* and others, this transfer of antibodies may protect against the development of clinical signs or may reduce the chances of infection. However, it does not avoid an animal becoming a carrier (e.g. in the tonsil tissue) without developing the disease itself (Baums et al., 2010; Blanco et al., 2004; Chiers et al., 2002; Solano-Aguilar et al., 1999).

There are two critical factors when piglets are moved to nurseries:

- animals from different litters are mixed; and
- maternally derived antibodies (MDA) fall below a level which protects against disease.

The mixing of infected animals, both with animals losing MDA and with animals of different statuses, favours transmission.

This combination of factors generates the typical endemic patterns shown by several diseases. Enteric diseases, for example, usually appear in nurseries at about the end of the first week after transfer of piglets to nurseries. A key reason is the loss of enteric IgAs ingested with colostrum and milk from the sow (the half-life of IgA was calculated to be around 3 days) (Curtis and Bourne, 1973). Other reasons include the presence of carrier pigs (e.g. carrying enterotoxigenic or enteropathogenic *Escherichia coli*) or the presence of the pathogen in facilities (e.g. rotaviruses).

In the case of other diseases such as influenza, some piglets are already infected when they leave their mothers and maintain a low level of transmission of the virus (probably to animals with the lowest MDA levels) until the point when most animals have no passive protection and disease incidence increases (Chamba-Pardo et al., 2019). Figure 2 depicts the prototypical patterns of immunity and infection in endemic herds.

The half-life of systemically absorbed colostral antibodies has been estimated in different studies. For example, Polo et al. (2012) calculated the

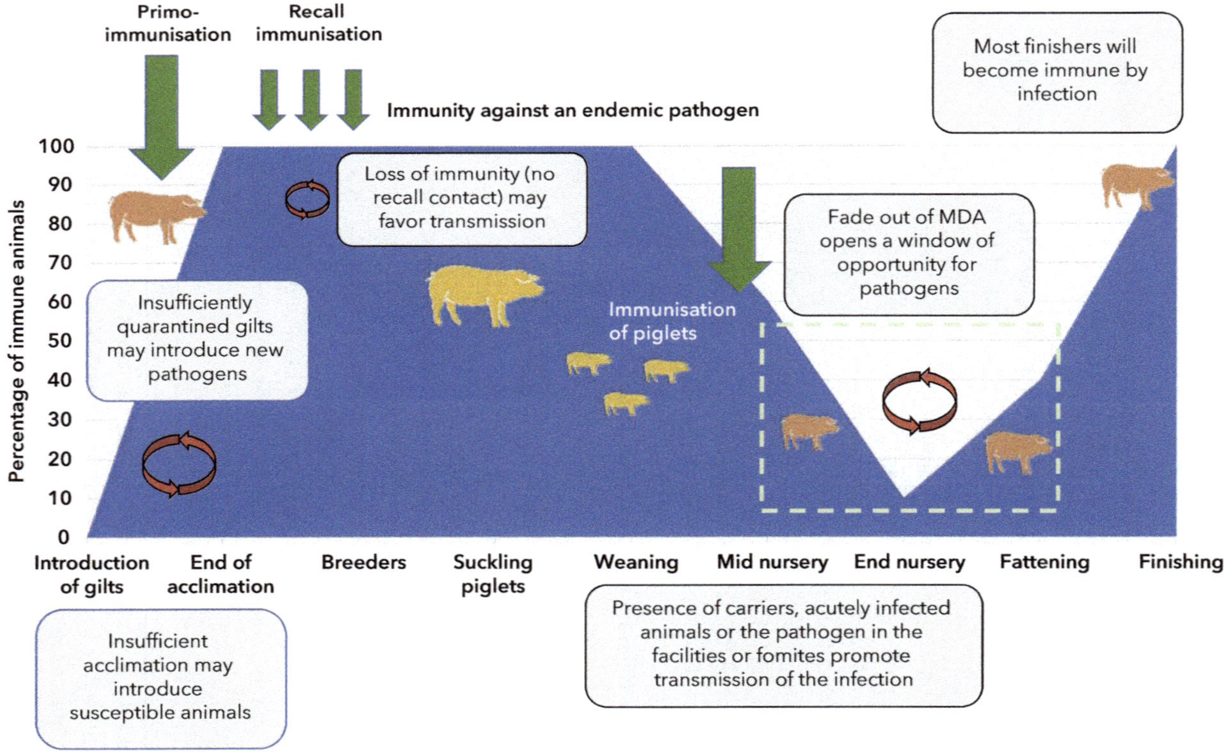

Figure 2 Graphical depiction of the evolution of immunity in the herd and the critical points for transmission and prevention. The introduction of external naïve gilts into will determine to a large extent the epidemiology of endemic infections in the herd. A successful immunisation of gilts during the acclimation period will prevent the infection of the newcomers upon their introduction in the breeding herd and will allow an adequate transfer of MDA to their offspring. Periodic recall vaccination of breeder sows will be required to maintain optimal levels of immunity in the sows' stock. The better the immunisation plan for sows, the higher the levels of MDA in piglets and the longer the duration of passive immunity. In most cases, the decline of MDA will take place at some moment during the nursery phase or at early growing-fattening periods. This opens a window of opportunity for the infection to be transmitted among weaners/growers. Vaccination must contribute to shorten or even to close that window.

half-life of IgG absorbed from a colostrum supplement as ranging from 3.0 days for swine influenza A virus (SwIAV)-specific antibodies, to 17.7 days, for porcine circovirus 2 (PCV2). Other studies have estimated half-life of MDA antibodies for PPV to be between 19.7 and 29 days, between 1 and 3 weeks for *Actinobacillus pleuropneumoniae*, 79 days for classical swine fever virus (CSFV), 21 days for pseudorabies virus (PRV) and 11.9 days for PRRSV (Müller et al., 2005; Paul et al., 1982; Polo et al., 2012; Vandeputte et al, 2001; Vigre et al., 2003). These estimations showed not only differences for different pathogens but also individual differences. These data may help to explain why the duration of protection derived from MDA is different for different agents.

Although a precise epidemiologic picture for all common diseases of pigs is lacking - and for many of them the determinants are yet to be understood - it is clear that some of the main forces driving endemic disease are common for several pathogens. This makes it possible for us to better understand the potential approaches we can adopt for use of vaccines in these situations.

3 Target populations and the dynamics of infection in endemically infected pig herds

The analysis of factors driving the endemic state of infectious agents in farms suggests that control of infections can be achieved by targeting the key populations at the right moment. Introduction of gilts into the stock of sows should only be done after an adequate immunisation/acclimation programme. In practical terms, this means that gilts should be immunised as soon as possible, must be isolated long enough to permit the development of solid immunity and, if possible, should be tested to confirm the development of such specific immunity. Depending on the strength and duration of immunity, booster or periodic recall immunisations may be needed during the acclimation period to maintain optimal immunisation (Bernaerdt et al., 2021).

If the first goal is achieved, that is, the gilts and sows have solid immunity, then the next target is the piglets on the farm. The main question here is to find the optimal moment for vaccination. Ideally, vaccination should take place at a point which avoids blocking by MDA, but at the same time, allows the development of adaptive immunity before MDA protection has fully disappeared. This is a very difficult task. To understand how to tackle this problem, we must first visualise the target population in terms of the proportion of immune (via MDA or previous infection), infected and susceptible piglets.

Let us first examine what happens in the case of passive immunity. In a large farm where no intervention is undertaken, we can expect to have a subpopulation of naïve sows (probably the younger ones) and a subpopulation of immune sows. The subpopulation of immune sows may have a more or less homogeneous immunity, or alternatively, may have different levels of immunity depending on several factors (e.g. age). This means we will find piglets both with no MDA against the pathogen and piglets with MDA. Within this latter group, different levels of immunity can be expected based on the immunity of sows.

For common pathogens on farms, older sows often have higher levels of protective antibodies or stronger cell-mediated responses than younger sows, because of repeated contact with the agent. This means that offspring of older sows have correspondingly higher levels of MDA. For example, Gelberg et al. (1991) reported that rotavirus-specific IgA titres in colostrum and milk increased significantly in third-parity sows compared to younger breeders. Rautiainen et al. (2000) indicated that in *Mycoplasma hyopneumoniae*-infected herds, older sows were more likely to have anti-Mycoplasma antibodies in colostrum than younger sows. Piñeiro et al. (2019) showed that, at weaning, the levels of serum IgG in piglets suckled by multiparous sows were significantly higher than those of piglets suckled by primiparous sows.

An additional factor is that not all animals within each litter have the same intake of colostrum or milk. Usually, the bigger the piglets, the better their colostrum or milk intake. In very large litters, where there is often a significant disparity in piglet size, we can find significant differences in the levels of MDA (reviewed by Blavi et al., 2021). Cross-fostering may also have an impact on MDA levels within a litter.

In a farm with a good immunisation programme for gilts and sows, we can assume that most females transfer some level of MDA to the offspring. However, different protocols for vaccination of sows may produce different levels of MDA in them. In general, it is assumed that blanket vaccination produces a more homogeneous immunity in sows compared to vaccination at a given moment of the reproductive cycle. However, pre-farrowing vaccination programmes may help to boost levels of antibodies in colostrum and milk, thus increasing the transfer of MDA in all animals (Poonsuk and Zimmerman, 2018).

The duration of MDA generally only extends for some weeks and rarely exceeds 2–3 months (with some exceptions such as PPV). However, since the duration of the MDA in the piglet depends on the amount of colostrum ingested and on the levels of specific antibodies in the colostrum, factors that increase antibody titres in sows (e.g. repeated vaccination or contact with the antigen, pre-farrowing vaccination etc.) may modify this duration. The following example illustrates this. Let us imagine that protective levels of antibodies against a given pathogen are 1:8 (that is 3 $\log_2$ in a viral neutralisation test) and the half-life of those antibodies is 7 days. If our population is homogeneous and we reach a normal (Gaussian) distribution of MDA titres at day 3 of life (average titre of 128 -8 $\log_2$- and a standard deviation of 2 $\log_2$), by the fifth week of life, 50% of the piglets would have lost protective titres of MDA, and from that point onwards, most of the population would be unprotected. Although the example is a simplification of a more complex reality, it provides a general picture of what would happen in a farm. Obviously, in a scenario where non-immune sows were present, the point where 50% of piglets would be unprotected would be reached sooner.

The second element that must be examined is transmission of the infection. This can be very different for different agents. In endemic states, it is relatively common to find acutely infected or carrier sows in maternity units. These females would transmit the infection to all or some of their piglets, which, in turn, would transmit the infection to other pen mates or to other piglets once they were transferred to the nursery. Animals with MDA are less likely to be infected and, depending on the pathogen, the infection in presence of MDA may result in lower shedding (and therefore in lower potential transmission). As a result, the reproduction rate (R) of the infection is expected to be lower in maternal units than in the nurseries when susceptible animals are in the majority. Table 1 shows some R values for different pig pathogens.

In a population where no intervention is made, we can again expect that a proportion of piglets has no MDA and that the remaining population becomes susceptible as MDA titres fall below the protection threshold. In most cases, this produces a 'drop-by-drop' pattern in which incidence follows a generally consistent pattern and cases will accumulate relatively slowly for several weeks. In contrast, in a more homogeneous population, most animals are either protected by MDA or have lost MDA. In this scenario, a substantial proportion of piglets may become susceptible simultaneously, resulting in an abrupt rise in incidence of disease.

The combination of these two elements (MDA and transmission at early ages) are the major drivers of the dynamics of infection in nurseries and determine the optimal moment for vaccination (Martelli et al., 2016). The higher the transmissibility of the pathogen, the higher the probability of early transmission. The higher the proportion of non-immune sows, the higher the number of non-immune piglets from birth and the higher the chances for early transmission. The earlier the transmission happens, the sooner vaccination must be applied and the higher the interference of MDA in vaccination.

4 The aims of vaccination in endemically infected farms and potential vaccine efficacy

The general aim of vaccination is to prevent infection or disease. In practical terms, vaccination may serve different purposes, ranging from prevention of infection to reduction of the economic impact of the disease. The aims of vaccination can be seen from different points of view: prevention of infection, prevention of clinical disease or just prevention of the most severe cases of disease (reviewed by Pollard and Bijker, 2021; Rose and Andraud, 2017).

Table 1 Estimated basic reproduction rates (R0) or transmission rates (β) for different swine pathogens.

Agent	References	R_0 or β	Study conditions
Porcine reproductive and respiratory syndrome virus	Pileri et al. (2017)	R_0: 3-7	Suckling piglets-weaners and fatteners in two farrow-to-finish farms
	Pileri et al. (2015)	R_0: 2.78	Naive
		R_0: 0.53	Vaccinated + heterologous challenge
	Rose et al. (2015)	R_0: 5.42	Naive
		R_0: 0.3	Vaccinated + heterologous challenge
Swine influenza A virus	Pileri et al. (2017)	R_0: 1.5-6.4	Suckling piglets-weaners and fatteners in two farrow-to-finish farms
	Allerson et al. (2013)	R_0: 10.4	Naïve
		R_0: 7.1	MDA+ heterologous challenge
		R_0: 0.8	MDA+ homologous challenge
	Cador et al. (2016)	R_0: 14.8	Naïve
		R_0: 5.8	MDA+ heterologous challenge
	Romagosa et al. (2012)	R_0: 10.6	Naïve
		R_0: 1	Heterologous vaccination
		R_0: 0	Homologous vaccination
Classical swine fever virus	Klinkenberg et al. (2002)	R_0: 7.7-100	Naïve weaners
		R_0: 3.4-15.5	Naïve slaughter-age pigs
	Weesendorp et al. (2011)	R_0: 23.1	Low-level shedder pigs
		R_0: 143	High-level shedder pigs
	Weesendorp et al. (2014)	B: 0.33/day	Air transmission
		B: 0.3/day	Adjacent-pen transmission
		B: 6.1/day	Within-pen transmission
Pseudorabies virus	de Jong and Kimman (1994)	R_0: 10	Naïve
		R_0: 0.5	Vaccinated
	Stegeman et al. (1995)	R_0: 3.4	Single vaccination
Mycoplasma hyopneumoniae	Betlach et al. (2020)	R_0: 1.5	Double vaccination
		B: 0.73/week	Naïve gilts

Some vaccines produce full protection against infection: vaccinated animals do not get an infection when exposed to pathogens. In this case, the vaccine produces a benefit in sanitary and economic terms and may also help to eradicate the infection from the herd if applied wisely. Unfortunately, vaccines for use in pigs rarely reach such a level of protection. However, in humans, the smallpox vaccine provides full immunity for a period of up to 5 years and lifelong protection against the disease (Belongia and Naleway, 2003). In cattle, the rinderpest vaccine produced long-term full immunity (Plowright, 1984) and permitted global eradication of the infection.

Other vaccines do not fully prevent the infection, but infected animals do not develop clinical signs or lesions and are able to clear the infection rapidly. Usually, in these cases, if a vaccinated animal is infected, its potential for transmission is lower than that of a non-vaccinated pig. For example, it has been shown that under experimental conditions, vaccination against PRRSV may help to reduce transmission of the infection (Pileri et al., 2015; Rose et al., 2015). Romagosa et al. (2011) showed that vaccination against SwIAV in pigs may contribute to a reduced transmission rate. For PCV2, vaccination also reduces the chances for infection and reduces shedding if infection occurs (Madson and Oppriessnig, 2011). For these vaccines, vaccination of the herd improves health, reduces direct losses, and may also contribute to eradication in combination with other measures.

In other cases, vaccinated animals can be infected but only develop very mild disease while still being able to transmit infection. For example, Villarreal et al. (2011) showed that a classical inactivated vaccine against *Mycoplasma hyopneumoniae* was unable to reduce transmission when administered to 3-week-old piglets. In this case, vaccination has a mainly economic impact and is unable to significantly contribute to the eradication of the disease. Finally, for some vaccines, the main effect is just prevention of death.

A second approach is the role of vaccination in maintaining production. Vaccination can have different purposes, for example:

- protecting incoming gilts from pathogens present in the farm;
- maintaining the immunity of already vaccinated sows throughout their productive lives; and
- promoting the transfer of MDA to piglets or preventing infection in piglets before (or as soon as) they lose MDA.

Vaccination can even be used in emergency situations to stop an outbreak of disease (Olanratmanee et al., 2013; Li et al., 2013).

A 'perfect vaccine' (if such a vaccine can be produced) should induce full sterilising and lifelong immunity in the whole population. Very few vaccines have been close to this ideal, with the probable exception of the smallpox vaccine in humans (Belongia and Naleway, 2003) and the rinderpest vaccine in animals (Plowright, 1984). In an ideal vaccine, none of the vaccinated individuals should get infected when exposed to any of the pathogen variants, even if exposed to high doses. Protection should be achieved with a single dose of the vaccine, in a very short period (days) and should last for life. Additionally, needleless administration would be preferable (Beverley, 2002). Most current vaccines cannot achieve this. In some cases, vaccination programmes require more than one dose, and immunity fades after several months.

5 The immunological basis of vaccination

Vaccination is based on the phenomenon of immunological memory. Immunological memory is the capacity of the immune system to produce a faster and stronger response against a pathogen that has been encountered before. This simple idea corresponds to a very complex process that includes the sensing of the pathogen by the innate immune system, antigenic processing and presentation to lymphocytes, activation of specific T-cell clones, cooperation of T- and B-cells, generation of an effector response (antibodies, cytotoxicity etc.) and generation of memory cells in charge of recall responses (reviewed by Beverley, 2002).

A pathogen contains many antigens, that is, many molecules that can induce an immune response in the host. Each of these molecules may, in turn, contain many epitopes, that is, the specific parts of the molecule that are recognised by the immune system. The immune response generated against a given epitope may result in the effective neutralisation/elimination of the pathogen, may not provide protection or may even be harmful to the host (Zinkernagel, 1993).

The first step in the generation of an immune response is the sensing of the pathogen by the immune system. This action is carried out by a wide group of receptors collectively designated as pattern recognition receptors (PRRs). These PRRs recognise molecular patterns that are common in groups of pathogens but are not present in the host. Examples of this could be the receptors for lipopolysaccharides (a common molecule in the cell wall of Gram-negative bacteria but absent in vertebrates) or for double-stranded RNA (present in some viruses). There are a large number of PRRs that can recognise all types of pathogens. This initial sensing results in the production and release of several mediators that send signals to other cells in the immune system, promoting their activation or migration to the site of the infection. The most studied mediators are the cytokines (designated as chemokines when their main function is to act as chemoattractants). Different antigens may activate different PRRs and may thus induce the release of different cytokines that condition the subsequent response (reviewed by Kumar et al., 2011).

The second step is antigenic processing and presentation. This action is carried out by antigen-presenting cells (mostly dendritic cells, macrophages and B cells) and happens in the lymph nodes or other secondary lymphoid tissues (reviewed by Janeway et al., 2001). Antigenic presentation requires the capture of the antigen, its processing and the presentation of an epitope in the major histocompatibility complex (MHC). There are two main types of MHC:

- MHC-I which is present in all cells; and
- MHC-II which is expressed constitutively only in antigen presenting cells.

As a general rule, antigenic presentation happens in MHC-I when the pathogen is replicating inside a cell. Pathogens or antigens in the extracellular space are presented in MHC-II. A particular case is that of pathogens replicating in vesicles inside the cell (e.g. *Mycobacterium tuberculosis*), which are mostly presented in MHC-II. Although this is a general rule, cross presentation may occur (i.e. presentation in MHC-I of antigens captured from the extracellular space).

The antigen presented in MHC is recognised by a specific T-cell (it is important to remember that a given T-cell recognises only one epitope). For this presentation to be effective, it is necessary to secure the collaboration of so-called co-stimulatory molecules on the T-cell and on the antigen-presenting cell surface. Presentation in MHC-I requires expression of CD8 in T-cells and presentation in MHC-II requires the expression of CD4 on the T-cells (reviewed by Rudolph et al., 2006).

This means that three elements are needed for a full and strong response:

- adequate sensing of the pathogen and cytokine milieu;
- presentation in MHC; and
- collaboration of co-stimulatory molecules during antigen presentation.

If all these requirements are fulfilled, the T-cell recognising a given epitope is activated, and proliferates and polarizes. This means that it commits to the production of specific cytokines that drive the immune response towards enhanced pro-inflammatory responses, enhanced antibody responses or enhanced cytotoxic responses. To produce antibodies, T- and B-cells must cooperate. In this case, B-cells capture antigens from the extracellular space through surface immunoglobulin, process the antigen, and then present the antigen in MHC-II to a T-cell that (upon recognition) sends several 'help' signals to the B-cell, inducing its commitment to developing a plasma cell (reviewed by Mitchison, 2004).

Once the primary response is mounted, some T- and B-cells become long-lived memory cells. At present, no less than five types of memory T-cells are recognised. Some of them circulate in blood while others reside in tissues. Several phenotypic subtypes of B-cells are also recognised (reviewed by Ratajczak et al., 2018). The drivers of the duration of the immunological memory are not well understood, although in general, natural infection or use of live-attenuated microorganisms tends to produce longer immunity than inactivated/non-replicating vaccines. Figure 3 summarises the chronology of the immune response.

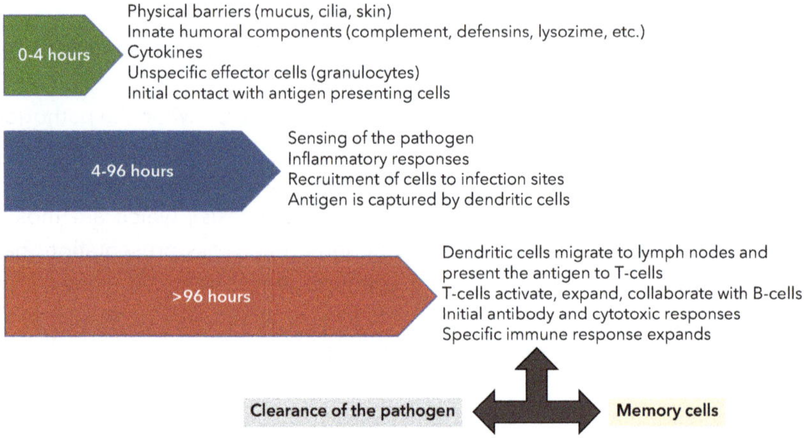

Figure 3 Chronology of the development of the immune response against a specific pathogen or antigen. The first barriers to infection are physicochemical elements such as the skin, the mucus, or the cilia of the respiratory epithelia. Pre-formed substances such as lysozyme or defensins or phagocytes present in the body surfaces may block the pathogen before it establishes the infection. If those first barriers fail, a second phase dominated by the innate response will occur. The pathogen will be sensed by the cells of the innate immunity and signals will be released to attract cells to the site of infection and to send messages to other cells. The dendritic cells will uptake the antigens and will migrate to the lymph nodes to execute antigen presentation to T-cells. In a third phase, the antigen will be recognised by specific T-cell clones that will activate, expand, and cooperate with B-cells. The development of these specific responses will eventually lead to the clearance of the infection and will generate memory cells that will react faster and stronger upon a second encounter with the same antigen.

The development of an immune response against the pathogen and the generation of memory cells are not enough to achieve protection. The immune response generated must be protective. To achieve this, the relevant epitopes of the pathogen must be targeted and a protective effector response must be produced. For example, generation of non-neutralising antibodies against a virus with a lack of cytotoxic responses would not clear the infection, despite having a strong immune response against the non-neutralising epitopes of the virus, such as occurs after PRRSV infection (reviewed by Mateu and Díaz, 2008).

In summary, the central element in a vaccine being effective is its capacity to induce memory responses. However, this memory response must target relevant epitopes of the pathogen and must comprise effective effector responses. To achieve this, the vaccine must induce the right response, and, in turn, this is determined by the right antigenic presentation and cytokine milieu. The type of antigen, the place where it is delivered and the induction of cytokines during this process are crucial in determining the efficacy of the vaccine and the duration of immunity (reviewed by Morel and Turner, 2010).

In a well-immunised animal, initial protection lies in the presence of preformed antibodies. For many (but not all) pathogens, entrance into the body occurs through the mucosa, which means that preformed IgA has a major role in blocking the pathogen at the point of entry. Preformed IgG plays a similar role in blood or in extracellular fluid. If blocking is successful, infection does not take place. However, preformed immunity is often not able to fully block the infection but reduces the number of infectious particles that are capable of infecting the host. This results in the need to receive a higher dose to cause infection, making transmission from an infected animal to another less likely, or causing less severe infection. If replication of the pathogen occurs, the presence of recall antigens boosts memory responses (antibody production and cytotoxic responses when previously developed). These memory responses may be detectable for 1–3 days after reencountering the pathogen. In a primary response, specific effector responses are rarely detected before days 5–7 after the onset of the infection, and in many cases, 7–14 days are needed for these responses to appear (reviewed by Pollard and Bijker, 2021).

Some conclusions can be inferred from this general picture. Firstly, for pathogens with very short incubation periods (e.g. influenza or enteric viruses), the presence of preformed antibodies in the mucosal surface is critical for achieving full protection, as the time needed to mount a recall response can be longer than the incubation period. Secondly, for most viruses, the development of cytotoxic responses is necessary to remove infected cells if fully protective titres of antibodies (sterilising immunity) are not achieved, resulting in potential infection occurring in vaccinated animals.

6 Vaccine types and development

6.1 Types of vaccines

One of the simplest ways to produce a vaccine is to grow the etiological agent of the disease, inactivate it and mix it with an adjuvant. In 1885, Pasteur was able to produce an efficacious rabies vaccine by using desiccated spinal cords of rabies-infected rabbits well before the discovery of viruses. A second classical way to produce a vaccine is to create an attenuated form of the pathogen, usually by serial passages in culture media in the laboratory. Using this technique, Pasteur was able to produce efficacious attenuated vaccines against fowl cholera (1879) and anthrax (1881). These two basic approaches are still in use, with most available vaccines made of either inactivated or attenuated agents (reviewed by Plotkin, 2014). Current technologies mean that attenuated vaccines for some pathogens can now also be made by deleting or modifying genes. Table 2 shows some examples of currently licensed or marketed vaccines for pigs.

Table 2 Some examples of different types of vaccines currently licensed for use in different countries.

Type of vaccine	Examples	Comments
Attenuated/ modified life vaccines	Pseudorabies virus Porcine reproductive and respiratory syndrome virus (1 and 2) Classical swine fever virus Swine influenza A virus (H1N1 plus H3N2) Rotavirus A (G5 and G9) *Lawsonia intracellularis* *Salmonella choleraesuis/ typhimurium*	✓ They produce an immune response close to the natural infection. ✓ Residual virulence or restriction of use in some individuals may exist. ✓ In some cases, recombination or reassortment may occur between the vaccine agent and the wild-type pathogen.
Inactivated/killed vaccines	Porcine parvovirus Porcine reproductive and respiratory syndrome virus (1 and 2) Swine influenza A virus Porcine epidemic diarrhoea virus *Erysipelothrix rhusiopathiae* *Mycoplasma hyopneumoniae* *Bordetella bronchiseptica* *Lawsonia intracellularis* *Actinobacillus pleuropneumoniae* *Glaeserella parasuis* *Leptospira* spp.	✓ Generally, very safe (no replication). ✓ Need for a high antigen concentration. ✓ Usually need an adjuvant. ✓ Two doses can be needed to establish first immunisation.
Toxoid vaccines	*Escherichia coli* *Clostridium perfringens* *Actinobacillus pleuropneumoniae* *Pasteurela multocida*	✓ Toxins modified to remove potential cause of damage. ✓ Generally, only induce antibodies.
Recombinant proteins (subunits)	Classical swine fever virus Porcine circovirus 2 Porcine parvovirus	✓ Useful when protection is determined by a single target in the pathogen.
Vectored vaccines	No vaccines with live vectors marketed for pigs at present Baculovirus-produced vaccine proteins for several agents	✓ May induce B- and T-cell responses. ✓ If vector is common in the target species, immunity against the vector may exist in the population.
Virus-like particles (VLP)	No vaccines marketed for pigs at present	✓ VLPs are self-adjuvanted. ✓ Difficult application for complex antigens.
Nucleic acids		
DNA	No vaccines licensed for pigs at present	✓ Usually require several doses. ✓ Need adjuvants. ✓ Potential integration in the host DNA.
RNA	Swine Influenza A virus Porcine epidemic diarrhoea virus	✓ Induce strong immune responses. ✓ They do not integrate in the host genome.

Examples of inactivated vaccines commonly used in pigs can be found for viral and bacterial pathogens such as PPV, swIAV, *Mycoplasma hyopneumoniae or Erysypelothrix rhusiopathiae*. Other vaccines for pigs made of attenuated microorganisms have been developed for some viral agents such as PRV or PRRSV. In some cases, the same agent can be found in both inactivated and attenuated vaccines. Examples range from viral (e.g. PRRSV) to bacterial (*Lawsonia intracellularis*) vaccines.

A third type of vaccines are those made of toxins that have been treated or modified to remove their potential to cause damage (called toxoids). A classic example is the tetanus vaccine that is made of inactive tetanus toxin. The use of vaccines containing toxoids for pigs is common, for example, for *Escherichia coli* (Johansen et al., 1997) and *Clostridium perfringens* (Kennedy et al., 1977).

In other cases, a vaccine is made of what are called subunits, that is, purified or recombinant proteins or purified lipopolysaccharides. There is one vaccine made of the recombinant E2 protein of the CSFV that is registered for use in pigs in Europe (EMA, 2015). One particular variant of this approach is the use of a vector to produce the protein of interest which is then injected in the animal. This is the case, for example, for baculovirus-vectored vaccines for PCV2 (Blanchard et al., 2003) or PPV (Martínez et al., 1992). Baculoviruses are viruses of insects that can be engineered to express foreign proteins and can be produced at scale. This feature has been used, for example, to insert the ORF2 (capsid) protein of PCV2, produce it in volume and, after adding an adjuvant, produce a highly efficient vaccine (Fort et al., 2009).

Virus-like particles (VLP) are self-assembled structures made of structural viral proteins that mimic the structure (often the capsid) of the native virus. In some cases, the VLP can be the antigen itself. In others, the VLP can be the vehicle with the antigen inserted by genetic engineering. VLPs offer the advantage of being non-replicating structures mimicking a virus and often their structures may have adjuvant properties. However, when the VLP itself is not the antigen of interest but only the vehicle for inserting the antigen, some limitations become apparent. The introduction of large foreign peptides in a VLP structure may result in its disassembly. On the other hand, the way in which the antigen is recognised by the immune system within the VLP is not necessarily the same as the way the same antigen is recognised in the original pathogen (reviewed by Jennings and Bachmann, 2008; and Donaldson et al., 2018). At present, there are no vaccines using this technology specially targeting pigs.

Vectored vaccines are vaccines that use a safe microorganism (e.g. a bacterium or a virus) that has been modified to express one or more antigens of the pathogen of choice. Upon inoculation, the vector microorganism replicates and expresses the antigens of interest in a way similar to what would happen in a natural infection. The use of a virus producing a harmless infection can protect very effectively against a second more virulent agent, as long as the relevant epitopes of the pathogens are expressed in the right way. At present, no vectored vaccines are in use for pigs, although many experimental vaccines are described in the literature. While vectored vaccines are advantageous because they only use relevant antigens of a pathogen in a way that mimics natural infection, they also pose some problems. The choice of the vector is a main issue. The vector must replicate in the target species for the vaccine to work. An obvious choice would be a virus or bacterium that is well adapted to that species, where replication causes no harm or in the worst case, a very mild infection. However, the use of a common microorganism as a vector can face an insurmountable obstacle: most animals may be already immune to that vector and can thus block its replication. For pathogens which depend on several mechanisms involving several proteins for virulence, it can also be technically difficult to express all those proteins in the vector.

Another type of vaccine is made of nucleic acids. When administered under appropriate conditions, nucleic acids may induce expression of the proteins they code in the vaccinated host (reviewed by Robinson and Pertmer, 2000). DNA vaccines were the first to be developed in the 1990s. This approach was used to develop experimental vaccines against several swine pathogens (Pirzadeh and Dea, 1998; Kamstrup et al., 2004; Olsen, 2000). However, none of them reached the market, mainly because of biosafety issues related to administration of foreign DNA and its potential integration in the host DNA (Nichols et al., 1995; Faurez et al., 2010).

A more recent advancement is the development of technology to produce RNA vaccines. The main advantage of RNA vaccines is that they do not replicate and cannot integrate in the genome. They also provide a very versatile platform, allowing the production of different variants of the original pathogen very rapidly. The development of the severe acute respiratory syndrome coronavirus 2 (SARS-CoV-2) RNA vaccines in just 2 months after the availability of the virus sequence is an example of its potential (reviewed by Dolgin, 2021). There are already RNA platforms providing the possibility of producing 'a la carte' vaccines for swIAV or porcine epidemic diarrhoea virus (PDEV).

6.2 Adjuvants

An adjuvant is commonly used to enhance the immunogenicity of a vaccine. Table 3 provides a list of adjuvants used in commercialised veterinary vaccines. Until very recently, the most common adjuvant was aluminium hydroxide (also called alum). Aluminium hydroxide was introduced as an adjuvant in the 1920s and was the main adjuvant used, through to the 1990s (reviewed by Marrack et al., 2009). Mineral oil adjuvants started to attract attention in the 1960s and were subsequently commercialised (reviewed by McKercher and Graves, 1976).

Table 3 Some of the adjuvants currently used in licensed vaccines.

Adjuvant	Antigens	Comments (references in parenthesis)
Mineral salts (Alum)	Swine influenza A virus Porcine circovirus 2 Porcine parvovirus *Actinobacillus pleuropneumoniae* *Bordetella bronchiseptica* *Clostridium perfringens* toxoids *Escherichia coli* toxoids *Erysipelothrix rhusiopathiae* *Lawsonia intracellularis* *Mannheimia haemolytica* *Mycoplasma hyopneumoniae* *Pasteurella multocida* toxoids	✓ Aluminium salts. ✓ Cell recruitment (neutrophils, monocytes, macrophage) (Awate et al., 2013; de Gregorio et al., 2008). ✓ Improve antigen uptake (Charerntantanakul, 2020; de Gregorio et al., 2008). ✓ Increase inflammatory responses (IL-1β and IL-18) (Awate et al., 2013) and the consequent expansion and differentiation of naïve CD4 T cells into Th2 cells (production of antibodies) (Apostólico et al., 2013; Awate et al., 2013; Coffman et al., 2010; Ghimire, 2015).
Mineral, animal and vegetable oils	Porcine epidemic diarrhoea virus *Escherichia coli* *Clostridium perfringens* *Mycoplasma hyopneumoniae*	✓ Can be found formulated as emulsions: Water-in-oil, oil-in-water, and water-in-oil-in-water. ✓ More effective than aluminium compounds but less safe (Beverley, 2002). ✓ Enhance mainly antibodies production; some of them also increase cytotoxic responses (Beverley, 2002; de Gregorio et al., 2013).
Paraffin	Porcine parvovirus *Mycoplasma hyopneumoniae*	✓ The mineral oil most included in veterinary vaccines.
• Oil-in-water emulsions	Swine influenza A virus	✓ Strong activator of cell recruitment (Awate et al., 2013; Charerntantanakul, 2020; Coffman et al., 2010). ✓ Promote antigen uptake (Charerntantanakul, 2020; Coffman et al., 2010)
LPS	*Mycoplasma hyopneumoniae*	✓ Major components of the membrane of Gram-negative bacteria. ✓ More toxic and pyrogenic than other adjuvants (Charerntantanakul, 2020). ✓ TLR4 ligand (Apostólico et al., 2013; Awate et al., 2013). ✓ Activation of macrophages and dendritic cells (Apostólico et al., 2013; Awate et al., 2013). ✓ Increase inflammatory responses (Apostólico et al., 2013; Awate et al., 2013). ✓ Mainly promote cytotoxic responses.

(*Continued*)

Table 3 (*Continued*)

Adjuvant	Antigens	Comments (references in parenthesis)
Carbomer	Swine influenza A virus Porcine circovirus 2 Porcine reproductive and respiratory syndrome virus *Glaesserella parasuis* *Mycoplasma hyopneumoniae*	✓ Hydrophilic synthetic polymers of acrylic acid. ✓ Promote recruitment of immune cells (Beverley, 2002). ✓ Promote antibody production (Beverley, 2002). ✓ Promote lymphocyte proliferation and early IFN-γ secreting cells (Beverley, 2002).
Levamisole	*Mycoplasma hyopneumoniae*	✓ Anthelmintic. ✓ Promote antibody production (Beverley, 2002). ✓ Promotes lymphocyte proliferation and T cell and macrophage activation (Beverley, 2002).
Ginseng	Porcine parvovirus *Bordetella bronchiseptica* *Clostridium perfringens* toxoids *Clostridium novyi* toxoids *Erysipelothrix rhusiopathiae* *Pasteurella multocida*	✓ Glycosidal saponins (ginsenosides) include different chemical complexes. ✓ Specific ginsenosides (Rb1 fraction) balance Th1 and Th2 immune response[8]Activation of macrophages (Block and Mead, 2003). ✓ Increase NK activity (Block and Mead, 2003). ✓ Increase cytokine responses (IFN-γ, IL-2, IL-4, IL-10 and TNF-α) (Block and Mead, 2003). ✓ Enhance antibodies production (Beverley, 2002; Block and Mead, 2003).
α-tocopherol	Porcine circovirus 2 Porcine parvovirus Porcine reproductive and respiratory syndrome virus *Actinobacillus pleuropneumoniae* *Bordetella bronchiseptica* *Clostridium perfringens* toxoids *Escherichia coli* toxoids *Erysipelothrix rhusiopathiae* *Glaesserella parasuis* *Lawsonia intracellularis* *Leptospira* spp. *Mycoplasma hyopneumoniae* *Pasteurella multocida*	✓ Vitamin E. ✓ Increase the antigen loading in monocytes (Beverley, 2002; Charerntantanakul, 2020). ✓ Promotes antibody production (Beverley, 2002; Charerntantanakul, 2020).

They are used as emulsions (with different formulations such as oil-in-water or water-in-oil-in-water).

The precise mechanism of action of classical adjuvants was not known until recently, and even now, some of them are only partially understood. For more than 60 years, for example, it was accepted as a dogma that the main effect of alum was producing a sort of 'depot effect' that allowed a slow and sustained release of the antigen, thus inducing a stronger immune response. Although it is possible that aluminium salts might produce this depot effect, it is now clear that alum also induces the release of some cytokines (mainly IL-1β) and chemokines. This suggests that it probably acts through interaction with a nucleotide-binding oligomerisation domain (NOD)-like receptor, resulting in enhanced antibody production and stronger Th2 polarisation (de Gregorio et al., 2008). However, there is still debate about the exact mechanisms of aluminium hydroxide/alum as an adjuvant.

Water-in-oil adjuvants and their variations may also act by inducing the release of cytokines - to a greater extent than alum - and by promoting danger signals that activate the immune system (O'Hagan et al., 2013). They also stimulate the recruitment of antigen-presenting cells to the site of inoculation. Squalene, in particular, is able to induce enhanced CD8+ responses by different mechanisms, including apoptosis-related signalling (Kim et al., 2020).

α-tocopherol is another commonly used adjuvant for pig vaccines. It is thought that α-tocopherol enhances the uptake of antigen through antigen-presenting cells and induces the release of several chemokines, thus increasing the magnitude of the immune response (Morel et al., 2011).

Combinations of different co-stimulatory molecules have also been used. The idea is to combine synergistically different substances targeting different receptors or triggering different innate responses. For example, the combination of saponin and MPL (lipid A) in a liposome has been shown to be very efficient for promoting CD4+T-cell-mediated immune responses, making it an ideal adjuvant for vaccines against viruses and intracellular agents (Didierlaurent et al., 2017).

Newer adjuvants are being introduced in both human and veterinary vaccines. For example, carbopol, an anionic polymer of acrylic acid, is being used as an adjuvant in a live-attenuated PRRSV vaccine. It has been shown that the addition of carbopol enhances vaccine-specific CD8 responses in a PRRSV vaccine (Mair et al., 2015). Other polymeric adjuvants are also commercially available.

The discovery of PRRs, the identification of their ligands, the understanding of their signalling pathways and the resulting cytokine/chemokine release now make it possible to design or choose adjuvants targeting the precise receptor, thus producing the right cytokines to induce the immune response of choice. For example, an adjuvant targeting a receptor for the release of IL-12 theoretically results in the induction of a stronger cell-mediated response (Ullrich et al., 2020). Nevertheless, the final result of immunisation with a combination of a given antigen and a given adjuvant can only be fully determined in an *in vivo* model. The same adjuvant also may have very different adjuvant capabilities in different species.

7 Limitations of vaccines and ways to overcome them

By definition, an endemic state implies that the agent of the disease is present in the population and that a fairly constant pattern of infection is observed. On one hand, the actual presence of the pathogen in the population is a problem, since immunity against it has to be solidly established before contact with the pathogen. For pathogens present in sows that transmit the infection to the offspring early in life, this is quite difficult since it implies the use of vaccines in the presence of MDA.

A current trend in pig preventive medicine is the establishment of early vaccination protocols, with some vaccines being licensed for use as early as the first week of age. This early vaccination has several advantages:

- If efficacious, it establishes an immunised population before transfer of piglets to nurseries, where most agents spread when animals lose MDA.
- Handling of animals for vaccination is easier at earlier ages.
- It can help to reduce transmission of infection from sows to their offspring.

Nevertheless, some circumstances may limit the practical use of vaccines in pig farms (Table 4).

Table 4 Limitations in the use of vaccines. The table shows some of the most common limitations to the practical use of vaccines in pig farms.

✓ Cost of a dose. ✓ Difficulty of administration. ✓ Storage and handling (for example, very low temperatures). ✓ Antigen can only be produced in limited amounts (low antigenic doses). ✓ Several doses needed to induce solid and durable immunity. ✓ Establishment of effective immunity requires several weeks. ✓ Short-lived immunity. ✓ Antigenic variability in the pathogen population (non-universal protection). ✓ Pathogen has several escape mechanisms. ✓ Lack of differentiating (DIVA) vaccines.	✓ Presence of maternally derived antibodies (MDA). ✓ Genetic/environmental factors may decrease the magnitude of the immune response induced by the vaccine. ✓ Infection with other concurrent pathogens impairing the immune response. ✓ Infection with microorganisms of very short incubation period, which implies need of an early vaccination (presence of MDA) and/or vaccination of already infected animals.

The presence of high levels of MDA may interfere with the development of immunity after vaccination. The presence of MDA in pigs involves IgG absorbed systemically from the colostrum. Lactogenic immunity in pigs (mediated by IgA) is important in protection against intestinal pathogens, but since very few oral vaccines are available, we will focus on MDA blocking by IgG.

The precise mechanisms by which MDA block vaccination of neonates are not well understood, although all types of vaccine can be affected. In the vaccination of children against measles, for example, it has been shown that even low titres of non-neutralising antibodies are able to block seroconversion after vaccination. Aaby et al. (2014) showed that mortality and attack rates in the presence of MDA were lower than in seronegative children, suggesting that imprinting of the immune system by MDA could also play a role in generating more balanced responses after vaccination. Several studies have shown that vaccination in the presence of MDA did not avoid development of T-cell responses (Gans et al., 2001, 2004).

One mechanism involved may simply be the formation of antigen-antibody complexes that are removed by macrophages, without leading to effective antigenic presentation. However, there is no direct evidence for this hypothesis. For live-attenuated viral vaccines, another common explanation for this blocking effect is the potential neutralisation of the virus, thus preventing its replication. Although this might be partially true, non-neutralising antibodies also can produce this effect (Weidinger et al., 2002). In addition, antibodies also block protein or inactivated vaccines, an indication that other mechanisms must be involved. Kim et al. (2011) found no evidence for the hypothesis that MDA might inhibit seroconversion after vaccination by masking vaccine epitopes from the immune system. Instead, they showed that a low-concentration cocktail of antibodies against different epitopes was more efficient than a single antibody at a high concentration, a fact which suggested that epitope masking is not the mechanism. They provided evidence that the MDA-antigen complex may crosslink the B-cell receptor and the Fc gamma receptor IIb (FcγRIIB), resulting in the inhibition of the proliferation of B-cells. This can be understood as a signal telling B-cells that antigen-specific antibodies are already present and thus no more antibodies must be produced. Other mechanisms have been proposed but lack clear evidence.

One classical approach to overcome this problem of blocking by MDA is to establish a double vaccination protocol. The first dose of the vaccine is administered at an early age, and then, 3 or 4 weeks later, a second dose is administered. If the first is not blocked, the second acts as a booster. In the worst case, if the first is blocked, the second probably immunises the animal because of the decline in MDA. This approach is widely used in dogs and cats, for example. For farm animals, the use of two doses has some limitations.

Firstly, it implies handling animals twice and doubling the cost of vaccination. Secondly, levels of MDA in a large population can be quite diverse, depending on the immunisation schedule and the composition of sow stock.

There are alternatives to blanket vaccination of all animals in a group, as can be seen for poultry. In the case of infectious bursal disease, serological tests (ELISA) can be used to estimate the optimal moment for vaccination in the presence of MDA. This requires knowing the half-life of MDA, the age of animals at sampling, the breakthrough titre of the vaccine and the distribution of MDA in the target population. For many pig diseases, there is a lack of quantitative information of this kind to make it possible to precisely balance titres of antibodies with the point at which vaccines may be blocked. This is in part a consequence of a lack of knowledge of the precise correlates of protection.

A second approach to overcome the presence of high levels of MDA is the use of less attenuated vaccines or a higher titre (or antigenic mass) vaccine. These approaches have been successfully used in several species. For example, a higher titre vaccine for canine parvovirus has been proven to overcome MDA as not all viral particles are neutralised. In poultry, the use of less attenuated strains may help to overcome MDA interference. However, these two approaches also have problems. By increasing the vaccine titre or by using less attenuated strains, the risk of residual virulence or undesired side effects increases (as well as the cost). Shedding of the vaccine strain can also increase, and this creates a potential issue of recombination of the vaccine strain with field strains, particularly for viruses. To our knowledge, these approaches have not been used systematically in pigs. However, some vaccines use the principle of enhanced replication to allow vaccination at very early ages.

Another approach to avoid MDA interference is the use of alternative routes of vaccination. The most common alternative routes are mucosal delivery and intradermal administration. One classic example of mucosal delivery of a vaccine in pigs is intranasal administration of PRV live-attenuated gE-deleted vaccines. From the 1970s, it was shown that intranasal vaccination was effective in clinically protecting piglets and reducing shedding, even when moderate-to-high titres of MDA were present at the moment of vaccination. However, protection started to decline after 2 months and was inefficient at 4 months after vaccination. A second dose intranasal or intramuscular dose is necessary to protect pigs until they reach market age (Arellano et al., 1992; McCaw and Xu, 1993; Mengeling et al., 1992; van Oirschot, 1987, 1991; Vannier, 1986). While this approach provides protection in the presence of MDA, it produces relatively short-lived immunity. Vaccination of 1-day-old piglets has not been found sufficiently protective.

An intranasal vaccine for influenza A containing live-attenuated virus has recently been marketed in the USA. The vaccine has been proven effective when administered to lactating piglets. However, reassortment/recombinant viruses containing wild-type viruses and vaccine viruses have been detected from respiratory clinical cases (Sharma et al., 2020), suggesting that these vaccines should only be used in the absence of SwIAV infection. In contrast, use of intranasal vaccines for other agents, such as PRRSV, has not been successful until now.

A second method of mucosal delivery of a vaccine is oral administration. At present, oral vaccines for *Lawsonia intracellularis*, *Salmonella* spp. and *Escherichia coli* (all live-attenuated) are being marketed in several countries. The interference of colostral/lactogenic immunity with these vaccines has not been well studied.

The second alternative route of administration is the intradermal route. This route of administration was explored from the 1960s in children, showing the potential efficacy of intradermal vaccination using reduced doses of the antigen. The rationale for intradermal vaccination is that skin is very rich in antigen-presenting cells. Dermal dendritic cells are thought to be particularly efficient in taking up antigens and migrating to the draining lymph nodes for antigen presentation.

The intradermal route has several advantages for pigs, since needles are not required, allowing gentler handling of animals. The chances for iatrogenic transmission are also substantially reduced. This approach has been used in vaccines for PRRSV, PCV2, *Mycoplasma hyopneumoniae* or *Lawsonia intracellularis* (Beffort et al., 2017; Jacobs et al., 2020; Martelli et al., 2007, 2009, 2014; Opriessnig et al., 2008; Stadler et al., 2018; Tassis et al., 2012). However, more studies are needed to establish whether this route can avoid MDA blocking.

Other alternatives, which are yet to be fully explored and commercialised, include the use of nanoparticles containing the antigen, the use of formulations avoiding interaction of the antigen with MDA or the use of mRNA vaccines. There is some evidence that induction of IFN-α responses may counterbalance blocking by MDA (Kim and Niewiesk, 2014). Since IFN-α is released in response to toll-like receptors (TLRs), adding appropriate adjuvants with the potential to stimulate TLRs may help to counteract blocking by MDA. Some of these approaches have been used in other species, particularly chickens. For example, the use of a complex of antibodies and an infectious bursal disease VP2 vaccine can overcome blocking by MDA (Sedeik et al., 2019; García et al., 2021).

A second important obstacle to vaccination protocols in farms with endemic disease is the time needed to establish effective immunity. Since the pathogen is already circulating in the population, animals can be exposed as soon as they enter the main stock. Gilts can be vaccinated in quarantine/adaptation units while in isolation from the main population. This is difficult for piglets since they can be exposed to the pathogen very early in life.

The onset of effective immunity can be seen as the moment after vaccination in which the animal achieves the desired goal of the vaccination programme (whether this goal is sterilising immunity or just prevention of production losses in the animal infected). Relatively little can be done to speed the onset of the immunity in terms of decreasing the time needed to mount an immune response. Biologically, it is extremely difficult or even impossible to achieve a significant response in less than 5–7 days after infection or vaccination, since the essential steps (antigen uptake, processing and presentation, clonal expansion of reactive lymphocyte clones, development of effector responses and later generation of memory clones) require this time to occur.

However, while the time needed to mount the immune response is difficult to shorten, we can modify the immune response to increase its intensity, achieving the required titres of antibodies or the required intensity of the cell-mediated response sooner. This suggests methods for directing the antigen to dendritic cells or the use of adjuvants or co-adjuvants to induce the release of cytokines and chemokines. The transfer of these potential measures to actual vaccines for use in the field is still under investigation.

Considering the current limitations of vaccines, the improvement of internal biosecurity is a much-needed addition to any vaccination plan. Internal biosecurity can be understood as the whole set of actions aimed to decrease, and eventually stop, the transmission of infection within the farm. If we consider that the two main groups of susceptible animals in a farm are gilts purchased from external sources and piglets as soon as they lose MDA, focus must be put on these two groups.

Gilts must be isolated before entering sow stock to avoid the introduction of new pathogens into the herd. This phase, usually designated as a quarantine, must be followed by an 'acclimation period' in which gilts are vaccinated against, or exposed to, pathogens present in the farm. The main objective here is to produce strong immunity before introduction of gilts in the breeder herd.

To be efficacious, immunisation should be undertaken simultaneously for all gilts (providing another reason to favour the use of vaccines over 'natural' exposure to the pathogen). This decreases the chances of leaving unprotected subpopulations that could be infected and maintain circulation of the pathogen. Homogeneous immunisation of gilts also reduces variability of MDA in their offspring and thus makes it easier to design vaccination programmes for piglets. However, unless vaccination produces sterilising immunity, there is still a chance for incoming gilts to get infected when entering the breeding stock.

The organisation of production batches in an all-in/all-out system and a well-managed hygiene programme minimise exposure to pathogens, thus enhancing the impact of vaccination. For piglets, the application of internal biosecurity measures is of capital importance, particularly for pathogens that are easily transmissible or when vaccines do not fully prevent infection. The most effective measures, beyond batch management and the all-in/all-out systems, are to limit cross-fostering and batch mixing, together with the application of strict hygiene measures, particularly regarding the use of needles or for all procedures relating to piglets.

8 Conclusion and future trends

Vaccination is the most convenient and the safest way to prevent infectious diseases in a farm. It can be planned strategically to anticipate and prevent disease rather than waiting for disease to occur. In farms with endemic disease, the application of a vaccine must consider the epidemiology of the pathogens against which vaccines are being administered, since different pathogens may have different patterns of transmission. Moreover, it is important to understand what level of protection is to be achieved with each vaccine. In some cases, vaccination can be aimed to eradicate disease altogether. In other cases, vaccination may just be needed to decrease the economic impact of the disease or to reduce transmission. Setting realistic goals helps to obtain the maximum benefit from vaccination. Target populations and their epidemiological characteristics must also be properly identified. Better understanding of the epidemiology of endemic pig diseases would contribute substantially to identifying patterns and determinants of transmission and the critical windows to undertake vaccination.

Vaccination goals and programmes for gilts, sows and piglets or fatteners can be substantially different. In the case of young piglets, the presence of MDA at the moment of vaccination is a major obstacle to achieving optimal herd immunisation. Identifying the levels of MDA that critically interfere with the development of immunity after vaccination is important. The development of tests allowing monitoring of MDA and the identification of adequate time-points for vaccination would be very helpful in optimising vaccination protocols for piglets. Exploration of new formulations or delivery methods would help to clarify whether alternative routes of administration may really help to overcome MDA blocking.

For more than a 100 years, all available vaccines have fallen in one of three categories: inactivated, attenuated or toxoids. This situation is now changing rapidly. Scientific and technological advances have allowed the development of several new types of vaccines, among which mRNA vaccines are the most promising. The COVID-19 pandemic will surely boost the development and production of these types of vaccines for veterinary as well as human use. However, a major obstacle is the lack of precise knowledge of the epitopes inducing effective immunity and of protection correlates. More immunological study of pathogens is necessary, with particular emphasis on escape mechanisms for which effective vaccines have not been developed.

A revolution in adjuvants is also taking place. Adjuvants designed to produce a pre-designed response in the host are being developed. This makes possible a future with 'à la carte' vaccines containing just one or a few relevant epitopes of a pathogen plus an adjuvant driving the desired immune response. There is a need to explore new methods for vaccine production as well as characterise the immune response of pigs to new adjuvants. This will require better knowledge of basic immunology, including pathogen receptors in dendritic cells or other antigen-presenting cells, signalling pathways of key cytokines and chemokines etc.

However, beyond the likely introduction of a new generation of vaccines, we must stick to well-established principles that remain valid:

- Choose the right time for vaccination.
- Immunise all animals in the target group together.
- Minimise exposure to a pathogen as much as possible, particularly if the vaccine does not produce sterilising immunity.

Separation of vaccinated animals from infected animals is important. Combining diagnostics, monitoring, biosecurity and vaccination is a winning strategy for any farm.

9 Where to look for further information

The subject of this chapter encompasses several areas of knowledge. We divided the information of this section according to these areas.

I. Vaccines and vaccinology:
   - A recent and comprehensive review on the current knowledge about vaccinology and the basis for vaccine efficacy can be found in Pollard, A. J. and Bijker, E. M. 2021. A guide to vaccinology: from basic principles to new developments. *Nat. Rev. Immunol.* 21, 83–100. doi: 10.1038/s41577-020-00479-7.
   - This book is one of the reference texts for vaccinology. Siegrist, C. A. 2017. Vaccine immunology. In: *Plotkin's Vaccines* (7th edn), Plotkin, S., Orenstein, W., Offit, P. and Edwards, K. M. (Eds). Amsterdam, The Netherlands: Elsevier.

II. Adjuvants: In the last few years, the research about adjuvant focuse on the development of more specific products aimed at producing specific effects on antigen-presenting cells, T or B cells. This paper is a review on the history and current status of this research. Shah, R. R., Hassett, K. J. and Brito, L. A. 2017. Overview of vaccine adjuvants:

Introduction, history, and current status. *Methods Mol. Biol.* 1494, 1–13. doi: 10.1007/978-1-4939-6445-1_1.

III. Interference of maternally derived antibodies with vaccination: This is an excellent and comprehensive review on the mechanism by which maternally-derived antibodies may block vaccines, Niewiesk, S. 2014. Maternal antibodies: Clinical significance, mechanism of interference with immune responses, and possible vaccination strategies. *Front. Immunol.* 5, 446. doi: 10.3389/fimmu.2014.00446.

IV. Future prospects for pig vaccines: The technologies for the production of new generation vaccines are mature enough to be applied not only to human but to animals disease as well. In this review, the authors present and discuss the potential application of these technologies for swine diseases. Opriessnig, T., Mattei, A. A., Karuppannan, A. K. and Halbur, P. G. 2021. Future perspectives on swine viral vaccines: Where are we headed? *Porcine Health Manag.*, 7:1. doi: 10.1186/s40813-020-00179-7.

V. Current trends on infectious diseases of swine: Infectious diseases are dynamic and the importance of some agents decreased while some others emerge. In this paper, the authors analyze the available data and formulate hypothesis about the future trends in the infectious diseases of pigs. VanderWaal, K. and Deen, J. 2018. Global trends in infectious diseases of swine. *Proc. Natl. Acad. Sci. U.S.A.* 115, 11495–11500. doi: 10.1073/pnas.1806068115.

VI. Sensing of pathogens by the innate immunity and interaction with the adaptive immunity: The adaptive immune response is the ultimate responsible for fighting infectious agents. However, the actions executed by the adaptive immunity are strongly influenced by the way in which the innate immunity senses a pathogen. This review explains how this happens. and Medzhitov, R. 2015. Control of adaptive immunity by the innate immune system. *Nat. Immunol.* 16, 343–353. doi: 10.1038/ni.3123.

VII. Immune response to pathogens: This is a reference book encompassing all major aspects of the immune response. Murphy, K., Weaver, C. and Janeway, C. 2017. *Janeway's Immunobiology* (9th edn). New York: Garland Science.

VIII. Transmission and maintenance of disease: This is the reference textbook for veterinary epidemiology. In this chapter, the authors present all the basic concepts on the transmission and maintenance of infectious agents in the population. Thrusfield, M. 2018. Chapter 6: The transmission and maintenance of infection. In: *Veterinary Epidemiology* (4th edn), Thrusfield, M. and Christley, R. (Eds)., Hoboken, NJ: Wiley-Blackwell.

IX. The concept of R0 and its relationship with vaccination: One of the expected effects of vaccination is reducing the transmission of the infection, contributing thus to stop the spread of the disease.

Transmission is often assessed by the calculation of a parameter called the basic reproduction number. In this review, the authors explain what this number represents and how is calculated. Delamater, P. L., Street, E. J., Leslie, T. F., Yang, Y. T. and Jacobsen, K. H. 2019. Complexity of the basic reproduction number (R_0). *Emerg. Infect. Dis.* 25, 1-4.

10 References

Aaby, P., Martins, C. L., Garly, M. L., Andersen, A., Fisker, A. B., Claesson, M. H., Ravn, H., Rodrigues, A., Whittle, H. C. and Benn, C. S. (2014). Measles vaccination in the presence or absence of maternal measles antibody: Impact on child survival. *Clin. Infect. Dis.* 59(4), 484–92.

Allerson, M. W., Cardona, C. J. and Torremorell, M. (2013). Indirect transmission of influenza A virus between pig populations under two different biosecurity settings. *PLoS One* 8(6), e67293. doi: 10.1371/journal.pone.0067293.

Apostólico, Jde S., Lunardelli, V. A., Coirada, F. C., Boscardin, S. B. and Rosa, D. S. (2016). Adjuvants: Classification, modus operandi, and licensing. *J. Immunol. Res.* 2016, 1459394.

Arellano, P. E., Morrison, R. B., Molitor, T. W. and Thawley, D. G. (1992). High- and low-challenge exposure doses used to compare intranasal and intramuscular administration of pseudorabies virus vaccine in passively immune pigs. *J. Am. Vet. Med. Assoc.* 200(5), 685–91.

Awate, S., Babiuk, L. A. and Mutwiri, G. (2013). Mechanisms of action of adjuvants. *Front. Immunol.* 4, 114.

Baums, C. G., Brüggemann, C., Kock, C., Beineke, A., Waldmann, K. H. and Valentin-Weigand, P. (2010). Immunogenicity of an autogenous *Streptococcus suis* bacterin in preparturient sows and their piglets in relation to protection after weaning. *Clin. Vaccine Immunol.* 17(10), 1589–97.

Beffort, L., Weiß, C., Fiebig, K., Jolie, R., Ritzmann, M. and Eddicks, M. (2017). Field study on the safety and efficacy of intradermal versus intramuscular vaccination against *Mycoplasma hyopneumoniae*. *Vet. Rec.* 181(13), 348.

Belongia, E. A. and Naleway, A. L. (2003). Smallpox vaccine: The good, the bad, and the ugly. *Clin. Med. Res.* 1(2), 87–92.

Bernaerdt, E., Dewulf, J., Verhulst, R., Bonckaert, C. and Maes, D. (2021). Purchasing policy, quarantine and acclimation practices of breeding gilts in Belgian pig farms. *Proc. Health Manag.* 7, 25.

Betlach, A. M., Valeris-Chacin, R., Singer, R. S., Allerson, M. and Pieters, M. (2020). Natural transmission and detection of *Mycoplasma hyopneumoniae* in a naïve gilt population. *Vet Microbiol.* 248, 108819. doi: 10.1016/j.vetmic.2020.108819.

Beverley, P. C. (2002). Immunology of vaccination. *Br. Med. Bull.* 62, 15–28.

Blanchard, P., Mahé, D., Cariolet, R., Keranflec'h, A., Baudouard, M. A., Cordioli, P., Albina, E. and Jestin, A. (2003). Protection of swine against post-weaning multisystemic wasting syndrome (PMWS) by porcine circovirus type 2 (PCV2) proteins. *Vaccine* 21(31), 4565–75.

Blanco, I., Galina-Pantoja, L., Oliveira, S., Pijoan, C., Sánchez, C. and Canals, A. (2004). Comparison between *Haemophilus parasuis* infection in colostrums-deprived and sow-reared piglets. *Vet. Microbiol.* 103(1–2), 21–7.

Blavi, L., Solà-Oriol, D., Llonch, P., López-Vergé, S., Martín-Orúe, S. M. and Pérez, J. F. (2021). Management and feeding strategies in early life to increase piglet performance and welfare around weaning: A review. *Animals (Basel)* 11(2), 302.

Block, K. I. and Mead, M. N. (2003). Immune system effects of echinacea, ginseng, and astragalus: A review. *Integr. Cancer Ther.* 2(3), 247–67.

Cador, C., Hervé, S., Andraud, M., Gorin, S., Paboeuf, F., Barbier, N., Quéguiner, S., Deblanc, C., Simon, G. and Rose, N. (2016). Maternally-derived antibodies do not prevent transmission of swine influenza A virus between pigs. *Vet Res.* 47(1), 86. doi: 10.1186/s13567-016-0365-6. Erratum in: Vet Res. 2016;47(1):95.

Chamba-Pardo, F. O., Wayne, S., Culhane, M. R., Perez, A., Allerson, M. and Torremorell, M. (2019). Effect of strain-specific maternally-derived antibodies on influenza A virus infection dynamics in nursery pigs. *PLoS ONE* 14(1), e0210700.

Charerntantanakul, W. (2020). Adjuvants for swine vaccines: mechanisms of actions and adjuvant effects. *Vaccine* 38(43), 6659–81.

Chiers, K., Donné, E., Van Overbeke, I., Ducatelle, R. and Haesebrouck, F. (2002). *Actinobacillus pleuropneumoniae* infections in closed swine herds: Infection patterns and serological profiles. *Vet. Microbiol.* 85(4), 343–52.

Coffman, R. L., Sher, A. and Seder, R. A. (2010). Vaccine adjuvants: Putting innate immunity to work. *Immunity* 33(4), 492–503.

Curtis, J. and Bourne, F. J. (1973). Half-lives of immunoglobulins IgG, IgA and IgM in the serum of new-born pigs. *Immunology* 24(1), 147–55.

de Gregorio, E., Caproni, E. and Ulmer, J. B. (2013). Vaccine adjuvants: Mode of action. *Front. Immunol.* 4, 214.

de Gregorio, E., Tritto, E. and Rappuoli, R. (2008). Alum adjuvanticity: Unraveling a century old mystery. *Eur. J. Immunol.* 38(8), 2068–71.

De Jong, M. C. and Kimman, T. G. (1994). Experimental quantification of vaccine-induced reduction in virus transmission. *Vaccine* 12(8), 761–6. doi: 10.1016/0264-410x(94)90229-1.

Didierlaurent, A. M., Laupèze, B., Di Pasquale, A., Hergli, N., Collignon, C. and Garçon, N. (2017). Adjuvant system AS01: Helping to overcome the challenges of modern vaccines. *Expert Rev. Vaccines* 16(1), 55–63.

Dolgin, E. (2021). The tangled history of mRNA vaccines. *Nature* 597(7876), 318–24.

Donaldson, B., Lateef, Z., Walker, G. F., Young, S. L. and Ward, V. K. (2018). Virus-like particle vaccines: Immunology and formulation for clinical translation. *Expert Rev. Vaccines* 17(9), 833–49.

European Medicine Agency (EMA). (2015). Suvaxyn CSF marker. Available at: https://www.ema.europa.eu/en/medicines/veterinary/EPAR/suvaxyn-csf-marker.

Faurez, F., Dory, D., Le Moigne, V., Gravier, R. and Jestin, A. (2010). Biosafety of DNA vaccines: New generation of DNA vectors and current knowledge on the fate of plasmids after injection. *Vaccine* 28(23), 3888–95.

Fort, M., Sibila, M., Pérez-Martín, E., Nofrarías, M., Mateu, E. and Segalés, J. (2009). One dose of a porcine circovirus 2 (PCV2) sub-unit vaccine administered to 3-week-old conventional piglets elicits cell-mediated immunity and significantly reduces PCV2 viremia in an experimental model. *Vaccine* 27(30), 4031–7.

Gans, H. A., Yasukawa, L., Rinki, M., DeHovitz, R., Forghani, B., Beeler, J., Audet, S., Maldonado, Y. and Arvin, A. M. (2001). Immune responses to measles and mumps vaccination of infants at 6, 9, and 12 months. *J. Infect. Dis.* 184(7), 817–26.

Gans, H. A., Yasukawa, L. L., Alderson, A., Rinki, M., DeHovitz, R., Beeler, J., Audet, S., Maldonado, Y. and Arvin, A. M. (2004). Humoral and cell-mediated immune responses to an early 2-dose measles vaccination regimen in the United States. *J. Infect. Dis.* 190(1), 83–90.

García, C., Soriano, J. M., Cortés, V., Sevilla-Navarro, S., Marin, C., Balaguer, J. L. and Catalá-Gregori, P. (2021). Monitoring serologic response to single in ovo vaccination with an immune complex vaccine against infectious bursal disease in broilers. *Poult. Sci.* 100(4), 100999.

Gelberg, H. B., Patterson, J. S. and Woode, G. N. (1991). A longitudinal study of rotavirus antibody titers in swine in a closed specific pathogen-free herd. *Vet. Microbiol.* 28(3), 231–42.

Ghimire, T. R. (2015). The mechanisms of action of vaccines containing aluminium adjuvants: An in vitro vs in vivo paradigm. *Springerplus* 4, 181.

Jacobs, A. A. C., Harks, F., Pauwels, R., Cao, Q., Holtslag, H., Pel, S. and Segers, R. P. A. M. (2020). Efficacy of a novel intradermal *Lawsonia intracellularis* vaccine in pigs against experimental infection and under field conditions. *Proc Health Manag.* 6, 25.

Janeway, C. A., Jr., Travers, P., Walport, M. and Shlomchik, M. (2001). The major histocompatibility complex and its functions. In: *Immunobiology: The Immune System in Health and Disease* (5th edn.). New York: Garland Publishing Science.

Jennings, G. T. and Bachmann, M. F. (2008). The coming of age of virus-like particle vaccines. *Biol. Chem.* 389(5), 521–36.

Johansen, M., Andresen, L. O., Jorsal, S. E., Thomsen, L. K., Waddell, T. E. and Gyles, C. L. (1997). Prevention of edema disease in pigs by vaccination with verotoxin 2e toxoid. *Can. J. Vet. Res.* 61(4), 280–5.

Kamstrup, S., Barfoed, A. M., Frimann, T. H., Ladekjaer-Mikkelsen, A. S. and Bøtner, A. (2004). Immunisation against PCV2 structural protein by DNA vaccination of mice. *Vaccine* 22(11–12), 1358–61.

Kennedy, K. K., Norris, S. J., Beckenhauer, W. H. and Hogg, A. (1977). Vaccination of pregnant sows with Clostridium perfringens type C toxoid. *Vet. Med. Small Anim. Clin.* 72(6), 1047–9.

Kim, D., Huey, D., Oglesbee, M. and Niewiesk, S. (2011). Insights into the regulatory mechanism controlling the inhibition of vaccine-induced seroconversion by maternal antibodies. *Blood* 117(23), 6143–51.

Kim, D. and Niewiesk, S. (2014). Synergistic induction of interferon a through TLR-3 and TLR-9 agonists stimulates immune responses against measles virus in neonatal cotton rats. *Vaccine* 32(2), 265–70.

Kim, E. H., Woodruff, M. C., Grigoryan, L., Maier, B., Lee, S. H., Mandal, P., Cortese, M., Natrajan, M. S., Ravindran, R., Ma, H., Merad, M., Gitlin, A. D., Mocarski, E. S., Jacob, J. and Pulendran, B. (2020). Squalene emulsion-based vaccine adjuvants stimulate CD8 T cell, but not antibody responses, through a RIPK3-dependent pathway. *eLife* 9, e52687.

Klinkenberg, D., de Bree, J., Laevens, H. and de Jong, M. C. (2002). Within- and between-pen transmission of Classical Swine Fever Virus: a new method to estimate the basic reproduction ratio from transmission experiments. *Epidemiol. Infect.* 128(2), 293–9. doi: 10.1017/s0950268801006537.

Kumar, H., Kawai, T. and Akira, S. (2011). Pathogen recognition by the innate immune system. *Int. Rev. Immunol.* 30(1), 16–34.

Li, X., Qiu, L., Yang, Z., Dang, R. and Wang, X. (2013). Emergency vaccination alleviates highly pathogenic porcine reproductive and respiratory syndrome virus infection after contact exposure. *BMC Vet. Res.* 9, 26.

Madson, D. M. and Oppriessnig, T. (2011). Effect of porcine circovirus type 2 (PCV2) infection on reproduction: Disease, vertical transmission, diagnostics and vaccination. *Anim. Health Res. Rev.* 12(1), 47-65.

Mair, K. H., Koinig, H., Gerner, W., Höhne, A., Bretthauer, J., Kroll, J. J., Roof, M. B., Saalmüller, A., Stadler, K. and Libanova, R. (2015). Carbopol improves the early cellular immune responses induced by the modified-life vaccine Ingelvac PRRS® MLV. *Vet. Microbiol.* 176(3-4), 352-7.

Marrack, P., McKee, A. S. and Munks, M. W. (2009). Towards an understanding of the adjuvant action of aluminium. *Nat. Rev. Immunol.* 9(4), 287-93.

Martelli, P., Cordioli, P., Alborali, L. G., Gozio, S., De Angelis, E., Ferrari, L., Lombardi, G. and Borghetti, P. (2007). Protection and immune response in pigs intradermally vaccinated against porcine reproductive and respiratory syndrome (PRRS) and subsequently exposed to a heterologous European (Italian cluster) field strain. *Vaccine* 25(17), 3400-8.

Martelli, P., Gozio, S., Ferrari, L., Rosina, S., De Angelis, E., Quintavalla, C., Bottarelli, E. and Borghetti, P. (2009). Efficacy of a modified live porcine reproductive and respiratory syndrome virus (PRRSV) vaccine in pigs naturally exposed to a heterologous European (Italian cluster) field strain: clinical protection and cell-mediated immunity. *Vaccine* 27(28), 3788-99.

Martelli, P., Saleri, R., Cavalli, V., De Angelis, E., Ferrari, L., Benetti, M., Ferrarini, G., Merialdi, G. and Borghetti, P. (2014). Systemic and local immune response in pigs intradermally and intramuscularly injected with inactivated *Mycoplasma hyopneumoniae* vaccines. *Vet. Microbiol.* 168(2-4), 357-64.

Martelli, P., Saleri, R., Ferrarini, G., De Angelis, E., Cavalli, V., Benetti, M., Ferrari, L., Canelli, E., Bonilauri, P., Arioli, E., Caleffi, A., Nathues, H. and Borghetti, P. (2016). Impact of maternally derived immunity on piglets' immune response and protection against porcine circovirus type 2 (PCV2) after vaccination against PCV2 at different age. *BMC Vet. Res.* 12, 77.

Martínez, C., Dalsgaard, K., López de Turiso, J. A., Cortés, E., Vela, C. and Casal, J. I. (1992). Production of porcine parvovirus empty capsids with high immunogenic activity. *Vaccine* 10(10), 684-90.

Mateu, E. and Diaz, I. (2008). The challenge of PRRS immunology. *Vet. J.* 177(3), 345-51. doi: 10.1016/j.tvjl.2007.05.022.

McCaw, M. B. and Xu, J. (1993). Protection against pseudorabies virus infection by intranasal vaccination of newborn pigs. *Am. J. Vet. Res.* 54(4), 527-33.

McKercher, P. D. and Graves, J. H. (1976). A review of the current status of oil adjuvants in foot-and-mouth disease vaccines. *Dev. Biol. Stand.* 35, 107-12.

Mengeling, W. L., Lager, K. M., Volz, D. M. and Brockmeier, S. L. (1992). Effect of various vaccination procedures on shedding, latency, and reactivation of attenuated and virulent pseudorabies virus in swine. *Am. J. Vet. Res.* 53(11), 2164-73.

Mitchison, N. A. (2004). T-cell-B-cell cooperation. *Nat. Rev. Immunol.* 4(4), 308-12.

Morel, P. A. and Turner, M. S. (2010). Designing the optimal vaccine: The importance of cytokines and dendritic cells. *Open Vaccine J.* 3, 7-17.

Morel, S., Didierlaurent, A., Bourguignon, P., Delhaye, S., Baras, B., Jacob, V., Planty, C., Elouahabi, A., Harvengt, P., Carlsen, H., Kielland, A., Chomez, P., Garçon, N. and Van Mechelen, M. (2011). Adjuvant system AS03 containing α-tocopherol modulates innate immune response and leads to improved adaptive immunity. *Vaccine* 29(13), 2461–73.

Muierhead, M. R., Alexander, T. J. L. and Carr, J. (2013). *Managing Pig Health: A Reference for the Farm* (2nd edn., vol. 5M). Essex, England: Books Ltd.

Mukherjee, R. (2020). Global efforts on vaccines for COVID-19: Since, sooner or later, we all will catch the coronavirus. *J. Biosci.* 45(1), 68.

Müller, T., Teuffert, J., Staubach, C., Selhorst, T. and Depner, K. R. (2005). Long-term studies on maternal immunity for Aujeszky's disease and classical swine fever in wild boar piglets. *J. Vet. Med. B Infect. Dis. Vet. Public Health* 52(10), 432–6.

Nichols, W. W., Ledwith, B. J., Manam, S. V. and Troilo, P. J. (1995). Potential DNA vaccine integration into host cell genome. *Ann. N. Y. Acad. Sci.* 772, 30–9.

Noguera, M., Vela, A., Kraft, C., Chevalier, M., Goutebroze, S., de Paz, X., Kunze, M., Rathkjen, P., Schacht, E. and Garcia-Morante, B. (2021). Effects of three commercial vaccines against porcine parvovirus 1 in pregnant gilts. *Vaccine* 39(29), 3997–4005.

O'Hagan, D. T., Ott, G. S., Nest, G. V., Rappuoli, R. and Giudice, G. D. (2013). The history of MF59(®) adjuvant: A phoenix that arose from the ashes. *Expert Rev. Vaccines* 12(1), 13–30.

Olanratmanee, E. O., Ayudhya, S. N. N., Thanawongnuwech, R., Kunavongkrit, A. and Tummaruk, P. (2013). Reproductive parameters following a PRRS outbreak where a whole-herd PRRS MLV vaccination strategy was instituted post-outbreak. *Trop. Anim. Health Prod.* 45(5), 1099–106. doi: 10.1007/s11250-012-0332-9.

Olsen, C. W. (2000). DNA vaccination against influenza viruses: A review with emphasis on equine and swine influenza. *Vet. Microbiol.* 74(1–2), 149–64.

Opriessnig, T., Madson, D. M., Prickett, J. R., Kuhar, D., Lunney, J. K., Elsener, J. and Halbur, P. G. (2008). Effect of porcine circovirus type 2 (PCV2) vaccination on porcine reproductive and respiratory syndrome virus (PRRSV) and PCV2 coinfection. *Vet. Microbiol.* 131(1–2), 103–14.

Opriessnig, T., Mattei, A. A., Karuppannan, A. K. and Halbur, P. G. (2021). Future perspectives on swine viral vaccines: Where are we headed? *Porcine Health Manag.* 7(1), 1.

Paul, P. S., Mengeling, W. L. and Pirtle, E. C. (1982). Duration and biological half-life of passively acquired colostral antibodies to porcine parvovirus. *Am. J. Vet. Res.* 43(8), 1376–9.

Pileri, E., Gibert, E., Soldevila, F., García-Saenz, A., Pujols, J., Diaz, I., Darwich, L., Casal, J., Martín, M. and Mateu, E. (2015). Vaccination with a genotype 1 modified live vaccine against porcine reproductive and respiratory syndrome virus significantly reduces viremia, viral shedding and transmission of the virus in a quasi-natural experimental model. *Vet. Microbiol.* 175(1), 7–16.

Pileri, E., Martín-Valls, G. E., Díaz, I., Allepuz, A., Simon-Grifé, M., García-Saenz, A., Casal, J. and Mateu, E. (2017). Estimation of the transmission parameters for swine influenza and porcine reproductive and respiratory syndrome viruses in pigs from weaning to slaughter under natural conditions. *Prev. Vet. Med.* 138, 147–155. doi: 10.1016/j.prevetmed.2017.01.008.

Pileri, E. and Mateu, E. (2016). Review on the transmission porcine reproductive and respiratory syndrome virus between pigs and farms and impact on vaccination. *Vet. Res.* 47(1), 108.

Piñeiro, C., Manso, A., Manzanilla, E. G. and Morales, J. (2019). Influence of sows' parity on performance and humoral immune response of the offspring. *Porcine Health Manag.* 5, 1.

Pirzadeh, B. and Dea, S. (1998). Immune response in pigs vaccinated with plasmid DNA encoding ORF5 of porcine reproductive and respiratory syndrome virus. *J. Gen. Virol.* 79(Pt 5):989–99. doi: 10.1099/0022-1317-79-5-989.

Plotkin, S. (2014). History of vaccination. *Proc. Natl. Acad. Sci. U. S. A.* 111(34), 12283–7.

Plowright, W. (1984). The duration of immunity in cattle following inoculation of rinderpest cell culture vaccine. *J. Hyg. (Lond)* 92(3), 285–96.

Pollard, A. J. and Bijker, E. M. (2021). A guide to vaccinology: from basic principles to new developments. *Nat. Rev. Immunol.* 21(2), 83–100. doi: 10.1038/s41577-020-00479-7. Epub 2020 Dec 22. Erratum in: Nat Rev Immunol. 2021 Jan 5.

Polo, J., Campbell, J. M., Crenshaw, J., Rodríguez, C., Pujol, N., Navarro, N. and Pujols, J. (2012). Half-life of porcine antibodies absorbed from a colostrum supplement containing porcine immunoglobulins. *J. Anim. Sci., Suppl.* 90(Suppl 4), 308–10.

Poonsuk, K. and Zimmerman, J. (2018). Historical and contemporary aspects of maternal immunity in swine. *Anim. Health Res. Rev.* 19(1), 31–45.

Ratajczak, W., Niedźwiedzka-Rystwej, P., Tokarz-Deptuła, B. and Deptuła, W. (2018). Immunological memory cells. *Cent. Eur J. Immunol.* 43(2), 194–203.

Rautiainen, E., Tuovinen, V. and Levonen, K. (2000). Monitoring antibodies to *Mycoplasma hyopneumoniae* in sow colostrum–A tool to document freedom of infection. *Acta Vet. Scand.* 41(3), 213–25.

Robinson, H. L. and Pertmer, T. M. (2000). DNA vaccines for viral infections: Basic studies and applications. *Adv. Virus Res.* 55, 1–74.

Romagosa, A., Allerson, M., Gramer, M., Joo, H. S., Deen, J., Detmer, S. and Torremorell, M. (2011). Vaccination of influenza A virus decreases transmission rates in pigs. *Vet. Res.* 42(1), 120.

Romagosa, A., Gramer, M., Joo, H. S. and Torremorell, M. (2012). Sensitivity of oral fluids for detecting influenza A virus in populations of vaccinated and non-vaccinated pigs. *Influenza Other Respir. Viruses* 6(2):110–8. doi: 10.1111/j.1750-2659.2011.00276.x. Epub 2011 Jul 21.

Rose, N. and Andraud, M. (2017). The use of vaccines to control pathogen spread in pig populations. *Porcine Health Manag.* 3, 8.

Rose, N., Renson, P., Andraud, M., Paboeuf, F., Le Potier, M. F. and Bourry, O. (2015). Porcine reproductive and respiratory syndrome virus (PRRSv) modified-live vaccine reduces virus transmission in experimental conditions. *Vaccine* 33(21), 2493–9.

Rudolph, M. G., Stanfield, R. L. and Wilson, I. A. (2006). How TCRs bind MHCs, peptides, and coreceptors. *Annu. Rev. Immunol.* 24, 419–66.

Sedeik, M. E., El-Shall, N. A., Awad, A. M., Abd El-Hack, M. E., Alowaimer, A. N. and Swelum, A. A. (2019). Comparative evaluation of HVT-IBD vector, immune complex, and live IBD vaccines against vvIBDV in commercial broiler chickens with high maternally derived antibodies. *Animals (Basel)* 9(3), 72.

Sharma, A., Zeller, M. A., Li, G., Harmon, K. M., Zhang, J., Hoang, H., Anderson, T. K., Vincent, A. L. and Gauger, P. C. (2020). Detection of live attenuated influenza vaccine virus and evidence of reassortment in the U.S. swine population. *J. Vet. Diagn. Invest.* 32(2), 301–11.

Sibila, M., Pieters, M., Molitor, T., Maes, D., Haesebrouck, F. and Segalés, J. (2009). Current perspectives on the diagnosis and epidemiology of Mycoplasma hyopneumoniae infection. *Vet. J.* 181(3), 221–31.

Solano-Aguilar, G. I., Pijoan, C., Rapp-Gabrielson, V., Collins, J., Carvalho, L. F. and Winkelman, N. (1999). Protective role of maternal antibodies against *Haemophilus parasuis* infection. *Am. J. Vet. Res.* 60(1), 81–7.

Stadler, J., Naderer, L., Beffort, L., Ritzmann, M., Emrich, D., Hermanns, W., Fiebig, K., Saalmüller, A., Gerner, W., Glatthaar-Saalmüller, B. and Ladinig, A. (2018). Safety and immune responses after intradermal application of Porcilis PRRS in either the neck or the perianal region. *PLoS ONE* 13(9), e0203560.

Stegeman, A., Van Nes, A., de Jong, M. C. and Bolder, F. W. (1995). Assessment of the effectiveness of vaccination against pseudorabies in finishing pigs. *Am. J. Vet. Res.* 56(5):573–8.

Tassis, P. D., Papatsiros, V. G., Nell, T., Maes, D., Alexopoulos, C., Kyriakis, S. C. and Tzika, E. D. (2012). Clinical evaluation of intradermal vaccination against porcine enzootic pneumonia (*Mycoplasma hyopneumoniae*). *Vet. Rec.* 170(10), 261.

Ullrich, K. A., Schulze, L. L., Paap, E. M., Müller, T. M., Neurath, M. F. and Zundler, S. (2020). Immunology of IL-12: An update on functional activities and implications for disease. *EXCLI J.* 19, 1563–89. doi: 10.17179/excli2020-3104.

Vandeputte, J., Too, H. L., Ng, F. K., Chen, C., Chai, K. K. and Liao, G. A. (2001). Adsorption of colostral antibodies against classical swine fever, persistence of maternal antibodies, and effect on response to vaccination in baby pigs. *Am. J. Vet. Res.* 62(11), 1805–11.

Vannier, P. (1986). Vaccination against Aujeszky's disease by different routes using live attenuated and inactivated virus vaccines in pigs with or without passive immunity. *Zentralbl. Veterinarmed. B* 33(9), 704–12.

van Oirschot, J. T. (1987). Intranasal vaccination of pigs against Aujeszky's disease: Comparison with one or two doses of attenuated vaccines in pigs with high maternal antibody titres. *Res. Vet. Sci.* 42(1), 12–6.

van Oirschot, J. T. (1991). Intranasal vaccination of pigs against Aujeszky's disease: Protective immunity at 2 weeks, 2 months and 4 months after vaccination in pigs with maternal antibodies. *Vet. Microbiol.* 27(2), 103–13.

Vigre, H., Ersbøll, A. K. and Sørensen, V. (2003). Decay of acquired colostral antibodies to *Actinobacillus pleuropneumoniae* in pigs. *J. Vet. Med. B Infect. Dis. Vet. Public Health* 50(9), 430–5.

Villarreal, I., Meyns, T., Dewulf, J., Vranckx, K., Calus, D., Pasmans, F., Haesebrouck, F. and Maes, D. (2011). The effect of vaccination on the transmission of *Mycoplasma hyopneumoniae* in pigs under field conditions. *Vet. J.* 188(1), 48–52.

Weesendorp, E., Backer, J. and Loeffen, W. (2014). Quantification of different classical swine fever virus transmission routes within a single compartment. *Vet. Microbiol.* 174(3–4), 353–361. doi: 10.1016/j.vetmic.2014.10.022.

Weesendorp, E., Loeffen, W., Stegeman, A. and de Vos, C. (2011). Time-dependent infection probability of classical swine fever via excretions and secretions. *Prev. Vet. Med.* 98(2–3), 152–64. doi: 10.1016/j.prevetmed.2010.11.010.

Weidinger, G., Ohlmann, M., Schlereth, B., Sutter, G. and Niewiesk, S. (2002). Vaccination with recombinant modified vaccinia virus Ankara protects against measles virus infection in the mouse and cotton rat model. *Vaccine* 19(20–22), 2764–8.

Zinkernagel, R. M. (1993). Protection and damage by antiviral immunity. *Harvey Lect.* 89, 29–51.

Chapter 13

Advances in developing vaccines for emerging diseases in pigs

Francisco Javier Martinez-Lobo, University of Lleida, Spain

1 Introduction

One of the main threats facing the pig industry is the emergence of new pathogens that are challenging veterinary services around the world (Perfumo et al., 2020). Emerging infectious diseases (EIDs) are a constant threat to producers and consumers, leading to a significant loss of national and especially international trade opportunities. EIDs are also a threat to human health, as the collected data support the idea that the emergence of animal diseases is the main source of the re-emergence of zoonoses worldwide.

There are various complex factors involved in the rise of emerging animal diseases in recent years, which can be summarized by three main points: (1) natural processes of natural selection of pathogens, (2) changes in the environment and in human and animal populations and (3) an increased ability to detect and diagnose new pathogens (Table 1).

In the first case, the rapid evolution of certain pathogens in the face of immune pressure generated by vaccines or natural infection, along with the rapid evolution of bacteria with antimicrobial resistance in recent years, must be taken into account. Secondly, it is important to highlight the impact of human behavior on the environment. Global warming may lead to an increase in vector-borne diseases beyond their historical geographical limits. In addition, the loss of ecological biodiversity that alters the ecosystem makes it possible

http://dx.doi.org/10.19103/AS.2022.0103.14

Table 1 Factors involved in the emergence/re-emergence of swine diseases

Factors	Mechanisms	Example of agents
Pathogens	• Evolving pathogens by mutation • Immune pressure selection • Antimicrobial pressure selection	• MRSA[a]
Host	• Genetic selection of susceptible animals	• PRRSV
Human behavior	• Globalization • Deforestation and change of land use • Climate change	• Nipah virus
Animal production systems	• Wildlife-livestock interface on extensive free-ranging pig • Intensive systems where pathogens can infect and spread to a lot of susceptible animals • Use of antimicrobials	• *Mycobacterium bovis* • Influenza Viruses
Improved diagnosis	• Improvement of veterinary services • Surveillance programs • Development of new diagnostic tools	• New sapelovirus

[a] Methicillin-resistant *Staphylococcus aureus*.

for pathogens previously circumscribed to very restricted areas to emerge. In this sense, globalization and international trade play a key role in the spread of these diseases from their area of origin, since the risk of introducing new diseases is proportional to the trade of raw materials and fresh and processed foods. It is imperative to note that air traffic makes it possible for humans to reach any part of the world in less than a day, which is a period of time below the incubation time of most animal diseases (Malik et al., 2020). These diseases are more likely to be acknowledged in developed countries due to their early detection and surveillance systems; however, they might emerge in developing countries with vast tropical areas and varied usage of land (Allen et al., 2017). Land use and severe deforestation of the planet have triggered complex ecological mechanisms, leading to conditions for a spillover, such as the Nipah virus in 1999, which occurred following direct contact between fruit bats and pigs when the former lost a fundamental part of their habitat to the latter (McLean and Graham, 2019).

Finally, increased and better detection of new pathogens is fundamentally due to a heightened awareness of veterinary services to detect clinical signs compatible with EIDs due to recent events, including the emergence of African swine fever (ASF) in Europe and Asia, or Porcine epidemic diarrhea (PED) in the US, which have led to strict passive surveillance measures being established in many territories. Moreover, in recent times, concerns regarding EID have led to training veterinarians in the detection of unusual clinical signs, their reporting and even the creation of specific organizations, such as the Swine

Health Information Center (SHIC) in the US, which includes both Surveillance and Discovery of Emerging Disease and Responding to Emerging Disease in its work plan. Diagnostic capacity has also increased due to the development and generalization of new molecular diagnostic tools, which can be implemented in a short space of time and can detect small amounts of genetic material in clinical samples, leading to the identification of new pathogens that were previously harder to discover.

All these factors have led to several new pig diseases being identified in the last 30 years, together with the re-emergence of others believed to be under control. Global efforts to control emerging/re-emerging diseases have focused mainly on early detection by veterinary services and transparent notification of the diagnosis to national and international agencies, which can implement appropriate measures depending on the nature of the disease in question. Early detection is essential to prevent the spread of the disease through trade, along with the restriction of movement and the slaughter of infected animals, which can be more effective from an economic point of view (Carpenter et al., 2011).

However, it should be taken into consideration that these measures might prove impossible to implement due to a lack of economic resources for their development and compensation for losses to farmers. In this sense, effective treatments and vaccines are necessary to control emerging diseases, as the Coronavirus infectious disease 2019 (COVID-19) pandemic has shown. Therefore, the development of new vaccines against emerging/re-emerging pig diseases and the development of new platforms for rapid and effective vaccine design are crucial in the global context, as described earlier.

2 Emerging and re-emerging pig diseases

2.1 Definition

According to the Terrestrial Animal Health Code (2021) of the World Organization for Animal Health (OIE), an emerging disease is defined as 'a new occurrence in an animal of a disease, infection or infestation, causing a significant impact on animal or public health resulting from a change of a known pathogenic agent or its spread to a new geographic area or species; or a previously unrecognized pathogenic agent or disease diagnosed for the first time'. Re-emerging diseases are those that reappear after they have been in significant decline (controlled or eradicated). On the other hand, Transboundary Animal Diseases (TAD) are 'highly contagious epidemic diseases that can spread extremely rapidly, irrespective of national borders. They cause high rates of death and disease in animals, thereby having serious socio-economic and sometimes public health consequences while constituting a constant threat to the livelihoods of livestock farmers' (Food and Agricultural Organization of United Nations, Transboundary animal diseases, www.fao.org/emergencies/emergency-types/transboundary-animal-diseases/). In addition, some countries have other definitions, such as Foreign Animal Diseases. As per the earlier result, the inclusion of a pig disease in the previous categories is not easy, since the definitions can be interpreted in different ways and can also be contradictory. In this regard, a disease can be endemic in one region, not causing severe outbreaks, and be considered exotic in another, generating a health and commercial alarm. In addition, there may be highly geographically localized diseases that can cause substantial economic damage to public health. Therefore, a disease can be considered as emerging or re-emerging in pigs if it has an economic, sanitary or public health impact in a given moment, regardless of its origin and epidemiological situation.

2.2 Main emerging infectious diseases

Table 2 demonstrates the main emerging/re-emerging diseases that affect pigs, based on their main characteristics, including those diseases caused by agents that fundamentally affect production parameters, such as Porcine Reproductive and Respiratory Syndrome (PRRSV) and Porcine Circovirus 2 (PCV-2). Since their first appearance in the pig population at the beginning of the 1990s, these diseases have caused millions of losses to the industry, together with a decrease in animal welfare and an increase in the use of antimicrobials due to the increased incidence of secondary infections in infected animals with respect to their immunomodulatory role.

Another notable case of disease that affects production parameters is PED, whose etiological agent is PED virus (PEDV), belonging to the genus *Alphacoronavirus*. The disease was identified in Europe in the late 1970s, with a clinical sign of diarrhea in animals of all ages with high mortality in neonates, similar to that described for the Transmissible gastroenteritis virus (TGEV) infection. The disease spread rapidly in the 1980s in both Europe and Asia, although clinically it gradually disappeared in Europe in the 2000s, with the only significant case reported in Italy in 2005 (Pensaert and Martelli, 2016). The scenario changed drastically from winter 2010 onwards, when a series of severe outbreaks of diarrhea in piglets was reported in China, leading to the death of 1 000 000 piglets. These outbreaks were associated with the appearance of highly virulent strains that spread from China to other Asian countries and were introduced into the US in 2013, with economic loss attributed to PEDV infection accounting for over 10% of the total losses in the pig industry in a 1-year period (Jung et al., 2020). In 2014, PED outbreaks were reported in Europe, although the sequencing of these European strains identified them as INDEL strains (with insertions or deletions in the S protein) of Asian origin, whose virulence

Table 2 Main emerging/re-emerging diseases of swine

Diseases	Agent	Productive parameters	Commercial vaccine	Treatment	Zoonosis	Geographic distribution	Main clinical signs[a]	Hosts/Reservoirs
Porcine reproductive and respiratory syndrome	PRRSV	Yes	Yes	No	No	Worldwide	Repr and Resp	Swine
Porcine circovirus diseases	PCV-2	Yes	Yes	No	No	Worldwide	Repr and Syst	Swine
African swine fever	ASFV	Yes	No	No	No	Africa, Asia, Europe and Central America	Syst	Swine/Wild boars, bushpigs, …
Classical swine fever	CSFV	Yes	Yes	No	No	Africa, Asia, Europe and Central America	Repr and Syst	Swine
Porcine epidemic diarrhea	PEDV	Yes	Yes	No	No	Worldwide	Diarrhea	Swine
Swine acute diarrhea syndrome	SADS-CoV	Yes	No	No	No	Asia	Diarrhea	Swine
Vesicular disease	Senecavirus A	Yes	No	No	No	North America	Cut and Syst	Swine
Rotavirosis	Rotavirus	Yes	Yes	No		Worldwide	Diarrhea	Mammals
Swine influenza	SIV	Yes	Yes	No	Yes	Worldwide	Resp	Humans and birds
Nipah infection	NiV	?	No	No	Yes	Southeast Asia	Resp and Neur	Fruit bats
Japanese encephalitis	JEV	Yes	Yes	No	Yes	Southeast Asia and Western Pacific	Repr and Neur	Birds and human

(*Continued*)

Table 2 (*Continued*)

Diseases	Agent	Productive parameters	Commercial vaccine	Treatment	Zoonosis	Geographic distribution	Main clinical signs[a]	Hosts/Reservoirs
Hepatitis E	HeV							
Sapelovirus infection	Sapelovirus	Yes	No	No	No	North America	Cut and Neur	Swine
Bocavirus inection	Bocavirus	?	No	No	No	Worldwide	?	Swine
Torque teno sus virus infetions	TTV	?	No	No	No	Worldwide	?	Swine
Reston Ebola	Ebola reston virus	?	No	No	No?	Asia (Philippines and China)	?	Swine and nonhuman primates
PCV-3 infection	PCV-3	?	No	No	No	Worldwide	Repr and Syst	Swine
Atypical pestivirus infection	APTV	?	No	No	No	Worldwide	Repr	Swine
Cisticercosis	*T. solium*	No	No	Benzimidazol	Yes	Asia, Africa and Central and South America	Asymptomatic	Swine and humans
Streptococcial infection	*S. suis*	Yes	No	Amoxicilin	Yes	Worldwide	Syst and Neur	Swine
Clostridiosis	*C. difficile*	Yes	No	Not recommended	Yes	Worldwide	Diarrhea	Swine and humans

[a] Clinical signs. Repr: reproductive; Resp: Respiratory; Syst: systemic; Cut: cutaneous; Neur: neurological.

is lower than the highly virulent no-INDEL strains that caused the acute disease outbreaks in China and the US. Due to these enormous economic losses and the effects of PEDV on swine health, the number of scientific publications on the disease (VanderWaal and Deen, 2018) significantly increased, as well as the development of inactivated and replication-deficient vector (alphavirus) vaccines licensed in the US.

Another example of a major emerging disease is the spread of ASF virus (ASFV) from Africa to across the world. Following the eradication of the virus in Europe after its original entry into Spain and Portugal in 1957–1960, a strain of genotype II was detected in Georgia in 2007, which spread through the Caucasus to countries in Eastern Europe and reached the European Union in 2014 (affecting wild boars in most cases). Furthermore, in 2018, the virus reached China, spreading further to Asian countries and reaching Oceania in 2019 and America in 2021 when it was introduced on Hispaniola Island (first introduced in the Dominican Republic and later in Haiti).

Another large group of agents causing EIDs are those with zoonotic potential, including swine influenza viruses (SIV), Japanese encephalitis virus (JEV) and Nipah virus (NiV) infection. SIV are influenza A viruses, belonging to the family *Orthomyxoviridae*. Hemagglutinin (HA) and neuraminidase (NA) are located on the surface of the virus, which are responsible for the viral entry (HA) and the efficient viral release from infected cells (NA). In addition, these surface proteins are highly variable and are the major targets of the host humoral immune response (Ma and Richt, 2010). Swine influenza is a respiratory disease that produces signs that vary from moderate to severe in growing animals, depending on the endemic or epidemic presentation of the disease. In addition to affecting the health and welfare of animals, SIV continues to be considered one of the main threats to global human health as producer of a pandemic, until the emergence of the new betacoronavirus as the Middle East respiratory syndrome-related coronavirus (MERS-CoV), Severe acute respiratory syndrome coronavirus (SARS-CoV) and Severe acute respiratory syndrome coronavirus 2 (SARS-CoV-2). In the ecology of influenza viruses, pigs have traditionally been attributed the role of 'mixing vessel', where new pandemic viruses are generated by reassortment, as pigs can become infected with human and avian viruses. However, the current data do not allow this role of the pig as a mixing vessel to be simplified. The ecology of influenza A viruses is complex, and although pigs can maintain many lineages of viruses circulating in the population, their susceptibility to stable infections by avian viruses does not appear to be greater than that of humans (Nelson and Vincent, 2015). However, the role of the pig is relevant in the emergence of new viruses, such as the pandemic virus pdmH1N1 in the pig population, found in 2009. The arrival of pdm H1N1 marked a significant change in the European epidemiology of the virus. This is because, in Europe, all subtypes affecting pigs possessed internal genes from the 1979 avian virus and H1pdmN1pdm caused genomic rearrangement requiring genotyping of isolates for characterization (Sosa-Portugal et al., 2021). Moreover, in some cases, vaccines were updated with the hemagglutinin of this virus due to the appearance of emerging strains. In this regard, the majority of vaccines against SIV are inactivated, as their efficacy is dependent on the antigenic similarity of the strains that circulate with those included in the vaccine. On the other hand, the immune response elicited by live attenuated viruses is excellent; however, a risk of reversion to virulence and/or recombination still exists, and therefore research into effective and safe vaccines against SIV remains a priority.

JEV is an arthropod-borne virus that belongs to the genus *Flavivirus* in the family *Flaviviridae*. It is the cause of Japanese encephalitis, whose zoonotic nature is much more significant than its role in animal health. This is because approximately half of the world's population lives in endemic areas, and JEV is most frequently found in viral encephalitis in Asia. The pig is a natural vertebrate host that amplifies the infection and maintains the cycle of the virus together with birds and the invertebrate vector *Culex* spp. The clinical significance in pigs is not high, as it is fundamentally asymptomatic, and sometimes causes reproductive failure in pregnant sows with the presence of abortions and medium-sized mummified pigs, leading to infrequent vaccination. However, in the Japanese encephalitis endemic region, vaccination against JEV is the most effective measure to control JEV infection in humans and domestic pigs. Modified live vaccines are used, although the current circulating virus change in genotype from G1 to G3 may impair the efficacy of these vaccines (Nah et al., 2015). In humans, various vector vaccines have been developed using an attenuated strain of yellow fever virus; however, no recombinant vaccines have been developed or tested in animals, i.e. horses or pigs, to date.

Finally, NiV is considered a priority pathogen on the World Health Organization (WHO) R&D Plan list and is known to be the sixth notifiable disease for OIE. Although the natural reservoir of the virus is the fruit bats of the *Pteropus* genus, the pig plays an amplifying role in infection and is the source of transmission to humans, where the virus can produce a lethality rate of approximately 70%. In pigs, clinical signs vary according to age and consist of respiratory and neurological alterations, with reproductive affectation in some cases in the form of abortions. Sporadic human outbreaks have been observed from 1998 to 2021 in Malaysia, Bangladesh and India. The possibility of developing attenuated vaccines with such lethal viruses is remote, since reversion to virulence is extremely risky for hosts. Inactivated vaccines of the G protein of the Hendra virus, closely related phylogenetically to NiV and some recombinant and vector attenuated vaccines have been developed experimentally; however, no commercial vaccines have been developed to date (McLean and Graham, 2019).

Among the emerging bacterial diseases in pigs, other diseases can also be considered that pose a risk due to their high resistance to antimicrobials. Bacteria, such as Methicillin-resistant *Staphylococcus aureus* (MRSA), β-lactamase producing *Escherichia coli* and multiple resistant *Streptococcus suis*, have been described in pigs. The increase in *S. suis* infection is related to restrictions regarding the use of preventive antibiotics together with the appearance of several strains resistant to various antibiotics, including tetracyclines, macrolides, β-lactams, aminoglycosydes, trimethoprim-sulfamethoxazole, chloramphenicol and fluoroquinolones (Hernandez-Garcia et al, 2017). In this scenario, vaccination could be a solution; however, no commercial vaccines are to be found in most pork-producing countries. This fact leads to the use of autogenous bacterin vaccines, whose protective effectiveness of *S. suis* vaccines is mostly strain-specific and often unpredictable due to differences between the health status of herds (e.g. co-infection with viruses that induce immunomodulation, such as PRRSV or PCV-2) and an incorrect diagnosis of the *S. suis* serotypes implicated in the diseases (Rieckmann et al., 2020). Several experimental vaccines have been developed to control the condition. Live attenuated vaccine candidates are developed by using auxotrophic mutants for aromatic amino acids or mutants lacking virulence factors. By contrast, subunit vaccines based on proteins constitute a promising approach, since they are considered safer than live attenuated vaccines. However, despite cases where they are effective against homologous challenges, no protection is generally achieved against different pathogenic circulating serotypes/strains. Therefore, control of the streptococcal swine disease by vaccination is one of the biggest challenges facing swine researchers.

In summary, due to increasing viral emerging diseases and the lack of therapeutic options in several bacterial infections due to multi-resistant agents to antibiotics, the development of effective and safe vaccines via new scientific and technological approaches is essential for swine health in the future.

3 Vaccine platforms and new technologies to develop vaccines against emerging swine diseases

It has been known since ancient times that the best way to protect a population against a disease is by obtaining a good average immunity. Hence, since the eighteenth century, attempts have been made to stimulate a protective immune response in individuals without causing disease, a notion that was applied for the first time to the control of smallpox, leading to the concept of vaccinations. The first developed vaccines were based on relatively simple approaches, such as the inactivation of pathogens to prevent their pathogenic action, but with the aim of preserving their immunogenicity, or attenuation through serial passages in heterologous systems (*in vivo* or *in vitro*) with the same objective as in the case of inactivated vaccines. With advances in scientific development, new alternatives and platforms have been established for the effective immunization of a population. Consequently, there is currently a wide range of possibilities that allow the best option to be selected based on the characteristics of the pathogens to be controlled.

These possibilities can be classified according to different criteria, such as the chronological development and composition of the immunogen. Firstly, a chronological classification of the manufacture and design of vaccine products can be performed, considering their evolution over time has grown with an increased knowledge of microbiology, immunology and molecular biology (Aida et al., 2021). Thus, first-generation vaccines, which were first to appear historically, included inactivated vaccines and attenuated or modified live vaccines (MLV). Subsequently, second-generation vaccines were developed, consisting of subunits or synthetic proteins, where heterologous expression systems are used for the production of antigens of interest in vaccinology. With the rise of molecular biology, third-generation vaccines were developed, using modified genetic material to express antigens of interest, such as DNA vaccines that use plasmids. In addition, recombinant viral vectors were also generated in which coding regions of antigens of interest were inserted. Finally, the appearance of mRNA vaccines can be distinguished as a milestone in the development of vaccines against infectious diseases, with the appearance of vaccines against SARS-CoV-2.

Alternatively, vaccines can be classified according to their composition, i.e. if they have the complete pathogen, either in attenuated, inactivated or recombinant form; subunits of the pathogen produced in various heterologous, prokaryotic or eukaryotic overexpression systems, or produced by a recombinant viral vector; and finally those that contain nucleic acids, both in the form of DNA plasmids and mRNA (reviewed in Opriessnigg et al., 2021). Following this last classification, some of the main vaccine generation platforms developed or under development for emerging swine viral diseases are summarized in Tables 3 and 4.

The main platforms today are described in the following sections, along with their potential advantages and disadvantages.

3.1 New attenuation strategies for live attenuated swine viral vaccines

3.1.1 Classical live attenuated swine viral vaccines: pros and cons

The advantages and disadvantages of live attenuated vaccines (LAVs) have already been stated in numerous articles (Opriessnigg et al., 2021; Aida

Table 3 Main commercial vaccines against emerging/re-emerging swine diseases

Vaccine platform	Etiological agent	Methodology	Advantages	Disadvantages
Live attenuated	PRRSV	Classical attenuation (cell culture passages)	Induction of strong immune response in seronegative animals	Reversion to virulence/Recombination Lack of full cross-protection
	JEV	Classical attenuation (cell culture passages)	Efficacious in experimental and natural infection	Genotype shift phenomenon (G3 to G1) in Asia
	SIV	Reverse genetics	Nasal application: local immunity induction	Reassortment of the live attenuated vaccine virus with SIV field strains
	Rotavirus	Classical attenuation (cell culture passages)	High efficacy	Several serotypes
Inactivated	PCV-2	Chemical inactivation	High efficacy	No DIVA vaccine
	PRRSV	Chemical inactivation	High safety	Recommended only as booster in PRRSV positive animals
	SIV	Chemical inactivation	High safety	Frequent antigenic drift and/or shift of the HA
	JEV	Chemical inactivation (mouse brain-derived)	High safety	Genotype shift phenomenon (G3 to G1) in Asia
	PEDV	Commercial vaccine	High safety	Poor efficacy in negative animals
Chimeric	PCV-2	Inactivated Chimeric PCV1-2	High safety Inclusion of different genotypes	Possible interference of maternal-derived antibody
	PRRSV	Reverse genetics	Safety	Recombination Lack of full cross-protection
Subunit	PCV-2	Baculovirus expression vector	High safety and efficacy	Possible interference of maternal-derived antibody

(*Continued*)

Table 3 (*Continued*)

Vaccine platform	Etiological agent	Methodology	Advantages	Disadvantages
	CSFV	Baculovirus expression vector	High safety and efficacy DIVA vaccine	
Vectors	PCV-2	*M. hyopneumoniae* expressing the PCV2 capsid protein	High safety profile High levels of expression of target genes	Possible interference of maternal-derived antibody
		Replication-deficient alphavirus	High safety profile High levels of expression of target genes	Possible interference of maternal-derived antibody
	PEDV	Replication-deficient alphavirus	High safety profile High levels of expression of target genes	Poor efficacy in seronegative animals
	SIV	Replication-deficient alphavirus	High safety profile High levels of expression of target genes	Frequent antigenic drift and/or shift of the HA

Table 4 Experimental vaccines against emerging/re-emerging swine diseases

Vaccine platform	Agent	Methodology/ Development	Advantages	Disadvantages	References
Live attenuated	ASFV	Recombinant live attenuated viruses (LAVs)	High efficacy against challenge	LAVs induce slightly viremia and nasal shedding	Bosch-Camós et al. (2020), Borca et al. (2020)
Inactivated	JEV	Inactivation with BEA	Updated with new genotypes High safety	Possible genotype shifting	Nah et al. (2015)
Chimeric	PRRSV	Shuffling genome	Inclusion of different strains in the chimeric full-length PRRSV clone	No information about protection correlates	Tian et al. (2017a)
Subunit	ASFV	Synthetic peptides	Circumvent problems of live vaccines	Lack of efficacy	Ivanov et al. (2011)
	SIV	Baculovirus expression system	Safety Production without live viruses and embryonated eggs	Difficult purification of HA Frequent antigenic drift and/or shift of the HA	Ma and Richt (2010)
	PEDV	Expression of S1 protein in HEK-293 T cells	Safety Production without live viruses	Risk of endogenous viruses contamination Poor efficacy in negative animals	Makadiya et al. (2016), Legastelois et al. (2017)
	CSFV	Expression of E2 protein in HEK-293 T and CHO cells	Safety Production without live viruses	Risk of endogenous viruses contamination Low yield Time to develop protection immunity	Chen et al. (2021)
		Nicotiana benthamiana plant expression system		Difficulties in extracting and purifying vaccine quality proteins	Legastelois et al. (2017)
	PCV-2	Virus-like particles	High efficacy and safety	Possible interference of maternal-derived antibody	He et al. (2020)

(*Continued*)

Table 4 (*Continued*)

Vaccine platform	Agent	Methodology/ Development	Advantages	Disadvantages	References
	PRRSV	Virus-like particles	DIVA vaccine	Poor efficacy	Murthy et al. (2015)
	NiV	Hendra virus (HeV) soluble G protein (sG)	Vaccine already developed for horses	No protection	McLean and Graham (2019)
Vectors	ASFV	Recombinant adenovirus vector	Production without live viruses DIVA vaccine	Poor protection	Lokhandwala et al. (2019)
		Modified Vaccinia Ankara (MVA) vector		No protection	Netherton et al. (2019)
	PCV-2	PRRSV vector	High efficacy	Disadvantages of MLVs PRRS vaccines	Tian et al. (2017b)
	NiV	Rabies vector	High NA Immunization against both NiV and rabies virus	No challenge performed	Shuai et al. (2020)
		Canarypox-virus-based vectorr	Protection against virulent NiV challenge	Viral RNA detected in challenged pigs	Weingartl et al. (2006)
		Newcastle disease virus vector	Induction of long-lasting NiV NA	No challenge performed	Kong et al. (2012)
		ChAdOx1 (replication-deficient simian adenoviral)	Potential immunogenicity and thermostability	Possible immunity against vector	McLean and Graham (2019)
Nucleic acid	ASFV	DNA fusion with Ub and DNA expression library	Production without live viruses	Partial protection	Argilaguet et al. (2012)
	PRRSV	DNA ORF5 mosaic sequences	Broad protection against some heterologous isolates	ORF5 variability and no clear correlation with protection	Cui et al. (2019)
	SIV	DNA Conserved HA peptides	High safety High mucosal IgA response	Frequent antigenic drift and/or shift of the HA	Sisteré-Oró et al. (2019)

et al., 2021). The application of these vaccines induces long-term humoral and cellular immunity with a low number of doses. In addition, they mimic the natural infection of the pathogen; hence, they do not require adjuvants. However, their conservation is complex, usually requiring refrigerated storage conditions, which can complicate their application in and transport to remote areas, greatly increasing the cost of the doses.

The main utility of this type of vaccine is for viruses whose replication in the host is essential for the induction of a protective immune response or where the mimicry of viral replication could induce broader protection. Thus, in the case of SIV, the live swine influenza attenuated vaccine not only induced strong systemic and mucosal antibody responses against homologous strains but also induced cross-clade protective immune responses against heterologous strains (Jang and Seong, 2012; Abente et al., 2018). In addition, this type of vaccine is necessary for pathogens where the molecular bases of protection are unknown, and therefore, it is very difficult to select fragments that can be included in subunit vaccines or vectors. One of the clearest examples would be that of ASFV, where the best results to date are those achieved through the use of LAVs, whether natural or recombinant, as opposed to inactivated vaccines that, independently of the inactivation method and the adjuvant used, have proven to be ineffective. Another example of the need for live vaccines is PRRSV since it meets the two previous requirements. First, the replication of the virus in seronegative animals is necessary to induce an adequate response, since the application of inactivated vaccines in these animals is ineffective, and thus is only recommended as a booster in sows. Second, although neutralizing antibodies (NA) and IFN-gamma-secreting cells can play a significant role in protection against the virus (Loving et al., 2015), the components of the immune response relevant to protection have not been established.

In contrast, the safety of these vaccines is not always adequate due to their inherent ability to replicate in the host, as these vaccines can be transmitted to susceptible pigs, revert to virulence and recombine with other infective viruses.

Transmission to susceptible pigs and the possibility of them being carriers of the disease is a major drawback of various MLV vaccines, as they can keep the pathogen circulating in the population for prolonged periods of time. A paradigmatic example of this phenomenon is the use of live ASFV vaccines in the field, which have been shown to have detrimental effects on control programs as they can produce carrier pigs that act as reservoirs of the disease (Bosch-Camós et al., 2020). These chronically infected pigs can complicate the eradication of the disease in areas with control and eradication programs due to difficulties in identification, as seen in Spain (Arias and Sánchez-Vizcaíno, 2002). In addition, these LAVs can be excreted and transmitted among the susceptible population, increasing the risk of accumulating mutations that may eventually produce virulent virus strains. This is known as reversion to virulence, which implies that an attenuated virus can revert to its wild-type form or mutate to a new isolate, potentially resulting in the new virus being infectious or pathogenic. It is one of the most controversial aspects of this type of vaccine, both in animal health and in human medicine. In this sense, the sequencing of numerous PRRSV field isolates has shown the existence of viruses with nucleotide sequences to be very similar to those of the vaccines. In fact, the use of PRRSV-2 vaccines in Denmark in 1996 produced a series of reproductive disease outbreaks in sows from vaccinated and non-vaccinated farms, since the attenuated vaccine virus had spread to the non-vaccinated sows, changed genetically and reverted to virulence (Nielsen et al., 2002).

Recently, the use of MLV vaccines derived from highly pathogenicity PRRSV strains (HP-PRRSV) in China has led to the isolation of field strains from vaccines that have recovered their pathogenic capacity to cause severe disease in inoculated animals (Jiang et al., 2015).

Lastly, there is great concern surrounding recombination events due to the overuse of MLV vaccines. In the case of PRRSV, where the MLV vaccines do not induce sterilizing immunity and replicate in the host for prolonged periods of time (Martinez-Lobo et al., 2012), the possibility of generating recombination of live attenuated viruses when two or more strains co-infect swine cells is high. Thus, numerous recombinant viruses have been described between field isolates and MLV, both in the case of PRRSV-2 and PRRSV-1 (Bian et al., 2017; Eclercy et al., 2019). Moreover, recombination has been detected between two MLV vaccines, causing an outbreak in a boar stud in Denmark (Kvisgaard et al., 2017). Due to this fact, the European Medicines Agency (EMA) had to intervene temporarily and form the conclusion that 'the benefits of modified live PRRS virus vaccines continue to outweigh the risk'. The Committee for Veterinary Medicinal Products (CVMP) finally recommended that 'certain warnings should be included in the product information in order to limit the opportunity for modified live PRRSVs to circulate and to reduce the risk and the frequency of recombination between PRRS viruses including PRRSV vaccine strains'. It must be noted that this problem is also observed in other RNA viruses, such as SIV, in which vaccination with live attenuated vaccines can lead to the appearance of new recombinants, since there is evidence that reassortment occurred between the vaccine virus and endemic field strains circulating in US swine (Sharma et al., 2020).

Finally, the development of these vaccines is generally slow since it is necessary to carry out several serial passages in systems heterologous to the host, such as other animal species or, more frequently, stable cell cultures. The number of passages required for attenuation is unpredictable and dependent on the virus species. In addition, virus strains in the process of attenuation must be tested periodically until the exact passage is observed where no clinical signs are produced in the animals following inoculation, yet they continue to maintain their immunological properties. Due to this long duration, LAVs are not the best system against a newly emerging disease, since the period of time for their development exceeds the reasonable time for control of the epidemic (although as will be discussed later, it will depend very much on the nature of the particular pathogen and the basic knowledge of its immunobiology).

Due to the issues regarding classically attenuated vaccines mentioned earlier, various approaches have been carried out to develop LAVs more quickly and safely against different diseases.

3.1.2 Reverse genetics

Reverse genetics has been fundamental to elucidating genes involved in attenuation by examining changes to phenotypes (of cells or organisms) caused by genetically engineering specific nucleic acid sequences. In this sense, reverse genetics, together with the possibility of deleting specific genes using recombination technologies by genome editing using the CRISPR-Cas9 system or modifying these genes by site direct mutagenesis, various strategies to produce attenuated strains more precisely and effectively can be developed.

First, modified live chimeric vaccines can be built, based on two or more different viruses or viral strains. One of the viruses, used as a genetic backbone, is an attenuated strain (obtained by classical attenuation or genetic engineering), whereas the other is a virulent strain, a potent cross immunity inducer or an updated virus circulating in the population. Usually, some structural proteins of the attenuated virus are replaced with counterparts of virulent viruses. This reverse-genetics approach is used to generate live attenuated influenza vaccines (LAIVs) using plasmids, encoding six internal genes of an attenuated donor strain and two HA and NA genes of circulating isolates that are co-transfected into cells to obtain the desired vaccine. Using this technology, LAIVs can be rapidly generated by co-transfection instead of the classical co-infection method, which is both laborious and time-consuming (Jang and Seong, 2012). Recently, a TGEV-PEDV chimeric virus has been generated by replacing the ectodomain of the S protein of an attenuated TGEV with that of a G2 PEDV strain (Pascual-Iglesias et al., 2019). In addition, the commercialization of a PRRSV vaccine can be highlighted, in which a backbone of a strain attenuated by passage in cell culture and the structural proteins of a highly pathogenic strain known as MN-1-8-4 have been used. Finally, the high genetic diversity of some viruses, such as PRRSV, presents a special challenge in vaccine development. Reverse genetics through the mosaic vaccine approach offers one potential solution (Corey and McElrath, 2010). Several field sequences are analyzed to effectively address the genetic diversity of a viral population. The selected sequences can be randomly recombined by DNA shuffling technology to expand the antigenic coverage of a viral vaccine (Locher et al., 2005). On the other hand, computational algorithms can be used to determine an artificial sequence that is located at the center of the phylogentic tree, usually known as centralized immunogen, which can be used to generate a candidate vaccine against high-variable viruses (Vu et al., 2017).

With information on the function of the modified genes in relation to the virulence or pathogenesis of the virus, recombinant viruses can be designed and recovered. These recombinant strains include the desired mutations, which, in many cases, affect the replication capacity of the virus in the host. This approach is more rational and usually safer than the classical methodology or the use of naturally attenuated isolates. One of the most characteristic examples would be to obtain LAVs by deleting ASFV genes. Numerous studies have been carried out to determine the genes that are involved in the virulence of the virus and the safety and efficacy of the generated deletion mutants as vaccine candidates. In recent years, important advances have been made and interesting candidates have been obtained from various research groups. An example of this would be the generation of LAVs by deleting the CD2v gene, known as BA71ΔCD2 (belonging to Genotype I) that conferred protection against homologous and heterologous viruses (Genotype I and II) (Monteagudo et al., 2017). This LAV lacks the non-haemadsorbaton capacity that confers an attenuation phenotype, a fact that also occurs in naturally attenuated isolates, such as ASFV Lv/17/WB/Rie1 strain, a non-HAD attenuated genotype II ASFV isolated in Europe, in which a point mutation generates a truncated form of the CD2 protein (Gallardo et al., 2019). Finally, another interesting LAV is a recombinant virus developed in Plum Island Animal Disease Center (USA), known as ASFV-G-ΔI177L, which lacks the I177L gene and confers sterilizing protection against the 2007 Georgia isolate (Borca et al., 2020).

Gene deletion or modification has also been used in other porcine viruses, such as PEDV, where this strategy can be used to generate LAVs that can replicate efficiently in enterocytes. This is a fundamental fact to achieve good lactogenic immunity in sows and protect neonatal piglets through colostrum, without causing tissue damage in the intestine. Thus, certain PEDV recombinants have been constructed by modifying the non-structural proteins nsp16, nsp15 and ORF3, or by inactivating the endocytosis signal YxxF motif in the cytoplasmic tail of the S protein (Hou and Wang, 2019). In order for these mutants to remain as stable as possible after replicating in the host, it has been suggested that they should contain small deletions/mutations for each targeted gene instead of deleting entire genes to increase genetic stability and introduce at least two distinct mutations into a separate gene (Hou and Wang, 2019). As with PEDV recombinants, the genes in which modifications are made to generate LAVs are usually non-essential for virus replication and do not play a role in virus immunogenicity as, on many occasions, these genes are encoding modulators of innate immune responses and virus replication. A typical example of this type of mutation is the non-structural protein 1 (NS1) truncated mutant influenza LAVs. This protein is not incorporated into the virion, nor does it interfere with the virus replication cycle; however, it plays an important role as an antagonist of interferon (IFN)-mediated antiviral responses, which gives it an attenuated phenotype in the host (Jang and Seong, 2012). However, in type I IFN-free production systems, such as embryonated eggs, it continues to replicate at high viral titers and is, therefore, interesting from an industrial point of view. In the case of pigs, there is a commercial vaccine that includes two subtypes of LAVS NS1 truncated, H1N1 and H3N2, whose intranasal application induces to generate strong, lasting immunity. However, reassortment of the LAV virus with SIV field strains could restore wild-type NS1 donated from these wild-type SIV strains (Sharma et al., 2020).

3.1.3 Codon pair deoptimization

Another promising strategy is the recoding of viral genomes, which has made it possible to generate new, safer and more stable attenuated vaccines through codon pair deoptimization (CPD). CPD uses suboptimal codon pairs to achieve attenuation of recoded viruses. These new viruses replicate at the host less efficiently than the wild-type strain due to increased mRNA decay and a reduced translation efficiency and protein production of recoded genes. However, the total amino acid identity of the virulent parent strain is completely preserved, inducing an immune response similar to those achieved against wild-type infection. This technique is applicable to a diverse variety of viruses (Coleman et al., 2008) and is also known to be quite rapid, leading to vaccine development in a short amount of time with the help of only sequences, circumventing the issue of slow development of attenuated vaccines in epidemic events. Finally, CPD can help to obtain more stable LAVs that do not revert to virulence as easily as traditional vaccines, since CPD rearranges several positions of existing synonymous codons, making it difficult for these mutations to change in the future. Although no commercial vaccines for pigs have been developed with this technology, experimentally attenuated isolates of Foot-and-mouth disease virus (FMDV) have been achieved (Diaz-San Segundo et al., 2015). Furthermore, attenuation of PRRSV by CPD has also been achieved in several non-structural proteins such as NSP1, NSP2, NSP9 or in envelope GP5 (Ni et al., 2014; Gao et al., 2015; Evenson et al., 2016), being highly stable after several passages *in vivo* and *in vitro*.

3.1.4 Ribavirin-screened viruses

The stability of the viral genome during several passages may be due to a higher fidelity in its RNA-dependent RNA polymerase (RdRp). Therefore, ribavirin, a mutagenic nucleoside analogue, has been used to select viruses with high-fidelity RdRp, improving their stability and acquiring an attenuated phenotype, possibly due to the decrease in quasi-species during the replication period. As in the previous case, these mutants have been obtained for FMDV and PRRSV, demonstrating their increased genetic and phenotypic stability even after three sequential passages in pigs (Khatun et al., 2016).

3.2 Strategies for improving inactivated and subunit vaccines

In contrast to live attenuated vaccines, the safety of inactivated vaccines is very high. This is because reversion to virulence or transmission to susceptible pigs is not possible as they are dead microorganisms (Burrel et al., 2016). In addition, storage is easier than in the case of live attenuated vaccines, being stable for very long periods of time in refrigeration. Obtaining these vaccines is rapid, since the attenuation of the etiological agent is not required; however, the cost of production is higher as a large amount of antigen is required, along with other components in the preparation. Another drawback of inactivated vaccines is that their induced immunity is usually only humoral, especially IgG, and of shorter duration than that obtained by MLV vaccines, which makes it necessary to administer several booster doses (Pollard and Bijker, 2021). This type of vaccine is usually used in agents where attenuation is technically difficult to be conducted, as in the case of autogenous vaccines against bacteria, e.g. *S. suis* and *E. coli*, or the risk of reversion or recombination is considered high, as in the case of SIV, where most of the vaccines applied are inactivated strains of the subtypes that affect pigs, such as H1N1, H3N2, H2N1 and pdmH1N1.

Subunit vaccines, where one or several antigens of a pathogen expressed in a heterologous system are used, are just as safe as inactivated vaccines; however, their manufacturing process is cheaper and production control much simpler. In addition, due to the use of a purified antigen, the possibility of side effects is less than in traditional vaccines. Numerous subunit vaccine production systems have been developed, including the use of prokaryotic systems, such as *Bacillus subtilis*, or eukaryotic yeast systems, such as *Sacharomyces cerevisiae* or *Pichia pastoris*, where several antigens have been expressed, especially protein S of PEDV. Systems have also been developed that use mammalian cells, plants and transgenic animals to express different antigens of PEDV, PRRSV and Classical swine fever virus (CSFV) (reviewed in Opriessnigg et al., 2021), although none have been used commercially. Conversely, the baculovirus expression vector system (BEVS) platform has become an established manufacturing technology for the production of swine viral vaccines as it offers several advantages, including high levels of protein expression with simple glycosylation and post-translational modifications, manufacturing speed, flexible product design, inherent safety and scalability (Felberbaum, 2015). As can be observed in Table 3, most PCV-2 commercial vaccines are expressed using this platform. Interestingly, these commercial vaccines contain a certain percentage of virus-like particles (VLPs), a self-assembled macromolecule composed of viral structural proteins that lack genetic materials. These VLPs hold several advantages over other particulate subunit vaccines. VLPs include a morphological resemblance to their parent virus, a highly repetitive immunogenic surface structure and the retention of cell uptake and immune processing pathways associated with their parent virus. Furthermore, VLPs are completely safe due to their incapability of infection or replication (Donaldson et al., 2018). For the aforementioned reasons, VLPs are a promising platform for the development of new vaccines against some emerging diseases, such as PRRS or PCV-3 infection (Murthy et al., 2015; Wang et al., 2020).

3.2.1 Adjuvants

The lack of replication in the host of both inactivated vaccines and subunit vaccines implies that, in most cases, they are not very immunogenic and, therefore, the use of adjuvants is necessary to enhance the immunogenicity of vaccines. These adjuvants can have two fundamental mechanisms of action - acting as a vehicle or stimulating the immune system. They can involve the recruitment and activation of antigen-presenting cells (APC) and T cells, an enhancement of APC functions, isotype switching of immunoglobulin (Ig) and adifferentiation of T cells (Cox and Coulter, 1997; Gerdts et al., 2006). In this regard, the design and use of new adjuvants may be essential for the development of non-replicative vaccines against emerging pig diseases, thereby increasing their efficacy. Several examples of adjuvants include oil-in-water emulsion, water-in-oil emulsion, several cytokins, pathogen-associated molecular patterns (PAMPs), saponins, etc. These adjuvants are used in both commercial and experimental vaccines that have been extensively reviewed in the scientific literature (Charerntantanakul, 2020). Focusing on the emerging diseases described in Section 2, some examples can be cited in which adjuvants, in addition to an antigen, have experimentally shown improvement over conventional vaccines. Thus, the use of ISA201 VG has induced a greater neutralizing antibody response against PEDV (Li et al., 2017), while the use of various cytokines has improved the poor immune response obtained on many occasions in experimental PRRSV vaccines (Charerntantanakul, 2020).

3.2.2 Subunit vaccine carriers

Not only can adjuvants help in the improvement of the immunogenicity of synthetic subunit antigens, but the use of vaccine carriers could improve the antigen presentation acting as depot effect, delivering antigens to the draining lymph nodes and targeting dendritic cells. Several carriers have been developed to perform this role, including liposomes, archeosomes, virosomes, virus-like particles (VLPs), polymeric and inorganic microparticles, and nanoparticles (Vartak and Sucheck, 2016). Additionally, some carriers act as adjuvants, increasing the efficacy of the vaccine.

These platforms appear to be highly successful in disease control as they induce strong and long-lasting immunity in the host. Thus, a platform consisting of the receptor binding domain (RBD) fragment of the SARS-CoV-2 RBD sortase-A-conjugated nanoparticle of ferritin induced higher neutralizing antibody titers than the SARS-CoV-2 mRNA vaccine and achieved sterilizing immunity in macaques (Saunders et al., 2021). This type of approach may have a place for coronaviruses that affect pigs, especially PEDV.

3.3 Viral vectors

Although viral vectors can be included within subunit vaccines in some classifications, we will treat them in a separate section due to their special characteristics. Replication-competent viral vectors exist; however, most platforms using viral vectors use replication-defective mutant viral strains. The use of these vectors, as in the case of subunit vaccines, is entirely safe as they only allow a single round of protein expression. This implies that the transmission to naïve pigs and reversion to virulence and recombination is not possible. In the same way, the use of this platform allows a rapid response to EIDs, since once the sequence is obtained, the vaccine can be developed in just a few weeks (Dudek and Knipe, 2006). Lastly, viral vectors are often compatible with differentiate infected from vaccinated animals (DIVA) vaccines, as the serological response to a protein is detectable against viral proteins absent in the vector vaccine (Baron et al., 2018).

The great advantage of replication-deficient vectors over subunit vaccines is that they have the ability to mimic infection and express gene products within cells to ensure that antigens can be efficiently presented by MHC-class pathways I and class II. Furthermore, they can stimulate PAMPs, leading to the innate immune response. Unlike inactivated or subunit vaccines, adjuvants are not required in the formulations of viral vector vaccines, therefore adverse reactions to the vaccines due to these adjuvants (that is, granulomas and local inflammation) are rare.

3.3.1 Poxvirus

Poxviruses are strong candidates for vaccine vectors because they grow to high titers, are highly stable when lyophilized and are capable of accepting large transgene sequences (Dudek and Knipe, 2006). The Modified Vaccinia Ankara (MVA), a strain of vaccine virus obtained by passages in chicken cells, has been used as an experimental vaccine vector for some swine EIDs. Thus, MVA-vectored pools of antigens use as a booster conferred protection against fatal AFSV infection (Goatley et al., 2020).

3.3.2 Adenoviruses

Adenoviruses are perhaps the most widely used replication-deficient vectors. Their use is safe, they are very stable and are capable of reaching high titers *in vitro*. In addition, adenoviruses can induce expression of transgenes at high levels in the host and can be administered by various routes. Thus, animals vaccinated orally with a recombinant adenovirus vaccine encoding the heat-labile enterotoxin B (LTB) and the core neutralizing epitope (COE) of PEDV (rAd-LTB-COE) induced humoral as well as cell-mediated immune responses in piglets (Do et al., 2020).

3.3.3 Alphaviruses

Alphaviral replicons are built by replacing structural genes with heterologous genes, abolishing their replication ability due to the loss of structural genes (Dudek and Knipe, 2006). Furthermore, these non-replicating viruses have dendritic cell tropism and induce high expression levels of target genes (10–20% of cell protein). A commercial Venezuelan equine encephalitis virus platform has been developed that is used to produce customized non-replicant vaccines against PEDV, SIV, PCV-2 (Opriesnigg et al., 2014; Sandbulte et al., 2015)

4 Nucleic acid vaccines

DNA vaccines offer several advantages over traditional vaccines. DNA vaccines are extremely safe, cost-effective and very versatile. Although many experimental DNA vaccines have been developed, no commercial vaccines have been licensed for swine veterinary practice. However, the use of plasmids as viral protein vectors is efficient and commercially affordable, as in the case of viruses affecting Atlantic salmon (Aida et al., 2021). In the case of swine EIDs, some experimental vaccines have been developed to improve their efficacy against PRRSV. Thus, plasmids could improve the immune response, including interferon proteins or other cytokines. In this sense, a DNA vaccine co-expressing GP3 and GP5 of PRRSV with interferon α/γ induced enhanced immune responses provides both immediate and long-lasting protection against HP-PRRSV challenge in pigs (Du et al., 2012). Moreover, these plasmids could be easily modified to design Mosaic Vaccines that are usually developed for high-variable viruses such as HIV (Cui et al., 2019).

Finally, mRNA-based vaccines are promising due to the success of SARS-CoV-2 vaccines. The ability to mimic many aspects of natural RNA viral infections can potentially enhance the immune host response (Pardi et al., 2018). The use of mRNA vaccines using lipid nanoparticles to deliver spike-PEDV encoding mRNA should be explored.

5 Conclusion and future trends

EIDs are a significant issue in animal health and, based on the evidence, climate change, antimicrobial resistance and deforestation will increase in the coming decades. For this reason, the use of vaccines is a fundamental tool to control and reduce health, social and economic consequences.

Technological advances have made it possible to generate new vaccines against these diseases, as well as the possibility that the development of a vaccine against an emerging disease is an engineering exercise. The availability of the sequence of the pathogen or of the gene of interest would be sufficient to efficiently develop and produce safe, effective and economically viable vaccines.

However, the main advantage of these systems compared to the classical development of vaccines is the rapid response to EIDs or new strains causing endemic diseases. From the sequencing of the pathogen to the development of the vaccine candidate, it takes only a few weeks to be included in pre-clinical trials. In addition, the methodology for the production of third- or fourth-generation vaccines is transferable between different pathogens, which previously did not occur in the classic attenuation or inactivation vaccines specific to each pathogen. Moreover, the use of high biosafety laboratories is not necessary as it works with sequences and genes of interest of the etiological agents and not with these living agents, which facilitates their development, production and testing (Excler et al., 2021). In addition, these platforms can be modeled with prototype pathogens, including prominent members of various virus families. Although the number of viral species may increase progressively, the families and groups seem to have reached a plateau; therefore, with sufficient knowledge about families, we can have a head start in vaccine development ahead of the emergence of the next crisis (Woolhouse, 2008).

Although these third-generation platforms may be valid for the development of safe and effective vaccines against EIDs, a few important factors should be taken into account. First, the biological and immunological properties of the pathogen in particular must be considered, since they are fundamental to the design of safe and effective vaccine candidates. In this sense, the development of a highly effective vaccine against PCV-2 has been rapid and carried out using various platforms and approaches, such as from inactivated vaccines, subunits and vectors. However, although the genetic diversity of the virus has increased and is greater than previously thought, vaccines are still quite effective against the disease (Franzo and Segales, 2020). By contrast, many aspects of the immunopathology of another DNA virus, ASFV, are unknown, due to the size of its genome and its complexity, especially the protective immune responses and the antigens against which these responses are induced. This implies that, to date, a safe and effective vaccine has not been obtained, although there are promising results using LAVs (Bosch-Camós et al., 2020). Finally, as already mentioned, PRRSV can practically locate all the previous difficulties to develop a universal vaccine against the pathogen, from its enormous genomic diversity to its ability to evade the host's immune response. Therefore, in addition to technological developments in vaccination platforms, new adjuvants and application devices, prior knowledge is required for the application of these technologies, as shown during the COVID-19 pandemic, in which previous studies by Karikó and Weissman were essential for the development of vaccines against SARS-Cov-2 (Nair, 2021). In this sense, veterinary science should be sufficiently funded to ensure the achievement of the following milestones:

1 Implementation of intensive epidemiological surveillance programs for the appearance of new pathogens in domestic animals and wildlife, including antimicrobial resistance.
2 Basic studies in the pathogenesis and immunology of animal diseases, including new technologies that are already applied in human medicine, such as the isolation of monoclonal antibodies of single B cells, definition of antibody lineages, structure-based vaccine (determination of structure of fusion proteins), structural analysis of antigenic sites and so on.
3 Using prototype pathogens in the previous development of vaccines and obtaining efficacy and safety data in bioassays in the target species.

Considering the aforementioned issues, the design of vaccines against emerging pathogens will require different efforts and timeframes depending on their nature and the epidemiological, social, economic and political conditions.

6 Where to look for further information

6.1 Further reading

- An excellent source of information regarding viral emerging diseases, including a One Health approach, is *Emerging and Transboundary Animal Viruses* by Y. S. Malik, R. K. Singh and M. P. Yadav (2020). 1st Edition. Springer.
- *Transboundary and Emerging Diseases* journal provides up-to-date information regarding EID prevention and management, vaccine development and rapid communication of new outbreaks. This international journal will be of vital interest to scientists and practitioners working in the field of infectious diseases, including veterinarians, animal scientists, agricultural scientists, policymakers, wildlife workers and conservationists. It focuses on novel science-based approaches to controlling transboundary and emerging diseases.
- A good introduction to swine EIDs is *Emerging Swine Diseases* by M. Martín, J. Segalés, L. Darwich, E. Mateu and J. Casal (2019). 1st Edition. SERVET. This book describes in a practical and concise manner the emerging and reemerging diseases of swine of economic or zoonotic importance.
- Very interesting reviews of state-of-the-art vaccines against swine diseases are included in 'Future perspectives on swine viral vaccines: where are we headed?', *Porcine Health Manag*, 7(1):1 by Opriessnig et al. (2021); and 'Novel vaccine technologies in veterinary medicine: a herald to human medicine vaccines', *Front Vet Sci*, 8:654289 by Aida et al. (2021).
- Good reviews of novel technologies in human medicine for developing emerging diseases vaccines are to be found in 'Vaccine development for emerging infectious diseases', *Nat Med.* 27(4):591-600 by Excler et al. (2021) and 'Emerging viral diseases from a vaccinology perspective: preparing for the next pandemic,' *Nat Immunol.* 19(1):20–28 by Graham and Sullivan (2018).

6.2 Organizations

- The World Organization for Animal Health (OIE) portal gives easy access to resources and information (outbreaks) on swine EIDs.
- The Swine Health Information Center communicates outbreaks of swine diseases with new or unexpected etiologies in EE.UU.
- The FAO Emergency Centre for Transboundary Animal Diseases (ECTAD) facilitates the Global Health Security Agenda (GHSA) and Emerging Pandemic Threats (EPT) programs, funded by the United States Agency for International Development.

7 References

Abente, E. J., Rajao, D. S., Santos, J., Kaplan, B. S., Nicholson, T. L., Brockmeier, S. L., Gauger, P. C., Perez, D. R. and Vincent, A. L. (2018). Comparison of adjuvanted-whole inactivated virus and live-attenuated virus vaccines against challenge with contemporary, antigenically distinct H3N2 influenza A viruses, *J. Virol.* 92(22):e01323-18. doi: 10.1128/JVI.01323-18.

Aida, V., Pliasas, V. C., Neasham, P. J., North, J. F., McWhorter, K. L., Glover, S. R. and Kyriakis, C. S. (2021). Novel vaccine technologies in veterinary medicine: a herald to human medicine vaccines, *Front. Vet. Sci.* 8:654289. doi: 10.3389/fvets.2021.654289.

Allen, T., Murray, K. A., Zambrana-Torrelio, C., Morse, S. S., Rondinini, C., Di Marco, M., Breit, N., Olival, K. J. and Daszak, P. (2017). Global hotspots and correlates of emerging zoonotic diseases, *Nat. Commun.* 8(1):1124. doi: 10.1038/s41467-017-00923-8.

Argilaguet, J. M., Pérez-Martín, E., Nofrarías, M., Gallardo, C., Accensi, F., Lacasta, A., Mora, M., Ballester, M., Galindo-Cardiel, I., López-Soria, S., Escribano, J. M., Reche, P. A. and Rodríguez, F. (2012). DNA vaccination partially protects against African swine fever virus lethal challenge in the absence of antibodies, *PLoS ONE* 7(9):e40942. doi: 10.1371/journal.pone.0040942.

Arias, M. and Sánchez-Vizcaíno, J. M. (2002). African swine fever eradication: the Spanish Model. In: *Trends in Emerging Viral Infections of Swine* (1st edn.). Ames: IA State Press (vol. 2002). pp. 133–139.

Baron, M. D., Iqbal, M. and Nair, V. (2018). Recent advances in viral vectors in veterinary vaccinology, *Curr. Opin. Virol.* 29:1–7. doi: 10.1016/j.coviro.2018.02.002.

Bian, T., Sun, Y., Hao, M., Zhou, L., Ge, X., Guo, X., Han, J. and Yang, H. (2017). A recombinant type 2 porcine reproductive and respiratory syndrome virus between NADC30-like and a MLV-like: genetic characterization and pathogenicity for piglets, *Infect. Genet. Evol.* 54:279–286. doi: 10.1016/j.meegid.2017.07.016.

Borca, M. V., Ramirez-Medina, E., Silva, E., Vuono, E., Rai, A., Pruitt, S., Holinka, L. G., Velazquez-Salinas, L., Zhu, J. and Gladue, D. P. (2020). Development of a highly effective African swine fever virus vaccine by deletion of the I177L gene results in sterile immunity against the current epidemic Eurasia strain, *J. Virol.* 94(7):e02017-19. doi: 10.1128/JVI.02017-19.

Bosch-Camós, L., López, E. and Rodriguez, F. (2020). African swine fever vaccines: a promising work still in progress, *Porcine Health Manag.* 6:17. doi: 10.1186/s40813-020-00154-2.

Burrel, C. J., Howard, C. L. and Murphy, F. A. (2016). *Fenner and White's Medical Virology* 5th edn. The Netherlands: Elsevier.

Carpenter, T. E., O'Brien, J. M., Hagerman, A. D. and McCarl, B. A. (2011). Epidemic and economic impacts of delayed detection of foot-and-mouth disease: a case study of a simulated outbreak in California, *J. Vet. Diagn. Invest.* 23(1):26-33. doi: 10.1177/104063871102300104.

Charerntantanakul, W. (2020). Adjuvants for swine vaccines: mechanisms of actions and adjuvant effects, *Vaccine* 38(43):6659–6681. doi: 10.1016/j.vaccine.2020.08.054.

Chen, J. Y., Wu, C. M., Chen, Z. W., Liao, C. M., Deng, M. C., Chia, M. Y., Huang, C. and Chien, M. S. (2021). Evaluation of classical swine fever E2 (CSF-E2) subunit vaccine efficacy in the prevention of virus transmission and impact of maternal derived antibody interference in field farm applications, *Porcine Health Manag.* 7(1):9. doi: 10.1186/s40813-020-00188-6.

Coleman, J. R., Papamichail, D., Skiena, S., Futcher, B., Wimmer, E. and Mueller, S. (2008). Virus attenuation by genome-scale changes in codon pair bias, *Science* 320(5884):1784–1787.

Corey, L. and McElrath, M. J. (2010). HIV vaccines: mosaic approach to virus diversity, *Nat. Med.* 16(3):268–270. doi: 10.1038/nm0310-268.

Cox, J. C. and Coulter, A. R. (1997). Adjuvants–a classification and review of their modes of action, *Vaccine* 15(3):248–256. doi: 10.1016/s0264-410x(96)00183-1.

Cui, J., O'Connell, C. M., Costa, A., Pan, Y., Smyth, J. A., Verardi, P. H., Burgess, D. J., Van Kruiningen, H. J. and Garmendia, A. E. (2019). A PRRSV GP5-Mosaic vaccine: protection of pigs from challenge and ex vivo detection of IFNgamma responses against several genotype 2 strains, *PLoS ONE* 14(1):e0208801. doi: 10.1371/journal.pone.0208801.

Diaz-San Segundo, F., Medina, G. N., Ramirez-Medina, E., Velazquez-Salinas, L., Koster, M., Grubman, M. J. and de los Santos, T. (2015). Synonymous deoptimization of foot-and-mouth disease virus causes attenuation in vivo while inducing a strong neutralizing antibody response, *J. Virol.* 90(3):1298–1310. doi: 10.1128/JVI.02167-15.

Do, V. T., Jang, J., Park, J., Dao, H. T., Kim, K. and Hahn, T. W. (2020). Recombinant adenovirus carrying a core neutralizing epitope of porcine epidemic diarrhea virus and heat-labile enterotoxin B of Escherichia coli as a mucosal vaccine, *Arch. Virol.* 165(3):609–618. doi: 10.1007/s00705-019-04492-7.

Donaldson, B., Lateef, Z., Walker, G. F., Young, S. L. and Ward, V. K. (2018). Virus-like particle vaccines: immunology and formulation for clinical translation, *Expert Rev. Vaccines* 17(9):833–849. doi: 10.1080/14760584.2018.1516552.

Du, Y., Qi, J., Lu, Y., Wu, J., Yoo, D., Liu, X., Zhang, X., Li, J., Sun, W., Cong, X., Shi, J. and Wang, J. (2012). Evaluation of a DNA vaccine candidate co-expressing GP3 and GP5 of porcine reproductive and respiratory syndrome virus (PRRSV) with interferon alpha/gamma in immediate and long-lasting protection against HP-PRRSV challenge, *Virus Genes* 45(3):474–487. doi: 10.1007/s11262-012-0790-1.

Dudek, T. and Knipe, D. M. (2006). Replication-defective viruses as vaccines and vaccine vectors, *Virology* 344(1):230–239. doi: 10.1016/j.virol.2005.09.020.

Eclercy, J., Renson, P., Lebret, A., Hirchaud, E., Normand, V., Andraud, M., Paboeuf, F., Blanchard, Y., Rose, N. and Bourry, O. (2019). A field recombinant strain derived from two Type 1 porcine reproductive and respiratory syndrome virus (PRRSV-1) modified live vaccines shows increased viremia and transmission in SPF pigs, *Viruses* 11(3):296. doi: 10.3390/v11030296.

Evenson, D., Gerber, P. F., Xiao, C. T., Halbur, P. G., Wang, C., Tian, D., Ni, Y. Y., Meng, X. J. and Opriessnig, T. (2016). A porcine reproductive and respiratory syndrome virus candidate vaccine based on the synthetic attenuated virus engineering approach is attenuated and effective in protecting against homologous virus challenge, *Vaccine* 34(46):5546–5553. doi: 10.1016/j.vaccine.2016.09.049.

Excler, J. L., Saville, M., Berkley, S. and Kim, J. H. (2021). Vaccine development for emerging infectious diseases, *Nat. Med.* 27(4):591–600. doi: 10.1038/s41591-021-01301-0.

Felberbaum, R. S. (2015). The baculovirus expression vector system: A commercial manufacturing platform for viral vaccines and gene therapy vectors, *Biotechnol. J.* 10(5):702–714. doi: 10.1002/biot.201400438.

Franzo, G. and Segalés, J. (2020). Porcine Circovirus 2 genotypes, immunity and vaccines: multiple genotypes but one single serotype, *Pathogens* 9(12):1049. doi: 10.3390/pathogens9121049.

Gallardo, C., Soler, A., Rodze, I., Nieto, R., Cano-Gómez, C., Fernandez-Pinero, J. and Arias, M. (2019). Attenuated and non-haemadsorbing (non-HAD) genotype II African swine fever virus (ASFV) isolated in Europe, Latvia 2017, *Transbound. Emerg. Dis.* 66(3):1399–1404. doi: 10.1111/tbed.13132.

Gao, L., Wang, L., Huang, C., Yang, L., Guo, X. K., Yu, Z., Liu, Y., Yang, P. and Feng, W. H. (2015). HP-PRRSV is attenuated by de-optimization of codon pair bias in its RNA-dependent RNA polymerase nsp9 gene, *Virology* 485:135–144. doi: 10.1016/j.virol.2015.07.012.

Gerdts, V., Mutwiri, G. K., Tikoo, S. K. and Babiuk, L. A. (2006). Mucosal delivery of vaccines in domestic animals', *Vet. Res.* 37(3):487–510. doi: 10.1051/vetres:2006012.

Goatley, L. C., Reis, A. L., Portugal, R., Goldswain, H., Shimmon, G. L., Hargreaves, Z., Ho, C. S., Montoya, M., Sánchez-Cordón, P. J., Taylor, G., Dixon, L. K. and Netherton, C. L. (2020). A pool of eight virally vectored African swine fever antigens protect pigs Against fatal disease, *Vaccines (Basel)* 8(2):234. doi: 10.3390/vaccines8020234.

He, Q., Cao, Z., Wang, P., Lu, Q., Zheng, H. and Sun, J. (2020). Efficient application of a baculovirus-silkworm larvae expression system for obtaining porcine circovirus type 2 virus-like particles for a vaccine, *Arch. Virol.* 165(10):2301–2309. doi: 10.1007/s00705-020-04754-9.

Hernandez-Garcia, J., Wang, J., Restif, O., Holmes, M. A., Mather, A. E., Weinert, L. A., Wileman, T. M., Thomson, J. R., Langford, P. R., Wren, B. W., Rycroft, A., Maskell, D. J., Tucker, A. W. and BRADP1T Consortium (2017). Patterns of antimicrobial resistance in Streptococcus suis isolates from pigs with or without streptococcal disease in England between 2009 and 2014, *Vet. Microbiol.* 207:117–124. doi: 10.1016/j.vetmic.2017.06.002.

Hou, Y. and Wang, Q. (2019). Emerging highly virulent porcine epidemic diarrhea virus: molecular mechanisms of attenuation and rational design of live attenuated vaccines, *Int. J. Mol. Sci.* 20(21):20(21):5478. doi: 10.3390/ijms20215478.

Ivanov, V., Efremov, E. E., Novikov, B. V., Balyshev, V. M., Tsibanov, S. Zh, Kalinovsky, T., Kolbasov, D. V., Niedzwiecki, A. and Rath, M. (2011). Vaccination with viral protein-mimicking peptides postpones mortality in domestic pigs infected by African swine fever virus, *Mol. Med. Rep.* 4(3):395–401. doi: 10.3892/mmr.2011.454.

Jang, Y. H. and Seong, B. L. (2012). Principles underlying rational design of live attenuated influenza vaccines, *Clin. Exp. Vaccin. Res.* 1(1):35–49. doi: 10.7774/cevr.2012.1.1.35.

Jiang, Y. F., Xia, T. Q., Zhou, Y. J., Yu, L. X., Yang, S., Huang, Q. F., Li, L. W., Gao, F., Qu, Z. H., Tong, W. and Tong, G. Z. (2015). Characterization of three porcine reproductive and respiratory syndrome virus isolates from a single swine farm bearing strong homology to a vaccine strain, Vet Microbiol. 179(3–4):242–249. doi: 10.1016/j.vetmic.2015.06.015.

Jung, K., Saif, L. J. and Wang, Q. (2020). Porcine epidemic diarrhea virus (PEDV): an update on etiology, transmission, pathogenesis, and prevention and control, *Virus Res.* 286:198045. doi: 10.1016/j.virusres.2020.198045.

Khatun, A., Shabir, N., Seo, B. J., Kim, B. S., Yoon, K. J. and Kim, W. I. (2016). The attenuation phenotype of a ribavirin-resistant porcine reproductive and respiratory syndrome virus is maintained during sequential passages in pigs, *J. Virol.* 90(9):4454–4468. doi: 10.1128/JVI.02836-15.

Kong, D., Wen, Z., Su, H., Ge, J., Chen, W., Wang, X., Wu, C., Yang, C., Chen, H. and Bu, Z. (2012). Newcastle disease virus-vectored Nipah encephalitis vaccines induce B and

T cell responses in mice and long-lasting neutralizing antibodies in pigs, *Virology* 432(2):327–335. doi: 10.1016/j.virol.2012.06.001.

Kvisgaard, L. K., Larsen, L. E., Hjulsager, C. K., Bøtner, A., Rathkjen, P. H., Heegaard, P. M. H., Bisgaard, N. P., Nielsen, J. and Hansen, M. S. (2017). Genetic and biological characterization of a Porcine Reproductive and Respiratory Syndrome Virus 2 (PRRSV-2) causing significant clinical disease in the field, *Vet. Microbiol.* 211:74–83. doi: 10.1016/j.vetmic.2017.10.001.

Legastelois, I., Buffin, S., Peubez, I., Mignon, C., Sodoyer, R. and Werle, B. (2017). Non-conventional expression systems for the production of vaccine proteins and immunotherapeutic molecules, *Hum. Vaccin. Immunother.* 13(4):947–961. doi: 10.1080/21645515.2016.1260795. DNA and Cell BiologyVol. 24, No. 4 Original Papers.

Li, B., Du, L., Yu, Z., Sun, B., Xu, X., Fan, B., Guo, R., Yuan, W. and He, K. (2017). Poly (d,l-lactide-co-glycolide) nanoparticle-entrapped vaccine induces a protective immune response against porcine epidemic diarrhea virus infection in piglets, Vaccine 35(50):7010–7017. doi: 10.1016/j.vaccine.2017.10.054.

Locher, C. P., Paidhungat, M., Whalen, R. G. and Punnonen, J. (2005). DNA shuffling and screening strategies for improving vaccine efficacy, *DNA Cell Biol.* 24(4):256–263. doi: 10.1089/dna.2005.24.256.

Lokhandwala, S., Petrovan, V., Popescu, L., Sangewar, N., Elijah, C., Stoian, A., Olcha, M., Ennen, L., Bray, J., Bishop, R. P., Waghela, S. D., Sheahan, M., Rowland, R. R. R. and Mwangi, W. (2019). Adenovirus-vectored African swine fever virus antigen cocktails are immunogenic but not protective against intranasal challenge with Georgia 2007/1 isolate, *Vet. Microbiol.* 235:10–20. doi: 10.1016/j.vetmic.2019.06.006.

Loving, C. L., Osorio, F. A., Murtaugh, M. P. and Zuckermann, F. A. (2015). Innate and adaptive immunity against porcine reproductive and respiratory syndrome virus, *Vet. Immunol. Immunopathol.* 167(1–2):1–14. doi: 10.1016/j.vetimm.2015.07.003.

Ma, W. and Richt, J. A. (2010). Swine influenza vaccines: current status and future perspectives, *Anim. Health Res. Rev.* 11(1):81–96. doi: 10.1017/S146625231000006X. Epub 2010 May 13.

Makadiya, N., Brownlie, R., van den Hurk, J., Berube, N., Allan, B., Gerdts, V. and Zakhartchouk, A. (2016). S1 domain of the porcine epidemic diarrhea virus spike protein as a vaccine antigen, *Virol. J.* 13:57. doi: 10.1186/s12985-016-0512-8.

Malik, Y. S., Singh, R. K. and Yadav, M. P. (2020). *Emerging and Transboundary Animal Viruses (Livestock Diseases and Management)* 1st edn. Luxemburg: Springer.

Martínez-Lobo, F. J., de Lome, L. C., Díez-Fuertes, F., Segalés, J., García-Artiga, C., Simarro, I., Castro, J. M. and Prieto, C. (2012). Safety of Porcine Reproductive and Respiratory Syndrome Modified Live Virus (MLV) vaccine strains in a young pig infection model, *Vet. Res.* 44(1):115. doi: 10.1186/1297-9716-44-115.

McLean, R. K. and Graham, S. P. (2019). Vaccine development for Nipah virus infection in pigs', *Front. Vet. Sci.* 6:16. doi: 10.3389/fvets.2019.00016.

Monteagudo, P. L., Lacasta, A., López, E., Bosch, L., Collado, J., Pina-Pedrero, S., Correa-Fiz, F., Accensi, F., Navas, M. J., Vidal, E., Bustos, M. J., Rodríguez, J. M., Gallei, A., Nikolin, V., Salas, M. L. and Rodríguez, F. (2017). BA71ΔCD2: a new recombinant live attenuated African swine fever virus with cross-protective capabilities, *J. Virol.* 91(21):e01058-17. doi: 10.1128/JVI.01058-17.

Murthy, A. M., Ni, Y., Meng, X. G. and Zhang, C. (2015). Production and evaluation of virus-like particles displaying immunogenic epitopes of porcine reproductive and respiratory syndrome virus (PRRSV), *Int. J. Mol. Sci.* 16(4):8382–8396. doi: 10.3390/ijms16048382.

Nah, J. J., Yang, D. K., Kim, H. H. and Song, J. Y. (2015). The present and future of veterinary vaccines for Japanese encephalitis in Korea, *Clin. Exp. Vaccin. Res.* 4(2):130–136. doi: 10.7774/cevr.2015.4.2.130.

Nair, P. (2021). QnAs with Katalin Karikó, *Proc. Natl Acad. Sci. U. S. A.* 118(51):e2119757118. doi: 10.1073/pnas.2119757118.

Nelson, M. I.. and Vincent, A. L. (2015). Reverse zoonosis of influenza to swine: new perspectives on the human-animal interface, *Trends Microbiol.* 23(3):142–153. doi: 10.1016/j.tim.2014.12.002.

Netherton, C. L., Goatley, L. C., Reis, A. L., Portugal, R., Nash, R. H., Morgan, S. B., Gault, L., Nieto, R., Norlin, V., Gallardo, C., Ho, C. S., Sánchez-Cordón, P. J., Taylor, G. and Dixon, L. K. (2019). Identification and immunogenicity of African swine fever virus antigens, *Front. Immunol.* 10(10):1318. doi: 10.3389/fimmu.2019.01318.

Ni, Y. Y., Zhao, Z., Opriessnig, T., Subramaniam, S., Zhou, L., Cao, D., Cao, Q., Yang, H. and Meng, X. J. (2014). Computer-aided codon-pairs deoptimization of the major envelope GP5 gene attenuates porcine reproductive and respiratory syndrome virus, *Virology* 450–451:132–139. doi: 10.1016/j.virol.2013.12.009.

Nielsen, J., Bøtner, A., Bille-Hansen, V., Oleksiewicz, M. B. and Storgaard, T. (2002). Experimental inoculation of late term pregnant sows with a field isolate of porcine reproductive and respiratory syndrome vaccine-derived virus, *Vet. Microbiol.* 84(1-2):1–13. doi: 10.1016/s0378-1135(01)00450-3.

Opriessnig, T., Gerber, P. F., Xiao, C. T., Mogler, M. and Halbur, P. G. (2014). A commercial vaccine based on PCV2a and an experimental vaccine based on a variant mPCV2b are both effective in protecting pigs against challenge with a 2013 U.S. variant mPCV2b strain, *Vaccine* 32(2):230–237. doi: 10.1016/j.vaccine.2013.11.010.

Opriessnig, T., Mattei, A. A., Karuppannan, A. K. and Halbur, P. G. (2021). Future perspectives on swine viral vaccines: where are we headed?, *Porcine Health Manag.* 7(1):1. doi: 10.1186/s40813-020-00179-7.

Pardi, N., Hogan, M. J., Porter, F. W. and Weissman, D. (2018). mRNA vaccines - a new era in vaccinology, *Nat. Rev. Drug Discov.* 17(4):261–279. doi: 10.1038/nrd.2017.243.

Pascual-Iglesias, A., Sanchez, C. M., Penzes, Z., Sola, I., Enjuanes, L. and Zuñiga, S. (2019). Recombinant chimeric transmissible gastroenteritis virus (TGEV) - porcine epidemic diarrhea virus (PEDV) virus provides protection against virulent PEDV, *Viruses* 11(8):682. doi: 10.3390/v11080682.

Pensaert, M. B. and Martelli, P. (2016). Porcine epidemic diarrhea: A retrospect from Europe and matters of debate, *Virus Res.* 226:1–6. doi: 10.1016/j.virusres.2016.05.030.

Perfumo, C. J., Pereda, A., Jongkaewwattana, A., Chen, Z., Perez, D. R. and Ma, J. (2020). Editorial: emerging swine viruses, *Front. Vet. Sci.* 7(7):132. doi: 10.3389/fvets.2020.00132.

Pollard, A. J. and Bijker, E. M. (2021). A guide to vaccinology: from basic principles to new developments, *Nat. Rev. Immunol.* 21(2):83–100. doi: 10.1038/s41577-020-00479-7. Rieckmann,

Rieckmann, K., Pendzialek, S. M., Vahlenkamp, T. and Baums, C. G. (2020). A critical review speculating on the protective efficacies of autogenous Streptococcus suis bacterins as used in Europe, *Porcine Health Manag.* 6:12. doi: 10.1186/s40813-020-00150-6.

Sandbulte, M. R., Spickler, A. R., Zaabel, P. K. and Roth, J. A. (2015). Optimal use of vaccines for control of influenza A virus in swine, *Vaccines (Basel)* 3(1):22–73. doi: 10.3390/vaccines3010022.

Saunders, K. O., Lee, E., Parks, R., Martinez, D. R., Li, D., Chen, H., Edwards, R. J., Gobeil, S., Barr, M., Mansouri, K., Alam, S. M., Sutherland, L. L., Cai, F., Sanzone, A. M., Berry, M., Manne, K., Bock, K. W., Minai, M., Nagata, B. M., Kapingidza, A. B., Azoitei, M., Tse, L. V., Scobey, T. D., Spreng, R. L., Rountree, R. W., DeMarco, C. T., Denny, T. N., Woods, C. W., Petzold, E. W., Tang, J., Oguin, T. H., Sempowski, G. D., Gagne, M., Douek, D. C., Tomai, M. A., Fox, C. B., Seder, R., Wiehe, K., Weissman, D., Pardi, N., Golding, H., Khurana, S., Acharya, P., Andersen, H., Lewis, M. G., Moore, I. N., Montefiori, D. C., Baric, R. S. and Haynes, B. F. (2021). Neutralizing antibody vaccine for pandemic and pre-emergent coronaviruses, *Nature* 594(7864):553–559. doi: 10.1038/s41586-021-03594-0.

Sharma, A., Zeller, M. A., Li, G., Harmon, K. M., Zhang, J., Hoang, H., Anderson, T. K., Vincent, A. L. and Gauger, P. C. (2020). Detection of live attenuated influenza vaccine virus and evidence of reassortment in the U.S. swine population, *J. Vet. Diagn. Invest.* 32(2):301–311. doi: 10.1177/1040638720907918.

Shuai, L., Ge, J., Wen, Z., Wang, J., Wang, X. and Bu, Z. (2020). Immune responses in mice and pigs after oral vaccination with rabies virus vectored Nipah disease vaccines, *Vet. Microbiol.* 241:108549. doi: 10.1016/j.vetmic.2019.108549.

Sisteré-Oró, M., López-Serrano, S., Veljkovic, V., Pina-Pedrero, S., Vergara-Alert, J., Córdoba, L., Pérez-Maillo, M., Pleguezuelos, P., Vidal, E., Segalés, J., Nielsen, J., Fomsgaard, A. and Darji, A. (2019). DNA vaccine based on conserved HA-peptides induces strong immune response and rapidly clears influenza virus infection from vaccinated pigs, *PLoS ONE* 14(9):e0222201. doi: 10.1371/journal.pone.0222201.Sosa

Sosa-Portugal, S., Cortey, M., Tello, M., Casanovas, C., Mesonero-Escuredo, S., Barrabés, S., Pineda, P., Wacheck, S., Martín-Valls, G. and Mateu, E. (2021). Diversity of influenza A viruses retrieved from respiratory disease outbreaks and subclinically infected herds in Spain (2017–2019), *Transbound. Emerg. Dis.* 68(2):519–530. doi: 10.1111/tbed.13709. Epub 2020 Jul 14.

Tian, D., Cao, D., Lynn Heffron, C., Yugo, D. M., Rogers, A. J., Overend, C., Matzinger, S. R., Subramaniam, S., Opriessnig, T., LeRoith, T. and Meng, X. J. (2017a). Enhancing heterologous protection in pigs vaccinated with chimeric porcine reproductive and respiratory syndrome virus containing the full-length sequences of shuffled structural genes of multiple heterologous strains, *Vaccine* 35(18):2427–2434. doi: 10.1016/j.vaccine.2017.03.046.

Tian, D., Sooryanarain, H., Matzinger, S. R., Gauger, P. C., Karuppannan, A. K., Elankumaran, S., Opriessnig, T. and Meng, X. J. (2017b). Protective efficacy of a virus-vectored multicomponent vaccine against porcine reproductive and respiratory syndrome virus, porcine circovirus type 2 and swine influenza virus, *J. Gen. Virol.* 98(12):3026–3036. doi: 10.1099/jgv.0.000964.

VanderWaal, K. and Deen, J. (2018). Global trends in infectious diseases of swine, *Proc. Natl Acad. Sci. U. S. A.* 115(45):11495–11500. doi: 10.1073/pnas.1806068115.

Vartak, A. and Sucheck, S. J. (2016). Recent advances in subunit vaccine carriers, *Vaccines (Basel)* 4(2):12. doi: 10.3390/vaccines4020012.

Vu, H. L. X., Pattnaik, A. K. and Osorio, F. A. (2017). Strategies to broaden the cross-protective efficacy of vaccines against porcine reproductive and respiratory syndrome virus, *Vet. Microbiol.* 206:29–34. doi: 10.1016/j.vetmic.2016.09.014.

Wang, Y., Wang, G., Duan, W. T., Sun, M. X., Wang, M. H., Wang, S. H., Cai, X. H. and Tu, Y. B. (2020). Self-assembly into virus-like particles of the recombinant capsid protein of porcine circovirus type 3 and its application on antibodies detection, *AMB Express* 10(1):3. doi: 10.1186/s13568-019-0940-0.

Weingartl, H. M., Berhane, Y., Caswell, J. L., Loosmore, S., Audonnet, J. C., Roth, J. A. and Czub, M. (2006). Recombinant Nipah virus vaccines protect pigs against challenge, *J. Virol.* 80(16):7929–7938. doi: 10.1128/JVI.00263-06.

Woolhouse, M. E. (2008). Epidemiology: emerging diseases go global, *Nature* 451(7181):898–899. doi: 10.1038/451898a.

Part 4

Optimising health through the life cycle

Chapter 14

Optimizing the health of gilts and sows during pregnancy and parturition

S. Björkman, C. Oliviero and O. A. T. Peltoniemi, University of Helsinki, Finland

1 Introduction

Current knowledge of factors affecting gilts' and sows' health during and after puberty and during pregnancy, parturition and after weaning is reviewed. Megatrends such as climate change, urbanization and globalization will eventually have a large impact on how pigs are kept and how healthy they will be in the long run. The current housing and breeding of gilts and sows worldwide need to be critically reviewed in light of megatrends and climatic changes. Because these changes are not likely to disappear, they must be considered when setting goals for breeding and designing new buildings in pig production.

Resilience of pig breeds should be a breeding goal from the point of view of health and welfare (Peltoniemi et al., 2019; Knap and Doeschl-Wilson, 2021).

http://dx.doi.org/10.19103/AS.2022.0103.15

While the present breeds are geared to produce high numbers of piglets in a litter, they might not be able to cope with emerging pathogens and megatrends such as the warming climate conditions. Also, technology is applied to improve the production environment; however, the technology used should be sustainable from the point of view of the environment and animal health and welfare.

One of the associated trends seen in the pig industry is a concomitant increase in litter size and prolongation of parturition. In the 1990s, the average duration of farrowing was 2–4 h (Madec and Leon, 1992; Jackson, 1995). Since 1990, there has been a linear increase in both (1) litter size from about 10 piglets in 1990 to close to 20 piglets in 2019 and (2) duration of farrowing from 2 h to 7–8 h (a conclusion based on 20 studies on duration of farrowing; Oliviero et al., 2019). While the described tendency is subject to differences between breeds and management (i.e. farrowing crate vs. free farrowing), the overall tendency is rather convincing. It seems that the average interval between piglets has not increased very much during the last 20 years, meaning that the extended duration of farrowing appears as an outcome of intensive breeding for prolificacy in the pig (Oliviero et al., 2019). In this review, we address the physiology behind these trends, focusing on potential management solutions to alleviate issues relating to the transfer of immunity and welfare of the sow.

2 Biosecurity issues, vaccination and deworming

At the level of an operational unit, external and internal pathogens circulate and create a threat to general health, leg health, continuation of pregnancy, successful parturition and fertility. A focus on external and internal biosecurity is essential (Neumann and Hall, 2019). It is therefore critical that gilt and sow units apply a high level of internal and external biosecurity, including a proper application of quarantine procedures when new breeding animals are bought in and are about to enter the herd (Alarcon et al., 2021).

Sows and gilts require routine vaccination and deworming programmes (Neumann and Hall, 2019). Even in well-managed indoor herds with high management standards, such as batch farrowing, all-in all-out systems and high biosecurity levels, helminth infections are still very prevalent. *Ascaris suum* is the most common helminth present on commercial farms. Sows and gilts should be dewormed about 2 weeks before moving to farrowing crates or pens. Fenbendazole is a widely used anthelmintic drug effective against the different life stages of the parasite (adults, eggs and larvae). Washing sows and gilts immediately prior to transfer to the farrowing rooms is also recommended. Depending on the country, area or farm health status, different vaccination protocols are established to maintain adequate protection against the main viral and bacterial diseases in gilts and sows. Protection against leptospira, erysipelas and parvovirus should be given either pre-farrowing or at weaning, depending on the optimal timing for the particular herd. Erysipelas can cause arthritis and death losses in the herd with a marked economic impact. Leptospirosis and erysipelas can also be zoonotic diseases. Parvovirus is extremely common in most swine herds (Oravainen et al., 2005). Protection of sows during gestation against these diseases (e.g. leptospirosis, erysipelas and parvovirus infections) can be obtained by vaccination of breeding animals at 6–7 months of age and 3–4 weeks later. Boosters would follow prior to insemination or every 6 months. Combined vaccines are available on the market.

Once or twice a year, vaccinating for influenza can boost herd protection. If influenza has been a problem in piglets historically, vaccinating sows pre-farrowing should be considered. Other pre-farrowing vaccinations may include either an *Escherichia coli* vaccine or an *E. coli* and rotavirus combination vaccine to help reduce the risk of scours in neonate piglets (Neumann and Hall, 2019).

If the herd has been at risk of contracting porcine reproductive and respiratory syndrome (PRRS), atrophic rhinitis or Clostridium, a pre-emptive vaccination should be given to sows to protect them and their litters (Zimmerman et al., 2019). A porcine circovirus 2 booster vaccination should also be administered to replacement gilts prior to breeding, as they are the most susceptible to reproductive issues. Some veterinarians suggest vaccinating previously exposed herds with a porcine epidemic diarrhoea (PED) vaccine.

At the time of vaccination, sows must be healthy. If they have a fever or show any other clinical signs of illness, the vaccine should be delayed for a few days. In case of attenuated bacterial vaccines, sows should not be treated with antibiotics 3 days before or after administration of the vaccine, as the antibiotic may kill the vaccine bacteria needed to stimulate immune responses.

Health of prepubertal gilts is affected by management of developing immunity, leg and udder health (Oravainen et al., 2005). Typically, immunity triggered by a vaccine requires repeated vaccinations to achieve adequate immunity against a pathogen (Oravainen et al., 2005). In real life, making staff members of a piggery operation aware of the basic concepts of use of a vaccine and immunity created by the vaccine is an important part of everyday work. If the staff is not sufficiently motivated and educated, the risk of misuse/improper use of vaccines will rise.

3 Production diseases causing fertility problems and reduced reproductive performance

Diseases of the mammary gland and urogenital tract such as mastitis, endometritis and cystitis are often related to intensive production systems, where breeding standard, hygiene and animal welfare are suboptimal, bringing both neonates and their dams to the edge of their metabolic and immunological competence. These so-called production diseases are usually connected, being either a cause or a result of each other. The postpartum dysgalactia syndrome (PDS) is probably the most important disease complex leading to high economic losses. These economic losses are due to an increased rate of piglet mortality as well as reduced number of piglets weaned. Further, these diseases are linked to reduced development and growth of the piglets as well as reduced fertility and longevity of sows. Bacteria involved are often non-specific and opportunistic, inhabiting the lower female reproductive tract. If immunity is compromised or the microbial balance disturbed, these bacteria can cause disease. Therefore, their incidence rate can vary a lot and can be very high depending on production system.

3.1 Vulvar discharge syndrome

The vulvar discharge syndrome (VDS) can cause considerable economic losses because of the small litter size at parturition and subsequent high piglet mortality and impaired sows' fertility (Dee, 1992). The source of pathological discharge may be reproductive or urinary tract disease, especially cystitis or endometritis (Althouse et al., 2019). The incidence of these diseases is on the rise because of an increase in the intensity of the production, mostly because of increased litter size and prolonged parturition (Björkman et al., 2018b).

The organisms underlying the signs and lesions of these infections are often similar (Dee, 1992), non-specific and opportunistic. Both gram-negative (e.g. *E. coli spp., Klebsiella spp., Proteus spp.*) and gram-positive bacteria (*Streptococcus spp., Staphylococcus spp., Enterococcus spp., A. suis)*) are usually involved (Biksi et al., 2002; Grahofer et al., 2014; Bellino et al., 2013; Boers et al., 2019)). The bacteria inhabit the normal physiological microbiota on the skin and mucosa of the lower female reproductive tract (Muirhead, 1986; D'Allaire et al., 1987; Dial and MacLachlan, 1988; de Winter et al., 1995; Glock and Bilkei, 2005; Oravainen et al., 2008; Kauffold, 2008; Tummaruk et al., 2010). They can ascend to the urinary and upper reproductive tract or enter into the bloodstream causing systemic effects whenever the infectious pressure is high and/or immunity of the sow is low (Oravainen et al., 2006; Bellino et al., 2013). Sows' immunity can be challenged during parturition, stressful situations, mycotoxin exposure and/or hormonal down regulation (Althouse et al., 2019). Especially hygiene, husbandry and management around the time of parturition and breeding are most important. Risk factors for puerperal endometritis are prolonged parturition, obstetrical intervention, impaired expulsion of placenta and birth of more than two stillborn piglets (Björkman et al., 2018b; Grahofer et al., 2019). Risk factors for non-puerperal endometritis are wrong time of insemination, more than two inseminations and confinement in crates, poor hygiene and lack of enrichment material (Oravainen et al., 2006, 2008). Further, any factor that causes a reduction in the frequency of urination (decreased water intake, decreased urination, incomplete emptying of bladder, etc.) should be considered important in the development of cystitis (Drolet, 2019). Therefore, the availability of water and water flow as well as water quality in terms of bacteria and physical/ chemical properties need to be frequently controlled (Drolet, 2019). Further, gilts and sows need to be in good body condition, stress-free and in good health (e.g. leg health) (Drolet, 2019).

If VDS is suspected in gilts or sows, for example, because of increased birth complications, increased incidence of postpartum diseases or decreased fertility, proper diagnostics with a systematic approach needs to be applied. The diagnostic procedure addressing herd fertility problems is comprehensive and must include a critical assessment of the tripartite contributors to herd reproductive performance: (1) farrowing personnel, (2) breeding personnel and (3) sow and semen management (Althouse et al., 2019). Diagnosticians must be aware of the multitude of factors and their inter-relatedness that might be causative of the problem, including facility type, the building environment, seasonality, nutrition, sanitation/hygiene and disease-related occurrences or interventions. A presumptive diagnosis can be based on the clinical signs. Further diagnostic methods have recently been reviewed by Grahofer et al. (2020) and Björkman and Grahofer (2020).

Both cystitis and endometritis are usually subclinical and systemic reactions are rare (Althouse et al., 2019). A body temperature of more than 39.5°C, together with clinical signs such as abnormal general behaviour (i.e. lethargy and apathy), reduced feed intake (anorexia), frequent urination and/ or abnormal vaginal discharge, is indicative of VDS (Stiehler et al., 2015; Grahofer et al., 2019). The vulvar discharge of affected sows varies extensively, depending on the pathogenic microorganism, duration of infection and the stage of the oestrous cycle (Dial and MacLachlan, 1988). Physiological vaginal discharge can be observed immediately after parturition, insemination and shortly before oestrus (Muirhead, 1986; Meredith, 1991; de Winter et al., 1992; Almond et al., 2006). Increased volumes of vaginal discharge are associated with endometritis, but there is no significance between the occurrence of endometritis and the colour of the vaginal discharge (Muirhead, 1986). In animals with cystitis, the urine is often reddish-brown in colour (haematuria) with a strong door (pyuria). Presumptive diagnosis of VDS in live animals is best achieved when an abnormal amount of discharge and/or frequent micturition of blood stained and cloudy urine are observed (Drolet, 2019).

Confirmatory diagnosis of VDS can be achieved by ultrasonography (Kauffold et al., 2010) and macroscopic, microscopic and chemical and bacteriologic examination of urine and/or vaginal discharge. Commonly used methods to confirm cystitis in live animals are urinalysis and urine culture (Gmeiner, 2007). For urinalysis, it is preferred to collect spontaneous midstream urine in a transparent tube. The best time to collect urine is in the morning before feeding (Kraft et al., 2005). A red or brown colour, cloudy or turbid appearance and/or a strong ammoniac or putrid odour could indicate the presence of bacteria in the urine (Tolstrup, 2017). After the macroscopic evaluation, the urine sample can be centrifuged and the sediment microscopically evaluated at ×400 magnification (Kraft et al., 2005). Erythrocytes, leukocytes and epithelial cells are counted. A sample is considered positive when there are more than five white blood cells per visual field (Bellino et al., 2013). Furthermore, the presence of transitional epithelial cells and bacteria, and a specific gravity of the urine higher than 1.020, can be indicative of cystitis (Gmeiner, 2007; Tolstrup, 2017). After urinalysis, dipslides can be used for bacteriological evaluation. They are placed into urine for about 10 s and the bacterial growth is evaluated approximately 18–24 h later. The threshold for a urinary tract infection is 10 × 5 colony forming units (cfu)/mL urine (Kraft et al., 2005). Dipslides can also be submitted to the laboratory for specification of the bacteria and antimicrobial susceptibility testing. Bacteriological evaluation allows to determine the exact bacteria and antibiotic sensitivities. It should always be done if antibiotics are planned to be used in the treatment in order to decrease the risk of creating antibiotic resistance.

The most common method to diagnose endometritis in live animals is ultrasonography. In evaluation of the structure of the uterus, the parameters of fluid echogenicity, echotexture and size are measured in order to provide a comprehensive diagnosis (Kauffold and Althouse, 2007; Björkman et al., 2018b; Grahofer et al., 219; Meile et al., 2019). In a sow with acute endometritis, the uterus size as well as the echotexture are increased (Kauffold and Althouse, 2007). However, the days postpartum and the parity should be taken into account when evaluating the uterine parameters (Kauffold and Althouse, 2007; Björkman et al., 2018b). In addition, fluid echogenicity in the uterus can be used as an indicator for an exudative inflammation of the uterus (Kauffold and Althouse, 2007). For further diagnostic tests, cytology and microbiology can be used. The number and type of inflammatory cells in the endometrium per visual field in the ×400 magnification of the light microscopy gets used to classify into acute and chronic endometritis. In sows with acute endometritis, more than 20 neutrophilic granulocytes can be detected in a field (de Winter et al., 1995). In comparison, chronic endometritis is defined as the presence of more than 20 lymphocytes, plasma cells or histiocytes in a field (de Winter et al., 1995). In sows without vaginal discharge, the endometrium contains a low number of neutrophilic and eosinophilic granulocytes as well as plasma cells (Kaeoket et al., 2001; Oravainen et al., 2008). Still, it needs to be considered that only about half of the sows with endometritis show vaginal discharge. Therefore, not only in sows with clinical signs but also in subclinical sows where endometritis is suspected, ultrasonography of the uterus is indicated. Further, as with cystitis, also in case of endometritis, a bacteriological examination should be performed if the use of antibiotics is needed in order to minimize the risk of creating antibiotic resistance. The uterine swab used to collect vaginal cytology can also be used for microbiological examination and antimicrobial susceptibility testing. A speculum (with a double-guarded swab) should be used to obtain a representative sample from the uterus and to avoid contamination of the bacterial flora from the vagina (Oravainen et al., 2008; Grahofer et al., 2017).

A highly valuable tool for the accurate diagnosis of reproductive disease and disorders such as the VDS is post slaughter gross pathological examination (de Jong et al., 2014). The genital organs, together with the urinary tract and bladder, are collected from specified sows on the slaughter line at the abattoir and submitted for pathological examination. In a study conducted with 502 sows on 7 farms, uterine exudate was detected in 18% with *E. coli* being the most common finding (de Jong et al., 2014).

3.2 Postpartum dysgalactia syndrome, mastitis and severe mammary oedema

Postparturient disorders in sows represent an economically important disease complex in modern pig production worldwide (Kemper, 2020). The most important disease complex is PDS. Mastitis, endometritis and cystitis have been linked as either a cause or sign of postpartum dysgalactia syndrome (PDS) (Wendt et al., 1990; Biksi et al., 2002; Kemper, 2020). PDS is characterized by insufficient colostrum and milk production by the sow during the first days of lactation (Maes et al., 2010). As a consequence, colostrum and milk intake by piglets is reduced and therefore their mortality increased (Maes et al., 2010). Further, PDS negatively affects subsequent reproductive health of the sow (Hoy, 2006).

PDS in sows has a complex multifactorial pathophysiology, and diagnosis is therefore challenging. Whenever there is an increased incidence rate in piglet mortality or disease, or a decrease in piglet weight gain and growth rate, PDS should be suspected. This is because piglets are in their first days of life totally reliant on the sow for access to colostrum and milk which is important for their growth, development and immunity. Every alteration in both milk yield and composition has major impact on the piglets. Nevertheless, other piglet-related causes need to be excluded. Causes for piglet mortality have been reviewed by Edwards and Baxter (2015).

Besides diseases of the uterus and urinary bladder, also diseases of the mammary gland can be either a cause or result of PDS. The mammary gland, unlike the urogenital tract, is located outside the body and is therefore easily accessible for examination. The diagnosis of mastitis can be done by adspection and palpation of the mammary gland. Mastitis can be a local process, restricted to one or several glands, but can also affect all mammary complexes (Gerjets and Kemper, 2009). Infected glands typically show signs of inflammation such as severe oedema and skin congestion, therefore appearing swollen, firm and warm (Gerjets and Kemper, 2009). With many glands involved, sows develop fever and lose their appetite (Gerjets and Kemper, 2009). If needed, ultrasonography can be used in the diagnosis. Affected mammary glands provide heterogeneous and hyperechoic images (Baer and Bilkei, 2005).

As with cystitis and endometritis, usually non-specific and opportunistic organisms such as *E. coli*, *Staphylococcus spp.* and *Streptococcus spp.* are involved. The current hypothesis is that interactions between endotoxins produced by gram-negative bacteria in the gut, mammary gland and/or urogenital tract, and alterations in the immune and endocrine functions, play a central role in the development of PDS (Maes et al., 2010). This is supported by a study where periparturient sows were challenged with lipopolysaccharide (LPS) endotoxin of *E. coli* and in which sows generated symptoms similar to PDS (Nachreiner and Ginther, 1974). *E. coli* originates from the environment or can already be present in the urogenital tract of pre-partum sows or enter during or after parturition.

If there is only one or a few glands involved, local treatment such as milking the affected gland(s) during nursing or after injection of oxytocin may be sufficient. If there are multiple glands involved, and the sow shows general signs, systemic treatment with antibiotics needs to be considered. Again, in order to prevent antibiotic resistance, it is essential to take a milk sample for identification of the bacteria and antimicrobial susceptibility testing. How to take a milk sample has been described, for example, by Gerjets and Kemper (2009) and Kaiser et al (2020). Shortly, after intramuscular (IM) injection of oxytocin (20–30 I.U.), any milk present in the nipples (ductus papillaris) is removed before sampling. The teats are cleaned with, for example, 5% chlorhexidine gluconate soap solution and disinfected with, for example, 70% isopropyl alcohol. The first streams of secretion from each teat are discarded and milk samples are milked on a transport medium.

As with cystitis and endometritis, prevention needs to focus on nutrition, housing and management. Especially high levels of animal welfare and hygiene are important in order to minimize stress and infectious pressure. Risk factors for PDS are therefore the use of the unslatted floors, no washing of sows and no use of disinfectants in the farrowing and breeding rooms (Hultén et al., 2004). Considering that parturition itself decreases immunity and causes significant inflammatory changes (Kaiser et al., 2018b), all other factors affecting immunity and endocrinology need to be kept at a minimum level. Especially, stress needs to be reduced as much as possible. Elevated cortisol levels in sows with PDS were found by Kaiser et al. (2018b). Stress can be due to restricted space in farrowing crates, lack of nest-building material, high ambient temperature and abrupt change from group housing during gestation to restraint in crates a few days before farrowing (Maes et al., 2010; Papadopoulos et al., 2010).

Another disease of the mammary gland which has been lately associated with PDS is severe mammary oedema (SME, Björkman et al., 2017a, 2018a; Kaiser et al., 2020). SME can also reduce colostrum quality (Björkman et al., 2017a, 2018a). One possible explanation is that sows that will suffer from PDS may already have before and during parturition a compromised cardiovascular system and vascular homeostasis. Kaiser et al. (2020) reported increased concentrations of chromogranin A (CgA) in sows with PDS. CgA is a stress marker of the sympathetic nervous system known to regulate vascular homeostasis. An increase in CgA can lead to sympathetic activation, leading to tachycardia and hypertension, as well as increased capillary wall permeability (Kaiser et al., 2020). This leads to increased vascular leakage which is central to oedema (Kaiser et al., 2020). This could also be an explanation for reduced colostrum quality in sows with SME. Before parturition, tight junctions of the capillary endothelium close in order to prevent milk components, for example, IgG, from the mammary gland alveoli to leak back to the bloodstream. This closure of tight junction may be impaired in sows with PDS.

Thus, determining the state of the cardiovascular system can be one way of diagnosing PDS. Besides an increase in heart rate (tachycardia), a prolonged capillary refill time in the vulvar mucosa (>2 s) has been reported by Kaiser et al. (2020). The degree of mammary gland oedema can be graded and SME can be diagnosed visually or via ultrasound (Björkman et al., 2017a, 2018a). At visual inspection, sows with severe udder oedema have dimpled skin with persistent marks on the floor. Further, teats are swollen and mammary glands are indistinct (Björkman et al., 2017a, 2018a). At palpation, glands feel doughy and pressure leaves an indentation (Kaiser et al., 2020). Ultrasound of the mammary glands shows thickened dermal and subdermal tissues, hyperechoic lobuloalveolar tissue with enlarged blood vessels and severe shadowing (Björkman et al., 2017a, 2018a; Björkman and Grahofer, 2020).

4 Infectious diseases causing fertility problems and reduced reproductive performance

From the physiological and immunological viewpoints, pregnancy in the pig can be divided into three phases. The first trimester involves recognition of the existence of embryos and the establishment of pregnancy, rendering the phase sensitive to environmental effects such as stress (see below), seasonality, especially light and temperature, and transmission of pathogens followed by early disruption of pregnancy. The middle phase is considered more stable, culminating in acquiring immune competence by the foetuses at around day 70 of pregnancy. Therefore, viral infections prior to this milestone can lead to mummification of embryos and prolonged length of pregnancy. The last trimester is characterized by foetal immunocompetence and therefore a safer phase that requires a high level of pathogenicity of the causative agent in order to be able to induce an abortion, as may happen with pathogens such as PRRS and *Leptospira* sp.

4.1 Viral causes for abortions

4.1.1 Porcine parvovirus (PPV) and picornavirus

In the scientific literature, porcine parvovirus (PPV) is a classic example of a viral pig disease that causes reproductive failure in the sow, which is manifested usually without visible abortion. PPV has also been used as an example of the temporal relationship between the acquisition of immunocompetence and the development of embryos and foetuses (Mengeling et al., 2000). During the embryonic period of development, infection is likely to cause death and resorption of developing early embryos. In the foetal period, if the infection occurs before day 70 of pregnancy, the foetuses lack the ability to respond and therefore become mummified. However, if infected after immunocompetence has been acquired, then the consequences are less serious, resulting in a slight weakness of the newborn piglets or, in some cases, no clinical signs at all.

PPV disease is globally widespread, with herd prevalence in most countries/studies exceeding 70% (Nash, 1990; Oravainen et al., 2005). The primary clinical symptoms are reported to return to oestrus and birth of dead born, mummified foetuses. However, abortion storms have been described in conditions where naïve female herd faces the virus for the first time (Brown et al., 1980). Thus for most producers, it is a disease with which they have to cope, because eradication is not feasible. The control programme involves vaccination of all breeding animals in the herd: gilts, sows and boars. As PPV is known as the 'gilt disease', it is essential that gilts are vaccinated twice, before they are inseminated for the first time. Vaccine-induced immunity is boosted in a stepwise manner, so that sows that have farrowed at least once clearly have higher antibody titers than unbred gilts (Oravainen et al., 2005). Therefore, in practice, signs of PPV may be seen in cases where vaccination of replacement gilts has not been carried out, or as a result of a vaccination failure. Common vaccination failures can be due to the use of out-of-date vaccines, an inappropriate injection site (neck instead of the base of the ear), storage problems (frozen/overheating of vaccine during storage) and improper vaccination technique. It may also be that some animals turn out to be lacking the ability to respond because of the remaining maternally derived immunity (can last until 7 months for PPV) that inhibit vaccine responses

In addition to PPV, different genera of picornaviruses are known to cause various reproductive problems in pigs. Encephalomyocarditis virus (EMCV) has been found and isolated from a number of mammalian species (rodents regarded as the main host) and is known for reproductive pathogenicity with virus-excreting pigs being a risk for other pigs. Therefore, pig-to-pig contact between healthy and diseased or dead pigs is considered as the main route of transmission; however, transplacental transmission followed by abortions may occur. Presence of rodents is considered as a risk factor for the spread of EMCV, even in a piggery setting, serving as a reservoir for infection. High mortality in weaners after suffering from general systemic symptoms is described with myocardial failure/myocarditis as a specific clinical sign. In breeding animals, abortion, increased number of mummified and stillborn foetuses have been described as the main clinical findings (Zimmerman et al., 2019). Control of EMCV is based on strict rodent control in endemic areas. Furthermore, low-level exposure of the virus through manure to naïve pigs may trigger some immunity. Vaccination is also available, being efficient and of value for control in endemic areas.

Some strains of porcine enteric picornaviruses (porcine teschovirus, PTV, strains PTV-1,-3,-6) have been associated with reproductive symptoms such as abortion, stillbirth, mummified foetuses, embryonic death and infertility.

4.1.2 Porcine circo virus (PCV 2/3)

Porcine circo virus (PCV 2 and PCV3) infections have been reported to cause reproductive disease including abortions in sow herds in a number of countries, including South Korea and Spain (Chae, 2005; Maldonado et al., 2005; Saporiti et al., 2021). The virus is described as infecting the conceptuses following transplacental transmission. Embryonic or foetal death and loss of the whole litter may occur at any stage of pregnancy (Chae, 2005). In addition to abortion, an increased number of stillborn and mummified foetuses, and weak born piglets, may be observed (Segales, 2012; Table 1). The oro-nasal route of transmission is considered to be the most common route of infection; horizontal transmission between pen mates and from one pen to another may be very efficient. However, transmission through semen may also occur, although it has not yet been proven to be a significant route. Herds suffering from the epizootic phase of infection associated with reduced farrowing rates and increased stillborn rates may benefit from vaccination programmes that have been developed (Segales, 2012). PCV3, since its first recovery in 2015, has received considerable research attention. It is apparent that in addition to having the ability to cause systemic disease in growing pigs, it also can cause a reproductive disease as described, which however needs further research.

4.1.3 Porcine reproductive respiratory syndrome (PRRS)

PRRS occurs causing severe illness with pyrexia, inappetence, listlessness, bleeding and swelling of the ear and, as the name implies, respiratory

Table 1 Clinical manifestations, gross lesions, microscopic lesions and diagnosis in PCV2 reproductive disease (PCV2-RD)

Clinical manifestation	Gross lesions	Microscopic lesions	Diagnosis
Abortions	foetal mummification foetal hepatic enlargement foetal cardiac hypertrophy ascites in foetuses	non-suppurative to necrotizing or fibrous myocarditis of foetuses hepatic congestion pneumonia in foetuses	1 late abortion 2 myocarditis 3 high/moderate PCV present in heart tissue by qPCR
Regular/ irregular returns to oestrus			1 regular/irregular return to oestrus PCV 2 2 seroconversion of PCV2 antibodies following the return to oestrus

Source: Modified from Segales (2012).

disease with bronchopneumonia and pleurisy in growing pigs. In gilts and sows, it affects the reproductive system and causes late-term interruption of pregnancy, mummified foetuses, stillbirths and weak born piglets (Hopper et al., 1992; Zimmerman et al., 2019). However, the transplacental infection may occasionally also occur earlier during pregnancy. During the acute phase of the disease, up to 1–3% of litters may be lost and mortality of sows is elevated by a few percentage points and is associated with pulmonary oedema and cystitis/nephritis. Late-term abortion occurs on days 100–118 and anything of the following may be encountered: normal piglets, a lot of variation in piglet birth weight, increased intrapartum death rate (piglets looking fresh), autolytic piglets and partially or completely mummified piglets (Zimmerman et al., 2019).

Routes for shedding of the virus include saliva, nasal secretions, urine and faeces. In semen, the virus may be shed for up to 43 days after exposure. It has also been reported that foetuses may die from hypoxia due to arteritis of the umbilical vessels (Lager and Halbur, 1996). Exposure of PRRS naïve replacement gilts to viral-infected tissues confirmed positive resulted in pre-breeding infection of gilts followed by sufficient immunity to protect the subsequent pregnancy (Menard et al., 2007). Lesions found in reproductive tissues include microscopic lesions in the uterus, oedematous myometrium and endometrium, with lymphohistiocytic perivascular cuffs. Prevention and control strategies involve internal and external biosecurity measures to maintain negative status in herds and areas that are free of the disease on one hand and keep new PRRSV variants out in those herds and areas that are endemic in terms of their PRRS status. In the latter situation, control is largely based on the application of internal biosecurity measures such as all-in all-out animal flow or partial depopulation. Application of vaccination programmes of breeding animals at adequate age to induce immunity prior to pregnancy belongs to the core of control measures in endemically infected herds. Eradication programmes based on total and partial depopulation/repopulation, test and removal have been described (Corzo et al., 2010). In addition, breeding for more resilient animals against PRRS has been recently proposed as an alternative strategy for the future (Knap and Doeschl-Wilson, 2021).

4.1.4 Aujzesky's disease (pseudorabies)

Another viral disease of pigs, mainly affecting the reproductive system, is Aujeszky's disease, due to infection with a herpes virus that may cause severe/fatal disease in cattle and other species such as sheep, cats, dogs, mice and rats (Pensart and Kluge, 1989). Herd prevalence of pseudorabies in Western countries is usually low, in fact, most EU countries are free of the disease, but once infected, a herd with a continuous pig production system may suffer from it for a long time. The route of transmission of the virus is usually by inhalation or ingestion. It may also be transmitted by coitus, although there is some argument as to whether true venereal transmission occurs. Aujeszky's disease is characterized by nervous and respiratory signs, associated with a rise in temperature and often death in young piglets. Depending on the timing, infection in breeding adults may result in embryonic death, resorption of foetuses, abortion or stillbirth (Mettenleiter et al., 2019). In adult boars and sows, the clinical signs of this disease are seldom severe and usually consist of pyrexia, depression and anorexia that lasts for up to a week. Of great significance to the breeding herd is the fact that the virus causes embryonic death, foetal mummification and stillbirths.

4.1.5 Other viral agents

Several other viruses have been associated with reproductive disorders in the pig such as ASF and CSF (Almond et al., 2006). However, they are more likely to exert their effect indirectly on reproduction by causing systemic illness in the sow or gilt, such as severe pyrexia rather than attacking the reproductive system directly.

4.2 Bacterial infections

4.2.1 Leptospirosis

Leptospirosis is known to be a widely spread disease compromising reproduction in the pig population worldwide. *Leptospira interrogans serovar pomona* is the most widely recognized pathogen, but serovars *tarassovi*, *bratislava* and *icterohaemorrhagiae* may also be pathogenic in pigs (Arent and Ellis, 2019). Leptospira has a predilection site for the kidneys where they persist and multiply, and they are then secreted via the urine into the environment intermittently for a period of up to 2 years. Rodents are often thought to be the most significant reservoir species, maintaining the infection in the production environment. During the acute phase of the disease, mild general symptoms such as fever and anorexia may be seen; however, this phase often passes unrecognized. During the chronic phase of the infection, abortion rates of sows increase considerably, especially during the second half of gestation (Ellis, 2006; Arent and Ellis, 2019). Furthermore, stillbirth rates may increase and newborn piglets may lack viability (Nagy, 1993). Diagnosis is based on serological testing microscopic agglutination test, MAT and fluorescence antibody testing, FAT and ELISA and attempts to demonstrate the presence of leptospires in tissues and body fluids by qPCR and DNA sequencing (OIE, 2008; Ellis, 1990; Arent et al., 2016; Arent and Ellis, 2019). Prevention is based on rodent control and better hygiene, particularly, improved urine drainage from the floor surface. In addition, antimicrobials (tetracycline and streptomycin) can be used for treatment, and vaccines against specific serotypes are available for prevention.

4.2.2 Brucellosis

Brucellosis, due to infection with a number of *Brucella* species, has been recognized as a disease associated with abortion in the pig since the early 1900s. Although they may not be highly pathogenic for the pig, the zoonotic nature of the disease presents a risk for humans handling pigs, and interspecies transmission between other domestic animal species makes it an important infectious disease (Olsen et al., 2019). The most important *Brucella* species in the pig is *B. suis*; however, other species such as *B. neotomae, B. ovis, B. canis, B. abortus* and *B. melitensis* have also been identified (Olsen et al., 2019).

The most significant route of transmission is direct contact between animals, and infection is usually acquired through the oro-nasal or genital route. Adult pigs that are infected usually lack general systemic signs, such as fever and loss of appetite. Abortion may occur at any stage of pregnancy; however, other clinical signs relating to reproduction such as stillbirth and infertility may be more significant from the reproductive viewpoint (Olsen et al., 2019).

As sows usually eliminate the bacteria within 30 days after infection, a period of reproductive rest of two oestrous cycles duration (42 days) may be enough to ensure that the infection has been eliminated, and subsequent AI may be warranted. Other signs of the disease include subfertility, orchitis, posterior paralysis and lameness. Boars may be unable to eliminate the bacteria, and these may act as carriers within a herd. Antimicrobial therapy is often ineffective, and an eradication programme may offer the best outcome.

4.3 Other bacterial diseases and parasites

Chlamydia, specifically *C. abortus*, has been reported in a few studies in conjunction with abortion, weak neonates and irregular returns to oestrus in the pig (Camenisch et al., 2004; Broes et al., 2019). Bacteria causing an acute systemic response resulting in pyrexia present a risk to pregnant sows and gilts. For example, endotoxemia caused by gram-negative bacteria, or generalized, acute clinical signs caused by *Erysipelothrix rhusiopathiae* can cause abortion at any stage of pregnancy (Opriessnig and Coutinho, 2019). The better the immunity of the herd towards these pathogens, the lower the risk of reproductive disorders. Parasites may also cause reproductive problems including abortions. The connection between *Toxoplasma gonadii* and late-term abortions is well established (Lindsay et al., 2019). The cat serves as a definite host of the infection, whereas the pig serves as an intermediate host by contact with infected other intermediate hosts such as mice and rats (Lindsay et al., 2019). Infected pork is a source of Toxoplasma infection for companion animals and humans.

5 Non-infectious factors causing fertility problems and reduced reproductive performance

5.1 Group housing, stress and social interaction

In group housing, sows can interact with each other, which can lead to positive but also negative social interactions. Aggression between sows is the most prominent form of negative social interaction and can affect sows' health, welfare and reproduction (Spoolder and Vermeer, 2015). Group housing has been linked with a decrease in pregnancy rate, farrowing rate and litter size (Munsterhjelm et al., 2008). Whenever mixing sows, establishing a dominance hierarchy and competition over resources such as space and food will lead to acute and eventual chronic stress. Therefore, appropriate management of the animals, the feeding system, the size and design of the pen, group size and the skills and attitude of the stock person are important to maximize reproductive success (Spoolder and Vermeer, 2015).

It is well documented that the embryonic period of the pregnancy is more vulnerable for loss of pregnancy than the subsequent foetal period (Tast et al., 2002). Stress negatively affects embryo survival, in particular, during early pregnancy, around the period of attachment of the embryos to the uterine wall (days 11–16), and the period shortly thereafter when maternal recognition of pregnancy takes place (Spoolder and Vermeer, 2015). Thus, it seems wise to avoid stressful events in early pregnancy, especially in weeks 2–3 following insemination, to avoid a decrease in productivity. Recognition among gilts and sows appears to be based on familiarity gained when reared or housed together, and sows can recognize each other even after several weeks of separation (Spoolder et al., 1996; Stookey and Gonyou, 1998). Rank order fights are initiated when animals do not recognize each other or when they dispute each other's position in the hierarchy (Spoolder and Vermeer, 2015). The social hierarchy is usually established within 2 days after the introduction of new animals (Eliasson-Selling et al., 2000), but full integration can take at least 3–4 weeks (Moore et al., 1993; Spoolder and Vermeer, 2015). After social hierarchy has been established, aggressions usually decrease dramatically (Bokma et al., 1984; Luescher et al., 1990). Therefore, it is advisable that sows are mixed before they become pregnant and that sows know each other before mixing. Best practice would be to rear gilts together from an early age on and to keep familiar groups of gilts/sows in stable groups. This ensured that they can establish a social hierarchy long before they are inseminated for the first time. After lactation, it would be good to keep the sows in the same group and keep the group as stable as possible. Further, mixing of familiar sows should occur right after weaning. Most important is that the rank is established before fertilization takes place. This decreases the risk related to aggression during the embryonic period.

Another benefit of mixing sows right after weaning and before insemination is that social contacts between females are known to advance and induce the onset of oestrus, alleviating oestrus detection (Eliasson-Selling et al., 2000; Peltoniemi et al., 2016). Group-housed sows at AI can show oestrus freely compared to sows restricted to stalls (Eliasson-Selling, 2000) which may improve the possibilities for observing oestrus behaviour and reduce the risk of inseminating a sow that is not in standing oestrus (Einarsson et al., 2014). However, there is some evidence to suggest that social interactions may suppress oestrus signs in subordinate sows (Tsuma et al., 1996). It is advisable to separate individual subordinate sows or a small group of sows from the rest of the group for heat detection and artificial insemination. This can be done, that is, by taking the sow/small group of sows out of the pen and moving them to the boar.

Further, it should be noted that social interaction during the establishment of a dominance order (fighting) and/or during the exhibition of oestrus behaviour (mating) are temporary behaviours of sows and require more space per sow as compared to behaviour expressed by pregnant sows (Spoolder and Vermeer, 2015). All animals need space to rest, eat, defecate and explore the environment (static and behavioural space) but also to perform social behaviours (interaction space) (Spoolder and Vermeer, 2015). Grouping sows in a situation where there is insufficient space will increase aggressive behaviours and induce chronic stress (Barnett et al., 2001; Remience et al., 2008). The housing system should therefore be designed in a way that allows sows to show proper social behaviour, and that gives subordinate sows a possibility to escape from confrontation with another sow. It is also important to make sure that all sows can eat undisturbed. Regarding space allowance, a number of studies suggest that as a general rule sows should be provided with 2.5–3.5 m^2 per sow in order to reach a reasonably high farrowing rate (Kemp and Soede, 2012). In addition, feeding enough bulk and fibre and providing the pen with areas to escape help to alleviate aggressive behaviour problems of housing (Peltoniemi et al., 1999). Furthermore, it is regularly reported that having a teaser boar in the pregnant sow group calms the group down and reduces aggression among sows (Peltoniemi et al., 2016). For the group size, small groups may be more natural for sows. However, large dynamic groups of up to 300 sows may provide the farmer with satisfactory fertility results given that there is enough space in the group and if they are otherwise well managed.

Floor type in a pregnant sow unit appears as an important determinant of health (reviewed by Maes et al., 2016). Use of unbedded, slatted floors increases the incidence of lameness in group housing, probably due to claw trauma/infections rather than joint problems (Maes et al., 2016). Furthermore, in case of use of slats, use of concrete instead of plastic building material seems as an increased risk factor for lameness; however, plastic slats can be slippery. Slip-resistance, abrasiveness, surface profile and hardness are regarded as four important floor quality factors contributing to injuries of legs and claws (Maes et al., 2016).

5.2 Seasonal infertility

Seasonal infertility is defined as decreased fertility of the gilt and sow in summer and autumn. The signs of subfertility include: reduced pregnancy rate, prolonged weaning to oestrus interval (lower rate of oestrus expression) and delayed puberty of gilts (Fig. 1). In feral wild pigs, decreasing day length in late summer and autumn would provide a physiological cue, to indicate that it is not an optimum time for breeding, since the piglets would be born in the mid-winter. Thus in nature that would most likely mean death of the newborn piglets and loss of the whole litter. The feral wild pigs are seasonally anoestrous, with the onset of the anoestrous period being under the control of the photoperiod (Tast et al., 2001). However, the end of the anoestrus period and return to cyclical ovarian activity is dependent upon other environmental cues like the increased availability of food (Mauget, 1982). In the domesticated environment, cues such as increased availability of food and, in-housed animals, a comfortable microclimate are absent, and these cues for seasonal breeding are removed.

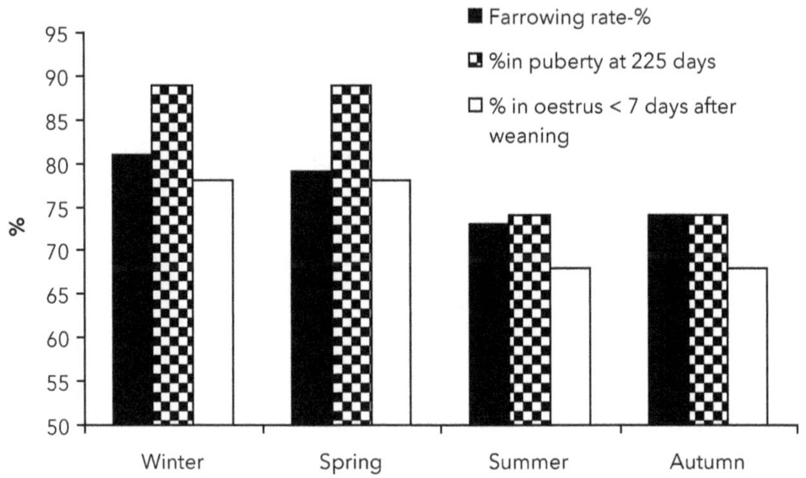

Figure 1 Major manifestations of seasonal infertility in the pig presented as percentages for each reproductive parameter. Data from Peltoniemi and Virolainen (2006).

Therefore, variable responses to decreasing daylight in the breeding activities of the female domestic pig population are frequently observed in late summer and autumn.

Seasonal infertility may be avoided by improving the reproductive management of the herd. This includes the use of a relevant light programme in which 16 h of light and 8 h of dark are provided in the breeding unit. This recommendation is combined with advice to explore and, if needed, increase the light intensity to a level of at least 200 lux at the level of the eyes of the sow. Feeding of pregnant, group-housed females during late summer and early autumn is also important. Increasing feed availability at this time is an effective method to decrease the stress of sows around the time of implantation of embryos and thereby alleviate early embryonic loss and return to oestrus. Moreover, the stress of gilts and sows prior to insemination and during pregnancy is reduced by grouping them correctly according to size and age. Furthermore, genetic selection, reduced stocking density in summer-autumn, cooling systems applied especially during lactation in a hot production environment and gonadotropin treatments may also be found useful in attempts to reduce seasonal effects on fertility (Peltoniemi and Virolainen, 2006).

Much of the reduction in farrowing rate in summer-autumn has been attributed to a disruption of early pregnancy rather than to conception failure (Love et al., 1993; Peltoniemi et al., 2000). These sows were mated, they conceived apparently well and pregnancy was established. The embryos were present for some time, but quite early on, pregnancy was interrupted and the sows returned to oestrus after prolonged oestrus to oestrus interval (Tast et al., 2002).

The diagnosis regarding seasonal disruption of pregnancy may be challenging since there is no diagnostic test for that. The diagnosis is by large based on the calendar, in other words, timing of the problem in late summer/ early autumn, and exclusion of other factors.

In the prevention of seasonal disruption of pregnancy, it is often beneficial to increase feeding for the duration of embryonic period of pregnancy (Peltoniemi et al., 1997). Other interventions that may prove beneficial include breaking down of the social group according to age and size, providing more space for the sows, including a boar in a group of pregnant sows, and taking care of lighting in the breeding unit (Peltoniemi and Virolainen, 2006; Auvigne et al., 2010).

5.3 Mycotoxins

Mycotoxins are fungal secondary metabolites produced by either field moulds (e.g. *Fusarium spp.*) or storage moulds (e.g. *Aspergillus spp.*) and *Penicillium spp.* (Eskola et al., 2020). In Europe, relevant mycotoxins are mainly derived from *Fusarium spp.* and include zearalenone (ZEA), trichothecenes (e.g. DON and T2) and Fumonisins (e.g. FB1 and FB2) (Eskola et al., 2020). Aflatoxins, a mycotoxin produced by the storage mould *Aspergillus spp.*, are usually abundant in warm and humid regions of the world, for example, in South America, Africa and South Asia (Eskola et al., 2020). Nevertheless, due to climate change, aflatoxins will also become one of the major mycotoxins gilts and sows are exposed to in Europe.

Mycotoxins can exert different effects depending on the production stage the gilts and sows are in. Exposure during rearing and sexual maturation can cause fertility problems, while exposure during gestation and lactation can also cause direct and indirect effects on foetal and neonatal development. Direct effects occur by transplacental and lactational transmission of mycotoxins from the dam to the foetuses and neonates (Dänicke et al., 2007; Goyarts et al., 2007; Sayyari et al., 2018; Benthem de Grave et al., 2021). Increased neonatal mortality, decreased neonatal vitality, reduced growth and impaired immune and organ functions are the results (Ensley and Radke, 2019; Kanora and Maes, 2010). Indirect effects occur through the impaired reproductive performance of exposed dams, for example, through reduced feed consumption, reduced colostrum and mild production, poor colostrum and milk quality, decreased maternal behaviour and increased restlessness and aggression (Ensley and Radke, 2019; Kanora and Maes, 2010). Thus, in gestating and lactating gilts and sows, mycotoxins can have various effects on the dams, their placentae and their foetuses and neonates depending on the type of mycotoxin, the dose and duration of exposure, and at what stage of gestation and lactation the exposure occurs.

The mechanisms of how mycotoxins exert their effects on sows' and gilts' reproduction as well as on foetal and neonatal development are not fully understood. It is known that mycotoxins have endocrine disruptive properties due to their oestrogenic activities interfering with the hypothalamic-pituitary-gonadal axis and impairing the blood-testes barrier in males, blood-milk barrier in dams and intestinal barrier in neonates (Ensley and Radke, 2019; Kanora and Maes, 2010). For instance, mycotoxins can lead to delayed maturity, pseudopregnancy, hyperoestrogenism syndrome, cystic ovarian degeneration, reduced litter size in female, delayed maturity and reduced testosterone, testis size and spermatogenesis in male. Further, they exert their toxicity at the molecular and genetic level (Ensley and Radke, 2019; Kanora and Maes, 2010). For instance, mycotoxins affect follicular growth, granulosa cell function and oocyte maturation in-vitro due to epigenetic modification, DNS damage and oxidative stress (Ensley and Radke, 2019; Kanora and Maes, 2010).

Mycotoxicosis is not easy to diagnose, but should be investigated whenever gilts, sows and boars show a reduction in their reproductive performance and piglets in their growth performance. Nevertheless, mycotoxicosis rarely causes a herd problem and a thorough anamnesis and history of the farm's reproductive management are important (Kanora and Maes, 2010). If mycotoxicosis is still suspected, the farmer needs to be advised on prevention, especially hygiene, and monitoring and control systems. Quality control programmes and regular check-ups of the quality of feed ingredients, including possible levels of mycotoxins present in the feedstuff are the base for prevention (Kanora and Maes, 2010). Further, advise on managing the critical conditions that facilitate the growth of fungi (e.g. dust, hygiene, temperature and moisture) in silos and feeders needs to be given and regular cleaning and disinfection needs to be encouraged (Kanora and Maes, 2010). Besides preventive measures, modern monitoring systems (Lawlor and Lynch, 2001; Berthiller et al., 2007) and other tools can be used for control. The other tools include toxin binders, acidifiers and dis-activators of the different mycotoxins present in feed and or feedstuffs (Dänicke et al., 2004; Völkl et al., 2004).

5.4 Lactation weight loss and metabolic condition at breeding

Management of sows' and gilts' metabolic conditions before breeding is important for their subsequent reproductive performance and piglets' growth performance after subsequent parturition (Soede and Kemp, 2015). For gilts, it is important that gilt development is optimized and that breeding occurs when gilts have reached an optimal body weight of 145–160 kg and an optimal body condition of 12–18 mm back-fat depth (Rozeboom, 2015). For sows, it is important that the energy and protein loss during previous lactation and the metabolic state at weaning are optimized because the following weaning-to-oestrus interval is too short to correct for a severe negative energy balance and to improve their subsequent reproductive performance (Kemp et al., 2018). The mechanism of action how the negative energy balance is affecting the next generation of growing follicles and thereby embryonic development appears IGF-1-dependent (Han et al., 2020, 2021).

During lactation, both the suckling intensity of piglets and the negative energy and protein balance inhibit follicles from growing. If lactation energy and protein loss are severe, follicle development is suppressed to such an extent that subsequent fertility of sows is impaired. This can result in negative effects on both the sows' reproductive performance and the piglet's growth performance. Sows can have a decrease in estrus rate, ovulation rate, pregnancy rate or litter size (Soede and Kemp, 2015). Lactation energy and protein loss are reflected in their body weight loss duration lactation. In multiparous sows, around 10–12% of body weight loss can be considered severe (Thaker and Bilkei, 2005), whereas in primiparous sows even bodyweight losses of less than 10% can be considered severe (Hoving et al., 2012). This is because young sows have higher energy requirement at breeding, not only for their maintenance and reproductive performance but also for their own body growth (Hoving et al., 2012). Further, primiparous sows have lower capabilities to meet these energy demands because they have a lower feed intake compared to older sows (Hoving et al., 2011).

The post-weaning follicular period is very short. Sows' metabolic condition needs to be optimized before weaning by preventing moderate (>5%) lactation weight loss in primiparous sows and severe (>10%) lactation weight loss in multiparous sows. This can be achieved by optimizing feed composition and intake during gestation, transition and lactation as well as by providing a stress-free and animal-friendly environment (Soede and Kemp, 2015). Another option is to reduce the suckling stimulus and thereby decrease the metabolic burden, for example, by split weaning or intermittent suckling (Soede and Kemp, 2015). Nevertheless, the benefit of decreasing the suckling stimulus on the metabolic state is rather marginal. The same applies to improving follicular development post-weaning by means of optimizing feed intake and composition during the weaning-to-oestrus interval because of its short duration. The only feasible management strategy would be to extend the weaning-to-oestrus interval by skipping a heat or by postponing the oestrus using a progesterone analogue for at least 10 days (Soede and Kemp, 2015). These management strategies should be considered in primiparous sows with moderate weight loss and in multiparous sows with severe weight.

Another possibility for those sows is to extend the lactation length up to 5–6 weeks. After the first 2–4 weeks of lactation, sows get into a positive energy balance as feed allowance reaches the maximum (Feyera and Theil, 2017). However, this positive energy balance in the remaining lactation may be too short to compensate for the body condition loss during lactation in those sows (Eissen et al., 2000). Allowing them 1–2 more weeks of lactation with maximum feed allowance may get them into better metabolic shape at weaning. Extending the lactation length can be combined with reducing the metabolic burden, for example, split weaning and intermittent suckling. If possible, special diets, for example, as dietary supplements/top-dressings, that stimulate IGF-1 production could be provided during this additional time (Han et al., 2020, 2021). Other factors that can affect post-weaning follicular development and subsequent reproductive performance are stress, social interactions between sows, boar contact, quality of oestrus detection and timing of insemination (Soede and Kemp, 2015).

6 Parturition

6.1 Physiology of farrowing

Physiology of farrowing is very complex, and several hormones act and interact to regulate the farrowing process. In the latest stages of pregnancy, progesterone, LH, oestrogens, cortisol, prolactin, relaxin and prostaglandins become the main actors that regulate all the physical events leading to parturition. All these hormones, regulated by various internal clocks, interact and influence each other in a very refined manner (Anderson, 2000). Behavioural expression is also very important for the farrowing process. Thorsen et al., 2017 Sows express an innate hormone-driven nest-building behaviour that signals the beginning of parturition (Algers and Uvnäs-Moberg, 2007). However, the environment can also have an important, though indirect, influence on the pattern of these hormones. A restricted environment can also prevent this nest-building behaviour, especially in the modern swine production system (Damm et al., 2003). This refined, delicate hormonal process clearly requires very well-defined premises in order to develop properly. In addition to housing, nutrition and disease may also affect, either directly or indirectly, the release or activity of farrowing hormones that interfere with the physiological process.

6.2 Hormones in pregnancy and parturition

During early pregnancy, after implantation of the embryos and the development activity of the corpora lutea, a higher and constant level of progesterone dominates the hormonal pattern (Meulen et al., 1988). This preserves the progress of pregnancy, and for almost two-thirds of its length, the level of progesterone circulating in the blood will remain high. After the first trimester of pregnancy, most of the other reproductive hormones (oxytocin, prostaglandins, relaxin and prolactin) show very basal levels or very little pulsating activity (Fig. 2A) for the remainder of the pregnancy. This very stable hormonal activity changes completely 48–24 h before the beginning of parturition. The quick drop in progesterone concentration produces a cascade effect on almost all the other hormones, which throughout the pregnancy remained quite stable. The level of prostaglandin peaks, the oxytocin concentration increases and begins to exhibit high pulsating activity (Gilbert et al., 1994), the prolactin concentration gradually increases, and oestrogens, after peaking quickly, gradually drop to basal levels (Ellendorff et al., 1979; Kindahl et al., 1982; Anderson, 2000) (Fig. 2B).

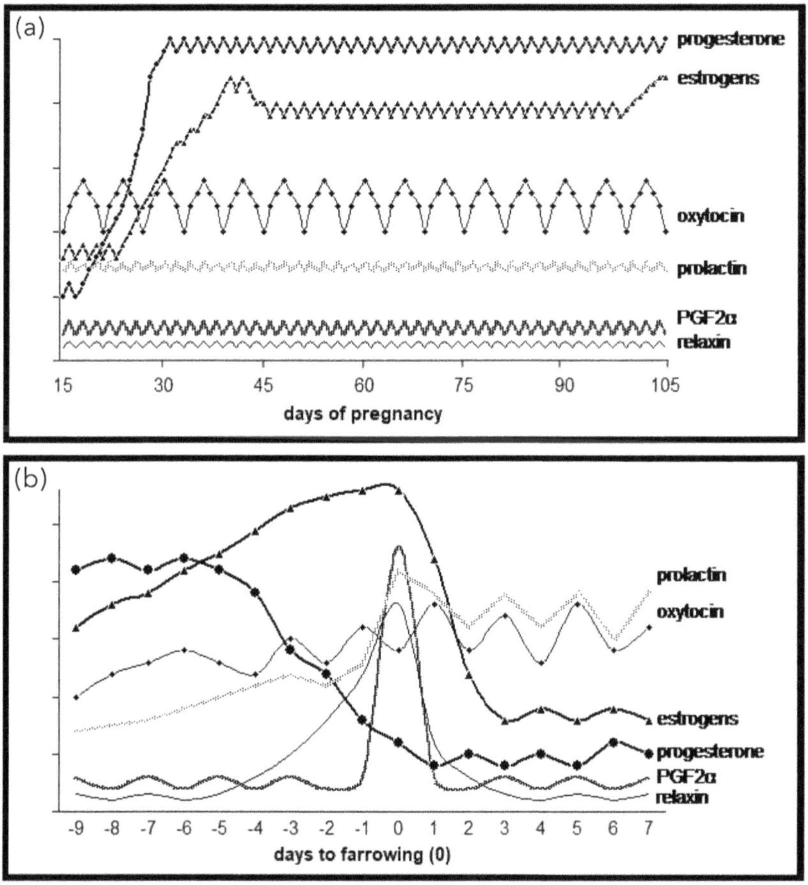

Figure 2 A schematic description of the level of reproductive hormones during pregnancy in the sow (modified from Anderson, 2000). Soon after early pregnancy, reproductive hormones show very stable levels for almost the entire pregnancy (a). Hormone levels undergo major changes only a few days before farrowing (b).

It is remarkable how such large hormonal changes occur in such a very limited amount of time.

6.3 Parturition behaviour and activity

Before and during farrowing, the intense hormonal activity described above is responsible not only for inducing the parturition event but also for triggering visible behavioural changes. Activities such as rooting, pawing, turning and walking increase considerably 24 h prior to farrowing (Hartsock and Barczewski, 1997; Oliviero et al., 2008b) and characterize the nest-building behaviour. One of the triggering factors of nest-building behaviour has been found to be a rise in prolactin (Castren et al., 1994), induced by a decrease in progesterone and an increase in prostaglandin (Algers and Uvnäs-Moberg, 2007). The external influence of the environment is also important for the expression of nest-building behaviour. The availability of proper nest-building material seems to speed up this process (Damm et al., 2000).

Most housing systems are based on the confinement of the farrowing sow in crates, where the sow has very limited movement and where bedding or any other nest-building substrate is often absent. In these particular conditions, the nest-building behaviour triggered by endogenous hormonal activity cannot find proper expression. In the absence of a nest-building substrate, confined sows express prolonged and unsuccessful nest-building behaviour (Damm et al., 2003). The lack of opportunity to express appropriate nest-building behaviour can lead to an increase in cortisol and ACTH (Jarvis et al., 1997), which indicates a stressful condition. Gustafsson et al. (1999) found that domestic sows were able to build nests identical to those of wild boars, even after several previous farrowing experiences in confined crates without bedding. This innate behaviour is therefore a clear indicator of impending farrowing and occurs independently of the housing or the bedding material available.

The benefits for sows farrowing in an open crate are significant and show how the welfare of the animals often correlates positively with economic benefit, as the lower number of stillborn piglets demonstrates (Oliviero, 2010). Adding roughage some days before farrowing would provide the sow with a substrate to better express its nest-building behaviour (Yun et al., 2014), and in the case of straw, a possible source of fibre which can alleviate the state of constipation that arises around farrowing (Oliviero et al., 2009).

6.4 Duration of farrowing and possible complications

As discussed in the introduction, the duration of farrowing has profoundly increased while the litter size has doubled over the past three decades (Fig. 3).

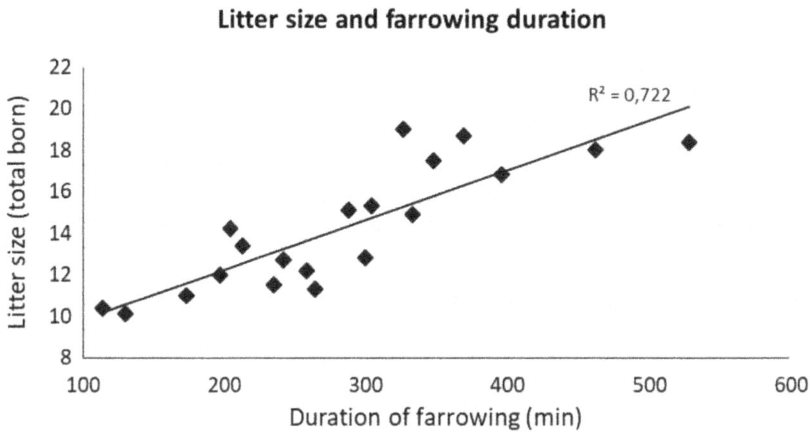

Figure 3 Relationship between litter size and the average duration of farrowing in 20 studies from 1992 to 2018 (adapted from Oliviero et al., 2019, Reproduction in Domestic Animals, Wiley-Blackwell).

The increasing litter size presents an immunological challenge for the sow and especially the piglets (Oliviero et al., 2013, 2019). The last 20–30% of the foetuses to be born likely miss out on access to good-quality colostrum that declines by 50% already by the sixth hour after the birth of the first piglet (Le Dividich et al., 2017). On the other hand, they also have less time to suckle colostrum due to decreased window of opportunity for colostrum intake, increased competition for teats and reduced birth weight. This all may show up later in the emergence of diseases in the growing phase of piglets/fattening pigs.

The metabolic challenge related to hyper-prolific sow production model begins in the growing phase of gilts and goes beyond farrowing and lactation. The sow is supposed to eat enough to meet the requirement of growing litters prior to farrowing, which may cause some of the problems seen around farrowing (Oliviero et al., 2008a, 2009). In the early part of lactation, sows with large litters loose more energy while producing milk than what they can consume, ending up in a negative energy balance (NEB) (Hoving et al., 2012; Costermans et al., 2020).

The growing litter size and intensity of production have triggered welfare concerns to the public. This seems to happen regardless of whether those concerns would be warranted or not.

7 Low state systemic inflammation involved in parturition and postpartum dysgalactia syndrome

Nest building and the phases of farrowing are orchestrated by responding to changes in reproductive hormones. It is well established that a decline in progesterone and peak in prostaglandin F2alpha triggers nest-building behaviour while oxytocin rise at the beginning of expulsion phase marks the session of nest building (Algers and Uvnäs-Moberg, 2007). Prostaglandin F2alpha peak also induces CL regression with a concomitant decline in progesterone, making uterine contractions and parturition possible. Oxytocin is mainly in charge of uterine contractions during the expulsion phase of parturition and letdown of colostrum and milk, while prolactin will promote mammary gland development to the extent that initiation of milk production after parturition will become possible (Taverne and van der Weijden, 2008; Farmer and Quesnel, 2009; Farmer, 2016).

It has also been described in the literature and also shown by our group that allowing the sow to build up a nest prior to farrowing will increase oxytocin release and shorten the duration of farrowing (Castren et al., 1993; Oliviero et al., 2008a). Other ways of shortening the duration of farrowing include increasing fibre in the feedstuff and encouraging water intake (Oliviero et al., 2008a, 2010). However, even after applying the best management interventions prior to farrowing, the duration of farrowing of modern hyper-prolific sows is extended four to five hold as described (Kaiser et al., 2018a; Oliviero et al., 2019; Yun et al., 2019).

Prolonged farrowing increases the risk for reduced quality and quantity of colostrum intake, intrapartum hypoxia of foetuses (Peltoniemi et al., 2020), retained placentae (Björkman et al., 2017b), uterine inflammation and PDS (Björkman et al., 2018b) and, likely, reduced development of next generation of follicles fertility (Oliviero et al., 2013; Peltoniemi et al., 2020).

Moreover, during the periparturient period, biological mechanisms coordinate the mobilization of body reserves in order to support foetal growth and milk production; insulin concentrations are reduced, and the response of hormone-sensitive lipase in adipose tissue (e.g. low insulin, high growth hormone and catecholamines, or high glucocorticoid concentrations) is greater to facilitate lipid mobilization. This periparturient period is also characterized by a low state of inflammation encompassing an increase in hepatic production of positive acute-phase proteins (APP) and a decrease in the production of negative APP (Petersen et al., 2004; Kaiser et al., 2018a). The evolving low state periparturient inflammation may be underway to be more profound in the hyper-prolific sows as litter size increases (Kaiser et al., 2018a). Also, the use of non-steroidal anti-inflammatory drugs (NSAIDs) seems useful in the control of periparturient inflammatory process like PDS (Schoos et al., 2020). However, this may be elucidated in further studies in the future.

It has been rather well described in the literature that these responses are mediated by the pro-inflammatory cytokines interleukin (IL)-6, IL-1β, and tumour necrosis factor-α (TNF-α) (Kaiser et al., 2018a). Additionally, evidence in the dairy cow indicates that oxidative stress also occurs during this period and is driven by the imbalance between the production of reactive oxygen metabolites (ROM), reactive nitrogen species (RNS) and the neutralizing capacity of antioxidant mechanisms in tissues and blood (Coleman et al., 2020).

The extent and duration of the inflammatory process will determine whether or not the condition is ending up as a clinical disease. However, it is noteworthy that in the hyper-prolific sows lines as those typical of Denmark and Belgium, the incidence of sows contracting a systemic disease postpartum is as high as >30% (Larsen and Thorup, 2006; Papadopoulos et al., 2010). Moreover, it is obvious that even in those sows staying healthy as far as clinical symptoms, the inflammatory process is heavily present as indicated by means of those markers described above (Kaiser et al., 2018a).

Typically, within 2–3 days postpartum, the process of inflammation may develop into endotoxemia, which involves the release of the inflammation markers described. Endotoxemia is associated with clinical symptoms indicating a systemic response to infectious agents such as coliform bacteria – and PDS (Bäckström et al., 1984; Peltoniemi et al., 2016; Kemper, 2020). The condition comes with acute general symptoms such as inappetite, lethargy and fever (Peltoniemi et al., 2016), followed by local symptoms that usually affect either the uterus (Björkman et al., 2018b) or the udder (Farmer et al., 2019) or both of them.

After parturition, concomitant with the process of inflammation, the sow undergoes metabolic stress due to loss of body reserves in favour of milk produced for large litters. This change rate is highest during the first 10 days of lactation. One of the major mediators of metabolic stress is IGF-1, which is also seen as an indicator of fertility. Low IGF-1 levels are associated with the increment of inflammation, presence of metabolic stress and reduced fertility. IGF-1 is also regarded as one of the most important factors driving follicle development (Han et al., 2019, 2020). The role of extracellular vesicles, although proposed as being key players in follicle development and the cross-talk between the mother and the embryo, in this inflammatory process and its effect on follicle development, however, remains less explored (Almiñana et al., 2017).

In summary, in hyper-prolific sows, the physiological process of farrowing is prolonged, making the system vulnerable in terms of increased rate of inflammation and emerging infectious uterine and mammary disease. In fact, recent evidence shows that even in sows staying without symptoms, there seems to be a considerable degree of 'silent inflammation' in the body. In an increased proportion of sows, however, postpartum disease PDS is detected and hopefully treated. The consequences of inflammation, regardless of clinical symptoms, include reduced quantity and quality of piglet colostrum intake and milk intake during early lactation.

8 Transfer of immunity

At birth, piglets have no circulating immunoglobulin from the mother, due to the epitheliochorial placenta, not allowing transfer of such large-sized molecules during the foetal period. Therefore, piglets need to acquire immunoglobulin passively from colostrum to gain adequate immune protection, before they will be able to start producing their own immunoglobulin from 3–4 weeks of age onwards (Rooke and Bland, 2002; Alexopoulos et al., 2018).

8.1 Emergence of large litters creates a challenge for transfer of immunity

As discussed previously, the average litter size has increased from 11 piglets to 14 piglets, with peaks of 16 piglets in some countries (Kemp et al., 2018; Baumgartner, 2020). With hyper-prolific sow lines, however, litters up to 18–20 piglets have become common (Björkman et al., 2017b; Kemp et al., 2018; Schoos et al., 2020). This poses a direct challenge during lactation, since most sows have an average of only 14–16 teats (Labroue et al., 2001). In a study with no litter balancing nor providing any kind of direct help to sow and piglets, it was found that a sow can wean successfully no more than 10–11 piglets, independent of how large the litter was (Andersen et al., 2011). Large litters also have a direct effect on the piglets' characteristics at birth. The larger the litter size, the lower is the piglets' average birthweight and the higher is the within litter weight variation (Quesnel et al., 2008; Beaulieu et al., 2010; Smit et al., 2013; Matheson et al., 2018). Competition for colostrum intake is increased with more piglets born than the available teats at the sow's udder, further increasing the weight variation within the litter (Vasdal and Andersen, 2012; Declerck et al., 2017). Consequently, being born in a large litter with a low birth weight decreases the piglet vitality, which can retard the access to the udder for adequate colostrum intake (Hoy et al., 1994; Islas-Fabila et al., 2018; Manjarin et al., 2018). Colostrum intake significantly influences piglets' performance as well as their survival, not only prior to weaning but also in the long term. As colostrum yield is reported to be independent of litter size, sufficient colostrum intake per piglet is crucial, especially in hyper-prolific sow (Declerck et al., 2016; Balzani et al, 2018).

Colostrum yield and its fat content are largely affected by different sow and litter factors. Pig producers should consider colostrum intake, yield and composition in their management to maximize production and reproduction potential (Declerck et al., 2015).

8.2 Transfer of local immunity in the gastrointestinal tract and gut closure

Sow's milk has a constant presence of secretory IgA (sIgA) which ensures good protection of the piglets' gut. This localized protection allows piglets to develop gradually their own immune response mechanisms, as long as they are able to suckle sufficient milk (Salmon et al., 2009). Other types of immunoglobulin, like IgG, are abundant in colostrum (Carbera et al., 2013; Theil et al., 2014). Sow colostrum has a high content of IgG (30–70g/l) and other bioactive compounds like growth factors and enzymes (Oliviero et al., 2019).

One limiting factor for piglets on the absorption of IgG from colostrum is the closure of the tight junctions in the gut occurring 24–36 h after birth, making piglets unable to absorb such large molecules. For this reason, the main cause of death in newborn piglets is the impossibility to obtain a sufficient amount of colostrum (Quesnel et al., 2012). A minimum of 250 g of colostrum is considered the minimum requirement to reduce mortality and allow proper growth (Quesnel et al., 2012).

8.3 Colostrum quantity and challenges due to competition for functional teats

Another limiting factor is the ability of the sow to produce adequate amount of colostrum to sufficiently supply large litters. According to Lessard et al. (2018) birth weight and colostrum intake can influence the genes' expression of immunity and oxidative stress in piglets' intestinal tissue. Low birth weight piglets had less intestinal antigen-presenting cells and altered development of B cells, when compared to piglets with higher birth weight (Lessard et al., 2018).

In large litters, immunity is also affected by social stress conditions (competition for colostrum and milk intake, crowding and regrouping). Psychosocial stress may alter both innate and adaptive immune responses, such as leukocyte distribution, cytokine secretion, lymphocyte proliferation, antibody production and immune responses to viral infection or vaccination (Gimsa et al., 2018). Additionally, social stress may lead to increased cortisol levels, promoting dysregulation of inflammatory processes and glucocorticoid resistance of lymphocytes and therefore predispose to gastrointestinal diseases (Gimsa et al., 2018).

There is an association between large litter size and increased pre-weaning mortality (Baxter et al., 2013; Rutherford et al., 2013), of which one visual example is shown in Fig. 4. Prolonged farrowing duration and low birth weight seen in large litters can explain the causes of this correlation (Oliviero et al., 2019).

8.4 Problems evolving due to prolonged process of farrowing

Prolonged parturition due to large litters can affect the level of absorbed IgG in piglet plasma with a linear decrease of 0.4 g/L for each piglet born (Kielland

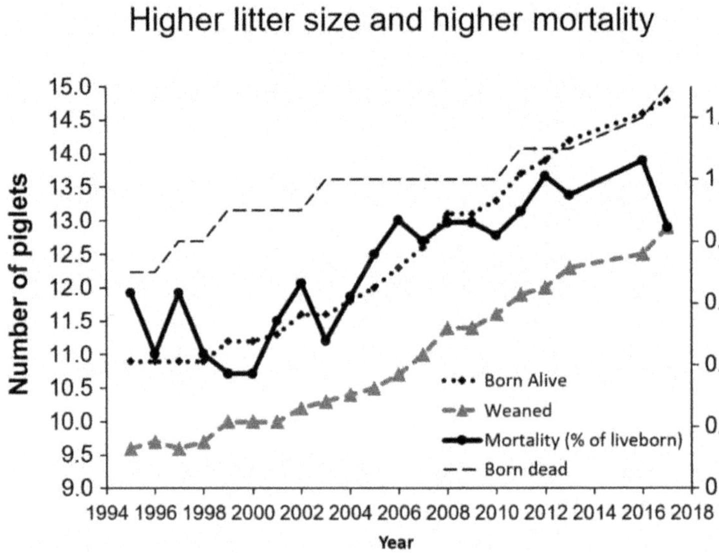

Figure 4 Increased mortality with increased litter size in the Netherlands (adapted from AgroVision B.V. The Netherlands, 2017).

et al., 2015). Many studies report a negative correlation between litter size and piglet birth weight (Quesnel et al., 2008; Beaulieu et al., 2010; Kapell et al., 2011; Smit et al., 2013). A decline in average birth weight relates to large litters with 35–43 g lower birth weight for each additional pig born (Quiniou et al., 2002; Beaulieu et al., 2010; Smit et al., 2013). The lower birth weight directly reduces colostrum intake, increasing the risk of mortality (Ferrari et al., 2014; Declerck et al., 2016; Le Dividich et al., 2017).

Piglets born with higher birth weight had higher serum IgG concentrations (Nguyen et al., 2013). Similarly, higher colostrum intake at birth increased the IgG content in serum of piglets at 24 h after birth (Moreira et al., 2017) and also at 10 and 20 days of age (Ferrari et al., 2014).

Immunoglobulin content in colostrum declines rapidly after the first piglets are born (Le Dividich et al., 2017), therefore prolonged duration of farrowing and the higher competition in large litters can reduce not only the possibility to suckle sufficient colostrum but also retard the time of access to the udder. Delaying the intake of colostrum affected negatively the piglets' immunoglobulin absorption and the maturation of their intestinal villi, with possible long-term detrimental effects on the digestion process (Cabrera et al., 2012). Due to the rapid decrease of IgG content in colostrum, Klobasa et al. (1987) found that birth order affected the amount of immunoglobulin absorbed by piglets. The latest piglets born in the litters had the lower plasma IgG level. Correspondingly, a 4% decrease in plasma IgG concentration was found in piglets of smaller birth weight (Cabrera et al., 2012). It is clear that the farrowing-to-suckling interval is fundamental in the acquisition of adequate IgG by piglets. After the start of parturition, a delay of 4 h in up taking of colostrum reduced the amount of plasma proteins from 24 h up to 12 days in piglets. It is therefore extremely important to consider also the time of birth of piglets in relation to the start of farrowing, when planning successful strategies to boost colostrum intake in large litters, like for instance split suckling (Oliviero et al., 2019).

9 Microbiota involvement during pregnancy, parturition and lactation

The gut microbiota has a constant shift over time, and in sows, both diversity and abundance of gut microbial population increase from pregnancy until the end of lactation (Ji et al., 2019). A more diverse gut microbiota can provide a more complete metabolism and functionality in sows, allowing a sufficient supply of nutrients for foetal growth and development (Ji et al., 2019). A recent study found that at farrowing, the most predominant phyla were Firmicutes, Bacteroidetes, Proteobacteria, Actinobacteria and Candidatus (Hasan et al., 2018a). The Firmicutes represent the most abundant proportion of the total population, followed by Bacteroides.

These two phyla accounted for approximately 98% of all bacteria present. These results are in line with the one published by Kim et al. (2011). However, Bacteroides increase linearly from the start to the end of pregnancy (Ji et al., 2019). Jost et al. (2014) reported that Firmicutes exhibited no detectable changes over perinatal period.

It seems that in sows, the increase in body weight or in the back-fat thickness in the different production phases is associated with the higher abundance of Firmicutes or the Firmicutes to Bacteroides ratio (Kim et al., 2011; Ji et al., 2019). In terms of phyla, the abundance of Tenericutes, Fibrobacteres and Cyanobacteria has been shown to increase with the progression of the pregnancy (Ji et al., 2019). Tenericutes have been connected with better intestinal cells' integrity and Fibrobacteres by having the potential to metabolize non-soluble polysaccharides, such as cellulose, hemicellulose or pectin (Ji et al., 2019).

In sows at late gestation, Romboutsia was the dominant genus, followed by Clostridium *sensu stricto*, Lactobacillus, Oscillibacter, Intestinimonas, Sporobacter, Christensenella, Barnesiella, Flavonifractor, Terrisporobacter, Acidaminobacter, Lachnospiracea incertae sedis and Turicibacter, with other genera being much less than 1% (Hasan et al., 2018a).

Another important factor responsible to modulate gut microbiota composition is diet, possibly having functional effects. Functional ingredients added to the diet, like hydrolysed yeast (Hasan et al., 2018a) and resin acid-enriched composition (Hasan et al., 2018b), probiotics (Menegat et al., 2019) and prebiotics (Tan et al., 2016; Li et al., 2020), showed the ability to modify the microbial gut population in sows. Other compounds of the diet like protein and fibre can modulate the gut microbial population both in gestating sows and in weaning piglets. Fibres can be classified as insoluble (like whole-wheat flour and wheat bran) and soluble (oats and barley). Fibre showed the most interesting properties when provided during late pregnancy, reducing Proteobacteria and increasing Ruminococcaceae, Oscillospira and Eubacterium (Tan et al., 2016; Li et al., 2020).

After dietary soluble fibre supplementation in late pregnancy sows, the genus Eubacterium increases, promoting propionate release (Xu et al., 2020). These microbiota are capable to ferment indigestible carbohydrates, producing short-chain fatty acids (SCFA) that can be an important energy source for the sow. Butyrate is considered to have gut health-promoting features, acting as the main energy source for colonocytes and having anti-inflammatory properties at gut level (Sassone-Corsi and Raffatellu, 2015).

With an increased production of SCFAs, there is more energy available at intestinal level, contributing to the high energetic demands of hyper-prolific sows. Therefore, in gestating sows, it could be favourable to promote fibre degrading microbiota in the gut. When dietary supplementation can modify substrate availability and gut physiological processes (fermentation, pH), due to the increase of beneficial bacteria, this also implies a reduction in pathogenic bacteria (Liu et al., 2008). After supplementation of yeast hydrolysate to sows during pregnancy, beneficial and fermentative bacteria (Roseburia, Paraprevotella and Eubacterium) increased, while some opportunistic pathogens like Desulfovibrio, Escherichia/Shigella and Helicobacter, belonging to the phylum Proteobacteria, were reduced (Hasan et al., 2018a).

Proteobacteria are usually not very abundant within a normal gut microbial community of adult animals (Eckburg et al., 2005). However, their expansion may indicate a dysbiosis in connection with gut inflammation (Shin et al., 2015; Litvak et al., 2017), like irritable bowel syndrome or inflammatory bowel disease (Morgan et al., 2012), and with increased inflammatory responses of women in late pregnancy (Koren et al., 2012).

For instance, Hasan et al. (2018a) found that some positive sow's productive and physiological performances (high colostrum yield, high colostrum proteins content, high colostrum IgG content, normal blood progesterone level and normal farrowing duration) were positively correlated to the gut bacterial families *Lactobacillaceae, Ruminococcaceae and Prevotellaceae*, the last two being bacteria capable to metabolize indigestible polysaccharides from dietary fibre.

Conversely, unfavourable productive and physiological performances of the sow (low colostrum yield, low colostrum proteins content, low colostrum

IgG content, high level of blood progesterone and long farrowing duration) were positively correlated with the gut bacterial families *Erysipelotrichaceae, Clostridiaceae, Streptococcaceae, Enterobacteriaceae, Desulfovibrionaceae and Bacteroidaceae*, some of these being pig pathogens bacteria or part of the phylum Proteobacteria.

10 Conclusion and clinical implications

Health challenges during pregnancy and parturition in the pig include a variety of internal and external factors that threaten the general health of the gilt and sow on the one hand and specific reproductive processes on the other hand. Physical quantities of the production environment, including spacing, flooring, feeding systems and the use of bedding materials provide the gilt and sow with a variety of physical environments that by large define possibilities for a healthy life. Internal and external farm and biosecurity determine how vulnerable gilts and sows are for pathogens threatening pregnancy and parturition. Season and chronic social stress in sow groups mainly disrupt pregnancies during the first half of pregnancy while abortions occurring during the last trimester of pregnancy more often point to a viral or bacterial cause. We conclude that breeding for ever-increasing litter size has prolonged the duration of farrowing by four to five fold over the past three decades, creating problems for transfer of immunity and thereby the health of the piglets. Mechanisms involved are numerous, including at least hormonal, behavioural and immunological alterations to what has been considered as normal and healthy. Transfer of immunity is affected by changes in microbiota that seem to be coupled with management of parturition and carry over to the general health of growing piglets. Breeding for resilience regarding health and welfare has been recommended for future studies and increased resilience against specific pathogens, and environmental changes may indeed pay off in the long term.

11 Where to look for further information

For online search engines, we recommend using keywords included in the subtitles of the present chapter. Key text books covering the areas discussed would include Diseases of Swine (the 11th edition, 2019, i.e. Neumann et al. 2019 in the list of references), Veterinary Reproduction and Obstetrics (the 10th edition, 2019, edited by Noakes, Parkinson and England, Elsevier) and The Gestating and Lactating sow (edited by Farmer, 2015) by Wageningen Academic Publishers. Regarding seasonal infertility, we recommend a case study in a Spanish herd, available online at: https://www.pig333.com/articles/diagnostic-darkness-recurring-fertility-decline-in-summer_14574/.

12 References

Alarcon, L. V., Allepuz, A. and Mateu, E. (2021). Biosecurity in pig farms: A review. *Porcine Health Management* 7(1):5. DOI: 10.1186/s40813-020-00181-z.

Alexopoulos, J. G., Lines, D. S., Hallett, S. and Plush, K. J. (2018). A review of success factors for piglet fostering in lactation. *Animals: An Open Access Journal From MDPI* 8(3):38. DOI: 10.3390/ani8030038.

Algers, B. and Uvnäs-Moberg, K. (2007). Maternal behavior in pigs. *Hormones and Behavior* 52(1):78–85.

Almiñana, C., Corbin, E., Tsikis, G., Alcântara-Neto, A. S., Labas, V., Reynaud, K., Galio, L., Uzbekov, R., Garanina, A. S., Druart, X. and Mermillod, P. (2017). Oviduct extracellular vesicles protein content and their role during oviduct-embryo cross-talk. *Reproduction* 154(3):153–168. DOI: 10.1530/REP-17-0054.

Almond, G. W., Flowers, W. L., Batista, L. and D'Allaire, S. (2006). Diseases of the reproductive system. In: Straw, B., Zimmermann, J. J., D'Allaire, S. and Taylor, D. J. (Eds) *Diseases of Swine* (9th edn.). Ames, IA: Blackwell Publishing, pp. 113–147.

Althouse, G. C., Kauffold, J. and Rossow, S. (2019). Diseases of the reproductive system. *Diseases of Swine*, pp. 373–392.

Andersen, I. L., Nevdal, E. and Bøe, K. E. (2011). Maternal investment, sibling competition, and offspring survival with increasing litter size and parity in pigs (Sus scrofa). *Behavioral Ecology and Sociobiology* 65(6):1159–1167.

Anderson, L. L. (2000). Reproductive cycle of pigs. In: Hafez, E. S. E. and Hafez, B. (Eds) *Reproduction in Farmanimals* (VII edn.). Lippincott, pp. 189–190.

Arent, Z., Frizzell, C., Gilmore, C., Allen, A. and Ellis, W. A. (2016). Leptospira interrogans serovars Bratislava and Muenchen animal infections: Implications for epidemiology and control. *Veterinary Microbiology* 190:19–26.

Arent, Z. J. and Ellis, W. A. (2019). Leptospirosis. In: Zimmerman, J. J., Karriker, L. A., Ramirez, A., Schwartz, K. J., Stevenson, G. W. and Zhang, J. (Eds) *Diseases of Swine* (11th edn.). Glasgow, Great Britain: Wiley Blackwell, pp. 854–862.

Auvigne, V., Leneveu, P., Jehannin, C., Peltoniemi, O. and Sallé, E. (2010). Seasonal infertility in sows: An observational field study to analyze relative roles of heat stress and photoperiod. *Theriogenology* 74(1):60–66.

Bäckström, L., Morkoc, A. C., Connor, J., Larson, R. and Price, W. (1984). Clinical study of mastitis-metritis-agalactia in sows in Illinois. *Journal of the American Veterinary Medical Association* 185(1):70–73.

Baer, C. and Bilkei, G. (2005). Ultrasonographic and gross pathological findings in the mammary glands of weaned sows having suffered recidiving mastitis metritis agalactia. *Reproduction in Domestic Animals* 40(6):544–547. DOI: 10.1111/j.1439-0531.2005.00629.x.

Balzani, A., Cordell, H. J. and Edwards, S. A. (2016). Relationship of sow udder morphology with piglet suckling behavior and teat access. *Theriogenology* 86(8):1913–1920. DOI: 10.1016/j.theriogenology.2016.06.007.

Barnett, J. L., Hemsworth, P. H., Cronin, G. M., Jongman, E. C. and Hutson, G. D. (2001). A review of the welfare issues for sows and piglets in relation to housing. *Australian Journal of Agricultural Research* 52(1):1–28.

Baumgartner, J. (2020). Pig industry in CH, CZ, DE, DK, NL, NO, SE, UK, AT and EU [Internet]. Available at: https://www.vetmeduni.ac.at/fileadmin/v/tierhaltung/I-.

Baxter, E. M., Rutherford, K. M. D., D'Eath, R. B., Arnott, G., Turner, S. P., Sandøe, P., Moustsen, V., Thorup, F., Edwards, S. A. and Lawrence, A. B. (2013). The welfare implications of large litter size in the domestic pig II: Management factors. *Animal Welfare* 22(2):219-238.

Beaulieu, A. D., Aalhus, J. L., Williams, N. H. and Patience, J. F. (2010). Impact of piglet birth weight, birth order, and litter size on subsequent growth performance, carcass quality, muscle composition, and eating quality of pork. *Journal of Animal Science* 88(8):2767-2778.

Becker, H. A., Kurtz, R. and Mickwitz, G. (1985). Chronische Harnwegsinfektionen beim Schwein, Diagnose und Therapie. *Praktische Tierarzt* 12:1006-1011.

Bellino, C., Gianella, P., Grattarola, C., Miniscalco, B., Tursi, M., Dondo, A., D'Angelo, A. and Cagnasso, A. (2013). Urinary tract infections in sows in Italy: Accuracy of urinalysis and urine culture against histological findings. *Veterinary Record* 172(7):183.

Benthem de Grave, X., Saltzmann, J., Laurain, J., Rodriguez, M. A., Molist, F., Dänicke, S. and Santos, R. R. (2021). Transmission of zearalenone, deoxynivalenol, and their derivatives from sows to piglets during lactation. *Toxins* 13(1):37.

Berner, H. (1981). Untersuchungen zum Vorkommen von Harnwegsinfektionen beim Schwein, 1. Mitteilung: Harnwegsinfektionen bei Muttersauen in Ferkelerzeugerbetrieben. *Tierärztl. Umsch.* 36:162-171.

Berthiller, F., Sulyok, M., Krska, R. and Schuhmacher, R. (2007). Chromatographic methods for the simultaneous determination of mycotoxins and their conjugates in cereals. *International Journal of Food Microbiology* 119(1-2):33-37.

Biksi, I., Takacs, N., Vetesi, F., Fodor, L., Szenci, O. and Fenyo, E. (2002). Association between endometritis and urocystitis in culled sows. *Acta Veterinaria Hungarica* 50(4):413-423.

Björkman, S. and Grahofer, A. (2020). Tools and protocols for managing hyperprolific sows at parturition: Optimizingpiglet survival and sows' reproductive health. In: *Animal Reproduction in Veterinary Medicine*. London: IntechOpen. DOI: 10.5772/intechopen.91337.

Björkman, S., Oliviero, C., Kauffold, J., Soede, N. M. and Peltoniemi, O. A. T. (2018a). Prolonged parturition and impaired placenta expulsion increase the risk of postpartum metritis and delay uterine involution in sows. *Theriogenology* 106:87-92.

Björkman, S., Grahofer, A., Han, T., Oliviero, C. and Peltoniemi, O. A. T. (2018b). Severe udder edema as a cause of reduced colostrum quality and milk production in sows - A case report. In: 10th European Symposium *of* Porcine Health Management, Barcelona, Spain (vols. 9-11), pp. 110-111.

Björkman, S., Oliviero, C., Rajala-Schultz, P. J., Soede, N. M. and Peltoniemi, O. A. T. (2017a). The effect of litter size, parity and farrowing duration on placenta expulsion and retention in sows. *Theriogenology* 92:36-44.

Björkman, S., Oliviero, C., Hasan, S. M. K. and Peltoniemi, O. A. T. (2017b). Poster presentations. *Reproduction in Domestic Animals* 52:66-146. DOI: 10.1111/rda.13026.

Boers, A., Taylor, D. J. and Martineau, G.-P. (2019). Miscellaneous bacterial infections. In: Zimmerman, J. J., Karriker, L. A., Ramirez, A., Schwartz, K. J., Stevenson, G. W. and Zhang, J. (Eds) *Diseases of Swine* (11th edn.). Glasgow, Great Britain: Wiley Blackwell (vol. 2019), pp. 981-1001.

Bokma, S., Kersjes, G. J., Unshelm, J., Putten, G. V. and Zeeb, K. (1984). The introduction of pregnant sows in an established group. In: Unshelm, J. (Ed.) *Proceedings of the*

International Congress on Applied Ethology in Farm Animals, Skara, Sweden, pp. 166–169.

Broes, A., Taylor, D. J. and Martineau, G. P. (2019). Miscellaneous bacterial infections. *Diseases of Swine*, pp. 981–1001.

Brown,T. T., Jr., Paul, P. S. and Mengeling, W. L. (1980). Response of conventionally raised weanling pigs to experimental infection with a virulent strain of porcine parvovirus. *American Journalof Veterinary Research* 41(8):1221–1224.

Cabrera, R., Lin, X., Ashwell, M., Moeser, A. and Odle, J. (2013). Early postnatal kinetics of colostral immunoglobulin G absorption in fed and fasted piglets and developmental expression of the intestinal immunoglobulin G receptor. *Journal of Animal Science* 91(1):211–218.

Cabrera, R. A., Lin, X., Campbell, J. M., Moeser, A. J. and Odle, J. (2012). Influence of birth order, birth weight, colostrum and serum immunoglobulin G on neonatal piglet survival. *Journal of Animal Science and Biotechnology* 3(1):42.

Camenisch, U., Lu, Z. H., Vaughan, L., Corboz, L., Zimmermann, D. R., Wittenbrink, M. M., Pospischil, A. and Sydler, T. (2004). Diagnostic investigation into the role of Chlamydiae in cases of increased rates of return to oestrus in pigs. *Veterinary Record* 155(19):593–596. DOI: 10.1136/vr.155.19.593.

Castrén, H., Algers, B., de Passille, A.-M., Rushen, J. and Uvnas-Moberg, K. (1993). Preparturient variation in progesterone, prolactin, oxytocin and somatostatin in relation to nest-building in sows. *Applied Animal Behaviour Science* 38(2):91–102.

Castrén, H., Algers, B., De Passillé, A.-M., Rushen, J. and Uvnäs-Moberg, K. (1994). Nest-building in sows in relation to hormone release. *Applied Animal Behaviour Science* 40(1):74–75.

Chae, C. (2005). A review of porcine circovirus 2-associated syndromes and diseases. *Veterinary Journal* 169(3):326–336.

Coleman, D. N., Lopreiato, V., Alharthi, A. and Loor, J. J. (2020). Amino acids and the regulation of oxidative stress and immune function in dairy cattle. *Journal of Animal Science* 98(Suppl 1):S175–S193. DOI:10.1093/jas/skaa138.

Corzo, C. A., Mondaca, E., Wayne, S., Torremorell, M., Dee, S., Davies, P. and Morrison, R. B. (2010). Control and elimination of porcine reproductive and respiratory syndrome virus. *Virus Research* 154(1–2):185–192. DOI: 10.1016/j.virusres.2010.08.016.

Costermans, N. G. J., Teerds, K. J., Middelkoop, A., Roelen, B. A. J., Schoevers, E. J., van Tol, H. T. A., Laurenssen, B., Koopmanschap, R. E., Zhao, Y., Blokland, M., van Tricht, F., Zak, L., Keijer, J., Kemp, B. and Soede, N. M. (2020). Consequences of negative energy balance on follicular development and oocyte quality in primiparous sows. *Biology of Reproduction* 102(2):388–398. DOI: 10.1093/biolre/ioz175.

D'Allaire, S., Stein, T. E. and Leman, A. D. (1987). Culling patterns in selected Minnesota swine breeding herds. *Canadian Journalof Veterinary Research* 51(4):506–512.

Damm, B. I., Lisborg, L., Vestergaard, K. S. and Vanicek, J. (2003). Nest-building behavioural disturbances and heart rate in farrowing sows kept in crates and Schmid pens. *Livestock Production Science* 80(3):175–187.

Damm, B. I., Vestergaard, K. S., Schrøder-Petersen, D. L. and Ladewig, J. (2000). The effects of branches on prepartum nest-building in gilts with access to straw. *Applied Animal Behaviour Science* 69(2):113–124.

Dänicke, S., Brüssow, K. P., Goyarts, T., Valenta, H., Ueberschär, K. H. and Tiemann, U. (2007). On the transfer of the Fusarium toxins deoxynivalenol (DON) and zearalenone

(ZON) from the sow to the full-term piglet during the last third of gestation. *Food and Chemical Toxicology* 45(9):1565–1574.

Dänicke, S., Valenta, H., Döll, S., Ganter, M. and Flachowsky, G. (2004). On the effectiveness of a detoxifying agent in preventing fusario-toxicosis in fattening pigs. *Animal Feed Science and Technology* 114(1-4):141–157.

Declerck, I., Dewulf, J., Piepers, S., Decaluwé, R. and Maes, D. (2015). Sow and litter factors influencing colostrum yield and nutritional composition. *Journalof Animal Science* 93(3):1309–1317. DOI: 10.2527/jas.2014-8282.

Declerck, I., Dewulf, J., Sarrazin, S. and Maes, D. (2016). Long-term effects of colostrum intake in piglet mortality and performance. *Journalof Animal Science* 94(4):1633–1643.

Declerck, I., Sarrazin, S., Dewulf, J. and Maes, D. (2017). Sow and piglet factors determining variation of colostrum intake between and within litters. *Animal* 11(8):1336–1343.

Dee, S. A. (1992). Porcine urogenital disease. The Veterinary Clinics of North America. *Food Animal Practice* 8(3):641–660. DOI: 10.1016/s0749-0720(15)30709-x.

de Jong, E., Appeltant, R., Cools, A., Beek, J., Boyen, F., Chiers, K. and Maes, D. (2014). Slaughterhouse examination of culled sows in commercial pig herds. *Livestock Science* 167:362–369. DOI: 10.1016/j.livsci.2014.07.001.

De Winter, P. J. J., Verdonck, M., de Kruif, A., de Vriese, L. A. and Haesebrouck, F. (1992). Endometritis and vaginal discharge in the sow. *Animal Reproduction Science* 28(1-4):51–58.

De Winter, P. J. J., Verdonck, M., De Kruif, A., Devriese, L. A. and Haesebrouck, F. (1995). Bacterial endometritis and vaginal discharge in the sow: Prevalence of different bacterial species and experimental reproduction of the syndrome. *Animal Reproduction Science* 37(3-4):325–335.

Dial, G. D. and MacLachlan, N. J. (1988). Urogenital infections of swine. Part I. Clinical manifestations and pathogenesis. *Compendium on Continuing Education for the Practicing Veterinarian* 10:63–68.

Drolet, R. (2019). Urinary system. In: Straw, B. E., Zimmerman, J. J., D'Allaire, S. and Taylor, D. J. (Eds) *Diseases of Swine* (11th edn.). Iowa: Blackwell Publishing.

Eckburg, P. B., Bik, E. M., Bernstein, C. N., Purdom, E., Dethlefsen, L., Sargent, M., Gill, S. R., Nelson, K. E. and Relman, D. A. (2005). Diversity of the human intestinal microbial flora. *Science* 308(5728):1635–1638. DOI: 10.1126/science.1110591.

Edwards, S. A., & Baxter, E. M. (2015). Piglet mortality: causes and prevention. In *The Gestating and Lactating Sow*. Wageningen Academic Publishers, Wageningen, The Netherlands. pp. 649–653.

Einarsson, S., Sjunneson, Y., Hulten, F., Eliasson-Selling, L., Dalin, A. M., Lundeheim, N. and Magnusson, U. A. (2014). 25 years experience of group-housed sows-reproduction in animal welfare-friendly systems. *Acta Veterinaria Scandinavica* 56:37.

Eissen, J. J., Kanis, E. and Kemp, B. (2000). Sow factors affecting voluntary feed intake during lactation. *Livestock Production Science* 64(2-3):147–165.

Eliasson-Selling, L., Hofmo, P. O. and Narum, M. (2000). The art of oestrous-control in loose housed sows. In: Cargill, C. and McOrist, S. (Eds) *Proceedings of the International Pig Veterinary Society Congres*, 17–21 September 2000. Melbourne. Causal Productions Pty, Limited (vol. 2000), p. 395.

Ellendorff, F., Taverne, M., Elsaesser, F., Forsling, M., Parvizi, N., Naaktgeboren, C. and Smidt, D. (1979). Endocrinology of parturition in the pig. *Animal Reproduction Science* 2(1-3):323–334.

Ellis, W. A. (1990). Leptospirosis. In: *OIE Manual of Recommended Diagnostic Techniques and Requirements for Biological Products for List A and B Diseases*, Paris, 27:1-11, sec. 7.

Ellis, W. A. (2006). Leptospirosis. In: Straw, B. , Zimmermann, J. J. , D'Allaire, S. and Taylor, D. J. (Eds) *Diseases of Swine* (9th edn.). London: Blackwell Publishing, pp. 691-700.

Ensley, S. M. and Radke, S. L. (2019). Mycotoxins in grains and feeds. In: Zimmerman, J. J., Karriker, L. A., Ramirez, A., Schwartz, K. J., Stevenson, G. W. and Zhang, J. (Eds) *Diseases of Swine* (11th edn.). Glasgow, Great Britain: Wiley Blackwell, pp. 1055-1071.

Eskola, M., Kos, G., Elliott, C. T., Hajšlová, J., Mayar, S. and Krska, R. (2020). Worldwide contamination of food-crops with mycotoxins: Validity of the widely cited 'FAO estimate'of 25%. *Critical Reviews in Food Science and Nutrition* 60(16):2773-2789.

Farmer, C. (2016). Altering prolactin concentrations in sows. *Domestic Animal Endocrinology* 56(Suppl):S155-S164. DOI: 10.1016/j.domaniend.2015.11.005.

Farmer, C., Maes, D. and Peltoniemi, O. A. T. (2019). The mammary system. In: Chwartz, K., Zimmerman, J., Karriker, L., Ramirez, A., Stevenson, G. and Zhang, J. (Eds) *Diseases of Swine* (11th edn.). London: Wiley Blackwell, pp. 313-338.

Farmer, C. and Quesnel, H. (2009). Nutritional, hormonal, and environmental effects on colostrum in sows. *Journal of Animal Science* 87(13)(Suppl):56-64. Available at: http://jas.fass.org/cgi/content/full/87/13_suppl/56.

Ferrari, C. V., Sbardella, P. E., Bernardi, M. L., Coutinho, M. L., Vaz, I. S., Wentz, I. and Bortolozzo, F. P. (2014). Effect of birth weight and colostrum intake on mortality and performance of piglets after cross-fostering in sows of different parities. *Preventive Veterinary Medicine* 114(3-4):259-266.

Feyera, T. and Theil, P. K. (2017). Energy and lysine requirements and balances of sows during transition and lactation: A factorial approach. *Livestock Science* 201:50-57.

Gerjets, I. and Kemper, N. (2009). Coliform mastitis in sows: A review. *Journal of Swine Health and Production* 17(2):97-105.

Gilbert, C. L., Goode, J. A. and MacGrath, T. J. (1994). Pulsatile secrection of oxytocin during parturition in the pig: Temporal relationship with foetal expulsion. *Journalof Physiology* 475(1):129-137.

Gimsa, U., Tuchscherer, M. and Kanitz, E. (2018). Psychosocial stress and immunity-what can we learn From pig studies? *Frontiers in Behavioral Neuroscience* 12:64. DOI: 10.3389/fnbeh.2018.00064.

Glock, X. T. P. and Bilkei, G. (2005). The effect of postparturient urogenital diseases on the lifetime reproductive performance of sows. *Canadian Veterinary Journal* 46(12):1103-1107.

Gmeiner, K. (2007). *Ultrasonographische Charakterisierung der gesunden und kranken Harnblase bei der Sau*. Doctoral Dissertation, University of Leipzig.

Goyarts, T., Dänicke, S., Brüssow, K. P., Valenta, H., Ueberschär, K. H. and Tiemann, U. (2007). On the transfer of the Fusarium toxins deoxynivalenol (DON) and zearalenone (ZON) from sows to their fetuses during days 35-70 of gestation. *Toxicology Letters* 171(1-2):38-49.

Grahofer, A., Bill, R. and Nathues, H. (2017). Vaginal discharge following artificial insemination of sows in a multisite sow pool system. In: *Proceedings of the 21nd Annual ESDAR Conference*.

Grahofer, A., Björkman, S. and Peltoniemi, O. (2020). Diagnosis of endometritis and cystitis in sows: Use of biomarkers. *Journal of Animal Science* 98(Suppl 1):S107-S116.

Grahofer, A., Meile, A. and Nathues, H. (2019). Detection and evaluation of puerperal disorders in sows after farrowing. In: *Proceedings of theFirst Symposium of the European College of Animal Reproduction*, Vienna, Austria.

Grahofer, A., Sipos, S., Fischer, L., Entenfellner, F. and Sipos, W. (2014). Relationship between bacteriological and chemic-analytical urinalysis from sows with reproductive disorders. In: *Proceedings of the Sixth European Symposium of Porcine Health Management* (vol. 126).

Gustafsson, M., Jensen, P., de Jonge, F. H., Illman, G. and Špinka, M. (1999). Maternal behaviour of domestic sows and crosses between domestic sows and wild boar. *Applied Animal Behaviour Science* 65(1):29–42.

Han, T., Björkman, S., Soede, N. M., Oliviero, C. and Peltoniemi, O. (2019). Effect of IGF-1 level at weaning on subsequent luteal developement and progesterone production in primiparous sows. In: *Abstract Book 11th European Symposium of Porcine Health Management; 22-24 May. Utrecht*. The Netherlands: ESPHM, p. 82.

Han, T., Björkman, S., Soede, N. M., Oliviero, C. and Peltoniemi, O. A. T. (2020). IGF-1 concentration patterns and their relationship with follicle development after weaning in young sows fed different pre-mating diets. *Animal* 14(7):1493-1501. DOI: 10.1017/S1751731120000063.

Han, T., Björkman, S., Soede, N. M., Oliviero, C. and Peltoniemi, O. A. T. (2021). IGF-1 concentrations after weaning in young sows fed different pre-mating diets are positively associated with piglet mean birth weight at subsequent farrowing. *Animal* 15(1):100029. DOI: 10.1016/j.animal.2020.100029.

Hartsock, T. G. and Barczewski, R. A. (1997). Prepartum behaviour in swine: Effects of pen size. *Journal of Animal Science* 75(11):2899–2904.

Hasan, S., Junnikkala, S., Peltoniemi, O., Paulin, L., Lyyski, A., Vuorenmaa, J. and Oliviero, C. (2018a). Dietary supplementation with yeast hydrolysate in pregnancy influences colostrum yield and gut microbiota of sows and piglets after birth. *PLoS ONE* 13(5):e0197586. DOI: 10.1371/journal.pone.0197586.

Hasan, S., Saha, S., Junnikkala, S., Orro, T., Peltoniemi, O. and Oliviero, C. (2018b). Late gestation diet supplementation of resin acid-enriched composition increases sow colostrum immunoglobulin G content, piglet colostrum intake and improve sow gut microbiota. *Animal* 27: 1–8. DOI: 10.1017/S1751731118003518.

Hermansson, I., Einarsson, S., Larsson, K. and Backstrom, L. (1978). On the agalactia post partum in the sow. A clinical study. *Nordisk Veterinaermedicin* 30(11):465–473.

Hopper, S. A., White, M. E. and Twiddy, N. (1992). An outbreak of blue-eared pig disease (porcine reproductive and respiratory syndrome) in four pig herds in Great Britain. *Veterinary Record* 131(7):140–144. DOI: 10.1136/vr.131.7.140.

Hoving, L. L., Soede, N. M., Feitsma, H. and Kemp, B. (2012). Lactation weight loss in primiparous sows: Consequences for embryo survival and progesterone and relations with metabolic profiles. *Reproduction in Domestic Animals* 47(6):1009–1016.

Hoving, L. L., Soede, N. M., Graat, E. A. M., Feitsma, H. and Kemp, B. (2011). Reproductive performance of second parity sows: Relations with subsequent reproduction. *Livestock Science* 140(1–3):124–130.

Hoy, S. (2006). The impact of puerperal diseases in sows on their fertility and health up to next farrowing. *Animal Science* 82(5):701–704. DOI: 10.1079/ASC200670.

Hoy, S., Lutter, C., Wähner, M. and Puppe, B. (1994). The effect of birth weight on the early postnatal vitality of piglets. *DTW. Deutsche Tierarztliche Wochenschrift* 101(10):393-396.

Hultén, F., Persson, A., Eliasson-Selling, L., Heldmer, E., Lindberg, M., Sjögren, U., Kugelberg, C. and Ehlorsson, C. J. (2004). Evaluation of environmental and management-related risk factors associated with chronic mastitis in sows. *American Journal of Veterinary Research* 65(10):1398–1403. DOI: 10.2460/ajvr.2004.65.1398.

Islas-Fabila, P., Mota-Rojas, D., Martínez-Burnes, J., Mora-Medina, P., González-Lozano, M., Roldan-Santiago, P., Greenwell-Beare, V., González-Hernández, M., Vega-Manríquez, X. and Orozco-Gregorio, H. (2018). Physiological and metabolic responses in newborn piglets associated with the birth order. *Animal Reproduction Science* 197:247–256.

Jackson, P. G. G. (1995). *Handbook of Veterinary Obstetrics* (vol. 1995, 2nd edn.). London: Saunders, pp. 221–222.

Jarvis, S., Lawrence, A. B., McLean, K. A., Deans, L. A., Chirnside, J. and Calvert, S. K. (1997). The effect of environment on behavioural activity, ACTH, beta-endorphin and cortisol in prefarrowing gilts. *Animal Science* 65(3):465–472.

Ji, Y. J., Li, H., Xie, P. F., Li, Z. H., Li, H. W., Yin, Y. L., Blachier, F. and Kong, X. F. (2019). Stages of pregnancy and weaning influence the gut microbiota diversity and function in sows. *Journal of Applied Microbiology* 127(3):867–879. DOI: 10.1111/jam.14344.

Jost, T., Lacroix, C., Braegger, C. and Chassard, C. (2014). Stability of the maternal gut microbiota during late pregnancy and early lactation. *Current Microbiology* 68(4):419–427.

Kaeoket, K., Persson, E. and Dalin, A. M. (2001). The sow endometrium at different stages of the oestrus cycle: Studies on morphological changes and infiltration by cells of the immune system. *Animal Reproduction Science* 65(1–2):95–114.

Kaiser, M., Jacobson, M., Andersen, P. H., Bækbo, P., Cerón, J. J., Dahl, J., Escribano, D. and Jacobsen, S. (2018a). Inflammatory markers before and after farrowing in healthy sows and in sows affected with postpartum dysgalactia syndrome. *BMC Veterinary Research* 14(1):83. DOI: 10.1186/s12917-018-1382-7.

Kaiser, M., Jacobsen, S., Andersen, P. H., Bækbo, P., Cerón, J. J., Dahl, J., Escribano, D., Theil, P. K. and Jacobson, M. (2018b). Hormonal and metabolic indicators before and after farrowing in sows affected with postpartum dysgalactia syndrome. *BMC Veterinary Research* 14(1):334. DOI: 10.1186/s12917-018-1649-z.

Kaiser, M., Jacobson, M., Bækbo, P., Dahl, J., Jacobsen, S., Guo, Y. Z., Larsen, T. and Andersen, P. H. (2020). Lack of evidence of mastitis as a causal factor for postpartum dysgalactia syndrome in sows. *Translational Animal Science* 4(1):250–263.

Kanora, A. and Maes, D. (2010). The role of mycotoxins in pig reproduction: A review. *Veterinarni Medicina* 54(12):565–576.

Kapell, D. N. R. G., Ashworth, C. J., Knap, P. W. and Roehe, R. (2011). Genetic parameters for piglet survival, litter size and birth weight or its variation within litter in sire and dam lines using Bayesian analysis. *Livestock Science* 135(2–3):215–224.

Kauffold, J. (2008). Nichtpuerperale Uterusentzündungen beim Schwein. *Tierärztliche Praxis Ausgabe G* 36(3):189–198.

Kauffold, J. and Althouse, G. C. (2007). An update on the use of B-mode ultrasonography in female pig reproduction. *Theriogenology* 67(5):901–911.

Kauffold, J., Gmeiner, K., Sobiraj, A., Richter, A., Failing, K. and Wendt, M. (2010). Ultrasonographic characterization of the urinary bladder in sows with and without urinary tract infection. *Veterinary Journal* 183(1):103–108.

Kemp, B., Da Silva, C. L. A. and Soede, N. M. (2018). Recent advances in pig reproduction: Focus on impact of genetic selection for female fertility. *Reproduction in Domestic Animals* 53(Suppl 2):28–36.

Kemp, B. and Soede, N. M. (2012). Reproductive issues in welfare-friendly housing systems in Pig husbandry: A review. *Reproduction in Domestic Animals* 47(Suppl 5):51–57.

Kemper, N. (2020). Update on postpartum dysgalactia syndrome in sows. *Journal of Animal Science* 98(Suppl 1):S117–S125. DOI: 10.1093/jas/skaa135.

Kielland, C., Rootwelt, V., Reksen, O. and Framstad, T. (2015). The association between immunoglobulin G in sow colostrum and piglet plasma. *Journal of Animal Science* 93(9):4453–4462.

Kim, H. B., Borewicz, K., White, B. A. and Singer, R. S. (2011). Longitudinal investigation of the age-related bacterial diversity in the feces of commercial pigs. *Veterinary Microbiology* 153(1–2):124–133.

Kindahl, H., Alonso, R., Cort, N. and Einarsson, S. (1982). Release of prostaglandin F2α during parturition in the sow. *Zentralblatt Fur Veterinarmedizin. Reihe A* 29(7):504–510.

Klobasa, F., Werhahn, E. and Butler, J. E. (1987). Composition of sow milk during lactation. *Journalof Animal Science* 64(5):1458–1466.

Knap, P. W. and Doeschl Wilson, A. (2021). Control and elimination of porcine reproductive and respiratory syndrome virus. *Virus Research* 154(1–2):185–192. DOI: 10.1016/j.virusres.2010.08.016.

Koren, O., Goodrich, J. K., Cullender, T. C., Spor, A., Laitinen, K., Bäckhed, H. K., Gonzalez, A., Werner, J. J., Angenent, L. T., Knight, R., Bäckhed, F., Isolauri, E., Salminen, S. and Ley, R. E. (2012). Host remodeling of the gut microbiome and metabolic changes during pregnancy. *Cell* 150(3):470–480. DOI: 10.1016/j.cell.2012.07.008.

Kraft, W., Dürr, U. M., Fürll, M., Bostedt, H. and Heinritzi, K. (2005). Harnapparat. In: Kraft, W. and Dürr, U. M. (Eds) *Klinische Labordiagnostik in der Tiermedizin* (vol. 6). Aufl. Stuttgart: Verlag Schattauer. S, pp. 186–217.

Labroue, F., Caugant, A., Ligonesche, B. and Gaudré, D. (2001). Étude de l'évolution des tétines dápparence douteuse chez la cochette au cours de sa carrier. *Journées de la Recherche Porcine en France* 33:145–150.

Lager, K. M. and Halbur, P. G. (1996). Gross and microscopic lesions in porcine fetuses infectedwith porcine reproductive and respiratory syndrome virus. *Journal of Veterinary Diagnostic Investigation* 8(3):275–282.

Larsen, I. and Thorup, F. (2006). The diagnosis of MMA. *The International Pig Veterinary Society* 52:256.

Lawlor, P. G. and Lynch, P. B. (2001). Mycotoxins in pig feeds-1: Source of toxins, prevention and management of mycotoxicosis. *Irish Veterinary Journal* 54(3):117–120.

Le Dividich, J., Charneca, R. and Thomas, F. (2017). Relationship between birth order, birth weight, colostrum intake, acquisition of passive immunity and pre-weaning mortality of piglets. *Spanish Journal of Agricultural Research* 15(2):e0603. DOI: 10.5424/sjar/2017152-9921.

Lessard, M., Blais, M., Beaudoin, F., Deschene, K., Lo Verso, L. L., Bissonnette, N., Lauzon, K. and Guay, F. (2018). Piglet weight gain during the first two weeks of lactation influences the immune system development. *Veterinary Immunology and Immunopathology* 206:25–34. DOI: 10.1016/j.vetimm.2018.11.005.

Li, H., Liu, Z., Lyu, H., Gu, X., Song, Z., He, X. and Fan, Z. (2020). Effects of dietary insulin during late gestation on sow physiology, farrowing duration and piglet performance. *Animal Reproduction Science* 219:106531.

Liebhold, M., Wendt, M., Kaup, F. J. and Drommer, W. (1995). Clinical and light and electron microscopical findings in sows with cystitis. *Veterinary Record* 137(6):141-144.

Lindsay, D. S., Dubey, J. P. and Santin-Duran, M. (2019). Coccidia and other protozoa. In: Zimmerman, J. J., Karriker, L. A., Ramirez, A., Schwartz, K. J., Stevenson, G. W. and Zhang, J. (Eds) *Diseases of Swine* (11th edn.). Glasgow, Great Britain: Wiley Blackwell (vol. 2019), pp. 1015–1027.

Litvak, Y., Byndloss, M. X., Tsolis, R. M. and Bäumler, A. J. (2017). Dysbiotic proteobacteria expansion: A microbial signature of epithelial dysfunction. *Current Opinion in Microbiology* 39:1–6. DOI: 10.1016/j.mib.2017.07.003.

Liu, P., Piao, X. S., Kim, S. W., Wang, L., Shen, Y. B., Lee, H. S. and Li, S. Y. (2008). Effects of chito-oligosaccharide supplementation on the growth performance, nutrient digestibility, intestinal morphology, and fecal shedding of and in weaning pigs. *Journal of Animal Science* 86(10):2609–2618.

Love, R. J., Evans, G. and Klupiec, C. (1993). Seasonal effects on fertility in gilts and sows. *Journal of Reproduction and Fertility Supplement* 48:191–206.

Luescher, U. A., Friendship, R. M. and McKeown, D. B. (1990). Evaluation of methods to reduce fighting among regrouped gilts. *Canadian Journal of Animal Science* 70(2):363–370.

Madec, F. and Leon, E. (1992). Farrowing disorders in the sow: A field study. *Zentralblatt Fur Veterinarmedizin. Reihe A* 39(6):433–444.

Maes, D., Papadopoulos, G., Cools, A. and Janssens, G. P. (2010). Postpartum dysgalactia in sows: Pathophysiology and risk factors. *Tierärztliche Praxis Ausgabe G: Großtiere/Nutztiere* 38(1):15–20.

Maes, D., Pluym, L. and Peltoniemi, O. (2016). Impact of group housing of pregnant sows on health. *Porcine Health Management* 2(17):17. DOI:10.1186/s40813-016-0032-3.

Maldonado, J., Segales, J., Martınez-Puig, D., Calsamiglia, M., Riera, P., Domingo, M. and Artigas, C. (2005). Identification of viral pathogens in aborted fetuses and stillborn piglets from cases of swine reproductive failure in Spain. *Veterinary Journal* 169(3):454–456.

Manjarin, R., Montano, Y. A., Kirkwood, R. N., Bennet, D. C. and Petrovski, K. R. (2018). Effect of piglet separation from dam at birth on colostrum uptake. *Canadian Journal of Veterinary Research* 82(3):239–242.

Matheson, S. M., Walling, G. A. and Edwards, S. A. (2018). Genetic selection against intrauterine growth retardation in piglets: A problem at the piglet level with a solution at the sow level. *Genetics, Selection, Evolution* 50(1):46.

Mauget, R. (1982). Seasonality of reproduction in the wild boar. In: Foxcroft, J. and Cole, R. (Eds) *Control of Pig Reproduction*. London: Butterworths, pp. 509-526.

Meile, A., Nathues, H. and Grahofer, A. (2019). Association between uterine involution in sows and reproductive performance in their next gestation. In: *Proceedings of the 17th International Conference on Production Diseases in Farm Animals*.

Menard, J., Batista, L. and D'Allaire, S. D. (2007). Gilt exposure to homologous Porcine Reproductive and Respiratory Syndrome (PRRS) virus strain during acclimatization as a tool for PRRS control. PMWS, PRRS and Swine influenza associated diseases. In: *5th international symposium on emerging and re-emerging pig diseases*, Krakow, Poland (vol. 1), pp. 24th–27th.

Menegat, M. B., DeRouchey, J. M., Woodworth, J. C., Dritz, S. S., Tokach, M. D. and Goodband, R. D. (2019). Effects of Bacillus subtilis C-3102 on sow and progeny performance, fecal consistency, and fecal microbes during gestation, lactation, and nursery periods. *Journal of Animal Science* 97(9):3920–3937. DOI: 10.1093/jas/skz236.

Mengeling, W. L., Lager, K. M. and Vorwald, A. C. (2000). The effect of porcine parvovirus and porcine reproductive and respiratory syndrome virus on porcine reproductive performance. *Animal Reproduction Science* 60–61:199–210.

Meredith, M. J. (1991). Non specific bacterial infections of the genital tract in female pigs. *Pig Veternery Journal* 27:110–121.

Mettenleiter, T. C., Ehlers, B., Müller, T., Yoon, K. J. and Teifke, J. P. (2019). Herpesviruses. In: Zimmerman, J. J., Karriker, L. A., Ramirez, A., Schwartz, K. J., Stevenson, G. W. and Zhang, J. (Eds) *Diseases of Swine* (11th edn.). Glasgow, Great Britain: Wiley Blackwell (vol. 2019), pp. 549–575.

Meulen, J., Helmond, F. A., Oudenaarden, C. P. J. and Van der Meulen, J. (1988). Effect of flushing of blastocysts on days 10–13 on the life-span of the corpora lutea in the pig. *Journal of Reproduction and Fertility* 84(1):157–162.

Moore, A. S., Gonyou, H. W. and Ghent, A. W. (1993). Integration of newly introduced and resident sows following grouping. *Applied Animal Behaviour Science* 38(3–4):257–267.

Moreira, L. P., Menegat, M. B., Barros, G. P., Bernardi, M. L., Wentz, I. and Bortolozzo, F. P. (2017). Effects of colostrum, and protein and energy supplementation on survival and performance of low-birth-weight piglets. *Livestock Science* 202:188–193.

Morgan, X. C., Tickle, T. L., Sokol, H., Gevers, D., Devaney, K. L., Ward, D. V., Reyes, J. A., Shah, S. A., LeLeiko, N., Snapper, S. B., Bousvaros, A., Korzenik, J., Sands, B. E., Xavier, R. J. and Huttenhower, C. (2012). Dysfunction of the intestinal microbiome in inflammatory bowel disease and treatment. *Genome Biology* 13(9):R79. DOI: 10.1186/gb-2012-13-9-r79.

Muirhead, M. R. (1986). Epidemiology and control of vaginal discharges in the sow after service. *The Veterinary Record* 119(10):233–235.

Munsterhjelm, C., Valros, A., Heinonen, M., Hälli, O. and Peltoniemi, O. A. T. (2008). Housing during early pregnancy affects fertility and behaviour of sows. *Reproduction in Domestic Animals* 43(5):584–591.

Nachreiner, R. F. and Ginther, O. J. (1974). Induction of agalactia by administration of endotoxin (Escherichia coli) in swine. *American Journal of Veterinary Research* 35(5):619–622.

Nagy, G. (1993). Comparative pathogenicity study of *Leptospira interrogans* serovar Pomona strains. *Acta Veterinaria Hungarica* 41(3–4):315–324.

Nash, W. A. (1990). Porcine parvovirus survey. *Veterinary Record* 126(7):175–176.

Neumann, E. J. and Hall, W. F. (2019). Disease control, prevention, and elimination. In: Zimmerman, J. J., Karriker, L. A., Ramirez, A., Schwartz, K. J., Stevenson, G. W. and Zhang, J. (Eds) *Diseases of Swine* (11th edn.). Glasgow, Great Britain: Wiley Blackwell (vol. 2019), pp. 123–157.

Nguyen, K., Cassar, G., Friendship, R. M. and Hodgins, D. (2013). An investigation of the impacts of induced parturition, birth weight, birth order, litter size, and sow parity on piglet serum concentrations of immunoglobulin G. *Journal of Swine Health and Production* 21:139–143.

OIE. (2008). Leptospirosis. In: *Manual of Diagnostic Tests and Vaccines for Terrestrial Animals* (6th edn.). Paris: Office International Des Epizooties (vol. 1), pp. 251–264.

Oliviero, C. (2010). Successful farrowing. PhD thesis, University of Helsinki. Available at: https://helda.helsinki.fi/bitstream/handle/10138/18971/successf.pdf?sequence=1&isAllowed=y.

Oliviero, C., Pastell, M., Heinonen, M., Heikkonen, J., Valros, A., Ahokas, J., Vainio, O. and Peltoniemi, O. A. T. (2008a). Using movement sensors to detect the onset of farrowing. *Biosystems Engineering* 100(2):281–285.

Oliviero, C., Heinonen, M., Valros, A., Halli, O. and Peltoniemi, O. A. T. (2008b). Effect of the environment on the physiology of the sow during late pregnancy, farrowing and early lactation. *Animal Reproduction Science* 105(3–4):365–377.

Oliviero, C., Heinonen, M., Valros, A. and Peltoniemi, O. A. T. (2010). Environmental and sow-related factors affecting the duration of farrowing. *Animal Reproduction Science* 119(1–2):85–91.

Oliviero, C., Junnikkala, S. and Peltoniemi, O. A. T. (2019). The challenge of large litters on the immune system of the sow and the piglets. *Reproduction in Domestic Animals* 54(Suppl 3):12–21. DOI: 10.1111/rda.13463.

Oliviero, C., Kokkonen, T., Heinonen, M., Sankari, S. and Peltoniemi, O. A. T. (2009). Feeding sows a high-fibre diet around farrowing and early lactation: Impact on intestinal activity, energy balance-related parameters and litter performance. *Research in Veterinary Science* 86(2):314–319.

Oliviero, C., Kothe, S., Heinonen, M., Valros, A. and Peltoniemi, O. A. T. (2013). Prolonged duration of farrowing is associated with subsequent decreased fertility in sows. *Theriogenology* 79(7):1095–1099.

Olsen, S. C., Nol, P. and Samartino, L. (2019). Brucellosis. In: Zimmerman, J. J., Karriker, L. A., Ramirez, A., Schwartz, K. J., Stevenson, G. W. and Zhang, J. (Eds) *Diseases of Swine* (11th edn.). Glasgow, Great Britain: Wiley Blackwell (vol. 2019), pp. 778–791.

Opriessnig, T. and Coutinho, T. A. (2019). Erysipelas. In: Zimmerman, J. J., Karriker, L. A., Ramirez, A., Schwartz, K. J., Stevenson, G. W. and Zhang, J. (Eds) *Diseases of Swine* (11th edn.). Glasgow, Great Britain: Wiley Blackwell (vol. 2019), pp. 635–843.

Oravainen, J., Heinonen, M., Seppä-Lassila, L., Orro, T., Tast, A., Virolainen, J. V. and Peltoniemi, O. A. T. (2006). Factors affecting fertility in loosely housed sows and gilts: Vulvar discharge syndrome, environment and acute-phase proteins. *Reproduction in Domestic Animals* 41(6):549–554.

Oravainen, J., Heinonen, M., Tast, A., Virolainen, J. V. and Peltoniemi, O. A. T. (2005). High porcine parvovirus antibodies in sow herds: Prevalence and associated factors. *Reproduction in Domestic Animals* 40(1):57–61.

Oravainen, J., Heinonen, M., Tast, A., Virolainen, J. V. and Peltoniemi, O. A. T. (2008). Vulvar discharge syndrome in loosely housed Finnish pigs: Prevalence and evaluation of vaginoscopy, bacteriology and cytology. *Reproduction in Domestic Animals* 43(1):42–47. DOI: 10.1111/j.1439-0531.2007.00852.x.

Papadopoulos, G. A., Vanderhaeghe, C., Janssens, G. P., Dewulf, J. and Maes, D. G. (2010). Risk factors associated with postpartum dysgalactia syndrome in sows. *Veterinary Journal* 184(2):167–171. DOI: 10.1016/j.tvjl.2009.01.010.

Peltoniemi, O. A. T., Björkman, S., Oropeza-Moe, M. and Oliviero, C. (2019). Developments of reproductive management and biotechnology in the pig. *Animal Reproduction* 16(3):524–538. DOI: 10.21451/1984-3143-AR2019-0055.

Peltoniemi, O. A. T., Oliviero, C., Yun, J., Grahofer, A. and Björkman, S. (2020). Management practices to optimize the parturition process in the hyperprolific sow. *Journal of Animal Science* 98(Suppl 1):S96–S106.

Peltoniemi, O. A. T., Love, R. J., Heinonen, M., Tuovinen, V. and Saloniemi, H. (1999). Seasonal and management effects on fertility of the sow: A descriptive study. *Animal Reproduction Science* 55(1):47–61.

Peltoniemi, O. A. T., Björkman, S. and Maes, D. (2016). Reproduction of group-housed sows. *Porcine Health Management* 2:15. DOI: 10.1186/s40813-016-0033-2.

Peltoniemi, O. A. T., Love, R. J., Klupiec, C. and Evans, G. (1997). Effect of feed restriction and season on LH and prolactin secretion, adrenal response, insulin and FFA in group housed pregnant gilts. *Animal Reproduction Science* 49(2–3):179–190.

Peltoniemi, O. A. T., Tast, A. and Love, R. J. (2000). Factors effecting reproduction in the pig:seasonal effects and restricted feeding of the pregnant gilt and sow. *Animal Reproduction Science* 60–61:173–184.

Peltoniemi, O. A. T. and Virolainen, J. V. (2006). Seasonality of reproduction in gilts and sows. *Society of Reproduction and Fertility Supplement* 62:205–218.

Pensart, M. B. and Kluge, J. P. (1989). Pseudorabies virus (Aujeszky's disease). In: Pensart, M. B. (Ed.) *Virus Infections in Porcines*. New York: Elsevier, pp. 39–64.

Petersen, H. H., Nielsen, J. P. and Heegaard, P. M. (2004). Application of acute phase protein measurements in veterinary clinical chemistry. *Veterinary Research* 35(2):163–187. DOI: 10.1051/vetres:2004002.

Quesnel, H., Brossard, L., Valancogne, A. and Quiniou, N. (2008). Influence of some sow characteristics on within-litter variation of piglet birth weight. *Animal* 2(12):1842–1849.

Quesnel, H., Farmer, C. and Devillers, N. (2012). Colostrum intake: Influence on piglet performance and factors of variation. *Livestock Science* 146(2–3):105–114.

Quiniou, N., Dagorn, J. and Gaudré, D. (2002). Variation of piglets' birth weight and consequences on subsequent performance. *Livestock Production Science* 78(1):63–70.

Remience, V., Wavreille, J., Canart, B., Meunier-Salaün, M. C., Prunier, A., Bartiaux-Thill, N., Nicks, B. and Vandenheede, M. (2008). Effects of space allowance on the welfare of dry sows kept in dynamic groups and fed with an electronic sow feeder. *Applied Animal Behaviour Science* 112(3–4):284–296.

Rooke, J. A. and Bland, I. M. (2002). The acquisition of passive immunity in the new-born piglet. *Livestock Production Science* 78(1):13–23.

Rozeboom, D. W. (2015). Conditioning of the gilt for optimal reproductive performance. In: *The Gestating and Lactating Sow*. Editor: Farmer C. Wageningen Academic Publishers, Wageningen, The Netherlands. pp. 649–653.

Rutherford, K. M., Baxter, E. M., D'Eath, R. B., Turner, S. P., Arnott, G., Roehe, R., Ask, B., Sandøe, P., Moustsen, V. A., Thorup, F., Edwards, S. A., Berg, P. and Lawrence, A. B. (2013). The welfare implications of large litter size in the domestic pig I: Biological factors. *Animal Welfare* 22(2):199–218.

Salmon, H., Berri, M., Gerdts, V. and Meurens, F. (2009). Humoral and cellular factors of maternal immunity in swine. *Developmental and Comparative Immunology* 33(3):384–393. DOI: 10.1016/j.dci.2008.07.007.

Saporiti, V., Valls, L., Maldonado, J., Perez, M., Correa-Fiz, F., Segalés, J. and Sibila, M. (2021). Porcine Circovirus 3 detection in aborted fetuses and stillborn piglets from swine reproductive failure cases. *Viruses* 13(2):264. DOI: 10.3390/v13020264.

Sassone-Corsi, M. and Raffatellu, M. (2015). No vacancy: How beneficial microbes cooperate with immunity to provide colonization resistance to pathogens. *Journal of Immunology* 194(9):4081–4087.

Sayyari, A., Uhlig, S., Fæste, C. K., Framstad, T. and Sivertsen, T. (2018). Transfer of deoxynivalenol (DON) through placenta, colostrum and milk from sows to their offspring during late gestation and lactation. *Toxins* 10(12):517.

Schoos, A., Chantziaras, I., Vandenabeele, J., Biebaut, E., Meyer, E., Cools, A., Devreese, M. and Maes, D. (2020). Prophylactic use of meloxicam and paracetamol in peripartal sows suffering From postpartum Dysgalactia syndrome. *Frontiers in Veterinary Science* 7:603719. DOI: 10.3389/fvets.2020.603719.

Segalés, J. (2012). Porcine circovirus type 2 (PCV2) infections: Clinical signs, pathology and laboratory diagnosis. *Virus Research* 164(1-2):10-19.

Shin, N. R., Whon, T. W. and Bae, J. W. (2015). Proteobacteria: Microbial signature of dysbiosis in gut microbiota. *Trends in Biotechnology* 33(9):496-503. DOI: 10.1016/j.tibtech.2015.06.011.

Sipos, W., Grahofer, A., Fischer, L., Entenfellner, F. and Sipos, S. (2014). Keimspektrum des Urogenitaltraktes von Sauen mit Fertilitätsstörungen. *Wien Tierarztl Monat* 101: 9-10, 214-220.

Smit, M. N., Spencer, J. D., Almeida, F. R., Patterson, J. L., Chiarini-Garcia, H., Dyck, M. K. and Foxcroft, G. R. (2013). Consequences of a low litter birth weight phenotype for postnatal lean growth performance and neonatal testicular morphology in the pig. *Animal* 7(10):1681-1689. DOI: 10.1017/S1751731113001249.

Soede, N. M. and Kemp, B. (2015). Best practices in the lactating and weaned sow to optimize reproductive physiology and performance. In: *The Gestating and Lactating Sow*. Editor: Farmer C. Wageningen Academic Publishers, Wageningen, The Netherlands. pp. 649-653.

Spoolder, H. A., Burbidge, J. A., Edwards, S. A., Lawrence, A. B. and Simmins, P. H. (1996). Social recognition in gilts mixed into a dynamic group of 30 sows. In: *Proceedings of the British Society of Animal Science* (vol. 1996). Cambridge: Cambridge University Press, p. 37.

Spoolder, H. A. and Vermeer, H. M. (2015). Gestation group housing of sows. In: *The Gestating and Lactating Sow*. Editor: Farmer C. Wageningen Academic Publishers, Wageningen, The Netherlands. pp. 649-653.

Stiehler, T., Heuwieser, W., Pfuetzner, A. and Burfeind, O. (2015). The course of rectal and vaginal temperature in early postpartum sows. *Journal of Swine Health and Production* 23:72-83.

Stookey, J. M. and Gonyou, H. W. (1998). Recognition in swine: Recognition through familiarity or genetic relatedness? *Applied Animal Behaviour Science* 55(3-4):291-305.

Tan, C., Wei, H., Ao, J., Long, G. and Peng, J. (2016). Inclusion of konjac Flour in the gestation diet changes the gut microbiota, alleviates oxidative stress, and improves insulin sensitivity in sows. *Applied and Environmental Microbiology* 82(19):5899-5909.

Tast, A., Halli, O., Ahlstrom, S., Andersson, H., Love, R. J. and Peltoniemi, O. A. T. (2001). Seasonal alterations in circadian melatonin rhythms of the European wild boar and domestic gilt. *Journal of Pineal Research* 30(1):43-49.

Tast, A., Peltoniemi, O. A. T., Virolainen, J. V. and Love, R. J. (2002). Early disruption of pregnancy as a manifestation of seasonal infertility in pigs. *Animal Reproduction Science* 74(1-2):75-86.

Taverne, M. A. M. and van der Weijden, G. C. (2008). Parturition in domestic animals: Targets for future research. *Reproduction in Domestic Animals* 43(Suppl 5):36-42.

Thaker, M. Y. and Bilkei, G. (2005). Lactation weight loss influences subsequent reproductive performance of sows. *Animal Reproduction Science* 88(3-4):309-318.
Theil, P. K., Lauridsen, C. and Quesnel, H. (2014). Neonatal piglet survival: Impact of sow nutrition around parturition on fetal glycogen deposition and production and composition of colostrum and transient milk. *Animal* 8(7):1021-1030.
Thorsen, C. K., Schild, S. A., Rangstrup-Christensen, L., Bilde, T. and Pedersen, L. J. (2017). The effect of farrowing duration on maternal behavior of hyper-prolific sows in organic outdoor production. *Livestock Science* 204:92-97.
Tolstrup, L. K. (2017). *Cystitis in Sows: Prevalence, Diagnosis and Reproductive Effect*. Doctoral dissertation, University of Copenhagen.
Tsuma, V. T., Einarsson, S., Madej, A., Kindahl, H. and Lundeheim, N. (1996). Effect of food deprivation during early pregnancy on endocrine changes in primiparous sows. *Animal Reproduction Science* 41(3-4):267-278.
Tummaruk, P., Kesdangsakonwut, S., Prapasarakul, N. and Kaeoket, K. (2010). Endometritis in gilts: Reproductive data, bacterial culture, histopathology, and infiltration of immune cells in the endometrium. *Comparative Clinical Pathology* 19(6):575-584.
Vasdal, G. and Andersen, I. L. (2012). A note on teat accessibility and sow parity–Consequences for newborn piglets. *Livestock Science* 146(1):91-94. DOI: 10.1016/j.livsci.2012.02.005.
Völkl, A., Vogler, B., Schollenberger, M. and Karlovsky, P. (2004). Microbial detoxification of mycotoxin deoxynivalenol. *Journal of Basic Microbiology* 44(2):147-156.
Waldmann, K. H. and Wendt, M. (2001). *Lehrbuch der Schweinekrankheiten* (4th edn.). Stuttgart: Parey Verlag, p. 608.
Weiss, E. (2007). Harnorgane. In: Dahme, E. and Weiss, E. (Eds) *Grundriß der speziellen pathologischen Anatomie der Haustiere*. Enke Verlag: Stuttgart (vol. 6). Aufl., S., pp. 197-199.
Wendt, M., Liebhold, M., Kaup, F., Amtsberg, G. and Bollwahn, W. (1990). Corynebacterium-suis-Infektion beim Schwein, 1. Mitteilung: Klinische Diagnose unter besonderer Berücksichtigung von Harnuntersuchung und Zystoskopie. *Tierärztl Prax* 18(4):353-357.
Winkler, I. (1987). *Zytodiagnostik des puerperalen Zervikalfluors beim Schwein*. Hannover.
Xu, C., Peng, C., Zhang, X. and Peng, J. (2020). Inclusion of soluble fiber in the gestation diet changes the gut microbiota, affects plasma propionate and odd-chain fatty acids levels, and improves insulin sensitivity in sows. *International Journalof Molecular Sciences* 21(2):635.
Yun, J., Ollila, A., Valros, A., Larenza Menzies, P., Heinonen, M., Oliviero, C. and Peltoniemi, O. A. T. (2019). Behavioural alterations in piglets after surgical castration: Effects of analgesia and anaesthesia. *Research in Veterinary Science* 125:36-42. DOI: 10.1016/j.rvsc.2019.05.009.
Yun, J., Swan, K. M., Vienola, K., Kim, Y. Y., Oliviero, C., Peltoniemi, O. A. T. and Valros, A. (2014). Farrowing environment has an impact on sow metabolic status and piglet colostrum intake in early lactation. *Livestock Science* 163:120-125. DOI: 10.1016/j.livsci.2014.02.014.
Zimmerman, J. J., Dee, S. A., Holtkamp, D. J., Murtaugh, M. P., Stadejek, T., Stevenson, G. W., Torremorell, M., Yang, H. and Zhang, J. (2019). Porcine reproductive and respiratory syndrome viruses (Porcine arteriviruses). In: Zimmerman, J. J., Karriker, L. A., Ramirez, A., Schwartz, K. J., Stevenson, G. W. and Zhang, J. (Eds) *Diseases of Swine* (11th edn.). Glasgow, Great Britain: Wiley Blackwell (vol. 2019), pp. 685-708.

Chapter 15

Optimising the health of weaned piglets

Andrea Luppi, Istituto Zooprofilattico Sperimentale della Lombardia e dell'Emilia Romagna (IZSLER), Italy

1 Introduction

The immediate post-weaning period is one of the most stressful phases in a pig's life in which piglets are usually exposed to environmental, social and psychological stressors such as abrupt separation from their mother, mixing with other litters in a usually new environment and switching from highly digestible (liquid) milk to a less-digestible plant-based dry diet containing complex protein and carbohydrate, including various antinutritional factors (Lallès et al., 2007). All these elements have direct or indirect effects on health and overall growth performance.

At weaning several factors influence the health and performance of pigs (Fig. 1), determining a suboptimal weight gain, predisposing weaned pigs to the disease, such as episodes of diarrhoea, and increased morbidity and mortality (Pluske et al., 1997) and reduced growth performance (Vente-Spreeuwenberg et al., 2003).

The immediate post-weaning period is also characterised by the immaturity of the digestive system, predisposing the piglet to diseases (Jayaraman and

http://dx.doi.org/10.19103/AS.2022.0103.17

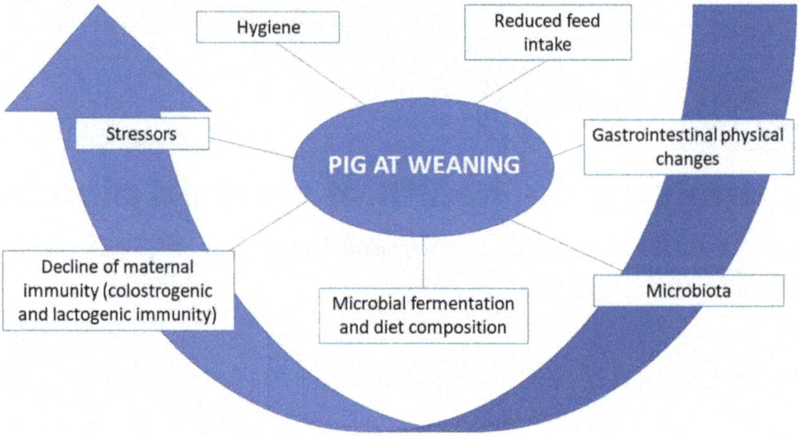

Figure 1 Main factors influencing health and performance of pigs at weaning (Pluske et al., 1997; Jayaraman and Nyachoti, 2017).

Nyachoti, 2017). In this phase, the immaturity of the intestinal immune system (complicated by the lack of a previous stimulation by pathogens) may reduce the ability of pigs to mount an appropriate immunological response to pathogens (Heo et al., 2013).

Piglets are born agammaglobulinemic and do not become fully immunologically competent until about 4 weeks of age (Langel et al., 2016). Maternal immunity, composed of colostrogenic and lactogenic immunity, is able to modulate the immune response and provide specific immunity in the newborn (Poonsuk and Zimmerman, 2018). More specifically the maternal immune components are characterised by: (1) circulating antibodies derived from colostrum; (2) mucosal antibodies from colostrum and milk; and (3) immune cells provided in mammary secretions (Poonsuk and Zimmerman, 2018). At weaning, in general terms, the protection given by the colostrogenic and lactogenic immunity ends. The severity of this condition is dependent on how much the immune system has developed during the pre-weaning period (Bailey et al., 2005) exposing piglets to several infections and possible diseases.

Weaning weight is an important factor able to influence the pig's growth and feed efficiency from weaning to market. Colostrum intake affects weaning weight showing long-term effects on piglets' growth from 3 weeks of age until after weaning (Devillers et al., 2011). Extending weaning age is a very effective way of improving weaning weight (Pinilla et al., 2008). This is very important considering that the weaning age imparted by a swine production system does not match with the natural situation, where weaning is a gradual process that takes place between 2 and 3 months of age (Allwin et al., 2016).

Another important factor associated with high variability in pig weight at weaning is the within-litter variation of piglet birth weight (Quesnel et al., 2008).

An increase in litter size in hyperprofilic sows was shown to be genetically associated with a decrease in the mean piglet birth weight and an increase in the within-litter variability of birth weight (Wolf et al., 2008).

Improving husbandry practices in the swine barn, in order to reduce the negative impact of the weaning transition period, include nutrition strategies, management practices, maintenance of hygienic standards, disease prevention protocols and animal welfare considerations (e.g. formation of groups, social stressors and ambient temperature for piglets) (Jayaraman and Nyachoti, 2017).

2 Transition from weaning: factors influencing health and performance of pigs

2.1 Reduced feed intake

The period of transition from weaning and the first post-weaning period are characterised by a marked reduction in voluntary feed intake. Lallès et al. (2007) reported that although 50% of weaned piglets consume their first meal within 24 h post-weaning, 10% have not eaten until 48 h. Thus, energy requirements for maintenance are only met 3 days post-weaning, and it can take 8–14 days for piglets to recover their pre-weaning level of energy intake. Available evidences indicate that reduced feed intake is a major contributing factor to the abruptly reduced intestinal villus height, affecting gut morphology, barrier function and the overall efficiency of nutrient capture and utilisation (Pluske et al., 1997; Vente-Spreeuwenberg et al., 2001; Vente-Spreeuwenberg et al., 2003). Feed intake and its reduction can predispose pigs to enteric diseases and is influenced by various factors (Jayaraman and Nyachoti, 2017) (Fig. 2).

In swine husbandry, practising encouraging feed intake and reducing weaning stresses are important factors for maintaining performance, the integrity of the small intestinal structure immediately after weaning (Moeser et al., 2017) and reducing mortality and morbidity (Collins et al., 2017). During the post-weaning phase psychological stressors, such as mixing with other littermates (Khafipour et al., 2014), crowding stress, and environmental stressors, such as hygienic standards, air quality and change in ambient temperature, act to drastically reduce feed intake (Campbell et al., 2013). Providing adequate feeder space is critical to weaned pig performance because limited feeder space can increase competition at the feeder, compromising feed intake and leading to reduced growth rates (Averos et al., 2012). He et al. (2016) evaluated the effect of restricted feeder space (2 spaces per pen vs. 5 spaces per pen; the area of each feeding space was 15 cm × 15 cm) on the growth performance of weaned pigs. The authors indicated that limited feeder space is associated with an increased risk of mortality or slow growth in weaned pigs.

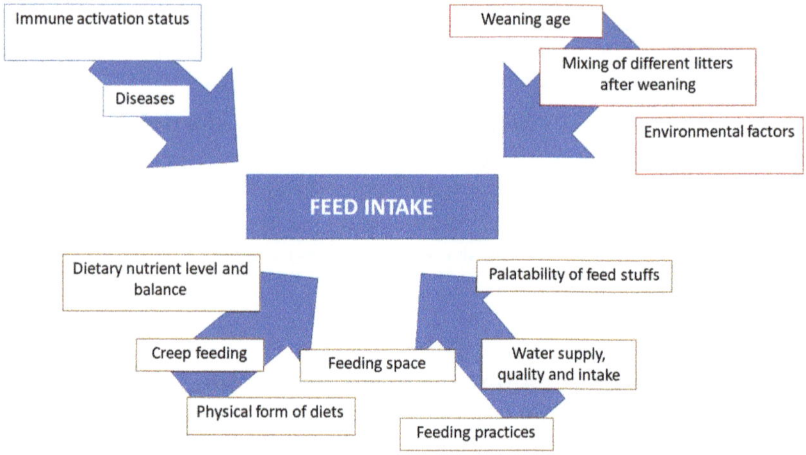

Figure 2 Main factors affecting feed intake at weaning (Jayaraman and Nyachoti, 2017).

Water intake might influence feed intake. Dybkjær et al. (2006) reported that a low feed intake may be due to insufficient drinking activity, as solid feed intake must be accompanied by water intake.

There is generally a negative relationship between inadequate thermal comfort (air temperature, relative humidity and ventilation rate) and feed intake (Noblet et al., 2001). Recommended air temperature ranges at the animal level and minimum ventilation rates for pigs of various sizes and ages are reported in Tables 1 and 2.

Table 1 Recommended air temperature ranges for pigs of various sizes and ages (Zulovich, 2012)

Pig's category	Optimal temperature (°C)	Range (°C)
Litter-newborn	35	32-35
Litter-3 weeks old	27	24-29
Nursery 5-14 kg	27	24-29
Nursery 14-23 kg	24	21-27
Nursery 23-34 kg	18	16-21

Table 2 Target minimum ventilation rates for pigs of various sizes (modified from Zulovich, 2012)

Pig's category	Ventilation rates
Litter	0.564 m^3/per minute-crate (20 cfm/crate)
Nursery 4-14 kg	0.0282-0.0564 m^3/per minute-pig (1-2 cfm/pig)
Nursery 14-23 kg	0.084-0.114 m^3/per minute-pig (3-4 cfm/pig)
Nursery 23-34 kg	0.144-0.198 m^3/per minute-pig (5-7 cfm/pig)

cfm, cubic feet per minute.

About the change in ambient temperature, fluctuations of ±4°C during the first week after weaning have been demonstrated to increase post-weaning scours (Brumm, 2019).

2.2 Gastrointestinal physical changes

Weaned pigs have a higher gastric pH value and this may be partly due to a lower acid secretion capacity of the stomach at weaning along with the reduction in lactic acid production from lactose. The high gastric pH value after weaning may contribute, in part, to the susceptibility of piglets to enteric infections at this time (Heo et al., 2013), considering that exposure to low pH values (i.e. 3.0–4.0) is bactericidal for many pathogenic bacteria, including *Escherichia coli*.

Contributing factors that increase gastric pH at weaning are:

- lower (hydrochloric acid) HCl production,
- abrupt change in diet,
- overeating after anorexia,
- protein level,
- dietary electrolytes balance,
- dietary concentration of lactose or lactogenic carbohydrates,
- lower secretion of saliva.

Weaning also reduces gastric motility, with a reduction in stomach emptying rate in pigs after weaning compared with suckling pigs. Martinez et al. (2004) reported that the activation of stress gene, corticotrophin-releasing factor receptor 2, has been implicated in the inhibition of gastric motility, while Moeser et al. (2007) described its up-regulation in the jejunum of weaned pigs. Gastric stasis may contribute to the development of post-weaning diarrhoea (PWD) in piglets by allowing the proliferation of pathogenic bacteria (Heo et al., 2013).

During the post-weaning period, changes in intestinal tissue, including villus and crypt architecture can be associated with depressed activities of many brush-border digestive enzymes and reduced absorptive capacity (Hampson and Kidder, 1986; Nabuurs et al., 1993; Vente-Spreeuwenberg et al., 2004). For this reason, activities of the brush-border enzymes in weaned pigs have been used as indicators of maturation and digestive capacity of the small intestine (Heo et al., 2013). The major effects of an immature digestive system in weaned pigs include the reduced activity of digestive enzymes, changes in intestinal morphology (Boudry et al., 2004) and reduced digestion in the small intestine.

2.3 Microbiota

An important topic involving piglets in the first weeks of life and in the weaning transition period is about the intestinal microbiota. The gut is sterile at birth and is then colonised by bacteria from the mother and the environment, starting with lactic acid bacteria, enterobacteria and streptococci. After the introduction of solid feed, obligate anaerobes increase in number and diversity until an adult-type pattern is achieved (Lallès et al., 2007). In a short period of time, the intestinal microbiota must ultimately develop from a simple unstable community into a complex and stable community, thus generating a tight 'colonisation resistance' or 'competitive exclusion' (Lallès et al., 2007). Microbial changes have been reported in the first 7–14 days after weaning when the number of lactobacilli diminished regardless of the weaning age (Franklin et al., 2002). It was also observed that lactobacilli numbers decreased whereas coliform and *E. coli* numbers increased in the various regions of the gastrointestinal tract at 28 days after weaning (Konstantinov et al., 2006).

One of the most important factors that affect the total population and diversity of the intestinal microbiota is the diet composition (Castillo et al., 2007; Metzler et al., 2009). Jeaurond et al. (2008) reported that the counts of *Clostridium* spp. in the large intestine were low in pigs fed higher fermentable carbohydrates (such as sugar beet pulp) and tended to increase by increasing dietary levels of fermentable protein.

2.4 Microbial fermentation and diet composition

It is recognised that microbial fermentation within the gastrointestinal tract is very important for the pig (Williams et al., 2001). Fermentation of carbohydrates, specifically resistant starches and dietary fibre, leads to the production of mainly short-chain volatile fatty acids (SCFAs) (acetic, propionic and butyric acids) and the use of NH_3 and other nitrogenous compounds, which are required for microbial growth (Lallès et al., 2007). When carbohydrates are in short supply relative to the available protein of non-degradable and endogenous origin, protein will be used as an energy source for fermentation, resulting in the end products NH_3, branched-chain fatty acids (BCFAs) and potentially toxic end products (Williams et al., 2001). These compounds include amines, volatile phenols and indoles, contributing to the increase in the incidence of diarrhoea at weaning in pigs (Lallès et al., 2007; Riaz Rajoka et al., 2017).

In recent years, the recommended levels of amino acids have increased and the National Research Council (NRC) reports crude protein (CP) concentrations of 20–23% (200–230 g/kg) in the diet of weaned pigs (NRC, 2012). Veldkamp and Vernooij (2021) reported that to maximise piglet growth, it is recommended that levels of CP be at 20–23% in pre-starter offered to piglets, before and just after weaning, and 18–20% in starter diets. There is evidence that feeding with such a high protein diet immediately after weaning could cause protein maldigestion as the digestive system of weaner pigs is insufficiently developed to fully digest and absorb dietary proteins. Consequently, increasing amounts of undigested CP materials are present in the large intestine where it is fermented by resident micro-organisms and could encourage the growth of nitrogen utilising bacteria (Heo et al., 2013; O'Doherty et al., 2017). This contributes to increased production of toxic by-products such as BCFAs, indole, phenols, ammonia and biogenic amines in the gastrointestinal tract (Pluske et al., 2002). Reducing dietary CP is a strategy to reduce the incidence of diarrhoea (Wu et al., 2015). A low-protein diet (<180 g/kg) in the post-weaning period reduces protein fermentation in the gastrointestinal tract (Nyachoti et al., 2006). Opapeju et al. (2009; 2010) showed as diets containing 170 g/kg CP can reduce inflammatory responses and the number of enterotoxigenic *E. coli* (ETEC) in the small and large intestine. Lordelo et al. (2008) reported that a low-protein diet (170 g/kg CP) did not compromise the growth of piglets if they receive similar digestible energy contents and adequate levels of essential amino acids (i.e. lysine, methionine, tryptophan and threonine) including valine and isoleucine. Rattigan et al. (2020) in a recent study showed as dietary CP can be reduced to 18% without compromising growth performance if the diet is correctly supplemented with the essential amino acids within the recommended range for maximising growth performance of weaned pigs. The reduction of CP in the diet with the aforementioned modalities and levels is a strategy currently adopted in many farms.

Dietary fibre has been reported to improve gut health and decrease diarrhoea incidence in pigs (Molist et al., 2009). In particular feeding weaned pigs high-insoluble-fibre diets might better protect them against pathogenic bacteria by increasing the villus length (McRorie and McKeown, 2017). The impact of dietary fibre on piglets' nutrition might be determined by the properties of fibre and/or fibre sources (Lindberg, 2014). For example, the fibre in the wheat bran diet was adapted by piglets and acted as prebiotics (Chen et al., 2013) increasing feed intake and the development of the gastrointestinal tract. A wheat bran diet has also been shown to decrease the amount of pathogenic *E. coli* in the faeces and reduce the incidence of PWD (Molist et al., 2009).

The electrolyte imbalance in post-weaning diets exhibited alterations in motility, changes in paracellular permeability, loss of absorption surface, a change in electrolyte fluxes in post-weaning piglets and, finally, induced PWD (Saravanan et al., 2013). Guzman-Pino et al. (2015) reported that the balance between cation (Na^+) and anions (K^+/Cl^-) and the acid or alkaline load from the diet may strongly alter the acid-base status and growth performance of weaned piglets, while Gao et al. (2019) described as an excess of chloride

Table 3 Minimum mineral levels in nursery diets (NRC, 2012)

	Nursery pigs - weight	
	5-7 kg/BW	7-11 kg/BW
Minerals	Requirements (% or mg/kg of diet)	
Calcium (%)	0.85	0.80
STTD phosphorus (%)	0.45	0.40
ATTD phosphorus (%)	0.41	0.36
Total phosphorus (%)	0.70	0.65
Sodium (%)	0.40	0.35
Chloride (%)	0.50	0.45
Magnesium (%)	0.04	0.04
Potassium (%)	0.30	0.28
Copper (mg/kg)	6	6
Iodine (mg/kg)	0.14	0.14
Iron (mg/kg)	100	100
Manganese (mg/kg)	4	4
Selenium (mg/kg)	0.30	0.30
Zinc (mg/kg)	100	100

STTD, standardised total tract digestible; ATTD, apparent total tract digestibility. The ATTD of phosphorus is calculated based on the phosphorus content in the diet and feaces. Because the phosphorus in the faeces includes phosphorus originating from the body as well as phosphorus derived from the diet, the ATTD of phosphorus is less than the real phosphorus digestibility. The major issue in the use of ATTD of phosphorus is that the values may not be always additive in mixed diets (NRC, 2012). The STTD of phosphorus takes into account the basal endogenous loss of phosphorus originating from the body (Almeida and Stein, 2010).

ions induces a negative dietary electrolyte balance and reduces the growth performance of weaned pigs.

Maintaining an appropriate dietary electrolyte balance (dEB) is critical for pigs to achieve optimal growth performance. Traditionally, the optimal dEB for pigs is reported to be approximately 250 mEq/kg (NRC, 2012).

Jones et al. (2019) reported that increasing dietary dEB up to 243 mEq/kg in nursery diets (from 0 to 8 days after weaning) improved the growth performance of weanling pigs. Minerals requirement estimates for nursery pigs (NRC, 2012) are reported in Table 3.

2.5 Decline of maternal immunity: colostrogenic and lactogenic immunity

Newborn piglet is immunodeficient at birth and, after exposure to infectious agents, takes 7-10 days for a primary antibody or cell-mediated immune response to develop. During this time, resistance to infection depends on the actions of the innate defence mechanisms and the antibody or immune components passively transferred from the sow to the piglet through colostrum and milk (Chase and Lunney, 2019). More specifically the maternal immune components are characterised by (1) circulating antibodies derived from colostrum; (2) mucosal antibodies from colostrum and milk; and (3) immune cells provided in mammary secretions (Poonsuk and Zimmerman, 2018). From 3 days of age until the end of lactation, IgA is the predominant Ig found in sow milk. The majority of milk Ig is synthesised in the mammary gland, whereas colostral Ig is mostly derived from serum, although this varies by Ig class (Chase and Lunney, 2019). At weaning the protection given by the colostrogenic and lactogenic immunity ends. The severity of this condition is dependent on how much the immune system was developed during the pre-weaning period (Bailey et al., 2005) also considering that the weaning age imparted by a swine production system does not match with the immune system capacity for optimal disease prevention (Chase and Lunney, 2019). The development of immuno-competence, namely the ability to mount appropriate responses to antigens, including the ability to generate tolerance to food and commensal bacterial antigens as well as active immune responses to pathogens, is an absolute requirement for optimum growth and performance (Bailey et al., 2001).

Maternal antibody half-life in pigs ranges from 11.3 days to 20 days. Specific examples for half-life for swine maternal antibodies include 16.2 days for Porcine Reproductive and Respiratory Syndrome virus (PRRSV) (Yoon et al., 1995), 14 days for Swine Influenza A viruses (swIAV) (Fleck and Behrens, 2002), 19 days for Porcine Circovirus Type 2 (PCV2) (Opriessnig et al., 2004), 11.3 days for Pseudorabies virus (PRV) (Müller et al., 2005), 11 days for classical swine fever virus (CSFV) (Müller et al., 2005), 20 days for Parvovirus (PPV) (Paul et al., 1982) and 15.8 days for *Mycoplasma hyopneumoniae* (Morris et al., 1994).

Measured by a serological test, maternal antibody levels against different etiological agents fall below the test cut-off level at different times in the pig's life (Table 4).

To measure and know the level of maternal immunity for different pathogens is of fundamental importance to set the vaccination interventions. The timing of vaccination has to consider when the level of maternal antibodies is low enough for an active immune response to progress sufficiently to provide vaccine immunity. Antibody levels often decay to a level still high enough to block responses to vaccine but not high enough to resist a field infection, which creates a window of opportunity for infecting organisms. For these reasons one of the major challenges in developing an active immune response in young pigs has been the interference from maternal immunity (Rose and Andraud, 2017).

Table 4 Maternal immunity duration of different etiological agents in pig diseases

Disease/etiological agent	Maternal antibody persistence (weeks)	Reference
Aujeszky's disease (SHV-1)	14–15	Thomas et al. (2019)
Porcine respiratory and reproductive syndrome (PRRSV)	6	Silva et al. (2015)
Swine influenza (swIAV)	4–14	Van Reeth and Vincent (2019)
Circovirosis (PCV2)	4.9–11.1	Opriessnig et al. (2004)
Parvovirosis (PPV)	20	Truyen and Streck (2019)
Rotavirosis (RVs)	2	Shepherd et al. (2019)
Classical swine fever (CSFV)	8–12	Kirkland et al. (2019)
Encephalomyocarditis (EMCV)	8	Alexandersen et al. (2019)
Porcine cytomegalovirus (PCMV)	8	Thomas et al. (2019)
Foot and mouth disease (FMDV)	8–12	Alexandersen et al. (2019)
Vomiting and wasting disease (pHEV)	4–18	Saif et al. (2019)
PED (PEDV)	4	Saif et al. (2019)
Hepatitis E (HEV)	7-9	Meng et al. (2019)
Porcine lymphotropic herpesviruses (PLHVs)	3	Thomas et al. (2019)
Mycoplasmosis (*Mycoplasma hyopneumoniae*)	4-8	Maes et al. (1996)
Pleuropneumonia (*Actinobacillus pleuropneumoniae*)	5–12	Gottschalk and Broes (2019)
Glässer's disease (*Glaesserella parasuis*)	3-6	Aragon et al. (2019)
Erysipelas (*Erysipelthrix rushiopathiae*)	8	Opriessnig and Coutinho (2019)
Streptococcosis (*Streptococcus suis*)	6	Gottschalk and Segura (2019)
Bordetellosis (*Bordetella bronchiseptica*)	3-4	Brockmeier et al. (2019)
Enteric colibacillosis (Enterotoxigenic *Escherichia coli*)	3-4	Melkebeek et al. (2013)
Oedema disease (Shiga toxin-producing *E. coli*)	3-4	Melkebeek et al. (2013)
Salmonellosis (*Salmonella* Choleraesuis and *Salmonella* monophasic and biphasic Typhimurium)	Up to 10	Cevallos-Almeida et al. (2019)
Ileitis (*Lawsonia intracellularis*)	3-5	Karuppannan and Opriessnig (2018)

2.6 Stressors

At weaning, the piglet is exposed to various stressors (such as removal of sow's milk, mixing with other littermates and change in ambient temperature) and pathogenic agents, mingling with unfamiliar animals, unfavourable environmental changes, contaminated air and low biosecurity measures. The combined and additive effects of these stressors and pathogenic threats accelerate the detrimental effects on the growth performance of pigs (Hyun et al., 1998).

Khafipour et al. (2014) reported that crowding stress had a direct impact on immune response in pigs as indicated by an elevated serum cytokine and plasma cortisol. Crowding stress leads also to an increase in intestinal pH, which inhibits beneficial bacteria and may create favourable conditions for ETEC colonisation and other harmful bacteria, resulting in diarrhoea (Ohland and MacNaughton, 2010).

2.7 Hygiene

Under poor sanitary conditions, piglets had depressed growth performance. The stimulation of the immune system and the consequent inflammatory response interferes with growth because of competition for nutrients, in particular amino acids, between structural tissues and immune function (Le Floc'h et al., 2009; Jayaraman et al., 2015). About amino acids, the optimal standardised ileal digestible threonine:lysine for weaned pigs, raised under unclean sanitary conditions, should be higher compared with weaned pigs raised under clean sanitary conditions (Jayaraman et al., 2015). Maintaining high standards of cleanliness in the nursery is critical for the optimal performance of piglets partly because of its direct or indirect effects on gut health and function. It was reported that piglets raised under unclean conditions had shorter villous height and less crypt depth compared to those raised under clean conditions (Jayaraman and Nyachoti, 2017).

Jayaraman et al. (2017) showed in a room with unclean sanitary conditions higher NH_3 (26.65 ppm vs. 18.17 ppm) and H_2S (0.099 ppm vs. 0.010 ppm) compared to a room with clean sanitary conditions. The deterioration of air quality can cause distress, drop in feed intake and predispose the respiratory tract to infectious agents.

3 Transition from weaning: intervention and prevention strategies

3.1 Creep feeding

In commercial swine production, creep feeding during the suckling period prior to weaning has been a common husbandry practice because it increases the weaning weight of piglets and leads to a smooth transition period for the piglets from sow's milk to the dry feed (Dong and Pluske, 2007; Cabrera et al., 2013). Previous studies demonstrated that creep feed intake during the suckling period has a positive effect on post-weaning feed intake (Bruininx et al., 2002; Cabrera et al., 2013), and it is assumed that nursery piglets offered creep feed prompts them to get adapted to solid feed (Dong and Pluske, 2007). This approach promotes gut development and therefore helps piglets more easily manage a dietary change after weaning (Jayaraman and Nyachoti, 2017). With respect to gut health, creep feeding helps to maintain nutrient supply after weaning, and consequently prevents villous atrophy, thereby reducing the chances of PWD in piglets (Pluske et al., 1996). Lallès et al. (2007) showed as *E. coli* infection is lower in piglets consuming an optimal amount of creep feed as compared with those consuming either no creep feed or high amounts before weaning. This finding illustrates the importance of voluntary feed intake regulation at this period.

3.2 Amino acids

Some traditionally classified dispensable amino acids, such as arginine, glutamine, glutamate and proline, play important roles in intestinal health. Supplementation of 0.4–0.8% L-arginine in a pre-weaning diet enhanced intestinal growth and development in the early post-weaning period (Yang et al., 2016). The administration of proline improved mucosal proliferation, intestinal morphology and tight junction and potassium channel protein expression in early-weaned piglets (Wang et al., 2015). Several mechanisms are highly involved in the benefits of glutamine or glutamine dipeptides on intestinal health (Wu et al., 2004, 2016; Wang et al., 2008). A growing evidence has revealed that the supplementation of indispensable amino acids, such as tryptophan and sulphur amino acids has positive effects on the intestinal health of weaned pigs by regulating host physiology, metabolism, oxidative status and immunity (Zong et al., 2018).

3.3 Phytochemicals and antimicrobial peptides

Dietary supplementation of phytochemicals, such as garlic, pepper, cinnamon, fennel, oregano, thyme, ginger, enhanced disease resistance (antibacterial and anti-viral effects, immune-regulatory activities and antioxidant action) and growth performance. These benefits were likely driven by improved gut health, such as improved intestinal barrier integrity (Zou et al., 2016).

Antimicrobial peptides, also known as host defence peptides, have been considered as potential alternatives to antibiotics in livestock (Robinson et al., 2018), possessing a strong and large-spectrum activity against gram-negative and gram-positive bacteria, fungi, parasites and viruses (Hancock, 2001). There are two ways to incorporate the benefits of antimicrobial peptides into animal health and nutrition. One is direct supplementation of exogenous antimicrobial peptides to animal feed, while the other one is to use dietary supplements/ingredients to stimulate the secretion of endogenous antimicrobial peptides by the host (Robinson et al., 2018). For example, Wang et al. (2006) reported that the supplementation of recombinant lactoferrin increased gut morphology and growth performance of piglets. Considering that the majority of exogenous antimicrobial peptides would be digested in the upper gastrointestinal tract, the stimulation of endogenous antimicrobial peptides secretion by nutritional manipulation may be a better approach (Xiong et al., 2019). Robinson et al. (2018) described the modulation of endogenous antimicrobial peptides synthesis in pigs by dietary compounds such as butyrate and vitamin D.

3.4 Organic acids and short-chain fatty acids

At weaning, natural acidification of the stomach via HCl secretion is reduced owing to immaturity of the digestive system and abrupt change in diet from milk to solid diets. Compared with the mature pig with a pH range of 2.0–3.0, the gastric pH in suckling and weanling piglets ranged between 2.6 and 5.0 (Heo et al., 2013). It is important to maintain a low gastric pH to optimise nutrient digestion and prevent pathogen overgrowth. Dietary addition of organic acids such as citric, fumaric, lactic and formic to weaned pig diets improved growth performance and health (Tsiloyiannis et al., 2001).

The SCFAs are a major fuel source for colonocytes and are essential for maintaining the normal metabolism of the colon mucosa. Other benefits of SCFAs can be summarised as below:

1 Modulate of the expression of genes involved in gut motility (Guilloteau et al., 2010),
2 Modulate the host defence (Guilloteau et al., 2010),
3 Modulate the inflammatory responses (Guilloteau et al., 2010),
4 Could stimulate the formation of intestinal barrier (Feng et al., 2018),
5 Protect intestinal barrier disruption (Feng et al., 2018),
6 May affect the composition of gut microbiota (Wen et al., 2012).

Among SCFAs, butyric acid has attracted significant research attention due to its importance in maintaining gut health in both humans and animals and has been shown to possess strong antimicrobial activity against both gram-positive and gram-negative pathogenic bacteria and effects on regulation of inflammatory responses and immunity (Xiong et al., 2019).

3.5 Probiotics and prebiotics

Probiotics are live microbial feed supplements. Commonly referred to as probiotics are *Bacillus*, yeast and lactic acid-producing bacteria such as *Lactobacillus*, *Bifidobacterium* and *Enterococcus* (Stein and Kil, 2006). Probiotics act inhibiting pathogen adhesion by steric hindrance or competitive exclusion, producing microbicidal substances such as bacteriosins and organic acids (Li et al., 2003) and increasing the production of SCFAs (Sakata et al., 2003).

Prebiotics, such as fructooligosaccharides and transgalactooligosaccharides, but also dietary fibre and resistant starch, are selectively fermented ingredients that allow specific changes, in both the composition and activity of microbiota, that confer benefits upon host well-being and health (Gibson et al., 2004; Heo et al., 2013).

3.6 Biosecurity

Academically, biosecurity can be defined as the application of measures aimed to reduce the probability of introduction and spread of pathogens (Barceló and Marco, 1988), to avoid transmission, either between farms or within the farm. A correct application of biosecurity measures implies a knowledge of the epidemiology of the diseases to be avoided, particularly of the routes of transmission, the stability of the agent in the environment and the role of fomites and vectors (Alarcón et al., 2021).

Biosecurity includes bio-exclusion, bio-containment and bio-management (Levis and Baker, 2011). Bio-exclusion (also known as external biosecurity) is preventing the entry of undesirable pathogens into the farm. Most pig producers concentrate on bio-exclusion as these types of practices are vital indicators for the chances of pathogen introduction, particularly in regards to the PRRSV (Bottoms et al., 2013).

Here are some very important rules that should be applied in all stages of production and in particular in the transition period from weaning.

Fomites carried by people (boots, clothes, etc.) or even the people itself, through contaminated skin, can spread various pig pathogens such as PRRSV, PEDV, TGEV and ETEC. Cleaning and disinfection of clothing, vehicles and equipment before and after contact with animals is essential. Boot baths should be provided at the entrance of the farm (Alarcón et al., 2021).

The control of visitors and fomites is a major focus, as both can bring pathogens to the farm (FAO, 2010). Visitors to farms should always be asked whether they have recently been to potentially contaminated places (FAO, 2010). The shower, before entering the clean area of the farm, should be compulsory. In the farm a delimitation between clean and dirty areas, perimeter fence and parking area outside the farm are important. The aim of these measures is to avoid vehicles delivering feedstuff, collecting dead animals or slurry entering the farm (Alarcón et al., 2021).

The transport of animals can be critical for the introduction in the farm of infectious agents. Trucks destined for the transport of replacements must not be used for the transport of animals to the slaughterhouse. The same truck should not pick up animals on different farms (Alarcón et al., 2021). The vehicles, especially those used to transport pigs, should be thoroughly cleaned before returning or visiting other farms (FAO, 2010). Bio-containment, avoiding the spread of pathogens from herd to herd, is mainly associated with external biosecurity.

Bio-management (also known as internal biosecurity) is characterised by the efforts to control infectious diseases already present in the farm population. Proper cleanliness and disinfection of the pig rooms, all-in/all-out and vaccinations, are the key components of internal biosecurity. In particular, the goal of all-in/all-out management system is to break the disease cycle by preventing the sharing of air-space of pigs carrying clinical disease with pigs susceptible to the infection by that disease (Jayaraman and Nyachoti, 2017). An important measure in preventing diseases in recently weaned pigs is the workflow, following the pig flow, from younger to older. Considering measures related to cleaning and disinfection, it is important to highlight that pens should be cleaned first by removing organic debris, then washed with soapy water, and after rinsing and drying they should be disinfected. As already reported for the external biosecurity, the personnel play a very important role in preventing the spread of the diseases among pigs. In particular, the respect of the assigned areas of work, the correct use of clothes, footwear, gloves and periodic hand washing and correct management of foot baths should be considered key factors (Alarcón et al., 2021). Assessing biosecurity includes measuring the potential routes for disease transmission. The most common biosecurity assessment has been the creation of scores. Most of these scores are based on values attributed to the biosecurity practices by expert opinion panels. Some of the scoring systems evaluate measures that are common to the transmission of different types of infectious agents while others are disease specific (Alarcón et al., 2021).

3.7 Vaccination

The aim of any vaccination is to prevent future disease, while, in many cases, it will not prevent future infection. For this reason, the main expected impact resides in a significant decrease of the clinical signs associated with the infection that the animals are vaccinated against. For several vaccines, a decrease in pathogen shedding in vaccinated animals has also been documented, with a reduction of the pathogen diffusion in the population (Rose and Andraud, 2017).

The transition from weaning is a challenging period from an immunological point of view. Maternal immunity tends to decrease from birth to weaning and, for many agents, it may no longer be present. At the same time, for some

Table 5 Examples of vaccines against the main pig diseases

Disease[a]	Target	Type of vaccine	Administration route
Swine erysipelas	• Piglets Sows, gilts and boars	Inactivated or attenuated	Parenteral
Colibacillosis	• Piglets Sows and gilts	Attenuated/Live non-pathogenic Inactivated	Oral Parenteral
Mycoplasmal pneumonia	• Piglets	Inactivated	Parenteral
Atrophic rhinitis	• Sows	Inactivated	Parenteral
Pleuropneumonia	• Pigs Sows, gilts and boars	Inactivated: 1. killed organisms (bacterins); 2. subunit toxin-based vaccines; 3. bacterins + toxins	Parenteral
Ileitis	• Piglets	Attenuated Inactivated	Oral Parenteral
Glässer's disease	• Piglets Sows and gilts	Commercial inactivated vaccines	Parenteral
Salmonellosis	• Piglets	Attenuated *S.* Choleraesuis/*S.* Typhimurium, alone or combined	Oral
Streptococcosis (*Streptococcus suis*)	• Piglets Sows and gilts	Commercial inactivated vaccines	Parenteral
Parvovirosis	• Sows	Inactivated	Parenteral
PCVAD	• Pigs Sows and gilts	Inactivated or subunit vaccines	Parenteral
PRRS	• Pigs Sows, gilts and boars	Attenuated/Modified-live vaccine or inactivated	Parenteral
Porcine epidemic diarrhoea (PED)	• Sows and gilts	Inactivated or modified-live vaccine	Parenteral
Swine Influenza	• Sows and gilts	Inactivated	Parenteral
Rotavirosis	• Piglets Sows and gilts	Attenuated/Modified-live vaccine	Oral/Parenteral
Aujeszky's disease	• Pigs Sows, gilts and boars	Attenuated or inactivated	Parenteral

[a]Refer to the leaflets of the different vaccines for the vaccination scheme.

diseases, the period just before or after weaning is characterised by the need to vaccinate pigs. The timing of vaccination must take into account the onset of the disease, for which there is the need to vaccinate pigs and the possible interference of maternal immunity capable of interfering with the piglets' own immune response to vaccines. Vaccination of sows and gilts may influence the duration of maternal immunity and may interfere with vaccination efficacy in young piglets.

In Table 5, examples of pig and sow vaccination programmes have been reported, considering different strategies of protection involving pigs in the period of transition from weaning or soon after.

4 Weaning transition and infectious agents

The transition from weaning and the following 2–3 weeks represent a complicated health period in the pig's life. In this period, enteric, respiratory and systemic diseases are common. Multiple pathogens are often involved, and some of them can cause severe immunosuppression and severely worsen the incidence and severity of many diseases. This is the case of PRRSV that induces poor innate and adaptive immune responses (Done and Paton, 1995) and PCV2 viraemic pigs with an increased risk of secondary infections due to immune suppression or immunomodulation (Segalés et al., 2004; Shen et al., 2012).

Among agents causing enteric diseases, PWD caused by enterotoxigenic *E. coli* (ETEC) is one of the most important diarrhoeal diseases worldwide. ETEC PWD is also an example of a disease influenced by predisposing risk factors associated with the transition from weaning. In this period, several agents can be involved in the aetiology of the porcine respiratory disease complex (Opriessnig et al., 2011), with huge differences between countries, regions and farms. Typical of the transition from weaning and the weeks soon after this period is the systemic forms caused by *Streptococcus suis* and *Glaesserella parasuis,* recognising predisposing factors such as concomitant infections, poor hygiene, stress and low biosecurity measures.

5 Viruses acting in the period of transition from weaning as immunosuppressive agents

5.1 Porcine reproductive and respiratory syndrome virus (PRRSV)

5.1.1 Impact and epidemiological characteristics

The PRRS virus causes viraemia, which typically leads to pyrexia, pneumonia with abnormal respiratory behaviour and reduced average daily weight gain (ADWG) (Martelli et al., 2009). Numerous studies have shown that pigs similarly infected with PRRSV elicit diverse response profiles. While some manage to clear the virus within a few weeks, others experience a prolonged infection with a rebound (Go et al., 2019). In an experimental study, Islam et al. (2013) demonstrated that around 20% of pigs exhibited viraemia rebound within 6 weeks postinfection, confirming that this phenomenon is common and could be due to a genetic factor, that would lead to variable immune responses (Boddicker et al., 2012). PRRSV seroconversion and viraemia typically occur during the nursery phase of production. It is a common practice for veterinarians and producers to adjust the timing (delay or move earlier) of the administration of common vaccinations (Drew, 2000) to avoid vaccine failure associated with PRRSV-induced immunosuppression. Overall, the immune responses against PRRSV are ineffective in resolving the infection completely, resulting in prolonged viraemia and persistent infection in lung and lymphoid tissues, potentiating the effects of other swine pathogens (Murtaugh et al., 2002).

It is important to know the main epidemiological characteristics of PRRSV in order to tackle the effect of PRRSV infection in pigs, implementing appropriate measures of control (Table 6).

5.1.2 Control measures

A key aspect to any control programme is the management of the gilt pool (Dee et al., 1994). The introduction of naïve replacement gilts into an infected breeding population results in recurrent episodes of reproductive failure and both horizontal and vertical transmission of the virus from the dam to offspring. To properly acclimatise incoming replacement gilts, many production systems use designated facilities for gilt development units (Dee et al., 1997). Replacement gilts enter these facilities at a young age (at weaning or at 25 kg/BW) and this allows them adequate time (4–6 months) for the development of protective immunity through the use of modified-live vaccines and/or exposure to farm-specific wild-type virus.

The selection of efficacious products and protocols for sanitising livestock transport vehicles and other surfaces in the farms when PRRSV is involved is extremely important in its control. PRRSV is inactivated by lipid solvents, for example chloroform and ether, and it is highly unstable in solutions containing low concentrations of detergents, which disrupt the envelope with concomitant release of the non-infectious core particles and loss of infectivity (Snijder and Meulenberg, 2001). At 'room temperature', complete inactivation of the virus is accomplished with chlorine (0.03%) in 10 min, iodine (0.0075%) in 1 min and a quaternary ammonium compound (0.0063%) in 1 min (Shirai et al., 2000). Decontamination protocols involving drying, thermo-assisted drying and foaming disinfectants containing glutaraldehyde and quaternary ammonium

Table 6 The main epidemiological characteristics of PRRSV influencing risk factors and measures of control

Main topics	Main subcategory	Rationale/specifications
Stability in the environment	PRRSV thermal stability and disinfection	Inactivation[a]: • solvents, heat, drying, or at a pH below 5 or above 7 • iodine (0.0075%) • quaternary ammonium compounds (0.0063%) • chlorine (0.03%) PRRSV is stable at: • −70°C and −20°C (months or years)
Routes of shedding	• Oral and nasal secretion • Semen • Urine • Faeces • Mammary secretion	Shedding of PRRSV in semen reported for up to 101 days postinfection[b]
Persistent infection	The most significant epidemiological feature of PRRSV infection	Detection of infectious virus for 100-165 DPI[b]
Transmission	Route of transmission: • Intranasal • Intramuscular • Oral • Intrauterine • Vaginal	Dose required to infect one-half of the exposed animals[b]: • Intranasal: $1 \times 10^{5.3}$ $TCID_{50}$ • Oral: $1 \times 10^{4.0}$ $TCID_{50}$ • Exposure via artificial insemination: $1 \times 10^{4.5}$ $TCID_{50}$
	Indirect transmission: • equipment • instruments • clothing • water • food • aerosols • arthropod vectors	Meteorological conditions supportive of airborne transport and transmission[b]: • low temperatures • moderate levels of relative humidity • rising barometric pressure slow directional winds
Vertical transmission	• Foetal death • Birth of weak infected pigs • Birth of infected pigs that may appear normal	PRRSVs cross the placenta efficiently only in the last trimester of pregnancy[a]
Introduction in a herd and transmission between herds	• Infected pigs • Virus-contaminated semen • Aerosols in herd-to-herd transmission • Lack of biosecurity measures • Infected transports • Insects • Herd density • Proximity to infected herds • Personnel • Equipment • Wild boar	• Airborne transport of PRRSV reported out to 4.7 km[c] • Wild boars: Infection confirmed in different European countries[d] • Houseflies: mechanically harbour the virus for up to 48 hours. PRRSV replication not supported[e]

[a] Pileri and Mateu (2017).
[b] Zimmerman et al. (2019).
[c] Dee et al. (2009).
[d] Bonilauri et al. (2006); Reiner et al. (2009); Vilcek et al. (2015); Roic et al. (2012); Albina et al. (2000); Oslage et al. (1994); Baroch et al. (2015).
[e] Schurrer et al. (2005).

chloride compounds are effective at inactivating PRRSV in farrowing rooms and transport vehicles in cold and warm climates (Dee et al., 2004). PRRSV ultraviolet light exposure (10 min) completely inactivated the virus on common farm surfaces and materials (Dee et al., 2011).

Treatment of vehicles by washing at high temperature (80°C), followed by phenol disinfection and overnight drying, was effective for complete sanitation of trailers. Alternatively, the use of a thermo-assisted drying and decontamination system or glutaraldehyde fumigation had equivalent efficacy to overnight drying for the complete trailer decontamination (Dee et al., 2005).

5.1.3 Vaccination

Vaccination of sows and piglets is one of the strategies commonly used for controlling PRRS; however, none of the vaccines commercially available has been fully effective in preventing the spread of the virus within a herd (Shirai et al., 2000; Pileri and Mateu, 2016) (Fig. 3). The most common PRRS vaccines can be broadly categorised into modified-live virus (MLV) vaccines or inactivated (IV) vaccines. PRRS MLV vaccines are attenuated live vaccines, which have shown effective protection against homologous and partial protection against heterologous PRRSV strains (Shirai et al., 2000; Chae, 2021), reducing clinical signs, severity and the duration of viraemia and virus shedding (Pileri and Mateu, 2016).

MLVs are able to replicate in the host, to develop viraemia for up to 4 weeks after immunisation, leading to the spread of vaccine virus to naïve animals (Charerntantanakul, 2012) and inducing an immune response similar

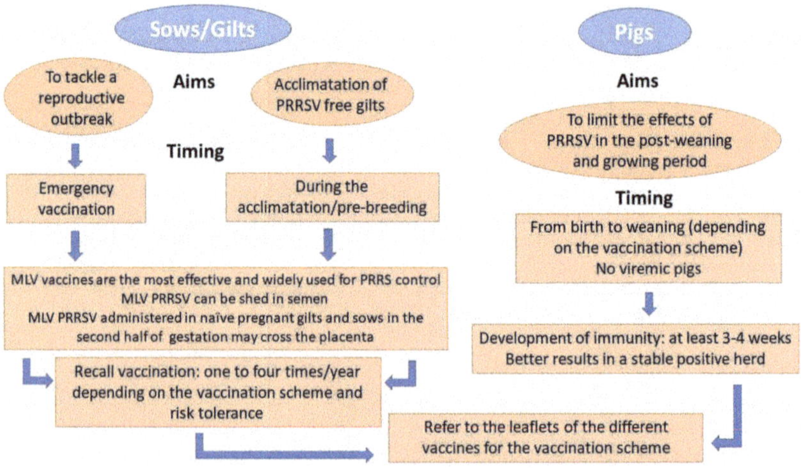

Figure 3 General considerations on PRRSV vaccination strategies (Zimmerman et al., 2019; Chae, 2021).

to that induced by mildly virulent PRRSV isolates. As mentioned earlier, the virological and clinical protection afforded by MLV vaccination is considered partial against a heterologous PRRSV strain. However, in this case, vaccinated pigs can show a viraemia of shorter duration compared to non-vaccinated pigs (Charerntantanakul, 2012; Roca et al., 2012).

The limited protective immunity and immunogenicity to many circulating PRRSV strains have raised major concerns regarding vaccine effectiveness (Shirai et al., 2000; Klinge et al., 2009). The potential efficacy of MLV vaccines in the field can be estimated:

- indirectly, by assessing the biological parameters related to transmission (i.e. duration of viraemia and shedding of the virus, number of chronically infected pigs, etc.),
- directly, by determining the reproduction rate of PRRSV (Pileri and Mateu, 2016) as most vaccines reduce pathogen transmission not only by offering protective immunity but also by reducing host infectivity and duration of infection (Bitsouni et al., 2019).

In this way, MLV vaccines, despite being far from perfect, could achieve drastic reduction in the occurrence and severity of PRRS outbreaks in commercial pig populations, and even help to eliminate the disease (Bitsouni et al., 2019).

PRRS inactivated vaccines show high safety, but the protective immunity that they provide against either homologous or heterologous PRRSV strains is extremely limited, and they often fail to significantly reduce clinical signs, viraemia and duration of shedding in naïve animals (Christianson et al., 1993; Albina et al., 1994). Consequently, MLV vaccines are the predominating vaccines in the field nowadays (Chae, 2021).

5.2 Porcine Circovirus type 2 (PCV2)

5.2.1 Impact and epidemiological characteristics

The viraemia is a characteristic feature of PCV2 infection. In conventional farms, viraemic animals are often present during the nursery and fattening stage. It has been seen that the presence of viraemic animals due to PCV2 occurs with a higher frequency and earlier in farms with PCV2 systemic disease. This condition is very rarely observed in pigs younger than 4 weeks of age (Segalés and Domingo, 2002), suggesting that certain levels of maternal immunity protect against the development of PCV2 systemic disease (Calsamiglia et al., 2007). Following PCV2 infection an important proportion of the animals can show intermittent and/or long-lasting viraemia that can go on until the 28th week of age (Rodríguez-Arrioja et al., 2002). A high PCV2 viraemia has already been shown to be associated with the development of porcine circovirus-associated diseases (Olvera et al., 2004; Meerts et al., 2006). Lopez-Soria et al. (2014) showed that the amount of PCV2 in blood had a negative effect on the ADWG, reporting that the higher the PCV2 load the lower the ADWG.

5.2.2 Vaccination

The most effective tool to tackle the production losses related to the PCV2 infection is to induce a passive or an active immunity that minimises the circulation of the virus in the farm. Most current vaccines worldwide are based on PCV2a strains (Segalés, 2015) and their efficacy against mixed infection of PCV2a and PCV2b and the recently emerged PCV2d is not so clear (Opriessnig et al., 2014; Hou et al., 2019). Different strategies of vaccination are available in sows and piglets (Fig. 4): (1) vaccination of sows for offspring and/or reproductive disease protection; (2) piglet vaccination as the way to control porcine circovirus-associated diseases in the farm and (3) vaccination in both sows and piglets. In this last approach, it is important to consider the possible interference of maternally derived antibodies (MDA) (Segalés, 2020). Martelli et al. (2016) concluded that some interference of MDA with the induction of an efficient immune response could be considered in case of vaccination of sows and piglets, even if, in a condition of high levels of MDA, piglets vaccination at 4, 6 or 8 weeks of age confers a protective immune response with a long-lasting (until 34 weeks of age) antibody response.

The reduction of PCV2 viraemia by the vaccine plays a critical role in controlling PCV2 infection (Chae, 2012) reducing significantly the percentage of viraemic piglets, as well as the viral load in the blood and the virus excretion (Seo et al., 2014).

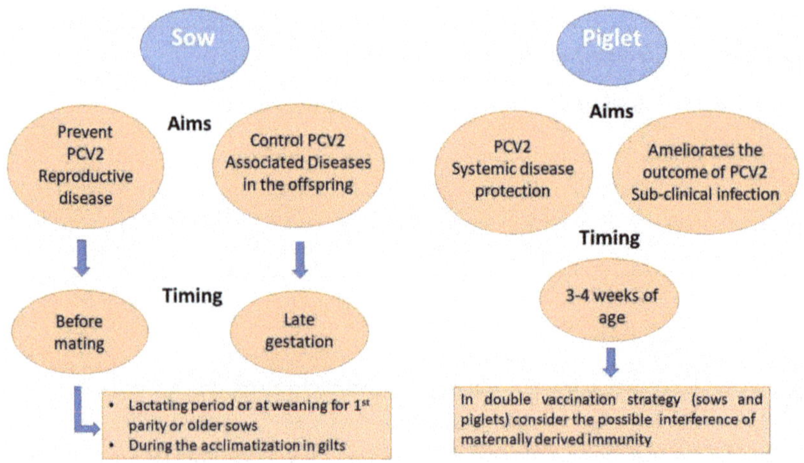

Figure 4 PCV2 vaccination strategies (Segalés, 2020).

Vaccination failure has been noticed in the last few years, showing an increase of PCV2 systemic disease in farms with vaccinated piglets. In these cases probably, the vaccination at weaning might not provide a sufficient vaccine-elicited immune response before natural infection occurs and a proportion of animals may develop PCV2 systemic disease and not just a PCV2 sub-clinical infection (Segalés, 2020). This condition is probably mostly associated with an inadequate management of the vaccine and timing of application (Segalés, 2020):

- too early, with potential interference with maternally derived immunity,
- too late, in case of early infections (too close to natural infection),
- in diseased animals (for example in PRRSV viraemic pigs).

6 Example of diseases recognising risk factors and prevention strategies in the transition from weaning

A large number of infectious agents are capable of causing disease in young weaners. *E. coli* is the major cause of PWD, enterotoxemia and mortality. Other infectious agents can determine severe episodes of diarrhoea in recently weaned pigs (Table 7), in some cases, recognising similar risk factors and measures of control.

6.1 Enterotoxigenic E. coli post-weaning diarrhoea: a multifactorial disease

The most important cause of PWD is associated with the proliferation of enterotoxigenic *E. coli* (ETEC) in the small intestine of pigs, in particular those

Table 7 Etiological agents of post-weaning diarrhoea (modified from Martelli, 2013)

Disease/etiological agent	Age	Lethality
Colibacillosis *Escherichia coli* (ETEC, EPEC)	Most commonly post-weaning until 45–50 days	Can reach 25%
Swine dysentery *Brachyspira hyodysenteriae*	6 weeks old to adult; frequent in the growing-fattening periods	Variable, usually low
Salmonellosis (*Salmonella* Typhimurium)	All ages, usually after weaning, mostly in the growing-fattening periods	Low
PED and TGE Coronavirus PEDV TGEV	All	Can be high; less severe than in neonates
Rotaviral enteritis Rotavirus	From 1 to 5 weeks	Low, < 20%
Proliferative enteropathy *Lawsonia intracellularis*	Post-weaning (5 weeks old to young adults)	Low

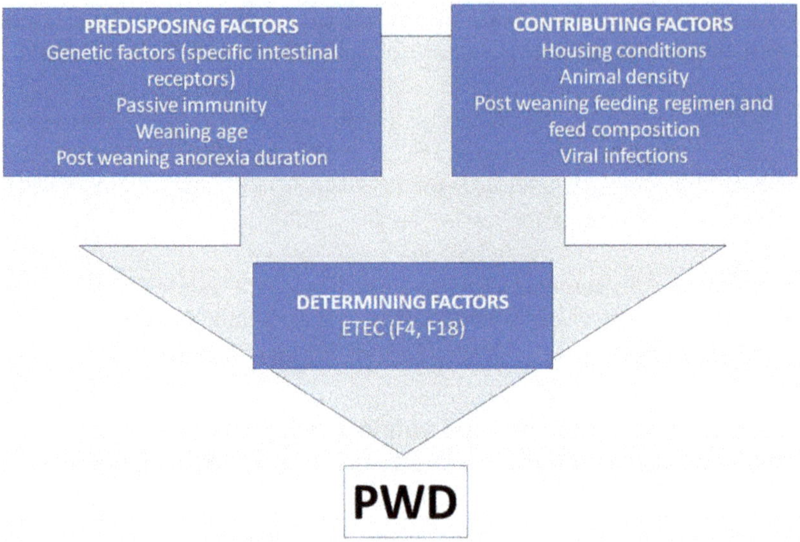

Figure 5 The multifactorial genesis of PWD in pig involves interaction between predisposing, contributing and determining factors (modified from Rhouma et al., 2017).

that express fimbrial adhesins F4 (K88) or F18 (determining factors). However, ETEC PWD shows a multifactorial genesis involving the interaction between predisposing, contributing and determining factors (Rhouma et al., 2017) (Fig. 5). Predisposing factors are elements that increase the risk of developing a disease (Racine et al., 2016). A contributing factor is a condition that influences the effect and severity of the infection. In other terms, outbreak contributing factors are conditions that enable or amplify an outbreak (Brown et al., 2017). The determining factors are considered as the causal factors of post-weaning colibacillosis such as ETEC (Rhouma et al., 2017).

6.1.1 Enterotoxigenic E. coli post-weaning diarrhoea: predisposing factors

The pig's susceptibility to developing an ETEC F4 and F18 infection depends on the presence and amount of F4 and F18 receptors, respectively, found in the brush border membrane of the small intestine, allowing the ETEC to adhere and develop diarrhoea (Jørgensen et al., 2003). The locus encoding the intestinal receptor responsible for adhesion of F4 ETEC was mapped to the q41 region on pig chromosome 13 (SSC13) (Fontanesi et al., 2012) and among the candidate genes found in this region (MUC4, MUC13, TFRC and MUC20), the MUC13 and MUC4 are the most likely responsible genes (Zhang et al., 2008). In addition tyrosine kinase non-receptor 2 (ACK1) and UDP-GlcNAc:betaGal beta-1,3-*N*-acetylglucosaminyltransferase 5 (B3GNT5) genes have been proposed as genetic markers for pig ETEC resistance/susceptibility (Wang et al., 2007; Ouyang et al., 2012). Regarding pig resistance to ETEC F18 infection, two main single nucleotide polymorphisms (SNPs) localised on alpha (1,2)-fucosyltransferase (FUT1) and bactericidal/permeability-increasing protein (BPI) (Meijerink et al., 2000; Liu et al., 2013) genes, respectively, have been proposed and greater consensus has been attained for the SNP located on FUT1.

Piglets are born with an immature mucosal immune system, which develops over the first few weeks of life following a programmed sequence, without reaching a complete maturity at weaning ages which are common on commercial farms (Stokes, 2017). At weaning the piglet's mucosal immune system is required to distinguish between antigens and recognise and respond appropriately to potential pathogens (Stokes, 2017). The relative immaturity of the intestinal immune system at weaning age (characterised by the lack of a previous stimulation by pathogens) may reduce the ability of the weaned pig to mount an appropriate immunological response to pathogens (Heo et al., 2013). This condition is coupled with the sow milk removal, and consequent discontinuation of nutritive intake of the IgA present in this milk contributes to increase the susceptibility of pigs to microbial infections (Rhouma et al., 2017).

Weaning age is an important ETEC PWD predisposing factor. Weaning pigs at an early age, before 21 days, increase the occurrence of PWD, encouraging them to move away from this practice and as a result wean pigs no earlier than 26 days of age (Rhouma et al., 2017).

Feed intake is usually reduced initially after weaning and this is strongly correlated with the risk of disease occurrence over the post-weaning period (Pluske et al., 1997; Vente-Spreeuwenberg et al., 2003). This condition promotes the creation of an ideal environment for the multiplication of bacteria such as pathogenic *E. coli* and allows toxins and bacteria to cross the epithelium as a result of intestinal inflammation (Campbell et al., 2013).

6.1.2 Enterotoxigenic E. coli post-weaning diarrhoea: contributing factors

Among the contributing factors, housing conditions are very important; in particular, it is essential to provide the correct environmental temperature, 26–28°C, to maintain pigs in their thermo-neutral zone, avoiding drafts and removing moisture and gases using adequate ventilation (Rhouma et al., 2017). Automatic temperature control was associated with a decreased risk of PWD (Laine et al., 2008).

Mixing piglets from different farms is a common practice in pig husbandry, resulting in fighting with aggressive interactions, occurring typically during the first few hours after grouping (Coutellier et al., 2007).

Post-weaning feeding regimen and feed composition play a role in the development of ETEC PWD. In Fig. 6, have been reported the effects of different nutrients on the incidence of PWD.

Resistant starch improves colon health thanks to the fermentation in the large intestine by colonic microflora and then is digested into hydrogen, methane and SCFAs, such as acetic, propionic and butyric acid (Alsaffar, 2010) (Fig. 7). The higher proportion of resistant starch showed a reduction in ileal and caecal digesta pH (Gao et al., 2019) improving the resistance to infective diseases (Mateos et al., 2007).

Diets characterised by high crude fibre (10–17%) and low in nutrients, particularly CP reduce proliferation and excretion of pathogenic *E. coli*. Considering that the apparent digestibility of CP at the terminal ileum is 75–85%, diets containing 220–250 g CP/kg, result in a considerable amount of protein entering the large bowel. As an example, a 7 kg pig eating 300 g/day of a 220 g CP/kg diet could lose 33–55 g protein daily to its hindgut (Pluske et al., 2002). Interestingly the activity of major gut proteases (trypsin and chymotrypsin) was stimulated after weaning by an increase in the level of CP in the weaning diet not exceeding 200 g/kg. This would explain data showing a decrease in the apparent ileal digestibility of nitrogen in pigs fed diets containing more than 225 g CP/kg (Li et al., 1993). Protein is fermented rapidly by the microbiota with the production of biogenic amines such as putrescine, cadaverine and tryptamine and gases (e.g. NH_3) that have been implicated in the clinical expression of ETEC PWD (Pluske et al., 2002).

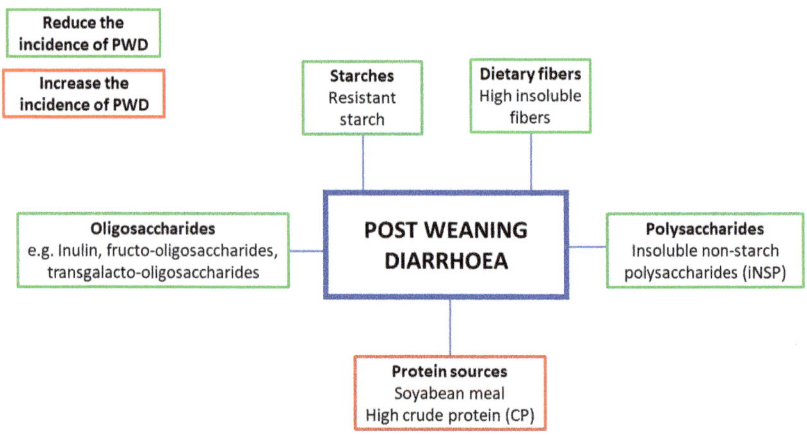

Figure 6 Effects of different nutrients on the incidence of PWD (Gerritsen et al., 2012; Rhouma et al., 2017; Liu et al., 2018; Gao et al., 2019).

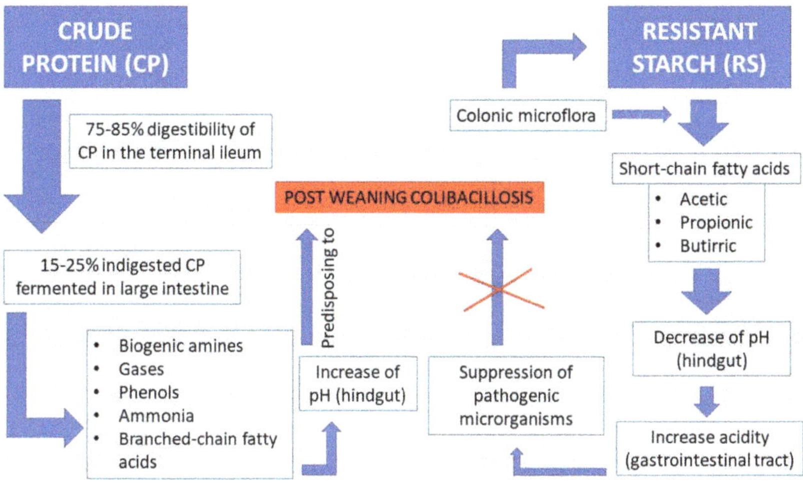

Figure 7 Role of crude protein and resistant starch in predisposing and preventing pig to PWD (Gao et al., 2019).

Apart from the feed's composition, the management of its administration also plays an important role. Interestingly it was shown that the prevalence of PWD was higher on farms that fed weaned piglets only twice a day with a restricted amount of feed, than on farms that provided more than two meals per day with or without feed restriction (Laine et al., 2008).

Considering the effect of other pathogens and in particular viral agents, as contributing factors, the isolation of ETEC strains in weaned pigs can often be accompanied by rotavirus or by the coronavirus of porcine epidemic diarrhoea (PED) (Nagy et al., 1996). Rotavirus is an important etiological agent and contributing factor of PWD, with different prevalence. In the USA, Canada and Mexico, rotavirus type A was detected in diarrhoeic pigs with the prevalence of 84% and 61%, from 21 to 55 days and >55 days old pigs respectively (Marthaler et al., 2014), while in Italy the 71.5% of pigs suffering from diarrhoea, from 28 to 84 days old, were positive for rotavirus type A (Martella et al., 2007). Co-infection with rotavirus and ETEC results in more severe disease, compared to the outcome of the infection with a single etiological agent (Tzipori et al., 1980). PRRSv infection can predispose pigs to ETEC infection, enteric colibacillosis and *E. coli* septicaemia mainly through an impairment of the immune response of piglets (Rhouma et al., 2017).

6.1.3 Strategies to control enterotoxigenic E. coli post-weaning diarrhoea

Several strategies can be implemented to prevent and control ETEC PWD, mainly based on feeding strategies, as reported above (see also Chapter 2), use

Table 8 Benefits and limitations of the major preventive strategies for the control of ETEC PWD in pigs (modified from Rhouma et al., 2017)

Strategies	Benefit	Rationale	Strategy limitations
Achieve high post-weaning feed intake/increase weaning age	Reduce the risk for PWD	Minimise post-weaning anorexia	Economically not convenient
Avoid some ingredients (such as soybean meal)	Reduces the severity and frequency of PWD	Soybean meal reduces: • duodenal specific activities intestinal enzymes • intestinal digestibility. Contradictory results have been reported	Growth retardation
Feeds with decreased protein content	Decrease *Escherichia coli* colonisation and to minimise PWD prevalence	High protein diet cause protein maldigestion	Growth retardation
Stocking density reduction	Decrease the occurrence of PWD and other diseases	Reduce stress factors	Significant costs
Diagnostic tools and appropriate diagnostic path	Avoid the inappropriate use of antimicrobials and improve the prevention	Correct diagnostic criteria to confirm ETEC PWD	Need of specialised laboratories
Vaccination	Specific protection against ETEC F4 and F18	Commercial F4 and F18 live avirulent vaccines are available	Interference with the lactogenic immunity
Selection of pigs genetically resistant to F4 and/or F18	Very effective approach	Pigs resistant to ETEC F4 and/or F18 do not express intestinal receptors for these fimbrial types	Expensive process Lack of techniques for a large-scale selection
Zinc Oxide (ZnO) administration	• Inhibition of bacterial adhesion to the intestinal mucosa • Growth rate stimulation • Integrity of intestinal mucosa • Modulate immune functions	ZnO at the levels of 2400–3000 ppm in pig feed was effective in reducing PWD	• Soil heavy metal contamination • Bacterial resistance • Co-resistance • No longer available from June 2022

Strategies	Benefit	Rationale	Strategy limitations
Addition of organic acid to reduce gastric pH	• Modulate immune function • Lower pH in the stomach • Improve growth performance • Reduce PWD	Exposure to low pH values (i.e. 3.0–4.0) is bactericidal for many pathogenic bacteria	Exact modes of action are still unknown Antimicrobial activities is different between acids
Addition of prebiotics and probiotics	• Improve intestinal health • Improve growth performance • Reduce ETEC attachment to intestinal mucosa • Reduced diarrhoea	Different mechanisms: • Inhibition of bacterial toxins production • Inhibition of the adhesion of pathogens to the intestinal mucosa	Sometimes contradictory studies and results on their effectiveness
Dehydrated porcine plasma administration	Improve growth performance Reduce incidence and severity of diarrhoea	Reduction of inflammatory cytokines expression Maintenance of mucosal integrity, and enhancement of specific antibody defense	High cost Rigorous control during the preparation process

of antimicrobials, prevention of predisposing viral infections, ETEC vaccination and use of zinc oxide (ZnO) (Table 8).

Antimicrobial treatments have been used administered in feed or water to treat and to prevent ETEC PWD (Luppi, 2017). Nowadays a prophylactic use of antimicrobials is no longer acceptable and preventive husbandry, active immunoprophylaxis and dietary preventive measures are the main tools for controlling PWD in weaned pigs.

To protect newly weaned piglets, an active intestinal mucosal immune response, in which the production of antigen-specific secretory IgA plays an important role, is required (McGhee et al., 1992). By vaccinating the oral route is the most logical route to obtain this type of response (Melkebeek et al., 2013). Several oral vaccines have been successfully implemented in weaned pigs, including subunit vaccines as well as live oral vaccines. A bivalent F4/F18 live vaccine comprising a F4 and a F18 non-pathogenic strain has been recently licensed in several countries for immunisation of pigs against post-weaning enteric colibacillosis (Fairbrother and Nadeau, 2019).

The addition of zinc oxide at the levels of 2400–3000 ppm in pig feed was demonstrated to reduce the incidence of PWD and mortality, improving growth performance in weaned pigs (Roselli et al., 2003). On 26 June 2017, the European Commission adopted a decision to withdraw all marketing authorisations for veterinary medicinal products, listed in Annex I of the decision, containing zinc oxide administered orally to food-producing species. The Member States have had the possibility to defer the withdrawal of the marketing authorisations for up to five years from this date, so the decision must be implemented no later than 26 June 2022.

6.2 Streptococcosis

6.2.1 Epidemiology and predisposing factors

S. suis is a major porcine pathogen causing high morbidity worldwide, including well-managed herds with high hygiene standards. Transmission of virulent strains between herds usually occurs by the movement of healthy carrier animals. The introduction of carrier pigs harbouring virulent strains (breeding gilts, boars and weaners) into a noninfected recipient herd may result in the subsequent onset of disease in weaners and/or growing pigs. Sows infect their piglets during farrowing via contamination from vaginal colonisation and probably through the respiratory route (Segura et al., 2016). Even though different serotypes and strains within the same serotype are present in a herd, a single strain usually causes most diseases (Marois et al., 2007). Horizontal transmission is important, especially during outbreaks, when diseased animals shed higher numbers of bacteria, thereby increasing transmission by direct contact or aerosol (Cloutier et al., 2003). The oral route

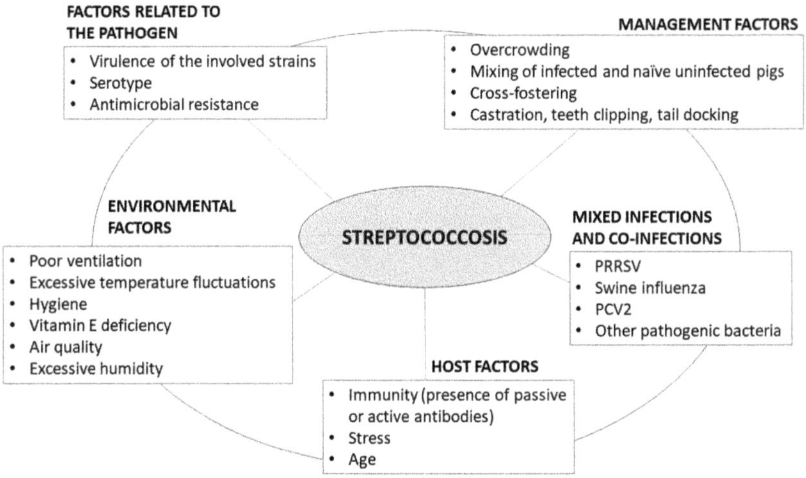

Figure 8 Main factors influencing and modulating the development of streptococcosis in pigs (modified from Obradovic et al., 2021).

of infection is not yet been proven, and recent studies suggested that *S. suis* is not able to survive in feed pellet or crumb feed, with or without formic acid or in the stomach contents (Warneboldt et al., 2016). *S. suis* may also be transmitted via fomites (Dee and Corey, 1993) and flies (Enright et al., 1987), although the importance of such vectors has still to be confirmed. In addition to the virulence of involved strains, other factors can impact the development of the disease such as overcrowding, poor ventilation and climatic conditions, poor hygiene status, high air pollution load and other stressors (Vötsch et al., 2018) (Fig. 8). Host-specific factors, such as age, genetic background and immunosuppression, influence the disease's development. Weaning piglets are most susceptible since protective maternal antibodies decline (Cloutier et al., 2003).

6.2.2 Control and preventive measures

Beta-lactam antibiotics (penicillin, ampicillin, amoxicillin and ceftiofur), gentamicin, tiamulin and a combination of trimethoprim and sulphonamide are the most useful antibacterial products for parenteral treatment (Gottschalk and Segura, 2019). Vaccines used in the field to prevent *S. suis* disease are either autogenous or (a very few) commercial bacterins, and results have been inconsistent (Segura, 2015). Possible reasons for the failure of autogenous vaccines include degradation of protective antigens caused by heat or formalin processing, inadequate bacterial antigen concentrations, the adjuvant included in the formulation, low production of antibodies and/or production of antibodies to antigens not associated with protection (Segura, 2015;

Rieckmann et al., 2020) and multiple *S. suis* serotypes circulating in the farm (Segura et al., 2020). For these reasons, it is of fundamental importance to obtain clear information about the serotypes/strains involved in the clinical cases, following appropriate diagnostic procedures (Fig. 9). In herds infected with multiple strains or serotypes of *S. suis*, multivalent vaccines or vaccines that provide a strong degree of cross-immunity are needed to provide adequate control of infection (Segura, 2015).

Even within the same serotype, protection obtained with a commercial bacterin may be difficult to predict due to high phenotypic variation within strains of the same serotype, based on available data for serotype 2 (Segura, 2015). Vaccination of sows and gilts has also been described as effective (Swildens et al., 2007). Most vaccination-challenge studies have been carried out with piglets. Because streptococcosis is most often observed at 6–10 weeks of age, due to maternal antibody decline, the first of two doses of bacterin need to be given at approximately 3–4 weeks of age and interference by maternal antibodies must be considered (Lapointe et al., 2002). Sow vaccination is less costly and labour intensive, thus representing an economical alternative to piglet vaccination. However, without subsequent active vaccination of the piglets to lengthen protection, this strategy could lead to highly susceptible growers (Baums et al., 2010). Preventing measures based on feed additives showed encouraging results. Correa-Fiz et al. (2020), in a comparative study about the prevention of streptococcosis, tested the effects of feed supplementation with medium-chain fatty acids and a natural anti-inflammatory. In this study clinical signs compatible with *S. suis* obtained with this supplementation were at the same level as those obtained with amoxicillin, concluding that feed additives may constitute an alternative to metaphylaxis.

6.3 Glässer disease

6.3.1 Epidemiology and predisposing factors

G. parasuis is a member of the normal respiratory microbiota, ubiquitous in swine herds worldwide, colonising the upper respiratory tract of piglets soon after birth through contact with the sow, which is the main reservoir and transmitter of *G. parasuis* to the piglets. Glässer's disease is commonly developed after weaning due to a reduction in maternal antibodies, mix of litters, strains variations, and other stress factors (Costa-Hurtado and Aragon, 2013; Costa-Hurtado et al., 2020). Basically, all pig farms are positive to *G. parasuis*, but not all farms develop the disease (Mahmmod et al., 2020) and this is usually due to specific strains, sometimes following a predisposition of the host due to distress factors and/or co-infections (Aragon et al., 2012), such as PRRSV and PCV2 (Pereira et al., 2017).

Among factors affecting Glässer's disease epidemiology, Mahmmod et al. (2020) showed as multi-site production type increases the risk of Glässer's disease in comparison to the one-site production type (farrow-to-finish). This could be due to stress factors caused by animal handling, transportation and relocation from one environment and management conditions to another. For these reasons, efforts should be done to achieve management practices that minimise stressful situations, particularly on the weaning phase (Costa-Hurtado et al., 2020).

Correa-Fiz et al. (2016) reported that the nasal microbiota composition was associated with the clinical status of the farm of origin of the piglets, leading to different susceptibilities to invasive infection by *G. parasuis*. Some microbial populations, such as *Chitinophagaceae* and *Corynebacteriaceae* families, may facilitate the colonisation of virulent *G. parasuis* strains, which are inhibited by *Enterobacteriaceae* and *Peptostreptococcaceae*, meanwhile, members from the *Flavobacteriaceae* family could facilitate the colonisation of non-virulent *G. parasuis* strains (Mahmmod et al., 2020). The interactions reported could be a base for innovative non-antimicrobial alternatives for Glässer's disease control.

6.3.2 Control and preventive measures

Antibiotics are widely used to prevent and control the disease, but increasing pressure to reduce antibiotics usage, particularly for prophylaxis strategies, puts more emphasis on vaccination strategies to prevent systemic infection and mortality.

The early colonisation of pigs with virulent strains in the presence of maternal immunity actually controls the evolution of the infection, preventing the disease and mortality post-weaning (Oliveira et al., 2004), avoiding subsequent Glässer's disease in the farms (Correa-Fiz et al., 2016). However, contradictory results have been obtained on the influence of MDA on nasal colonisation of piglets (Cerdà-Cuéllar et al., 2010; Brean et al., 2016). Sow vaccination may be effective to protect piglets during lactation, but lasting protection may require vaccination of piglets after the farrowing phase, together with actions to ensure early colonisation. There is not a complete agreement on the interference of the MDA with piglet vaccination against *G. parasuis*. While some studies report no effective interference when vaccinating piglets with two doses (Oh et al., 2013), other studies report an interference effect after sow vaccination that precluded the effectiveness of vaccination in piglets (Pomorska-Mol et al., 2011).

Commercially available bacterin vaccines are based on serovar 5, combined serovars 4 and 5, or combined serovars 1 and 6 for use in sows and pigs (Zhao et al., 2017). All these products show variable cross-protection to heterologous strains (Aragon et al., 2019). When a commercial vaccine fails

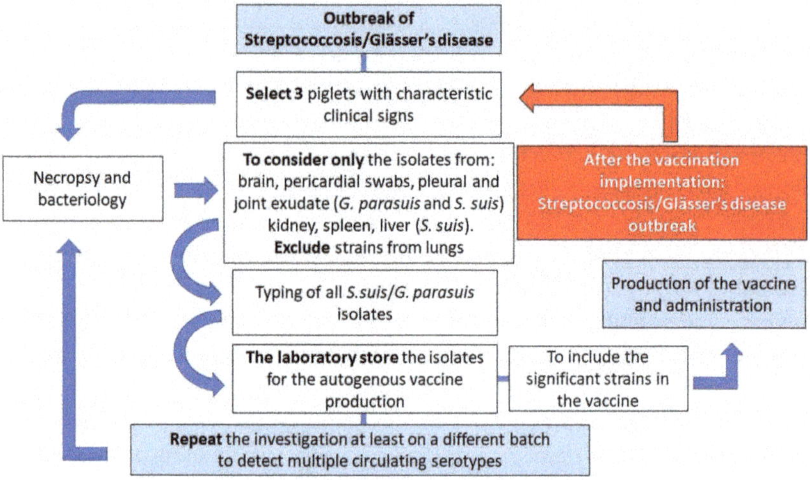

Figure 9 Steps and criteria for selecting strains of *Streptococcus suis* or *Glaesserella parasuis* for inclusion in an autogenous vaccine.

to protect vaccinated herds, the most appropriate solution, in a short term, is to develop autogenous vaccines, which need to be developed rationally. It is recommended to include in the formulation, strains necessarily typed and characterised, isolated from systemic sites such as meninges, pericardium and joints (Oliveira and Pijoan, 2004; McOrist et al., 2009), avoiding isolates from the respiratory tract (Fig. 9).

7 Conclusion and future trends

The weaning transition period, as described in this chapter, is a very stressful phase in pig's life in which several factors act in a synergistic way, predisposing piglets to poor health and reducing their performance.

Measures aimed at improving animal health and welfare in the period of transition from weaning, are of huge importance. Among these the stocking densities reduction, the improvement of piglet health at weaning, the diagnostic approach, and the vaccination strategies implementation are the most important. A very important challenging point is that veterinarians and farmers update the concept of control of the diseases intended as a control of a single disease, switching it in the control of an entire system, where we have complex interaction between pathogens and predisposing factors.

The next steps to improve the health of recently weaned pigs should focus the research on four main fields. The first is about the improvement of knowledge about the diagnostic approach, in order to reach the correct diagnosis and consequently correct measures of prevention and control. Even if this may seem a sufficiently developed topic, it is frequent to have in the practice, diagnostic conclusions not supported by correct criteria. The second concerns the study of the impact of pig's genetics on disease resistances and how this aspect modifies the epidemiology of the diseases themselves. The third aspect concerns the study of the microbiota and its impact on diseases, as well as the tools available to influence the bacterial populations that compose it. Last but not least, is the development and use of technology to improve the health and performance of pigs. The implementation of sensors to monitor parameters such as humidity, temperature, ammonia, CO_2 and dust are of fundamental importance to check environmental variables and their impact on pig health. Cameras to detect problems such as lameness or immobile animals, to identify aggressive behaviours such as fights or tail bites can be useful tools, as well as devices for measuring the animal temperature and individual feed consumption. All the data collected using these tools can be extremely valuable, but only if taken accurately, analysed correctly and presented in a way that allows veterinarians to make decisions.

8 Where to look for further information

8.1 Further reading

The following articles provide a good overview of the subject:

- Alarcón, L. V., Allepuz, A. and Mateu, E. (2021). 'Biosecurity in pig farms: a review', *Porcine Health Management* 7:5. https://doi.org/10.1186/s40813-020-00181-z.
- Drew, T. W. (2000). A review of evidence for immunosuppression due to porcine reproductive and respiratory syndrome virus. *Veterinary Research* 31:27–39.
- Gao, J., Yin, J., Xu, K., Li, T. and Yin, Y. (2019). 'What Is the Impact of Diet on Nutritional Diarrhea Associated with Gut Microbiota in Weaning Piglets: A System Review', *Hindawi BioMed Research International* Article ID 6916189, 1–14.
- Moeser, A. J., Pohl, C. S. and Rajput, M. (2017). 'Weaning stress and gastrointestinal barrier development: Implications for lifelong gut health in pigs', *Animal Nutrition* 3:313–321.
- Rhouma, M., Fairbrother, J. M., Beaudry, F. and Letellier, A. (2017). 'Post weaning diarrhea in pigs: risk factors and non-colistin-based control strategies', *Acta Veterinaria Scandinavica* 59:31.
- Rieckmann, K., Pendzialek, S. M., Vahlenkamp, T. and Baums, C. G. (2020). A critical review speculating on the protective efficacies of autogenous *Streptococcus suis* bacterins as used in Europe, *Porcine Health Management*, 6:12.

8.2 Key conferences held regularly every year or few years

- AllenD.LemanSwineConference.https://ccaps.umn.edu/allen-d-leman-swine-conference.
- American Association of Swine Veterinarians. https://www.aasv.org.
- European Symposium of Porcine Health Management (ESPHM). https://ecphm.org/members-residents-area/meetings-trainings.
- International pig veterinary society (IPVS). https://www.theipvs.com.

8.3 Organisations and websites worth visiting where can be found research in this subject

- EU PiG Innovation Group (EU PiG). EU PiG aims are to help pig producers find tried-and-tested best practice from fellow producers across Europe, sharing all new knowledge in one place, online - funded by the European Horizon 2020 programme. https://www.eupig.co.uk.
- National Agricultural Library – U.S. Department of Agriculture. https://www.nal.usda.gov/topics/swine.
- Pig333. This website contains articles by swine specialists on health, management, environment, biosecurity, welfare, genetics, facilities and markets, as well as news, pig prices, clinical cases, events. https://www.pig333.com.
- Prohealth. The Prohealth Consortium has expertise in veterinary science and epidemiology, physiology and immunology, genetics, nutrition, socio-economics, welfare and production science of pigs and poultry https://www.fp7-prohealth.eu.
- The pig site. An updated guide focusing main subjects on pig health and management. https://www.thepigsite.com.

9 References

Alarcón, L. V., Allepuz, A. and Mateu, E. (2021). Biosecurity in pig farms: a review, *Porcine Health Manag.* 7(1):5. https://doi.org/10.1186/s40813-020-00181-z.

Albina, E., Madec, F., Cariolet, R. and Torrison, J. (1994). Immune response and persistence of the reproductive and respiratory syndrome virus in infected pigs and farm units, *Vet. Rec.* 134(22):567–573.

Albina, E., Mesplède, A., Chenut, G., Le Potier, M. F., Bourbao, G., Le Gal, S. and Leforban, Y. (2000). A serological survey on classical swine fever (CSF), Aujeszky's disease (AD) and porcine reproductive and respiratory syndrome (PRRS) virus infections in French wild boars from 1991 to 1998, *Vet. Microbiol.* 77(1–2):43–57.

Alexandersen, S., Knowles, N. J., Belsham, G. J., Dekker, A., Nfon, C., Zhang, Z. and Koenen, F. (2019). Picornaviruses. In: Zimmerman, J. J., Karriker, L. A., Ramirez, A., Schwartz, K. J., Stevenson, G. W. and Zhang, J. (Eds), *Disease of Swine* (11th edn.), Wiley-Blackwell, Hoboken, NJ, pp. 641–684.

Allwin, B., Swaminathan, R., Mohanraj, A., Suhas, G. N., Vedaminckam, S., Gopal, S. and Kumar, M. (2016). The wild pig (Sus scrofa) behavior - a retrospective study, *J. Vet. Sci. Techno* 7(4):333.

Almeida, F. N. and Stein, H. H. (2010). Performance and phosphorus balance of pigs fed diets formulated on the basis of values for standardized total tract digestibility of phosphorus, *J. Anim. Sci.* 88(9):2968–2977.

Alsaffar, A. A. (2010). Effect of thermal processing and storage on digestibility of starch in whole wheat grains, *J. Cereal Sci.* 52(3):480–485.

Aragon, V., Segalés, J. and Oliveira, S. (2012). Glässer disease. In: Zimmerm, J., Karrker, L., Ramirez, A., Schwarz, K. and Stevenson, G. (Eds), *Diseases of Swine* (10th edn.), Wiley-Blackwell, Ames, IA; The Atrium, Southern Gate, Chichester: Oxford, pp. 760–770.

Aragon,V., Segalés, J. and Tucker, A. W. D. (2019). Glässer disease. In: Zimmerman, J. J., Karriker, L. A., Ramirez, A., Schwartz, K. J., Stevenson, G. W. and Zhang, J. (Eds), *Disease of Swine* (11th edn.), Wiley-Blackwell, Hoboken, NJ, pp. 844–853.

Averos, X., Brossard, L., Dourmad, J. Y., de Greef, K. H., Edwards, S. A. and Meunier-Salaun, M. C. (2012). Meta-analysis on the effects of the physical environment, animal traits, feeder and feed characteristics on the feeding behaviour and performance of growing finishing pigs, *Animal* 6(8):1275–1289.

Bailey, M., Plunkett, F. J., Rothkotter, H. J., Vega-Lopez, M. A., Haverson, K. and Stokes, C. R. (2001). Regulation of mucosal immune responses in effector sites, *Proc. Nutr. Soc.* 60(4):427–435.

Bailey, M., Haverson, K., Inman, C., Harris, C., Jones, P., Corfield, G., Miller, B. and Stokes, C. (2005). The influence of environment on development of the mucosal immune system, *Vet. Immunol. Immunopathol.* 108(1-2):189–198.

Barcelo, J. and Marco, E. (1988). On-farm biosecurity. In: Varley, M. A., Done, S., Thomson, J. and International Pig Veterinary Society (Eds), *Scientific Committee of the 15th IPVS Congress. Proceedings of the 15th IPVS Congress*, Nottingham University Press. Nottingham (United Kingdom), pp. 129–133.

Baroch, J. A., Gagnon, C. A., Lacouture, S. and Gottschalk, M. (2015). Exposure of feral swine (Sus scrofa) in the United States to selected pathogens, *Can. J. Vet. Res.* 79(1):74–78.

Baums, C. G., Brüggemann, C., Kock, C., Beineke, A., Waldmann, K. H. and Valentin-Weigand, P. (2010). Immunogenicity of an autogenous *Streptococcus suis* bacterin in preparturient sows and their piglets in relation to protection after weaning, *Clin. Vaccine Immunol.* 17(10):1589–1597.

Bitsouni, V., Lycett, S., Opriessnig, T. and Doeschl-Wilson, A. (2019). Predicting vaccine effectiveness in livestock populations: a theoretical framework applied to PRRS virus infections in pigs, *PLoS ONE* 14(8):e0220738. https://doi.org/ 10.1371/journal.pone .0220738.

Boddicker, N., Waide, E. H., Rowland, R. R. R., Lunney, J. K., Garrick, D. J., Reecy, J. M. and Dekkers, J. C. M. (2012). Evidence for a major QTL associated with host response to porcine reproductive and respiratory syndrome virus challenge, *J. Anim. Sci.* 90(6):1733–1746.

Bonilauri, P., Merialdi, G., Dottori, M. and Barbieri, I. (2006). Presence of PRRSV in wild boar in Italy, *Vet. Rec.* 158(3):107–108.

Bottoms, K., Poljak, Z., Dewey, C., Deardon, R., Holtkamp, D. and Friendship, R. (2013). Evaluation of external biosecurity practices on Southern Ontario sow farms, *Prev. Vet. Med.* 109(1-2):58–68.

Boudry, G., Péron, V., Le Huérou-Luron, I., Lallès, J. P. and Sève, B. (2004). Weaning induces both transient and long-lasting modifications of absorptive, secretory, and barrier properties of piglet intestine, *J. Nutr.* 134(9):2256–2262.

Brean, M., Abraham, S., Hebart, M. and Kirkwood, R. N. (2016). Influence of parity of birth and suckled sows on piglet nasal mucosal colonization with *Haemophilus parasuis*, *Can. Vet. J.* 57(12):1281–1283.

Brockmeier, S. L., Register, K. B., Nicholson, T. L. and Loving, C. L. (2019). Bordetellosis. In: Zimmerman, J. J., Karriker, L. A., Ramirez, A., Schwartz, K. J., Stevenson, G. W. and Zhang, J. (Eds), *Disease of Swine* (11th edn.), Wiley-Blackwell, Hoboken, NJ, pp. 767–777.

Brown, L. G., Hoover, E. R., Selman, C. A., Coleman, E. W. and Schurz Rogers, H. (2017). Outbreak characteristics associated with identification of contributing factors to foodborne illness outbreaks, *Epidemiol. Infect.* 145(11):2254–2262.

Bruininx, E. M., Binnendijk, G. P., van der Peet-Schwering, C. M., Schrama, J. W., den Hartog, L. A., Everts, H. and Beynen, A. C. (2002). Effect of creep feed consumption on individual feed intake characteristics and performance of group-housed weanling pigs, *J. Anim. Sci.* 80(6):1413–1418.

Brumm, M. C. (2019). Effect of environment on health. In: Zimmerman, J. J., Karriker, L. A., Ramirez, A., Schwartz, K. J., Stevenson, G. W. and Zhang, J. (Eds), *Disease of Swine* (11th edn.), Wiley-Blackwell, Hoboken, NJ, pp. 50–58.

Cabrera, R. A., Usry, J. L., Arrellano, C., Nogueira, E. T., Kutschenko, M., Moeser, A. J. and Odle, J. (2013). Effect of creep feeding and supplemental glutamine or glutamine plus glutamate (Amino gut) on pre- and post-weaning growth performance and intestinal health of piglets, *J. Anim. Sci. Biotechnol.* 4(1):29.

Calsamiglia, M., Fraile, L., Espinal, A., Cuxart, A., Seminati, C., Martín, M., Mateu, E., Domingo, M. and Segalés, J. (2007). Sow porcine circovirus type 2 (PCV2) status effect on litter mortality in postweaning multisystemic wasting syndrome (PMWS), *Res. Vet. Sci.* 82(3):299–304.

Campbell, J. M., Crenshaw, J. D. and Polo, J. (2013). The biological stress of early weaned piglets, *J. Anim. Sci. Biotechnol.* 4(1):19.

Castillo, M., Martin-Orue, S. M., Nofrarias, M., Manzanilla, E. G. and Gasa, J. (2007). Changes in caecal microbiota and mucosal morphology of weaned pigs, *Vet. Microbiol.* 124(3–4):239–247.

Cerdà-Cuéllar, M., Naranjo, J. F., Verge, A., Nofrarias, M., Cortey, M., Olvera, A., Segales, J. and Aragon, V. (2010). Sow vaccination modulates the colonization of piglets by *Haemophilus parasuis*, *Vet. Microbiol.* 145(3–4):315–320.

Cevallos-Almeida, M., Fablet, C., Houdayer, C., Dorenlor, V., Eono, F., Denis, M. and Kerouanton, A. (2019). Longitudinal study describing time to Salmonella seroconversion in piglets on three farrow-to-finish farms, *Vet. Rec. Open* 6(1):e000287. https://doi.org/10.1136/vetreco-2018-000287.

Chae, C. (2012). Commercial porcine circovirus type 2 vaccines: efficacy and clinical application, *Vet. J.* 194(2):151–157.

Chae, C. (2021). Commercial PRRS modified-live virus vaccines, *Vaccines* 9(2):185. https://doi.org/10.3390/vaccines9020185.

Charerntantanakul, W. (2012). Porcine reproductive and respiratory syndrome virus vaccines: immunogenicity, efficacy and safety aspects, *World J. Virol.* 1(1):23–30. https://doi.org/10.5501/wjv.v1.i1.23.

Chase, C. and Lunney, J. K. (2019). Immune system. In: Zimmerman, J. J., Karriker, L. A., Ramirez, A., Schwartz, K. J., Stevenson, G. W. and Zhang, J. (Eds), *Disease of Swine* (11th edn.), Wiley-Blackwell, Hoboken, NJ, pp. 264–291.

Chen, H., Mao, X., He, J., Yu, B., Huang, Z., Yu, J., Zheng, P. and Chen, D. (2013). Dietary fibre affects intestinal mucosal barrier function and regulates intestinal bacteria in weaning piglets, *Br. J. Nutr.* 110(10):1837–1848.

Christianson, W. T., Choi, C. S., Collins, J. E., Molitor, T. W., Morrison, R. B. and Joo, H. S. (1993). Pathogenesis of porcine reproductive and respiratory syndrome virus infection in mid-gestation sows and fetuses, *Can. J. Vet. Res.* 57(4):262–268.

Cloutier, G., D'Allaire, S., Martinez, G., Surprenant, C., Lacouture, S. and Gottschalk, M. (2003). Epidemiology of *Streptococcus suis* serotype 5 infection in a pig herd with and without clinical disease, *Vet. Microbiol.* 97(1–2):135–151. https://doi.org/10.1016/j.vetmic.2003.09.018. PMID: 14637045.

Collins, C. L., Pluske, J. R., Morrison, R. S., McDonald, T. N., Smits, R. J., Henman, D. J., Stensland, I. and Dunshea, F. R. (2017). Post-weaning and whole-of-life performance of pigs is determined by live weight at weaning and the complexity of the diet fed after weaning, *Anim. Nutr.* 3(4):372–379.

Correa-Fiz, F., Fraile, L. and Aragon, V. (2016). Piglet nasal microbiota at weaning may influence the development of Glasser's disease during the rearing period, *BMC Genomics* 17:404.

Correa-Fiz, F., Neila-Ibáñez, C., López-Soria, S., Napp, S., Martinez, B., Sobrevia, L., Tibble, S., Aragon, V. and MiguraGarcia, L. (2020). Feed additives for the control of post-weaning *Streptococcus suis* disease and the effect on the faecal and nasal microbiota, *Sci. Rep.* 10(1):20354. https://doi.org/10.1038/s41598-020-77313-6.

Costa-Hurtado, M. and Aragon, V. (2013). Advances in the quest for virulence factors of *Haemophilus parasuis*, *Vet. J.* 198(3):571–576.

Costa-Hurtado, M., Barba-Vidal, E., Maldonado, J. and Aragon, V. (2020). Update on Glässer's disease: how to control the disease under restrictive use of antimicrobials, *Vet. Microbiol.* 242:108595. ISSN 0378-1135. https://doi.org/10.1016/j.vetmic.2020.108595.

Coutellier, L., Arnould, C., Boissy, A., Orgeur, P., Prunier, A., Veissier, I. and Meunier-Salaün, M. (2007). Pig's responses to repeated social regrouping and relocation during the growing-finishing period, *Appl. Anim. Behav. Sci.* 105(1–3):102–114.

Dee, S. A. and Corey, M. M. (1993). The survival of *Streptococcus suis* on farm and veterinary equipment, *Swine Health Prod.* 1(1):17–20.

Dee, S. A., Joo, H. S. and Pijoan, C. (1994). Controlling the spread of PRRS virus in the breeding herd through management of the gilt pool, *Swine Health Prod.* 3(2):64–69.

Dee, S. A., Joo, H. S., Polson, D. D., Park, B. K., Pijoan, C., Molitor, T. W., Collins, J. E. and King, V. (1997). Evaluation of the effects of nursery depopulation on the persistence of porcine reproductive and respiratory syndrome virus and the productivity of 34 farms, *Vet. Rec.* 140(10):247–248.

Dee, S. A., Deen, J., Burns, D., Douthit, G. and Pijoan, C. (2004). An assessment of sanitation protocols for commercial transport vehicles contaminated with porcine reproductive and respiratory syndrome virus, *Can. J. Vet. Res.* 68(3):208–214.

Dee, S. A., Torremorell, M., Thompson, B., Deen, J. and Pijoan, C. (2005). An evaluation of thermo-assisted drying and decontamination for the elimination of porcine reproductive and respiratory syndrome virus from contaminated livestock transport vehicles, *Can. J. Vet. Res.* 69(1):58–63.

Dee, S., Otake, S., Oliveira, S. and Deen, J. (2009). Evidence of long distance airborne transport of porcine reproductive and respiratory syndrome virus and *Mycoplasma hyopneumoniae*, *Vet. Res.* 40(4):39.

Dee, S. A., Otake, S. and Deen, J. (2011). An evaluation of ultraviolet light (UV 254) as a means to inactivate porcine reproductive and respiratory syndrome virus on common farm surfaces and materials, *Vet. Microbiol.* 150(1-2):96–99.

Devillers, N., Le Dividich, J. and Prunier, A. (2011). Influence of colostrum intake on piglet survival and immunity, *Animal* 5(10):1605–1612.

Done, S. H. and Paton, D. J. (1995). Porcine reproductive and respiratory syndrome: clinical disease, pathology and immunosuppression, *Vet. Rec.* 136(2):32–35.

Dong, G. Z. and Pluske, J. R. (2007). The low feed intake in newly-weaned pigs: problems and possible solutions, *Asian. Australas J. Anim. Sci.* 20(3):440–452.

Drew, T. W. (2000). A review of evidence for immunosuppression due to porcine reproductive and respiratory syndrome virus, *Vet. Res.* 31(1):27–39.

Dybkjær, L., Jacobsen, A. P., Tøgersen, F. A. and Poulsen, H. D. (2006). Eating and drinking activity of newly weaned piglets: effects of individual characteristics, social mixing, and addition of extra zinc to the feed, *J. Anim. Sci.* 84(3):702–711.

Enright, M. R., Alexander, T. J. L. and Clifton Hadley, F. A. (1987). Role of houseflies (*Musca domestica*) in the epidemiology of *Streptococcus suis* type 2, *Vet. Rec.* 121(6):132–133.

Fairbrother, J. M. and Nadeau, É. (2019). Colibacillosis. In: Zimmerman, J. J., Karriker, L. A., Ramirez, A., Schwartz, K. J., Stevenson, G. W. and Zhang, J. (Eds), *Disease of Swine* (11th edn.), Wiley-Blackwell, Hoboken, NJ, pp. 807–834.

FAO (Food and Agriculture Organization) of the United Nations/World Organisation for Animal Health/World Bank (2010). Good practices for biosecurity in the pig sector - issues and options in developing and transition countries. FAO Animal Production and Health Paper No. 169. Rome, FAO.

Feng, Y., Wang, Y., Wang, P., Huang, Y. and Wang, F. (2018). Short-chain fatty acids manifest stimulative and protective effects on intestinal barrier function through the inhibition of NLRP3 inflammasom and autophagy, *Cell. Physiol. Biochem.* 49(1):190–205.

Fleck, R. and Behrens, A. (2002). Evaluation of a maternal antibody decay curve for H1N1 swine influenza virus using the hemagglutination inhibition and the IDEXX ELISA tests. *Proceedings of the American Association of Swine Veterinarians* 2002: 109–110.

Fontanesi, L., Bertolini, F., Dall'Olio, S., Buttazzoni, L., Gallo, M. and Russo, V. (2012). Analysis of association between the MUC4 g.8227C>G polymorphism and production traits in Italian heavy pigs using a selective genotyping approach, *Anim. Biotechnol.* 23(3):147–155.

Franklin, M. A., Mathew, A. G., Vickers, J. R. and Clift, R. A. (2002). Characterization of microbial populations and volatile fatty acid concentrations in the jejunum, ileum, and cecum of pigs weaned at 17 vs 24 days of age, *J. Anim. Sci.* 80(11):2904–2910.

Gao, J., Yin, J., Xu, K., Li, T. and Yin, Y. (2019). What is the impact of diet on nutritional diarrhea associated with gut microbiota in weaning piglets: a system review, *Biomed Res. Int.* 2019:6916189

Gerritsen, R., van der Aar, P. and Molist, F. (2012). Insoluble nonstarch polysaccharides in diets for weaned piglets, *J. Anim. Sci.* 90 (Suppl. 4):318–320.

Gibson, G. R., Probert, H. M., Loo, J. V., Rastall, R. A. and Roberfroid, M. B. (2004). Dietary modulation of the human colonic microbiota: updating the concept of prebiotics, *Nutr. Res. Rev.* 17(2):259-275.

Go, N., Touzeau, S., Islam, Z., Belloc, C. and Doeschl-Wilson, A. (2019). How to prevent viremia rebound? Evidence from a PRRSv data-supported model of immune response, *BMC Syst. Biol.* 13(1):15. https://doi.org/10.1186/s12918-018-0666-7.

Gottschalk, M. and Broes, A. (2019). Actinobacillosis. In: Zimmerman, J. J., Karriker, L. A., Ramirez, A., Schwartz, K. J., Stevenson, G. W. and Zhang, J. (Eds), *Disease of Swine* (11th edn.), Wiley-Blackwell, Hoboken, NJ, pp. 749-766.

Gottschalk, M. and Segura, M. (2019). Streptococcosis. In: Zimmerman, J. J., Karriker, L. A., Ramirez, A., Schwartz, K. J., Stevenson, G. W. and Zhang, J. (Eds), *Disease of Swine* (11th edn.), Wiley-Blackwell, Hoboken, NJ, pp. 934-950.

Guilloteau, P., Martin, L., Eeckhaut, V., Ducatelle, R., Zabielski, R. and Van Immerseel, F. (2010). From the gut to the peripheral tissues: the multiple effects of butyrate, *Nutr. Res. Rev.* 23(2):366-384.

Guzman-Pino, S. A., Sola-Oriol, D., Davin, R., Manzanilla, E. G. and Perez, J. F. (2015). Influence of dietary electrolyte balance on feed preference and growth performance of post-weaned piglets, *J. Anim. Sci.* 93(6):2840-2848.

Hampson, D. J. and Kidder, D. E. (1986). Influence of creep feeding and weaning on brush border enzyme activities in the piglet small intestine, *Res. Vet. Sci.* 40(1):24-31.

Hancock, R. E. (2001). Cationic peptides: effectors in innate immunty and novel antimicrobials, *Lancet Infect. Dis.* 1(3):156-164.

He, Y., Deen, J., Shurson, G. C., Wang, L., Chen, C., Keisler, D. H. and Li, Y. Z. (2016). Identifying factors contributing to slow growth in pigs, *J. Anim. Sci.* 94(5):2103-2116.

Heo, J. M., Opapeju, F. O., Pluske, J. R., Kim, J. C., Hampson, D. J. and Nyachoti, C. M. (2013). Gastrointestinal health and function in weaned pigs: a review of feeding strategies to control post-weaning diarrhoea without using in-feed antimicrobial compounds, *J. Anim. Physiol. Anim. Nutr. (Berl)* 97(2):207-237.

Hou, Z., Wang, H., Feng, Y., Li, Q. and Li, J. (2019). A candidate DNA vaccine encoding a fusion protein of porcine complement C3d-P28 and ORF2 of porcine circovirus type 2 induces cross-protective immunity against PCV2b and PCV2d in pigs, *Virol. J.* 16:57.

Hyun, Y., Ellis, M., Riskowski, G. and Johnson, R. W. (1998). Growth performance of pigs subjected to multiple concurrent environmental stressors, *J. Anim. Sci.* 76(3):721-727.

Islam, Z. U., Bishop, S. C., Savill, N. J., Rowland, R. R., Lunney, J. K., Trible, B. and Doeschl-Wilson, A. B. (2013). Quantitative analysis of porcine reproductive and respiratory syndrome (PRRS) viremia profiles from experimental infection: a statistical modelling approach, *PLoS ONE* 8(12):e83567.

Jayaraman, B., Htoo, J. K. and Nyachoti, C. M. (2015). Effects of dietary threonine:lysine ratioes and sanitary conditions on performance, plasma urea nitrogen, plasma-free threonine and lysine of weaned pigs, *Anim Nutr.* 1(4):283-288.

Jayaraman, B., Htoo, J. K. and Nyachoti, C. M. (2017). Effects of different dietary tryptophan: lysine ratios and sanitary conditions on growth performance, plasma urea nitrogen, serum haptoglobin and ileal histomorphology of weaned pigs, *Anim. Sci. J.* 88(5):763-771.

Jayaraman, B. and Nyachoti, C. M. (2017). Husbandry practices and gut health outcomes in weaned piglets: a review, *Anim. Nutr.* 3(3):205-211.

Jeaurond, E. A., Rademacher, M., Pluske, J. R., Zhu, C. H. and de Lange, C. F. M. (2008). Impact of feeding fermentable proteins and carbohydrates on growth performance, gut health and gastrointestinal function of newly weaned pigs, *Can. J. Anim. Sci.* 88(2):271–281.

Jones, A. M., Wu, F., Woodworth, J. C., Dritz, S. S., Tokach, M. D., DeRouchey, J. M. and Goodband, R. D. (2019). Evaluation of dietary electrolyte balance on nursery pig performance, *Transl. Anim. Sci.* 3(1):378–383.

Jørgensen, C. B., Cirera, S., Anderson, S. I., Archibald, A. L., Raudsepp, T., Chowdhary, B., Edfors-Lilja, I., Andersson, L. and Fredholm, M. (2003). Linkage and comparative mapping of the locus controlling susceptibility towards E. coli F4ab/ac diarrhoea in pigs, *Cytogenet. Genome Res.* 102(1–4):157–162.

Karuppannan, A. K. and Opriessnig, T. (2018). *Lawsonia intracellularis*: revisiting the disease ecology and control of this fastidious pathogen in pigs, *Front. Vet. Sci.* 5:181.

Khafipour, E., Munyaka, P. M., Nyachoti, C. M., Krause, D. O. and Rodriguez-Lecompte, J. C. (2014). Effect of crowding stress and Escherichia coli K88+ challenge in nursery pigs supplemented with anti-Escherichia coli K88+ probiotics, *J. Anim. Sci.* 92(5):2017–2029.

Kirkland, P. D., Le Potier, M. F. and Finlaison, D. (2019). Pestiviruses. In: Zimmerman, J. J., Karriker, L. A., Ramirez, A., Schwartz, K. J., Stevenson, G. W. and Zhang, J. (Eds), *Disease of Swine* (11th edn.), Wiley-Blackwell, Hoboken, NJ, pp. 622–640.

Klinge, K. L., Vaughn, E. M., Roof, M. B., Bautista, E. M. and Murtaugh, M. P. (2009). Age-dependent resistance to porcine reproductive and respiratory syndrome virus replication in swine, *Virol. J.* 6:177.

Konstantinov, S. R., Awati, A. A., Williams, B. A., Miller, B. G., Jones, P., Stokes, C. R., Akkermans, A. D., Smidt, H. and de Vos, W. M. (2006). Post-natal development of the porcine microbiota composition and activities, *Environ. Microbiol.* 8(7):1191–1199.

Laine, T. M., Lyytikäinen, T., Yliaho, M. and Anttila, M. (2008). Risk factors for post-weaning diarrhoea on piglet producing farms in Finland, *Acta Vet. Scand.* 50(21):21.

Lallès, J. P., Bosi, P., Smidt, H. and Stoke, C. R. (2007). Nutritional management of gut health in pigs around weaning, *Proc. Nutr. Soc.* 66(2):260–268.

Langel, S. N., Paim, F. C., Lager, K. M., Vlasova, A. N. and Saif, L. J. (2016). Lactogenic immunity and vaccines for porcine epidemic diarrhea virus (PEDV): historical and current concepts, *Virus Res.* 226:93–107.

Lapointe, L., D'Allaire, S., Lebrun, A., Lacouture, S. and Gottschalk, M. (2002). Antibody response to an autogenous vaccine and serologic profile for *Streptococcus suis* capsular type 1/2, *Can. J. Vet. Res.* 66(1):8–14.

Le Floc'h, N., Lebellago, L., Matte, J.J., Melchior, D. and Seve, B. (2009). The effect of sanitary status degradation and dietary tryptophan content on growth rate and tryptophan metabolism in weaning pigs, *J. Anim. Sci.* 87:1686–1694.

Levis, D. G. and Baker, R. B. (2011). Biosecurity of pigs and farm security. *University of Nebraska Lincoln Extension Publications.* Available at: http://extesnion.unl.edu/publications.

Li, S., Sauer, W. C. and Fan, M. Z. (1993). The effect of dietary crude protein level on ileal and faecal amino acid digestibility in early-weaned pigs, *J. Anim. Physiol. Nutr.* 70(1–5):117–128.

Li, D. F., Jiang, J. Y. and Ma, Y. X. (2003). Early weaning diets and feed additives'. In: Xu, R. J. and Cranwell, P. D. (Eds). *The Neonatal Pig: Gastrointestinal Physiology and Nutrition*, Thrumptom, Nottingham. Nottingham University Press, p. 247.

Lindberg, J. E. (2014). Fiber effects in nutrition and gut health in pigs, *J. Anim. Sci. Biotechnol.* 5(1):15.

Liu, L., Wang, J., Zhao, Q., Zi, C., Wu, Z., Su, X., Huo, Y., Zhu, G., Wu, S. and Bao, W. (2013). Genetic variation in exon 10 of the BPI gene is associated with Escherichia coli F18 susceptibility in Sutai piglets, *Gene* 523(1):70–75.

Liu, Y., Espinosa, C. D., Abelilla, J. J., Casas, G. A., Lagos, L. V., Lee, S. A., Kwon, W. B., Mathai, J. K., Navarro, D. M. D. L., Hans, N. W. and Stein, H. H. (2018). Non-antibiotic feed additives in diets for pigs: a review, *Anim. Nutr.* 4(2):113–125.

López-Soria, S., Sibila, M., Nofrarías, M., Calsamiglia, M., Manzanilla, E. G., Ramírez-Mendoza, H., Mínguez, A., Serrano, J. M., Marín, O., Joisel, F., Charreyre, C. and Segalés, J. (2014). Effect of porcine circovirus type 2 (PCV2) load in serum on average daily weight gain during the postweaning period, *Vet. Microbiol.* 174(3–4):296–301.

Lordelo, M. M., Gaspar, A. M., Le Bellego, L. and Freire, J. P. B. (2008). Isoleucine and valine supplementation of a low-protein corn-wheat-soybean meal-based diet for piglets: growth performance and nitrogen balance, *J. Anim. Sci.* 86(11):2936–2941.

Luppi, A. (2017). Swine enteric colibacillosis: diagnosis, therapy and antimicrobial resistance, *Porcine Health Manag.* 3:16.

Maes, D., Verdonck, M., Deluyker, H. and de Kruif, A. (1996). Enzootic pneumonia in pigs, *Vet. Q.* 18(3):104–109.

Mahmmod, Y. S., Correa-Fiz, F. and Aragon, V. (2020). Variations in association of nasal microbiota with virulent and non-virulent strains of *Glaesserella* (*Haemophilus*) *parasuis* in weaning piglets, *Vet. Res.* 51(1):7. https://doi.org/10.1186/s13567-020-0738-8.

Marois, C., Le Devendec, L., Gottschalk, M. and Kobisch, M. (2007). Detection and molecular typing of *Streptococcus suis* in tonsils from live pigs in France, *Can. J. Vet. Res.* 71(1):14–22.

Martella, V., Banyai, K., Lorusso, E., Bellacicco, A. L., Decaro, N., Camero, M., Bozzo, G., Moschidou, P., Arista, S., Pezzotti, G., Lavazza, A. and Buonavoglia, C. (2007). Prevalence of group C rotaviruses in weaning and post-weaning pigs with enteritis, *Vet. Microbiol.* 123(1–3):26–33.

Martelli, P., Gozio, S., Ferrari, L., Rosina, S., De Angelis, E., Quintavalla, C., Bottarelli, E. and Borghetti, P. (2009). Efficacy of a modified live porcine reproductive and respiratory syndrome virus (PRRSV) vaccine in pigs naturally exposed to a heterologous European (Italian cluster) field strain: clinical protection and cell-mediated immunity, *Vaccine* 27(28):3788–3799. https://doi.org/10.1016/j.vaccine.2009.03.028 PMID: 19442420.

Martelli, P. (2013). Tabelle diagnosi differenziale. In: Martelli, P. (Ed.). *Le patologie del maiale*. Point Veterinaire Italie, Milano, pp. 2–5.

Martelli, P., Saleri, R., Ferrarini, G., De Angelis, E., Cavalli, V., Benetti, M., Ferrari, L., Canelli, E., Bonilauri, P., Arioli, E., Caleffi, A., Nathues, H. and Borghetti, P. (2016). Impact of maternally derived immunity on piglets' immune response and protection against porcine circovirus type 2 (PCV2) after vaccination against PCV2 at different age, *BMC Vet. Res.* 12:77.

Marthaler, D., Rossow, K., Culhane, M., Goyal, S., Collins, J., Matthijnssens, J., Nelson, M. and Ciarlet, M. (2014). Widespread rotavirus H in commercially raised pigs, United States, *Emerg. Infect. Dis.* 20(7):1195–1198.

Martinez, V., Wang, L., Million, M., Rivier, J. and Taché, Y. (2004). Urocortins and the regulation of gastrointestinal motor function and visceral pain, *Peptides* 25(10):1733– 1744.

Mateos, G. G., López, E., Latorre, M. A., Vicente, B. and Lázaro, R. P. (2007). The effect of inclusion of oat hulls in piglet diet base on raw cooked rice and maize, *Anim. Feed Sci. Tech.* 135(1–2):100–112.

McGhee, J. R., Mestecky, J., Dertzbaugh, M. T., Eldridge, J. H., Hirasawa, M. and Kiyono, H. (1992). The mucosal immune system: from fundamental concepts to vaccine development, *Vaccine* 10(2):75-88.

McOrist, S., Bowles, R. and Blackall, P. (2009). Autogenous sow vaccination for Glasser's disease in weaner pigs in two large swine farm systems, *J. Swine Health Prod.* 17(2):90-96.

McRorie Jr., J. W. and McKeown, N. M. (2017). Understanding the physics of functional fibers in the gastrointestinal tract: an evidence-based approach to resolving enduring misconceptions about insoluble and soluble fiber, *J. Acad. Nutr. Diet.* 117(2):251-264.

Meerts, P., Misinzo, G., Lefebvre, D., Nielsen, J., Botner, A., Kristensen, C. S. and Nauwynck, H. J. (2006). Correlation between the presence of neutralizing antibodies against porcine circovirus 2 (PCV2) and protection against replication of the virus and development of PCV2-associated disease, *BMC Vet. Res.* 2:6.

Meijerink, E., Neuenschwander, S., Fries, R., Dinter, A., Bertschinger, H. U., Stranzinger, G. and Vögeli, P. (2000). A DNA polymorphism influencing (1,2) fucosyltransferase activity of the pig FUT1 enzyme determines susceptibility of small intestinal epithelium to Escherichia coli F18 adhesion, *Immunogenetics* 52(1-2):129-136.

Melkebeek, V., Goddeeris, B. M. and Cox, E. (2013). ETEC vaccination in pigs, *Vet. Immunol. Immunopathol.* 152(1-2):37-42.

Meng, X. J., Halbur, P. G. and Opriessnig, T. (2019). Hepatitis E. In: Zimmerman, J. J., Karriker, L. A., Ramirez, A., Schwartz, K. J., Stevenson, G. W. and Zhang, J. (Eds), *Disease of Swine* (11th edn.), Wiley-Blackwell, Hoboken, NJ, pp. 544-547.

Metzler, B. U., Vahjen, W., Baumgartel, T., Rodehutscord, M. and Mosenthin, R. (2009). Changes in bacterial populations in the ileum of pigs fed low-phosphorus diets supplemented with different sources of fermentable carbohydrates. *Anim. Feed Sci. Technol.* 148:69-89.

Moeser, A. J., Klok, C. V., Ryan, K. A.,Wooten, J. G., Little, D., Cook, V. L. and Blikslager, A. T. (2007). Stress signaling pathways activated by weaning mediate intestinal dysfunction in the pig, *Am. J. Physiol. Gastrointest. Liver Physiol.* 292(1):G173-G181.

Moeser, A. J., Pohl, C. S. and Rajput, M. (2017). Weaning stress and gastrointestinal barrier development: implications for lifelong gut health in pigs, *Anim. Nutr.* 3(4):313-321.

Molist, F., de Segura, A. G., Gasa, J., Hermes, R. G., Manzanilla, E. G., Anguita, M. and Pérez, J. F. (2009). Effects of the insoluble and soluble dietary fibre on the physicochemical properties of digesta and the microbial activity in early weaned piglets, *Anim. Feed Sci. Technol.* 149(3-4):346-353.

Morris, C. R., Gardner, I. A., Hietala, S. K., Carpenter, T. E., Anderson, R. J. and Parker, K. M. (1994). Persistence of passively acquired antibodies to *Mycoplasma hyopneumoniae* in a swine herd, *Prev. Vet. Med.* 21(1):29-41.

Müller, T., Teuffert, J., Staubach, C., Selhorst, T. and Depner, K. R. (2005). Long-term studies on maternal immunity for Aujeszky's disease and classical swine fever in wild boar piglets, *J. Vet. Med. B Infect. Dis. Vet. Public Health* 52(10):432-436.

Murtaugh, M. P., Xiao, Z. and Zuckermann, F. (2002). Immunological responses of swine to porcine reproductive and respiratory syndrome virus infection, *Viral Immunol.* 15(4):533-547.

Nabuurs, M. J. A., Hoogendoorn, A., van der Molden, E. J. and Van Osta, A. L. M. (1993). Villous height and crypt depth in weaned and unweaned pigs, reared under various circumstances in the Netherlands, *Res. Vet. Sci.* 55(1):78-84.

Nagy, B., Nagy, G., Meder, M. and Mocsari, E. (1996). Enterotoxigenic Escherichia coli, rotavirus, porcine epidemic diarrhoea virus, adenovirus and calici-like virus in porcine postweaning diarrhoea in Hungary, *Acta Vet. Hung.* 44(1):9–19.

Noblet, J., Le Dividich, J. and Van Milgen, J. (2001). Thermal environment and swine nutrition. In: Lewis, A. J., Southern, L. L. and Swine Nutrition (Eds), 2nd ed., CRC Press LLC Press, Boca Raton, Palm Beach, FL, pp. 519–544.

NRC (2012). *Nutrient Requirements of Swine* (11th edn.), National Academic Press.

Nyachoti, C. M., Omogbenigun, F. O., Rademacher, M. and Blank, G. (2006). Performance responses and indicators of gastrointestinal health in early-weaned pigs fed low-protein amino acid-supplemented diets, *J. Anim. Sci.* 84(1):125–134.

Obradovic, M. R., Segura, M., Segalés, J. and Gottschalk, M. (2021). Review of the speculative role of co-infections in *Streptococcus suis*-associated diseases in pigs, *Vet. Res.* 52(1):49. https://doi.org/10.1186/s13567-021-00918-w.

O'Doherty, J. V., Bouwhuis, M. A. and Sweeney, T. (2017). Novel marine polysaccharides and maternal nutrition to stimulate gut health and performance in post-weaned pigs, *Anim. Prod. Sci.* 57(12):2376–2385. https://doi.org/10.1071/AN17272.

Oh, Y., Han, K., Seo, H. W., Park, C. and Chae, C. (2013). Program of vaccination and antibiotic treatment to control polyserositis caused by *Haemophilus parasuis* under field conditions, *Can. J. Vet. Res.* 77(3):183–190.

Ohland, C. L. and MacNaughton, W. K. (2010). Probiotic bacteria and intestinal epithelial barrier function, *Am. J. Physiol. Gastrointest. Liver Physiol.* 298(6):G807–G819.

Oliveira, S. and Pijoan, C. (2004). *Haemophilus parasuis*: new trends on diagnosis, epidemiology and control, *Vet. Microbiol.* 99(1):1–12.

Olvera, A., Sibila, M., Calsamiglia, M., Segalés, J. and Domingo, M. (2004). Comparison of porcine circovirus type 2 load in serum quantified by a real time PCR in postweaning multisystemic wasting syndrome and porcine dermatitis and nephropathy syndrome naturally affected pigs, *J. Virol. Methods* 117(1):75–80.

Opapeju, F. O., Krause, D. O., Payne, R. L., Rademacher, M. and Nyachoti, C. M. (2009). Effect of dietary protein level on growth performance, indicators of enteric health, and gastrointestinal microbial ecology of weaned pigs induced with postweaning colibacillosis, *Journal of Animal Science* 87:2635–2643.

Opapeju, F. O., Rademacher, M., Payne, R. L., Krause, D. O. and Nyachoti, C. M. (2010). Inflammation-associated responses in piglets induced with post-weaning colibacillosis are influenced by dietary protein level, *Livest. Sci.* 131(1):58–64.

Opriessnig, T., Jones, D., McKeown, N., Meng, X.-J. and Halbur, P. G. (2004). Development of a mouse model for PCV2-associated diseases. *Proceedings of the Con. Res. Work Animal Dis., Chicago, Illinois* 85:143.

Opriessnig, T., Giménez-Lirola, L. G. and Halbur, P. G. (2011). Polymicrobial respiratory disease in pigs, *Anim. Health Res. Rev.* 12(2):133–148.

Opriessnig, T., Gerber, P. F., Xiao, C. T., Halbur, P. G., Matzinger, S. R. and Meng, X. J. (2014). Commercial PCV2a-based vaccines are effective in protecting naturally PCV2b-infected finisher pigs against experimental challenge with a 2012 mutant PCV2, *Vaccine* 32(34):4342–4348.

Opriessnig, T. and Coutinho, T. A. (2019). Erysipelas. In: Zimmerman, J. J., Karriker, L. A., Ramirez, A., Schwartz, K. J., Stevenson, G. W. and Zhang, J. (Eds), *Disease of Swine* (11th edn.), Wiley-Blackwell, Hoboken, NJ, pp. 835–843.

Oslage, U., Dahle, J., Müller, T., Kramer, M., Beier, D. and Liess, B. (1994). Prevalence of antibodies against the viruses of European swine fever, Aujeszky's disease and porcine reproductive and respiratory syndrome in wild boars in the federal states Sachsen-Anhalt and Brandenburg, *Dtsch Tierarztl Wochenschr.* 101:33–38.

Ouyang, J., Zeng, W., Ren, J., Yan, X., Zhang, Z., Yang, M., Han, P., Huang, X., Ai, H. and Huang, L. (2012). Association of B3GNT5 polymorphisms with susceptibility to ETEC F4ab/ac in the white Duroc × Erhualian intercross and 15 outbred pig breeds, *Biochem. Genet.* 50(1–2):19–33.

Paul, P. S., Mengeling, W. L. and Pirtle, E. C. (1982). Duration and biological half-life of passively acquired colostral antibodies to porcine parvovirus, *Am. J. Vet. Res.* 43(8):1376–1379.

Pereira, D. A., Dalla Costa, F. A., Ferroni, L. B., Moraes, C. N., Schocken-Iturrino, R. P. and Oliveira, L. G. (2017). The challenges with Glässer's disease in technified pig production, *Aust. J. Vet. Sci.* 49:63–69.

Pileri, E. and Mateu, E. (2016). Review on the transmission porcine reproductive and respiratory syndrome virus between pigs and farms and impact on vaccination, *Vet. Res.* 47(1):108.

Pinilla, J. C., Geiger, J., Kummer, R., Piva, J., Schott, R. and Williams, N. H. (2008). *Management Strategies to Maximize Weaning Weight*, American Association Of Swine Veterinarians, pp. 185–192.

Pluske, J. R., Williams, I. H. and Aherne, F. X. (1996). Maintenance of villous height and crypt depth in piglets by providing continuous nutrition after weaning, *Anim. Sci.* 62(1):131–144.

Pluske, J. R., Hampson, D. J. and Williams, I. H. (1997). Factors influencing the structure and function of the small intestine in the weaned pig: a review, *Livest. Prod. Sci.* 51(1–3):215–236.

Pluske, J. R., Pethick, D. W., Hopwood, D. E. and Hampson, D. J. (2002). Nutritional influences on some major enteric bacterial diseases of pigs, *Nutr. Res. Rev.* 15(2):333–371.

Pomorska-Mol, M., Markowska-Daniel, I., Rachubik, J. and Pejsak, Z. (2011). Effect of maternal antibodies and pig age on the antibody response after vaccination against Glässer's disease, *Vet. Res. Commun.* 35(6):337–343.

Poonsuk, K. and Zimmerman, J. (2018). Historical and contemporary aspects of maternal immunity in swine, *Anim. Health Res. Rev.* 19(1):31–45.

Quesnel, H., Brossard, L., Valancogne, A. and Quiniou, N. (2008). Influence of some sow characteristics on within-litter variation of piglet birth weight, *Animal* 2(12):1842–1849. HTTPS://doi.org/10.1017/S175173110800308X.

Racine, N. M., Riddell, R. R., Khan, M., Calic, M., Taddio, A. and Tablon, P. (2016). Systematic review: predisposing, precipitating, perpetuating, and present factors predicting anticipatory distress to painful medical procedures in children, *J. Pediatr. Psychol.* 41(2):159–181.

Rattigan, R., Sweeney, T., Maher, S., Ryan, M. T., Thornton, K. and O'Doherty, J. V. (2020). Effects of reducing dietary crude protein concentration and supplementation with either laminarin or zinc oxide on the growth performance and intestinal health of newly weaned pigs, *Anim. Feed Sci. Technol.* 270:114693. https://doi.org/10.1016/j.anifeedsci.2020.114693.

Reiner, G., Fresen, C., Bronnert, S. and Willems, H. (2009). Porcine reproductive and respiratory syndrome virus (PRRSV) infection in wild boars, *Vet. Microbiol.* 136(3–4):250–258.

Rhouma, M., Fairbrother, J. M., Beaudry, F. and Letellier, A. (2017). Post weaning diarrhea in pigs: risk factors and non-colistin-based control strategie's, *Acta Vet. Scand.* 59(1):31.

Riaz Rajoka, M. S., Shi, J., Mehwish, H. M., Zhu, J., Li, Q., Shao, D., Huang, Q. and Yang, H. (2017). Interaction between diet composition and gut microbiota and its impact on gastrointestinal tract health, *Food Sci. Hum. Wellness* 6(3):121–130.
Rieckmann, K., Pendzialek, S. M., Vahlenkamp, T. and Baums, C. G. (2020). A critical review speculating on the protective efficacies of autogenous *Streptococcus suis* bacterins as used in Europe, *Porcine Health Manag.* 6:12.
Robinson, K. M. X., Liu, Y., Qiao, S., Hou, Y. and Zhang, G. (2018). Dietary modulaton of endogenois host defense peptides synthesis as an alternative approach to infeed antibiotics, *Anim. Nutr.* 4:160–169.
Roca, M., Gimeno, M., Bruguera, S., Segalés, J., Díaz, I., Galindo-Cardiel, I. J., Martínez, E., Darwich, L., Fang, Y., Maldonado, J., March, R. and Mateu, E. (2012). Effect of challenge with a virulent genotype II strain of porcine reproductive and respiratoiry syndrome virus on piglets vaccinated with an attenuated genotype I strain vaccine, *Vet. J.* 193(1):92–96.
Rodríguez-Arrioja, G. M., Segalés, J., Calsamiglia, M., Resendes, A. R., Balasch, M., Plana-Durán, J., Casal, J. and Domingo, M. (2002). Dynamics of porcine circovirus type 2 infection in a herd of pigs with postweaning multisystemic wasting syndrome, *Am. J. Vet. Res.* 63(3):354–357.
Roic, B., Jemersic, L., Terzic, S., Keros, T., Balatinec, J. and Florijancic, T. (2012). Prevalence of antibodies to selected viral pathogens in wild boars (Sus scrofa) in Croatia in 2005–2006 and 2009–2010, *J. Wild Dis.* 48:131–137.
Rose, N. and Andraud, M. (2017). The use of vaccines to control pathogen spread in pig populations, *Porcine Health Manag.* 3:8 https://doi.org/10.1186/s40813-017-0053-6.
Roselli, M., Finamore, A., Garaguso, I., Britti, M. S. and Mengheri, E. (2003). Zinc oxide protects cultured enterocytes from the damage induced by *Escherichia coli*, *J. Nutr.* 133(12):4077–4082.
Saif, L. J., Wang, Q., Vlasova, A. N., Jung, K. and Xiao, S. (2019). Coronaviruses. In: Zimmerman, J. J., Karriker, L. A., Ramirez, A., Schwartz, K. J., Stevenson, G. W. and Zhang, J. (Eds), *Disease of Swine* (11th edn.), Wiley-Blackwell, Hoboken, NJ, pp. 488–523.
Sakata, T., Kojima, T., Fujieda, M., Takahashi, M. and Michibata, T. (2003). Influences of probiotic bacteria on organic acid production by pig caecal bacteria in vitro, *Proc. Nutr. Soc.* 62(1):73–80.
Saravanan, S., Geurden, I., Orozco, Z. G., Kaushik, S. J., Verreth, J. A. and Schrama, J. W. (2013). Dietary electrolyte balance affects the nutrient digestibility and maintenance energy expenditure of Nile tilapia, *Br. J. Nutr.* 110(11):1948–1957.
Schurrer, J. A., Dee, S. A., Moon, R. D., Murtaugh, M. P., Finnegan, C. P., Deen, J., Kleiboeker, S. B. and Pijoan, C. B. (2005). Retention of ingested porcine reproductive and respiratory syndrome virus in houseflies, *Am. J. Vet. Res.* 66(9):1517–1525.
Segalés, J. and Domingo, M. (2002). Postweaning multisystemic wasting syndrome (PMWS) in pigs: a review, *Vet. Q.* 24(3):109–124.
Segalés, J., Domingo, M., Chianini, F., Majo, N., Dominguez, J., Darwich, L. and Mateu, E. (2004). Immunosuppression in postweaning multisystemic wasting syndrome affected pigs, *Vet. Microbiol.* 98(2):151–158.
Segalés, J. (2015). Best practice and future challenges for vaccination against porcine circovirus type 2', *Vaccines* 14(3):473–487.
Segalés, J. (2020). Porcine circoviruses in 2020: what's new? In: Stingelin, G. M., Guilherme de Oliveira, L. and Montenegro Franceschini, V. (Eds) *Swine Health and Production: Updating, Innovation and Technology*, FUNEP Jaboticabal, São Paulo, pp. 4–9.

Segura, M. (2015). *Streptococcus suis* vaccines: candidate antigens and progress, *Expert Rev. Vaccines* 14(1):1-22.
Segura, M., Calzas, C., Grenier, D. and Gottschalk, M. (2016). Initial steps of the pathogenesis of the infection caused by *Streptococcus suis*: fighting against nonspecific defenses, *FEBS Lett.* 590(21):3772-3799.
Segura, M., Aragon, V., Brockmeier, S. L., Gebhart, C., de Gree, A., Kerdsin, A., O'Dea, M. A., Okura, M., Saléry, M., Schultsz, C., Valentin-Weigand, P., Weinert, L. A., Wells, J. M. and Gottschalk, M. (2020). Update on *Streptococcus suis* research and prevention in the era of antimicrobial restriction: 4th international workshop on *S. suis*. Pathogens, 9(5):374.
Seo, H. W., Park, C., Han, K. and Chae, C. (2014). Effect of porcine circovirus type 2 (PCV2) vaccination on PCV2-viremic piglets after experimental PCV2 challenge, *Vet. Res.* 45:13.
Shen, H. G., Loiacono, C. M., Halbur, P. G. and Opriessnig, T. (2012). Age-dependent susceptibility to porcine circovirus type 2 infections is likely associated with declining levels of maternal antibodies, *J. Swine Health Prod.* 20(1):17-24.
Shepherd, F. K., Freeman, M. J., Culhane, M. R. and Marthaler, D. G. (2019). Reoviruses (rotaviruses and Reoviruses). In: Zimmerman, J. J., Karriker, L. A., Ramirez, A., Schwartz, K. J., Stevenson, G. W. and Zhang, J. (Eds), *Disease of Swine* (11th edn.), Wiley-Blackwell, Hoboken, NJ, pp. 715-727.
Shirai, J., Kanno, T., Tsuchiya, Y., Mitsubayashi, S. and Seki, R. (2000). Effects of chlorine, iodine, and quaternary ammonium compound disinfectants on several exotic disease viruses, *J. Vet. Med. Sci.* 62(1):85-92.
Silva, J., Rocha, D., Cunha, I., Rui Sales, L., Neto, F., Fontes, M. C. and Simões, J. (2015). Serological profile of offspring on an intensive pig farm affected by porcine reproductive and respiratory syndrome, *Asian Pac. J. Reprod.* 4(4):317-321.
Snijder, E. J. and Meulenberg, J. M. (2001). Arteriviruses. In: Knipe, D. M., Howley, P. M., Griffin, D. E., et al. (Eds), *Fields Virology* (4th edn.), Lippincott Williams and Wilkins, Philadelphia, PA, pp. 1205-1220.
Stein, H. H. and Kil, D. Y. (2006). Reduced use of antibiotic growth promoters in diets fed to weanling pigs: dietary tools, part 2, *Anim. Biotechnol.* 17(2):217-231.
Stokes, C. R. (2017). The development and role of microbial-host interactions in gut mucosal immune development, *J. Anim. Sci. Biotechnol.* 8:12.
Swildens, B., Nielen, M., Wisselink, H. J., Verheijden, J. H. and Stegeman, J. A. (2007). Elimination of strains of *Streptococcus suis* serotype 2 from the tonsils of carrier sows by combined medication and vaccination, *Vet. Rec.* 160(18):619-621.
Thomas, C., Mettenleiter, B. E., Müller, T., Yoon, K. J. and Teifke, J. P. (2019). Herpesviruses. In: Zimmerman, J. J., Karriker, L. A., Ramirez, A., Schwartz, K. J., Stevenson, G. W. and Zhang, J. (Eds), *Disease of Swine* (11th edn.), Wiley-Blackwell, Hoboken, NJ, pp. 548-575.
Truyen, U. and Streck, A. F. (2019). *Parvoviruses*. In: Zimmerman, J. J., Karriker, L. A., Ramirez, A., Schwartz, K. J., Stevenson, G. W. and Zhang, J. (Eds), *Disease of Swine* (11th edn.), Wiley-Blackwell, Hoboken, NJ, pp. 611-621.
Tsiloyiannis, V. K., Kyriakis, S. C., Vlemmas, J. and Sarris, K. (2001). The effect of organic acids on the control of porcine post-weaning diarrhoea, *Res. Vet. Sci.* 70(3):287-293.
Tzipori, S., Chandler, D., Makin, T. and Smith, M. (1980). *Escherichia coli* and Rotavirus infections in four-week-old gnotobiotic piglets fed milk or dry food, *Aust. Vet. J.* 56(6):279-284.

Van Reeth, K. and Vincent, A. L. (2019). Influenza virus. In: Zimmerman, J. J., Karriker, L. A., Ramirez, A., Schwartz, K. J., Stevenson, G. W. and Zhang, J. (Eds), *Disease of Swine* (11th edn.), Wiley-Blackwell, Hoboken, NJ, pp. 576–593.

Veldkamp, T. and Vernooij, A. G. (2021). Use of insect products in pig diets, *J. Insects Food Feed* 0(0):1–14. https://doi.org/10.3920/JIFF2020.0091. in Press.

Vente-Spreeuwenberg, M. A. M., Verdonk, J. M. A. J., Gaskins, H. J. and Verstegen, M. W. A. (2001). Small intestine epithelial barrier function is compromised in pigs with low feed intake at weaning, *J. Anim. Nutr.* 131:1520–1527.

Vente-Spreeuwenberg, M. A. M., Verdonk, J. M. A. J., Verstegen, M. W. A. and Beynen, A. C. (2003). Villus height and gut development in weaned piglets receiving diets containing either glucose, lactose or starch, *Br. J. Nutr.* 90(5):907–913.

Vente-Spreeuwenberg, M. A. M., Verdonk, J. M. A. J., Bakker, G. C. M., Beynen, A. C. and Verstegen, M. W. A. (2004). Effect of dietary protein source on feed intake and small intestinal morphology in newly weaned piglets, *Livest. Prod. Sci.* 86(1–3):169–177.

Vilcek, S., Molnar, L., Vlasakova, M. and Jackova, A. (2015). The first detection of PRRSV in wild boars in Slovakia, *Berl. Munch. Tierarztl Wochenschr.* 128(1–2):31–33.

Vötsch, D., Willenborg, M., Weldearegay, Y. B. and Valentin-Weigand., P. (2018). *Streptococcus suis* – the "two faces" of a pathobiont in the porcine respiratory tract, *Front. Microbiol.* 9:480.

Wang, Y., Shan, T., Xu, Z., Liu, J. and Feng, J. (2006). Effect of lactoferrin on the growth performance, intestinal morphology, and expression of PR-39 and protegrin-1 genes in weaned piglets, *J. Anim. Sci.* 84(10):2636–2641.

Wang, Y., Ren, J., Lan, L., Yan, X., Huang, X., Peng, Q., Tang, H., Zhang, B., Ji, H. and Huang, L. (2007). Characterization of polymorphisms of transferrin receptor and their association with susceptibility to ETEC F4ab/ac in pigs, *J. Anim. Breed. Genet.* 124(4):225–229.

Wang, J., Chen, L., Li, P., Li, X., Zhou, H., Wang, F., Li, D., Yin, Y. and Wu, G. (2008). Gene expression is altered in piglet small intestine by weaning and dietary glutamine supplementation, *J. Nutr.* 138(6):1025–1032.

Wang, J., Li, G. R., Tan, B. E., Xiong, X., Kong, X. F., Xiao, D. F., Xu, L. W., Wu, M. M., Huang, B., Kim, S. W. and Yin, Y. L. (2015). Oral administration of putrescine and proline during the suckling period improve epithelial restitution after early weaning in piglets, *J. Anim. Sci.* 93(4):1679–1688.

Warneboldt, F., Sander, S. J., Beineke, A., Valentin-Weigand, P., Kamphues, J. and Baums, C. G. (2016). Clearance of *Streptococcus suis* in stomach contents of differently fed growing pigs, *Pathogens* 5(3):56.

Wen, Z.-S., Lu, J.-L. and Zou, X.-T. (2012). Effect of sodium butyrate on the intestinal morphology and DNA-binding activity of intestinal nuclear factor-κB in weanling pigs, *J. Anim. Vet. Adv.* 11(6):814–821.

Williams, B. A., Verstegen, M. W. A. and Tamminga, S. (2001). Fermentation in the monogastric large intestine: its relation to animal health, *Nutr. Res. Rev.* 14(2):207–228.

Wolf, J., Žáková, E. and Groeneveld, E. (2008). Within-litter variation of birth weight in hyperprolific Czech Large White sows and its relation to litter size traits, stillborn piglets and losses until weaning, *Livest. Sci.* 115(2-3):195–205.

Wu, G., Fang, Y. Z., Yang, S., Lupton, J. R. and Turner, N. D. (2004). Glutathione metabolism and its implications for health, *J. Nutr.* 134(3):489–492.

Wu, Y. P., Jiang, Z. Y., Zheng, C. T., Wang, L., Zhu, C., Yang, X. F., Wen, X. L. and Ma, X. Y. (2015). Effects of protein sources and levels in antibiotic-free diets on diarrhea, intestinal morphology, and expression of tight junctions in weaned piglets, *Anim. Nutr.* 1(3):170–176.

Wu, M., Xiao, H., Liu, G., Chen, S., Tan, B., Ren, W., Bazer, F. W., Wu, G. and Yin, Y. (2016). Glutamine promotes intestinal SIgA secretion through intestinal microbiota and IL-13, *Mol. Nutr. Food Res.* 60(7):1637–1648.

Xiong, X., Tan, B., Song, M., Ji, P., Kim, K., Yin, Y. and Liu, Y. (2019). Nutritional intervention for the intestinal development and health of weaned pigs, *Front. Vet. Sci.* 6:46.

Yang, X. F., Jiang, Z. Y., Gong, Y. L., Zheng, C. T., Hu, Y. J., Wang, L., Huang, L. and Ma, X. Y. (2016). Supplementation of pre-weaning diet with L-arginine has carry-over effect to improve intestinal development in young piglets, *Can. J. Anim. Sci.* 96(1):52–59.

Yoon, K. J., Zimmerman, J. J., Swenson, S. L., McGinley, M. J., Eernisse, K. A., Brevik, A., Rhinehart, L. L., Frey, M. L., Hill, H. T. and Platt, K. B. (1995). Characterization of the humoral immune response to porcine reproductive and respiratory syndrome (PRRS) virus infection, *J. Vet. Diagn. Invest.* 7(3):305–312.

Zhang, B., Ren, J., Yan, X. M., Huang, X., Ji, H. Y., Peng, Q. L., Zhang, Z. Y. and Huang, L. S. (2008). Investigation of the porcine MUC13 gene: isolation, expression, polymorphisms and their strong association with susceptibility to 282 enterotoxigenic *E. coli* F4ab/ac, *Anim. Genet.* 39(3):258–266.

Zhao, Z., Liu, H., Xue, Y., Chen, K., Liu, Z., Xue, Q. and Wang, C. (2017). Analysis of efficacy obtained with a trivalent inactivated *Haemophilus parasuis* serovars 4, 5, and 12 vaccine and commercial vaccines against Glässer's disease in piglets, *Can. J. Vet. Res.* 81(1):22-27.

Zimmerman, J. J., Dee, S. A., Holtkamp, D. J., Murtaugh, M. P., Stadejek, T., Stevenson, G. W., Torremorell, M., Yang, H. and Zhang, J. (2019). Porcine reproductive and respiratory syndrome viruses (porcine arteriviruses). In: Zimmerman, J. J., Karriker, L. A., Ramirez, A., Schwartz, K. J., Zhang, J. and Stevenson, G. W. (Eds), *Disease of Swine* (11th edn.), Wiley-Blackwell, Hoboken, NJ, pp. 685–708.

Zong, E., Huang, P., Zhang, W., Li, J., Li, Y., Ding, X., Xiong, X., Yin, Y. and Yang, H. (2018). The effects of dietary sulfur amino acids on growth performance, intestinal morphology, enzyme activity, and nutrient transporters in weaning piglets, *J. Anim. Sci.* 96(3):1130–1139.

Zou, Y., Xiang, Q., Wang, J., Peng, J. and Wei, H. (2016). Oregano essential oil improves intestinal morphology and expression of tight junction proteins associated with modulation of selected intestinal bacteria and immune status in a pig model, *BioMed Res. Int.* 2016:5436738. https://doi.org/10.1155/2016/5436738.

Zulovich, J. M. (2012). Effect of the environment on health. In: Zimmerman, J., Karrker, L., Ramirez, A., Schwarz, K. and Stevenson, G. (Eds), *Diseases of Swine* (10th edn.), Wiley-Blackwell, Ames, IA; The Atrium, Southern Gate, Chichester; Oxford, pp. 60–66.

Chapter 16

Optimising the health of finisher pigs

Edgar Garcia Manzanilla, Pig Development Department, Teagasc, The Irish Agriculture and Food Development Authority, Ireland

1 Introduction

The growing-finishing period is the most expensive part of the pig production cycle accounting for 60–75% of the total costs depending on the type of production (Calderón Díaz et al., 2019). It should also be the easiest part to manage of the pig production cycle. Thus, it offers a big opportunity to maximise profit. Unfortunately, the performance of pigs in the finisher stage depends on the previous management. These aspects have been discussed in previous chapters. However, there are still a few factors that are important to optimise the health and productive performance of growing-finisher pigs. This chapter points out these factors and provides some insights on how to monitor and improve them.

2 How to measure pig health in grower-finisher phase

Health of pigs can be measured in different ways, and sometimes it is not measured properly but in the simplest way. Indeed, when there is an outbreak of an infectious disease, it is easy to detect and describe it. However, herds with

http://dx.doi.org/10.19103/AS.2022.0103.18

no outbreaks still have differences due to subclinical disease, stress or deficient management. Among the common variables used to measure health in finisher pigs, we could find feed intake, growth, feed conversion rate (FCR), time to slaughter, mortality, antimicrobial use (AMU) and lesions at slaughter (Agostini et al., 2014, 2015; O'Neill et al., 2021; Pessoa et al., 2021a,b; Rodrigues da Costa et al., 2020). Serology and pathogen detection are also very useful but are not always related to clinical disease. Indeed, all these variables can be affected by genetics, nutrition and climate, but once those are reasonably homogeneous, all the above variables can be used to compare health between herds and batches. The most important health measures for growing-finishing pigs are briefly described below:

1 Anorexia/hyporexia is one of the first signs of disease in animals. Pigs are no exception to this rule, and they tend to reduce activity, including feed intake, when they have any health problem and often before showing any other clinical signs. Thus, feed intake is a good way to measure how pigs are doing, although it should be combined with other indicators when possible.
2 Growth is indeed affected in poor health scenarios by the drop in intake, and this is also a health indicator to consider. Growth can also be affected by factors other than feed intake like severe activation of the immune system. However, in such cases, there would be other clinical signs that would be better to identify the health issue.
3 Feed efficiency, the relationship between intake and growth, is often assumed to be affected by the health status of the herd. However, in many cases, feed efficiency is more an indicator of issues with facilities and management (e.g. feeder type, feed presentation form, split sex and mixing) than of health. To some extent, it makes sense because hiporexia in sick animals is accompanied by a decrease in growth but does not necessarily affect FCR unless there is a severe disease. This might be counterintuitive for nutritionists since maintenance requirements are still constant and FCR should be affected (NRC, 2012). However, many field studies fail to identify the increase in FCR due to poor health status.
4 Time to slaughter is not a measure commonly included in scientific literature, but it is often an easier and more commercial way to detect issues in the growth of animals. Again, for the same genetics, diet and climate, herds with worse health status normally have problems selling pigs at the planned dates. This, at the same time, affects the farms and creates a lack of space, increases density and leaves less time for cleaning and disinfection. In current commercial pig farms, a period of 21 weeks to 23 weeks from birth to slaughter should be considered appropriate.

Pig farms with health problems are often around 25-27 weeks or even more weeks to reach the slaughter (Rodrigues da Costa et al., 2020).

5 Mortality and AMU are variables more directly related to clinical health issues and can be used for syndromic surveillance of herds (Lopes Antunes et al., 2017). Farms differ in their culling policy, and we need to consider mortality in all phases, including slaughterhouse, when looking at finisher mortality. The breakdown of the mortality causes is often not available, but it is very useful information to identify the aetiology of the problems, whether it is an infectious agent or problems in the facilities and management, and put in place the necessary corrective measures.

Other variables that are useful to measure the health status of finishing pigs and are easy to perform regularly are coughing measurements (Pessoa et al., 2021b; Fig. 1), lesions on farm or in slaughterhouse (van Staaveren et al., 2017) and, in the near future, farm biomarkers (López-Martínez et al., 2022).

3 Types of farms

The first thing to consider when planning how to manage our finisher stage to optimise health is the type of production structure. The management needed differs greatly between, for example, a multi-site production system and a farrow-to-finish farm where the finisher stage is metres away from the nursery (Fig. 2). In principle, no system is better than others if properly managed. In fact, literature favours one system or the other depending on the health indicator used (Casal et al., 2007; Fablet et al., 2018; Jager et al., 2012). For the chapter's purpose, the pros and cons of each system are presented.

Pigs in a farrow-to-finish farm are generally less stressed because no transport among production phases is required; they just walk to the next building. These pigs are also exposed to the pathogen burden to which they have already been exposed through their mothers, in the environment or equipment. On the other hand, a farrow-to-finish farm needs to be managed in a stricter way, trying to keep the breeding (gestation + lactation), the nursery and the finishers as separated units, so any outbreaks can be controlled easier. A multi-site system is ideal from an infectious disease management point of view if it is well run. It allows for easier all-in-all-out management and establishes a physical distance between the different phases. However, this advantage can be lost when animals from different origins are mixed or when the facility is managed with a continuous flow of animals. An inherent problem of the multi-site system is the stress caused by handling, transportation and relocation from one environment and management conditions to another, which can be a triggering factor for diseases promoted by stress such as Glässer's disease or ileitis.

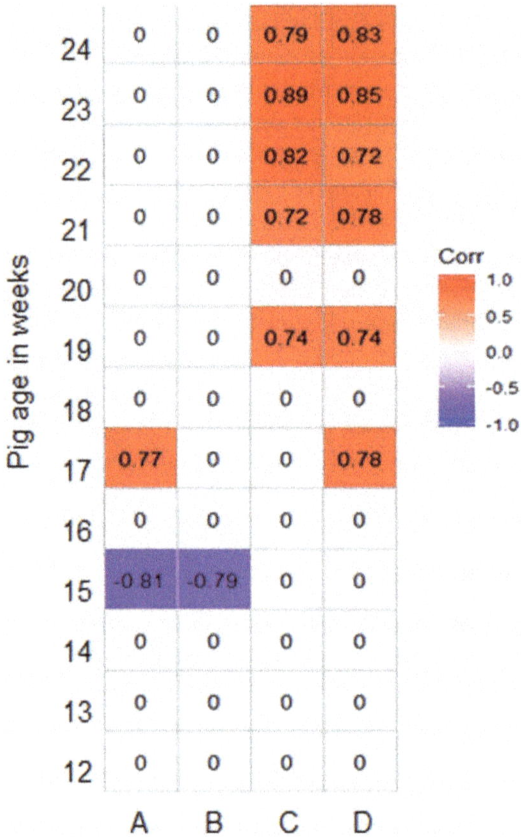

Figure 1 Lung lesions in slaughter can indicate disease as far as 8 weeks earlier in growing-finishing pigs. Automatic coughing monitors can detect disease on average 2 days before farm staff. In the figure, correlations between the prevalence of dorsocaudal (A) and cranial (B) pleurisy, scar lesions (C) and pneumonia (D) and the respiratory distress index measured automatically with coughing monitors along the growing-finishing stage are shown. Zeros correspond to non-significant results (Pessoa et al., 2021b).

4 Animal flow

In any type of farm, understanding how the animals flow through the system and segregating the different flows is one of the best ways to optimise health. Ideally, one batch of pigs should move all together (all-in-all-out), and those pigs brought to a hospital should never come back to the normal flow of pigs to avoid the spread of pathogens. The use of tags of different colours for each batch is probably the easiest way to keep batches separated and detect mixing. It is also easier to keep batches of pigs segregated and reduce disease incidence in farms working with batches at least 3 weeks apart from each other (Fablet et al., 2018). However, in many farms, the segregation between batches

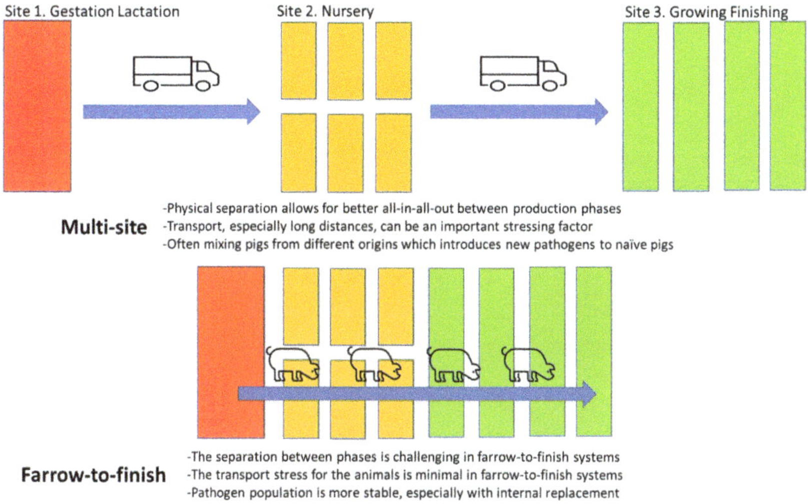

Figure 2 Pros and cons of the multi-site and farrow-to-finish pig systems.

is not respected. In Fig. 3, a real example of a farrow-to-finish farm is shown where one-weekly batch of 1096 pigs gets broken down into different flows as soon as the first week after weaning and along the production cycle (Rodrigues da Costa et al., 2021). In this example, pigs are divided at least into three main flows: pigs that move as planned, pigs that get delayed by 1 week and are mixed with the following batch and pigs that are delayed by several weeks and get mixed with several batches. Some of the pigs that are delayed by several weeks went through hospital rooms and went back to groups of healthy pigs. At slaughter, these delayed pigs were still in many cases sick and showed much higher levels of pleurisy, pericarditis, lameness and condemnations at slaughter than pigs that were not delayed (Calderón Díaz et al., 2017). Basically, the pathogens are being transmitted across different batches of naïve pigs. This may seem obvious to the veterinarian, but for a farm staff member with no training in infectious disease, this is the most normal thing to do. Farm staff need to be educated on the basics of infectious diseases and biosecurity.

Indeed, Fig. 3 is an extreme example where pigs are moved through four successive stages after weaning (nursery 1, nursery 2, growing and finishing). However, it serves the purpose of showing how wrong it can go and the need to avoid such practices. Probably the first thing to avoid is to design production cycles as complicated as this one. The more movements take place, the more stressed the pigs become and the more chances of mixing batches there are. Several epidemiological studies have shown strong associations between mixing growing-finishing pigs of different ages and higher incidence of diseases like enzootic pneumonia or pleurisy (Jager et al., 2012; Nathues et al., 2014).

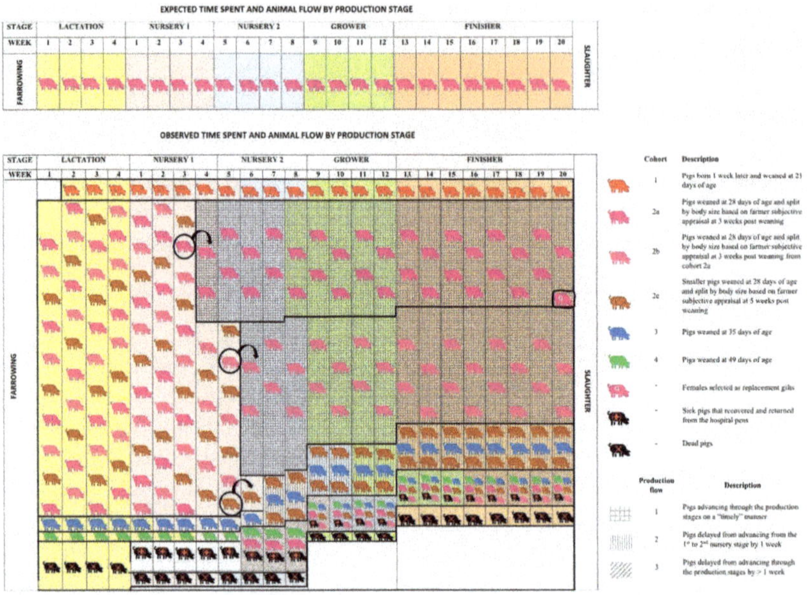

Figure 3 Expected vs. observed time spent and animal flow by production stage in a farrow-to-finish commercial pig farm where 1096 pigs were followed up from birth to slaughter to investigate management approaches for body weight variation and their possible implications for animal health (Rodrigues da Costa et al., 2021).

There are several causes of these different flows and mixing between ages. The two main causes are the natural variability of pig weights and the false impression that reorganising pigs by size helps reduce variability or improve performance (López-Vergé et al., 2018). Pigs' weights follow a normal distribution and, despite the positive correlation between birth and slaughter weight, reorganising pens by weight fails to reduce variability. In a recent unpublished study, 900 piglets were weighed individually and reorganised at weaning in 18 pens with average weights between 5 kg and 10 kg (weight range between pens: 5 kg) and weight ranges ((weight difference between the lower and higher weight in a pen)) within the pen of 300 g. By 16 weeks of age, the average weights of the pens ranged between 60 kg and 73 kg (13 kg), which is a 2.6-fold increase compared to weaning. However, the range of weights within pens was more than 20 kg, which is a 50-fold increase compared to weaning. The variability of weights at 16 weeks was the same in a control group of pigs not reorganised by weights. This proves the inefficacy of reorganising pigs to reduce variability. In fact, reorganising pigs by weight in the case of finishers has a negative effect, as proved by Camp Montoro et al. (2021, 2022). According to the results by Camp Montoro et al. (2021, 2022), mixing pigs in the growing-finishing stage has a more detrimental effect on growth than suboptimal space allowance and has an effect similar to using diets 10% deficient in lysine.

5 Stocking density

Stocking density or space allowance refers to the size of the area allocated to a pig and plays a critical role from a productive performance, animal health, welfare and economic standpoint (Thomas et al., 2017; Vermeer et al., 2017). Pig producers need to maximise production efficiency and minimise the housing cost, without compromising animal health. At the commercial scale, space allowance is adjusted by changing the number of pigs per pen. However, this method induces confounding on whether the pig is affected by space allowance, group size or feeder space. Flohr et al. (2018) stated that floor space is the main factor that influences grower-finisher pigs' performance; however, other factors can induce chronic stress by changes in the group structure and competition for resources. Kyriazakis and Whittemore (2006) came up with a formula where space allowance is expressed as:

$$\text{Space allowance}\left(m^2\right) = k \times BW^{0.667}$$

Table 1 Laying down minimum standards for the protection of nursery and grower-finisher pigs in the EU (Council of the European Union, 2008)

Body weight, kg	Space per pig, m^2	k-value[a]	Space needed to keep k at 0.034[b]
Not more than 10	0.15	0.032	0.16
>10 but not more than 20	0.20	0.027	0.25
>20 but not more than 30	0.30	0.031	0.33
>30 but not more than 50	0.40	0.029	0.46
>50 but not more than 85	0.55	0.028	0.65
>85 but not more than 110	0.65	0.028	0.78
>110	1.00	0.039[c]	0.87

[a] $k - \text{value} = \dfrac{\text{Space allowance}}{\text{Body weight}^{0.667}}$; *k*-value was calculated based on the final body weight of the study.

[b]Based on Gonyou et al. (2006).

[c]*k*-value calculated based on pigs going to slaughter with a target body weight of 130 kg.

where k represents the space allowance coefficient and $BW^{0.667}$ means the geometric conversion of body weight (BW) in kilogram to area. The relationship between k-value and feed intake is highly similar for all pigs and, below a certain critical value, the pig reduces the feed intake proportionally, which is probably related to stress. Gonyou et al. (2006) reported a critical k-value of 0.0336; however, recent studies reported higher k critical values (Carpenter et al., 2018; Kim et al., 2017; Thomas et al., 2017). These values may depend on the type of facilities, environmental enrichment, pig genetics and, indeed, health status. It is likely that farms with lower health status are more affected by high stocking density because most of the research is done in high-health farms.

Directive 2008/120/EC establishes the minimum standards for pigs in the European Union (EU) in m^2/pig (Table 1). For pig weights between 85 and 110 kg of BW, the minimum space is 0.65 m^2/pig, which is a k-value = 0.028. Such legal minimum is clearly insufficient from a productive point of view and can also lead to adverse social behaviours that may affect pig health by chronic stress with clinical disease and higher condemnations at slaughter (Pandolfi et al., 2018). Table 1 shows a calculation of the minimum space recommended following Gonyou et al.'s (2006) critical k-value.

6 The importance of transfer weight and the transition diet

Agostini et al. (2014, 2015) studied the effects of different factors on the performance and mortality of 688 batches of pigs from nine different companies in Spain. The most relevant factor that affected mortality in all the companies was body weight at transfer from nursery to the growing phase, ranging from 15 to 25 kg. This is considered expectable because higher weights are probably related to a better ability of the pig to take the challenge of a new transport, new stressors and a new diet. Although research in this area is scarce, it is normal practice to not introduce the pig to the new diet directly but to try a feed transition during the first week in growing stage from the weaner diet to the new growing diet. The weights mentioned above (15–25 kg) are not the same in all the countries, and the later the transition between phases happens, the better for the pig from a health point of view.

Water, often referred to as the forgotten nutrient, becomes a key element in this stage. Dybkjær et al. (2006) reported that a low feed intake may be due to insufficient drinking activity, as solid feed intake must be accompanied by sufficient water intake. This is not as important for wet feed systems, although a separated drinker is always recommended. In practice, it is quite common not installing or restricting the flow of these drinkers to avoid water waste, especially when pigs play with the water. This type of action is detrimental to the pig, and the solution should be to provide other toys to the pigs or look for alternative drinker designs. The importance of dietary electrolyte balance has been mentioned in previous chapters for weaners and should also be considered here (Guzmán-Pino et al., 2015). Problems in water supply affect electrolyte balance and make pigs more sensitive to disease.

7 General control of infectious diseases

Previous chapters tackled different infectious diseases in detail, and the intention of this chapter is to describe the general aspects to minimise their occurrence in commercial farms. Randomised studies in these areas are difficult to perform, but there are quite a few epidemiological approaches that indicate some areas of importance. The main risk factors observed in epidemiological studies looking at digestive and respiratory infectious diseases in growing-finishing pigs are related to environmental control and biosecurity.

Ventilation is one of the areas that keeps appearing as a key aspect associated with health. It is well known that respiratory issues and mortality are increased during the colder quarters of the year (Agostini et al., 2015; Fablet et al., 2012). Thus, it is not surprising that automatic control of ventilation is associated with better health indicators in several studies (Agostini et al., 2014, 2015; Chantziaras et al., 2020; Matheson et al., 2022). Recommendations on temperature ranges and ventilation rates are included in Chapter 17. The technology for environmental control has improved significantly in recent years. The classic control of ventilation by temperature can now be easily improved by including sensors for humidity, CO_2 or NH_3 (Fig. 4). These changes can have a significant impact on the occurrence of infectious diseases. As an example, a high mean CO_2 concentration has been associated with a higher incidence of pneumonia (Fablet et al., 2012).

Another factor that is associated with bad health in epidemiological studies is the number of origins of the animals. Mixing piglets from different origins in any phase of production, but especially in growing-finishing pigs, is detrimental (Agostini et al., 2014, 2015; Fablet et al., 2018; Jager et al., 2012).

Cleaning and disinfecting of grower and finisher accommodation between groups is often considered less important than in previous stages of the production cycle, but it is related to better performance (van der Meer et al., 2016, 2017) and lower incidence of lesions like pleurisy (Jager et al., 2012).

Co-infections have been receiving more and more attention in recent years. Receiving pigs from weaner facilities infected with porcine reproductive and respiratory syndrome virus or porcine circovirus 2 is associated with lower performance and more pathology in the finisher stage (Fablet et al., 2018; Magalhaes et al., 2022). Management in such cases needs to be done considering the pathogens present on the farm. A correct application of

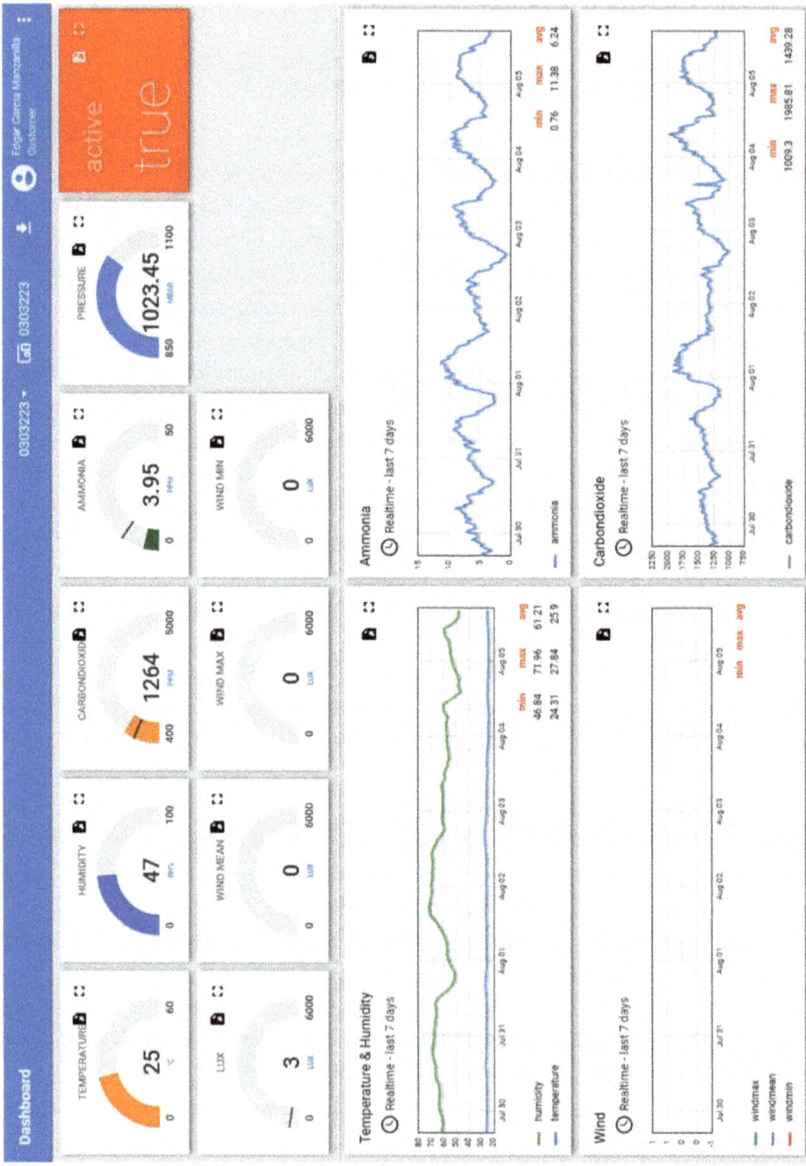

Figure 4 The current systems for environmental control allow the real-time monitoring of several parameters (e.g. temperature, humidity, NH_3 and CO_2) with access to any device and that can be used to automatically control the ventilation of the facilities.

biosecurity measures in these cases implies a knowledge of the epidemiology of the disease (Alarcón et al., 2021).

There are other aspects of management like the use of straw or other environmental enrichment (Matheson et al., 2022) or the herd size (Agostini et al., 2015), which are sometimes associated with the health status of the herd, but it is difficult to define whether they are positive or negative factors. In the case of herd size, there should not be any benefits related to the size if management is properly done. In the case of bedding and enrichment, in general, it should be beneficial. However, more studies are needed in this area.

8 Infectious vs. non-infectious diseases

Pig health management often focuses too much on infectious diseases. Probably, the main reason for this bias is the fact that more tools do exist to deal specifically with infectious diseases (antibiotics and vaccination, diagnostic analytical tests, etc.) than with non-infectious ones. However, growing-finishing pigs suffer from a series of diseases with no clear infectious component but with important prevalence and associated with high costs. Among these issues are stomach ulcers, hernias (inguinal or abdominal), prolapses, tail-biting lesions, lameness and arthritis (Heinonen et al., 2007; Teixeira et al., 2020). For many of these issues, there is not much literature available, and much more research is needed. The prevalence of most of these problems varies significantly among farms. For example, Gottardo et al. (2017) identified farm of origin as the main factor for gastric ulcers, with prevalence ranging from 0% to 36%. Robertson et al. (2002) also found the use of automated feeding systems, pelleted ration and changes in formulation and water from a dam rather than from a well or river as predisposing factors for stomach ulcers. Hernias, prolapses, lameness and arthritis are also typically problems observed in finishers more than in weaners, probably because of the size and weight of the animals. The heavy weight and fast growth of pigs in the grower-finisher stage exerts more physical pressure on these issues. Thus, whenever possible, these issues should be addressed at earlier stages to avoid their progress. Finally, a problem with no clear solution so far is tail biting (Taylor et al., 2010). Despite the extensive literature available in this area and the EU legislation ban on it, tail docking is so far the most effective measure to avoid tail biting in the current production systems. It is well known though that tail biting is associated with the occurrence of infectious diseases like abscessation in different parts of the body (Teixeira et al., 2016). Good health status is one of the preventive factors highlighted by Valros (2022), together with larger space allowance, partly slatted flooring, use of manipulable materials and use of long troughs for feeding. Obviously, all these improvements come with an extra cost that sometimes is difficult to justify (Niemi et al., 2021), and tail biting may still be a problem on some farms.

9 Conclusion

Based on the previous discussions, the general recommendations to optimise the health of pigs in growing-finishing stage include the following:

1. There is no good system or bad system. There is good management and bad management.
2. Bring the knowledge to the farm staff and make things easy for them. They take care of the details.
3. Define very well what health indicators you are using, why and how you will monitor them.
4. The health of growing-finishing pigs depends greatly on previous phases. Always keep in contact with those working with the breeding and the nursery farms and feedback information on the issues you are finding.
5. Put the animals to a good start bringing them into the growing stage with good weights and making a proper diet transition.
6. Avoid mixing pigs from different origins in the growing-finishing phase (or any phase).
7. Make the interval between batches as long as possible (ideally >3 weeks) and avoid mixing pigs of different ages to stop the transmission of infectious diseases between batches.
8. Avoid overstocking and always respect the ratio of pig:feeder space to reduce stress.
9. Always look for ways to improve environmental conditions for pigs and biosecurity. There is always room for improvement.

10 Where to look for further information

Grower-finisher pigs have been managed in the same way for more than 30 years. The development of new technologies and tools to manage their health will be key in the following years, especialy to anticipate health issues. However, research on new systems to raise pigs is needed to adapt pig farming to the needs of society and of an animal that, thanks to genetic selection, has little to do with what a pig was in the 90s.

If you want to obtain further information or stay up to date on the most recent advances in this area, please follow the links included in this section and subscribe to their regular publications.

- https://extension.umn.edu/swine/swine-health.
- https://www.asi.k-state.edu/research-and-extension/swine/.
- https://www.ipic.iastate.edu/.
- https://eaphm.org/.
- https://www.teagasc.ie/animals/pigs/.

11 References

Agostini, P. S., Fahey, A. G., Manzanilla, E. G., O'Doherty, J. V., de Blas, C. and Gasa, J. 2014. Management factors affecting mortality, feed intake and feed conversion ratio of grow-finishing pigs. *Animal* 8(8):1312–1318.

Agostini, P. S., Manzanilla, E. G., de Blas, C., Fahey, A. G., da Silva, C. A. and Gasa, J. 2015. Managing variability in decision making in swine growing-finishing units. *Irish Veterinary Journal* 68(1):20.

Alarcón, L. V., Allepuz, A. and Mateu, E. 2021. Biosecurity in pig farms: a review. *Porcine Health Management* 7(1):5.

Calderón Díaz, J. A., Diana, A., Boyle, L. A., Leonard, F. C., McElroy, M., McGettrick, S., Moriarty, J. and Manzanilla, E. G. 2017. Delaying pigs from the normal production flow is associated with health problems and poorer performance. *Porcine Health Management* 3:13.

Calderón Díaz, J. A., Shalloo, L., Niemi, J. K., Kyriazakis, I., McKeon, M., McCutcheon, G., Bohan, A. and Manzanilla, E. G. 2019. Description, evaluation, and validation of the Teagasc Pig Production Model. *Journal of Animal Science* 97(7):2803–2821.

Camp Montoro, J., Boyle, L. A., Solà-Oriol, D., Muns, R., Gasa, J. and Manzanilla, E. G. 2021. Effect of space allowance and mixing on growth performance and body lesions of grower-finisher pigs in pens with a single wet-dry feeder. *Porcine Health Management* 7(1):7.

Camp Montoro, J., Pessoa, J., Solà-Oriol, D., Muns, R., Gasa, J. and Manzanilla, E. G. 2022. Effect of phase feeding, space allowance and mixing on productive performance of grower-finisher pigs. *Animals* 12(3):390.

Carpenter, C. B., Holder, C. J., Wu, F., Woodworth, J. C., Derouchey, J. M., Tokach, M. D., Goodband, R. D. and Dritz, S. S. 2018. Effects of increasing space allowance by removing a pig or gate adjustment on finishing pig growth performance. *Journal of Animal Science* 96(7):2659–2664.

Casal, J., Mateu, E., Mejia, W. and Martin, M. 2007. Factors associated with routine mass antimicrobial usage in fattening pig units in a high pig-density area. *Veterinary Research* 38(3):481–492.

Chantziaras, I., De Meyer, D., Vrielinck, L., Van Limbergen, T., Pineiro, C., Dewulf, J., Kyriazakis, I. and Maes, D. 2020. Environment-, health-, performance- and welfare-related parameters in pig barns with natural and mechanical ventilation. *Preventive Veterinary Medicine* 183:105150.

Council of the European Union (2008). COUNCIL DIRECTIVE 2008/120/EC of 18 December 2008. *Laying down minimum standards for the protection of pigs.* 47/5-47/13. Available at: https://eur-lex.europa.eu/legal-content/EN/TXT/PDF/?uri=CELEX:32008L0120&rid=19.

Dybkjær, L., Jacobsen, A. P., Tøgersen, F. A. and Poulsen, H. D. 2006. Eating and drinking activity of newly weaned piglets: effects of individual characteristics, social mixing, and addition of extra zinc to the feed. *Journal of Animal Science* 84(3):702–711.

Fablet, C., Dorenlor, V., Eono, F., Eveno, E., Jolly, J. P., Portier, F., Bidan, F., Madec, F. and Rose, N. 2012. Noninfectious factors associated with pneumonia and pleuritis in slaughtered pigs from 143 farrow-to-finish pig farms. *Preventive Veterinary Medicine* 104(3–4):271–280.

Fablet, C., Rose, N., Grasland, B., Robert, N., Lewandowski, E. and Gosselin, M. 2018. Factors associated with the growing-finishing performances of swine herds: an exploratory study on serological and herd level indicators. *Porcine Health Management* 4:6.

Flohr, J. R., Dritz, S. S., Tokach, M. D., Woodworth, J. C., Derouchey, J. M. and Goodband, R. D. 2018. Development of equations to predict the influence of floor space on average daily gain, average daily feed intake and gain:feed ratio of finishing pigs. *Animal* 12(5):1022–1029.

Gonyou, H. W., Brumm, M. C., Bush, E., Deen, J., Edwards, S. A., Fangman, T., McGlone, J. J., Meunier-Salaun, M., Morrison, R. B., Spoolder, H., Sundberg, P. L. and Johnson, A. K. 2006. Application of broken-line analysis to assess floor space requirements of nursery and grower-finisher pigs expressed on an allometric basis. *Journal of Animal Science* 84(1):229–235.

Gottardo, F., Scollo, A., Contiero, B., Bottacini, M., Mazzoni, C. and Edwards, S. A. 2017. Prevalence and risk factors for gastric ulceration in pigs slaughtered at 170 kg. *Animal* 11(11):2010–2018.

Guzmán-Pino, S. A., Solà-Oriol, D., Davin, R., Manzanilla, E. G. and Pérez, J. F. 2015. Influence of dietary electrolyte balance on feed preference and growth performance of postweaned piglets. *Journal of Animal Science* 93(6):2840–2848.

Heinonen, M., Hakala, S., Hämeenoja, P., Murro, A., Kokkonen, T., Levonen, K. and Peltoniemi, O. A. T. 2007. Case-control study of factors associated with arthritis detected at slaughter in pigs from 49 farms. *Veterinary Record* 160(17):573–578.

Jager, H. C., McKinley, T. J., Wood, J. L. N., Pearce, G. P., Williamson, S., Strugnell, B., Done, S., Habernoll, H., Palzer, A. and Tucker, A. W. 2012. Factors associated with pleurisy in pigs: a case-control analysis of slaughter pig data for England and Wales. *PLoS ONE* 7(2):e29655.

Kim, K. H., Kim, K. S., Kim, J. E., Kim, D. W., Seol, K. H., Lee, S. H., Chae, B. J. and Kim, Y. H. 2017. The effect of optimal space allowance on growth performance and physiological responses of pigs at different stages of growth. *Animal* 11(3):478–485.

Kyriazakis, I. and Whittemore, C. 2006. *Whittemore's Science and Practice of Pig Production* (3rd edn.). Blackwell Publishing Ltd., Oxford.

Lopes Antunes, A. C., Ersbøll, A. K., Bihrmann, K. and Toft, N. 2017. Mortality in Danish Swine herds: spatio-temporal clusters and risk factors. *Preventive Veterinary Medicine* 145:41–48.

López-Martínez, M. J., Franco-Martínez, L., Martínez-Subiela, S. and Cerón, J. J. 2022. Biomarkers of sepsis in pigs, horses and cattle: from acute phase proteins to procalcitonin. *Animal Health Research Reviews* 23(1):82–99.

López-Vergé, S., Gasa, J., Farré, M., Coma, J., Bonet, J. and Solà-Oriol, D. 2018. Potential risk factors related to pig body weight variability from birth to slaughter in commercial conditions. *Translational Animal Science* 2(4):383–395.

Magalhaes, E. S., Zimmerman, J. J., Thomas, P., Moura, C. A. A., Trevisan, G., Holtkamp, D. J., Wang, C., Rademacher, C., Silva, G. S. and Linhares, D. C. L. 2022. Whole-herd risk factors associated with wean-to-finish mortality under the conditions of a Midwestern USA swine production system. *Preventive Veterinary Medicine* 198:105545.

Matheson, S. M., Edwards, S. A. and Kyriazakis, I. 2022. Farm characteristics affecting antibiotic consumption in pig farms in England. *Porcine Health Management* 8(1):7.

Nathues, H., Chang, Y. M., Wieland, B., Rechter, G., Spergser, J., Rosengarten, R., Kreienbrock, L. and Grosse Beilage, E. 2014. Herd-level risk factors for the seropositivity to Mycoplasma hyopneumoniae and the occurrence of enzootic pneumonia Among fattening pigs in areas of endemic infection and high pig density. *Transboundary and Emerging Diseases* 61(4):316–328.

Niemi, J. K., Edwards, S. A., Papanastasiou, D. K., Piette, D., Stygar, A. H., Wallenbeck, A. and Valros, A. 2021. Cost-effectiveness analysis of seven measures to reduce tail biting lesions in fattening pigs. *Frontiers in Veterinary Science* 8:682330.

NRC 2012. *Nutrient Requirements of Swine* (11th edn.). National Academic Press, Washington, DC.

O'Neill, L., Calderón Díaz, J. A., Rodrigues da Costa, M., Oakes, S., Leonard, F. C. and Manzanilla, E. G. 2021. Risk factors for antimicrobial use on Irish pig farms. *Animals* 11(10):2828.

Pandolfi, F., Edwards, S. A., Maes, D. and Kyriazakis, I. 2018. Connecting different data sources to assess the interconnections between biosecurity, health, welfare, and performance in commercial pig farms in Great Britain. *Frontiers in Veterinary Science* 5:41.

Pessoa, J., McAloon, C., Rodrigues da Costa, M., Manzanilla, E. G., Norton, T. and Boyle, L. A. 2021a. Adding value to food chain information: using data on pig welfare and antimicrobial use on-farm to predict meat inspection outcomes. *Porcine Health Management* 7(1):55.

Pessoa, J., Rodrigues da Costa, M., Manzanilla, E. G., Norton, T., McAloon, C. and Boyle, L. A. 2021b. Managing respiratory disease in finisher pigs: combining quantitative assessments of clinical signs and the prevalence of lung lesions at slaughter. *Preventive Veterinary Medicine* 186:105208.

Robertson, I. D., Accioly, J. M., Moore, K. M., Driesen, S. J., Pethicka, D. W. and Hampson, D. J. 2002. Risk factors for gastric ulcers in Australian pigs at slaughter. *Preventive Veterinary Medicine* 53(4):293–303.

Rodrigues da Costa, M., Fitzgerald, R. M., Manzanilla, E. G., O'Shea, H., Moriarty, J., McElroy, M. C. and Leonard, F. C. 2020. A cross-sectional survey on respiratory disease in a cohort of Irish pig farms. *Irish Veterinary Journal* 73(1):24.

Rodrigues da Costa, M., Manzanilla, E. G., Diana, A., van Staaveren, N., Torres-Pitarch, A., Boyle, L. A. and Calderón Díaz, J. A. 2021. Identifying challenges to manage body weight variation in pig farms implementing all-in-all-out management practices and their possible implications for animal health: a case study. *Porcine Health Management* 7(1):10.

Taylor, N. R., Main, D. C. J., Mendl, M. and Edwards, S. A. 2010. Tail biting: a new perspective. *Veterinary Journal* 186(2):137–147.

Teixeira, D. L., Harley, S., Hanlon, A., O'Connell, N. E., More, S. J., Manzanilla, E. G. and Boyle, L. A. 2016. Study on the association between tail lesion score, cold carcass weight, and viscera condemnations in slaughter pigs. *Frontiers in Veterinary Sciences* 3:24.

Teixeira, D. L., Salazar, L. C., Enriquez-Hidalgo, D. and Boyle, L. A. 2020. Assessment of animal-based pig welfare outcomes on farm and at the abattoir: a case study. *Frontiers in Veterinary Science* 7:576942.

Thomas, L. L., Goodband, R. D., Woodworth, J. C., Tokach, M. D., Derouchey, J. M. and Dritz, S. S. 2017. Effects of space allocation on finishing pig growth performance and carcass characteristics. *Translational Animal Science* 1(3):351-357.

Valros, A. 2022. Review: the tale of the Finnish pig tail - how to manage non-docked pigs? *Animal* 16 (Suppl. 2):100353.

van der Meer, Y., Gerrits, W. J. J., Jansman, A. J. M., Kemp, B. and Bolhuis, J. E. 2017. A link between damaging behaviour in pigs, sanitary conditions, and dietary protein and amino acid supply. *PLoS ONE* 12(5):e0174688.

van der Meer, Y., Lammers, A., Jansman, A. J. M., Rijnen, M. M. J. A., Hendriks, W. H. and Gerrits, W. J. J. 2016. Performance of pigs kept under different sanitary conditions affected by protein intake and amino acid supplementation. *Journal of Animal Science* 94(11):4704-4719.

van Staaveren, N., Doyle, B., Manzanilla, E. G., Calderón Díaz, J. A., Hanlon, A. and Boyle, L. A. 2017. Validation of carcass lesions as indicators for on-farm health and welfare of pigs. *Journal of Animal Science* 95(4):1528-1536.

Vermeer, H. M., Dirx-Kuijken, N. C. P. M. M. and Bracke, M. B. M. 2017. Exploration feeding and higher space allocation improve welfare of growing-finishing pigs. *Animals* 7(5):3-11.

Index

www.ingramcontent.com/pod-product-compliance
Ingram Content Group UK Ltd.
Pitfield, Milton Keynes, MK11 3LW, UK
UKHW021824100926
8556IPUK00003B/138

9 781786 768834